Fault Mechanics and Transport Properties of Rocks

This is Volume 51 in the
INTERNATIONAL GEOPHYSICS SERIES
A series of monographs and textbooks
Edited by RENATA DMOWSKA and JAMES R. HOLTON

A complete list of the books in this series appears at the end of this volume.

Fault Mechanics and Transport Properties of Rocks

A Festschrift in Honor of W. F. Brace

Edited by

Brian Evans
Department of Earth, Atmospheric and Planetary Sciences
Massachusetts Institute of Technology
Massachusetts
USA

and

Teng-fong Wong
Department of Earth and Space Sciences
State University of New York, Stony Brook
New York
USA

ACADEMIC PRESS
Harcourt Brace Jovanovich, Publishers
London San Diego New York Boston
Sydney Tokyo Toronto

ACADEMIC PRESS LIMITED
24/28 Oval Road
LONDON NW1 7DX

United States Edition published by
ACADEMIC PRESS INC.
San Diego, CA 92101

This book is printed on acid-free paper

A catalogue record for this book is available from the British Library

ISBN 0-12-243780-2

Typeset by P & R Typesetters Ltd, Salisbury, Wiltshire, UK
Printed in Great Britain at The University Press, Cambridge

Contents

Brittle Failure of Rocks

Permeability and Fluid Flow in Rocks

Fracture Characterization and Physical Properties of Rock

Implications of Rock Mechanics on Crustal Tectonics

Contributors

Yves Bernabé
Institut de Physique du Globe du Strasbourg, Universite Louis Pasteur, 5 Rue Descartes 67084 Strasbourg Cedex, France

Stephen C. Blair
Earth Sciences Dept., Lawrence Livermore National Laboratories, MS-L201, PO Box 808, Livermore, CA 94550, USA

James G. Berryman
Earth Sciences Dept., Lawrence Livermore National Laboratories, MS-L201, PO Box 808, Livermore, CA 94550, USA

E. G. Bombolakis
Department of Geology and Geophysics, Boston College, Chestnut Hill, Boston, MA 02167, USA

James D. Byerlee
US Geological Survey, Mail Stop 977, 345 Middlefield Road, Menlo Park, CA 94205, USA

Robert D. Cicerone
Earth Resources Laboratory, Department of Earth, Atmospheric and Planetary Sciences, Massachusetts Institute of Technology, Cambridge, MA 02139, USA

C. H. Cheng
Earth Resources Laboratory, Department of Earth, Atmospheric and Planetary Sciences, Massachusetts Institute of Technology, Cambridge, MA 02139, USA

Carlos A. Dengo
Exxon Production Research, PO Box 808, Houston, TX 77252-2189, USA

Terry Engelder
Department of Geosciences, Pennsylvania State University, 336 Deike, University Park, PA 16802, USA

Brian Evans
Department of Earth, Atmospheric and Planetary Sciences, Massachusetts Institute of Technology, Cambridge, MA 02139, USA

George J. Fischer
Research School of Earth Science, Australian National University, PO Box 4, Canberra, ACT 2601 Australia

Yaojun Gu
Department of Earth and Space Sciences, State University of New York, Stony Brook, NY 11794-2100, USA

Stephen H. Hickman
US Geological Survey, Mail Stop 977, 345 Middlefield Road, Menlo Park, CA 94025, USA

Nigel G. Higgs
Amoco Production Research, PO Box 3385, Tulsa, OK 74102, USA

David L. Kohlstedt
Department of Geology and Geophysics, University of Minnesota, Pillsbury Hall, Minneapolis, MN 55455, USA

Victor Kuksenko
A.F. Iofee Physical-Technical Institute, Russian Academy of Sciences, Polytechnicheska 26, St. Petersburg, 194021 Russia

Alfred Lacazette
Texaco Inc., Exploration and Production Technology Department, 3901 Briarpark, Houston, TX 77042, USA

David A. Lockner
US Geological Survey, Mail Stop 977, 345 Middlefield Road, Menlo Park, CA 94025, USA

John M. Logan
Center for Tectonophysics, Texas A&M University, College Station, TX 77843, USA

Randolph J. Martin
New England Research Associates, 76 Olcott Drive, White River Junction, VT 05001, USA

John Mecholsky, Jr
Department of Material Sciences and Engineering, University of Florida, Gainesville, FL 32611, USA

Peter H. Molnar
Department of Earth, Atmospheric and Planetary Sciences, Massachusetts Institute of Technology, Cambridge, MA 02139, USA

Carolyn A. Morrow
US Geological Survey, Mail Stop 977, 345 Middlefield Road, Menlo Park, CA 94205, USA

Amos M. Nur
Department of Geophysics, Stanford University, Stanford, CA 94305, USA

Mervyn S. Paterson
Research School of Earth Science, Australian National University, PO Box 4, Canberra ACT 2601, Australia

Alexander Ponomarev
Institute of Physics of the Earth, Russian Academy of Sciences, Bolshaya, Gruzinskay 10, Moscow D-242, 123810 Russia

Burlinda Gay Radney
Amoco Oil Co, Houston, Texas, USA

James R. Rice
Division of Applied Sciences, Harvard University, Pierce Hall, 29 Oxford St, Cambridge, MA 02138, USA

G. N. Riley, Jr
American Superconductor Corporation, 149 Grove Street, Watertown, MA 02172, USA

Christopher H. Scholz
Lamont-Doherty Geological Observatory, Columbia University, Palisades, NY 10964, USA

Paul A. Scott
BP Exploration Inc., Sage Plaza, 5151 San Felipe St., Houston, TX 77210, USA

A. Sidorin
Institute of Physics of the Earth, Russian Academy of Sciences, Bolshaya, Gruzinskay 10, Moscow D-242, 123810 Russia

Eve Sprunt
Mobil Research Development Corporation, PO Box 819047, Dallas, TX 75381-9047, USA

Nafi M. Toksöz
Earth Resources Laboratory, Department of Earth, Atmospheric and Planetary Sciences, Massachusetts Institute of Technology, Cambridge, MA 02139, USA

Jan Tullis
Department of Geological Sciences, Brown University, Providence, RI 02912, USA

Zai-zhang Wang
K. Brown and Associates Int., College Station, TX 78753, USA

Teng-fong Wong
Department of Earth and Space Sciences, State University of New York, Stony Brook, NY 11794, USA

Joseph S. Walder
US Geological Survey, Cascades Volcano Observatory, 5400 MacArthur Blvd., Vancouver, WA 98661, USA

Takashi Yanagidani
Department of Mineral Science and Technology, Kyoto University, Yoshida Houmachi, Sakyo 606, Kyoto, Japan

Richard A. Yund
Department of Geological Sciences, Brown University, Providence, RI 02912, USA

Yusheng Zhao
Department of Earth and Space Sciences, State University of New York, Stony Brook, NY 11794, USA

William F. Brace

Introduction: William F. Brace

W. F. Brace's career has several parallels with the institution from which all his degrees were granted. Both emphasized fundamental training in sciences, mathematics, physics, chemistry and the rigorous application of that training to applied problems. Both demand scrupulous attention to data collection, followed by aggressive interpretation of the implications of the data. Both flirted with Harvard. Early in this century, M.I.T. almost became Harvard's engineering school. During his doctoral research, Brace took courses and advice from J. B. Thompsen and others at Harvard; and he returned later to do postdoctoral work with Francis Birch.

Bill Brace was born in 1926 in Littleton, NH. His parents separated early in his life, and his mother supported the two of them by working as an accountant and secretary in Boston. Bill's elementary and secondary schooling were taken in the Boston area. He finished preparatory school in Danvers, a Boston suburb, and matriculated to M.I.T. in 1943 to study engineering. The war intervened and Bill served in the Navy from 1944–1946, resuming his undergraduate study to obtain an S.B. degree in Naval Architecture in 1946 and a degree in Civil Engineering in 1949.

Some seeds for the change from engineering to geology were sown in the early part of his undergraduate career, in lectures in introductory geology by F. K. Morris, in courses for engineering geology with W. J. Mead, and in laboratory work and field trips with H. W. Fairbairn. But, perhaps a larger motivation to become an Earth scientist came from his love of the outdoors. Drawn to the physical challenges of mountains as well as their intellectual challenges, Bill was an avid rock climber and hiker and continues to the present to walk and run through and over mountains.

Brace's dissertation was a structural petrology study of the deformation of rocks of the Green Mountain anticlinorium around Rutland, VT. The work included field examinations, thin section studies, and petrofabric investigations. He drew heavily on the methods of structural petrology pioneered by Bruno Sander in Austria, and taught to him by Harold Fairbairn. Eventually Brace spent part of 1953 and 1954 in Sander's laboratory as a Fulbright fellow, although he also found time to climb and ski in the Tyrolean Alps.

Paradoxically, it was Brace's detailed field work that motivated his transition into laboratory studies. The last paragraph of his thesis reads, 'The most obvious need today...is for experimental data on rock deformation. We know very little of the processes which contribute to permanent change in shape of rocks. The data presently being gathered [are] of limited value inasmuch as deformaton is being carried on under conditions of limited application in nature; the products of many of these tests doubtless bear little resemblance to natural rock fabric... At present there is little to limit the imagination of the structural geologist in the interpretation of rock fabric.'

After his fellowship with Sander, Brace began to establish a laboratory to provide fundamental data for structural geologists to use. To gain expertize in laboratory methods he arranged to work with Francis Birch in the Dunbar laboratory at Harvard. In addition to gaining insight in technique from Birch, Brace was relieved of teaching responsibilities and could concentrate on developing an experimental program. In a series of early papers he published his field work, and studied the energetics of anisotropic elastic crystals, indentation hardness of minerals, and methods to analyze finite strain.

The first major problem that Brace attacked while at the Dunbar laboratory, though, was obtaining a better understanding of the fracture strength of rocks. Workers of the day recognized that the treatment of brittle failure in tension would have to be amended to predict failure of rocks in compression. Among those trying to formulate a failure criterion was Frank McClintock of the Mechanical Engineering department at M.I.T. Both Brace and McClintock were consulting for a company investigating drilling technology, and it was there that Brace first met Joe Walsh, a young student of McClintock who had worked on an extension of the Griffith criterion. Soon Brace and Walsh (1962) collaborated on an adaptation of Obreimoff's experiments to determine fracture surface energy of some minerals. (Typically, Bill had begun with experiments designed to examine fundamental processes and constants.) That paper, still quoted in fracture mechanics papers today, marks the first in their long and fruitful collaboration. It was also during this time that Bill was introduced to Westerly granite, via Birch's investigation of its elastic wave properties, Griggs's tests on its strength at high temperature, and Fairbairn's determinations of its geochemical properties.

Guided by new experimental data and mechanical analyses, Brace and his first graduate student at M.I.T., Mike Bombolakis, examined crack growth in compression using a photo-elastic technique (Brace and Bombolakis, 1963). The configuration they investigated, the growth of branch or wing cracks from an inclined flaw, has become the basis for several new approaches to understanding compressional failure in rock, which have great promise for finally producing a robust description of brittle failure.

With the combination of engineering and geology studies at M.I.T., with laboratory training in Birch's laboratory, and with careful and systematic mechanical analyses done with Walsh, the fuel was in place. Brace's (1964) paper on brittle fracture of rocks was the initial report in a remarkable, sustained explosion of ideas, experiments, analyses, and techniques which took place over approximately a dozen years. In addition to providing the conceptual foundation for understanding brittle rock failure. Bill's work also provided a systematic framework for rationalizing attenuation and velocity of elastic waves, permeability, electrical resistivity, and frictional properties of rocks.

In his watershed study, Brace (1964) applied Griffith's theory to rocks and demonstrated the causal relation between shear fracture and stress-induced microcracking. This paper and accompanying discussion are notable in that they clearly document Brace's conviction that the search for a robust fracture criterion would require careful microstructural observations, systematic suites of experiments with new and improved testing equipment, and rigorous mathematical analyses. This approach flowered into a definitive study of the dilatancy which generally occurs before shear localization (Brace *et al.*, 1966) and into systematic observations of the micromechanics of fracture in compression (Brace, 1971; Sprunt and Brace, 1974; Montgomery and Brace, 1975; Brace, 1976; Tapponier and Brace, 1976). Combining the insight of the micromechanical studies with his knowledge of the effect of cracks on transport properties, he documented dilatancy hardening (Brace and Martin, 1968) and explored the possibility that changes in physical properties might be used to predict earthquakes (Brace, 1975; Myachkin *et al.*, 1975). Experiments of frictional sliding in Brace's laboratory showed for the first time that frictional instabilities could occur under crustal pressure conditions, and that the instabilities are sensitively dependent on pressure, temperature and machine stiffness, motivating Brace and Byerlee (1966) to formulate the paradigm that stick-slip is the physical mechanism for crustal earthquake rupture.

In pursuing this research, Brace was always conscious of the errors associated with the experiments and was willing to spend the time and energy to improve the apparatus or generate a new technique. To illuminate the crack and pore structure of deformed and intact rocks, he explored the use of the scanning electron microscope and the argon ion thinner (Brace *et al.*, 1971). To improve strain measurement, Brace investigated the effect of pressure on resistance strain gages (Brace, 1964). To further investigate instability, he built a new testing apparatus (Wawersik and Brace, 1971) with the help of Gene Robertson. The key component of the device, an extremely stiff thrust bearing on the axial-load generator, which Brace remembers designing on his dining room table at night, harks back to his training as a nautical engineer, since the drive shaft of every large ship terminates in an enormous bearing.

To broaden the conditions and mechanisms of failure, Brace designed and built an internally heated servo controlled vessel. With this machine and others, he and his colleagues explored the anelasticity of intact crustal rocks at high temperature (Goetze and Brace, 1972), sliding along faults at high temperature (Stesky *et al.*, 1974), and the rheology of partially molten rock (Arzi, 1978). He supported Chris Goetze in his efforts to understand plastic flow and encouraged Dave Kohlstedt, then at the Cavendish laboratory, to join Goetze in these studies.

Brace's interest in microstructure of cracks and pores dovetailed nicely with his desire to understand other physical properties. Here too, he developed measurement techniques necessary for rigorous analysis. He pioneered the pulse transient technique to resolve the very low permeability of crystalline rocks (Brace *et al.*, 1968), provided the first systematic measurements of electrical conductivity of water-saturated crystalline rocks under elevated pressures (Brace *et al.*, 1965), and measured resistivity during fracture and frictional sliding (Brace and Orange, 1968). In a synergistic collaboration with J. B. Walsh, Brace explored the dominant role of crack closure in controlling the elastic, thermal, electrical, and hydraulic transport properties of rocks. These experiments and observations motivated Walsh to formulate a unified explanation of the mechanical and transport properties of cracked rocks and their variation with pressure (e.g. Walsh and Brace, 1966; Walsh and Brace, 1984).

Throughout his career Brace did not lose sight of the goal: to apply laboratory and theoretical experience to natural Earth processes. His work contains ubiquitous insights into tectonic problems guided by interpretations of his data. He explored techniques for determining the strength of *in situ* rock (Pratt *et al.*, 1971, 1974) and the effect of size on rock strength (Pratt *et al.*, 1972) and integrated the laboratory data and field determinations of permeability to show that an interconnected pore space probably spans a significant portion of the Earth's crust (Brace, 1971, 1981). These benchmark analyses of crustal permeability have extensively influenced geophysicists, geochemists and engineering geologists. The rock friction tests of Brace, Byerlee, and others, and the high temperature creep data from Goetze at M.I.T. and from other labs, provide a basis for quantitative estimates of the strength profile through the continental crust and upper mantle. Although Brace and Kohlstedt's (1980) paper on lithospheric stress is probably the most cited of Brace's papers, one of the most important aspects of this paper is often overlooked: the detailed comparison of *in situ* stress with the predictions from laboratory data.

Much of Bill's energies in the last four years before retirement were consumed by the administrative duties associated with being the department head. During his administration the Earth sciences department underwent a merger with meteorology to become Earth, Atmospheric, and Planetary Sciences. Despite the seeming disparity between these programs, he accomplished the merger into a coherent, if not seamless department. During this period there were a number of difficult and important personnel transitions and appointments, not the least involving a diminution in effort in field structural geology and petrology. Brace retained an allegiance to and insight in this field and strengthened the department, building a extraordinary vigorous community of scholars, hiring Marcia McNutt, Kip Hodges, John Grotzinger, and Leigh Royden to join Clark Burchfiel and Peter Molnar.

Summary

Well-executed science is like a well-built piece of cabinetry. Each will be long-lasting, both have simple, clear, but elegant, lines. Like the joinery in a hidden dovetail, Brace's science contains thorough preparation and execution. He does not suffer half-measures; his endeavors are done correctly, whether they are woodworking, marathon running, rowing shells, playing the oboe, doing science, designing an apparatus or writing a paper. Returning to the question raised in his thesis, it seems clear that W. F. Brace's work will not limit the geologist's imagination; rather his efforts will both constrain and extend it.

Brace's distinguished career was recognized by election to the National Academy of Sciences (1971), and by selection as a Fellow of the Geological Society of America (1953), the American Geophysical Society (1963), and the American Academy of Arts and Sciences (1971). He served as associate editor and editor for several journals, as President and Vice-President of the Tectonophysics section of AGU (1963–1969), and on numerous visiting committees and review panels for national laboratories, universities and government agencies. From 1976 to 1988, he was the Cecil and Ida Green Professor of Geology at M.I.T. and served as the department head for seven years (1981–1988). The American Geophysical Society presented him with the Bucher Medal in 1987, and the U.S. National Committee on Rock Mechanics gave him a Distinguished Achievement Award in the same year.

Brace's former students and post-doctoral associates include Y. Bernabé, E. G. Bombolakis, J. D. Byerlee, K. Hadley, L. M. Jones, R. M. Martin, III, C. Morrow, A. S. Orange, B. W. Paulding, Jr., P.-Y. F. Robin, M. Rowan, C. Scholz, E. Sprunt, R. Stesky, P. Tapponier, W. Wawersik, and T.-f. Wong. In addition, there were many other research associates and students at M.I.T. with whom Bill interacted through scientific and technical help, financial support, and personal encouragement. These include A. Arzi, W. B. Durham, Y. D. Caristan, B. Evans, C. G. Goetze, S. H. Hickman, D. L. Kohlstedt, D. Lockner, C. Montgomery, A. Nur, and D. Olgaard. He co-authored papers with Jim Dieterich, John Handin, Ted Madden, Frank Press, Howard Pratt, Barry Raleigh, Gene Simmons, and Joe Walsh, and offered the hospitality of his lab to K. Mogi, M. S. Paterson, and others. His thoughtful analysis, careful data collection, thorough planning, and insightful scientific intuition set high standards for those who follow. Brace's research accomplishments will be long-lasting, but it is likely that his impact on Earth sciences will be felt most lastingly from the training, methods, and standards which he imparted to those with whom he worked.

The Brace Symposium and this Volume

William F. Brace's pioneering work has defined many important research areas and opened questions which are still being intensively investigated. To honor Brace, a symposium was held at the Massachusetts Institute of Technology on June 10–11, 1990. Twenty-six papers, on the theme of fault mechanics and transport properties in rocks, were presented by Bill's colleagues, friends, and former students. The papers in this volume, based on the oral presentations of the symposium are divided into four general topics in which Brace's work has had fundamental impact: brittle failure of crustal rocks, permeability and fluid flow, fracture characterization and physical properties of rocks, and implications of rock mechanics on crustal tectonics.

From earlier work by Brace and others, we now know that shear fracture develops through the interaction and coalescence of a multiplicity of micro cracks, but, in the past, real-time observation of this complex process has been hampered by dynamic instability. In the lead paper of this volume, Lockner *et al.* (Chapter 1) report the first successful attempt to use acoustic emission activity to control the growth of a quasi-static shear fault and simultaneously map the spatial and temporal evolution of micro-cracking. The results have important implications for the mechanics of shear localization and the energetics of shear faulting. Under elevated temperatures and pressures, shear localization may be inhibited causing rocks to fail by macroscopically ductile cataclastic flow. Tullis and Yund's paper (Chapter 4) details the micro mechanics of the transition from faulting to cataclastic flow and then to dislocation creep in feldspar aggregates. Tullis and Yund's work highlights the importance of cataclastic flow in feldspar deformation, with far reaching implications on the stress state of the lithosphere. Shear localization is also commonly associated with the fabrics of experimental and natural fault zones. Logan *et al.* (Chapter 2) systematically document the evolution of Riedel shears during frictional sliding. They establish the connection between the fabrics of experimental fault zone and the frictional sliding modes, and emphasize the important role of kinematic constraint in controlling the stress field in the fault zone and the development of boundary shear localization.

The anomalous frictional strength of clays such as montmorillonite has often been cited to explain the low shear stresses along the San Andreas fault zone as inferred from heat flow measurements. To extrapolate laboratory measurements to crustal conditions, one needs an appropriate law governing the dependence of frictional strength on the overburden and pore pressures. Addressing this question, Morrow *et al.* (Chapter 3) conclude that even if montmorillonite remains stable throughout the seismogenic layer, its frictional strength would still be higher than the stress values inferred from heat flow measurements – unless the pore pressure gradient is greater than hydrostatic. Although the frictional strength they discuss bounds the stress state for the onset of frictional sliding, it does not provide information as to the stability of sliding. The nonlinear dynamics of the transition from stable sliding to cyclic stick-slip is considered by Wong *et al.* (Chapter 5). Those authors demonstrate that the transition is through a route with period doubling and self-sustained oscillations, in agreement with theoretical predictions for the dynamic behavior of a spring-slider system with friction dependent on state and rate. Wong *et al.* also show that wear processes along a seismogenic fault zone will generally stabilize the faulting behavior.

The second section focuses on permeability and fluid flow. Three papers address permeability measurements in the laboratory. Fischer (Chapter 8) analyzes the advantages and disadvantages of the steady state flow method, the pulse transient method and the sinusoidal oscillation method, pointing out that the last method can simultaneously measure permeability and storage capacity. Reporting the first successful implementation of this innovative technique with elevated temperature and pressure, Fischer and Paterson (Chapter 9) draw important conclusions on the effects of stress and temperature on pore geometry. Although it is well known that permeability can be highly anisotropic in sedimentary basins, shear zones and accretionary complexes, the laboratory measurement of the permeability tensor is seldom attempted (Bernabé, Chapter 6). Through numerical simulations and model experiments, Bernabé provides important guidelines and suggestions for optimizing the experimental design.

Water is not the only fluid that percolates through the Earth: melts and two-phase mixtures of gas and water are also present. The influence of water and carbon dioxide on the kinetics of melt migration in an olivine system is reported by Riley and Kohlstedt (Chapter 11). The experimental data indicate that dissolved volatiles have only a modest influence on the infiltration rate and the volume of melt that a rock can absorb. In the transport of multiple phases through an interconnected pore space, it is difficult, if not impossible, to define permeability and relative permeability. Using digitized images of pore geometry, Blair and Berryman (Chapter 7) illustrates the complexity of this problem and address the limitations of the Kozeny-Carman model of equivalent channel in the context of multi-phase flow. The pore geometry of rocks is expected to be continually modified by diagenetic and tectonic processes. The lithification of sediments may involve cementation, diagenetic reaction or mechanical compaction. In their experimental study, Hickman and Evans (Chapter 10) consider the growth of grain contacts by solution transfer in halite. They formulate neck growth models to interpret their data, and conclude that this neck growth mechanism can be very important in the induration or sediments and the strength recovery of fault gouge.

Two papers in section III of this volume report innovative methods to characterize fractures in the laboratory and in the field. Sprunt (Chapter 15) used scanning X-ray computer tomography to characterize the spatial and temporal distribution of fluid during saturation. If saturation is not uniform, then complications may arise in the use of well-log data to infer hydrocarbon saturation. Toksöz *et al.* (Chapter 16) used vertical seismic profiling data of tube waves to characterize the geometric and hydraulic attributes of fractures which intersect a borehole. They demonstrate the power of the technique by applying it to two localities in New York and Massachusetts where borehole televiewer and core data are available for comparison.

The effect of hydrostatic and non hydrostatic stresses on remanent magnetization in Ralston diabase was investigated by Martin (Chapter 13) using an unique triaxial rig. He concluded that the decrease in intensity of magnetic remanence was independent of the relative orientation of the magnetic vector and the stress applied. The correlation between fracture toughness anisotropy and surface morphology of the Ithaca siltstone was investigated by Scott *et al.* (Chapter 14) using two different techniques (short bar and indentation). The fracture toughness increases with increasing fractal dimension. In a related study of fluid driven joint formation in the Appalachian Basin, Lacazette and Engelder (Chapter 12) consider compressibility-limited, flow-limited or infiltration-limited joint propagation. Their field observations agree best with the first conceptual model.

The last five papers in the volume discuss the application of rock mechanics data to large scale tectonics. Bombolakis (Chapter 17) considers thrust-sheet emplacement in a foreland belt. He develops a spring-slider model to analyze the role of elastic stiffness and fault damping on the thrusting process. In many different regions, oblique convergence seems to be partitioned into elements with nearly pure strike-slip or thrust-slip components – faults along the San Andreas system being prime examples. Molnar (Chapter 18) points out that the Brace–Goetze strength profile implies that the upper mantle is the strongest region, and that brittle deformation and crustal block motion are coupled to ductile flow in the upper mantile, rather than decoupled. Oblique-slip faults are likely to be unstable under finite deformation dictated by the vorticity of the underlying viscous substratum, and thus, partitioning of oblique-slip into pure strike-slip and thrust-slip should be preferred. In the brittle portion of crust, shear stress is limited by frictional strength in accordance with Byerlee's law. If the pore pressure is nearly hydrostatic, then shear heating along seismogenic systems such as the San Andreas, should induce a measurable heat flow anomaly. But, extensive heat flow measurements in the past two decades show no indication of any anomaly. In the symposium, Stephen H. Hickman presented a comprehensive summary of 'Field evidence for and against a weak San Andreas fault' subsequently published elsewhere.

Two groundbreaking ideas were orally proposed at the Brace symposium to resolve the apparent paradox between field measurements of heat flow and stress and laboratory friction data. In Chapter 20, Rice emphasizes two fundamental mechanical attributes of the San Andreas Fault which must be incorporated into any physical model: the fault is weak in an absolute sense in that it slips under shear stresses far smaller than inferred by Byerlee's law – assuming hydrostatic pore pressure – and it is also weak in a relative sense, in that the adjoining crust seems to be mechanically stronger – assuming loading with compression close to the fault-normal. Rice's unified hydromechanical model is probably the first which agrees with both the laboratory friction data and the field observations of *in situ* stress and heat-flow. His paper clarifies previous misconceptions on stress continuity along slipping fault zones, and spells out the minimal assumptions of the fluid source and transport properties necessary for the generation and maintenance of nearly lithostatic pore pressure along a mature fault zones. Concurrently with Rice's presentation, James D. Byerlee also presented a model for 'Friction, overpressure and fault-normal compression' at the Brace symposium, but which was published elsewhere. Byerlee appeals to non-Darcian behavior in granular gouge material to maintain nearly lithostatic pore pressure under an anomalous low pressure gradient. While either Rice's or Byerlee's model is capable of explaining many of the geophysical and geological observations, the extent to which they apply to a particular fault must be verified by further seismologic and hydrogeologic studies, some of which are well in progress.

The migration of fluids in the Earth has an enormous impact on tectonic processes. Conversely, as Brace's work profoundly illustrates, mechanical processes can influence the permeability and transport properties of the rock mass. Nur and Walder (Chapter 19) consider quantitative models of fluid transport in tectonically active regions. They argue that the ubiquitous evidence for high pore pressure at depth can be reconciled with the suggestions of high permeability at depth if cycles of pulsed fluid flow are followed by periods in which porosity and permeability are reduced. The modeling suggests that such cycles might be 10^3–10^5 years long. In the last chapter, Scholz (Chapter 21) traces the formulation and development of the stick-slip theory of earthquakes since Brace and Byerlee (1966) proposed the paradigm more than 25 years ago. Using a standard methodology from history of science, Scholz contrasts the responses to and interpretation of this paradigm in the rock mechanics and seismological communities. He contends that rock friction research has evolved to the point where a relatively complete and self-consistent description of the earthquake cycle is now possible. However, these advances in rock mechanics may still be somewhat esoteric to the average observation seismologist.

Bibliography: William F. Brace

Brace, W. F. (1953). *Ph.D. Dissertation, Dept. Geology and Geophysics: Rock Deformation in the Rutland VT Area*, M.I.T., Cambridge, MA. 164 pp.

Brace, W. F. (1953). *Geology of the Rutland Vermont Geol. Surv. Bull. #6*, Vermont Development Commission, Montpelier, VT. 124 pp.

Brace, W. F. (1954). Die Einregelung von (100) der Hellglimmer in alpinen Tektoniten. *Znz. math.-naturw. Klasse Oest. Akad. Wiss., Jahrgang 1954*, **9**, 79–85.

Brace, W. F. (1955). Quartzite pebble deformation in Central Vermont. *Am. J. Sci.* **253**, 129–45.

Boucot, A. J., Brace, W. F., and Demar, R. (1958). Distribution of brachiopod and pelecypod shells by currents. *J. Sed. Petrol.* **28**, 321–332.

Brace, W. F. (1958). Interaction of basement and mantle during folding near Rutland, Vermont. *Am. J. Sci.* **256**, 241–256.

Brace, W. F. (1960). Behavior of rock salt, limestone and anhydrite during indentation. *J. Geophys. Res.* **65**, 1773–1788.

Brace, W. F. (1960). Analysis of large two-dimensional strain in deformed rocks. In *Rept. Int. Geolog. Congr. 21st Session, Norden, 1960, Part XVIII. Structure of the Earth's Crust and Deformation of Rocks*, pp. 261–269. International Geological Congress, Copenhagen.

Brace, W. F. (1960). An extension of the Griffith theory of fracture to rocks. *J. Geophys. Res.* **65**, 3477–2480.

Brace, W. F. (1960). Orientation of anisotropic minerals in a stress field: Discussion. In *Rock Deformation, G. S. A. Memoir 79* (ed. D. T. Griggs and J. W. Handin), pp. 9–20. Geological Society of America, New York, NY.

Dulaney, E. N. and Brace, W. F. (1960). The velocity behavior of a growing crack. *J. Appl. Phys.* **31**, 2233–2236.

Brace, W. F. (1961). *Experimental Study of the Indentation of Rocks and Minerals*, MIT, Cambridge, Mass. 81 pp.

Brace, W. F. (1961). Mohr construction in the analysis of large geologic strain. *Bull Geol. Soc. Am.* **72**, 1059–1080.

Brace, W. F. (1961). Review of The Mechanics of Faulting, with Special Reference to the Fault-plane Work (A Symposium). *Trans. Am. Geophys. Un.* **42**, 39–40.

Brace, W. F. (1961). Dependence of fracture strength of rocks on grain size. *Bull. Penn. State Univ. Mineral Industries Experiment Station*, **76**, 99–103.

Brace, W. F. and Walsh, J. B. (1962). Some direct measurements of the surface energy of quartz and orthoclase. *Am. Mineralogist*, **47**, 1111–1122.

Dulaney, E. N. and Brace, W. F. (1962). Velocity behavior of a growing crack: Comments on a discussion by J.P. Berry. *J. Appl. Phys.* **33**, 227.

Brace, W. F. and Bombolakis, E. G. (1963). A note on brittle crack growth in compression. *J. Geophys. Res.* **68**, 3709–3713.

Brace, W. F. (1963). Behavior of quartz during indentation. *J. Geol.*, **71**, 581–595.

Brace, W. F. (1964). Brittle fracture of rocks. In *State of Stress in the Earth's Crust* (ed. W. R. Judd), pp. 110–178. Elsevier Publishing Co. New York.

Brace, W. F. (1964). Effect of pressure on electric-resistance strain gages. *Experimental Mechanics*, **4**, 212–216.

Brace, W. F. (1964). Indentation hardness of minerals and rocks. *N. Jb. Miner. Mh. Jahrgang*, 257–269.

Walsh, J. B. and Brace, W. F. (1964). A fracture criterion for brittle anisotropic rock. *J. Geophys. Res.* **69**, 3449–3456.

Brace, W. F. (1965). Some new measurements of linear compressibility of rocks. *J. Geophys. Res.* **70**, 391–398.

Brace, W. F. (1965). Relation of elastic properties of rock to fabric. *J. Geophys. Res.* **70**, 5657–5667.

Brace, W. F., Orange, A. S., and Madden, T. R. (1965). The effect of pressure on the electrical resistivity of water-saturated crystalline rocks. *J. Geophys. Res.* **70**, 5669–5678.

Simmons, G. and Brace, W. F. (1965). Comparison between static and dynamic measurement of compressibility of rocks. *J. Geophys. Res.* **70**, 5649–5656.

Walsh, J. B., Brace, W. F., and England, A. W. (1965). Effect of porosity on compressibility of glass. *J. Am. Ceram. Soc.* **48**, 605–608.

Brace, W. F. and Orange, A. S. (1966). Electrical resistivity changes in saturated rock under stress. *Science*, **153**, 1525–1526.

Brace, W. F., Paulding, Jr., B. W., and Scholz, C. (1966). Dilatancy in the fracture of crystalline rocks. *J. Geophys. Res.* **71**, 3939–3954.

Brace, W. F. and Byerlee, J. D. (1966). Stick-slip as a mechanism for earthquakes. *Science*, **153**, 990–992.

Brace, W. F. (1966). Laboratory studies of frictional sliding and of the effect of stress on the electrical resistivity of saturated crystalline rocks. In *Proc. of 2nd USA–Japan Conference on Research Related to Earthquake Prediction* (ed. R. Page), pp. 65–65. Nat. Sci. Found. and Jap. Soc. Promotion Sci. Wash. DC.

Press, F. and Brace, W. F. (1966). Earthquake prediction. *Science*, **152**, 1575–1584.

Walsh, J. B. and Brace, W. F. (1966). Elasticity of rock: A review of recent theoretical studies. *Felsmechanik Ingenieurgeologie*, **4**, 283–297.

Walsh, J. B. and Brace, W. F. (1966). Cracks and pores in rocks. In *Int. Congress Rock Mechanics, Lisbon*, pp. 643–646.

Brace, W. F. and Byerlee, James D. (1967). Recent experimental studies of brittle fracture of rocks. In *Failure and Breakage of Rock, Proc. Eighth Symp. on Rock Mechanics* (ed. C. Fairhurst), pp. 58–81. Am. Inst. Mining, New York, NY.

Brace, W. F. and Orange, A. S. (1968). Electrical resistivity changes in saturated rocks during fracture and frictional sliding. *J. Geophys. Res.* **73**, 1433–1445.

Brace, W. F., Walsh, J. B., and Frangos, W. T. (1968). Permeability of granite under high pressure. *J. Geophys. Res.* **73**, 2225–2236.

Brace, W. F. and Martin III, R. J. (1968). A test of the law of effective stress for crystalline rocks of low porosity. *Int. J. Rock Mech. Min. Sci. Geomech. Abstr.* **5**, 415–426.

Brace, W. F. and Orange, A. S. (1968). Further studies of the effect of pressure on electrical resistivity of rocks. *J. Geophys. Res.* **73**, 5407–5420.

Brace, W. F. (1968). Current laboratory studies pertaining to earthquake prediction. *Tectonophysics*, **6**, 75–87.

Byerlee, J. D. and Brace, W. F. (1968). Stick-slip, stable sliding and earthquakes, Effect of rock type, pressure, strain-rate, and stiffness. *J. Geophys. Res.* **73**, 6031–6037.

Brace, W. F. (1969). The mechanical effects of pore pressure on fracturing of rocks. In *Proc. Conf. Research in Tectonics, Ottawa, March, 1968, GSC Paper 68–52*, pp. 113–124. Geological Soc. Canada.

Brace, W. F., Scholz, C. H., and La Mori, P. N. (1969). Isothermal compressibility of kyanite, andalusite, and sillimanite from synthetic mineral aggregates. *J. Geophys. Res.* **74**, 2089–2098.

Brace, W. F. (1969). Laboratory studies pertaining to earthquakes. *Trans. New York Acad. Sci, Ser. II*, **31**, 892–906.

Byerlee, J. D. and Brace, W. F. (1969). High pressure mechanical instability in rocks. *Science*, **164**, 713–715.

Greenberg, R. J. and Brace, W. F. (1969). Archie's law for rocks modelled by simple networks. *J. Geophys. Res.* **74**, 2099–2102.

Brace, W. F., Ernst, W. G., and Wittels Kallberg, R. (1970). An experimental study of tectonic overpressure in Franciscan rocks. *Geol. Am. Soc. Bull.* **81**, 1325–1338.

Brace, W. F. and Byerlee, J. D. (1970). California earthquakes: Why only shallow focus? *Science*, **168**, 1573–1576.

Brace, W. F. (1971). Micro-mechanics in rock systems. In *Structure, Solid Mechanics, and Engineering Design, Proc. Southhampton Civil Eng. Materials Commission, April 1969* (ed. M. Te'eni), pp. 187–204. John Wiley-Interscience, New York, NY.

Brace, W. F. (1971). Resistivity of saturated crustal rocks to 40 km based on laboratory measurements. In *The Structure and Physical Properties of the Earth's Crust, Geophysical Monograph Series, Vol. 14* (ed. J. Heacock), pp. 243–255. Am. Geophys. Un. Washington, DC.

Brace, W. F. and Jones, A. H. (1971). Comparison of uniaxial deformation in shock and static loading of three rocks. *J. Geophys. Res.* **76**, 4913–4921.

Pratt, H. R., Brown, W. S., and Brace, W. F. (1971). In-situ determination of strength properties in a quartz diorite rock mass. In *Dynamic Rock Mechanics, Proc. 12th Symposium on Rock Mechanics, Rolla, 1970* (ed. G. B. Clark), pp. 27–43. Port City Press, Baltimore.

Wawersik, W. R. and Brace, W. F. (1971). Post-failure behavior of a granite and diabase. *Rock Mechanics*, **3**, 61–85.

Brace, W. F., Silver, E., Hadley, K., and Goetze, C. (1972). Cracks and pores: A closer look. *Science*, **178**, 162–164.

Brace, W. F. (1972). Laboratory studies of stick-slip and their application to earthquakes. *Tectonophysics*, **14**, 189–200.

Brace, W. F. (1972). Pore pressure in geophysics. In *Flow and Fracture of Rocks, the Griggs Volume, Geophys. Mon. Vol. 16* (ed. H. C. Heard, I. Y. Borg, N. L. Carter and C. B. Raleigh), pp. 265–273. Am. Geophys. Un. Washington, DC.

Brace, W. F. and Riley, D. (1972). Static uniaxial deformation of 15 rocks to 30 kb. *Int. J. Rock Mech. Min. Sci.* **9**, 271–288.

Byerlee, J. D. and Brace, W. F. (1972). Fault stability and pore pressure. *Bull. Seismol. Soc. Am.* **62**, 657–660.

Goetze, C. and Brace, W. F. (1972). Laboratory observations of high temperature rheology of rocks. *Tectonophysics*, **13**, 583–600.

Pratt, H. R., Black, A. D., Brown, W. S., and Brace, W. F. (1972). The effect of specimen size on the mechanical properties of unjointed diorite. *Int. J. Rock Mech. Min. Sci. Geomech. Abstr.* **9**, 513–529.

Walsh, J. B. and Brace, W. F. (1972). Elasticity of rock in uniaxial strain. *Int. J. Rock Mech. Min. Sci*, **9**, 7–15.

Stesky, R. M. and Brace, W. F. (1973). Estimation of frictional stress on the San Andreas Fault from laboratory measurements. In *Geological Sciences, Vol. XIIIProc. Conf.*

on Tectonic Problems of the San Andreas Fault System. (ed. R. L. Kovach and A. Nur), pp. 206–214. Stanford, CA.

Stesky, R. M. and Brace, W. F. (1973). Electrical conductivity of serpentinized rocks to 6 kb. *J. Geophys. Res.* **78**, 7618–7621.

Walsh, J. B. and Brace, W. F. (1973). Mechanics of rock deformation. In *ASME Annual Meeting, Proc. Rock Mechanics Symposium, AMD-Vol. 3*, Detroit.

Brace, W. F. (1974). Experimental studies of seismic behavior of rocks under crustal conditions. *Engineering Geol.* **8**, 109–127.

Pratt, H. R., Black, A. D., Brown, W. S., and Brace, W. F. (1974). A new technique for determining the deformation and frictional characteristics of in-situ rock. In *Field Testing and Instrumentation of Rock, ASTM STP 554* (ed. G. B. Clark), pp. 3–19. Am. Soc. for Testing and Materials, Baltimore.

Pratt, H. R., Black, A. D., and Brace, W. F. (1974). Friction and deformation of jointed quartz diorite. In *Advances in Rock Mechanics, Reports of Current Research, Proc. 3rd. Int. Cong. Rock Mechanics, Denver, Vol. II, Part A* (ed. U.S. National Comm. Rock Mechanics), pp. 306–310. Nat. Acad. Sci. Washington D.C.

Sprunt, E. and Brace, W. F. (1974). Direct observation of microcavities in crystalline rocks. *Int. J. Rock Mech. Min. Sci.* **11**, 139–150.

Sprunt, E. and Brace, W. F. (1974). Some permanent structural changes in rocks due to pressure and temperature. In *Advances in Rock Mechanics, Reports of Current Research, Proc. 3rd. Int. Cong. Rock-Mechanics, Denver, Vol. II, Part A* (ed. U.S. Nat. Comm. Rock Mechanics), pp. 524–529. Nat. Acad. Sci. Washington D.C.

Stesky, R. M., Brace, W. F., Riley, D. K., and Robin, P.-Y. F. (1974). Friction in faulted rock at high temperature and pressure. *Tectonophysics*, **23**, 177–203.

Brace, W. F. (1975). Dilatancy-related electrical resistivity changes in rocks. *Pure Appl. Geophys.* **113**, 207–217.

Brace, W. F. and Dieterich, J. H. (1975). Notes on earthquake related experimental studies in the USSR. *EOS, Transactions Am. Geophys. Un.* **56**, 221–223.

Brace, W. F. (1975). The physical basis for earthquake prediction. *Technology Review*, **77**, 844–848.

Brace, W. F. (1975). Earthquake research in China, Parts 3 and 9 of Report by members of American Seismological Delegation to China, October, 1974. *EOS, Trans. Am. Geophys. Un.* **56**, 844–848.

Montgomery, C. and Brace, W. F. (1975). Micropores in plagioclase. *Contrib. Mineral. Petrol.* **52**, 17–28.

Myachkin, V. I., Brace, W. F., Sobolev, G. A., and Dieterich, J. H. (1975). Two models for earthquake forerunners. *Pure Appl. Geophys.* **113**, 169–181.

Brace, W. F. (1976). Direct observation of dilatant voids in rock. In *The Effects of Voids on Material Deformation* (ed. S. C. Cowin and M. M. Caroll), pp. 1–12. Am. Soc. Mech. Engrs. New York, NY.

Tapponnier, P. and Brace, W. F. (1976). Development of stress-induced microcracks in Westerly granite. *Int. J. Rock Mech. Min. Sci. Geomech. Abstr.* **13**, 103–112.

Brace, W. F. (1977). Recent laboratory studies of earthquake mechanics and prediction. *J. Phys. Earth*, **25** (Suppl.), 185–202.

Brace, W. F. (1977). Permeability from resistivity and pore shape. *J. Geophys. Res.* **82**, 3343–3349.

Pratt, H. R., Swolfs, H. S., Brace, W. F., Black, A. D., and Handin, J. W. (1977). Elastic and transport properties of an in-situ jointed granite. *Int. J. Rock Mech. Min. Sci. Geomech. Abstr.* **14**, 35–45.

Brace, W. F. (1978). Volume changes during fracture and frictional sliding: A review. *Pure Appl. Geophys.* **116**, 603–614.

Brace, W. F. (1978). Note on permeability changes in geologic material due to stress. *Pure Appl. Geophys.* **116**, 627–633.

Brace, W. F. (1979). Measured crustal stress compared with laboratory strength of rocks. In *Proc. Res. Conf. on Intra-continental Earthquakes, Ohrid, Yugoslavia, Sept. 17–21, 1979*, (ed. J. Petrovski and C. R. Allen).

Wong, T.-f. and Brace, W. F. (1979). Thermal expansion of rocks: Some measurements at high pressure. *Tectonophysics*, **57**, 95–117.

Brace, W. F. (1980). Permeability of crystalline and argillaceous rocks. *Int. J. Rock Mech. Min. Sci. Geomech. Abstr.* **17**, 241–251.

Brace, W. F. and Kohlstedt, D. L. (1980). Limits on lithospheric stress imposed by laboratory experiments. *J. Geophys. Res.* **85**, 6248–6252.

Evans, B., Rowan, M., and Brace, W. F. (1980). Grain-size sensitive deformation of a stretched conglomerate from Plymouth, Vermont. *J. Struct. Geol.* **2**, 411–424.

Brace, W. F. (1981). The effect of size on mechanical properties of rock. *Geophys. Res. Lett.* **8**, 651–652.

Brace, W. F. (1981). A note on downhole detection of conductive fractures. In *Proc. Workshop XVII, Hydraulic Fracturing Stress Measurements, Monterrey, Calif. Dec. 2–5, 1981, USGS Open File Report 82-1075* (ed. M. D. Zoback and B. C. Haimson), pp. 624–634. U.S. Geolog. Surv. Menlo Park, CA.

Morrow, C., and Brace, W. F. (1981). Electrical resistivity changes in tuffs due to stress. *J. Geophys. Res.* **86**, 2929–2934.

Swolfs, H. S., Brechtel, C. E., Brace, W. F., and Pratt, H. R. (1981). Field-mechanical properties of a jointed sandstone. In *Mechanical Behavior of Crustal Rocks, The Handin Volume, Geophysics Monograph, Vol. 24* (ed. N. L. Carter, M. Friedman, J. M. Logan and D. W. Stearns), pp. 161–172. Am. Geophys. Union, Washington, DC.

Bernabé, Y., Brace, W. F., and Evans, B. (1982). Permeability, porosity and pore geometry of hot-pressed calcite. *Mech. Mater.* **1**, 173–183.

Olgaard, D. and Brace, W. F. (1983). The microstructure of gouge from a mining-induced seismic shear zone. *Int. J. Rock Mech. Min. Sci. Geomech. Abstr.* **20**, 11–19.

Voegele, M. D. and Brace, W. F. (1985). Measurement of permeability at elevated stresses and temperatures. In *Measurement of Rock Properties at Elevated Pressures and Temperatures ASTM STP 869* (ed. H. J. Pincus and E. R. Hoskins), pp. 3–23. Am. Soc. Test. Mater., Philadelphia, PA.

Brace, W. F. (1984). Permeability of crystalline rocks: New in-situ measurements. *J. Geophys. Res.* **89**, 4327–30.

Walsh, J. B. and Brace, W. F. (1984). The effect of pressure on porosity and the transport properties of rock. *J. Geophys. Res.* **89**, 9425–31.

Brace, W. F. (1987). Elasticity and rigidity of rock. In *Encyclopedia of Structural Geology and Plate Tectonics* (ed. C. K. Seyfert), pp. 208–213. Van Nostrand Reinhold Pub. New York.

Bernabé, Y. and Brace, W. F. (1990). Deformation and fracture of Berea Sandstone. In *The Brittle-Ductile Transition in Rocks, The Heard Volume, Geophysical Monograph 56* (ed. A. G. Duba, W. B. Durham, J. W. Handin and H. F. Wang), pp. 91–101. Am. Geophys. Union, Washington, DC.

Acknowledgments

That the conference held in Bill's honor was a resounding success was due to the generosity and hard work of the participants. The papers presented were uniformly outstanding; in fact, one contributor complained that the talks were so good and so continuous that he was unable to slip out to the men's room during the proceedings.

It is not surprising that much of the administrative work for the conference was handled by Madge Slavin. Madge was Bill's administrative assistant for 15 years. In addition to her capable and efficient administrative help, she is a practicing artist and contributed artwork for the cover design of the book and for advertising for the conference. The same year that Madge came to work for Bill, a young machinist, Derek Hirst, was added to the technical support staff. Since much of the equipment used in Bill's research was of home design. Hirst was intensely involved in the design, fabrication, and maintenance of the equipment. He still continues to turn illegible sketches from students and professors into working pieces of apparatus, and the machinery that Bill designed, though aging, still provides useful facilities for state of the art experiments.

Many others helped in staging the conference and in preparing the book. Renata Dmowska encouraged us to commit the proceedings to print, and gave useful advice during the planning of the book. Peter Molnar helped in organizing the conference and in introducing speakers. Georg Dresen, Joanne Fredrich, Dave Olgaard and Lisa dell'Angelo pitched in to deal with numerous last minute details and emergencies; Georg organized the fun run held after the conference. Paolo Harabagilio and Pierre Ihmli tended the registration desk and helped with audiovisual details. Teresa Hernandez and the late David Scott provided a great deal of support during the production of the book; Teresa, in particular, was cheerful and efficient throughout the editing process. Beverly Kozol-Tattlebaum gave much aid in organizing the reception and banquet. Tom Jordan, the head of Earth, Atmospheric and Planetary Sciences, strongly supported the conference with both effort and money from department headquarters. Some of the material for the biography was drawn from a two volume history of M.I.T. by Robert Shrock; Dick Lindzen took the photograph of Bill, and his wife, Nadine, was able to recover it.

Finally, we list below the reviewers of the papers submitted to the conference. Good reviewing is an unsung, but time-consuming task, and there is no doubt that the thoughtful and constructive comments of the folks listed below improved the scientific quality and the clarity of this volume. Thanks to all of you: Michael Batzle, Yves Bernabé, James Berryman, Mike Blanpied, E. G. Bombolakis, Brian Bonner, Clark Burchfiel, James D. Byerlee, Fred Chester, Anthony Dahlen, Christian David, Dan M. Davis, Georg Dresen, Jeremy Dunning, William B. Durham, Terry Engelder, Joanne Fredrich, Kate Hadley, Thomas Henyey, Stephen H. Hickman, Malcolm Johnson, Stephen H. Kirby, David Kohlstedt, Robert Kranz, David Lockner, John Logan, T. R. Madden, Randolph Martin, Peter Molnar, Diane Moore, Dale Morgan, Carolyn Morrow, Amos Nur, Jason Phipps-Morgan, Neil Ribe, James R. Rice, Pierre-Yves F. Robin, Dan Rothman, Leigh Royden, John Rudnicki, Ernest H. Rutter, Charles Sammis, T. Satoh, Chris H. Scholz, Norman Sleep, Eve Sprunt, Lisa Stewart, Larry Teufel, Nafi M. Toksöz, Jan Tullis and Chi-yuen Wang.

Brittle Failure of Rocks

Chapter 1

Observations of Quasistatic Fault Growth from Acoustic Emissions

D.A. Lockner[1], J.D. Byerlee[1], V. Kuksenko[2], A. Ponomarev[3], and A. Sidorin[3]

[1] *U.S. Geological Survey, 345 Middlefield Rd., Menlo Park, CA 94025, U.S.A.*

[2] *A.F. Ioffe Physical-Technical Institute, Russian Academy of Sciences, Polytechnicheska 26, St. Petersburg, Russia 194021.*

[3] *Institute of Physics of the Earth, Russian Academy of Sciences, Bolshaya Gruzinskay 10, Moscow D-242, Russia 123810.*

Abstract

The location of acoustic emission (AE) sources during deformation of rock has proven to be a useful non-destructive analytic technique. We present experimental results,based on AE observations, that show the nucleation and growth of macroscopic fault planes in granite and sandstone samples. By controlling axial stress to maintain constant AE rate rather than more conventional loading conditions such as constant strain rate or constant stress, we have been successful in retarding the failure process in triaxial experiments. As a result, the post-failure stress curve has been followed quasistatically, extending to minutes or hours the fault growth process that normally would occur violently in a fraction of a second. In addition to this novel control system, three-dimensional locations of AE events were determined by analyzing the relative arrival times of AE pulses recorded on a network of transducers attached to the sample. In this manner, as many as 40 000 AE events were located in the course of a single experiment.

While the details of fault formation varied from experiment to experiment, a number of features were consistently observed. In all three granite experiments, the fault plane nucleated abruptly at a point on the sample surface soon after reaching peak stress. Prior to fault nucleation, microcrack growth was distributed evenly throughout the granite samples. From the nucleation site, the fault plane grew across the sample, accompanied by a gradual drop in axial stress. AE locations showed that the fault propagated as a fracture front (process zone) with in-plane dimension of 10–50 mm. As the fracture front passed, the AE from a given region would drop to a low level. If allowed to progress to completion, stress eventually dropped to the frictional sliding strength. Sandstone samples showed somewhat

FAULT MECHANICS AND TRANSPORT PROPERTIES OF ROCKS
ISBN 0-12-243780-2

different response. In these experiments, a diffuse damage zone appeared prior to peak strength and gradually localized into an incipient fault plane. After passing through peak stress, this plane grew, as in the granite samples, to eventually bisect the sample.

1. Introduction

It is well known that brittle rocks, when loaded in compression, will commonly fail by the development of a large-scale fault. The growth of such a fault results in an overall weakening of the rock and is usually accompanied by a violent release of energy (earthquake). It is surprising that, while this mode of failure has been studied in detail for many years (Jaeger and Cook, 1984), the details of how faults nucleate and grow have remained an enigma for so long. This situation is due in part to the tendency for faults to grow at near shear wave velocity, making the detailed observation of fault propagation technically difficult. It is also due to the lack of an adequate theory of how faults can grow in-plane.

In the past, experiments designed to investigate the fracture development or 'post failure' stage in the brittle failure process have resorted to 'stiff' loading frames (Wawersik and Fairhurst, 1970; Wawersik and Brace, 1971; Wong, 1982a,b). By reducing the amount of elastic energy available to the sample during fault formation, this approach can greatly increase stability. Even so, many crystalline rocks such as Westerly granite store sufficient energy in the sample itself to drive unstable fault growth. In this case, the loading system must be fast enough to actively remove energy from the system as the fault grows. We present here an experimental technique in which the rate of acoustic emissions (AE) occurring in the sample is used to control axial stress. This system, similar to one employed by Terada et al. (1984), has proven capable of arresting unstable fault growth so that it can be observed under quasistatic growth conditions. Then, by recording the relative arrival times of AE signals at transducers mounted on the sample, we are able to locate the microcrack damage that occurs during the faulting process. In this manner, information concerning the complete nucleation and growth history is obtained.

When deformed samples of brittle rock are examined in thin section and with SEM (Hadley, 1975; Tapponnier and Brace, 1976), microcrack damage is found to be dominated by the growth of mode I tensile cracks that are subparallel to the maximum compressive stress direction. On the microscopic scale, mode I crack damage is consistently favored over shear failure modes, indicating that tensile fracture toughness is lower than shear fracture toughness in this class of materials. These observations are supported by elastic crack theory, which predicts that cracks will tend to grow in mode I towards the direction of maximum compressive stress (Lawn and Wilshaw, 1975). Since faulting is consistently observed on the macroscopic scale, considerable effort has gone into bridging the gap between prefaulting microscopic damage and the coalescence of damage needed to form a fault. Models for fault development typically assume that microcracks at first do grow parallel to the direction of maximum compression, and in so doing transfer stress from weakened zones to their surroundings (Madden, 1983). This process continues until the increased crack density ultimately causes cracks to interact, leading to a mechanical instability (e.g., Horii and Nemat-Nasser, 1985; Sammis and Ashby, 1986; Costin, 1987; Lockner and Madden,

1991a,b). Based on the observations presented here, Reches and Lockner (1990) develop a coherent analysis of the process by which microcracks organize themselves to form a fracture plane, while a similar treatment was recently presented by Du and Aydin (1991).

Considerable discussion has been devoted in the literature to the question of when, during the loading cycle of an initially intact rock, the incipient fracture plane can be recognized. Experimentally, researchers have undertaken the laborious task of loading samples to different stages of deformation and then counting microcrack damage in thin section or using SEM techniques (Wawersik, 1973; Hadley, 1975; Tapponnier and Brace, 1976). Using an argillaceous quartzite, Hallbauer et al. (1973) reported the localization of microcrack damage onto an incipient fracture plane as between 95% and 98% of (preceding) peak strength. By means of a holographic technique, Spetzler et al. (1981) reported localized surface dilatancy preceding peak strength by 2% in a pyrophyllite sample. Lockner and Byerlee (1977a) reported that localization of AE sources onto the eventual fault plane occurred in one triaxially loaded Weber sandstone sample at 95% failure strength. However, in a second sandstone sample and two Westerly granite samples, no localization occurred prior to failure. In this paper, we will show examples of faulting in initially intact granite in which nucleation occurs suddenly and after peak strength. In sandstone samples, however, microcrack localization occurs during early loading, presumably as the result of preexisting flaws in the sample. This aspect of the current experiments is analyzed in detail in Lockner and Byerlee (1992a,b) and Lockner et al. (1992). Providing a coherent explanation of these various observations is important if we are to fully understand the brittle fracture process.

An understanding of this fracture process in rock is of obvious importance to our understanding of earthquakes and to our ability to predict earthquakes. This problem can be divided into three classes: long-, intermediate- and short-term predictions. Successful prediction on these time scales has different utility to the public and also requires different types of observations and analysis. To date, long-term predictions have met with the most success by making use of, for example, historical earthquake data, trenching and long-term geodetic strain and tidal data. The success rate for short-term prediction, in which earthquakes are anticipated by a few days or less, has been far worse. This area of research has undergone repeated cycles of optimism and discouragement as various prediction schemes have gained attention and then failed under further testing. A recent example of optimism is the observation of ultra-low-frequency electromagnetic noise prior to the Loma Prieta earthquake of October 1989 (Fraser-Smith et al., 1990).

An important set of borehole strain observations has now been compiled that relates to short-term prediction. Sacks–Everton dilatometers have recorded variations at the nanostrain level or better in the near-field of a number of moderate earthquakes (Johnston et al., 1987; Linde and Johnston, 1989). In no case has precursory strain been observed over a timescale of days to seconds before any of these earthquakes. Owing to the high sensitivity of these instruments, the investigators conclude that any precursory strain that did occur on the eventual rupture plane must have been less than 0.1% of the coseismic strain release. Stated differently (Johnston, 1990), they conclude that short-term precursory strain preceding some of these moderate-sized earthquakes would have been observable if it had occurred on a fault patch larger than about 100 m in dimension. How much smaller the regions that underwent precursory strain actually were remains an open question. This important result has obvious implications for the problem of short-term earthquake prediction. Any plausible short-term prediction scheme that has been proposed can ultimately be related to strain changes in the earthquake source region. Knowledge of the volume of rock that can be expected to deform will guide us in our estimates of the magnitude of the precursory signals that can be detected remotely.

The experiments presented here are conducted on samples in which faults can grow to approximately 0.1 m in dimension. Thus, just as the borehole dilatometer observations can be used to place an upper bound on the nucleation patch size for earthquakes, we anticipate that our laboratory observations will allow us to place a lower bound on the nucleation size. In one sense, the initially intact samples that are deformed in these experiments could be considered an analog of a stuck patch or asperity on a yielding fault plane. Breakage of such an asperity could represent the initiation of an earthquake. While the samples used in the present experiments were chosen to be homogeneous, future experiments are planned to study the effects of preexisting faults as well as samples with strength heterogeneities. In the present experiments we are able to observe not only the nucleation of fractures, but we will also show that, for the first time, we can monitor the growth phase of fractures as they propagate quasistatically in-plane. Through this new technique, it should also be possible to analyze the energetics of fault propagation (Lockner et al., 1991) to determine the conditions in the earth that will lead to unstable earthquake rupture as well as the arresting phase that terminates earthquakes.

2. Experimental

2.1. Sample assembly and loading conditions

Cylindrical samples of Westerly granite and Berea sandstone (18% porosity) were prepared. Sandstone samples were cored perpendicular to bedding. Sample dimensions were 76.2 mm diameter by 190.5 mm length. Ends were ground parallel to ± 0.05 mm. Samples were jacketed in either a 0.38 mm walled copper sleeve or a 3.2 mm thick polyurethane tube (Figure 1). A set of six piezoelectric transducers (resonance at 0.6 MHz) were attached to the sample and used to monitor high-frequency acoustic emission (AE) occurring in the rock as it was deformed. In all experiments a constant confining pressure of 50 ± 0.2 MPa was maintained by computer control. Axial load on the sample was measured with an internal load cell to a precision of ± 0.2 MPa.

A pulse counter was connected to the amplified output of one of the piezoelectric transducers. This counter generated an output voltage that was proportional to the AE rate occurring in the sample. By including this device in the feedback loop for the axial load, experiments could be run at constant AE rate. A similar system was tested by Terada et al. (1984) in which they employed a computer to condition the AE signal instead of the simple analog device used here. At low differential stress, when there was little AE activity, we adjusted the loading system to provide an axial shortening rate between 0.1 and 1.0 $\mu m\ s^{-1}$. As the load increased, this rate would drop, requiring repeated adjustments during the early stages of each experiment. With increasing stress and therefore increasing AE activity, the control system was designed to maintain an approximately constant acoustic emission rate by reducing, or reversing when necessary, the axial strain rate. In this manner, we were able to monitor the growth of the fracture plane under quasistatic conditions; typically extending the fault formation phase so that it would last a matter of minutes or even hours. In the early stages of fault growth, it

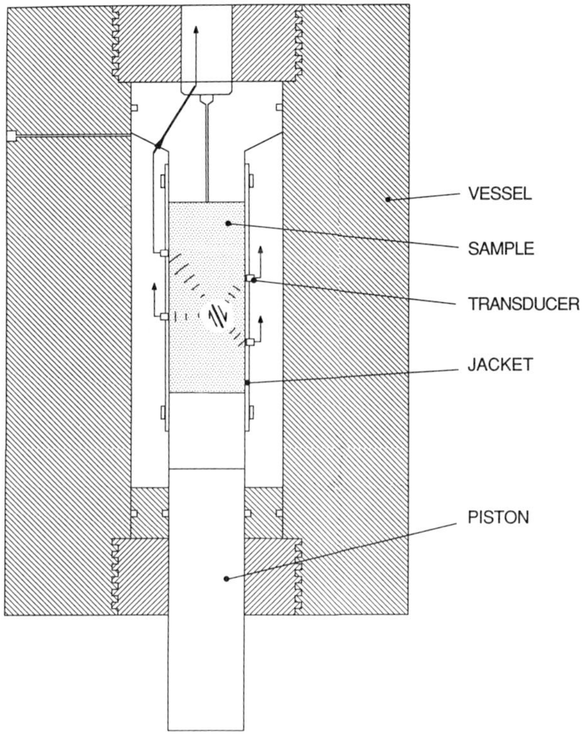

Figure 1. Schematic view of sample mounted in pressure vessel. Six piezoelectric transducers mounted on sample surface detected acoustic emissions during deformation. An additional set of four transducers was used to determine velocity field.

was necessary to drop stress rapidly. To accomplish this, axial load was controlled by a fast-acting hydraulic valve with an overall system response of approximately 0.2 Hz. The hydraulic valve was capable of lowering axial stress at a maximum rate of 6 MPa s^{-1}. This unloading rate was sufficient to prevent unstable fault growth in all experiments.

Axial shortening of the sample column was measured outside the pressure vessel with a DCDT displacement transducer. Accuracy and sensitivity were, respectively, $\pm 5 \times 10^{-3}$ mm and $\pm 0.3 \times 10^{-3}$ mm. In selected experiments circumferential strain was measured with a single 40 mm^2 foil strain gage applied to the sample mid-plane. In addition, axially oriented strain gages were mounted on some of the samples. Gage sensitivity was $\pm 1 \times 10^{-6}$ for short time intervals; however, thermal drift resulted in a long-term stability of $\pm 5 \times 10^{-6}$. In the two experiments (referred to as G1 and S1) that used copper jackets, gages were cemented on the jacket. In all other experiments, gages were cemented to the rock surface, underneath the polyurethane jacket.

2.2. Acoustic emission monitoring system

A set of six piezoelectric transducers were attached to the sample (shown schematically in Figure 1) to detect ultrasonic noise radiated during the sudden growth of microcracks or slip along existing crack surfaces. The transducers, which were 6.4 mm diameter, were mounted in brass holders whose bottom faces were machined to the radius of curvature

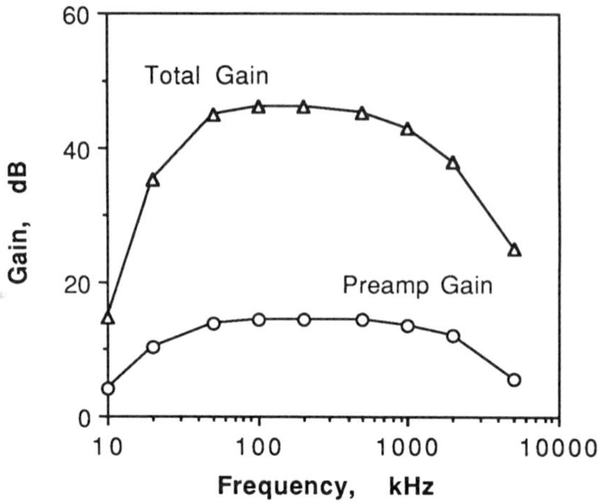

Figure 2. Frequency response of amplifiers used in data acquisition system. Resonant frequency of transducers was 600 kHz.

of the sample and were 2 mm thick. The holders were designed so that transducers were immersed in the silicone oil confining medium. In experiments using copper jackets, the transducer mounts were soldered directly to the jacket. In the remaining experiments, holes were drilled in the polyurethane jackets and transducer mounts were epoxied directly to the sample surface. Transducer signals were brought out of the pressure vessel to a set of wide-band preamplifiers (amplifier response is shown in Figure 2). Signals were amplified by 45 db and sent to a six-channel data acquisition and recording system.

The data acquisition system digitizes the relative arrival time of the AE signal at each of the six transducers, the absolute time of arrival of the signal at the transducer nearest the source, the sign (positive or negative) of the first maximum in the wave train at each transducer, the amplitude of the first maximum in the wave train at each transducer, and the maximum amplitude of the wave train seen by any of the transducers during an entire AE event. Arrival time, as determined by the system, is the time when the input signal first exceeds a preset threshold level and is recorded to ± 0.05 μs accuracy. A discussion of the location accuracy of this acquisition system appears in the appendix. Amplitudes are stored as integer values on a linear scale from 2 to 99 (over-range events are recorded as 99). Calibration of the system showed that this internal scale corresponds to a range, as referred to transducer output, of 1.0–17.8 mV, to provide a dynamic range of 1.25 decades. The acquisition system can buffer data from up to 32 AE events for a short-term repetition rate of 25 kHz. Data are stored for later analysis by writing blocks of 16 events to magnetic tape. Throughput rate for sustained AE activity is limited by the tape transfer rate to 300 events/s. In the experiments presented here, load on the sample was adjusted dynamically to maintain constant AE rate. This rate was set to <100 events/s, so that events would not be lost as a result of system saturation. Electromagnetic noise will occasionally trigger the acquisition system and will typically appear as coincident arrivals on two or more inputs. To eliminate this noise, the acquisition system automatically rejects events for which the apparent shortest travel time is identical for two stations. Additional details of the data acquisition system are given in Byerlee and Lockner (1977).

2.3. Location procedure

The primary use of the AE data is in determining the time and three-dimensional position of microseisms to determine where damage is occurring in the sample. To accomplish this task, we treat the six transducer array as a miniature seismic network and invert the relative arrival time data for hypocentral location and event time. In past experiments (e.g., Lockner and Byerlee, 1977a), we inverted arrival time data for velocity as well as time and position. With only six arrival times and five unknowns, there was little redundancy in that inversion procedure. The situation is actually even more serious owing to P-wave anisotropy that develops as the sample is deformed. At the large differential stresses that occur in these experiments, microcracks tend to open parallel to the maximum compressive principal stress direction (along the sample axis). These oriented cracks produce a severe velocity anisotropy (often exceeding 40%) which varies during the course of an experiment (Lockner et al., 1977; Lockner and Byerlee, 1980). In the present experiments, we added an independent set of four piezoelectric transducers to the sample assembly. By driving one transducer with an externally generated 100 V pulse, and measuring the transit time to the three receiving transducers, we were able to determine the P-wave velocity for declinations from the sample axis of 31°, 50° and 90°. From these measurements we determined the mean velocity field in the sample, assuming radial symmetry and an ellipsoidal field. Velocities were measured in this manner between five and eight times during each experiment to determine how the velocity field varied with time. By supplying the velocity to the inversion routine as a known parameter, we have significantly increased the accuracy of the source location determinations.

In seismic inversions, stations are often located on a plane (earth's surface), leading to a reduced resolution of hypocentral depth. In our case, however, transducers are close to AE sources and completely surround them, providing us with good spatial resolution in all directions and making the inversion routine quite robust. We employ a least-squares technique to estimate hypocentral locations by minimizing the travel time residuals. Since the technique is nonlinear, AE locations are determined iteratively by successively improving the hypocentral estimate. When accurate arrival times are provided, the inversion routine generally requires four to six iterations to come within 0.5 mm of the optimum solution. The procedure is similar to that described in Lockner and Byerlee (1980) with the modification that velocities are not solved for, since they are measured independently during each experiment. We will denote our estimate of the source parameters by $\mathbf{m} = [T_e, X_e, Y_e, Z_e]$, the time and space coordinates of the AE event. Next, parameters are defined that are related to the travel path to the ith transducer. These include

$(x, y, z)_i$	coordinates of ith station
d_i	distance from AE event to ith station
ϕ_i	the declination of the travel path relative to the sample axis
v_i	average P-wave velocity along travel path to ith station
t_i	observed arrival time at ith station
t_i^m	estimated arrival time $= T_e + d_i/v_i$ as predicted by model parameters
R_i	$t_i - t_i^m$, travel time residual.

The velocity field in the sample is assumed to be homogeneous. While this can be achieved during the early portions of the experiment by careful choice of sample, the progressive deformation of the sample invariably leads to velocity field heterogeneity. One problem is the mismatch in elastic modulus between the sample and steel end-plugs. The steel plugs tend to provide lateral support for the ends of the sample, retarding the dilatancy and therefore the velocity anisotropy in these regions. The contrast in modulus between rock and steel also creates zones of high shear stress which may influence the mode of failure in the sandstone experiments. However, in all three granite samples tested, end effects appeared to have little effect on fracture initiation since initiation consistently occurred in the central region of the sample. An additional source of velocity heterogeneity is the development of the fault plane itself. This effect will be most severe for ray paths traveling along the fault plane. Clearly, a detailed knowledge of the spatial velocity variation would be desirable. However, obtaining this information is beyond our present capabilities and, for now, we must be satisfied with the assumption of homogeneity. This assumption should be reasonable for events located in the central region of the sample but will lead to systematic location biases for events occurring near the sample ends. This bias will, however, have little effect on relative locations of events occurring near each other. This is an important distinction, since relative location accuracy is the most important for the present analysis. We make the further assumption that the velocity field can be represented by a radially symmetric ellipsoid. Given the loading symmetry, this is a reasonable first approximation, although the preceding discussion on heterogeneity suggests that there are also errors associated with this assumption. The velocity anisotropy is then expressed as $\zeta = v^{\text{transverse}}/v^{\text{axial}}$, and generally decreases from 1 to less than 0.7 as the sample is loaded. v_i is then expressed as

$$v_i/v^{\text{axial}} = \frac{\zeta}{(\sin^2\phi_i + \zeta^2\cos^2\phi_i)^{1/2}}$$

which is equivalent to

$$\frac{1}{v_i} = \left[\left(\frac{\sin\phi_i}{v^{\text{transverse}}}\right)^2 + \left(\frac{\cos\phi_i}{v^{\text{axial}}}\right)^2\right]^{1/2}$$

The model adjustment vector $\partial\mathbf{m}$ is calculated by a linearized least-squares procedure which minimizes $|\mathbf{R}|^2 = \mathbf{R}^{\mathrm{T}}\cdot\mathbf{R}$ where $\mathbf{R}^{\mathrm{T}} = \mathbf{R}$-transpose. Then the updated model at the $k+1$ step is given by $\mathbf{m}^{k+1} = \mathbf{m}^k + \partial\mathbf{m}^k$. The adjustment vector is determined by solving the normal equations

$$\mathbf{A}\,\partial\mathbf{m} = \mathbf{b}$$

$\mathbf{A}$ and $\mathbf{b}$ are constructed according to the least-squares formulation from

$$\mathbf{A} = \mathbf{P}^{\mathrm{T}}\mathbf{P}$$

and

$$\mathbf{b} = \mathbf{P}^{\mathrm{T}}\mathbf{R}$$

where $\mathbf{P}$ is the matrix of partial derivatives given by $P_{ij} = \partial R_i/\partial m_j$. In expanded form, elements of $\mathbf{P}$ are expressed as

$$\frac{\partial R_i}{\partial m_1} = 1, \qquad \frac{\partial R_i}{\partial m_2} \cong \frac{(X_e - x_i)v_i}{d_i(\zeta v^{\text{axial}})^2}$$

$$\frac{\partial R_i}{\partial m_3} \cong \frac{(Y_e - y_i)v_i}{d_i(\zeta v^{\text{axial}})^2}, \qquad \frac{\partial R_i}{\partial m_4} \cong \frac{(Z_e - z_i)v_i}{d_i(v^{\text{axial}})^2}$$

The $\pm 0.05\ \mu s$ arrival time resolution of the data acquisition system sets a limit of about ± 0.4 mm for the accuracy of determining source locations. However, in practice, other measurement errors further degrade the location accuracies. The primary source of error is a tendency to systematically pick arrivals late for low-amplitude emergent waveforms (Lockner and Byerlee, 1978; also see discussion in appendix), although large-amplitude events are much less affected by this problem. As a result, we estimate that location precision for large-amplitude AE events used in the present analysis is ± 3 mm. This estimate is confirmed by the tightness of clusters of AE event locations from events occurring on fault planes during the experiments. As mentioned earlier, errors in determining the true velocity field will result in a systematic spatial bias in hypocentral determinations. However, this bias will not have a significant effect on the relative locations of neighboring AE events.

3. Results

Differential stress ($\sigma_d = \sigma_{\text{axial}} - P_{\text{conf}}$) and strain components are plotted in Figure 3 for granite samples G1, G2, G3 and Figure 4 for sandstone samples S1 and S2. Horizontal axis in all plots is the calculated axial shortening of the sample, dz. Since remote axial displacement dz_r is measured outside the pressure vessel, it includes both the sample shortening and an additional component of elastic shortening of the steel piston. Thus the sample shortening is computed from

$$dz = dz_r - \frac{\sigma_d}{k}$$

where $k = 205 \pm 5$ MPa mm^{-1} for the sample assembly used in these experiments. In all three granite samples, fault nucleation was accompanied by an abrupt reversal in slope of the stress–displacement curve soon after peak stress. As the AE locations will show, the post-peak weakening curve corresponds to propagation of the macroscopic fault plane across the sample. In experiments G1, S1 and S2 (Figures 3a, 4a,b), the fracture surface cuts across the sample, just grazing the steel endplug. Once the fault plane has completely bisected the sample in these experiments, further deformation involves slip along the entire fault surface at $\sigma_d \sim 150$ MPa or a coefficient of friction of approximately 0.75. In experiments G2 and G3, however, the fracture plane initially formed at an angle of 20–25° and rotated, as it grew, to form an even smaller angle with the sample axis. As a result, in these two experiments the fault intersected the steel endplug, preventing complete loss of strength. In experiment G1 (Figure 3a), the shallow ramp halfway down the weakening curve occurred when the displacement control system was briefly turned off to conduct a velocity measurement. The sample underwent accelerating creep during this period. In all five experiments, the post-peak weakening curve has positive slope and indicates that all of these samples stored sufficient elastic energy to form a through-going fracture surface and would have fractured unstably during a conventional deformation experiment.

Details of the peak strength regions of experiments G1, G3 and S2 are shown in Figure 5 (plotted at 1 s intervals). The episodic stress drops and recoveries are the result of short bursts of AE activity. Since the displacement feedback system attempts to maintain constant AE rate, sudden acoustic bursts result in a rapid unloading of the system. In experiment G1, at least two episodes of incipient fault nucleation occurred

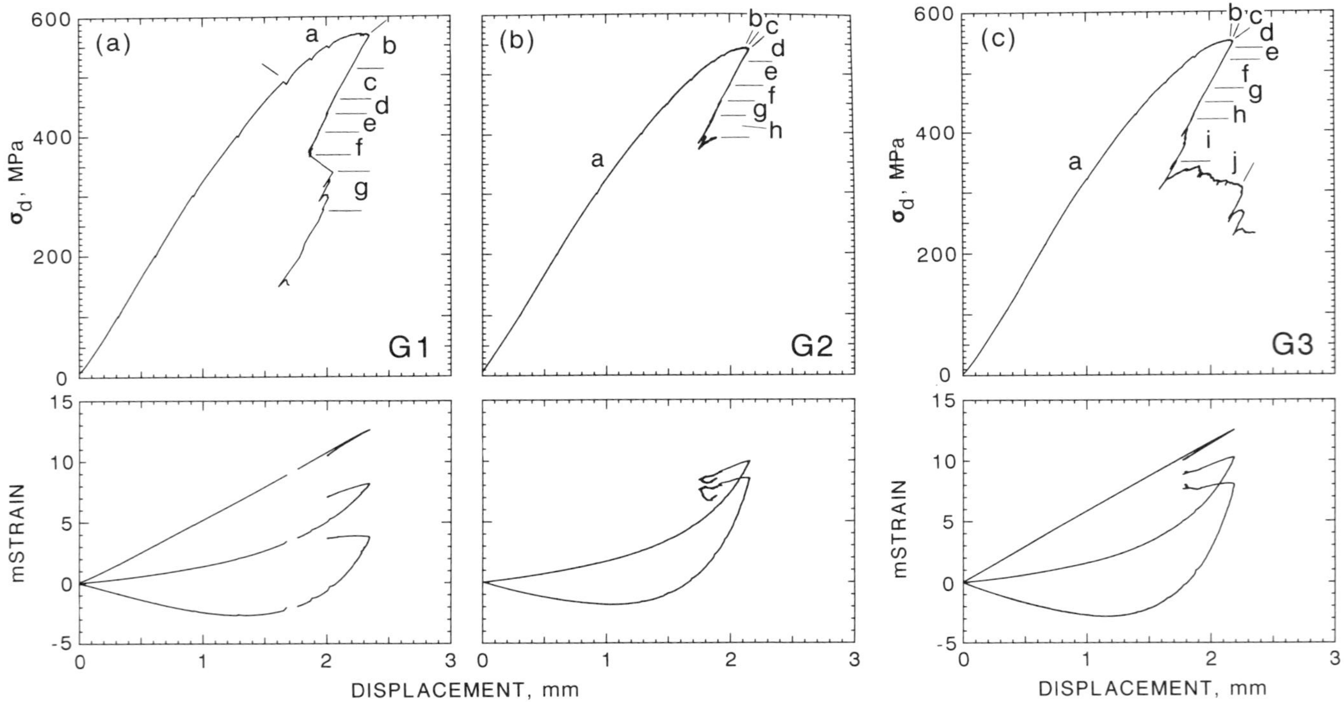

Figure 3. Differential stress and strain components plotted as a function of axial shortening of sample (corrected for elastic shortening of loading system). (a) Westerly granite sample G1; (b) granite sample G2; (c) granite sample G3. All experiments conducted at 50 MPa confining pressure. Regions indicated on stress curves correspond to AE location plots in Figures 7–9.

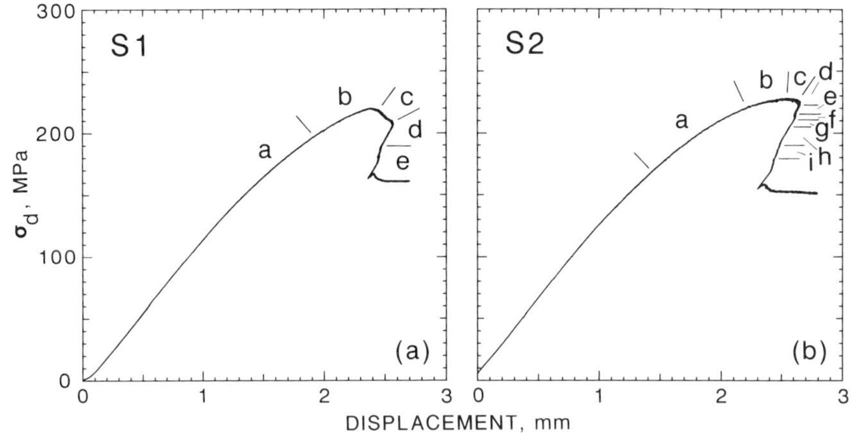

Figure 4. Differential stress plotted as a function of axial shortening of sample (corrected for elastic shortening of loading system). (a) Berea sandstone sample S1; (b) sandstone sample S2. $P_{conf} = 50$ MPa. Regions indicated on stress curves correspond to AE location plots in Figures 10 and 11.

that apparently were arrested before they could go unstable. Similar behavior can be seen for sample G3 which was conducted at a lower AE rate than G1. The sandstone samples also showed short-term stress variations due to bursts of AE events. However, as shown in Figure 5c, the character of the resulting stress–displacement curve differed from that of the granite runs. This difference is mainly the result of the larger number of AE events that occurred in the peak strength region of the sandstone experiments (see Table 1). Between 10 and 30 times as many events were recorded in this region in the sandstone experiments as in the granite. As a result the peak stress region was traversed more slowly in the sandstone experiments. The strength loss that occurred just after the reversal in slope of the stress–displacement curve was also more gradual in the sandstone experiments.

In Figure 6 we show two examples of the P-wave velocity structure used in the hypocentral location program. Solid symbols are measured velocities and open symbols are the inferred axial and transverse velocities. Both the granite and sandstone show an increase in v^{axial} with initial loading. With the onset of dilatancy about mid-way through the loading curve, v^{axial} begins to decrease in the granite but not in the sandstone. Following the onset of dilatancy, v^{trans} drops dramatically in both the granite and sandstone, leading to a pronounced velocity anisotropy by the time of fracture initiation. The granite data are in good agreement with earlier determinations of velocity variations due to deviatoric stress (Lockner et al., 1977). By the end of the fault formation process and the accompanying partial unloading of the sample, both v^{axial} and v^{trans} have decreased, although significant anisotropy remains. Return to isotropic loading results in a permanent decrease in P velocity of 10–15% as well as a dramatic loss of anisotropy for both granite and sandstone.

3.1. Fault formation in granite

In Figure 7 we present the locations of the largest-amplitude AE events recorded during experiment G1. Each set of events is plotted from three perspectives. In the bottom

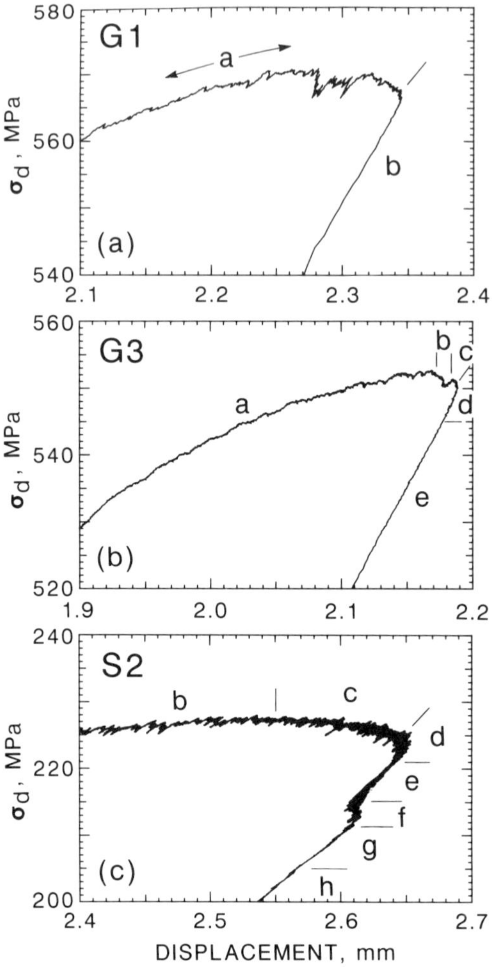

Figure 5. Detailed stress–displacement plots of peak stress regions in (a) exp. G1; (b) exp. G3; (c) exp. S2. Regions indicated on stress curves correspond to AE location plots in Figures 7, 9 and 11.

projection of each set the sample is viewed along strike, while in the middle plot the sample has been rotated 90° counterclockwise on its axis to view the fault plane face-on. In the upper plot, the same events are viewed looking down on the sample. Each dot represents one AE event. A summary of the data presented in Figures 7–11 is presented in Table 1. For reference, the projection of the surface trace of the final fault plane is shown in Figure 7a. Since the faults that developed in all experiments were curved to varying degrees, the surface traces drawn in this and subsequent plots can only be used as an approximate indication of the three-dimensional faults. The individual segments of experiment G1 shown in Figure 7a–g are indicated on the stress–displacement curve shown in Figures 3a and 5a.

AE events recorded during the dilatant stage through peak stress in sample G1 are plotted in Figure 7a. Note that events are distributed throughout the central portion of the sample, reflecting the uniform dilatancy occurring during this portion of the

Table 1. Summary of AE data

Sample	Figure	Number of events	Start of plot[a] interval (s)	Plot interval (s)	Initial faulting angle [deg]
G1	1.7a	867	−18 450	18 450	18 ± 3
	1.7b	205	0	100	
	1.7c	443	100	200	
	1.7d	965	300	500	
	1.7e	1 360	800	500	
	1.7f	618	1 300	1 000	
	1.7g	2 103	2 300	600	
		6 561			
G2	1.8a	235	−7 900	7 700	25 ± 3
	1.8b	16	−200	100	
	1.8c	20	−100	100	
	1.8d	72	0	100	
	1.8e	383	100	200	
	1.8f	412	300	400	
	1.8g	587	700	800	
	1.8h	2 565	1 500	9 000	
		4 290			
G3	1.9a	403	−10 800	10 600	22 ± 3
	1.9b	35	−200	100	
	1.9c	36	−100	100	
	1.9d	50	0	100	
	1.9e	159	100	100	
	1.9f	695	200	200	
	1.9g	786	400	1 000	
	1.9h	1 691	1 400	2 000	
	1.9i	5 102	3 400	4 000	
	1.9j	10 000	7 400	10 000	
		18 957			
S1	1.10a	1 529	−29 300	14 100	35 ± 3
	1.10b	3 808	−15 200	7 900	
	1.10c	7 574	−7 300	7 300	
	1.10d	1 790	0	2 300	
	1.10e	6 559	2 300	8 300	
		21 260			
S2	1.11a	830	−53 000	10 000	33 ± 3
	1.11b	2 431	−43 000	8 000	
	1.11c	3 592	−35 000	10 000	
	1.11d	2 733	−25 000	10 000	
	1.11e	3 020	−15 000	10 000	
	1.11f	2 934	−5 000	5 000	
	1.11g	1 992	0	1 000	
	1.11h	2 133	1 000	2 000	
	1.11i	3 534	3 000	3 000	
		23 199			

[a] Relative to time of fault nucleation as indicated by reversal of slope in stress-displacement plot. This time would correspond to the onset of tertiary creep in a constant-stress experiment.

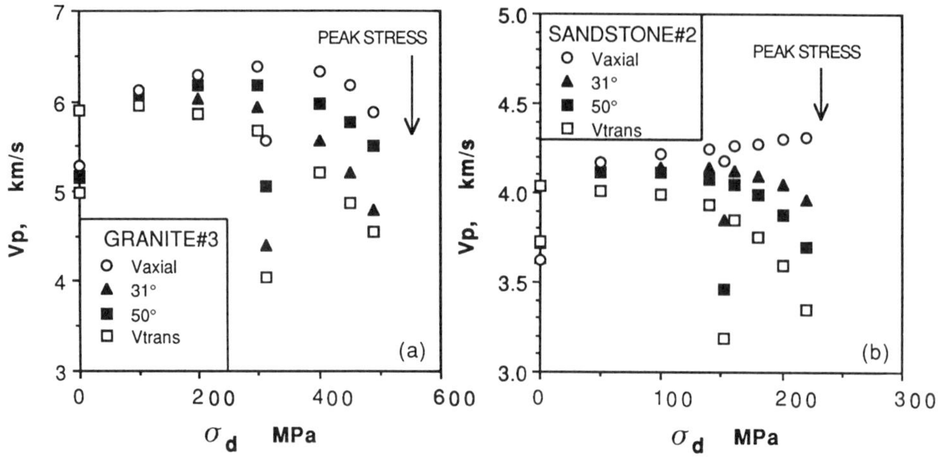

Figure 6. Average P-wave velocity structure used in hypocentral inversion routine for (a) sample G3 and (b) sample S2 throughout loading histories. Solid symbols represent measured velocities; open symbols represent calculated fast- and slow-direction velocities.

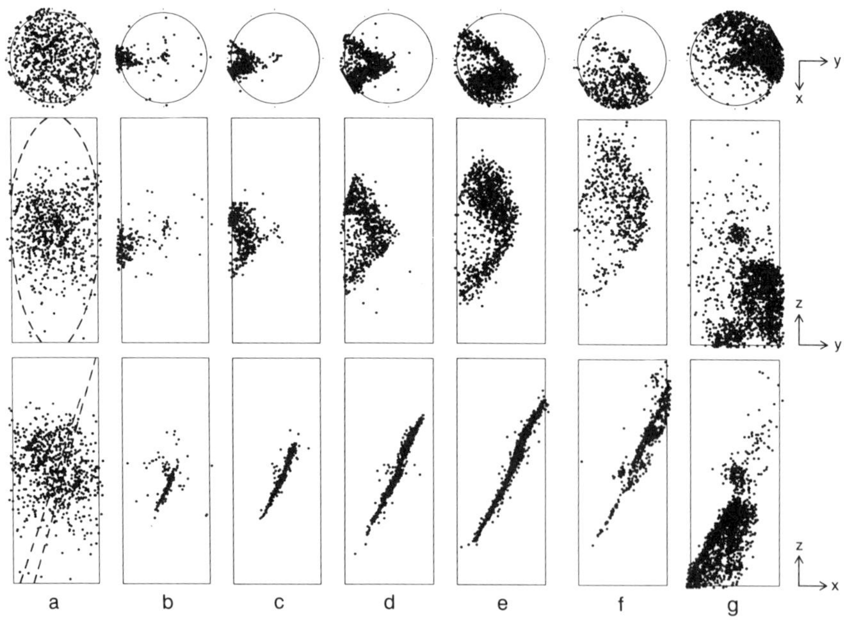

Figure 7. Plots of AE locations for sample G1. Bottom plot of each set is view looking along-strike of eventual fracture plane. Middle plot is face-on view of fault in which sample has been rotated counterclockwise 90°. Top plot is view looking down on sample. Projections of surface trace of eventual fault plane are shown in (a) for reference. Each point represents one AE event. Associated statistics are given in Table 1. Stress intervals for each plot are indicated in Figures 3a and 5a. Fault nucleation occurs in plot (b).

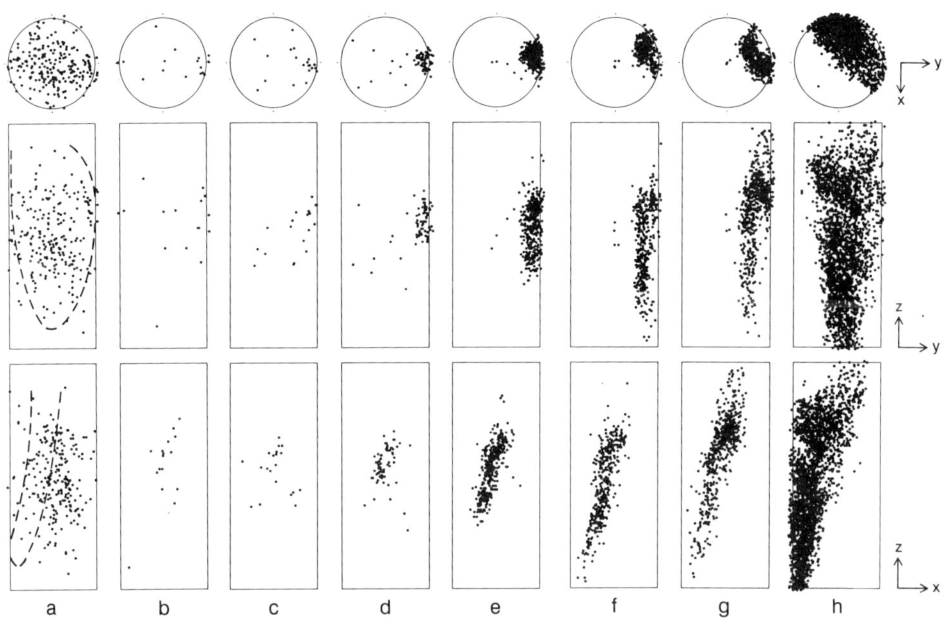

Figure 8. Plots of AE locations for sample G2 (explanation is similar to that for Figure 7). Stress intervals for each plot are indicated in Figure 3b. Fault nucleation occurs in plot (d). Owing to equipment problems, this experiment gave the least accurate locations.

experiment. The acoustic emission pattern shifts dramatically in Figure 7b in which AE events that occurred during the 100 s following fault nucleation are plotted. While a few events are still occurring throughout the volume of the sample, most events are now clustered near the surface of the sample in the nucleation zone. In the next 200 s (Figure 7c), distributed AE events have all but ceased and a distinct fracture front, or process zone, has developed that is growing out from the nucleation site. The fracture tends to grow upward and across the sample more easily than downward (Figure 7d,e,f), possibly reflecting the fact that it has started to graze the bottom endplug. A segment of data was lost following Figure 7f while we measured acoustic velocity in the sample (shown as shallow-sloping ramp in Figure 3a). The final AE events recorded (Figure 7g) show growth of the lower right corner of the fault plane, as well as numerous events off the fault due to the interference with the endplug. While the AE locations in the central region of the sample coincide with the observed fault plane, events near the ends of the sample show a systematic bias away from the observed fault. As discussed in the previous section, we attribute this discrepancy to errors in the velocity structure supplied to the inversion program. A 5% adjustment in the velocity structure will rotate the locus of AE events to agree with the observed fault surface. We are currently exploring ways of improving our velocity determinations to correct this problem.

Acoustic emission locations from experiment G2 are plotted in Figure 8 and referenced to the stress history in Figure 3b. A relatively complicated, steeply dipping fault formed in this experiment (seen from fault trace in Figure 8a). Similarly to experiment G1, events occurring through peak stress (Figure 8a,b) were distributed throughout the central region of the sample. However, AE activity in this case shifted to the fault nucleation region over an approximately 100s interval (Figure 8c). Owing to equipment

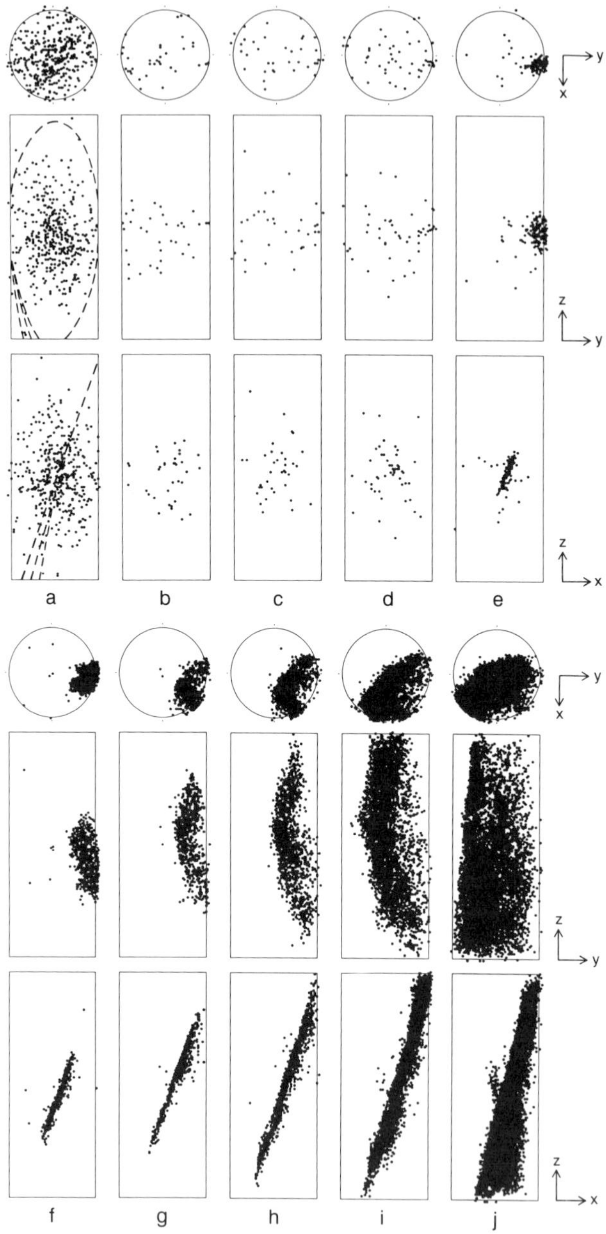

Figure 9. Plots of AE locations for sample G3 (explanation is similar to that for Figure 7). Stress intervals for each plot are indicated in Figures 3c and 5b. Fault nucleation occurs in plot (d).

problems, event locations are not as accurate in this experiment as in the other granite runs, although the fault plane is again seen to nucleate at the sample surface near the sample mid-plane (Figure 8d). The fault then grows away from the nucleation site across the sample (Figures 8e–h).

AE locations from experiment G3 are plotted in Figure 9, with stress history shown in Figures 3c and 5b. Experiment G3 was the last run of the series and, from the travel-time residuals (shown in the appendix), appears to have the most accurate locations of all the runs. The final fault surface (indicated in Figure 9a) again intersected one endplug, causing distributed damage to the bottom end of the sample during the late stages of faulting. As in the other granite runs, loading through peak stress produced distributed microcracking in the central region of the sample (Figure 9a,b). In the 100 s prior to nucleation (Figure 9c), there is a suggestion of increased AE activity in the nucleation zone, although this clustering becomes more organized in the next 100 s (Figure 9d). In the next 200 s (Figure 9e) the nascent fault plane has clearly formed, again on the sample surface near the mid-plane. In the remainder of the plots, the fault plane propagates across the sample, eventually bisecting it. During the late stages of fault growth, the fault tends to steepen. This feature was observed in all three granite experiments where the initial fault formed at an angle of 18–25° to the sample axis (Table 1) but the trace where the fault broke out on the opposite surface of the sample was consistently steeper, forming angles of 5–15°.

3.2. Fault formation in sandstone

A number of features of the standstone experiments differed from the granite runs. Peak stress, and therefore stress drop, was lower for the sandstone. The overall weakening of the granite samples during fault formation was 400–420 MPa, whereas for the sandstone, the drop in differential stress during faulting was 70–80 MPa. Inclination of fractures in the sandstone was approximately 32° as compared to 18–25° for the granite. In addition, the acoustic emissions that occurred during deformation were different for the two rock types. Little AE occurred in the granite samples below about 60% of peak stress, while in the sandstone samples significant AE was occurring by 30 to 40% peak stress. The overall AE activity was also more intense in the sandstone samples than in granite. We interpret the enhanced sandstone activity as the result of distributed grain crushing and sliding. The open pore structure in the Berea (18% porosity) results in stress concentrations at the grain contacts, making this rock susceptible to grain crushing and grain rearrangement when loaded.

The progression of faulting in the sandstone was also notably different from that in the granite. AE locations from sample S1 are shown in Figure 10 and can be compared to the stress history plotted in Figure 4a. A region of intense AE activity occurred in sample S1, beginning with the initial loading (Figure 10a). This activity was primarily in the core of the sample and was located on the eventual fracture plane. Owing to the contrast in modulus between the sandstone and the steel endplugs, conically-shaped zones of high shear stress develop at the ends of samples loaded in this configuration. The coincidence of early AE activity, shown in Figures 10a and 10b, with the zone of high shear stress may indicate that these features are related. In any case, by peak stress (Figure 10b) the region of intense activity has broadened and considerable uniform microcracking has been generated throughout the rest of the sample. For the next 7000 s (Figure 10c), the stress steadily drops by 10 MPa, indicating that the damage zone is gradually weakening the sample, and the stress must be decreased slightly to maintain constant AE rate. During this period, AE activity also migrates to the periphery of the damage zone (outward and downward). In some sense, the nucleation stage of the fracture has been completed by Figure 10d. While this process is not as distinct as in the granite samples, a change in the character of the AE event locations occurs between

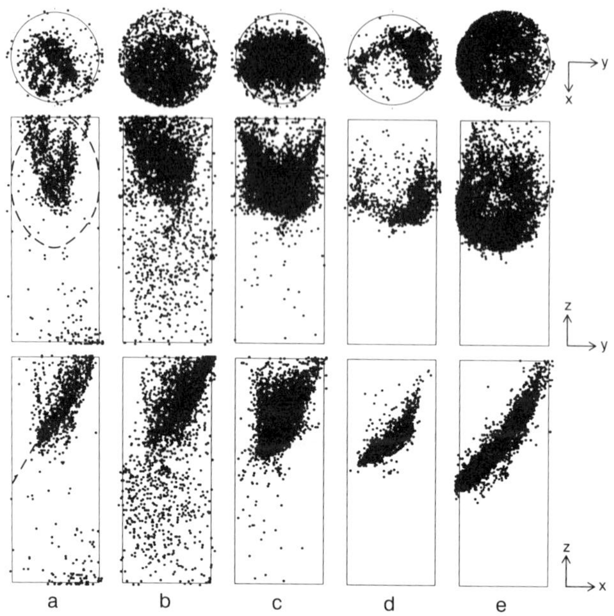

Figure 10. Plots of AE locations for sample S1 (explanation is similar to that for Figure 7). Stress intervals for each plot are indicated in Figure 4a. Unlike granite experiments, AE activity concentrated in region of fault from beginning of experiment; possibly indicating a flaw in sample. AE activity appears to organize into planar feature in (d). In one sense, this interval represents fault nucleation.

Figures 10c and 10d. In the along-strike views (bottom plots), the AE events become notably localized along the eventual fracture plane by Figure 10d. In the face-on views (middle plots), a narrow fracture front has developed by Figure 10d with most activity concentrated on the right-hand side of the fracture as it propagates out to the sample surface. This process continues in Figure 10e, where the toe and lower left side of the fault are formed. A more detailed view of the events shown in Figures 10d and 10e reveals that after the fracture breaks through to the surface on the right side of the sample, the fracture front systematically progresses down, around the toe and up the left side.

AE locations from sandstone sample S2 are shown in Figure 11 and can be compared to the stress history plotted in Figures 4b and 5c. The surface trace of the eventual fracture plane is indicated in Figure 11a. Also shown in Figure 11a are AE locations for events occurring during loading to 73% peak stress. In this case, two and possibly three clusters of microcracking occur during the early stages of loading. The most distinct cluster, in the central region of the sample, occurs on the eventual fault plane and within 10 mm of the sample surface. This cluster appears to be the nucleation site for the fracture plane. Visual inspection of the sample after the experiment showed nothing unusual about this area. The upper cluster of activity occurs in the core of the sample in the region of high shear stress discussed in the previous paragraph. Loading to peak stress (Figure 11b) results in continued AE activity in these same regions, although the activity at this stage is diffuse, extending over volumes with diameters of approximately 40 mm. Over the next 5.5 hours (Figures 11c,d), AE activity gradually shifts to the fault

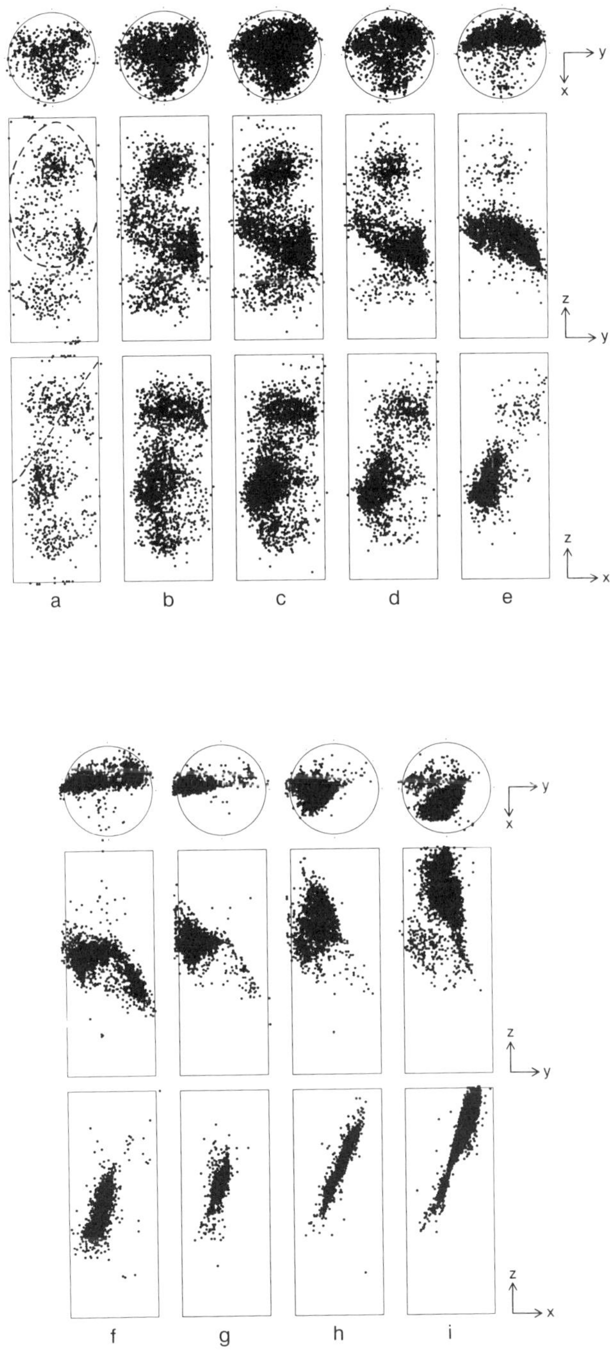

Figure 11. Plots of AE locations for sample S2 (explanation is similar to that for Figure 7). Stress intervals for each plot are indicated in Figures 4b and 5c. Similarly to sample S1, AE activity is localized in early loading stage of experiment, suggesting strength inhomogeneity in this sample. AE activity does not become well organized into planar feature until plots (e) and (f), which are well beyond peak strength.

nucleation site and decreases in intensity throughout the rest of the sample. By this stage the nucleation cluster has shrunk to roughly 30 mm in diameter, although it is somewhat elongated in the direction of the fault plane. In the next 2.8 hours (Figure 11e), the toe of the fault forms and microcracking outside of the nucleation zone has virtually ceased. As the stress continues to drop (Figure 11f,g), microcracking migrates to the lower left side of the fault. The width of the zone of AE activity, normal to the fault plane, is now 12 mm. At this stage, the fault propagates upward and across the sample as the remainder of the fault is formed.

4. Discussion

4.1. Prenucleation

In the preceding section we have shown how shear fractures can be controlled to propagate quasistatically in brittle rock. Then, by analyzing the distribution of impulsive microcrack events through their acoustic emissions, we are able to observe nucleation and growth of the fractures. This is the first time that such an approach has been taken to the study of fault formation. While some details change from one experiment to the next, a number of general features have been observed. In the granite experiments, early loading resulted in little AE activity. Not until the onset of dilatancy at 50–60% peak stress did significant AE activity occur. Microcracking, as determined by AE locations, remained uniformly distributed in the central portion of the granite samples until after peak stress. This result is consistent with our earlier granite experiments (Lockner and Byerlee, 1977a, 1980) and a more recent confined experiment by Hirata et al. (1987). On the other hand, Sondergeld and Estey (1981), Nishizawa et al. (1984), and Yanagidani et al. (1985) all reported clustering of AE sources near the sample surface in uniaxially deformed granite and andesite. To explain this difference, Sondergeld and Estey speculated that prefracture AE was of such low amplitude that our automatic detection system would systematically mislocate these events, and that we would therefore not detect event clustering. A discussion of the location errors for our acquisition system appears in the appendix along with a plot (Figure 13) of the frequency–magnitude distribution for AE events recorded in experiment G3. As seen in Figure 13, the *b*-value for prenucleation events is indeed larger than for postnucleation events, indicating a greater percentage of low-amplitude events in the prenucleation phase. However, we were still able to obtain high-quality locations for 474 events prior to fault nucleation (Figures 9a,b,c). These represent over four times the number of events located by Sondergeld and Estey. Even so, we see only subtle changes in event clustering prior to fault nucleation (Lockner and Byerlee, 1992b), and certainly nothing that is comparable to the clustering reported in the unconfined experiments. Consequently, we prefer the explanation proposed by Yanagidani et al. (1985), that the intense surface damage they observed was the result of stress corrosion cracking due to exposure of the free surface to water vapor. Still, the differences between our results and those of other researchers must be considered an open question and further study of this problem is required.

Related to this issue, a sampling problem that is inherent to AE studies should be kept in mind. While numerous studies have pointed out the close correlation between number of AE events and inelastic strain in the sample (e.g. Lockner and Byerlee, 1977c, 1980), the number of new microcracks detected by AE is a small fraction of the total microcrack population. As we show in the appendix, this condition holds for AE studies in general. As an example, we consider the AE detected during the initial loading stages in our granite experiments. During this period, 2000–5000 AE events were typically detected. For a grain size in Westerly granite of ~0.3 mm, this gives approximately one AE event for every 3000 to 8000 grains, or one AE event per 200 mm^3. We can make a rough comparison between AE event density and microcrack density in the following way. Madden (1984) analyzed crack data compiled by Hadley (1975) for deformation of Westerly granite. Hadley reported that the largest cracks showing significant increase in crack density during loading were 30–100 μm long, a result that we have independently confirmed (Lockner et al., 1992). She found that by peak stress, the areal density of these cracks had increased by 115 cracks/mm^2. Assuming that crack length and width were approximately the same, volumetric crack density of this size crack had increased during loading on the order of 3000 cracks/mm^3. Thus, cracking events detected by AE represent a small fraction of the total microcrack damage occurring in the rock. How representative the microcracks detected by AE are of the overall microcrack population is still unknown. For fracture studies, however, AE activity does appear to be a reliable indicator of zones of fault movement and crack damage. The excellent agreement between AE locations and observed faults in this study as well as faults and hydrofracture surfaces in Lockner and Byerlee (1977b) and Lockner et al. (1982) supports this conclusion.

4.2. Fault nucleation

The observation that in our granite experiments the fault consistently nucleated at the sample surface suggests that nucleation is controlled by conditions at the sample boundaries in the triaxial geometry. It is possible that a misalignment of the loading frame caused bending of the sample and a high-stress region on the surface. However, if this were the case, we would expect to see increased AE activity from this region during the entire loading cycle. Figures 7–9 show that this was not the case. An alternate explanation is based on the observation that fracture nucleation in brittle rock involves microcracking and the development of a dilatant zone. Since damaging earthquakes nucleate at depth in the earth, the dilating nucleation zone must expand against the surrounding rock mass. This surrounding rock acts to confine and support the dilatant zone. In the triaxial test geometry, however, the confining fluid has zero modulus (control system maintains constant pressure regardless of volume change). Thus, while a failing region on the rock surface will do work as it expands against the confining fluid, there will be no net transfer of stress as would be the case for a dilating zone in the interior of the sample. As a result, we would always expect the fault to nucleate at the rock surface. This interpretation is supported by bifurcation analysis of finite-sized samples in which failure also tends to occur on the sample surface. In this sense, triaxial rock tests provide a lower bound for the strength of rock deformed under more appropriate boundary conditions. We are currently considering methods of fabricating composite samples that will provide boundary conditions more appropriate for earthquake nucleation studies. Fault nucleation appeared to occur in the interior of at least the first sandstone experiment. However, it should be noted that the dilatancy needed for fault formation in a low-porosity granite is only a few percent. Berea sandstone contains about 18% porosity and also has a lower modulus than granite. As a result there should be room for grains to readjust locally with little transfer of load to the surrounding material.

In contrast to the granite experiments, AE localization occurred during the early stages of loading in both sandstone experiments. Berea is a bedded sandstone and consequently exhibits strength anisotropy. From the early localization of AE events in the sandstone experiments, we infer that Berea also has significant strength heterogeneity. This heterogeneity appears to be of sufficient magnitude to dictate where failure will initiate. One interesting feature of the sandstone experiments is that early AE clustering was relatively diffuse, filling volumes of 40 mm or more in linear dimension. Then, as the fracture surface formed, the zone of microcracking narrowed in the direction perpendicular to the fault to between 10 and 20 mm thickness. For comparison, the zone of crushed material that defined the surface trace of the fracture was less than 2 mm thick.

In the granite experiments, fault nucleation involved the occurrence of a relatively small number of large-amplitude events. This can best be seen in Figure 9d where about 10 events have occurred in the nucleation zone. With the addition of about 20 more events (included in Figure 9e), the nucleation patch that initially had no preferred orientation has established the narrow diagonal structure which defines the orientation of the final fault plane. In the granite samples, the nucleation zone appears to require a volume of approximately 2 cm^3. In the sandstone samples the nucleation zone, just before fault propagation, is approximately 9 cm^3. As defined by the AE locations, the nascent fault plane that forms in the granite samples has a half-disk shape (Figures 7b, 9e), with a thickness of no more than 3 mm and radius of approximately 15 mm. From this initial structure, the fault grows in-plane away from the nucleation site.

4.3. Fault growth

Once fault nucleation begins in the granite samples, it proceeds relatively quickly to establish the nascent fault plane. Following this phase of fault development is the growth phase, in which a band of AE activity develops, indicating a zone of intense microcracking, and propagates away from the nucleation site. Thus we are able to observe mixed modes II and III shear propagation as the fault grows across the sample. Growth in the direction perpendicular to the sample axis is pure mode III anti-plane shear while propagation towards the sample ends is nearly pure mode II in-plane shear. A relatively quiet region forms behind the advancing fracture front (Figures 12, 7c–e, 9g–i), indicating that intense grain crushing is restricted to the fracture front. Sliding must continue on the fault surface after the fracture front passes, furthering the process of grain comminution. However, the relatively few large-amplitude AE events in this region suggest that most large grains have been crushed during the passage of the fracture front. The localized zone of grain crushing at the tip of the advancing fault corresponds closely to the process zone discussed, for example, in the Rice (1980, 1984) model for shear crack propagation along an existing fault. In this model the process zone, having in-plane dimension w, is the region at the crack tip over which the shear stress drops from the peak value to the frictional strength level. If we assume that the breakdown zone is represented by the region of intense AE activity, samples G1 and G3 give $w_{\text{granite}} \approx 20–50$ mm (Figures 7, 9, and 12). Wong (1982a) reported estimates of w for room temperature experiments that ranged from 40 to 290 mm. Thus our estimates of w fall near the lower end of the range tabulated by Wong. In the sandstone samples, the breakdown zone that developed was never as well-defined as in the granite experiments. An approximate value of w, based on the AE patterns, is $w_{\text{sandstone}} \approx 60–90$ mm.

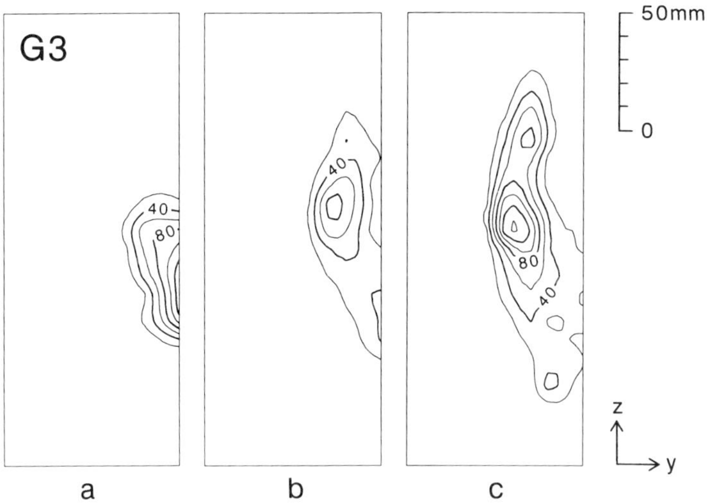

Figure 12. Density of AE events during progressive time intervals in Experiment G3. Fault plane is viewed face-on. Time intervals correspond to Figure 9f, g, and h. Contour interval is 20 events/cm^2. Width w of process zone is taken to correspond to width of band of AE activity.

The shear fracture energy, or energy release rate, $\mathscr{G}$ is a fundamental quantity used in models of shear fracture propagation. $\mathscr{G}$ represents the energy flux needed to extend the fracture. Following Rice (1980), Wong (1982a) compiled representative values of $\mathscr{G}$ from confined laboratory experiments. Values ranged from 0.3×10^4 to 7.3×10^4 J m^{-2} for granite and quartzite. The procedure for calculating $\mathscr{G}$ assumes that the sample is small enough so that the entire fault plane slides as a unit. While this is not the case for our experiments, we can still provide a rough estimate of $\mathscr{G}$. Sample G1 yields a value of $\mathscr{G}_{\text{granite}} = 1.3 \times 10^4$ J m^{-2} and sample S2 yields $\mathscr{G}_{\text{sandstone}} = 0.25 \times 10^4$ J m^{-2}, in good agreement with the earlier reported values. In a related paper (Lockner et al., 1991) we use the radiated acoustic energy to map the local energy release rate on the fault, much as seismic moment is used to infer stress drop heterogeneity for large earthquakes. The rapid drop in stress level that marks the transition from fault initiation to fault propagation indicates that it is easier for a fault to grow once its initial structure has formed. It may therefore be inappropriate to infer properties of the growth of fractures from experiments on small samples in which nucleation dominates. We intend to explore this possibility in future experiments.

The thickness, h, of the fracture normal to its plane is another important parameter for describing faults. Fracture energy for aggregates is generally significantly greater than for single crystals. The main reason for this increase is the development of a zone of damage that moves with the advancing fault tip. The development of this damage zone depends on mobilization of preexisting flaws which are abundant in multigrain aggregates. By impregnating fractured samples with dyed epoxy, the damaged region bordering the fault can easily be measured. In sample G2, the fault propagated toward the upper endplug and stopped about 25 mm from it (Figure 8). The damage zone at the arrested crack tip was $h = 1.0 \pm 0.2$ mm. Near the lower part of the fracture, where possibly 0.5 mm shear slip had occurred, damage zone thickness was approximately 1.5–3 mm. In sample G3, where more slip occurred (~ 1 mm), damage zone thickness ranged from 4 to 5 mm. Thin-section analysis (Lockner et al., 1992) showed that this damage was composed primarily of microcracks oriented parallel or subparallel to the maximum compressive stress direction. This damage was much more severe than microcrack damage in regions of the sample far from the fracture. The 5–6 mm width of the zone of intense AE activity (Figure 9f,g) suggests that the resolving power of the AE monitoring system is approaching this minimum dimension of the fractures.

5. Conclusions

Through this series of experiments we have demonstrated the ability to retard unstable fracture propagation so that quasistatic fault growth can be studied on a timescale of minutes to hours. This ability is based on the technique of controlling stress to maintain a constant acoustic emission rate. By combining this novel control system with an AE acquisition system capable of locating AE source hypocenters, we have observed a number of features of fracture formation. The brittle fracture process, under triaxial conditions, was found to involve three stages:

(1) *Distributed damage during loading of the sample to peak stress.* In granite samples, dilatancy and the associated microcracking were found to be distributed uniformly throughout the sample. No clustering of AE locations was observed that was comparable to clustering reported in previous unconfined experiments. In sandstone, however, AE clustering was observed from the earliest stages of loading in regions which ultimately developed into the fault plane. This clustering was initially diffuse but became more localized as the sample was loaded. It is interpreted as indicating the presence of significant weak zones or zones of stress concentration which acted as nucleation sites and determined the location of the fault plane.

(2) *Fault nucleation.* In granite, soon after peak stress, a zone of intense AE activity formed on the surface, near the sample mid-plane. The nucleation site, with volume approximately 2 cm^3, rapidly evolved into the nascent fault, narrowing into a half-disk shape that defined the position and orientation of the fracture. This nucleation process was accompanied by a rapid drop in stress. In the sandstone samples, this nucleation stage involved localization of the relatively diffuse AE activity into a roughly planar feature.

(3) *Fault propagation.* In both granite and sandstone samples, the newly formed fault grew by developing a characteristic zone of intense AE activity which is interpreted as a fracture front propagating across the sample. This process zone was 10–50 mm wide in granite and 60–90 mm wide in sandstone. The thickness of the process zone, normal to the fracture surface, was 1–5 mm in granite and approximately 10 mm in sandstone.

These results suggest interesting new avenues of research for earthquake studies. The nucleation patch size for the highly homogeneous Westerly granite was only a few cubic centimeters. Once the fault nucleated on this size scale, further growth in a compliant system such as the earth would be rapid and uncontrolled. Such a small nucleation zone would be very difficult to recognize remotely. On the other hand, the sandstone samples showed that a heterogeneous material would generate zones of intense activity well before catastrophic failure. The nucleation process in this case involved a concentration of this acoustic activity into a more and more compact region, as well as a gradual evolution towards a planar shape. Thus, if the heterogeneity that controls fault nucleation in the earth is on a large enough scale, the possibility exists that the occurrence and evolution of the nucleation zone could be identified remotely.

Acknowledgements

Many people have contributed to the success of these experiments. In particular, we thank T. Madden, Z. Reches, J. Rudnicki, T. Satoh, J. Savage and T.-f. Wong for their thoughtful and constructive comments.

References

Byerlee, J.D. and Lockner, D.A. (1977). Acoustic emission during fluid injection in rock. In *Proc. 1st Conf. on Acoustic Emission/Microseismic Activity in Geological Structures and Materials* (ed. H.R. Hardy and F.W. Leighton), pp. 87–98. Trans-Tech. Publications, Clausthal-Zellerfeld.

Costin, L.S. (1987). Deformation and failure. In *Fracture Mechanics of Rock* (ed. B.K. Atkinson), pp. 167–215. Academic Press, New York.

Du, Y. and Aydin, A. (1991). Interaction of multiple cracks and formation of echelon crack arrays. *Int. J. Numerical and Analytical Methods in Geomech.* **15**, 205–218.

Fraser-Smith, A.C., Bernardi, A., McGill, P.R., Ladd, M.E., Helliwell, R.A., and Villard, O.G., Jr. (1990). Low-frequency magnetic field measurements near the epicentre of the M_S 7.1 Loma Prieta earthquake. *Geophys. Res. Lett.* **17**, 1465–1468.

Gutenberg, B. and Richter, C.F. (1949). *Seismicity of the Earth.* Princeton University Press.

Hadley, K. (1975). Dilatancy: further studies in crystalline rocks. Ph.D. thesis, MIT, Cambridge, Mass.

Hallbauer, D.K., Wagner, H., and Cook, N.G.W. (1973). Some observations concerning the microscopic and mechanical behavior of quartzite specimens in stiff, triaxial compression tests. *Int. J. Rock Mech. Min. Sci.* **10**, 713–726.

Hirata, T., Satoh, T., and Ito, K. (1987). Fractal structure of spatial distribution of microfracturing in rock. *Geophys. J. R. Astron. Soc.* **90**, 369–374.

Horii, H. and Nemat-Nasser, S. (1985). Compression-induced microcrack growth in brittle solids: axial splitting and shear failure. *J. Geophys. Res.* **90**, 3105–3125.

House, L.S., Kranz, R.L., Nishizawa, O., and Satoh, T. (1989). Use of acoustic emissions to investigate stresses in laboratory rock samples. *EOS* (*Trans. Amer. Geophys. Union*), **70**, 1340.

Jaeger, J.C. and Cook, N.G.W. (1984) *Fundamentals of Rock Mechanics.* Chapman and Hall, New York.

Johnston, M.J.S. (1990). Review of continuous high precision borehole strain measurements near the locked and creeping sections of the San Andreas fault. *Abst. of NSF-USGS Workshop on Crustal Deformation Measurement and Earthquake Mechanics, Morro Bay, California, March 18, 1990.*

Johnston, M.J.S., Linde, A.T., Gladwin, M.T., and Borcherdt, R.D. (1987). Fault failure with moderate earthquakes. *Tectonophysics* **144**, 189–206.

Lawn, B.R. and Wilshaw, T.R. (1975). *Fracture of Brittle Solids.* Cambridge University Press.

Linde, A.T. and Johnston, M.J.S. (1989). Source parameters of the October 1, 1987 Whittier Narrows earthquake from crustal deformation data. *J. Geophys. Res.* **94**, 9633–9643.

Lockner, D.A. and Byerlee, J.D. (1977a). Acoustic emission and fault formation in rocks. In *Proc. 1st Conf. on Acoustic Emission/Microseismic Activity in Geol. Structures and Materials* (ed. H.R. Hardy and F.W. Leighton), pp. 99–107. Trans-Tech. Publications, Clausthal-Zellerfeld.

Lockner, D.A. and Byerlee, J.D. (1977b). Hydrofracture in Weber sandstone at high confining pressure and differential stress. *J. Geophys. Res.* **82**, 2018–2026.

Lockner, D.A. and Byerlee, J.D. (1977c). Acoustic emission and creep in rock at high confining pressure and differential stress. *Bull. Seismol. Soc. Am.* **67**, 247–258.

Lockner, D.A. and Byerlee, J.D. (1978). Velocity anomalies: an alternative explanation based on data from laboratory experiments. *Pure Appl. Geophys.* **116**, 765–772.

Lockner, D.A. and Byerlee, J.D. (1980). Development of fracture planes during creep in granite. In *2nd Conf. on Acoustic Emission/Microseismic Activity in Geological Structures and Materials* (ed. H.R. Hardy and F.W. Leighton), pp. 11–25. Trans-Tech. Publications. Clausthal-Zellerfeld.

Lockner, D.A. and Byerlee, J.D. (1992a). Fault growth and acoustic emissions in confined granite. *Appl. Mechanics Rev.* (in press).

Lockner, D.A. and Byerlee, J.D. (1992b). Precursory AE patterns leading to rock fracture. In *Proc. 5th Conf. on Acoustic Emission/Microseismic Activity in Geol. Structures and Materials* (ed. H.R. Hardy), 14 pp. Trans-Tech. Publications, Clausthal-Zellerfeld, Germany (in press).

Lockner, D.A. and Madden, T.R. (1991a). A multiple crack model of brittle fracture part I: non-time-dependent simulations. *J. Geophys. Res.* **96**, 19623–19642.

Lockner, D.A. and Madden, T.R. (1991b). A multiple crack model of brittle fracture part II: time-dependent simulations. *J. Geophys. Res.* **96**, 19643–19654.

Lockner, D.A., Recher, Z., and Moore, D.E. (1992). Microcrack interaction leading to shear fracture. In *Proc. 33rd U.S. Symposium on Rock Mechanics* (ed. W. Wawersik), A.A. Balkema, Rotterdam.

Lockner, D.A., Walsh, J.B., and Byerlee, J.D. (1977). Changes in seismic velocity and attenuation during deformation of granite. *J. Geophys. Res.* **82**, 5374–5378.

Lockner, D.A., Summers, R., Moore, D., and Byerlee, J.D. (1982). Laboratory measurements of reservoir rock from the Geysers geothermal field, California. *Int. J. Rock Mech. Min. Sci. & Geomech. Abstr.* **19**, 65–80.

Lockner, D.A., Byerlee, J.D., Kuksenko, V., Ponomarev, A., and Sidorin, A. (1991). Quasi-static fault growth and shear fracture energy in granite. *Nature* **350**, 39–42.

Madden, T.R. (1983). Microcrack connectivity in rocks: a renormalization group approach to the critical phenomena of conduction and failure in crystalline rocks. *J. Geophys. Res.* **88**, 585–592.

Nishizawa, O., Onai, K., and Kusunose, K. (1984). Hypocenter distribution and focal mechanism of AE events during two stress stage creep in Yugawara andesite. *Pure Appl. Geophys.* **122**, 36–52.

Reches, Z. and Lockner, D. (1990). Self-organized cracking – a mechanism for brittle faulting. *EOS, Amer. Geophys. Union Trans.* **71**, 1586.

Rice, J.R. (1980). The mechanics of earthquake rupture. In *Physics of the Earth's Interior* (Proc. Int'l School of Physics "E. Fermi", course 78, 1979) (eds. A.M. Dziewonski and E. Boschi). Italian Physical Society/North Holland.

Rice, J.R. (1984). Shear instability in relation to the constitutive description of fault slip. In *Proc. 1st Int. Congress on Rockbursts and Seismicity in Mines*, pp. 57–62. SAIMM, Johannesburg.

Sammis, C.G. and Ashby, M.F. (1986). The failure of brittle porous solids under compressive stress states. *Acta Metall.* **34**, 511–526.

Scholz, C.H. (1968). Experimental study of the fracturing process in brittle rocks. *J. Geophys. Res.* **73**, 1447–1454.

Sondergeld, C.H. and Estey, L.H. (1981). Acoustic emission study of microfracturing during the cyclic loading of Westerly granite. *J. Geophys. Res.* **86**, 2915–2924.

Spetzler, H., Sobolev, G., Sondergeld, C., Salov, B., Getting, I., and Koltsov, A. (1981). Surface deformation, crack formation, and acoustic velocity changes in pyrophyllite under polyaxial loading. *J. Geophys. Res.* **86**, 1070–1080.

Tapponnier, P. and Brace, W.F. (1976). Development of stress-induced microcracks in Westerly granite. *Int. J. Rock Mech. Min. Sci.* **13**, 103–112.

Terada, M., Yanagidani, T., and Ehara, S. (1984). A. E. rate controlled compression test of rocks. In *3rd Conf. on Acoustic Emission/Microseismic Activity in Geological Structures and Materials* (ed. H.R. Hardy and F.W. Leighton), pp. 159–171. Trans-Tech. Publications, Clausthal-Zellerfeld.

Wawersik, W.R. (1973). Time-dependent behavior of rock in uniaxial compression. In *Proc. 14th Symposium on Rock Mechanics*, pp. 85–106.
Wawersik, W.R. and Brace, W.F. (1971). Post-failure behavior of a granite and a diabase. *Rock Mech.* **3**, 61–85.
Wawersik, W.R. and Fairhurst, C. (1970). A study of brittle rock fracture in laboratory compression experiments. *Int. J. Rock Mech. Min. Sci.* **7**, 561–575.
Weeks, J.D., Lockner, D.A., and Byerlee, J.D. (1978). Change in *b*-values during movement on cut surfaces in granite. *Bull. Seismol. Soc. Am.* **68**, 333–341.
Wong, T.-f. (1982a). Shear fracture energy of Westerly granite from post-failure behavior. *J. Geophys. Res.* **87**, 990–1000.
Wong, T.-f. (1982b). Effects of temperature and pressure on failure and post-failure behavior of Westerly granite. *Mechanics of Materials* **1**, 3–17.
Yanagidani, T., Ehara, S., Nishizawa, O., Kusunose, K., and Terada, M. (1985). Localization of dilatancy in Ohshima granite under constant uniaxial stress. *J. Geophys. Res.* **90**, 6840–6858.

Appendix Evaluation of Location Accuracy

In this appendix we examine the resolving power of the AE acquisition system used in these experiments. Lockner and Byerlee (1978) demonstrated the susceptibility of an automatic arrival-time detector (or P-picker in seismological parlance) to systematically picking arrivals of small-amplitude, emergent signals too late. It is necessary, therefore, to consider the effect of this sampling bias on the location accuracy of the hypocentral inversion routine. This is an important issue, since the conclusions of our current experiments rely on our ability to locate accurately AE sources. In previous studies (Lockner and Byerlee, 1977b, 1980; Lockner et al., 1982) direct correspondence between AE locations and observed fault planes has demonstrated our ability to locate faults through inversion of AE arrival times. We now present a more quantitative assessment of our location accuracy.

A useful starting point in a discussion of location bias is through the use of an amplitude–frequency plot. Weeks et al. (1978) and Lockner et al. (1991) have shown that, as with earthquakes, AE events obey the well-known frequency–magnitude relation

$$N(M) = a - bM$$

(Gutenberg and Richter, 1949) where N is the number of earthquakes greater than magnitude M and a and b are constants. Using the complete data set obtained from experiment G3 as well as the subset of prenucleation events, we have plotted cumulative number of events as a function of log(amplitude) in Figure 13. In this plot, amplitudes have been adjusted for attenuation due to geometric spreading and are shown as the amplitude that would be expected 10 mm from the hypocenter. If we define the RMS travel time residual as $TT^{\text{rms}} = \sqrt{(\mathbf{R}^{\mathrm{T}} \cdot \mathbf{R}/6)}$, we can set an arbitrary cutoff for potentially 'successful' event locations of $TT^{\text{rms}} < 5\ \mu\text{s}$. On this criterion, the total number of events successfully located in this experiment was 45 713. Plotted in the upper left corner is the total number of events (165 117) that triggered the acquisition system at the preset threshold level. As seen in this plot, most AE events have small amplitudes. For example, we could have doubled the number of recorded events simply by lowering the threshold level by 30%. This fact, resulting from the nature of the frequency–magnitude relation, can be expected for any brittle rock deformation experiment. Of the events that did trigger the system, some 28% gave acceptable locations according to this initial criterion.

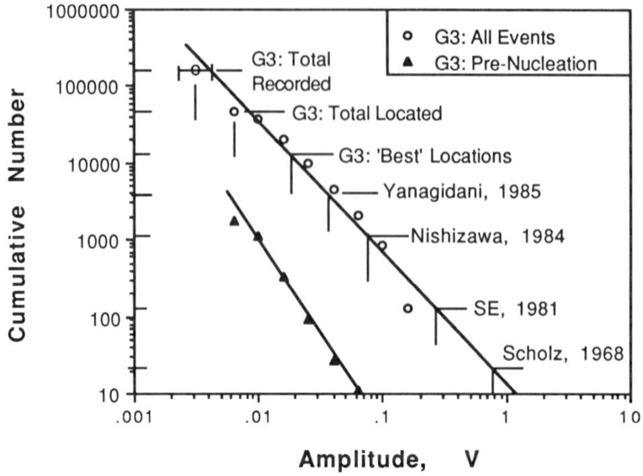

Figure 13. Cumulative number of AE events plotted vs. log (amplitude) for experiment G3. Open symbols, entire experiment; solid symbols, prenucleation events. Of 165117 events detected, 45713 were located with $TT^{rms} < 5\ \mu s$. Of these, approximately 19000 had adjusted amplitudes >15 mV and $TT^{rms} < 1\ \mu s$. These events were considered to have accurate locations and were used in the analysis. Numbers of events analyzed by other authors are plotted for comparison.

While this yield may at first seem low, it must be remembered that the location inversion routing requires good arrivals at all six transducers. If the first arrival is missed at any station owing, for example, to attenuation or to location of a station along a nodal plane of the radiation pattern, the location procedure will fail. Since most events have small amplitudes, this type of error will occur frequently. One way to improve the event yield is to have a denser array of stations, which we intend to incorporate in the next generation of acquisition system. To insure that we had strong first arrivals and therefore good timing accuracy, we culled the set of located events to approximately 19000 large-amplitude events. Thus, we ultimately used only 11% of the recorded events in our detailed analysis, although the amplitude and travel-time residual criteria used to select events were precisely defined. As we will show, this yield ratio of usable events compares very favorably to those reported for hand-picked location studies.

To compare our results to hand-picked arrival-time data from other studies, we will assume, for the sake of argument, that the number and distribution of AE events occurring in each deformation experiment are similar to the G3 results shown in Figure 13. An early and well-known AE study was published by Scholz (1968), in which he reported microcrack localization based on 22 selected events (shown near bottom of Figure 13). These represent a small fraction, indeed, of the thousands of events that occurred during that experiment. Sondergeld and Estey (1981) detected 3360 events of which they located 116, presumably choosing the largest-amplitude events for their analysis. Even so, these represent a significantly smaller sampling than do our results. Nishizawa et al. (1984) located over 1000 events and, after considerable man-hours of work, Yanagidani et al. (1985) located 3933 hypocenters. These last two studies begin to approach our results in terms of sheer number of locations. Sondergeld and Estey, Nishizawa, and Yanagidani were also interested in analyzing first motions to obtain fault-plane solutions. In this case, inspection of the waveform is advised, although with more sophisticated techniques, automatic determination of first motion polarity should be possible. A similar assessment

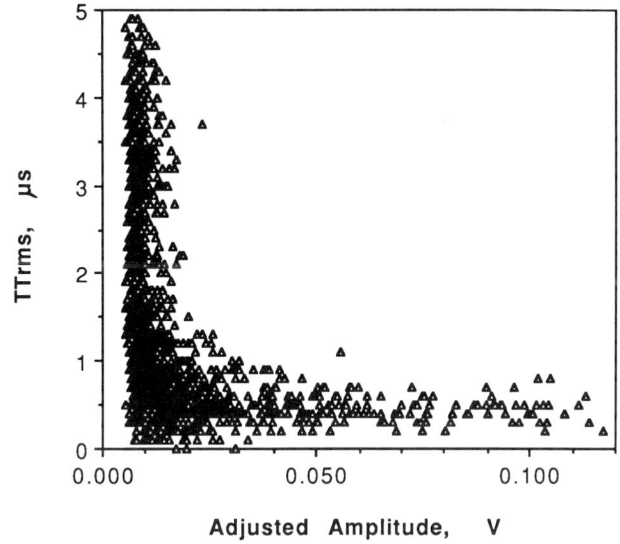

Figure 14. Distribution of TT^{rms} plotted as a function of event amplitude (adjusted to 10 mm from source) for 1600 events recorded in experiment G3. Above ~15 mV, residuals cluster below 1 μs, indicating that the first arrival is picked reliably for these events. Below 15 mV, residuals blow up, indicating that first arrivals are not picked reliably at all stations for these events.

can be made concerning the approximately 1000 events located during a recently reported triaxial test of granite by House et al. (1989).

To further analyze the performance of our automatic P-picker, we plot TT^{rms} as a function of amplitude in Figure 14. For adjusted amplitudes greater than 15 mV, travel-time residuals cluster below 1 μs. Smaller-amplitude events give much larger residuals. This result indicates that below 15 mV the P-picker frequently misses the first arrival at one or more stations. However, above this cutoff, first arrivals were picked reliably at all transducers and for a typical P-wave velocity of 5 km/s would result in location errors of approximately ±2 mm. This value is based on a formal maximum-likelihood estimate using actual transducer locations and events occurring in the central region of the sample. The fact that large-amplitude events give residuals that cluster at 0.4 μs indicates a small systematic error in locating events in this segment of the experiment. This error could be the result of, for example, a 5% error in determination of the velocity field, a systematic late pick of first arrivals at the most distant stations, or timing errors due to the finite size of the transducers (3.2 mm radius). We cannot eliminate any of these possible causes at this time. The systematic deviations of AE locations from the observed fault planes near the ends of the sample (Figures 7–11) are most likely the result of an improper determination of the velocity structure. We are currently investigating methods for improving these determinations. However, what is most important for the present analysis is the relative error in determining locations of neighboring events. The small scatter in event locations when viewed along-strike (Figures 7–9) confirms that this error is less than ±3 mm. The automated threshold detector system we have used does, in fact, perform well when compared to hand-picked arrival schemes and is certainly adequate for the present analysis. A more sophisticated system, employing current technology, can be expected to perform even better.

Chapter 2

Fabrics of Experimental Fault Zones: Their Development and Relationship to Mechanical Behavior

J.M. Logan[1], C.A. Dengo[2], N.G. Higgs[3], and Z.Z. Wang[4]

[1] *Center for Tectonophysics, Texas A & M University, College Station, TX 77843, U.S.A.*

[2] *Exxon Production Research, Houston, Texas, U.S.A.*

[3] *Amoco Production Research, Tulsa, Oklahoma, U.S.A.*

[4] *K. Brown and Associates, Inc., College Station, Texas, U.S.A.*

Abstract

The fracture array of simulated fault zones is shown to evolve in a predictable and reproducible manner, from a stepwise fashion to a steady-state condition. At low confining pressures and increasing shear strain the sequence is: (1) Homogeneous shearing by grain-to-grain movements. (2) R_2- and R_1-fractures initiate at about the same time but propagate only a few grain diameters. They are at relatively high angles to the gouge-forcing block interface and widely spaced. These first two stages are one primarily of gouge compaction characterized by strain hardening. (3) Extension of R_1s and coincident reorientation to lower angles closely paralleling the interface with the forcing blocks. P-fractures initiate. These occur from the ultimate strength through a strain softening stage. (4) Y-fractures form along which most of the displacement is accommodated, with the fracture array now close to steady state. Y's initially are close to one or both interfaces with the forcing blocks, but with increasing shear strain shift to the interior of the gouge. At this stage, sliding may change from stable slip to periodic oscillations, characteristic of stick–slip sliding.

The development of the fracture array is interpreted to be the result of a reorientation of the stress field across and within the gouge zone. Riedel shears form in response to Coulomb failure, but Y-fractures appear as a result of the kinematic constraint produced by the more rigid bounding blocks.

Modeling of the weak gouge zone within a stronger medium shows that the stress field may rotate to higher angles at the gouge boundaries. This is consistent with recent field observations. A significant implication is that without this recognition, laboratory values of frictional coefficients may be overestimated.

FAULT MECHANICS AND TRANSPORT PROPERTIES OF ROCKS
ISBN 0-12-243780-2

1. Introduction

The mechanical behavior of brittle fault zones has been of great interest in predicting earthquakes, evaluating the role of faults as permeability conduits and barriers to fluid flow, and for more accurate regional reconstruction of structural deformation. This motivation has produced considerable experimental effort to delineate the frictional response of simulated fault gouge material under controlled laboratory conditions. Despite the rewarding mechanical insights that have been achieved, characterization of the fault gouge has often received only minor attention, in part owing to the difficulties in quantifying cataclastic deformation. An understanding of the fabrics of brittle faults and their development is the aim of the present study.

The study of fabrics generated in controlled laboratory experiments serves a number of purposes: (1) it provides understanding of the physical processes operating during shear of a simulated fault zone; (2) it establishes a link between observed features of the fault zone and the mechanical response; (3) it provides comparison of the laboratory geometry with that of natural faults thereby validating the experiments; and (4) it yields methods for applying the laboratory results to the field.

2. Early Studies

Our first systematic observations of the fabric of experimental shear zones deformed in the brittle field were presented in 1978 (Logan et al., 1979). Our observations on fabric at that time are summarized in Figure 1 and Table 1. Salient observations from this early study are:

1. A repeatable fabric was produced under a wide range of confining and pore pressures, displacement rates, gouge thicknesses, and materials.

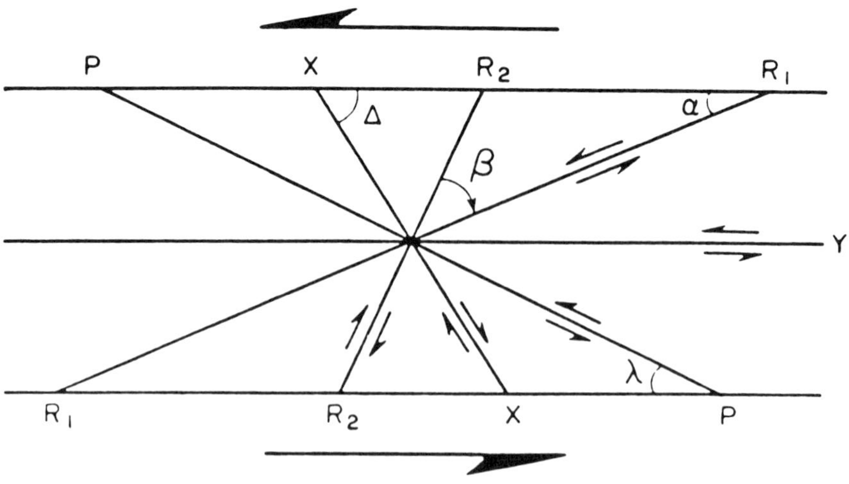

Figure 1. Fracture array and geometrical relationships defined by experiments on simulated fault gouge deformed in triaxial compression (Logan et al., 1979). Sense of shear is as shown; labeled angles as used in Table 1.

Table 1. Angular relations among gouge fractures

Specimen[a]	Gouge composition (%)[b]	P_c (MPa)	Average angle (degrees)[c]			
			α	β	δ	λ
G3	100 Q	50	17	55	70	17
CG6	95 Q, 5 A	100	14	52	76	18
G4	100 Q	250	18	53	72	19
CG7	85 Q, 15 A	100	18	53	74	19
CG5	25 Q, 75 A	100	16	57	71	18
G1	100 O	100	12	50	69	9
G4	100 O	300	13	52	74	15
		Averages	**15**	**53**	**72**	**16**

[a] Specimens are 35°-precut cylinders of Tennessee sandstone with 1 mm thick gouge deformed dry, at room temperature and a shortening rate of $2\text{–}10 \times 10^{-4}\,s^{-1}$.
[b] Q = quartz; A = anhydrite; O = orthoclase.
[c] For definition of angles see Figure 1.

2. R_1-, R_2-, P- and Y-fractures (Figure 1) are the most abundant elements, other orientations are generally less common.
3. All elements develop in reproducible abundance under given conditions, but relative proportions change as the conditions change; some may even be absent depending upon the conditions.
4. Although Y-fractures are close to the interface between the simulated gouge and the bounding rock, they form within the gouge proper. Thus, the sliding response reflects the gouge material (e.g., Figure 1). The bounding blocks influence the mechanical response through their elastic or inelastic response, but not by their frictional properties.

Independent recognition of R_1, R_2 and P fractures was reported by Byerlee and colleagues shortly thereafter (Byerlee et al., 1978) as well as by other workers using a wide range of experimental conditions (e.g., Rutter et al., 1986; Shimamoto, 1986; Moore et al., 1989; Power et al., 1989).

This same fabric array was recognized earlier in experiments on clay cakes (Cloos, 1928; Reidel, 1929; Morgenstern and Tchalenko, 1967), in shear-box experiments on sand and other granular materials (Mandl et al., 1977; Mandl, 1988), and in rock model experiments of an intact layer deformed by sliding of an underlying block (Bartlett et al., 1981). A common fracture geometry emerges from all of these studies even under a wide range of experimental conditions, suggesting that there is a generic relationship. The common factor may be that all have been shown to approximate simple shear conditions (Mandl, 1988; Chester and Logan, 1990) despite widely different boundary conditions.

A number of field studies of natural fault zones have documented many of these same fabric elements (e.g., Dengo, 1982; Welker, 1985; Rutter et al., 1986; Chester and Logan, 1987). These results suggest that the fabric array may be self-similar over a range of scales from microscopic to outcrop and within a variety of rock types.

We have followed our initial laboratory observations with (1) more detailed characterizations of the shear zones, (2) studies of the sequence of development of individual elements of the fracture array, (3) experiments of the relationship of the fracture array to mechanical response during shearing. The results are integrated here in an attempt to obtain a coherent understanding of the problem.

3. Shear Zone Fabric and Experimental Boundary Conditions

3.1. Fault zone boundaries

Natural fault zones often have surprisingly planar boundaries separating the gouge zone from the country rock; although the country rock is highly fractured, it generally has not undergone the extensive comminution and grain-size reduction observed in the gouge (e.g., Engelder, 1974; Dengo, 1982; Chester and Logan, 1985). Other faults, however, have irregular, curved interfaces. Dieterich (1981) found differences in velocity dependence of simulated granite gouge as a function of interface (sawcut) roughness. To what extend do nonplanar boundaries affect the fabric already described? If they change the fabric substantially, then clearly the prior laboratory results are of limited significance.

Air-dried, right circular cylinders of Tennessee sandstone, 8.4 cm long and 3.8 cm in diameter, were fractured in triaxial compression at a confining pressure of 30–50 MPa, room temperature, and a displacement rate of 10^{-4} cm s^{-1}. The fractured specimens were carefully separated in two; a calcite gouge layer about 1 mm thick was spread along the fracture, and the specimen was provided with end spacers to allow displacement along the fractured surface. Frictional tests were then conducted at confining pressures ranging from 40 to 140 MPa. Crushed Yule marble was used as the simulated gouge material as considerable data exists on its frictional behavior with saw cuts (Shimamoto and Logan, 1981). The fractured specimens when separated were found in some cases to have a zone of crushed quartz adjacent to the fractures. The effect of this zone on subsequent shearing after the calcite had been introduced and the subsequent fracture array was found to be small as shown in the results below.

The mechanical behavior agrees well with data on calcite gouge deformed with a sawcut boundary (Figure 2). A transition from stable sliding to stick–slip behavior with increasing confining pressure is present at about the same pressure for both geometries. The frictional levels agree qualitatively. The control that the calcite gouge has on the mechanical response is shown when similar experiments, without calcite gouge, are compared (Figure 2). Note that only stable sliding is found in these latter experiments and the character of the stress-shortening curves is different.

Petrographic studies of the specimens show that damage in prefractured specimens is severe and that this fracture is a zone rather than a sharp, planar surface as found in specimens with a precut (Figure 3, upper). Comminution has occurred in the quartz during the initial fracturing process, so in some places the calcite gouge is in contact with quartz gouge which in turn interfaces with the sandstone. The zone of deformation can in this case be considered as a dilatant fractured region as indicated by the penetration of the thin-section epoxy (Figure 3). Notice also that the surface is not planar.

The fracture array of the calcite gouge (Figure 4) is similar to that observed in sawcut specimens (Wang, 1989). There is extensive development of R_1-, R_2- and Y-fractures which are the loci of grain-size reduction. While fractures do occasionally extend into

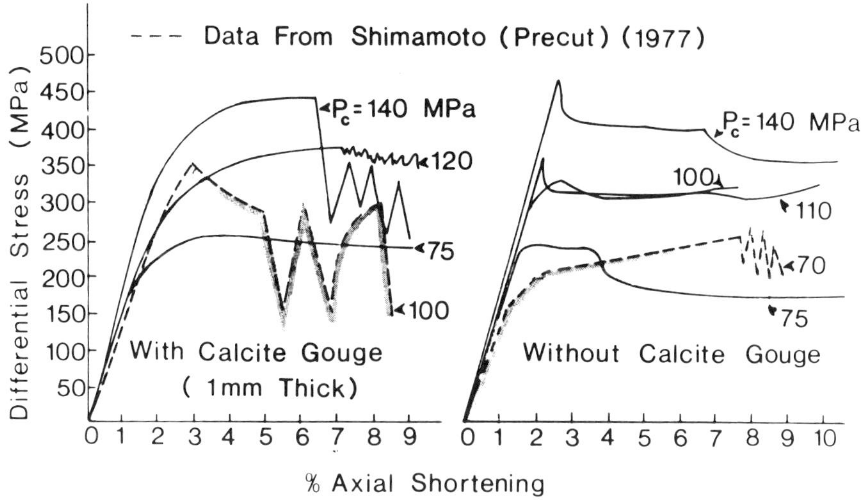

Figure 2. Stress-shortening curves for experiments on prefractured sandstone with (left) and without (right) simulated gouge of calcite. Dashed curves (Shimamoto, 1977) are for sawcut specimens. Confining pressures for each experiment as shown.

the granulated quartz, they are largely restricted to the weaker calcite. The R_1-fractures curve to become asymptotic with the Y-fractures (Figure 4a), a characteristic identical to that found in precut specimens. Although the interface between the calcite and the quartz is irregular, the Y-fractures are generally planar (Figure 4a). The result is for Y-fractures, at least locally, to make the displacement sites more planar, producing a lower energy requirement for shear. They are within the calcite (the weaker material), but close to the interface joining interior peaks of the contact. With increased displacement they leave pockets of calcite behind (Figure 4b) which become 'relics' from the time of formation of the fracture. This feature has also been observed in natural faults (Logan et al., 1979).

In summary, four facets emerge from this study of calcite gouge in sandstone. (1) The fabric produced with the irregular interfaces is not significantly different from that of precut specimens. Studies of gouge bounded by planar (sawcut) boundaries appear satisfactory for our problem, significantly increasing the ease of such experimental investigations. (2) The Y-fractures, which appear to localize most of the displacement in the shear direction, are more or less planar surfaces, at least over small distances, regardless of the irregular character of the interface. (3) The irregular interface does not significantly alter the mechanical response of these specimens. This suggests that overall frictional strength is related to the fabric of the gouge and not to the geometric character of the interfaces.

3.2. Sequence of formation

If the described fabric is pervasive in the experimental studies, is there a sequence of development of the fabric elements?

To investigate this, an experimental configuration was designed to allow for larger displacements than allowed by the traditional triaxial configuration. Our limit even with

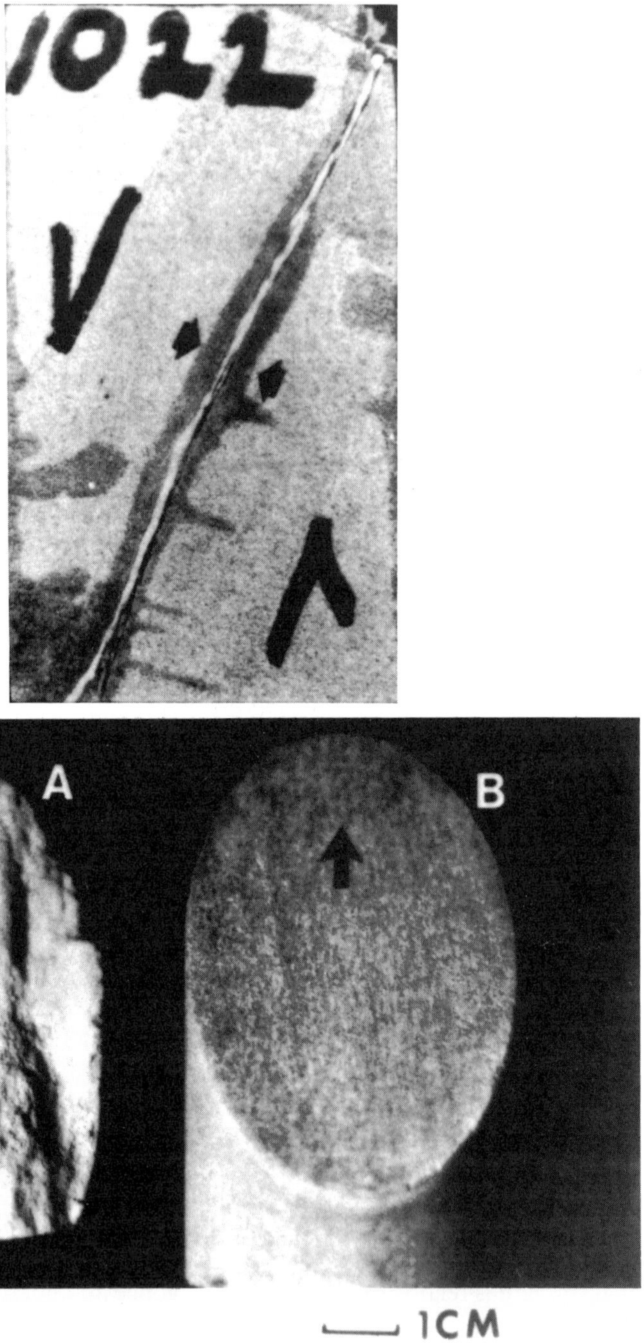

Figure 3. *Upper*: Photograph of a cut through part of a prefractured specimen. Calcite has been distributed along the fracture and sheared as shown. The darker band observed on both sides of the sliding surface, produced by a concentration of epoxy from the thin-section preparation, indicates the extent to which the deformation penetrated the country rock. *Lower*: Compare the sliding surfaces of a prefractured specimen (A) and a precut one (B). Both were deformed at a confining pressure of 50 MPa, room temperature, and a displacement rate of 10^{-4} cm s^{-1}.

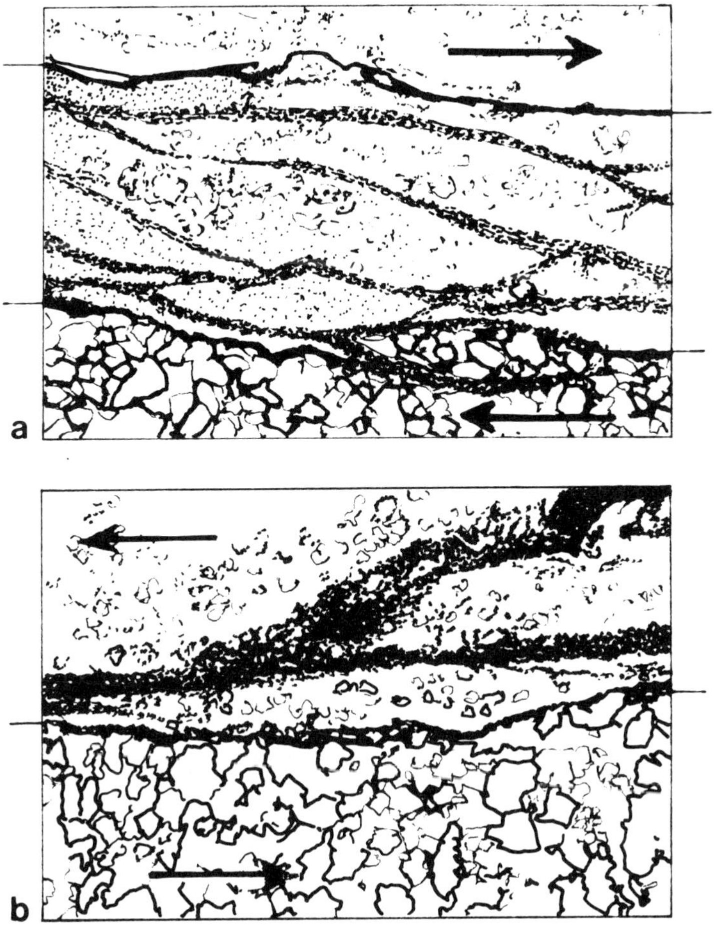

Figure 4. Drawings of deformed calcite gouge distributed along the fracture. Lines show interface of calcite with sandstone. Degree of stippling suggests the grain size with darker patterns indicating finer material. (a) Specimen with right-lateral displacement with gouge about 1 mm thick. Regularly spaced R_1-fractures are inclined at about 18° to the sandstone contact. Note the change in orientation of the R_1s to parallel the interface as Y-fractures in the lower part of the diagram. (b) Drawing of a portion of one interface in a different specimen that has undergone left-lateral displacement. Here a Y-fracture has separated the calcite in the depression from the main gouge mass, and left it as a relic. Sandstone is shown at the interface with the calcite. See text for further discussion.

large-diameter specimens is about 10 mm before the alignment of the specimen seriously affects the stress state (Shimamoto and Logan, 1981). A modification of the double shear block arrangement employed by Hoskins et al. (1968) was designed for use in a triaxial pressure vessel (Figure 5). This configuration allows displacement to 3 cm and has the additional benefits of constant normal stress and alignment during displacement.

A series of experiments, stopped at incremental displacements, thin-sectioned and studied petrographically, provide a view of the fabric development (Figure 6). The experiments utilize a direct shear configuration and are done with simulated calcite

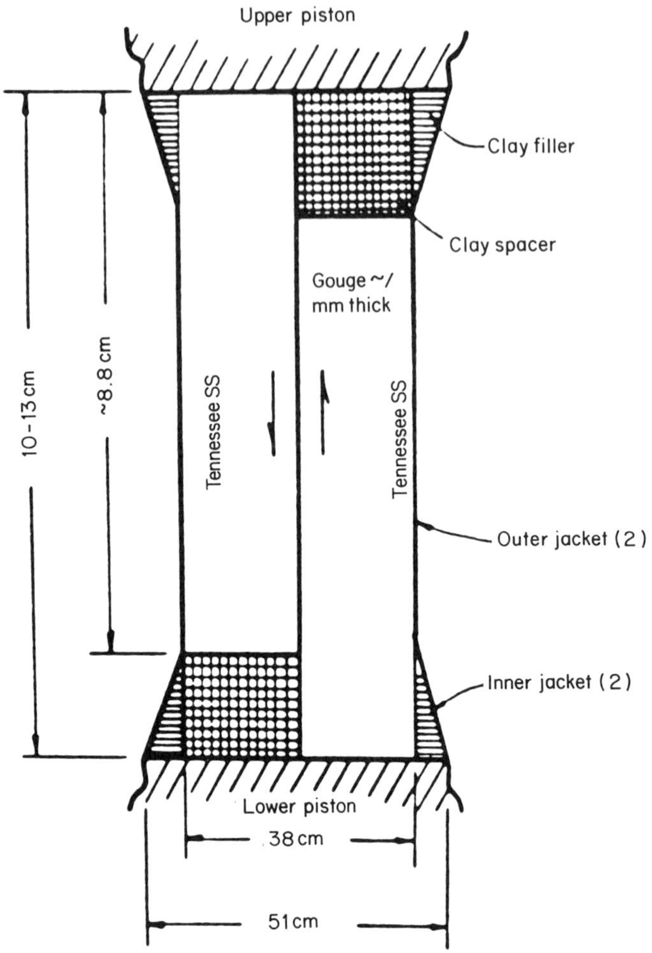

Figure 5. Diagram of direct shear assembly used in a triaxial pressure vessel. The assembly is jacketed in polyolefin to isolate the specimen from the confining pressure.

gouge, a constant normal stress of 40 MPa, room temperature, and a constant displacement rate of 10^{-4} cm s^{-1}. The resulting sequence is not unlike that observed by Tchalenko (1970) in his clay model studies. In the initial linear portion of the stress–displacement curves, no fractures are observed after unloading. Prior to the peak strength R_1- and R_2-fractures have developed across the grains, but have virtually no displacement along them (Figure 7). Fractures in the X orientation (Figure 1) are quite abundant, but there is a fanning of orientations from this to that of R_2. At about 4 mm of displacement, a drastic reduction in grain size has occurred throughout the gouge and the predominant fractures, R_1-fractures extend completely across the gouge zone and are transected at places by R_2-fractures. At 7 mm, R_1-fractures appear to extend by changing orientation to parallel the forcing block boundaries, taking a Y orientation. Incipient P-fractures are also visible. With further displacement, R_1s continue to extend and change orientation so that by 17 mm of displacement many have a segment parallel to the forcing block boundaries. P's have now propagated across the shear zone. New

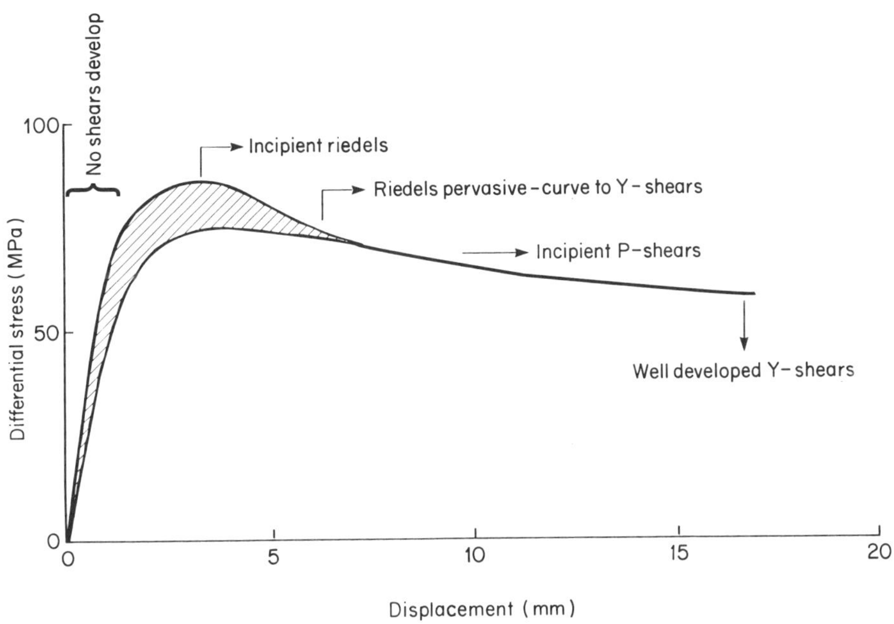

Figure 6. Composite of five direct shear triaxial experiments (with ~1 mm thick calcite gouge; $P_c = 40$ MPa; $T = 25°$C; $\mathring{d} = 10^{-4}$ cm s^{-1}, air dried). The stress–displacement curve is an average of the experiments, with the shaded portion indicating the spread in the data. The different stages in the evolution of the gouge are shown.

Y-fractures have formed distinct from extensions of R_1s. To about 25 mm displacement, the stress–strain curve shows steady-state behaviour with no new fabric elements present.

These results approximate those observed for other materials subjected to large shear displacements (e.g., Morgenstern and Tchalenko, 1967; Tchalenko, 1970; Mandl et al., 1977; Mandl, 1988), where the common stages of deformation are (1) a prepeak, strain hardening stage, (2) a post-peak, strain softening stage, and (3) a steady-state or residual stage.

3.3. Fracture array as a function of shear strain and displacement

It has been suggested that a change in mechanical response occurs with a change in gouge thickness (Byerlee and Summers, 1976). But if a correlation exists between the sequence of fracture development and changes in mechanical behavior, does the thickness of the fault zone become important or only the absolute displacement parallel to it? To address this, fracture arrays from two groups of experiments were studied. In the first, gouges of different thickness were deformed to different displacements so as to yield the same overall shear strain of about 1.5 (Table 2). In the second group the same thicknesses were used but now deformed to the same displacement (about 5.2 mm), thereby producing different shear strains.

The same shear geometry described above (Figure 5) was used with layers of simulated gouge of calcite, either 0.5, 1.0, 2.0 or 3.0 mm thick. The thickness desired for the gouge layer was obtained by calculating the weight of gouge material needed to fill the geometric

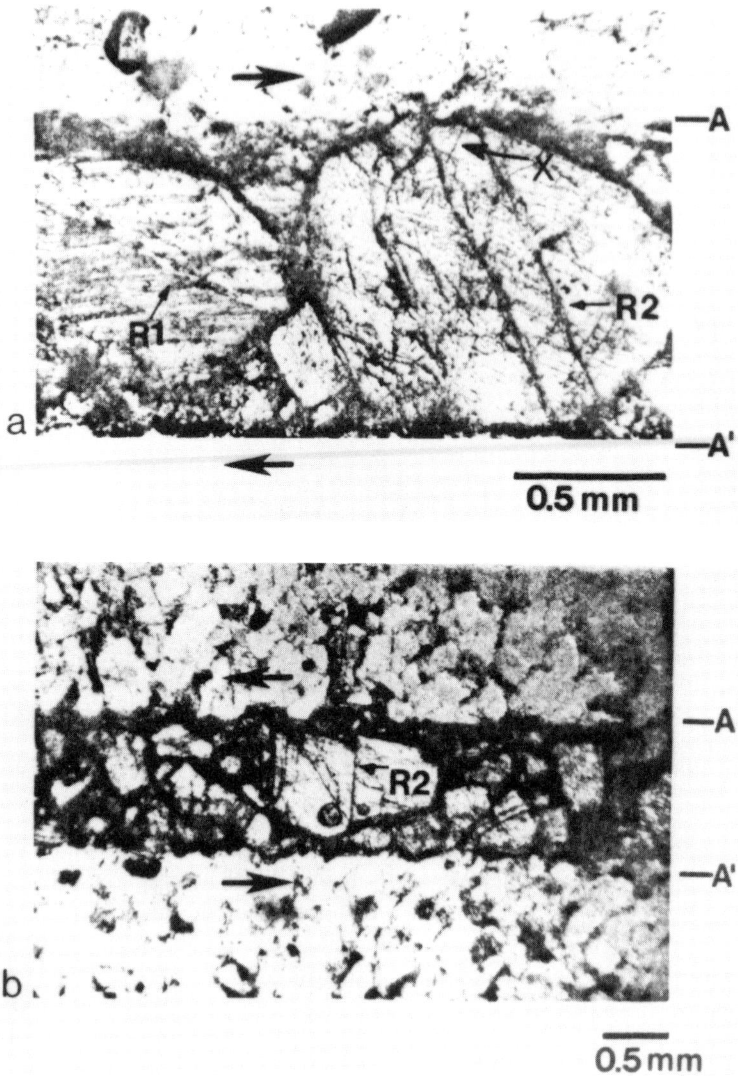

Figure 7. Photomicrographs (plane-polarized light) of deformed specimens of Tennessee sandstone with calcite gouge about 1 mm thick. Specimens deformed in direct shear at a constant normal stress of 40 MPa, room temperature, and a displacement rate of 10^{-4} cm s^{-1}. Sense of shear shown by bold arrows, and gouge interfaces by A–A′. (a) Specimen displaced to peak strength. Incipient R_1-, R_2-, and X-fractures are indicated. (b) Specimen displaced just past peak strength (about 3 mm). Displacement on a R_2-fracture is clear. X-fractures are also visible.

volume, using a density of 1.512 g cm^{-3}. The thickness of the deformed gouge, measured in thin section, was the value used for the calculation of shear strain.

3.3.1. Experiments to equal shear strain

Figure 8a shows the shear stress–shear strain curves for specimens deformed to a shear strain of 1.5. This value was chosen as a reference as it had been shown that R_1-fractures

Table 2. Summary of direct shear experiments to evaluate gouge fabric as a function of displacement and strain[a]

Experiment (specimen) number	Original gouge thickness (mm)	Deformed gouge thickness (mm)	Total shear displacement (mm)	Total shear strain[b] (γ)
Group 1				
11	2.00	1.60	2.8	1.7
12	0.50	0.47	0.7	1.6
13	3.00	2.60	3.4	1.3
14	1.00	0.85	1.2	1.5
Group 2				
20	0.50	0.44	5.2	11.2
21	0.50	0.42	5.4	12.7
22	2.00	1.61	5.0	3.1
23	1.00	0.80	5.3	6.7

[a] All experiments were conducted room dry, at 40 MPa confining pressure, a displacement rate of 10^{-4} cm s^{-1} and temperature of 25°C.
[b] Shear strain equals engineering shear strain.

werc well developed under this condition. Figure 8b shows that larger amounts of displacement are required in samples with a thicker gouge to reach the same shear stress as in samples with a thinner gouge layer, as expected.

Petrographic observations of all specimens show fabrics very similar to that already described except that Y-fractures are only incipiently developed. The only significant difference is that where a 3-mm gouge layer was used (not plotted in Figure 8), a lower shear strain resulted than in the other specimens for the same value of displacement and R_1-fractures are barely discernible and do not cross the gouge zone.

3.3.2. Experiments of equal displacement

The shear stress–shear strain behavior of all specimens is remarkably similar (Figure 9). Only the peak stress is slightly different, where a higher shear stress is generally needed to deform specimens with a thicker gouge.

Unlike the previous group of experiments where fracture arrays were similar, in this group the fabrics observed differ in all specimens. In specimens with a gouge layer 0.5 mm thick (Figure 10a), R_1-fractures are at very low angles, and almost parallel in some cases, to the forcing block interface. Anastomosing fractures, with an R_1, P, and particularly a Y orientation are observed. Y-fractures within the gouge and close to the interface dominate the fabric. In the specimen with a 1-mm gouge layer (Figure 10b), the R_1-fractures are abundant with some curvature toward the Y orientation. Y-fractures are also present but only close to the host–rock interface. The specimen with a gouge layer 2 mm thick (Figure 10c) shows the same fabric as found in the specimens deformed to a strain of 1.5. R_1-fractures dominate with R_2- and P-fractures present. Y-fractures are barely visible.

Another feature was found that was not previously observed owing to the larger shear strains achieved. In our initial characterization of the fabric, we found the deformation

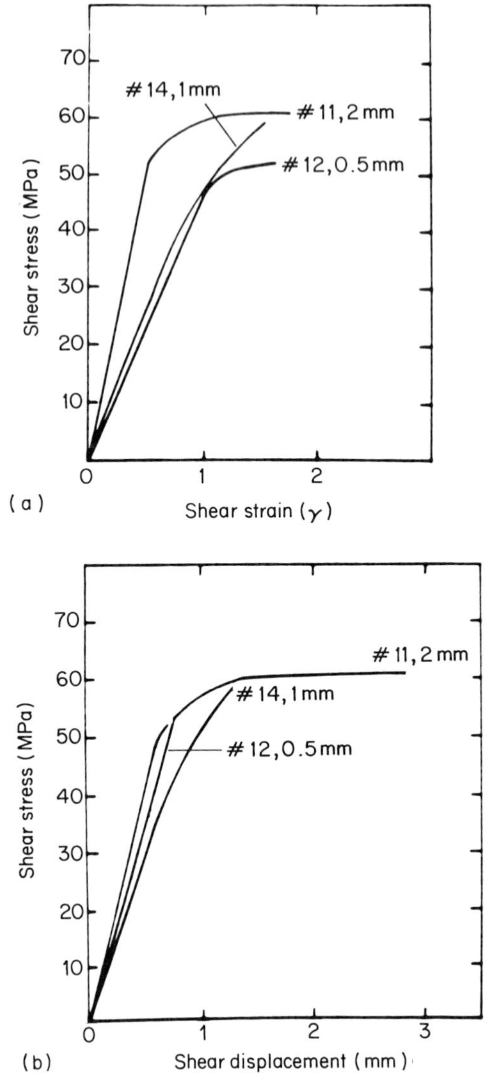

Figure 8. (a) Shear stress–shear strain curves for specimens deformed to approximately an equal shear strain (Table 2), and (b) shear stress–shear displacement curves for the same experiments. Experiment numbers and corresponding gouge thickness are shown. All experiments were run at 40 MPa confining pressure, room temperature, and a displacement rate of 10^{-4} cm s^{-1}.

to be very heterogeneous. Although grain size is reduced within the gouge, R_1, R_2, P, and Y are the loci of greater grain comminution, with the greatest decreases along the Y-fractures. The areas between these shear fractures are less deformed. Many of the grains are about the size of the starting material. With larger shear strains in these experiments, displacement along adjacent Y-fractures initiates new R_1 and R_2 sets confined between these Y's. These are analogous to the 'Riedels within Riedels' that have been observed in natural faults. Our integration of this development with our

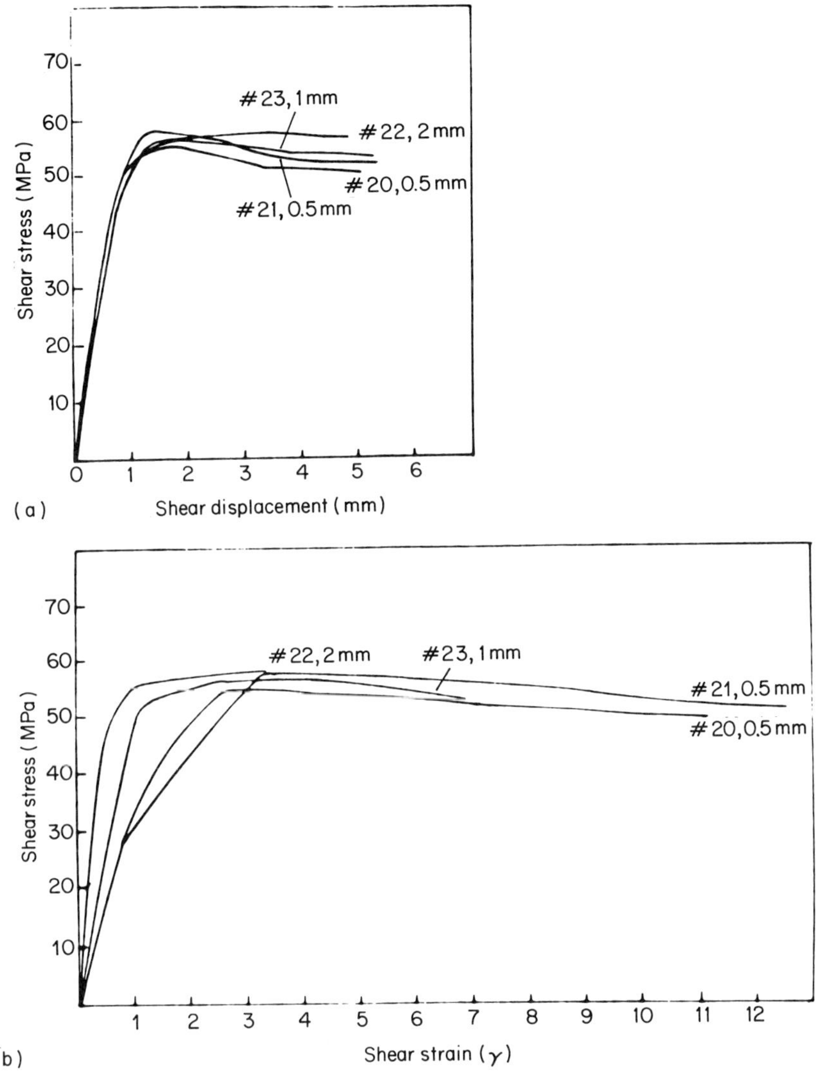

Figure 9. (a) Shear stress–shear displacement curves for specimens deformed to approximately equal shear displacements, and (b) shear stress–shear strain curves for the same experiments. Experiment numbers and corresponding gouge thickness are shown. All experiments were conducted at 40 MPa confining pressure, room temperature, and a displacement rate of 10^{-4} cm s^{-1}.

previous findings results in a composite picture of the evolution of the fabric in the brittle regime shown in Figure 11.

4. Fabric and Mechanical Behavior

Is the fabric evolution related to changes in mechanical response of the fault gouge? For instance, (1) Is slip partitioned among the fabric elements? (2) Is stable sliding or

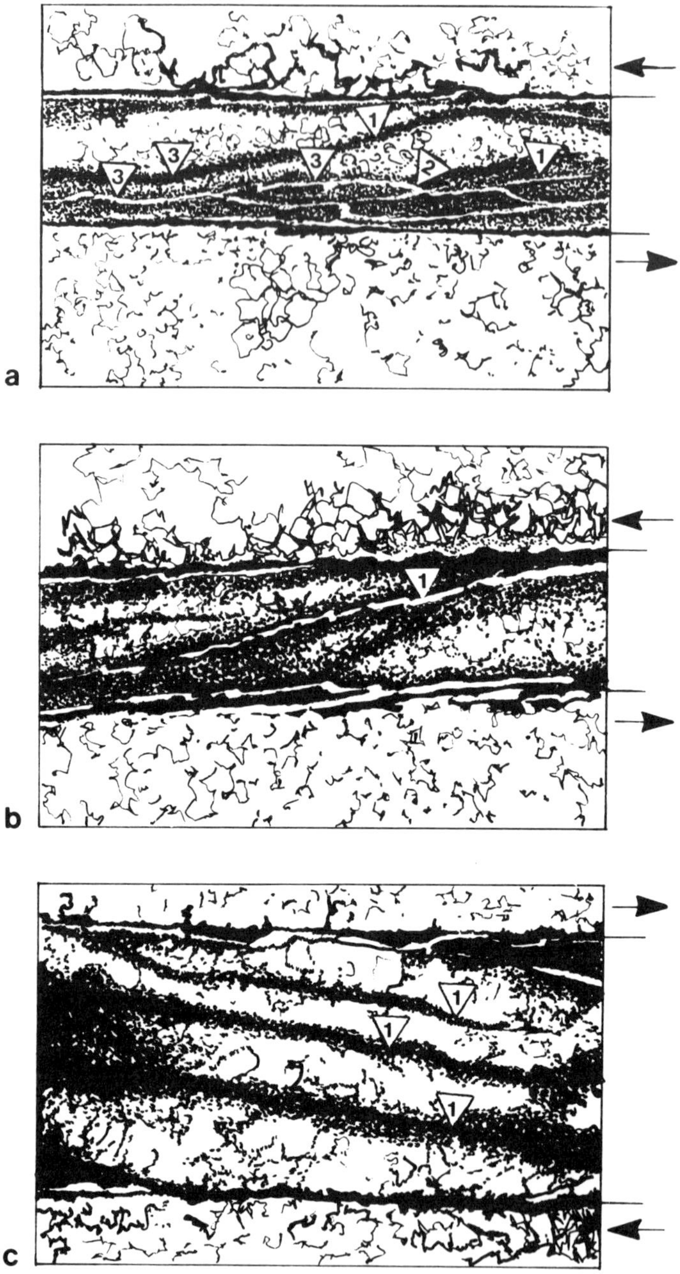

Figure 10. Drawing of deformed specimens of Tennessee sandstone with a layer of calcite gouge along a sawcut of differing thicknesses displaced to about 5.2 mm, but resulting in differing shear strains. The lines on the right mark the gouge interfaces and arrows show the sense of shear. (a) Specimen 21 with a thickness of 0.5 mm and a resulting shear strain of 12.7. The number 1 arrows point to R_1-fractures, number 2 to P-fractures and 3 to Y-fractures. (b) Specimen 23 with an initial gouge thickness of 1.0 mm and a resulting strain of 6.7. White arrow shows R_1-fractures, which change orientation toward lower angles close to the sawcut interface. (c) Specimen 22 having an initial thickness of 2 mm, and a resulting shear strain of 3.1. Arrows point to R_1s. Note absence of change in angle to these and lack of Y-fractures.

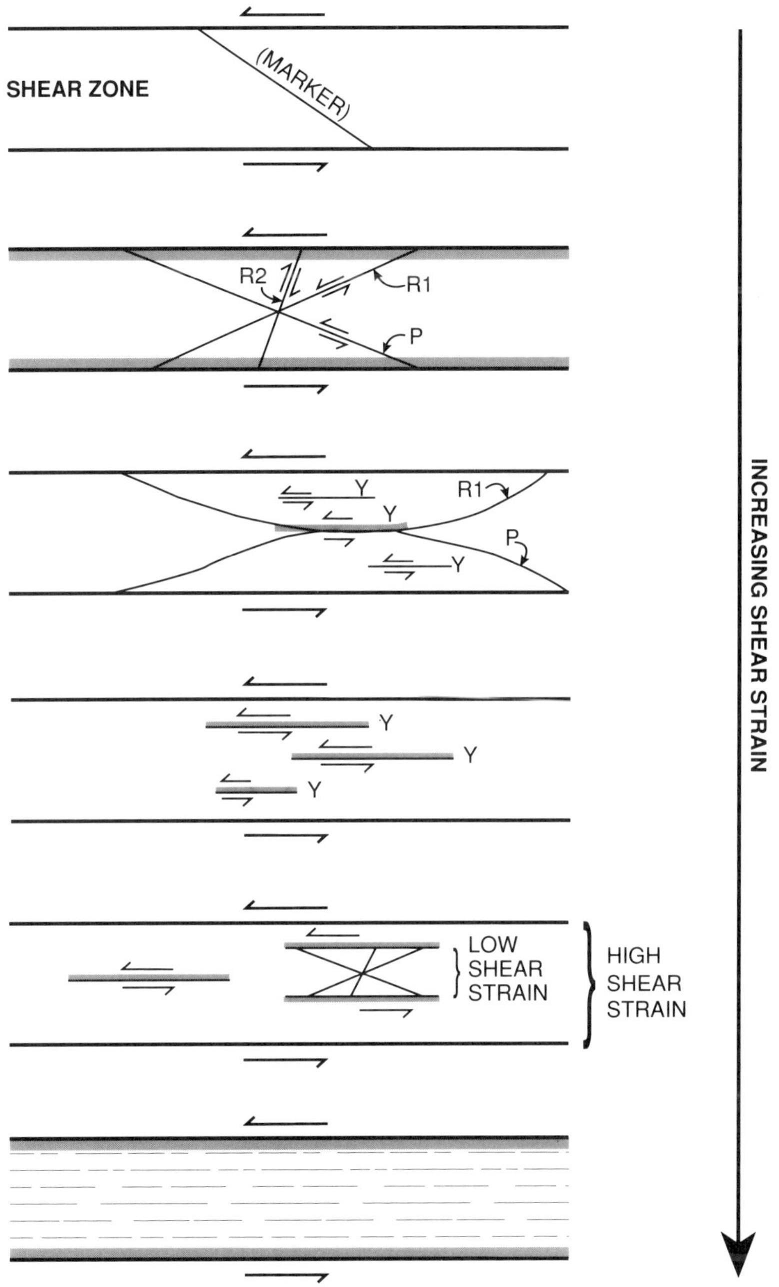

Figure 11. Evolution of shear zone fabric as elucidated from experiments. Fabric elements shown in black, with sites of major displacement shown by heavy lines. One or several Y-fractures may accommodate the displacement, but generally they are close to the interface with the country rock. Note that changes are a function of the shear strain and not of absolute displacement.

Figure 12. Boundary between black and white calcite simulated gouge which remains at high angles to the interface of Tennessee sandstone after precompaction at 150 MPa confining pressure. The upper part of the white calcite layer was lost in sectioning. The gouge layer is about 0.6 mm thick.

stick–slip sliding uniquely related to slip on selected fabric elements? (3) Are strain hardening or softening indicative of specific stages in the evolution? These questions are considered in this order.

The general difficulty of documenting the internal deformation of experimental shear zones has been the lack of internal markers. Displacement of grains has provided local sense of shear, but more detailed information and particularly magnitudes of displacements have not been available. To overcome this deficiency we used calcite coated with hydrocarbon, which gives it a black color (Wang, 1989). After determining through standard triaxial tests that its properties were not significantly different from uncoated grains, we formed color boundaries at high angles to the interface with the forcing blocks. Two boundaries were formed in each specimen. We utilized a sawcut geometry in triaxial compression configuration.

All specimens were compacted at 150 MPa confining pressure for 5 hours after which, the layer has a bulk density of about 1.57 g cm^{-3} and a porosity of about 42%. The color interface is still inclined at high angles but now has moved from about 90° to about 78–80° (Figure 12). With shear, the expected fracture array develops, but the fracture sets – R_1, R_2, P, and Y – cut the color interface so displacements along them can be measured (Figure 13). These can be used with values of the final thickness W, the angle the color interface makes with the sawcut, and the projected length L of the color interface on the sawcut interface to determine the partitioning of the displacement within the gouge.

The axial shortening of the specimen is measured with an external transducer and is corrected for apparatus distortion. The resulting corrected axial shortening δ is a

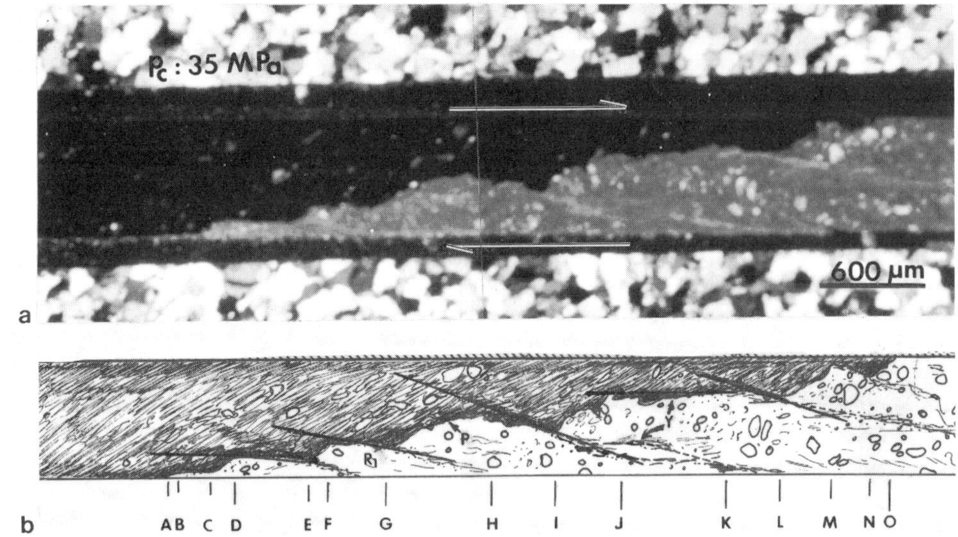

Figure 13. (a) Photomicrograph of deformed calcite boundary in specimen no. 3029 after 14.5 mm of shear displacement. (b) Sketch of photo shown indicating fracture elements. Sense of shear is right-lateral. See text for discussion of letters on bottom of sketch and analysis.

composite of (i) elastic shortening of the specimen, which has been determined to be negligible (Shimamoto and Logan, 1981); (ii) compaction of the gouge, which can be calculated knowing the initial and final thickness of the gouge layer; (iii) shear displacement along Y-fractures, which can be calculated by subtracting the other effects from the corrected shortening; and (iv) deformation involving the total gouge thickness reflected by changes in W/L. This fourth contribution is composed of displacement on R_1, R_2, and P-fractures and closely spaced distributed shearing within the gouge. This latter phenomenon was only recognized by using the internal markers.

A first set of tests was run at constant confining pressures of 30 and 50 MPa, but tests were stopped at predetermined displacements to document the partitioning of deformation with increasing shear strain (Figure 14). The second set was a series deformed to about the same shear strain but with confining pressures to 300 MPa to assess changes with increasing stress.

4.1. Fabric development and mechanical response at a constant confining pressure

At 30 and 50 MPa confining pressures, when the experiments are stopped and thin sections examined, the fracture development is very similar to that documented in our direct shear experiments. The black–white boundary reveals an additional stage, which was suggested in earlier experiments but which we were unable to document, and that is an early stage of homogeneous shear.

The black–white boundary is clearly cut and offset by R_1-fractures (Figure 13), but the displacement along these fractures is not sufficient to account for the rotation of

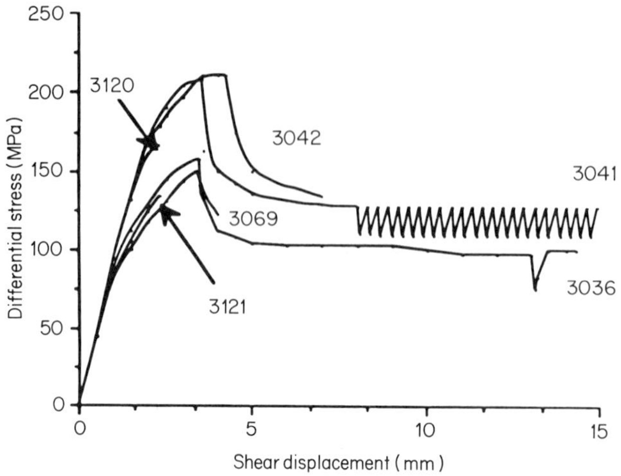

Figure 14. Stress–shear displacement curves for sequential tests to evaluate fabric changes as a function of shear strain. Shear displacement measured along 35° sawcut. Experiment number is shown and tests were done at constant confining pressures of 30 and 50 MPa. See Table 3 for data.

the boundary. Although the boundary between adjacent R_1-fractures is still often at high angles to the rock interface, in many places (e.g., between G and H, Figure 13) it can be found to have rotated on a scale smaller than R_1-fractures. The shearing appears to be by cataclastic shearing at the grain scale, hence the term 'homogeneous shearing.' As this does not disrupt R_1-fractures it is presumed to precede them, and occur in the earliest stages of deformation, probably coincident with compaction.

Utilizing the black–white boundary to assess partitioning of displacement, measurements are made using drawings similar to those in Figure 13 (Tables 3 and 4). The displacement of the black–white boundary along each fracture is projected on to a line parallel to the gouge–rock interface and the type of fracture (R_1, etc.) is noted for each segment. Displacements for a single fracture type are summed. For example,

$$R_1 = \sum(D - E + F - G + H - I + K - L + M - N)$$

These values are then compared to ones made for displacement along Y-fractures and those assigned to homogeneous shear. P-fractures generally cut the boundary because of the parallelism in orientation corroborating the late development of these fractures. R_2-fractures, while recognizable, were so small that displacements of the black–white boundary could not be determined accurately.

Reviewing the experiments at 30 MPa, we find that the ratio L/δ increases to a value approaching 1 (Table 3) and then decreases with increasing displacement. Unity indicates that the total displacement is entirely accommodated within the gouge, that is there is no displacement along Y-fractures. Simultaneously, the ratio of W/L changes from a maximum value of about 0.26, prior to the ultimate strength, and decreases with further displacement. This maximum value corresponds to the period of compaction, where R_1 and R_2 are dominant, and prior to any evidence of Y-fractures. This initial stage is also when the density shows the greatest increase (Figure 15). Table 4 quantifies the change

Table 3. Data compiled from experiments with black–white calcite

Experiment *number*	Thickness W (mm)	Internal shear L (mm)	W/L	Simple shear angle (degrees)	Externally measured displacement (mm)	L/δ	Density (g cm^{-3})	Density change[a] (%)
Experiments at a confining pressure of 30 MPa								
3121	0.5	2.0	0.3	70.0	2.3	0.86	1.84	17.1
3069	0.6	3.4	0.2	77.0	3.6	0.94	1.93	44.6
3036	0.6	5.0	0.1	81.0	15.0	0.33	1.97	44.3
3035	0.5	4.8	0.1	80.0	15.2	0.37	1.94	44.3
Experiments at a confining pressure of 50 MPa								
3120	0.45	1.6	0.28	65	1.6	1.00	1.92	22.2
3042	0.54	2.1	0.26	70	6.2	0.33	2.04	29.9
3040	0.53	2.2	0.24	69	12.8	0.17	2.07	31.8
3041	0.54	2	0.27	70	13.1	0.15	2.04	29.9

[a] Initial value taken as 1.57, that measured after precompaction at 150 MPa (sample number 3070, Table 5)

Table 4. Displacement partitioning as a function of shear strain at 30–35 MPa confining pressures

Experiment number	Confining pressure (MPa)	Total displacement (mm)	Displacement along R_1-shears (% of total)	Homogeneous shearing (% of total)	Displacement along Y-fractures (% of total)	Experiment terminated[a]: Slope of stress–strain curve
3121	30	2.3	8	92	0	Before ultimate strength Strain hardening
3069	30	3.6	42	50	8	After ultimate strength Strain softening
3029	35	14.5	36	30	34	Steady-state sliding
3036	30	15.0	35	27	38	Steady-state sliding

Note: Percentages have an error of about $\pm 5\%$.
[a] See Figure 14.

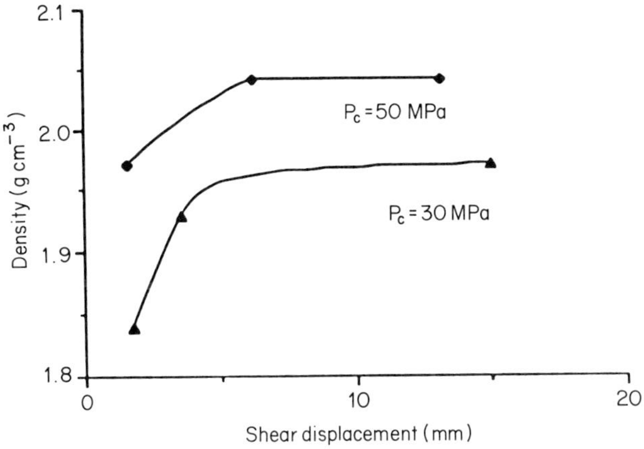

Figure 15. Gouge density versus shear displacement for experiments shown in Figure 14. Densities were measured at the end of each run.

in the relative amounts of axial displacement accommodated by R_1-, Y-fractures, and through homogeneous shearing.

Combining these observations provides an interpretation of fabric evolution. Initial loading causes compaction accompanied by grain to grain movements which we have termed homogeneous shearing. R_2- and R_1-fractures initiate but propagate only about one or two grain diameters. This is a period of strain hardening, little displacement parallel to the shear-zone boundaries, density increases of the gouge, and small rotation of the color interface. With further axial displacement of the specimen, compaction decreases as shear displacement increases. Strain softening occurs, R_1-fractures propagate and change orientation to parallel the rock interface to accommodate more boundary-parallel shear. P-fractures and incipient Y-fractures are formed. The gouge density increases at a slower rate and the amount of deformation accommodated within the gouge (exclusive of the contribution of the Y-fractures) reaches a maximum; L/δ approaches unity.

Additional displacement shows a continued decrease of L/δ as Y-fractures form and shear displacement shifts to them almost totally. Compaction and density of the gouge seems to change little nor does the ratio of W/L. Once displacement is localized along the Y-surfaces, conditions for unstable sliding appear to be present for stick–slip begins (Figure 14). Displacement during stick-slip seems to make the Y-fractures smoother.

Data from the experiments at 50 MPa suggest that the higher confining pressures accelerate the evolution, substituting for larger displacements at the lower confining pressures. This corroborates the findings on the influence of preconditioning the gouge layers by a stress cycle as found by Shimamoto (1985) and Chester (Chester and Logan, 1990). Here the effect appears to substitute the stress cycle for displacement in creating the Y-surfaces.

Additional documentation of this process is seen in experiments on halite simulated gouge (Figure 16). In this case, as with calcite, unstable sliding is not found until after the Y-fracture is formed (Figure 16c).

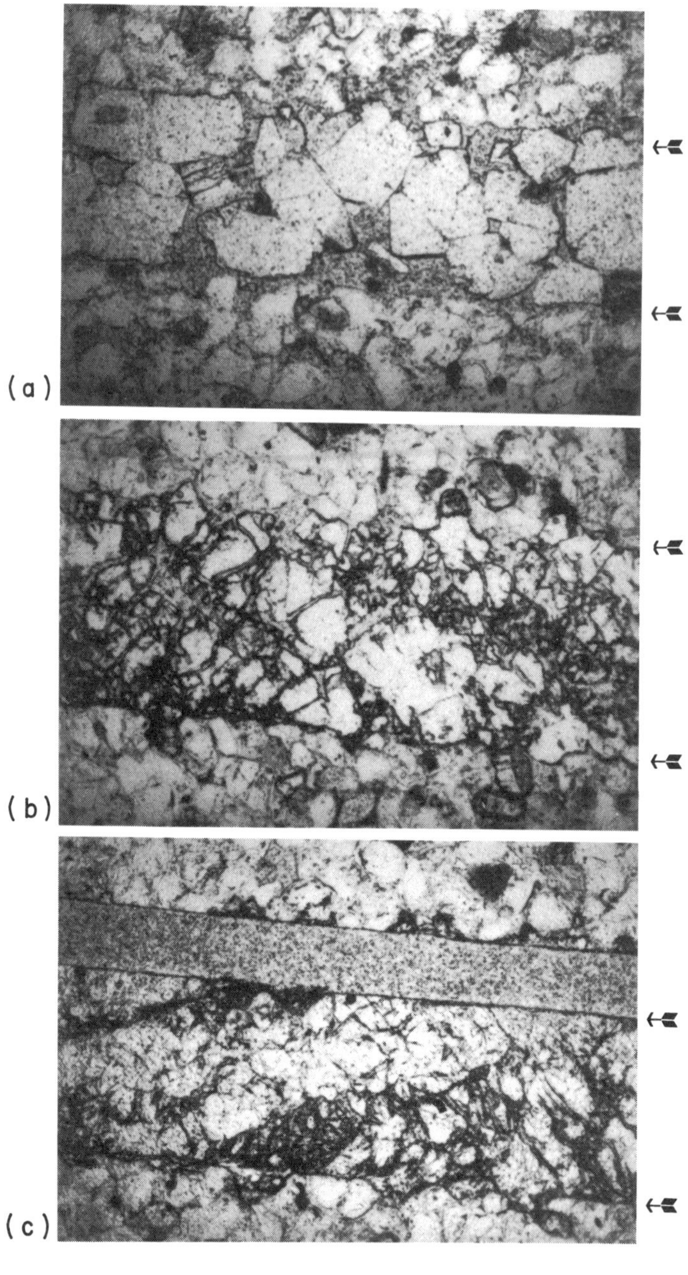

Figure 16. Photomicrographs of simulated gouge of halite. Halite–sandstone forcing block interface indicated by arrows. Gouge thickness 0.5 mm. Specimens deformed at a confining pressure of 70 MPa. (a) Subjected to confining pressure only. (b) Deformed in right-lateral shear to 1.7 mm of displacement. Note that R_1-shear zones are well developed. (c) Deformed to 4.1 mm of displacement in left-lateral displacement and the onset of stick–slip. Y-fracture now truncates R_1s. Upper contact between halite and sandstone filled with epoxy. (Courtesy F. Chester.)

4.2. Fabric development and mechanical behavior with increasing confining pressure

The second set of experiments at confining pressures to 300 MPa illustrates the interchangeability of normal stress for displacement in the fabric development. A change from stable sliding to stick–slip occurs in our apparatus at about 50 MPa confining pressure for the calcite. Above this pressure, the amount of displacement before the onset of stick–slip decreases with increasing confining pressure. Additionally, the ultimate strength appears to reach a limiting value at about 250 MPa.

Photomicrographs (Figure 17) illustrate the internal changes in fabric from localized deformation concentrated along widely spaced R_1s at low confining pressures to closely distributed shear within the gouge at higher pressure. At higher pressures the R_1s are more closely spaced, extend only a few grains, are not as long, and are much more numerous. They are also inclined at higher angles to the interface similar to the results found by Gamond (1983) and Naylor et al. (1986). The result is similar to that of homogeneous shearing recognized in the early stages of gouge evolution at low pressures. At high pressures, the largest and dominant fractures are still Y's which are close to one or both interfaces with the forcing block.

Using similar measurements as before (Table 5) an evaluation of the displacement partitioning can be made. With increasing confining pressure, L/δ increases and W/L decreases, both approaching more or less constant values at about 50 MPa confining pressure coincident with the transition to stick–slip. The deformation within the gouge (again not counting the contribution of slip on the Y-fractures), decreases as the confining pressure increases, and the Y-fractures are formed with less prior internal deformation. The density also increases to a maximum at about 50 MPa, where Y-fractures and stick–slip appear dominant (Figure 18).

In all cases displacement eventually shifts to the Y-surfaces, but at higher confining pressures less displacement is necessary to produce this shift than at low pressures. This is shown in Figure 19, by the amount of 'run in' displacement that is necessary before the fabric evolves and stick–slip develops.

4.3. The influence of an existing fault on the stress field

The deformation in a gouge layer that already has within it an R_1 fracture array has been modeled using the finite-element method. The finite-element mesh and boundary stress conditions that were selected to simulate triaxial loading are shown in Figure 20a. A confining pressure of 150 MPa was used, and an imposed differential stress parallel to the axis of the specimen. Initially, a linear elastic constitutive equation was used for all elements in the model. Three different portions of the model (forcing block, gouge and R_1-shear) were assigned different elastic moduli. The forcing block elements have value of $E = 5 \times 10^5$ MPa, the gouge elements one order of magnitude less, and the Riedel shear elements two orders of magnitude less. These values were selected not to simulate any specific rock or gouge type but to approximate estimated differences within an experiment.

These results (Figure 20b,c,d) show that in the vicinity of the shear zone maximum principal stress trajectories are reoriented to high angles at the shear zone boundary and also reoriented within the Riedel shear. In the latter it forms an angle of about 50°

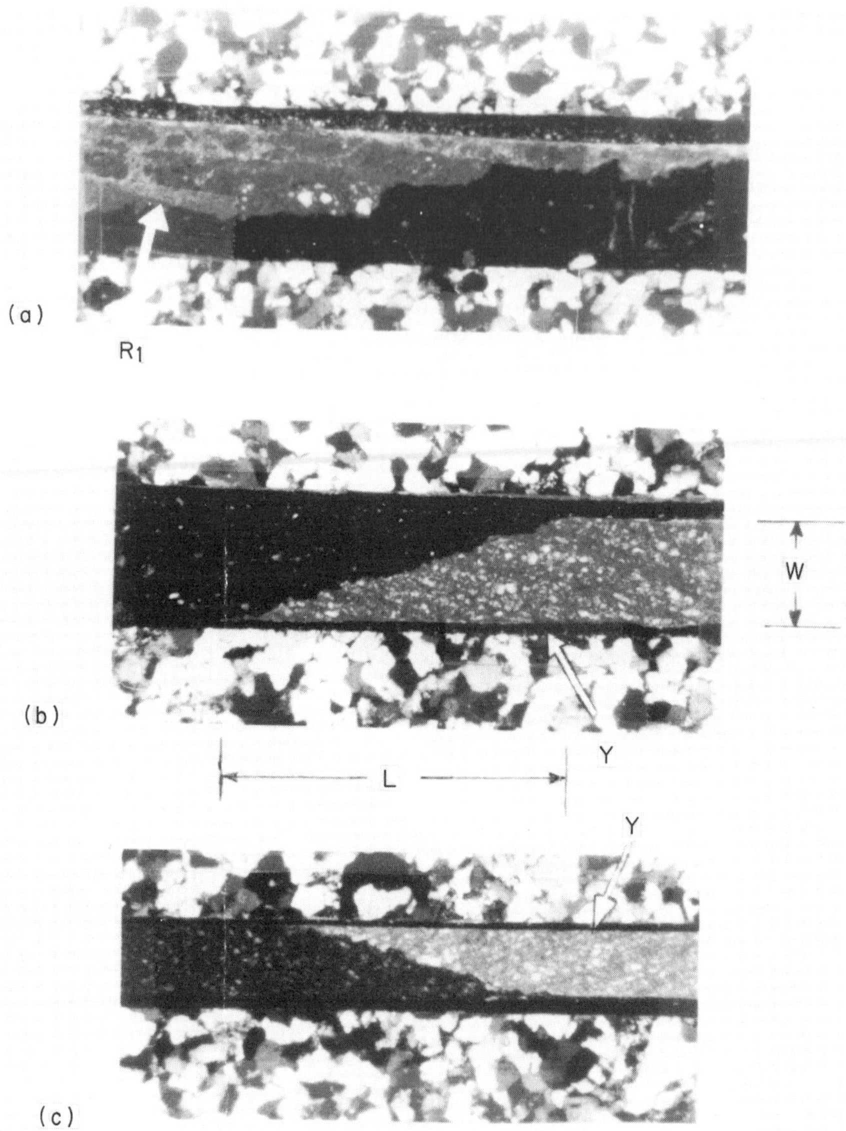

Figure 17. Photomicrographs of shear zone showing the boundary between black and white calcite at increasing confining pressures after each is shared about 12 mm in triaxial compression. *W* and *L*, as defined in the text, are shown in (b). Shear zones are about 0.55 mm thick except for (c) which is 0.33 mm. (a) Confining pressure 25 MPa, right-lateral shear. (b) Confining pressure 50 MPa, right-lateral shear. (c) Confining pressure 295 MPa, shear is left-lateral.

to the Riedel shear boundary. The reorientation at the shear zone boundary supports the contention that both across and within the shear zone the state of stress is not uniform.

Although these results are intriguing, there remains doubt of the restrictiveness of the assumed elastic behavior. Additionally these results could be explained simply by the effect of welded boundaries joining elements of greatly differing moduli. So it is

Table 5. Data for experiments with black–white calcite boundaries as a function of confining pressure

Sample number	P_c (MPa)	Thickness W (mm)	Internal shear L (mm)	Simple shear angle (degrees)	External shear δ (mm)	Steady state shear strength (MPa)	Density (g cm^{-3})
3070	150	0.68	0.15	12			1.57
3037	20	0.64	7.5	72	14.7	88	1.67
3027	25	0.54	6.0	78	12.5	96	1.85
3035	30	0.55	5.0	80	15.2	105	1.94
3036	30	0.54	4.8	81	15.0	120	1.97
3029	35	0.60	3.9	71	14.5	110	1.83
3071	40	0.54	5.0	70	12.9		1.97
3040	50	0.53	2.2	69	12.8	130	2.07
3041	52.5	0.54	2.0	70	13.1	128	2.04
3042	52.5	0.54	2.1	70	6.2	143	2.04
3072	100	0.34	1.3	73	5.0	200	2.35
3044	200	0.33	1.0	72	4.1	310	2.42
3046	250	0.38	1.1	73	4.9	380	2.4
3049	300	0.33	1.1	74	5.4	430	2.42

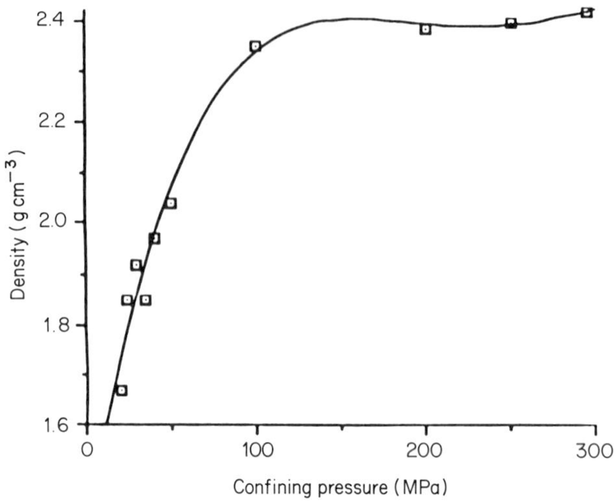

Figure 18. Gouge density measured at the end of the experiments shown in Figure 19 versus confining pressure. The transition from stable sliding to stick–slip occurs at about 50 MPa.

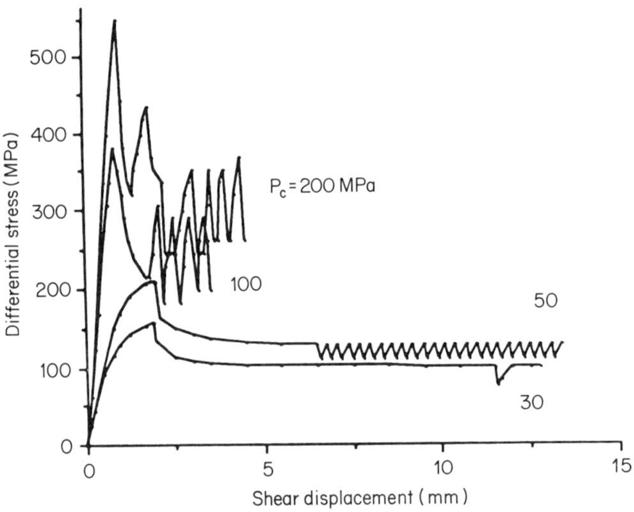

Figure 19. Stress–displacement curves at confining pressures shown. Displacement is that measured along the sawcut containing simulated gouge of calcite. Specimens were deformed at room temperature and constant displacement rates of 10^{-4} cm s^{-1}.

relevant to consider a shear model having an elastic–plastic constitutive law, especially as this rheology is more consistent with the experimental observations. For the forcing block elements, a linear elastic constitutive equation has been retained, with the same elastic modulus as previously described. Unlike the first model, this one ascribes to the shear zone and Riedel elements an elastic–plastic constitutive equation, with a von Mises yield criterion and an isotropic work hardening rule. The gouge and Riedel shear elements initially deform elastically as in the previous model. The plastic deformation

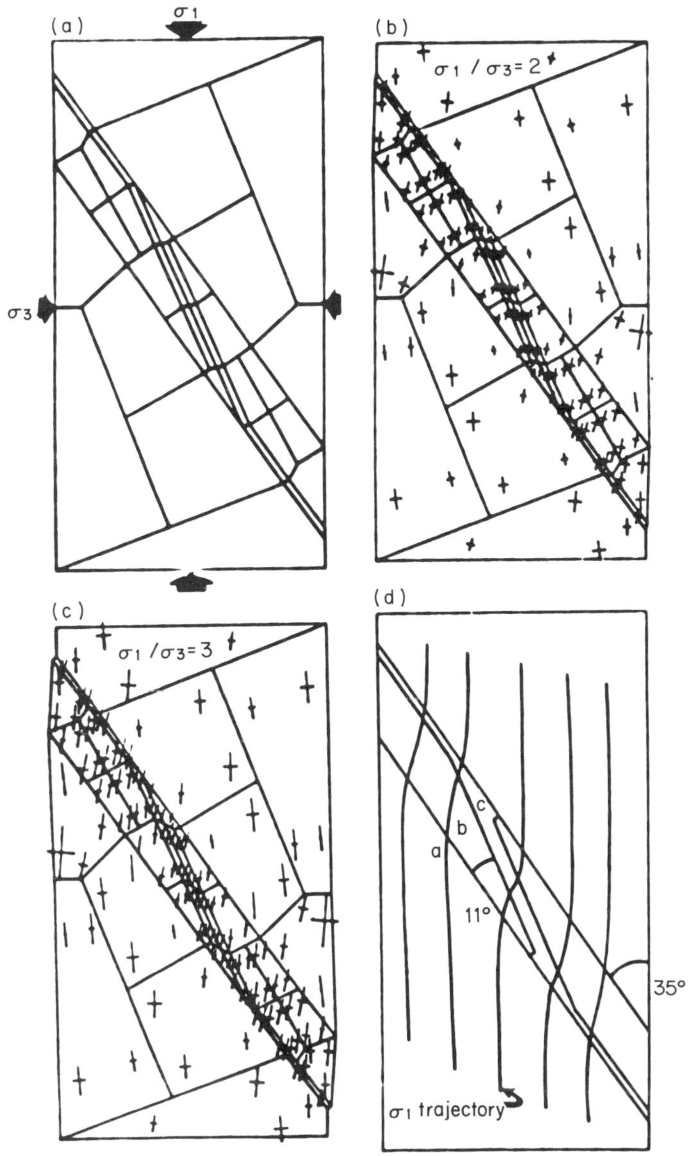

Figure 20. Finite-element results of gouge zone with a simulated Riedel shear using a linear elastic constitutive law. (a) The element size and arrangement used, with the far-field stresses. (b) Using a σ_1/σ_3 ratio of 2, the orientation of the principal stresses are shown, with their relative magnitudes scaled to the length of the lines. (c) Here the ratio $\sigma_1/\sigma_3 = 3$ is used at the boundaries of the specimen and the principal stresses are shown within the specimen. (d) Smoothed trajectories of σ_1 through the gouge zone containing the R_1.

follows a prescribed stress–strain path, where the first point is that of the yield point and the last point is equal to the desired amount of strain. Two stress–strain curves are used (Figure 21). The one with the highest yield point is used for the gouge elements, and corresponds to a curve obtained for experiments done at 150 MPa confining pressure,

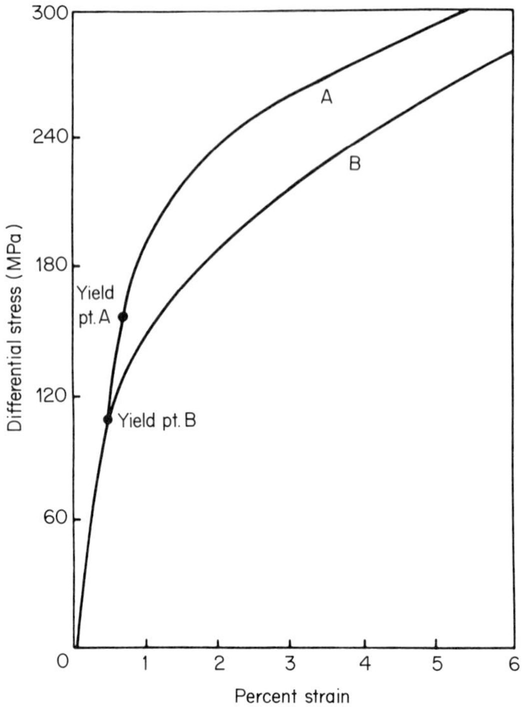

Figure 21. Differential stress–strain curves used to model the elastic–plastic elements. Curve A is ascribed to the gouge elements and curve B to the Riedel shear elements. The yield point for each is the first stress–strain point used for the program.

using a precut sample with quartz gouge. The curve with the lower yield point is ascribed to the Riedel shear elements. It was arbitrarily chosen to have a lower yield point, such that upon deformation the Riedel shear elements would yield before those surrounding it.

Figure 22a is a plot of the stress orientations and relative magnitudes for a ratio of the maximum and least principal stresses equal to 2.0. The stress trajectories of the maximum principal stress are shown in Figure 22c. Similar models with the stress ratio equal to 2.5 do not show appreciable changes from this one.

A stress reorientation appears near the host-rock–gouge interfaces, where the angle changes from 35° to 45°. Also, at the Riedel shear-gouge boundaries the orientation changes such that the maximum principal stress forms an angle of about 50° to such a boundary. It is worth noting that the larger angles are observed where the Riedel shear is surrounded by gouge, and not where it is in contact with the host-rock. A comparison between the linear elastic model and the elastic–plastic one shows: (1) in the elastic–plastic model the stress magnitudes are more constant than they are in the elastic one for the bounding block elements; (2) the stress reorientation at the shear zone boundaries and within the Reidel shears is observed for both models. The angular changes are about the same for both.

The changes that take place within the shear zone in our model appear to be independent of the exact constitutive properties of the material (i.e., elastic versus elastic–plastic). This suggests that the observed stress reorientation arises because of the deformation gradient at the shear zone and Riedel shear boundaries. Thus, the

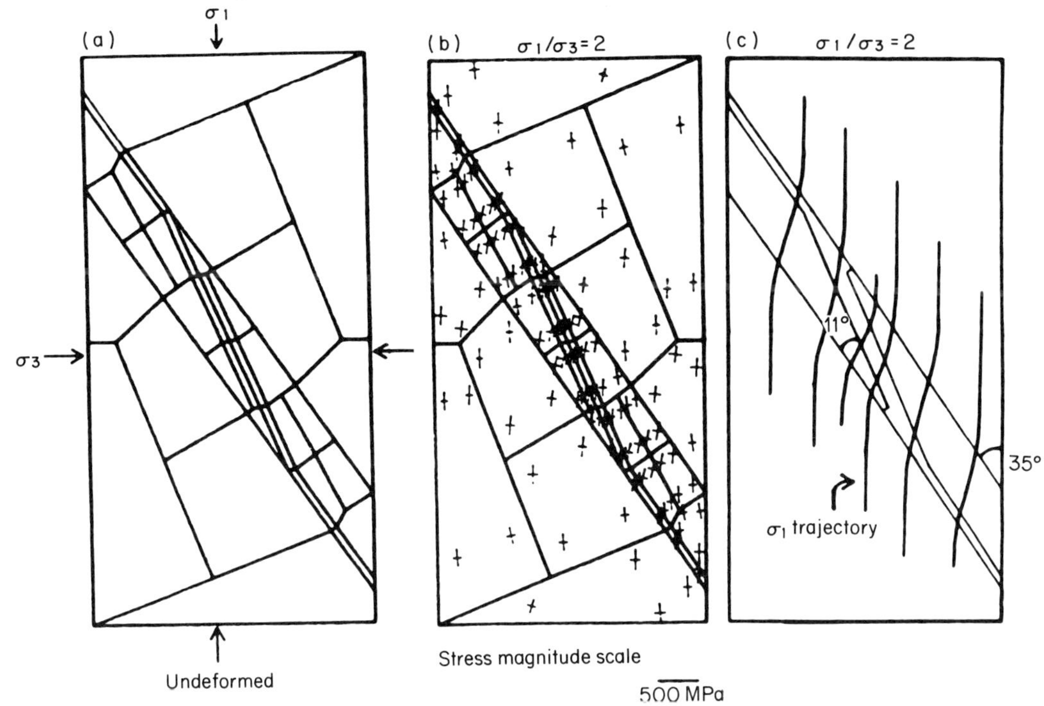

Figure 22. (a) The undeformed mesh and far-field boundary stress conditions of the elastic–plastic finite-element model. See text for details. (b) Using $\sigma_1/\sigma_3 = 2$, the orientation and scaled magnitudes of the principal stress are shown for the elastic–plastic solution. (c) σ_1 stress trajectories drawn for $\sigma_1/\sigma_3 = 2$.

contact of a highly deformable material with a less deformable material apparently causes the reorientation of the stress tensor.

5. Discussion

5.1. Geometric changes in fabric

One of the salient findings of our work is that the fabric of shear zones in laboratory experiments changes in a predictable and reproducible manner. The evolution of the fabric is linear and directional; that is, in general one stage follows another in a given sequence and leads to a final geometry. There is a variation found in this sequence at high confining pressures and normal stress. The R_1s are more closely spaced and are not as continuous. Their extension and curvature to approach the Y orientation is not as apparent, and the shift in major displacement from the R_1s to the Y-fractures appears to be more abrupt. The more uniform distribution of R_1-fractures is in marked contrast to the more wisely spaced ones found at the lower stress conditions. The differences in heterogeneity of deformation within the gouge as a function of confining pressure may allow at least qualitative retrospective estimates of the magnitude of normal stress conditions in natural fault zones.

The culmination of the evolution is a fracture array associated with 'steady-state' mechanical behavior. Here steady-state is used where upon returning to a previous sliding velocity the same frictional response is measured, regardless of the magnitude of intervening velocities (Shimamoto and Logan, 1986; Logan and Rauenzahn, 1987). This is not the case at conditions prior to this steady-state stage. In these cases one cannot obtain the same response when changes in velocity are made. Therefore, such mechanical data accumulated during the evolutionary period are not unique and are directly tied to the current fracture array. For this reason they cannot be directly compared to other experiments, unless the condition of the fracture array can be shown as similar. In contrast, data acquired once the gouge fracture array has evolved to a steady state can be compared to other results as long as it is known that fracture array has achieved steady state in all cases.

With this insight, care must be taken in comparing data not only from one experiment to that from another but from laboratory to laboratory where experimental procedure may vary widely. Although such caution is germane to the interpretation of laboratory data, it is not clear at the present time what condition is pertinent to appraise natural faults. Our current data on the long term mechanical response of a particular fault is limited, providing little insight into this question. Thus, an assessment of whether or not steady-state fracture arrays have been achieved on a particular segment would seem presently to rely upon characterization of the fabric. This step has not been attempted to our knowledge. Additionally, any a priori assumption is complicated by healing processes that are expected to take place during recurrence intervals when much of the fault is not moving. It may be argued that healing or partial healing may sufficiently change a steady-state fabric so that the evolution process starts over again with each renewed slip event.

An implication of the documented evolution towards a steady-state fracture condition in gouge where the displacement is localized along Y-fractures is the use of laboratory data derived from experiments of sliding rock by rock. When simulated gouge studies were initiated (Engelder et al., 1975) it was postulated that the frictional response might be fundamentally different from that of intact rock/rock sliding as had been used in prior frictional studies. It is now apparent, however, that steady-state sliding along Y-fractures involves the same general phenomenon of two relatively intact masses moving by each other as does rock/rock sliding. This has been empirically assumed in velocity stepping experiments when the results of gouge experiments produced similar results to rock/rock tests (e.g., Dieterich, 1981; Weeks and Tullis, 1985), but an explanation of these results is provided by studies of the gouge fabric.

5.2. Mechanics associated with formation of fabric elements

Although the exact mechanical development of the fabric elements has not been explicitly modeled to date, a general understanding of the mechanics seems to be emerging. Mandl (Mandl et al., 1977; Mandl, 1988) has argued, from experiments in direct shear on granular materials, that principal stresses within the shear zone, which acts in an elastic–plastic manner, will rotate as the differential stress is increased. At some critical stress level, one or both of the Riedel fractures (R_1 and R_2) will form in response to Coulomb failure. These will initially be at high angles to the gouge interface. As the stress field rotates, new Riedels will form at lower angles to the boundary. Studies in a ring shear apparatus found the rotation to reach a limit at 45° to the boundaries (Mandl et al., 1977). Here rotation stops, Rs cease forming and 'shear bands' (similar to our Y-fractures) form parallel to the interfaces. He argues that the relatively rigid boundaries (which restrict displacement along the Riedels) impose a kinematic constraint restricting displacement along R_1s, and results in the formation of the Y-fractures along directions of maximum shear stress. Our observations agree with this general interpretation, although they offer some additional insights.

First our results support the hypothesis of the stress field reorienting within the gouge zone. We find R_1-fractures at high angles that appear to extend themselves by reorienting to lower angles to the interface, in the extreme becoming parallel to the Y orientation. This change in orientation is clearly a function of increasing shear strain. This rotation is very evident in the direct shear experiments, where one can trace on a single fracture the change in angle.

Mandl has argued that P-fractures result from either a slight overrotation in the development of the Y-fractures or by the local stress increase caused by the 'transpressure' produced by en echelon R_1-fractures. Although direct evidence of either is not found, it is clear in our experiments that they develop contemporaneously with the formation of Y-fractures.

From this argument of stress rotation it is plausible that heterogeneous states of stress will potentially exist in the gouge. Some portions along Y-fractures may have followed the material stress–strain curve well beyond the ultimate strength while other domains may still be approaching the peak stress conditions. Our observation of the development of 'Riedels within Riedels' directly supports this contention. Additionally, some portions of the gouge may be strain softening while others are still strain hardening. It is then possible that although the dominant displacement of the gouge zone may be accommodated along Y-fractures displacement an extension of R_1- and P-fractures may also take place simultaneously to a minor extent. This appears to be in agreement with observations by Tullis and colleagues of R_1s developing after extensive displacement has occurred in their rotary shear machine (Weeks and Tullis, 1985; Tullis and Weeks, 1987). Generally however, we expect elements on either side of a Y-surface to be relics of prior deformation.

The imposition of kinematic control upon a changing stress field is supported by our comparison of sawcut and fracture interfaces. Regardless of the geometry, both provide rigid boundaries which ultimately lead to Y-fractures being formed. Additionally, the ubiquitous nature of the gouge fabric regardless of material – from pure montmorillonite (Logan and Rauenzahn, 1987) to quartz (Higgs, 1981) to calcite (Friedman and Higgs, 1981; Wang, 1989) to halite (Shimamoto, 1986; Chester and Logan, 1990) – supports the importance of the kinematic constraint.

Our observations further extend these ideas. The first is that the development of the fracture array is a function of shear strain. That is, formation of Y-fractures and accommodation of displacement along them occurs earlier in zones that have undergone higher shear strain regardless of the absolute displacement. More specifically, thinner gouge layers achieve steady-state fracture arrays before thicker ones, with a given displacement. This most likely reflects the increased kinematic constraint imposed by the closer boundaries in the thinner zones. Less adjustment can be made along early-formed fractures to accommodate any displacement parallel to the interfaces. There is less distance for reorientation of the R_1s, thus development of the Ys is forced at an earlier displacement. The implication is that in the laboratory, because thicker gouge layers will require either larger displacements or some preconditioning to achieve steady-state, care must be taken in comparing their results to other experiments.

Our second caveat is that although similar fabrics develop in all rock materials that we have tested, the amount of strain necessary to achieve steady state varies with gouge composition. Generally, weaker material such as halite and calcite achieve steady-state fabrics with less shear strain than stronger minerals such as quartz or feldspars. As would be expected, upon the application of stress the weaker minerals begins to compact sooner than the stronger ones and the gouge increases its bulk density faster. For a given initial thickness, this results in the layer becoming thinner more quickly with the weaker material, and the kinematic constraint is imposed earlier. The increase in bulk density which precedes formation of Y-fractures in the calcite experiments supports this. The implication, as just mentioned, is that the Y-surface also develops earlier. This fabric change is compatible with the differences in mechanical behavior. Shimamoto has pointed out the propensity for unstable sliding in gouges formed of minerals of intermediate hardness, whereas those of high hardness slide stably under moderate confining pressures (Shimamoto and Logan, 1981).

5.3. Existing faults and stress fields

Our finite-element models support the contention of highly heterogeneous stress fields existing within fault gouge. They also point to an explanation for the potential reorientation of the stress field at the boundaries of the fault zone, a phenomenon that has been observed in the measurements from the Cajon Pass well (Zoback et al., 1987) and for which explanations have recently been offered by others (Morrow et al., this volume, Chapter 3; Rice, this volume, Chapter 20).

These finite-element models also clarify the concentration of displacement at the host-rock–gouge interface during the fabric evolution. The differences in properties would tend to localize displacement at these sites. As deformation proceeds and the gouge becomes more indurated, the contrast in properties is expected to diminish. With this decrease in property contrast, Y-fractures would be expected to move to sites within the gouge (Logan et al., 1981).

An important implication for friction studies in the laboratory is that calculated values for coefficients of friction based on far-field measurements of stress may be significantly in error. In the worst cases considered in the models, this error may be as much 25%. Actual values for coefficients of friction would be consistently lower than those calculated without consideration of stress reorientation. This is especially pertinent as many of the existing questions about fault-zone behavior and strength involve the value of the coefficient of friction. The friction of natural faults has been estimated by using coefficients derived from laboratory experiments, and generally taken to be about 0.6 (Byerlee, 1978), but it now appears pertinent to review the experimental results with a new perspective.

Although these models are simplified, they emphasize the complex stress conditions that may be expected in fault zones, and the associated heterogeneity of deformation. Additionally, the fault zone itself, being a weaker segment within the crust, may alter the local stress field, an observation that is now receiving attention in light of recent field data.

6. Conclusions

This study has attempted to elucidate the development of fabrics in laboratory fault zones, their correlation to mechanical behavior, and their implications for stress conditions in and around fault zones. The following conclusions are supported by our work.

Laboratory fault zones develop a repeatable fabric, which evolves with increasing shear strain. The general character of this experimental fracture array has been sufficiently verified by field studies to suggest that it offers a viable model to assess natural brittle faults. The experimentally produced fracture array progresses in a stepwise fashion to a steady-state condition. Strain hardening initially occurs, during which compaction accompanies much of the deformation. Strain softening follows, during which the deformation shifts to accommodate displacement parallel to the shear zone boundaries. This culminates in a steady-state fracture array where the dominant deformation is by slip parallel to, but not necessarily at, the interfaces. Periodic instability, or stick–slip, appears to occur after the fracture array has reached the steady-state condition. As the mechanical response of the fault is dependent upon the state of the gouge, caution must be exercised in correlating the mechanical data without knowledge of the condition of the gouge.

The fracture array apparently develops in response to a stress reorientation within the gouge zone which is limited by the kinematic constraints imposed by the more rigid bounding blocks. Both the deformation and stress fields are highly heterogeneous within the gouge, especially at low to moderate normal stresses. Even when the interface with the gouge is irregular, this constraint acts to change Coulomb failure to kinematically controlled fractures. The importance of this kinematic control is exacerbated by weaker material and thinner gouge zones.

Finally, our models suggest that under some boundary conditions the presence of the weaker fault gouge may produce reorientation of the overall stress field in the vicinity of the fault. This is predicted to increase the angle of the maximum principal stress to the fault. An important implication for laboratory studies is that they may lead to a significant overestimation of the coefficient of friction accompanying shearing. The latter needs further investigation as it addresses a number of paradoxes of fault movement that remain without satisfactory explanation at the present.

Acknowledgements

We wish to acknowledge the significant improvements made in this chapter from the reviews of D. Moore and G. Dresen. Fred Chester did the experiments and contributed the photos in Figure 16. Many colleagues too numerous to mention have contributed ideas over the years. If ideas that originated with them are not acknowledged, it is merely because our memories get dim; our apologies. Not the least we want to recognize the stimulation, insights and friendships that we have enjoyed with Bill Brace over the years. It has made a difference in the quality of both our science and personal life. This work has been supported in part by the following USGS grants: 14-08-001-G-460, 14-08-001-17677, 14-08-001-19293, 14-08-001-21181, 14-08-001-G-820, 14-08-001-G-981, 14-08-001-G-1509; NSF grants: EAR-8200726 and EAR-8513651, and GRI contract 1509809T.

References

Bartlett, W.L., Friedman, M., and Logan, J.M. (1981). Experimental folding and faulting of rocks under confining pressure; Part IX. Wrench faults in limestone layers. *Tectonophysics* **79**, 255–277.

Byerlee, J.D. (1978). A review of rock mechanics studies in the United States pertinent to earthquake prediction. *Pure Appl. Geophys.* **116**, 586–602.

Byerlee, J.D. and Summers, R. (1976). A note on the effect of fault gouge thickness on fault stability. *Int. J. Rock Mech. Min. Sci. Geomech.* **13**, 35–36.

Byerlee, J.D., Mjachkin, V., Summers, R., and Voevoda, O. (1978). Structures developed in fault gouge during stable sliding and stick-slip. *Tectonophysics* **44**, 161–171.

Chester, F.M. and Logan, J.M. (1985). Implications for mechanical properties of brittle faults from observations of the Punchbowl Fault Zone, California. *Pure Appl. Geophys.* **24**, 89–106.

Chester, F.M. and Logan, J.M. (1987). Composite planar fabric of gouge from the Punchbowl Fault, California. *J. Struct. Geol.* **9**, 621–634.

Chester, F.M. and Logan, J.M. (1990). Frictional faulting in polycrystalline halite: correlation of microstructure, mechanisms of slip and constitutive behavior. *Am Geophys. Mon.* **56** (The Heard Volume), 49–56.

Cloos, H. (1928). Experimente zur inneren Tektonik. *Centralbl. Mineral. Geol. u. Pal.* **1929B**, 609–621.

Dengo, C.A. (1982). Structural analysis of the Polochic Fault Zone in Western Guatemala, Central America, Ph.D. diss. Texas A & M University.

Dieterich, J.H. (1981). Constitutive properties of faults with simulated gouge. *Am Geophys. Union Monogr.* **24**, 103–120.

Engelder, J.T. (1974). Cataclasis and the generation of fault gouge. *Geol. Soc. Am Bull.* **85**, 1515–1522.

Engelder, J.T., Logan, J.M., and Handin, J. (1975). The sliding characteristics of sandstone on quartz fault gouge. *Pure Appl. Geophys.* **113**, 69–86.

Friedman, M. and Higgs, N.G. (1981). Calcite fabrics in experimental shear zones. *Am Geophys. Union Mon.* **24**, 11–28.

Gamond, J.F. (1983). Displacement features associated with fault zones: A comparison between observed examples and experimental models. *J. Struct. Geol.* **5**, 33–45.

Higgs, N. (1981). Mechanical properties of ultrafine quartz, chlorite and bentonite in environments appropriate to upper-crustal earthquakes. Ph.D. diss. Texas A & M University.

Hoskins, E., Jaeger, J.C., and Rosengren, K.J. (1968). A medium scale direct friction experiment. *Int. J. Rock Mech. Min. Sci.* **5**, 143–154.

Logan, J.M. and Rauenzahn, K.A. (1987). Velocity-dependent behavior of mixed quartz-montmorillonite gouge. *Tectonophysics* **144**, 87–108.

Logan, J.M., Friedman, M., Higgs, M., Dengo, C., and Shimamoto, T. (1979). Experimental studies of simulated gouge and their application to studies of natural fault zones. *Proc. Conf. VIII, Analysis of Actual Fault Zones in Bedrock*, pp. 305–343. U.S. Geol. Survey, Menlo Park, CA.

Logan, J.M., Higgs, N., and Friedman, M. (1981). Laboratory studies of natural fault gouge from the U.S.G.S. No. 1 well in the San Andreas fault zone. *Am. Geophys. Union Monogr.* **24**, 121–134.

Mandl, G. (1988). *Mechanics of Tectonic Faulting.* Elsevier, New York.

Mandl, G., de Jong, L.N.J., and Maltha, A. (1977). Shear zones in granular materials – an experimental study of their structure and mechanical genesis. *Rock Mech.* **9**, 95–144.

Moore, D. E., Summers, R., and Byerlee, J. (1989). Sliding behavior and deformation textures of heated illite gouge. *J. Struct. Geol.* **11**, 329–342.

Morgenstern, N.R. and Tchalenko, J.S. (1967). Microscopic structures in Kaolin subjected to direct shear. *Geotechnique* **17**, 309–328.

Morrow, C.A. and Byerlee, J.D. (1989). Experimental studies of compaction and dilatancy during frictional sliding of faults containing gouge. *J. Struct. Geol.* **11**, 815–825.

Morrow, C., Radney, B., and Byerlee, J. (1992). This volume, Chapter 3.

Naylor, M.A., Mandl, G., and Sijpenstein, C.J.K. (1986). Fault geometries in basement-induced wrench faulting under different initial stress states. *J. Struct. Geol.* **8**, 737–752.

Power, W.L., Tullis, T.E., and Weeks, J.D. (1989). The relationship between slickensided surfaces in fine-grained quartz and the seismic cycle. *J. Struct. Geol.* **11**, 879–893.

Riedel, W. (1929). Zur mechanik geologischer Brucherscheinungen. *Centralbl. Mineral. Geol. u. Pal.* **1929B**, 354–368.

Rutter, E.H., Maddock, R.H., Hall, S.H., and White, S.H. (1986). Comparative microstructures of natural and experimentally produced clay bearing fault gouges. *Pure Appl. Geophys.* **124**, 3–29.

Shimamoto, T. (1977). Effects of fault gouge on the frictional properties of rocks: an experimental study. Ph.D. diss., Texas A & M University.

Shimamoto, T. (1985). Confining-pressure reduction experiment: A new method for measuring frictional strength over a wide range of normal stress. *Int. J. Rock Mech. Min. Sci. Geomech. Abstr.* **22**, 227–236.

Shimamoto, T. (1986). A transition between frictional slip and ductile flow for halite undergoing large shearing deformation at room temperature. *Science* **231**, 711–714.

Shimamoto, T. and Logan, J.M. (1981). Effects of simulated fault gouge on the sliding behavior of Tennessee sandstone: non-clay gouges. *J. Geophys. Res.* **86**, 2902–2914.

Shimamoto, T. and Logan, J.M. (1986). 'Velocity-dependent behavior of simulated halite shear zones: an analog for silicates.' *Am. Geophys. Union Monogr.* **37**, 49–64.

Tchalenko, J.S. (1970). Similarities between shear zones of different magnitude. *Bull. Geol. Soc. Am.* **81**, 1625–1640.

Tullus, T.E. and Weeks, J.D. (1987). Constitutive behavior and stability of frictional sliding of granite. *Pure Appl. Geophys.* **124**, 383–414.

Weeks, J.D. and Tullis, T.E. (1985). Frictional sliding of dolomite: a variation in constitutive behavior. *J. Geophys. Res.* **90**, 7821–7826.

Welker, M. (1985). Investigation of the structure, deformation mechanisms and mechanical behavior of a brittle–ductile shear zone along the insubric-tonal line in Northern Italy, M.S. Thesis. Texas A & M University.

Wang, Z.Z. (1989). Frictional sliding and the fabric developed in experimental shear zones. Ph.D. diss., Texas A & M University.

Zoback, M.D., Zoback, M.L., Mont, V.S., Suppe, J., Eaton, J.P., Healy, J.H., Oppenheimer, D., Reasdenberg, P., Jones, L., Raleigh, C.B., Wong, I.G., Scotti, O., and Wentworth, C. (1987). New evidence on the state of stress of the San Andreas Fault system. *Science* **238**, 1105–1112.

Chapter 3

Frictional Strength and the Effective Pressure Law of Montmorillonite and Illite Clays

C. Morrow, B. Radney[1]*, and J. Byerlee*
U.S. Geological Survey, Menlo Park, CA 94025, U.S.A.
[1] *Now at Amoco Oil Co., Houston, Texas, U.S.A.*

Abstract

Low-strength clay minerals are a common constituent of fault gouges, and are often cited as a possible explanation for the low ambient shear stresses along the San Andreas fault inferred from heat flow constraints and in situ stress measurements. Montmorillonite, the weakest of the clay minerals, undergoes a gradual phase transition to illite with depth. In order to compare the shear stresses supported by these two minerals with those thought to exist along the San Andreas, we have measured the frictional sliding behavior of pure montmorillonite, mixed montmorillonite/illite and pure illite as a function of effective pressure, simulating burial to seismogenic depths. Strength measurements verify that the effective pressure law for friction holds for these minerals under all conditions. That is, the measured stresses were a function of the effective pressure, $P_c - P_p$, independent of the choice of confining and pore pressure. This relation, common for many other rock types, was previously untested for these clays under most conditions. Results show that dry samples were consistently stronger than saturated samples, and that strength increased with increasing illite content. In addition, the coefficient of friction increased as a function of pressure for the montmorillonite gouge, but was independent of pressure for the illite gouge. This behavior may be explained by the presence of loosely bonded interlayer water in the montmorillonite, which is squeezed out at higher pressures, changing the frictional characteristics of the clay. The nonexpanding illite was not affected in this way. For the montmorillonite-to-illite compositional profile, an average shear stress of 60 MPa was determined for crustal conditions to 15 km, assuming a normal hydrostatic gradient. If montmorillonite remains stable at depth, the resulting average shear stress is reduced to 30 MPa. In either case, these values are above the 10–20 MPa shear stress limit along the San Andreas inferred from heat flow constraints. Strength may be reduced to in-situ levels if fluid pressures become greater than hydrostatic within the gouge zone.

FAULT MECHANICS AND TRANSPORT PROPERTIES OF ROCKS
ISBN 0-12-243780-2

1. Introduction

It has long been recognized that the mechanical properties of clay-rich materials are pertinent to the discussion of fault zone behavior. Many varieties of clay minerals are commonly found in fault gouges along the San Andreas and other major active faults. These minerals are thought to persist to seismogenic depths, on the basis of geochemical and geophysical evidence such as velocity and density profiles (Wu et al., 1975; Wang, 1984). Clay minerals are consistently weaker than natural rock-flour gouges composed of crushed granitic material (Morrow et al., 1984). Anomalously low frictional strength is often cited as a possible factor in the long-standing stress–heat flow debate. This debate has arisen because the strength of typical crustal rocks measured in the laboratory is inconsistent with the low strength of the San Andreas fault inferred from in-situ measurements and the lack of a frictionally generated heat flow anomaly (Lachenbruch and Sass, 1980). Conductive heat flow measurements in numerous shallow boreholes near the fault imply an average shear stress level of 10–20 MPa, at least an order of magnitude below that implied by the laboratory strength data for nonclay rock gouges. Although the stress–heat flow debate, in which gouge strength is only one aspect, has not been fully resolved, it has lead to numerous studies of the frictional properties of clays.

The expandable clay, montmorillonite, is by far the weakest of the clay minerals commonly found in fault zones and as such, is of particular interest to the discussion of fault strength. This clay typically undergoes a gradual dehydration and phase change to the stronger, nonexpandable clay, illite, with increasing temperature and pressure. The two minerals coexist as a mixed-layer phase at intermediate states of the diagenesis process. This phase transition is well documented in hydrothermally altered rocks, contact metamorphic environments, and sedimentary basins such as the Gulf Coast of North America. The temperatures and pressures at which diagenesis begins and ends depend on a number of factors including composition of the original clay, chemistry of available fluids, porosity and geologic environment (Freed and Pecor, 1989; Colten-Bradley, 1987; Meunier and Velde, 1989). Generally, the transition begins at temperatures less than 100°C, as in the case of the Gulf Coast. However, Wu et al. (1975), Wang (1984), and others suggest that in certain environments montmorillonite may be stable to 300–400°C, corresponding to seismogenic depths in the earth. If this weak mineral persists at depth, then its deformation behavior may be important to the understanding of fault dynamics.

Numerous frictional sliding studies of montmorillonite yield a wide range of reported strengths, largely because of the strong influence of fluids on this highly expandable clay. Reported shear stresses vary depending on whether the clay is in the dry or saturated state, the degree of saturation, and whether overpressured fluids have been unintentionally generated due to the low permeability of the clay (Summers and Byerlee, 1977; Byerlee, 1978; Wang et al., 1979; Wang and Mao, 1979; Morrow et al., 1982). As a result, the manner in which the strength of montmorillonite varies with effective pressure is not well known. For this reason, we have undertaken a series of frictional sliding experiments in which care has been taken to maintain pore pressure without generating overpressured fluids. In particular, we wish to study the important case in which pore pressure varies with depth according to the hydrostatic gradient. The choice of samples represents two different scenarios for the composition of a fault zone. The first assumes that montmorillonite persists to seismogenic depths. The strength of pure montmorillonite was measured under a series of confining and pore pressures simulating depths of burial from 0 to 15 km. The second case assumes that montmorillonite undergoes a phase change to illite with increasing pressure. For these tests, a mixture of montmorillonite and illite was used at intermediate pressures, representing a mixed-layer phase, and pure illite was tested at the higher pressures, simulating the completion of diagenesis. In this way the average shear stress values of the two compositional profiles can be compared with the 10–20 MPa shear stress level thought to exist along the San Andreas fault. The magnitude of these shear stresses will indicate whether or not low-strength clay minerals are sufficient to explain the existence of a weak fault.

2. Effective Pressure Law for Friction

In order to report the true strength of these gouges as a function of effective pressure, we need to determine the proper form of the effective pressure relation for friction. With the exception of experiments conducted on montmorillonite at 100 MPa by Wang and Mao (1979), this point has not been specifically tested for the clay minerals used in this study. Accordingly, an investigation of the effective pressure law forms the second major focus of this chapter.

The effective pressure is usually defined as

$$P_e = P_c - \alpha P_p \tag{1}$$

where P_e is the effective pressure, P_c is the confining pressure, α is a parameter that incorporates the physical properties of the rocks, and P_p is the pore pressure. This equation can also be written in terms of the effective stress, σ_e, where the effective stress is equivalently defined as

$$\sigma_e = \sigma - \alpha P_p \tag{2}$$

This form is relevant to the discussion of friction, where the applied normal stress on a slip surface is reduced by the fluid pressure. The parameter α in these equations has been analytically derived for certain elastic rock properties. See, for example, Nur and Byerlee (1971), Robin (1973) and Garg and Nur (1973), among others. It should be noted that the effective pressure need not vary linearly with confining and pore pressure, and that it may not have a simple analytic expression.

Generally, α is determined emperically for a particular rock property and rock type, yielding values between 0 and 1. The larger the value of α, the more dependent the effective pressure is on the fluid pressure. When $\alpha = 1$, eq. (1) reduces to the more common form

$$P_e = P_c - P_p \tag{3}$$

This expression has been found to be approximately correct for many properties of brittle rocks (Brace, 1972).

The effective pressure law for friction has been studied for a number of materials. Byerlee (1967) found that the effective pressure law (3) held for the sliding of granite at room temperature. Deviations from this law in crystalline rocks have been linked to dilatancy and strain rate-dependent pore pressure, where the true pore pressure in the rock may be different from the externally measured pore pressure (Brace and Martin, 1968; Garg and Nur, 1973). Moore et al. (1984a,b) studied the friction of granite, serpentinite, and clay gouges at a variety of temperatures. They found a deviation from the effective pressure law for the serpentine and sheet silicate gouges, possibly because pore fluids became trapped and overpressured during shearing. It is important to determine whether these gouges follow the effective pressure law under drained conditions, where over- or underpressured fluids are not generated, as was observed in the experiments noted above. In addition, expandable clays such as montmorillonite contain loosely bonded interlayer water whose role in relation to the effective pressure law is unknown. Does this bonded water interact with the free water in a way that may alter the effective pressure law? Does the strength depend on how many bonded layers are present, or on the bonding strength of each layer?

These questions must be answered before the true state of stress in the experiments can be determined. For this reason, the strengths of the three gouge types were studied under many combinations of confining and pore pressures, in addition to those representing a normal lithostatic-to-hydrostatic ratio, so that the effective pressure law could be fully investigated. We wish to address several main points: What are the shear strengths of these clays as a function of effective pressure? What is the form of the effective pressure law for friction that is required to describe these materials? Finally, is there some difference between the expandable and nonexpandable clays that may influence the frictional characteristics of the samples?

3. Sample Description and Procedure

The samples used in this study consisted of fine-grained particles of montmorillonite, illite, and a 50% by weight mixture of these two clays. The montmorillonite was a commercially obtained powder (Volclay SPV-200), with 65% of the grains finer than 200 mesh (74 μm). The illite gouge was obtained by crushing a natural shale from Fithian, Illinois.

The clay gouges were saturated with distilled water to form a thick paste using the same weight of clay and water in each case. A 1 mm-thick layer of the saturated gouge was placed between the surfaces of a 2.5 cm diameter cylindrical sample, cut at an angle of 30° as shown in Figure 1. The upper half of this sample was composed of Berea Sandstone, the lower half of fused silica, with a total length of 6.3 cm. The Berea Sandstone has a porosity of 18%, and a permeability of around 10^{-13} m^2 along the direction of bedding (Zoback and Byerlee, 1975). This permeability is about eight orders of magnitude higher than the permeability of the clay layer ($\sim 10^{-21}$ m^2; Morrow et al., 1984) at the pressures of these experiments. Fused silica was used in the bottom section because of the low permeability of the clay gouges, which would cause a long time delay in the saturation of the bottom section if it were composed of sandstone. In addition, pore volume reduction during deformation could create excess fluid pressures which become trapped beneath the low-permeability clay layer. The impermeable fused silica eliminates these problems by preventing fluids from entering the bottom section entirely, where the pore pressure cannot easily be monitored. The highly permeable Berea Sandstone on the upper half insured close communication between fluids in the rock and those in the fluid pressure generator and pore pressure transducers. Each rock–glass assembly was held together with a copper foil jacket (0.05 mm thick), and placed in a vacuum oven at 100°C for 24 hours to dry the samples. The copper foil tears easily during shearing, and does not affect the strength measurements. After drying, the samples were then placed in a polyurethane jacket and fastened to the steel endplugs

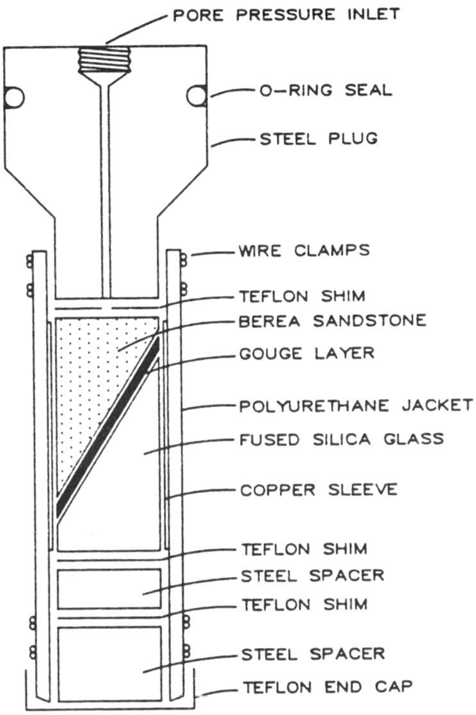

Figure 1. Sample assembly.

with wire clamps. Since the two halves of each sample must move slightly sideways during shearing, lubricated shims were added at the interfaces of the sample column to minimize frictional resistance in the horizontal direction.

All frictional sliding experiments followed the same deformation history, under a wide range of confining and pore pressures, to produce effective pressures of 1, 2, 20, 30, 50, 100, 150, 200, 300, and 400 MPa. All pressures and displacement rates were maintained by computer-controlled servomechanisms. Pressure measurements were accurate to 0.05 MPa, displacement to 10^{-6} m. At the start of each run, the pore fluid lines were evacuated to remove all water in the system. Confining pressure was applied and sliding initiated at an axial displacement rate of 0.866 $\mu m\ s^{-1}$. This corresponds to a shear displacement rate along the inclined gouge layer of 1 $\mu m\ s^{-1}$. After 2.5 mm of axial displacement, the piston was held fixed for 10 000 s. Distilled water was then introduced to the sample and pore pressure was raised to the desired level where it was left to equilibrate for the duration of the pause. Thereafter, sliding continued at the previous rate of 1 $\mu m\ s^{-1}$ to a maximum value of 7 mm. In some experiments the sample was left dry throughout. The purpose of the dry/wet procedure was to assure that the fluid in the gouge layer did not become overpressured owing to the initial loading of the sample. Most of the pore volume reduction occurs during this phase. If the permeability of the gouge layer is particularly low, then fluids may not be expelled fast enough to maintain the desired pore pressure. In that case, we could not determine true effective pressure acting on the sample. By introducing the fluid after the sample is already under shear stress, we are assured that excess water has not become trapped in the gouge.

Velocity-dependent strength behavior was investigated on the three gouge types with frictional sliding experiments similar to those described above. These tests are pertinent because laboratory experiments are typically conducted at displacement rates faster than those of active faults, and may yield strength results that differ slightly from true in-situ conditions. The samples for these tests were deformed in a dry state at a shear displacement rate of 1 $\mu m\ s^{-1}$ for 2.5 mm at a confining pressure of 300 MPa. Sliding paused for 10 000 s as before, and distilled water was added to achieve a pore pressure of 100 MPa, simulating a depth of burial of around 10 km. After saturation, sliding resumed at a rate of 0.01 $\mu m\ s^{-1}$, 100 times slower than the previous rate, to a displacement of 7 mm. The sliding rate was then increased to the original value of 1 $\mu m\ s^{-1}$ for the remaining 1 mm of sliding. The slower rate, equivalent to 31 cm/year, is an order of magnitude greater than the average slip rate along the San Andreas fault of around 2.5 cm/year. Any changes in the coefficient of friction during these experiments can be used to extrapolate the laboratory strength measurements to in-situ conditions.

A final set of experiments was designed to quantify the effects of elastic stiffness. Because the elastic stiffness of the Berea Sandstone and glass sample configuration depends on confining pressure, the slope of the initial loading curve and the displacement at which sliding initiates will vary between experiments. Samples for these stiffness tests were assembled with no gouge layer between the sandstone and glass surfaces. The samples were loaded and unloaded at the standard rate of 1 $\mu m\ s^{-1}$ several times to a maximum displacement of 1 mm. This computer-controlled procedure was repeated at all the confining pressures used in the frictional sliding experiments. The results of these tests were used to derive the true shear displacement for each sample, so that experiments at different pressures could be compared. As a part of the stiffness testing procedure, the seal friction of the machine was also determined by cycling the piston back and forth under pressure while not in contact with the sample. This friction is the force needed to slide the piston past the O-ring seal assembly. Since axial force is measured outside the pressure vessel, the true axial force on the sample must be corrected for this friction.

4. Results

4.1. Dry runs

Shear stress versus axial displacement for selected dry runs is shown in Figure 2. The stress data are corrected for the decreasing area of contact as the two sawcut surfaces slide past one another. This decreasing area results in an increase in the normal stress during shearing. Confining pressures of 30, 150, and 300 MPa represent depths of burial of 1, 5, and 10 km, respectively. This plot shows data for all three gouge types, with data for the illite and mixed clay at the higher pressures only, representing the greater depths at which they would be expected to occur in nature. Shear stresses rose rapidly during the initial elastic portion of the curves, then continued to rise more slowly as the samples began to slide. At 2.5 mm displacement, the stresses dropped slightly during the 10 000 s pause, owing to creep in the sample under the sustained load. This behavior is typical of granular materials under stress (Morrow and Byerlee, 1989). When sliding

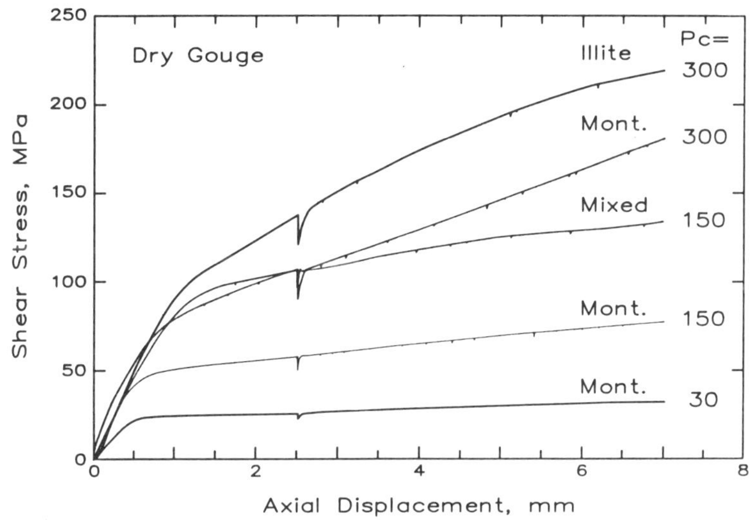

Figure 2. Shear stress as a function of axial displacement for vacuum-dried samples of pure montmorillonite, mixed montmorillonite/illite, and pure illite gouges. Confining pressures are indicated at right in MPa. The drop in shear stress at 2.5 mm corresponds to the 10 000 s pause in sliding.

was resumed, the shear stresses quickly rose to the previous level as if there had been no interruption. The relative strength of the gouges at a particular pressure was dependent on the amount of montmorillonite present, which is the weaker of the constituent minerals. In addition, the samples all strain hardened with displacement. The gradual increase in strength is a function of confining pressure, with more strain hardening observed at the higher pressures than at the lower. This observation is consistent with the strain hardening behavior of numerous other clay-rich fault gouges collected along the San Andreas and nearby faults (Morrow et al., 1982). Clearly, the reported strength of these gouges depends on how far the samples have slid if strain hardening behavior is observed. With the exception of the illite, whose strain hardening rate began to decrease near the end of the experiment, we have not slid far enough to observe any changes in the displacement-dependent behavior that may indicate how the samples will behave at large shear strains. Owing to the geometric limitations of the triaxial system, such large-strain behavior would be better observed in a rotary shear apparatus.

4.2. Saturated runs

Several dozen runs were conducted with saturated gouge at a variety of confining and pore pressures on the illite, montmorillonite, and mixed clays. Figure 3 shows the effects of saturation for the same gouges and confining pressures discussed in Figure 2. For these runs, distilled water was injected into the sample during the 10 000 s pause, and held at atmospheric pressure after the onset of shearing, so that the effective pressure was unchanged before and after the pause. Shear stresses increased during the dry portion as before. However, the strength increases were not entirely reproducible, particularly for the expandable montmorillonite, because of slight variations in the atmospheric humidity absorbed by the clay before the experiment. The introduction of

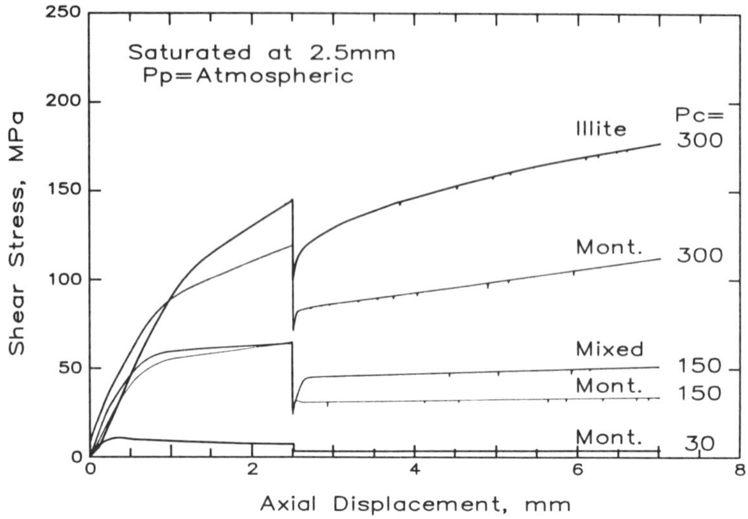

Figure 3. Shear stress as a function of axial displacement for the same gouges and stress conditions as shown in Figure 2. During the 10 000 s pause at 2.5 mm, distilled water was introduced at atmospheric pressure. Confining pressures are shown at right in MPa.

water during the pause equalized these differences, and also caused a reduction in the shear stress even more pronounced than the decrease observed during the pause in the dry runs. When sliding resumed, the samples did not regain the former shear stress levels, as was observed for the dry samples. In addition, the strain hardening rate was significantly reduced for samples at the highest pressure (300 MPa), from around 30 to 8 MPa mm^{-1} for the illite and from 16 to 7 MPa mm^{-1} for the montmorillonite. At lower pressures, the gouges slid at a steady-state shear stress level with little or no strain hardening. These tests verify that the presence of fluids decreased the strength and strain hardening of the gouges even though the applied pressure remained unchanged.

Petrographic observations of both the wet and dry samples show textural evidence for the strength differences. The deformation fabric of the dry samples for all three of the gouge types was much more developed than that of the wet samples at the same pressure. At 30 MPa, the dry sample contained boundary shear zones along both sawcut interfaces, in which clay platelets were aligned parallel to the direction of shearing. At 150 MPa this shear zone encompassed the entire width of the sample, and at 300 MPa kink bands within the fabric were prominent. In contrast, deformation in the saturated samples at the lower pressures was confined to a few boundary layers and Riedel shear bands, separated by large areas of undeformed gouge. Only at 300 MPa did the samples show evidence of deformation throughout the entire gouge layer.

The strength of the saturated gouge was further affected when pore pressure was increased from atmospheric to near lithostatic values. Figure 4 shows the shear stress as a function of axial displacement for montmorillonite at a constant confining pressure of 150 MPa. After the pause, fluids were added at pressures of 0.1, 50, 100, and 148 MPa, with a resulting decrease in strength for each increase in pore pressure. This behavior is to be expected, as the effective confining pressure acting on the sample is gradually reduced. This figure illustrates how readily the strength of the gouge can decrease to very low values in the natural environment if the fluid pressure is increased. The mixed

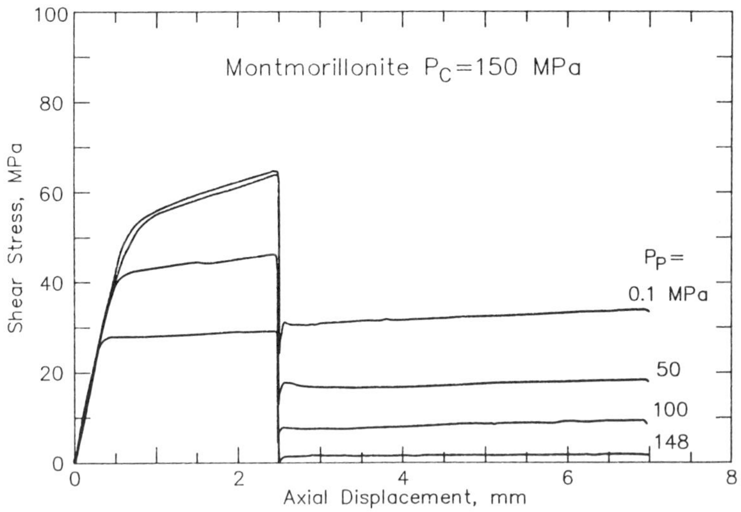

Figure 4. Shear stress as a function of axial displacement for the pure montmorillonite gouge at a confining pressure of 150 MPa. Increasing pore pressure reduces the strength.

clay and pure illite samples behaved in a similar fashion, although the drop in strength was less pronounced as montmorillonite content decreased. Note that the dry and wet portions of the strength curves are no longer directly comparable since they are not at the same stress state. Further discussions of strength will concentrate on the stress measurements taken after the pause at 2.5 mm. Figure 4 is an example of the numerous stress plots produced as a function of confining and pore pressure which were needed to investigate the strength of the clays as a function of effective pressure. These other plots have been omitted for brevity, as the stress values at different pressure conditions can be derived from the effective pressure analysis discussed below.

4.3. Velocity dependence

In order to determine whether these laboratory strength measurements are comparable to the in-situ conditions, a velocity step test was conducted for each of the gouges at confining and pore pressures of 300 and 100 MPa, respectively, representing a depth of burial of around 10 km. The resulting coefficients of friction are plotted as a function of displacement in Figure 5. The true change in strength of the gouge is masked at the first velocity step (2.5 mm) because of the decrease in strength due to saturation. Therefore, the relevant portion of this plot is at 7 mm, where the sliding velocity changes by a factor of 100 from 0.01 $\mu m\ s^{-1}$ to 1 $\mu m\ s^{-1}$, as measured along the inclined gouge layer. The coefficient of friction increased for all of the samples by about 0.025, indicating a velocity-strengthening behavior. This increase was 5–8% of the total coefficient of friction at 7 mm displacement.

The velocity dependence of friction can be defined following Logan and Rauenzahn (1987) as

$$a - b = \frac{d\mu}{d(\ln(\dot{\delta}))} \tag{4}$$

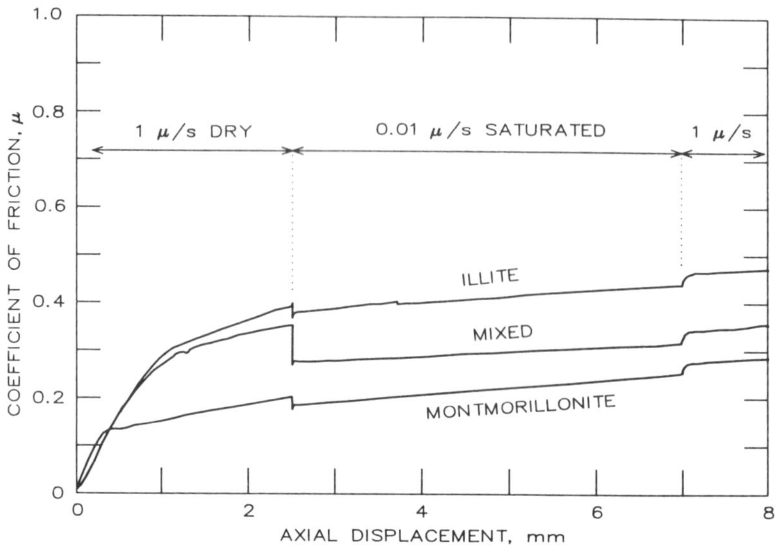

Figure 5. Coefficient of friction as a function of axial displacement at 300 MPa confining pressure and 100 MPa pore pressure. The sample was saturated at 2.5 mm; slip rates along the gouge layer are indicated above the curves.

where a is the initial response to the velocity change, b is the magnitude of the decay to the steady-state value, μ is the steady-state coefficient of friction, and $\dot{\delta}$ is the displacement rate. The $a - b$ values for these experiments were 0.005 for both the montmorillonite and illite samples, and 0.007 for the mixed clay. These values are consistent with the findings of Logan and Rauenzahn (1987) on experiments with pure montmorillonite. Therefore, the velocity dependence of friction does not appear to be affected by the expandable or non-expandable nature of the clays. If we assume an average slip rate on the San Andreas fault of 2.5 cm/year, or 7.9×10^{-7} mm s^{-1}, then the slower rate of the step test was around one order of magnitude higher than the San Andreas rate. We would therefore expect a further few percent reduction in shear strength for samples deformed at these rates. Such low-velocity experiments are not practical in the laboratory because of time constraints, and were not used to investigate the effective pressure law, but they are relevant to the discussion of the strength of natural faults.

4.4. True shear displacement

In order to test the effective pressure law for friction, the shear and normal stress values of the different samples must be compared at the same shear displacement. This is necessary because the elastic stiffness of the sample column is a function of confining pressure, and therefore the point at which sliding commences varies between experiments. From the load and displacement data of the gouge-free sample, elastic stiffnesses were calculated ranging from 271 to 317 MPa mm^{-1}. Stiffness increased with confining pressure at 300 MPa. At 400 MPa, grain crushing within the Berea Sandstone reduced the stiffness of the sample. These data were used to calculate the shear displacement

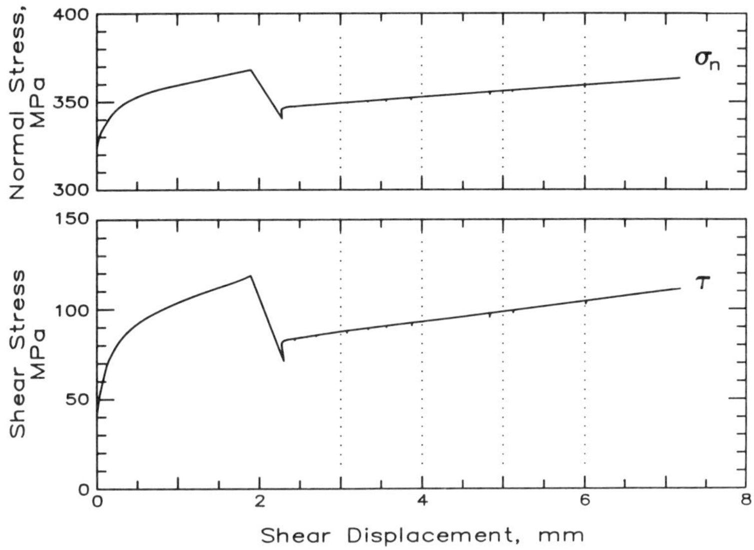

Figure 6. Shear and normal stress as a function of shear displacement for the montmorillonite gouge at a confining pressure of 300 MPa and atmospheric pore pressure. The data have been modified to normalize the elastic stiffness of the machine and rock cylinder under different pressures, and resolve the axial displacement along the inclined gouge layer.

along the gouge layer at each confining pressure, defined as

$$dl = \frac{dz}{\cos\theta} - \frac{\Delta\sigma}{K\cos\theta} \tag{5}$$

where

dz = axial displacement (mm)
$\Delta\sigma$ = differential stress (MPa), corrected for the decreasing area of contact during sliding
K = elastic stiffness of the gouge-free sample (MPa mm^{-1})

This calculation resolves the displacement along the gouge layer (first term) and shifts the data along the displacement axis to normalize the elastic portion of the deformation (second term). With this normalization, the shear stress values can be compared at the same displacement for experiments run at equivalent effective pressures but widely varying confining and pore pressures. This displacement correction does not account for compaction of the gouge during shearing; however, most of the compaction takes place during the initial loading of the sample.

Figure 6 illustrates the data normalization for the montmorillonite gouge at a confining pressure of 300 MPa and atmospheric pore pressure shown in Figure 3. Shear and normal stresses are plotted as a function of the shear displacement parallel to the fault. This plot shows the true deformation in the gouge layer, rather than the deformation observed external to the loading apparatus. As a consequence, the initial elastic loading curve is steeper, and the sloping trend during the 10 000 s pause reflects the creep and stress relaxation in the sample under the sustained load. Shear and normal stress values were picked at several values along the shear displacement axis, as shown by the dotted lines in Figure 6. This procedure was repeated for all experiments.

Shear stress values were plotted as a function of effective normal stress for each of the displacements using the simplest case of $\alpha = 1$ (eq. 1) as an initial reference point. If $\alpha = 1$ for these samples, we would expect to see a single trend line, with data for like effective pressures clustering together. For values other than 1, the data set would form a family of lines connecting points with the same pore pressure. The amount of spread

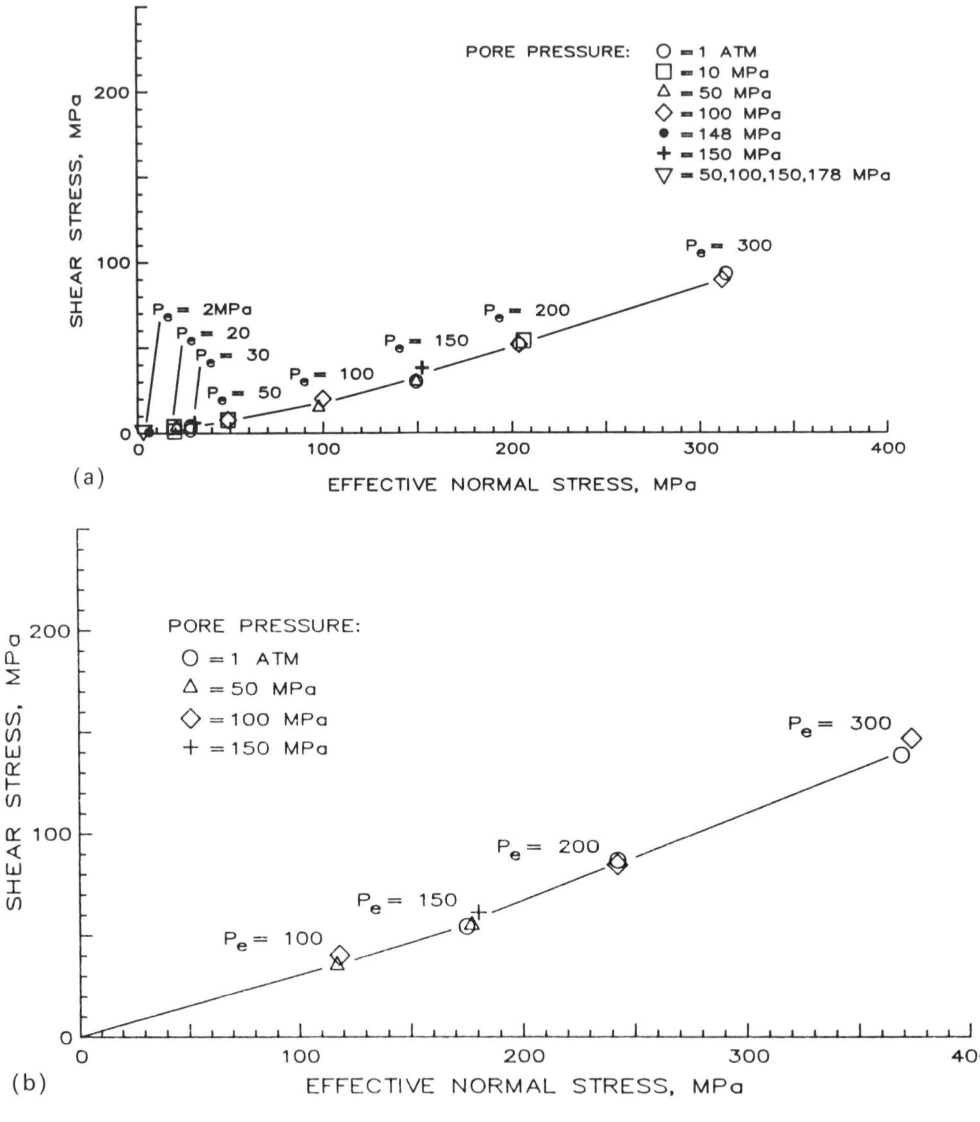

Figure 7. (a) Shear stress as a function of effective normal stress for the montmorillonite gouge at a shear displacement of 4 mm. Effective normal stress $P_c - \alpha P_p$ calculated with the coefficient $\alpha = 1$. Pore pressures are indicated by symbols at right. (b) Shear versus effective normal stress ($P_c - \alpha P_p$), for the mixed montmorillonite/illite gouge at shear displacement of 4 mm with the coefficient $\alpha = 1$.

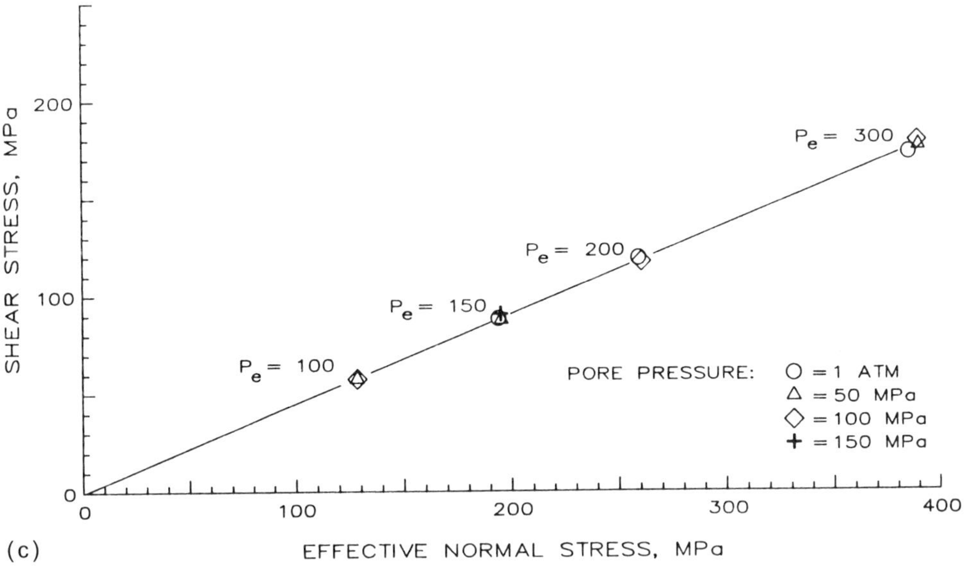

(c)

Figure 7. *Continued.* (c) Shear versus effective normal stress ($P_c - \alpha P_p$) for the illite gouge at shear displacement of 4 mm with the coefficient $\alpha = 1$.

in the lines would indicate the magnitude of the correction in α required to bring the data into a single line. Figures 7a–c show examples of these plots for the three gouge types at 4 mm shear displacement. Pore pressures are indicated by thc symbols, and confining pressures for each experiment can be determined by adding the pore pressure to the stated effective pressure. Data for all other shear displacements were nearly identical, indicating that the effective pressure relation is not strongly dependent on displacement. For pure montmorillonite (Figure 7a), the data points cluster together along a single line, regardless of the confining or pore pressures applied. Variations in α above and below a value of 1 cause an immediate divergence of the data points, showing that $\alpha = 1.00 \pm 0.01$ is the best fit. This plot verifies that the effective pressure law for the friction of montmorillonite holds under the pressure conditions tested; the effective pressure in eq. (1) can be written as $P_c - P_p$, and the effective normal stress as $\sigma_n - P_p$. Data for the mixed clay (Figure 7b) and pure illite samples (Figure 7c) also fall along a single trend line for all pressure conditions, indicating that the simple effective pressure law for friction, $P_c - P_p$, holds for these samples as well. This important conclusion verifies that the expandable nature of the montmorillonite does not influence the effective pressure law in a way that differs from the nonexpandable illite.

4.5. Friction

Additional aspects of the strength behavior of these gouges can be gleaned from other details of Figure 7. For instance, the shear and normal stress relation for montmorillonite does not follow a straight line. These data bend upward at effective pressures of around 50–100 MPa, so that the gouge becomes proportionally stronger at the higher pressures. This behavior was observed regardless of shear displacement. The mixed clay (Figure 7b) shows this trend to a lesser degree. Illite, on the other hand, shows no

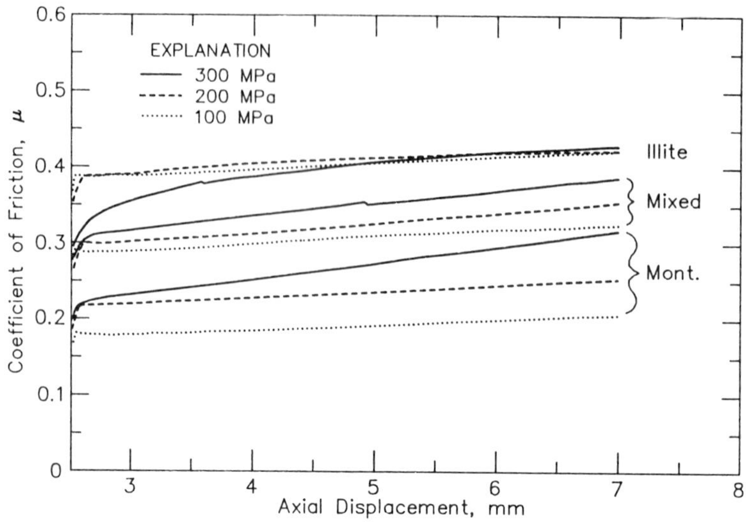

Figure 8. Coefficient of friction as a function of axial displacement for montmorillonite, mixed clay, and illite after sample saturation. Effective pressures indicated in MPa.

variation in the slope of the shear and normal stress relation with effective pressure (Figure 7c). These data follow a straight line that intersects the origin.

The pressure dependence of the shear stress shown in Figure 7 is more clearly demonstrated if we plot the coefficient of friction, $\tau/(\sigma_n)_e$, at selected effective pressures, as shown in Figure 8. The dry portions of the curves are not of interest and have been eliminated, so that displacement begins at 2.5 mm. The coefficient of friction of the montmorillonite is strongly dependent upon effective pressure as noted above, with a range in the friction values from around 0.2 to 0.3 by the end of the experiment for pressures between 100 and 300 MPa. The spread in the friction values diminishes with the addition of illite, as seen in the mixed clay samples. Finally, the coefficients of friction of the pure illite runs are nearly coincident at around 0.4 indicating little dependence on effective pressure, corresponding to the linear relation in Figure 7c. This systematic progression towards a constant coefficient of friction illustrates that as the illite content increases the samples become stronger and the pressure dependence of friction is reduced.

5. Discussion

Frictional sliding experiments conducted on the clay gouges, montmorillonite, and illite verify that the strength of these materials follows the simple effective pressure law. The measured shear stresses depended solely on the difference between the confining and pore pressures $P_c - P_p$, at all pressures and displacements tested. Frictional sliding studies that appear to deviate from this law for montmorillonite and illite are most likely the result of experimental conditions in which the true fluid pressure acting on the material is different from the externally measured pressure. This can easily be caused by the low permeability of these mineral aggregates, which may prevent changes in fluid pressure from equilibrating rapidly. Such a condition was observed by Wang et al. (1979), who reported an apparent pressure independence of strength in montmorillonite above 200 MPa, which was attributed to high fluid pressures trapped in the clay. It is important to remember that such samples still follow the effective pressure law, even though the true effective pressure acting in the sample is unknown.

One of the questions which we wished to address was whether the expandable nature of the montmorillonite would result in a different frictional behavior from that of the nonexpandable illite. Although the effective pressure law was not sensitive to this difference, the response of the two clays was different, as shown by the nonlinear increase in strength of the montmorillonite at higher effective pressures. This result can be understood if we consider the properties of the bonded water in the montmorillonite.

The hydration phases of montmorillonite have been extensively treated in the literature. Expandable clays can take on numerous layers of bonded water, which cause discrete increases in the *d*-spacing of the silicate layers (swelling). The interaction of this structural water with the silicate layers creates interlayer forces whose exact origin is a matter of debate (Viani et al., 1983). Bird (1984) measured the pressure stability fields of different hydrated phases, and the shear stresses required to cause slip on each phase at different normal stresses. He found that the shear strength of montmorillonite was greatest in the dehydrated state, and decreased with the addition of each water layer. Since the interlayer forces decay with distance, the first few layers had the most effect on strength. These water layers act as glide planes along which slip can more easily take place, accounting for the fact that expandable clays are weaker than nonexpandable ones. In addition, the water layers can be expelled with applied pressure. The strength of the hydrated phases is also a function of mineral composition. For instance, the hydrated phases of Na-montmorillonite are weaker than those of Ca-montmorillonite, because of the higher concentration of Ca cations in the Ca-montmorillonite. These additional cations form a stronger bond between the silicate layers, which in turn act on the hydrated phase. Since our samples were composed predominantly of Na-montmorillonite, the corresponding strengths for this composition are most relevant to our dicsussion. Bird (1984) found that for frictional sliding experiment on Na-rich samples, the phase change between the dehydrated and first water layer state occurred in a transitional pressure range of a few tens of MPa centered at an effective normal stress of around 100 MPa. This pressure was higher (~ 180 MPa) and more sharply defined for Ca-rich samples. Above that pressure, a state of dehydration exists in the mineral. The expelled water becomes free water that contributes to the pore pressure. These results are corroborated by the work of Israelachvili et al. (1988), who showed that the physical properties of molecularly thin liquid films are different than those of bulk liquid, and that the frictional force between thinly separated surfaces is a discrete function of the number of boundary layers.

The behavior of bonded water layers under pressure may explain the observed increases in the coefficient of friction of our montmorillonite samples with effective pressure. Shear and normal stress data exhibit a broad inflection point at around 50–150 MPa for the pure montmorillonite (Figure 7a). This inflection point is more clearly defined for the mixed clay (Figure 7b) at around 150 MPa, possibly because of the limited data at the lower pressures compared to the montmorillonite. The change in slope may represent the normal stresses at which the interlayer water is squeezed out under pressure. Above that pressure, the mineral structure does not change further, so that the slope of the shear vs. normal stress curve remains constant. Although we do not know the hydration state that existed during our experiments, the behavior of the samples was consistent with Bird's findings. The mixed gouges were less influenced by the expulsion of the interlayer water, as we might expect, because only half of the sample by weight was composed of an expandable clay. The coefficient of friction of the nonexpandable illite (Figure 7c) was independent of effective pressure.

These results have important implications for the state of stress at depth in fault zones, because they imply that at depths greater than a few kilometers, where lithostatic pressures are squeezing the water out of the clay structure, the relevant coefficient of friction of a montmorillonite gouge will be that of the dehydrated state rather than the expanded state. The true coefficient of friction at depth may be higher than values typically reported in the literature on the basis of measurements made at lower effective pressures. For instance, our measurements below 100 MPa effective pressure gave a coefficient of friction of around 0.2, whereas at 300 MPa the coefficient of friction increased to around 0.3. Although the strengthening of the gouge with pressure does not increase the coefficient of friction to levels near those of crystalline rocks, or even to that of illite ($\mu = 0.4$), this behavior needs to be taken into account if expandable clays are called upon to explain the low strength of faults inferred from heat flow measurements.

Several other details of frictional behavior must be considered when comparing these laboratory results to the in-situ environment. For instance, the velocity step tests show that the gouges may be several percent weaker if shearing takes place at rates comparable to the average San Andreas slip rate. Although this factor contributes to the 'weak gouge' explanation for the low strength of the San Andreas fault, the velocity effects alone are probably insufficient to lower the strengths to the necessary levels (Figure 5). Strain hardening has the opposite effect on strength, particularly at the higher effective pressures. Strength increased substantially with displacement for all of the gouge types studied. Since this scales with strain rather than total slip, the strain hardening may not be as significant for slip on thick natural faults as compared to laboratory experiments. The strain hardening behavior renders a definitive measurement of gouge strength difficult, because it is not clear from these triaxial tests at what point the samples will reach a steady-state stress level. Clearly, the strain hardening can not go on forever. At best, we can compare the relative strengths of the gouges at a given strain. Finally, these room-temperature experiments do not take into account the effects of increasing temperature with depth in the earth. Moore et al. (1989) found that the strength of heated illite gouges increased with temperature owing to greater degrees of induration of the particles. This might suggest that our laboratory measurements are an underestimate of the true in-situ strength. Other temperature effects that might be relevant in the in-situ environment include a change in deformation mechanism, change in velocity dependence, reactions to stronger or weaker products, and ductility.

Although the limitations of the laboratory data are apparent from the discussion above, these experimental results are nonetheless sufficiently relevant to address the final and most important question we wish to consider. That is, how do the measured shear stresses of the clay gouges compare with the average shear stresses to seismogenic depth? Two cases are considered, both with the condition that fluid pressures remain hydrostatic with depth. The first case assumes that montmorillonite undergoes a phase transformation to illite with increasing pressure. This view is generally supported by geologic evidence in other natural environments. For this model, the stress levels given in Figures 7a–c for montmorillonite at low pressure, mixed clay at mid-pressure, and illite at high pressure are plotted over a depth interval of 0–15 km (Figure 9, mixed symbols). Clearly the exact shape of this profile will depend on the depth range of the

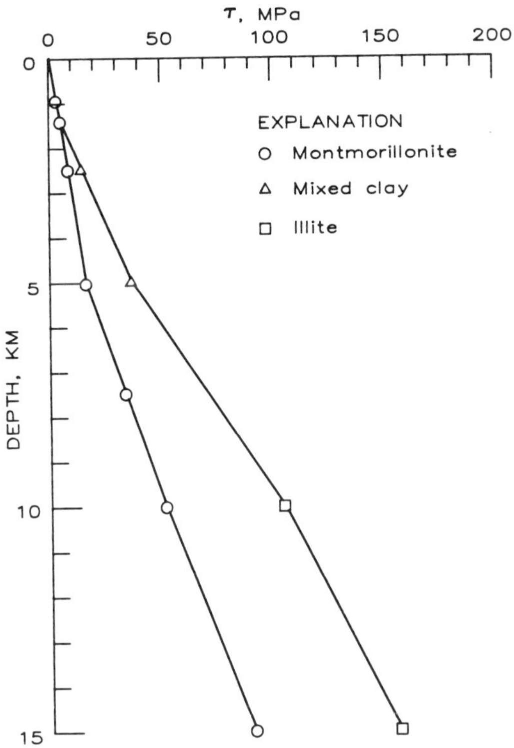

Figure 9. Selected shear stress data from Figure 7 as a function of depth for the case of montmorillonite undergoing a phase change to illite (mixed symbols), and the case of montmorillonite remaining stable with depth (round symbols). Temperature effects with depth are not considered in this plot.

mineral transition and the proportion of the endmember clays at intermediate depths. However, integrating the stress values from these experiments with depth yields an average shear stress of around 60 MPa in the crust. This value is higher than the 10–20 MPa prescribed by heat flow measurements, seismic stress drops and in-situ testing, so that this clay profile alone is insufficient to explain the existence of a weak fault.

The other case pertinent to the discussion of fault strength assumes that montmorillonite, the weaker of the two minerals, persists at depth. For the case of stable montmorillonite, Figure 7a is relevant over all pressures studied (Figure 9, round symbols). Integrating these measurements over the depth to the seismogenic zone gives an average shear stress value of around 30 MPa. This is still higher than the assumed in-situ stress state of 10–20 MPa. This result would not be anticipated if the laboratory stress measurements made at low effective pressures were extrapolated to higher pressures, because of the nonlinear relation between strength and pressure. The implication of these calculations is that the presence of low-strength clay minerals cannot be reconciled with the stress–heat flow paradox if the gouges are under normal hydrostatic pressures, no matter what the composition with depth, because the clay gouges are stronger than the prescribed state of stress on the fault. If strain hardening or temperature effects are considered, the average shear strength of the clay could be even higher still.

One possible explanation for the low strength of the fault is that the strength of clay is reduced to anomalously low values by high fluid pressures, which become trapped in the gouge zone owing to the low permeability of the clay. Such low strengths are commonly observed in undrained experiments. However, there is an important problem with this concept as an explanation for the weak fault model if fluid pressures are uniformly high within the gouge zone. Zoback et al. (1987) found that the principal stress direction along the San Andreas was nearly normal to the fault, based on evidence from hydraulic fracturing, borehole breakouts, focal mechanisms, and geologic features. If this interpretation is correct, then Mohr–Coulomb theory dictates that slip cannot occur on the fault even if fluid pressures reach the value of the minimum horizontal principal stress (at higher pressures hydraulic fracturing would occur in the country rock). For a typical coefficient of friction of 0.6, the Mohr–Coulomb construction shows that the minimum σ_1 angle necessary to cause slip is 31° from fault normal when high fluid pressures are invoked. For the friction values measured in this study ($\mu \sim 0.3$ for montmorillonite and $\mu = 0.4$ for illite), the σ_1 angles drop to 17° and 22° from fault normal, respectively. Although the stress orientation data reported by Zoback et al. (1987) show some variability, these reduced angles are still insufficiently low to explain the stress–heat flow paradox, especially in central California where the principal stress directions are closer to fault normal. The difficulties in explaining fault-normal compression may be overcome if a fluid pressure gradient exists within the gouge zone, as proposed by Byerlee (1990) and Rice (1990). Because the magnitudes of the fluid pressures are not well known at depth within the fault, the possibility of abnormally low effective pressures resulting from trapped fluids cannot be satisfactorily addressed at present. Clearly, the factors that control the stresses and constitutive properties along the San Andreas fault must be further explored in order to resolve the long standing stress–heat flow debate.

6. Conclusions

(1) The effective pressure law for friction holds under all pressure conditions for both the montmorillonite and illite gouges. The measured shear stresses were a function of the effective pressure, $P_c - P_p$, independent of the particular choice of confining and pore pressures.

(2) Pure montmorillonite gouge was the weakest of the gouge types studied. Shear strength increased with increasing illite content. Dry gouges were consistently stronger than the saturated gouges.

(3) The coefficient of friction of the montmorillonite gouge was dependent on effective pressure, increasing at higher pressures, whereas the illite gouge showed no dependence on effective pressure. We suggest that this behavior is caused by the presence of loosely bonded interlayer water in the montmorillonite that is squeezed out at the higher pressures, causing an increase in strength. The nonexpanding illite gouge is not influenced in this way.

(4) The average shear stresses of both the illite and montmorillonite gouges measured over pressures simulating depths to the seismogenic zone were greater than the average stresses along the San Andreas fault inferred from heat flow measurements and borehole stress measurements. Therefore, the low strength of these clay gouges under normal hydrostatic conditions is insufficient to account for the weak fault model.

References

Bird, P. (1984). Hydration-phase diagrams and friction of montmorillonite under laboratory and geologic conditions, with implications for shale compaction, slope stability, and strength of fault gouge. *Tectonophysics* **107**, 235–260.

Brace, W.F. (1972). Pore pressure in geophysics. In *Flow and Fracture of Rocks* (eds. H.C. Heard, I.Y. Borg, N.L. Carter, and C.B. Raleigh), pp. 265–273. American Geophysical Union Geophysical Monograph No. 16.

Brace, W.F. and Martin, R.J. (1968). A test of the effective stress law for crystalline rocks of low porosity. *Int. J. Rock Mech. Min. Sci.* **5**, 415–426.

Byerlee, J.D. (1967). Frictional characteristics of granite under high confining pressure. *J. Geophys. Res.* **72**(14), 3639–3648.

Byerlee, J. (1978). Friction of rocks. *Pure Appl. Geophys.* **116**, 615–626.

Byerlee, J. (1990). Friction, overpressure and fault normal compression. *Geophys. Res. Lett.* **17**(12), 2109–2112.

Colten-Bradley, V.A. (1987). Role of pressure in smectite dehydration–effects on geopressure and smectite-to-illite transformation. *Am. Assoc. Pet. Geol. Bull.* **71**(11), 1414–1427.

Freed, R.L. and Peacor, D.R. (1989). Variability in temperature of the smectite/illite reaction in gulf coast sediments. *Clay Minerals* **24**, 171–180.

Garg, S.K. and Nur, A. (1973). Effective stress law for fluid saturated rocks. *J. Geophys. Res.* **78**(26), 5911–5921.

Israelachvili, J.N., McGuiggan, P.M., and Homola, A.M. (1988). Dynamic properties of molecularly thin liquid films. *Science* **240**, 189–190.

Lachenbruch, A.H. and Sass, J.H. (1980). Heat flow and energetics of the San Andreas fault. *J. Geophys. Res.* **85**, 6185–6222.

Logan, J.M. and Rauenzahn, K.A. (1987). Frictional dependence of gouge mixtures of quartz and montmorillonite on velocity, composition and fabric. *Tectonophysics* **144**, 87–108.

Meunier, A. and Velde, B. (1989). Solid solutions in I/S mixed-layer minerals and illite. *Am. Mineral.* **74**, 1106–1112.

Moore, D.E., Ma Jin, Summers, R., and Byerlee, J. (1984a). The effects of strain rate and pore pressure on the strength of heated fault gouge. *EOS* **65**, 1078.

Moore, D.E., Ma Jin, Summers, R., and Byerlee, J. (1984b). The effect of pore pressure and heating time on the strength and stability of a serpentine gouge. In *Rock Mechanics in Productivity and Protection*, Proc. 25th Symp. on Rock Mech. (eds. F. Dowding and M. Singh), pp. 312–319.

Moore, D.E., Summers, R., and Byerlee, J. (1989). Sliding behavior and deformation textures of heated illite gouge. *J. Struct. Geol.* **11**(3), 329–342.

Morrow, C.A. and Byerlee, J.D. (1989). Experimental studies of compaction and dilatancy during frictional sliding on faults containing gouge. *J. Struct Geol.* **11**(7), 815–825.

Morrow, C.A., Shi, L.Q., and Byerlee, J.D. (1982). Strain hardening and strength of clay-rich fault gouge. *J. Geophys. Res.* **87**(B8), 6771–6780.

Morrow, C.A., Shi, L.Q., and Byerlee, J.D. (1984). Permeability of fault gouge under confining pressure and shear stress. *J. Geophys. Res.* **89**(B5), 3193–3200.

Nur, A. and Byerlee, J.D. (1971). An exact effective stress law for elastic deformation of rocks with fluids. *J. Geophys. Res.* **76**(26), 6414–6419.

Rice, J.R. (1990). Fault stress states, pore pressure distributions, and the weakness of the San Andreas fault. *EOS* **71**(43), 1652.

Robin, P.F. (1973). Note on effective pressure. *J. Geophys. Res.* **78**(14), 2434–2437.

Summers, R. and Byerlee, J. (1977). Summary of results of frictional sliding studies, at confining pressures up to 6.98 kb, in selected rock materials. *U.S. Geological Survey, Open File Report 77-142.*

Viani, B., Low, P., and Roth, C. (1983). Direct measurement of the relation between interlayer force and interlayer distance in the swelling of montmorillonite. *J. Colloid Interface Sci.* **96**(1), 229–244.

Wang, C.Y. and Mao, N. (1979). Shearing of saturated clays in rock joints at high confining pressures. *Geophys. Res. Lett.* **6**(11), 825–828.

Wang, C.Y., Mao, N., and Wu, F.T. (1979). The mechanical property of montmorillonite clay at high pressure and implications on fault behavior. *Geophys. Res. Lett.* **6**(6), 476–478.

Wang, C.Y., Mao, N., and Wu, F.T. (1980). Mechanical properties of clays at high pressure. *J. Geophys. Res.* **85**(B3), 1462–1468.

Wang, C.Y. (1984). On the constitution of the San Andreas fault zone in central California. *J. Geophys. Res.* **89**(B7), 5858–5866.

Wu, F.T., Blatter, L. and Robertson, H. (1975). Clay gouges in the San Andreas fault system and their possible implications. *Pure Appl. Geophys.* **113**, 87–95.

Wu, F.T. (1978). Mineralogy and physical nature of clay gouge. *Pure Appl. Geophys.* **116**, 655–689.

Zoback, M.D. and Byerlee, J.D. (1975). Permeability and effective stress. *Bull. Am. Assoc. Pet. Geo.* **59**(1), 154–158.

Zoback, M.D., Zoback, M.L., Mont, V.S., Suppe, J., Eaton, J.P., Healy, J.H., Oppenheimer, D., Reasenberg, P., Jones, L., Raleigh, C.B., Wong, I.G., Scotti, O., and Wentworth, C. (1987). New evidence on the state of stress of the San Andreas fault system. *Science* **238**, 1105–111.

Chapter 4

The Brittle–Ductile Transition in Feldspar Aggregates: An Experimental Study

J. Tullis and R. Yund

Department of Geological Sciences, Brown University, Providence, RI 02912, U.S.A.

Abstract

The micromechanical processes associated with the brittle-to-crystal plastic transition in anorthosite have been investigated. Axial compression experiments have been conducted on Bushveld anorthosite at a constant displacement rate of $10^{-5}\,s^{-1}$ over a wide range of pressures (250–1500 MPa), temperatures (25–1200°C), and total strains (1–60% shortening). A combination of optical and transmission electron microscopic observations shows that there is a progression with increasing pressure and temperature from brittle faulting, to macroscopically ductile cataclastic flow, to dislocation creep. The style of faulting changes from a wide zone ~30°C to σ_1 of obliquely connected more axial strands at lower pressures and temperatures, to a very sharp, planar surface at ~40°C to σ_1 at pressures >750 MPa or temperatures >500°C. The transition from faulting to distributed cataclastic flow depends on temperature, pressure, and water content; at pressures of 1000–1500 MPa it is fairly abrupt, occurring between 250 and 300°C. Unlike any other material described to date, this transition involves a change in the nature and distribution of the microcracks without any crystal plasticity. Cataclastic flow occurs by distributed cracking, primarily on cleavages, with a length and spacing down to <0.1 μm, and formation of micro-crush zones. Optically this is manifested as multiple and commonly conjugate grain-scale faulting, with a very strong and patchy undulatory extinction. The transition from cataclastic flow to dislocation creep is quite broad; at a pressure of 1000 MPa mechanical twins are first noted at 600°C, dislocations are first activated at 700°C, and steady-state dislocation creep with no cracking requires temperatures ≥1100°C.

Microstructures typical of cataclastic flow have been described in low-grade naturally deformed feldspathic rocks. The existence of a broad regime of cataclastic flow in feldspars is facilitated by the ease of cracking on the cleavages and the difficulty of dislocation glide and climb. The brittle–ductile transition in many portions of the crust where feldspars form the stress-supporting matrix may

FAULT MECHANICS AND TRANSPORT PROPERTIES OF ROCKS
ISBN 0-12-243780-2

reflect the transition from faulting to cataclastic flow, and the grain size reduction and fluid influx resulting from this deformation mechanism may promote strain weakening.

1. Introduction

The mechanisms and conditions associated with the macroscopic brittle-to-ductile transition in the crust are of great interest to geoscientists attempting to interpret the record preserved in naturally deformed rocks and to model inter- and intraplate deformation. The brittle–ductile transition is commonly idealized as a sharp change from brittle faulting and frictional sliding to dislocation creep; on a stress vs. depth plot this is represented by the intersection of Byerlee's law with a power-law creep curve (e.g., Brace and Kohlstedt, 1980). Many modeling studies have used an experimental flow law for 'wet' quartz to represent the behavior of the ductile portion of the crust, in part because this predicts a transition depth ($\sim$15 km) that agrees with the observed depth limit of earthquakes on the San Andreas fault. However, recently it has been realized that there is probably a broad region of complex deformation associated with the brittle–ductile transition, that quartz may not be the controlling phase, and that the depth limit of plate boundary earthquakes may be due to a change in the frictional properties of faults rather than to a change in the deformation behavior of intact rocks (e.g., Tse and Rice, 1986; Rutter and Brodie, 1991).

There have been relatively few studies of the grain-scale deformation mechanisms associated with the brittle-to-crystal plastic transition in geologic materials, and thus the terminology has been confusing. Cataclastic flow is a term originally used by Borg et al. (1960) in a study of low-temperature and pressure deformation of loose quartz sand. It is generally used to mean macroscopically ductile deformation that involves no crystal plasticity (Rutter, 1986). The term has been applied to the flow of loose granular aggregates at low temperatures and pressures (e.g., Mandl et al., 1977), to the deformation of the gouge within a fault zone (e.g., Chester et al., 1985), and to the distributed deformation of an initially intact rock at relatively low temperatures and moderate pressures (e.g., Heard, 1960). Deformation involving some distributed cracking and some dislocation glide has been termed semibrittle flow (Carter and Kirby, 1978).

Previous experimental work on the brittle–ductile transition in geologic materials indicates a difference between carbonates and silicates. Studies of calcite aggregates (Heard, 1960) showed a transition with increasing pressure at room temperature from sharp faults with stress drops, to wider fault zones and no stress drops, to macroscopically ductile deformation which still showed a pressure dependence of strength. No microstructural studies were made on these samples, but more recently Fredrich et al. (1989) used SEM and TEM to show that for calcite marble deformed at room temperature, increasing pressure induced a transition from longer, axial cracks that coalesced into a fault to a combination of shorter and more isotropic cracks and dislocation glide. In silicates, a transition from brittle to ductile behavior with increasing pressure at room temperature has been demonstrated in porous aggregates (e.g., basalts (Shimada, 1986); quartzites (Hadizadeh and Rutter, 1983)), as well as in serpentine-bearing dunite and gabbro (Byerlee, 1968). However, in porous quartzites the cataclastic flow is only transient, and faulting occurs at $\sim$25% strain after net compaction changes to net dilation (Hirth and Tullis, 1989). Similarly, nonporous and unaltered silicates show brittle faulting up to 2000–3000 MPa (e.g., granite (Schock et al., 1973); gabbro, granite, dunite, and eclogite (Shimada and Yukutake, 1983)). Thus pressure alone is not sufficient to cause a switch to ductile cataclastic flow at laboratory strain rates, contrary to earlier expectations (e.g., Paterson, 1978).

We have been conducting experimental studies of the brittle-to-crystal plastic transition in crustal rocks in order to determine the grain-scale deformation mechanisms responsible for the macroscopic behavior. Initial studies on Westerly granite (Tullis and Yund, 1977) showed a transition from faulting to distributed deformation with increasing temperature ($\sim 300°C$) at high pressure (1000 MPa), where crystal plasticity is extremely limited. Later studies on pure albite aggregates documented a gradual transition from dislocation creep to distributed cracking (cataclastic flow) with decreasing temperature (900–700°C) at high pressure (1000–1500 MPa) (Tullis and Yund, 1987). Recently, Hadizadeh and Tullis (1991, 1992) presented preliminary observations on the brittle–ductile transition in anorthosite deformed over a wide range of pressures and temperatures. In the present study we report the results of a detailed investigation of the grain-scale deformation mechanisms in these and additional anorthosite samples. We have focused our efforts on monomineralic aggregates of feldspar because its volumetric importance makes it likely to control the behavior of large portions of the crust, and because its two excellent cleavages coupled with its difficulty of dislocation creep (Tullis and Yund, 1985) should make it especially likely to undergo cataclastic flow. Parallel studies conducted on pure quartz aggregates (Hirth and Tullis, 1991) show important differences.

Our study shows that unlike any other material investigated to date, experimentally deformed feldspar aggregates exhibit a broad regime at intermediate temperatures and pressures where the deformation is macroscopically ductile and occurs by distributed cracking without any evidence of crystal plasticity. These results have important implications for the interpretation of microstructures in feldspathic rocks deformed at low to moderate metamorphic grades and for the changes in rheology that occur with increasing depth in the crust.

2. Experimental Procedures

2.1. Starting material

Most of our experiments were conducted on Bushveld anorthosite, kindly supplied by B. Hobbs, consisting of roughly 98% plagioclase ($An_{78}Ab_{21}Or_1$), 1% pyroxene, $<1\%$ opaques, with trace amounts of sericite(?) along some grain boundaries and healed cracks. The plagioclase grains are equant to slightly tabular, with grain diameters ranging from 100 to 900 μm and averaging $\sim 500\ \mu m$; they contain albite and pericline growth twins, and Huttenlocher exsolution lamellae. There is no obvious shape or crystallographic preferred orientation (see Figure 3a).

Most experiments were done on intact cylinders, 6.35 mm in diameter and 10–15 mm long. A few experiments were done on synthetic aggregates made from crushed and sieved anorthosite (as described in Tullis and Yund, 1987). Most samples were air-dried, although a few were given a prior vacuum heat treatment, and a few had 0.1–0.2 wt% water added and mechanically sealed inside the metal jacket.

Bytownite (An_{78}) undergoes a reaction within the range of temperatures and pressures

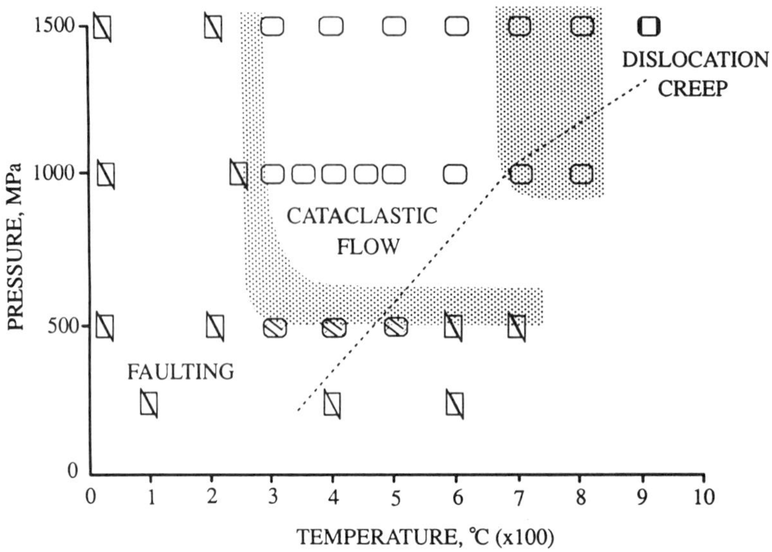

Figure 1. *P–T* diagram showing conditions of experiments performed on 'as-is' Bushveld anorthosite cores at a strain rate of $10^{-5}\,s^{-1}$, and location of regimes of faulting, cataclastic flow, and dislocation creep. Stippled regions indicate transitions between different regimes. First mechanical twins are noted at 600°C and first dislocations at 700°C. Plagioclase reacts with any available water to form zoisite to the left of the dashed line, which is drawn for $P_{water} = P_{total}$ (Goldsmith, 1982).

utilized in this study (Goldsmith, 1982): An-rich plagioclase + water = zoisite + kyanite + quartz + vapor + more Ab-rich plagioclase (dashed line on Figure 1). Lath-shaped grains of zoisite ranging up to ~20 μm long are present in samples deformed at ≥700°C at 1500 MPa, even though no water was added to the samples; presumably it results from traces of moisture originally present along grain boundaries, and/or decomposition of the sericite. The maximum amount of zoisite in any deformed sample is <1%. The zoisite only occurs in samples deformed in the transitional regime between cataclastic flow and dislocation creep; this transition has been well studied in albite aggregates (Tullis and Yund, 1987), which do not undergo this reaction.

2.2. Apparatus and sample assemblies

Samples were deformed in a modified Griggs apparatus, using a sample assembly similar to that illustrated in fig. 2 of Kronenberg and Tullis (1984), except that the sample was the same diameter as the pistons, with no inner nickel jacket. The confining medium was a sleeve of NaCl, thick enough (3.2 mm) to allow high sample strains (~50%) without impinging on the furnace. For most experiments soft-fired pyrophyllite was used between the furnace and the pressure vessel walls, but for comparison a few experiments were done with NaCl, and a few room-temperature experiments were done in an all-lead assembly. Alumina end pistons and a stepped furnace were used to achieve low thermal gradients and homogeneous sample strains. Samples were jacketed in thin-walled (0.15 mm) platinum, silver, or copper tubing. In the few experiments with a trace of added water, a disk of platinum foil between each end of the sample and the end pistons constituted an effective mechanical seal (see Tullis and Yund, 1980). Although the samples were not buffered with oxide pairs, tests with pieces of various metal foils between quartzite disks have shown that the effective buffering of our NaCl assembly is equivalent to Ni–NiO.

2.3. Experimental conditions

Samples were taken to pressure and temperature using a path designed to retain any original water (Kronenberg and Tullis, 1984). Approximately 6–8 hours were taken for pressurization (at 300°C). When at pressure, the piston was advanced at $10^{-4}\,s^{-1}$ until a slight increase in force defined the hit point; the piston was then backed off ~1–3 mm. After the temperature was raised to run conditions (over a time of 5–15 minutes), the motor was started at the desired rate. At the end of an experiment the motor was stopped; the temperature immediately decreased to 300°C (in 1–2 mins); the piston was retracted at $10^{-4}\,s^{-1}$, and then the pressure was slowly decreased while maintaining an end load of < 100 MPa to help reduce the number of extension fractures. The temperature was decreased stepwise with the pressure decrease from 500 to 250 MPa.

More than 50 experiments were performed in axial compression, at constant displacement rate; the conditions are shown on Figure 1. Almost all experiments were done at a strain rate of $2 \times 10^{-5}\,s^{-1}$, but a few were done at $2 \times 10^{-4}\,s^{-1}$ and $2 \times 10^{-6}\,s^{-1}$. The confining pressure ranged from 200 to 1400 MPa. (Recent calibrations of the outer pyrophyllite inner NaCl assemblies using germanates indicate that pressures are 24% lower than listed.) Temperatures ranged from 20 to 1200°C. The thermocouple temperature is known to within ±5°C, and the central 90% of the sample volume is within 10°C of this value; at a thermocouple temperature of 600°C the ends of the sample are at approximately 550°C (see Kronenberg and Tullis, 1984). Total shortening strains ranged from 2% to 60%.

2.4. Sample analysis and data reduction

Deformed samples were impregnated with epoxy, sawn in half longitudinally, and reimpregnated. A doubly polished epoxy thin section was made from one half, for optical observations, and a lakeside section from the other half. The latter was used to mount one to three transmission electron microscopc (TEM) grids and rings, which were subsequently ion-thinned. TEM observations were made using a Philips 420 machine operating at 120 kV. Weak-beam techniques were used to examine the strain contrast along cracks.

The choice of a confining medium which is solid but weak over the temperature range from 20 to 900°C is nontrivial. At a pressure of 200 MPa and a strain rate of $10^{-5}\,s^{-1}$, NaCl has a strength of ~200 MPa at 20°C and ~35 MPa at 400°C (Heard, 1972). Additional contributions to the externally measured sample strength may come from friction against the colder, moving upper piston, as well as the interface with the relatively rigid pyrophyllite outside the NaCl sleeve. We have done several comparison experiments in an assembly consisting entirely of NaCl except for the stepped graphite furnace, and find that the externally measured sample strengths are ~200 MPa lower for sample temperatures of 400°C.

We have also done comparison experiments in an all-lead assembly at room temperature and 1000 MPa, where samples always fault. The initial fracture stress was close to the same as that in the NaCl-sleeve assembly. However, samples in lead showed no work hardening; each stress peak of the subsequent stick–slip behavior was the same as the initial strength, whereas samples in NaCl showed an overall work hardening envelope to the stick–slip behavior.

Aside from strength contributions from the solid confining medium, there is a question of how to reduce stress–strain curves for the high-strain experiments. This is an important question, since there has been considerable debate as to whether cataclastic flow is capable of producing steady-state deformation. All of the stress–strain curves shown in this chapter have been reduced taking into account the measured apparatus distortion, and assuming that the sample volume remains constant, with the diameter changing so as to maintain a right circular cylinder. For a constant applied load this results in decreasing stress with time, owing to the larger cross-sectional area of the sample. Although the sample strain is usually quite homogeneous, owing to the very low temperature gradients, the piston which transmits the stress remains a constant area. If one assumes that only the portion of the sample directly between the pistons contributes to the measured strength, then the stress–strain curve would show more hardening than the ones we present. The truth probably lies somewhere between; the sample material that moves sideways, outside the direct line of force between the pistons, is subjected to slower strain rates and somewhat higher temperatures.

3. Results

3.1. Overview

The experimentally deformed anorthosite samples show a range of deformation behavior with increasing temperature and pressure, from brittle faulting to distributed cracking (which we have called cataclastic flow), to steady-state dislocation creep. The transition from faulting to cataclastic flow is fairly abrupt, whereas that from cataclastic flow to dislocation creep is quite broad. The deformation mechanisms have been identified on the basis of the mechanical behavior plus the optical and TEM microstructures. The pressure–temperature conditions of the different mechanism fields, for a strain rate of $2 \times 10^{-5}\ s^{-1}$, are shown in Figure 1. Representative stress–strain curves for confining pressures of 250, 500, 1000, and 1500 MPa are shown in Figure 2.

In this paper we will concentrate mostly on the regime of cataclastic flow, which has not been described in detail before, and the transition from faulting to cataclastic flow. We have previously described the transition from cataclastic flow to dislocation creep in albite aggregates (Tullis and Yund, 1987), and find the behavior of the anorthosite almost identical. At our laboratory strain rates, the transition from faulting to cataclastic flow requires temperatures $\geq 300°C$ as well as pressures ≥ 750 MPa (Hadizadeh and Tullis, 1986), and the transition from cataclastic flow to dislocation creep requires temperatures $> 1000°C$ and pressures ≥ 1000 MPa.

There is a major complication in extrapolating the experimentally determined regimes to nature. In general one expects the flow stress for faulting and cataclastic flow to depend strongly on pressure and relatively little on temperature, but that for dislocation creep to depend strongly on temperature and not at all on pressure. However, the experimental flow stress for dislocation creep of feldspar, just as that for quartz, varies inversely with pressure, apparently owing to a dependence on water fugacity (e.g., Tullis

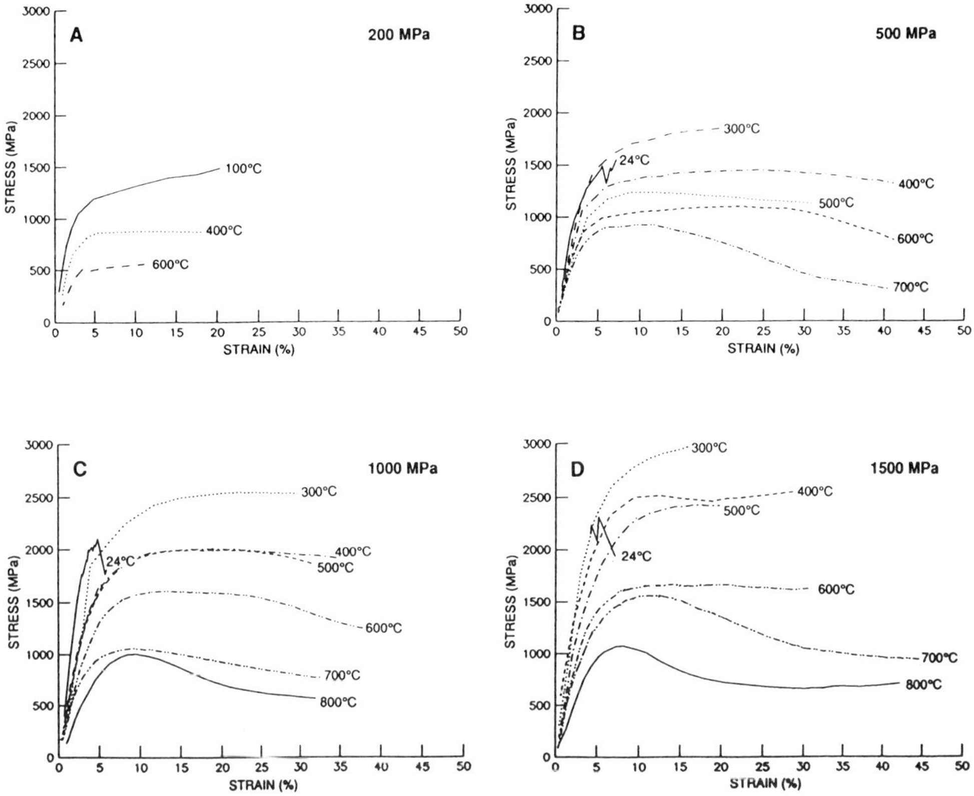

Figure 2. Representative stress–strain curves for intact cores of Bushveld anorthosite deformed at a strain rate of $10^{-5}\,s^{-1}$. (A) Confining pressure of 250 MPa. (B) Confining pressure of 500 MPa. (C) Confining pressure of 1000 MPa. (D) Confining pressure of 1500 MPa.

and Yund, 1989; Farver and Yund, 1990). At laboratory strain rates, steady-state dislocation creep requires pressures of 1000–1500 MPa, whereas in the crust at much slower strain rates it occurs at ~500 MPa. Obviously, cracking and frictional sliding have a direct dependence on pressure. It is possible that the crack propagation rates controlling cataclastic flow also have a positive dependence on water fugacity, and thus an inverse dependence on pressure, but this is unknown. Given these complications, we will focus most of our attention on samples deformed at moderately high pressures (1000–1500 MPa) and therefore high water fugacities. Ultimately our ability to accurately extrapolate laboratory data to the crust will depend on determining the appropriate water fugacity term in the flow laws for these deformation mechanisms.

In the sections below we describe the mechanical and microstructural results for samples of Bushveld anorthosite which show brittle faulting, cataclastic flow, and dislocation creep. We will first briefly describe the behavior of samples deformed 'as-is' at 200–500 MPa and $10^{-5}\,s^{-1}$ as a function of increasing temperature. We will then describe in more detail the behavior of samples deformed at 1000–1500 MPa, and compare these with results for other strain rates and water treatments where appropriate.

3.2. Pressures of 200–500 MPa

At pressures of 200–500 MPa, faulting extends to high temperatures, and the nature of the faults is different from those at higher pressures. There is no temperature interval of fully distributed cataclastic flow, nor of dislocation creep. The microstructures in these faulted samples are complicated by the relatively high strains imposed after faulting. In addition, the nonnegligible strength of the confining medium at the lower temperatures ($\leq 300°C$) has undoubtedly enhanced the amount of distributed deformation away from the fault. Therefore only brief descriptions are given of these samples.

At 200 MPa, samples deformed at 100, 400, and 600°C all developed faults at $\sim 30°$ to σ_1, although without a sudden stress drop (Figure 2a). The fault in the 400°C sample consists of three portions; the outer thirds are relatively sharp and lie at $\sim 25°$ to σ_1, and they connect to a central region of relatively short (3–5 grain diameters), en echelon, and steep ($\sim 16°$ to σ_1) segments (Figure 3b). The gouge thickness is variable along the length of the fault; there are thicker pockets at the intersections of the subaxial segments with the oblique connecting links. In the 600°C sample, a relatively sharp and planar through-going fault offsets an earlier conjugate fault.

At 500 MPa, samples exhibit transitional behavior. A sample deformed at 24°C faulted with audible, small stress drops. A sample deformed at 200°C developed a broad fault zone consisting of a number of anastamosing strands; this zone is offset by a later, conjugate fault. This sample showed slight work hardening. Samples deformed at 300 and 400°C developed a single, very broad fault zone consisting of a number of shorter, more axial strands. These samples underwent approximately steady-state flow out to 25–35% shortening, followed by gradual strain weakening. A sample deformed at 500°C shows broad, conjugate fault zones initiated at the sample corners, intersecting in a region of distributed deformation in the sample center. The faults formed at 24–500°C all developed at $\sim 30°$ to σ_1.

Samples deformed at 600 and 700°C and 500 MPa develop a single very sharp, through-going fault at $\sim 45°$ to σ_1 (Figure 3d), as previously observed in Westerly granite and Hale albite deformed at the same conditions (Tullis and Yund, 1977, 1980). The 600°C sample shows approximately steady-state flow out to 30% strain followed by gradual strain weakening, whereas the 700°C sample shows a more pronounced strain weakening after a strength maximum at $\sim 10\%$ strain (Figure 2b). The faults in these two higher-temperature samples are quite different in character from those in lower-temperature samples. They are narrow (100–200 μm wide, less than one grain diameter), with sharp, planar boundaries; they do not appear to originate from a number of en echelon, shorter, and more axial strands. There is a thin zone of very fine-grained material along the fault; TEM shows this to be fault gouge, with no evidence for dynamic recrystallization. The exposed ends of the 700°C fault showed shiny slickensides on the surfaces. Both samples show evidence of mechanical twinning.

3.3. Pressures of 1000–1500 MPa

At pressures of 1000–1500 MPa, there is a well-defined transition from faulting to cataclastic flow at ~ 250–300°C, and a broader and more continuous transition from cataclastic flow to dislocation creep at 700–900°C. We will describe these three regimes separately.

Figure 3. Optical photomicrographs, crossed nicols. For deformed samples σ_1 is vertical. (A) Undeformed Bushveld anorthosite. Note randomly oriented grains, original twins, some healed cracks. (B) Fault in sample deformed at 400°C and 250 MPa, formed by linking of several en echelon, more axial strands. (C) Fault in sample deformed at 200°C and 1000 MPa, consisting of single sharp plane. (D) Fault in sample deformed at 600°C and 500 MPa, consisting of thin planar zone of intensely sheared material.

3.3.1. Faulting regime

Samples of Bushveld anorthosite deformed at ≤250°C fault with audible, sudden stress drops. Stress–strain curves are shown in Figures 2c and d. The strength increases with increasing pressure.

At 24°C and 1000 and 1500 MPa, samples deformed in an NaCl-sleeve assembly show small stress drops superimposed on a strain hardening curve, whereas samples deformed in all-lead assemblies show the same initial yield strength but no strain hardening associated with the several subsequent large stress drops. The nature of the faulting also differs depending on the confining medium. Samples deformed in all-lead assemblies with up to seven major stress drops (20% shortening) develop a single fault with extensive offset. The fault has several splays enclosing less-deformed slivers, but in general it is a thin (~100–200 μm) zone of gouge. A zone approximately one grain wide on either side of the fault contains cracks and patchy undulatory extinction, but farther away the grains appear completely undeformed. In contrast, samples deformed in NaCl-sleeve assemblies commonly develop several separate through-going faults, and there are more cracks and grain-scale faults in grains away from the faults. This is undoubtedly due to the greater strength of the NaCl confining medium.

For all samples deformed at >24°C, an NaCl-sleeve or an all-NaCl assembly has been used. Samples deformed at 250°C at 1000 MPa (and 200°C at 1500 MPa) show three sharp and distinct through-going faults corresponding to three major stress drops. These faults and the zones of disruption on either side of them are extremely narrow (2–5 μm) where they follow favorably inclined cleavage planes and grain boundaries, and somewhat wider where they cut across grains (Figure 3c). The maximum thickness of the gouge zone is 40 μm, and the particle size ranges down to <0.01 μm. At all scales the fragments have sharp edges and commonly a tabular shape, suggesting breakage along cleavages. Both samples (shortened 12% and 20%, respectively) show minor grain-scale faulting and patchy undulatory extinction in regions away from the faults, perhaps reflecting the relatively high strain and the strength of the confining medium.

The faults formed at these high pressures do not appear to have formed by the linking of en echelon, shorter, and more axial strands, in contrast to faults developed at lower pressure. The high-pressure faults (in samples deformed in either the lead or the NaCl medium) are much sharper and more continuous, and form at a distinctly higher angle to σ_1 (~40°).

3.3.2. *Cataclastic flow regime*

Feldspar aggregates deformed at 300–600°C are macroscopically ductile. TEM observations have been crucial in determining the deformation processes operative in this regime. They show that there is no dislocation multiplication or glide over this entire temperature range. At 300–500°C, deformation has occurred solely by distributed microcracking with associated fragment rotation and comminution; at 600°C there is some mechanical twinning, although cracking remains dominant. We will refer to this as the cataclastic flow regime. Stress–strain curves for samples deformed in this regime are shown in Figures 2c and d; the strength increases with increasing pressure and decreases with increasing temperature.

Microstructural development At 1000–1500 MPa, samples deformed at 300–500°C develop very similar microstructures. We will first describe the optical and TEM microstructures developed with increasing strain in samples deformed at 300°C.

In a sample shortened 1% at 30°C and 1000 MPa, a few grain-scale faults are visible within some grains as discrete but small offsets of original twins (Figure 4a); occasionally two conjugate sets are activated. These faults are inclined at ~15–60° to

---

Figure 4. Optical photomicrographs (crossed nicols) of Bushveld anorthosite deformed in the cataclastic flow regime at 300°C and 1000 MPa (σ_1 is vertical). (A) Sample shortened 1%; grain-scale faults are few. (B) Sample shortened 10%; note multiple grain-scale faults (NW–SE) and patchy undulatory extinction. (C)–(E) Sample shortened 30%. (C) shows grain-scale faults on one cleavage plane which are more closely spaced than at lower strain. (D) shows conjugate grain-scale faults on two cleavage planes. (E) shows grain-scale shear zones of discrete width, with strong patchy undulatory extinction. (F) Sample shortened 50%; grain-scale shear zones on conjugate planes and intense patchy undulatory extinction.

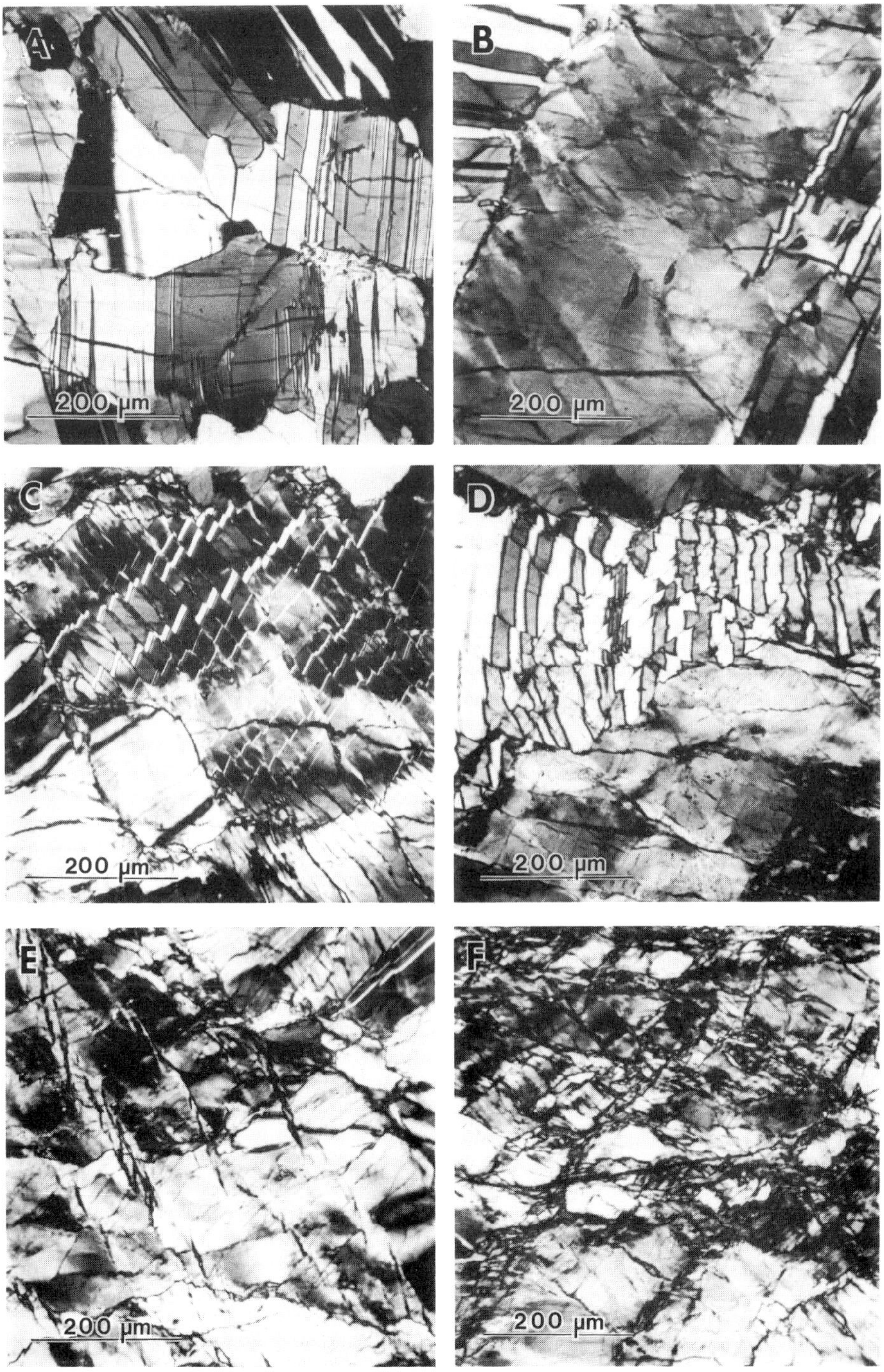
A
200 µm
B
200 µm
C
200 µm
D
200 µm
E
200 µm
F
200 µm

σ_1, and commonly follow cleavage planes. Very few original healed cracks have opened and been offset. A few of the faults cut across two or three grains, but typically they terminate within grains, in a small region of undulatory extinction. The faults are very sharp, with no optically visible gouge along them. Few if any axial cracks are present, even associated with the small grains of pyroxene or oxides, or the interior patches of sericite. TEM shows abundant microcracks within most grains, with a spacing of about 0.1–2 μm. These are mostly on the two cleavages, and so tend to form an irregular grid, although there are also oblique cracks connecting these two sets. We have never observed tensile wing cracks growing from inclined cleavage cracks.

In a sample shortened 10%, almost all grains contain multiple grain-scale faults with somewhat larger offsets and a spacing down to ~20 μm (Figure 4b). Most of these faults terminate within grains, in a small region of intense undulatory extinction; others appear to terminate at the grain boundary in a local zone of intense crushing. Where these grain-scale faults change orientation slightly, there are small regions of sharply different extinction, giving the grain a puckered appearance (Figure 4b). Only a few of the faults have open cracks along them, and they would not be obvious without the offset twins. In the regions between the grain-scale faults there is commonly a strong, patchy undulatory extinction (Figure 4b). A few grains contain irregular microfaults and crush zones, rather than parallel, planar faults; these grains have very strong and irregular undulatory extinction. TEM shows that the spacing of the cleavage microcracks ranges down to <0.1 μm, and that many of them begin and end in regions of 'good' crystal structure (Figure 5a). Some of the cracks are not open along their entire length, but show contrast where their sides are touching (Figure 5b). The operation of cracking on two nearly perpendicular planes quickly isolates blocky fragments (Figure 6b), which can then rotate slightly, undergoing further cracking at their corners. This very effectively produces a progressive grain size reduction. We also observe irregular regions and semiplanar zones (~0.1–1 μm across) which contain extremely finely crushed material (<0.01 μm) (right side of Figure 6a). These micro-crush zones have very sharp boundaries. The material within the crush zones undergoes extensive rotation, as shown by the strong ring pattern in the SAD (Figure 6a). However, this material is not amorphous, as observed in the feldspars of granite samples subjected to room-temperature frictional sliding (Yund et al., 1990).

In a sample shortened 30%, although a few grains remain relatively undeformed, almost all grains contain one or two (nearly perpendicular) sets of multiple grain-scale faults; the spacing of these faults ranges down to ~20 μm and the offset on each may be up to 15 μm (Figures 5c,d). The 'plates' or 'beams' isolated by the grain-scale faults (on one or two cleavages) are commonly strongly bent, producing a patchy undulatory extinction which may resemble subgrains. TEM shows that the spacing of cleavage cracks is commonly ~0.05 μm. Their average length is difficult to determine, but it is often just 2–3 μm, and it is common to observe a set of closely spaced cracks terminate in one region and a new set initiate only 1–2 μm farther on. Although there are no markers at the TEM scale to show the amount of offset, shearing motion on the cleavage cracks is indicated by the observations that the 'beams' are bent and that SAD patterns show rotation (as in left side of Figure 6a). Micro-crush zones are abundant, occurring on average every 2–3 μm. Many of the cracks in this sample exhibit strain contrast (as in Figure 5b), which gives the impression of a high density of dislocations. However, careful examination with weak-beam techniques indicates that there are no dislocations associated with the cracks. The strain contrast indicates that the cracks are not completely

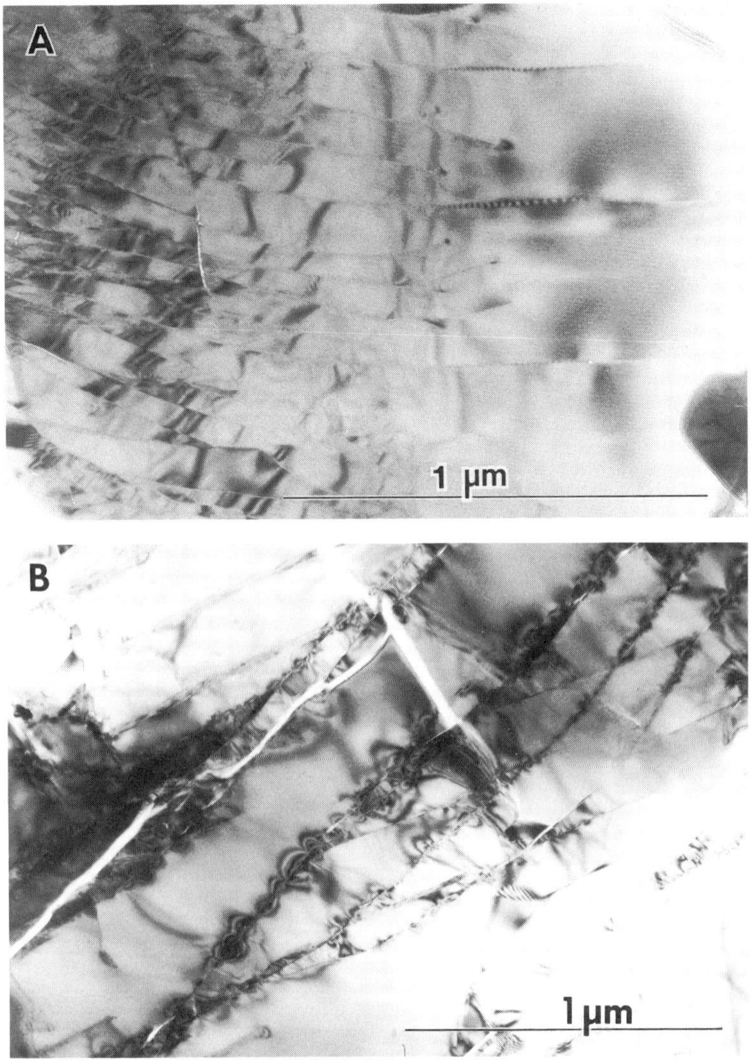

Figure 5. TEM photomicrographs (bright-field) from Bushveld anorthosite shortened 10% at 300°C and 1000 MPa. (A) Multiple cracks subparallel to (010) cleavage which terminate in an area of good structure; faint horizontal lines are the Huttenlocher exsolution lamellae. (B) Multiple subparallel cracks showing strong strain contrast.

open but that their sides are touching in many places. Although it is impossible to know exactly how the openness of the cracks might have changed during unloading of the sample and section preparation, cracks with strain contrast are never seen in samples from the faulting or dislocation creep regimes. The occurrence of numerous points of impingement along these cracks in the cataclastic flow regime means that their effective length is quite low.

In a sample shortened about 50%, original twins and grain boundaries are generally not distinguishable (Figure 4f). The finely crushed material resulting from penetrative microcracking and associated offset and rotation of the fragments produces, on the

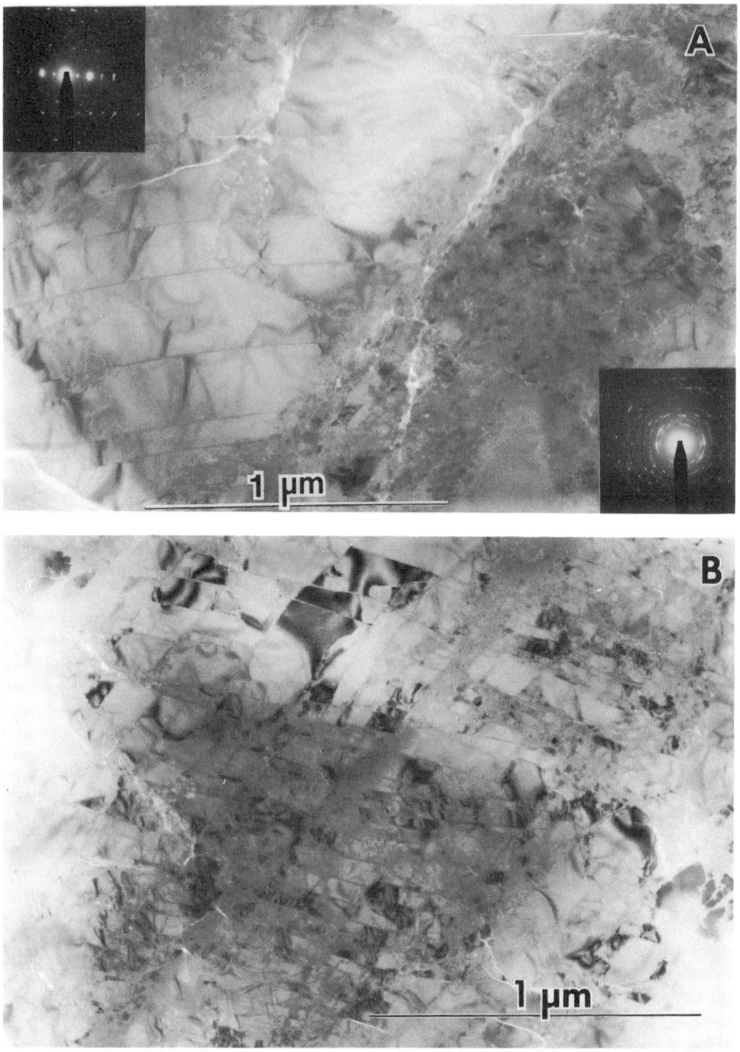

Figure 6. TEM photomicrographs (bright-field) from Bushveld anorthosite shortened 10% at 300°C and 1000 MPa. (A) Multiple subparallel cracks which terminate in a thin (~1 μm wide) micro-crush zone, within which the grain size is <0.1 μm. Streaking of spots in SAD from cracked region indicates some shear on the cracks; extensive streaking in SAD from crush zone reflects rotation of fragments. (B) Penetrative cracking on two orthogonal cleavage planes.

optical scale, strongly patchy undulatory extinction and regions which appear similar to subgrains or recrystallized grains (Figure 4f). In addition, this sample has developed a fairly strong crystallographic preferred orientation, as described previously in albite aggregates (Tullis and Yund, 1987), because multiple grain-scale faulting is geometrically similar to dislocation glide.

Samples deformed at 400–600°C show a similar microstructural progression with increasing strain as just described for 300°C. Because there is no dislocation

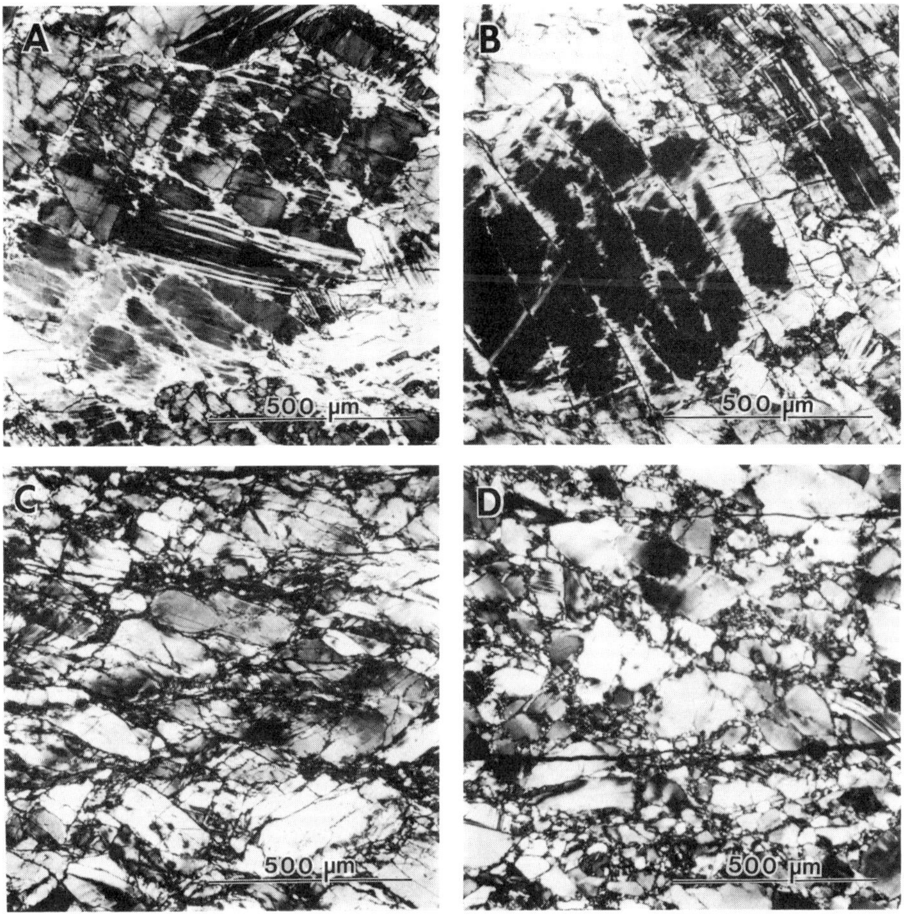

Figure 7. Optical photomicrographs (crossed nicols) of Bushveld anorthosite deformed in the transitional and dislocation creep regimes (σ_1 is vertical). (A) Sample deformed at 600°C and 1500 MPa; fuzzy white areas are regions of very fine-scale mechanical twinning. (B) Samples deformed at 700°C and 1500 MPa; note mechanical twins (white fuzzy patches) associated with the grain-scale faults. (C) Sample deformed at 800°C and 1500 MPa; dynamic recrystallization has occurred on grain-scale faults, isolating relatively undeformed 'lozenges.' (D) Sample deformed at 1000°C, $10^{-6}\ s^{-1}$ and 1500 MPa; dynamic recrystallization is more extensive, but grain interiors remain little strained.

multiplication or glide in this regime, it is clear that the optical-scale deformation bands and patchy undulatory extinction must be due to local rotations of the crystal structure by submicroscopic cracks and crush zones. This has been noted previously in experimentally deformed feldspars (Marshall and MacLaren, 1977; Tullis and Yund, 1977, 1987).

Samples deformed at 600°C show very similar microstructures, although they also undergo mechanical twinning. This twinning is visible optically as a distinctive type of patchy undulatory extinction between the grain-scale faults (Figure 7a). In TEM the twins are visible as very thin ($<0.1\ \mu m$) lamellae (Figure 8b). These samples also contain closely spaced conjugate cracks with strong strain contrast (Figure 8a).

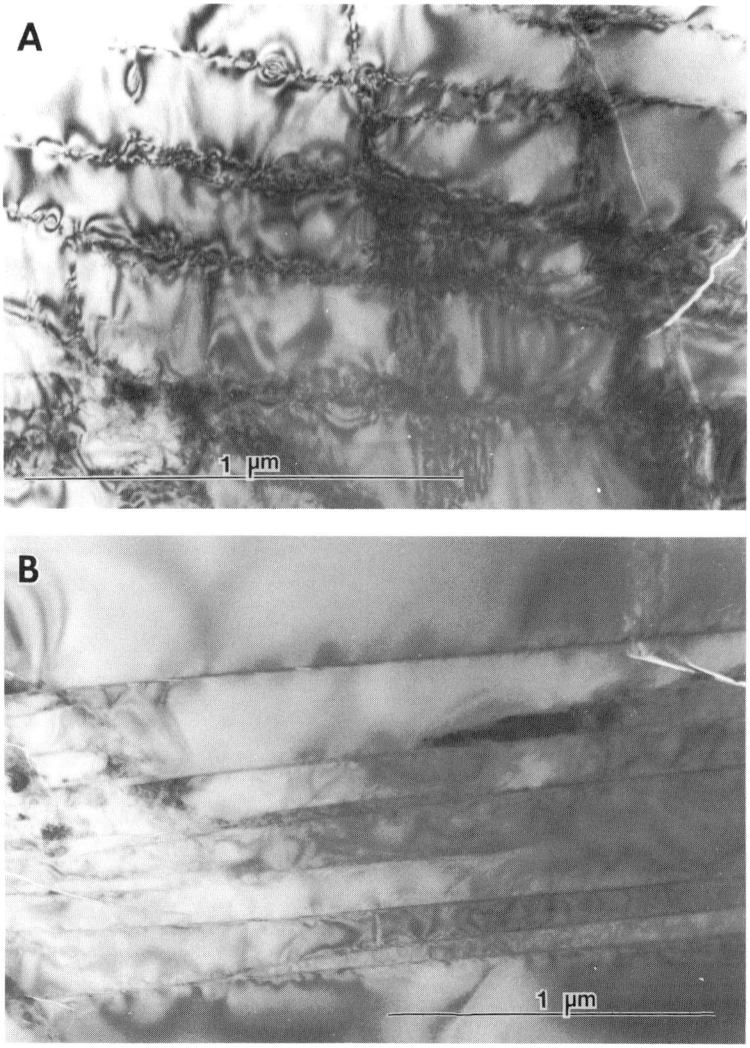

Figure 8. TEM photomicrographs (bright-field) from Bushveld anorthosite shortened 30% at 600°C and 1500 MPa. (A) Multiple, conjugate cracks showing strong strain contrast; no dislocations are present. (B) Deformation-induced mechanical twins.

The sample deformed at 600°C and 1000 MPa has a corner-to-corner ductile fault. It appears to have developed after about 25% strain, when strain weakening in the sample began (Figure 2c), as it is superimposed on distributed cataclastic flow. Except for its late initiation, the fault in this sample is similar to the ductile faults developed in samples deformed at 600–700°C and 500 MPa (Figure 3d). In contrast, samples deformed at 700°C at 1000 MPa, as well as those deformed at 600 and 700°C at 1500 MPa, show fully distributed deformation out to 30–40% shortening.

Transition from faulting to cataclastic flow In attempting to understand the deformation mechanism of cataclastic flow and especially the causes of the switch from faulting to distributed cracking, it is important to determine which cracks were produced during the deformation and which were produced during unloading and/or section preparation. We can never know this for sure; we can only compare the nature and distribution of cracks produced in samples deformed at different conditions, but treated identically afterward. All of the experimentally deformed feldspar aggregates contain extension cracks due to thermal quenching and decompression from high pressure. However, the nature and abundance of the cracks in samples deformed in the cataclastic flow regime is in striking contrast to those in samples deformed in either the faulting or the dislocation creep regimes. In the latter case one observes only occasional cracks at the TEM scale; these tend to be oriented normal to the compression direction and to be open with no strain contrast.

Our results indicate that there is a fundamental change in the nature of the cracking process at a temperature of about 300°C, at pressures of 1000–1500 MPa and a strain rate of $2 \times 10^{-5}\ s^{-1}$: at lower temperatures the cracks link up to form a through-going fault, whereas at higher temperatures they remain distributed. This change is not associated with dislocation activation. The fundamental nature of this transition is indicated in two ways. First, several samples deformed at 300–500°C developed faults at one or more corners, where the temperature was lower and stress concentrations were high due to the pistons, but in all cases these grade into ductile shear zones of finite width a short distance in towards the center of the sample, and they continue to widen until they intersect in a broad central zone or distributed cataclastic flow. In addition, one sample was originally loaded to failure at 24°C and 1000 MPa with the production of a through-going fault, then unloaded and reloaded at 400°C. Despite the presence of the fault, the deformation at 400°C was fully distributed cataclastic flow with a flow strength identical to that of an originally intact sample.

In order to test the effect of water on the transition from faulting to cataclastic flow, two samples were heated at 800°C under vacuum for 24 hours before being mechanically sealed in a silver jacket. This heat treatment should have completely dehydrated the sericite and also removed any water in fluid inclusions in the plagioclase (Kronenberg et al., 1989). One of these predried samples was deformed at 300°C, $2 \times 10^{-5}\ s^{-1}$, and 1000 MPa. Several undried samples deformed at these conditions showed fully distributed cataclastic flow. However, the dried sample faulted, with two audible and sudden stress drops. There are two separate although closely adjacent fault planes in the sample; they appear to be continuous and sharp. The other predried sample was deformed at 600°C, $2 \times 10^{-5}\ s^{-1}$, and 1000 MPa, where an undried sample showed distributed cataclastic flow out to $\sim 30\%$ shortening followed by development of a ductile fault. The dried sample did not show a sudden stress drop, but it was stronger than the undried sample, and the deformation was more localized into a ductile shear zone. Thus the presence or absence of trace amounts of water does seem to influence the nature of the cracking and crack interaction in feldspars.

Mechanical behavior Within the cataclastic flow regime, the yield strengths are greater than the confining pressure (Figure 2c,d), consistent with the dominance of microcracking. The strength increases with increasing pressure, decreasing temperature, and increasing strain rate.

One major question concerns whether cataclastic flow is a process capable of steady-state flow to high strains, or whether it is inherently a transient process, as found by Hirth and Tullis (1989) for porous quartzites deformed at room temperature. At

1000 MPa and $2 \times 10^{-5}\ s^{-1}$, samples deformed at 300 and 400°C show steady-state flow out to 25–30% strain, whereas samples deformed at 500 and 600°C show steady-state flow out to 20–25% strain followed by strain weakening (Figure 2c). The latter sample contains a corner-to-corner localized ductile shear zone superimposed on fairly high homogeneous strain, as described above. At 1500 MPa samples deformed at 400–600°C show steady state flow out to the highest strains tested, 20–30% (Figure 2d).

It is difficult to evaluate the achievement of steady-state flow in high-strain axial compression experiments because the geometry of the sample changes significantly and causes an inhomogeneous stress distribution. In an attempt to simulate a higher-strain experiment in the cataclastic flow regime, where the chief process appears to be progressive grain size reduction, we have utilized a synthetic aggregate so that the sample starts out with considerable finely crushed material. A sintered aggregate consisting of 95% grains $<10\ \mu m$ in diameter plus 5% grains $\sim 100\ \mu m$ in diameter was deformed at 400°C and 1000 MPa. It did not have a well-defined hit point, owing to the significant compaction strain, but at ~30% strain it achieved a flow strength essentially identical to that of an intact core, and it then showed slight strain softening out to 45% strain. The center of the sample developed a very strong foliation; both small and large grains are flattened by multiple grain-scale faulting. Discrete sharp faults of small offset propagate in toward the center of the sample from the sample corners, reflecting the 'punching' of the piston into the compacted sample. However, these faults die out toward the center which has undergone extensive cataclastic flow, testifying to the fundamental stability of this deformation mode.

The deformation mechanism of cataclastic flow shows a definite temperature dependence, which may primarily reflect the temperature dependence of frictional sliding of the feldspar fragments. Because many of the samples achieved approximately steady-state flow, it is possible to calculate an approximate value for the activation energy in this regime. Using experiments at 400, 500, and 600°C at $2 \times 10^{-5}\ s^{-1}$ and 1500 MPa, we find an activation energy of $\sim 80\ kJ\ mol^{-1}$. This is similar to the value of $130\ kJ\ mol^{-1}$ reported by Stesky (1977) for frictional sliding of granite between 300 and 500°C.

The temperature dependence of the yield strength in the cataclastic flow regime suggests that there is also a strain rate dependence. Two experiments were performed at 400°C and 1000 MPa in which the strain rate was varied between $10^{-4}\ s^{-1}$ and $10^{-6}\ s^{-1}$, allowing approximately steady-state flow to be achieved in each increment before changing the rate. When fitted to a power-law form of flow law, the results of these experiments yield a stress exponent $n \sim 90$.

3.4. Transition from cataclastic flow to dislocation creep

Samples deformed at 700, 800, and 900°C show decreasing amounts of grain-scale faulting and distributed cracking, and increasing amounts of dislocation glide and dynamic recrystallization, indicating a broad transition from cataclastic flow to dislocation creep. Samples deformed in this temperature interval show homogeneously distributed deformation, but they all undergo strain softening after an early stress peak (Figure 3c,d), and they all contain fine-grained zoisite which grew during the deformation.

Samples deformed at 700°C contain multiple grain-scale faults with small offsets; these appear to be less regular, less closely spaced, and less planar than those at lower

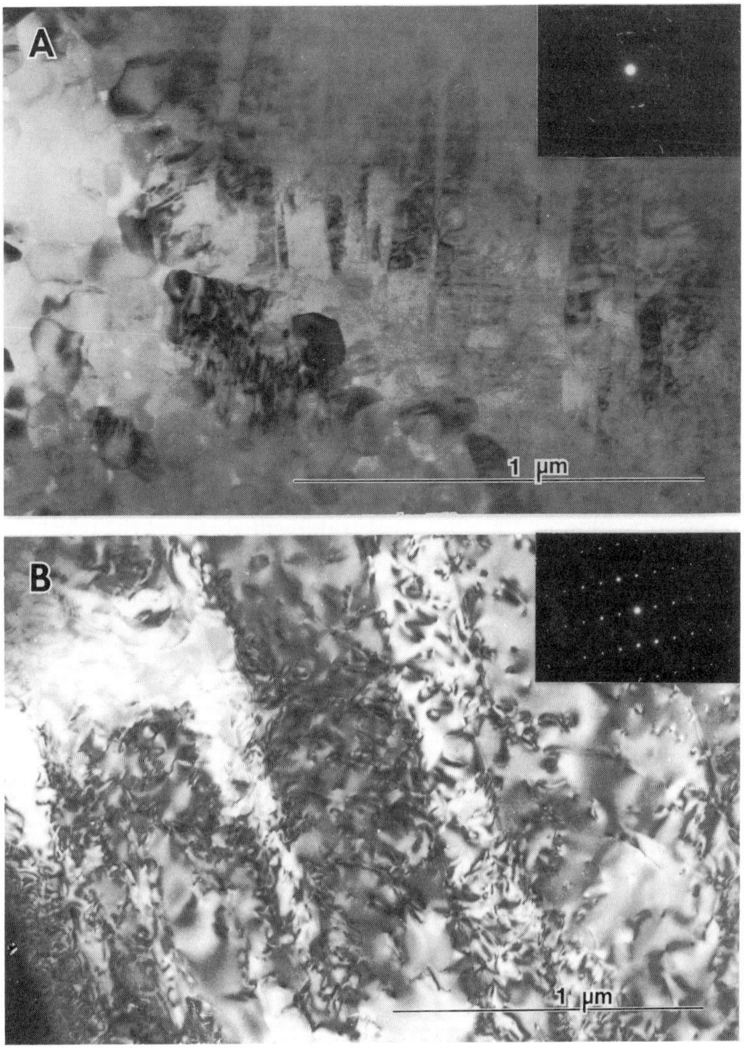

Figure 9. TEM photomicrographs (bright-field) of Bushveld anorthosite shortened 40% at 700°C and 1500 MPa. (A) Region with multiple cracks (NE) and high dislocation density, adjacent to former crush zone which has undergone some grain growth and shows evidence of dynamic recrystallization (SW). (B) Region with high dislocation density and no cracks; broad N–S bands are original twins.

temperatures, and the regions between the faults almost always have extreme undulatory extinction (Figure 7b). Many of the grain-scale faults appear as thin deformation bands, with a discrete, thin zone of material having a different extinction. Mechanical twinning appears to be associated with the grain-scale faults (Figure 7b). An irregular high density ($>10^{10}$ cm^{-2}) of tangled dislocations is present within most grains (Figure 9a,b), in addition to an irregular network of cracks (Figure 9a). Some crush zones appear to have undergone some grain growth and dynamic recrystallization (Figure 9a).

Samples deformed at 800 and 900°C show a different style of deformation; original grains have separated into lozenge-shaped fragments, with very fine recrystallized grains as well as some small zoisite grains along the boundaries (Figure 7a). The grain-scale faults are less planar than at lower temperatures, and they commonly diffuse out into a zone of undulatory extinction. In the regions between the grain-scale faults the undulatory extinction is smooth and sweeping, instead of sharp and patchy as it is at lower temperatures. TEM shows that the grains contain a more uniform high dislocation density, with an irregular cell structure. Dislocation segments are very straight, implying little or no climb, and many of these dislocations are dissociated. These samples show moderately high crack densities, even though the flow stress is significantly less than the confining pressure. This suggests that the local stress fields associated with the high dislocation densities have been important in nucleating the cracks. These samples also show evidence that some early-formed crush zones undergo later sintering, grain growth, and/or dynamic recrystallization (as in Figure 9a); in some planar zones the extremely small grains are polygonal, all about the same size, and larger ($\sim$0.1 μm) than seen in similar zones at temperatures $<700°$C.

Small (2–20 μm) grains of zoisite grow along original grain boundaries and grain-scale faults during deformation at 700–900°C. It is difficult to assess the effect of the zoisite on the deformation and to determine, for example, whether it has contributed to the strain weakening. A comparison with albite is useful, since albite also shows a broad transition from cataclastic flow to dislocation creep over approximately the same temperature interval at moderate to high pressure, without any accompanying phase changes (Tullis and Yund, 1987). Albite aggregates deformed at 700°C show quasisteady-state flow, whereas the anorthosite shows slight strain weakening, suggesting that the zoisite in the latter samples might be responsible. However, albite aggregates deformed at 800 and 900°C show strain weakening that is just as pronounced as that in the anorthosite, owing to the onset of grain boundary migration recrystallization (Tullis and Yund, 1985). Thus overall we do not believe that the nucleation and growth of the zoisite has significantly affected the deformation of the anorthosite samples.

3.5. Dislocation creep regime

At temperatures $>900°$C, anorthosite samples undergo dislocation creep. The lower-temperature portion of the dislocation creep regime is complicated by the occurrence of a minor amount of reaction to zoisite. However, careful optical and TEM analyses indicate that the deformation processes in this interval are essentially identical to those described previously in albite aggregates (Tullis and Yund, 1985, 1987).

Two samples were deformed in the dislocation creep regime at conditions where no zoisite develops (1000°C and 1200 MPa, 1200°C and 1500 MPa, both at $10^{-6}\,s^{-1}$). Both samples show early grain-scale faulting, with small offsets, on one or both cleavage planes; these divide most of the grains into smaller lozenge-shaped pieces. These lozenges contain clear, sweeping undulatory extinction, rather than the fine-scale patchy undulatory extinction characteristic of lower temperatures. They appear to have undergone very little internal strain, even for sample strains of 50–60% (Figure 7d). Very small (<5 μm) recrystallized grains have grown along the grain and lozenge boundaries and grain-scale faults (Figure 7d). TEM shows that the original grain interiors in the 1000°C sample contain a high density of tangled dislocations. The dislocation segments tend to be straight, and there are no subgrain boundaries, indicating that climb was very limited. The recrystallized grains are 1–3 μm in diameter, and many of them are dislocation-free, although others have a high density of dislocations.

It appears that in this lower-temperature portion of the dislocation creep regime, dislocation climb in bytownite is very limited and recovery occurs by grain boundary migration recrystallization, as previously described for albite (Tullis and Yund, 1985) and quartz (Hirth and Tullis, 1992). Because the recrystallized grains are initially strain-free, they are weaker than the work-hardened original grains and are able to undergo an increment of easy glide. When they in turn work harden, they are replaced by a new episode of grain boundary migration. Thus the recrystallized grains take up almost all of the sample strain.

In the 1200°C sample, TEM shows definite evidence of dislocation climb. The dislocation density is lower, some dislocation segments are curved, and subgrain boundaries are present. Most recrystallized grains appear to have formed by grain boundary migration rather than subgrain rotation. However, it appears that these conditions are transitional between dislocation creep regimes 1 and 2 (Hirth and Tullis, 1992), where the recovery mechanism switches from grain boundary migration recrystallization to dislocation climb.

4. Discussion

4.1. Faults: effect of pressure and temperature

There is a change in the nature of the faulting with increasing confining pressure (at temperatures <300°C). At 250 and 500 MPa through-going faults have an overall orientation ~30° to σ_1, but they appear to develop from the linking up of a series of en echelon shorter segments (3–6 grains in length) having a more axial orientation. This set of sinistral, left-stepping (or dextral, right-stepping) segments produces a complex zone of distributed deformation between the segments. At 1000 and 1500 MPa, in contrast, through-going faults consist of a continuous and very narrow zone; they occur at higher angles to σ_1 (40–45°); and there tend to be several per sample, corresponding to the multiple stress drops. These faults tend to follow grain boundaries or cleavage planes of several favorably inclined grains.

The nature of the fault gouge shows a corresponding change. At lower pressures the gouge zone is wider because it develops from more irregular surfaces, which break off more material as they slide. At higher pressures the fault is sharp and smooth from the start and so can slide with less gouge development. A related aspect is that there is a wider range of sizes of the fragments in the gouge formed at low pressures, whereas at high pressures all the fragments are very fine-grained.

Most previous experimental studies of faulting in rocks have been done at pressures <500 MPa. Early studies showed that new cracks are primarily axial in orientation (e.g., Tapponier and Brace, 1976). However, several studies have shown a change in the nature of new cracks with increasing pressure. First, the density of prefailure stress-induced cracks increases with increasing pressure (e.g., Wong and Biegel, 1985). Second, the anisotropy of the new cracks decreases with increasing pressure, even within the range 100–300 MPa (e.g., granite (Kranz, 1980); gabbro (Wong and Biegel, 1985); calcite (Fredrich et al., 1989)). Recently Shimada and Cho (1990) described a fundamental transition in the nature of brittle fractures in silicate rocks deformed at room temperature, at the pressure where compressive strength equals frictional strength. For granite and gabbro they observed that at ~800 MPa there was a transition from 'classical' faults to faults which are sharp and oriented at ~45° to σ_1, which have no concentration of mode I microcracks along them, and whose acoustic emission activity remains constant until failure. (However, Ismail and Murrell (1990) report that for similar rock types, at this same pressure where compressive strength equals frictional strength, there is a switch from fracture to cataclastic flow.) Hirth and Tullis (1991) have observed a change in the microstructural aspects of faults developed in nonporous quartzite deformed at room temperature, at ~750 MPa, almost identical to the transition described by Shimada and Cho (1990). Our qualitative observations on the anorthosite indicate a similar transition, also at ~750 MPa. Intriguingly, Thatcher and Hill (1991) have noted several examples of natural seismic faults at ~45° to σ_1.

Within the faulting regime at pressures >750 MPa there is no pronounced change in the nature of the faulting with increasing temperature (20–250°C). However, there does appear to be an increase in the distributed deformation away from the faults, which must be related to the transition to cataclastic flow. If this were due to the strength of the confining medium, it should diminish with increasing temperature.

The sharp faults with pronounced strain weakening which develop in samples deformed at 600–700°C and 500 MPa are a puzzle. They have the same sharp nature as those developed at low temperature but high pressure, and similarly tend to form at close to 45° to σ_1. Similar sharp faults have been observed at identical conditions in other feldspathic rocks, including fine-grained albite aggregates, aplite, and Westerly granite (Tullis and Yund, 1977; Shelton et al., 1981). However, they do not develop in quartzites (Hirth and Tullis, 1991). The transition from faulting to distributed deformation with increasing pressure at 700°C might seem to be explained by the well-known pressure dependence of water weakening in quartz and feldspar (e.g., Kronenberg and Tullis, 1984; Tullis and Yund, 1980). However, this cannot explain why the same sequence occurs at 600°C, because at this temperature dislocations are not activated at any pressure. It is at 600°C that the first mechanical twins are noted, but this occurs at all confining pressures, including 500 MPa where faults develop and 1500 MPa where deformation is fully distributed. These low-pressure, high-temperature sharp faults may represent a switch from mode I to mode II cracks (Shelton et al., 1981), but experiments to much lower sample strains would be necessary to test this possibility.

4.2. Transition from faulting to cataclastic flow

One of the important results of this study is the finding that increase of pressure alone, at low temperatures, is not sufficient to cause a transition from faulting to cataclastic flow. However, at modest temperatures (300–600°C) Bushveld anorthosite does show a transition from faulting to cataclastic flow with increasing pressure (at ~750 MPa). Similarly, at high pressures (1000–1500 MPa), anorthosite shows a transition from faulting to cataclastic flow with increasing temperature (250–300°C), at a strain rate of $10^{-5}\,s^{-1}$.

The transition from faulting to cataclastic flow might be influenced by the compliance of the apparatus and/or the temperature-dependent strength of the confining medium. Our machine is relatively compliant, which should tend to drive the sample in the post-failure regime into faulting. Similarly, the decreasing strength of the NaCl confining medium with increasing temperature should tend to favor localized over distributed deformation. The transition to distributed cracking occurs in spite of these factors. Testimony to the inherent stability of the distributed cracking ≥300°C is provided by several observations. For example, in samples deformed at 300–500°C (and 1000–1500 MPa) short fault segments commonly extend in from the sample corners, but these do not propagate through; they become diffuse toward the sample center. Similarly, a sample loaded to brittle failure at room temperature at 1000 MPa and subsequently re-loaded at 400°C showed fully distributed deformation. Thus the transition to cataclastic flow appears to indicate an inherent temperature- (and pressure-) dependent property of the cracking process.

The behavior of cracks under all-compressive loading has been treated in a number of recent theoretical and observational studies. Under such loading conditions cracks propagate stably and tend to grow into parallelism with σ_1 (e.g., Nemat-Nasser and Horii, 1982); the equilibrium length of such wing or wedge cracks decreases with increasing pressure (e.g., Wong, 1990). At lower pressures it is the interactions of the relatively long, axial cracks that lead to macroscopic shear fracture (e.g., Ashby and Hallam, 1986). Observations from several types of experimentally deformed rocks confirm the dominance of axial cracks at low pressures (e.g., Tapponier and Brace, 1976; Wong, 1982), and the decreasing crack length with increasing confining pressure (e.g., Fredrich et al., 1989). These studies also document a decreasing anisotropy of crack orientations with increasing confining pressure, suggesting an approach to distributed cataclastic flow (e.g., Wong and Biegel, 1985). However, this is not achieved in silicates at room temperature, for experimental strain rates. Brittle rocks with little or no porosity or alteration products, such as granites, granodiorites and gabbros, undergo brittle fracture to pressures as high as 3000 MPa (Byerlee, 1968; Schock et al., 1973; Tullis and Yund, 1977; Shimada and Yukutake, 1983). The Bushveld anorthosite shows similar behavior. Nonetheless, we infer that the reduction of crack length and anisotropy with increasing pressure observed in these other studies contributes to the observed requirement of a critical pressure for cataclastic flow in anorthosite. An additional factor contributing to the small effective length of the cracks in the cataclastically deformed anorthosite is the fact that many of them are partly closed, and load-bearing. Apparently, as the cracks propagate, the local stresses change and older portions can close again.

The requirement of a critical temperature for the achievement of fully distributed cataclastic flow must be related to the nature of the cracking process, since it does not activate mechanical twins or dislocations in the interval where the transition to distributed cataclastic flow occurs. Increased temperature would tend to increase the number of cracks nucleated, owing to bursting of fluid inclusions (e.g., Montgomery and Brace, 1975) and mismatch of thermal expansion between adjacent grains or between host and exsolved lamellae (e.g., Nur and Simmons, 1970). However, this factor is not believed to be very important. A sample preannealed at 1000 MPa and 600°C (conditions where deformed samples undergo cataclastic flow) for 2 hours and then deformed at room temperature faulted; the sample strength and the character of the fault are identical to those of samples deformed at the same conditions but which had not been preannealed. Increased temperature would tend to increase the crack propagation rate, at least in the presence of trace amounts of moisture, which are certainly present in our 'as-is' experiments (e.g., Atkinson and Meredith, 1987). It is not clear how this would favor distributed, isotropic cracking. However, the importance of water to the process was indicated by the fact that a sample that was predried (and presumably thermally cracked) developed a fault when deformed at conditions where 'as-is' samples showed cataclastic flow.

One factor of obvious importance is the existence in plagioclase of two perfect cleavage planes at approximately right angles. Cracking on these planes is obviously easier than on other planes, and hence may occur even when these planes are at high angles to σ_1. This would contribute to crack isotropy within the aggregate. Even small amounts of cracking on inclined, conjugate cleavages, especially if coupled with connecting axial cracks, would isolate small fragments which could then slide and rotate. A continuation of this process reduces the aggregate to fine 'rubble', and deformation will result from rotation and frictional sliding between the fragments. We do not have the low-strain experiments spanning the faulting to cataclastic flow transition with increasing temperature at high pressure which would be necessary to document whether the increase in temperature which promotes the transition does cause a greater isotropy of cracking. Nor are we aware of any theoretical treatment or other observations which suggest that temperature (and moisture) should have such an effect. This remains an unexplained observation.

4.3. Cataclastic flow: steady state?

This study has demonstrated the existence of an extensive regime of cataclastic flow in experimentally deformed feldspar aggregates. The results of previous exploratory studies of anorthosite deformation (e.g., Seifert, 1969; Hadizadeh and Tullis, 1991, 1992) are consistent with our findings. However, our identification of a regime of distributed cracking without any associated crystal plasticity has depended on detailed TEM analysis, not done previously.

One of the chief questions concerning the deformation regime of cataclastic flow is whether it is capable of producing steady-state flow. One of the difficulties in answering this question stems from the uncertainties in reducing stress–strain curves for high-strain experiments, discussed earlier. At 1000 MPa, samples appear to undergo steady-state flow to ~25% strain, although samples deformed at 500 and 600°C show strain weakening associated with localization at higher strains. At 1500 MPa, samples deformed at 400–600°C show steady-state flow out to 20–30% strain.

Deformation in this regime involves progressive comminution with frictional sliding of the fragments produced. It is therefore closely similar to what occurs within the gouge along faults. The end product of such deformation in room-temperature rotary shear of feldspathic rock (Westerly granite) is amorphous material (Yund et al., 1990). The frictional sliding of these samples was velocity neutral to slightly velocity weakening, consistent with the steady-state flow that we have observed in our much lower-strain experiments on anorthosite.

4.4. Transition to dislocation creep

With increasing temperature at high pressure there is a broad transition from cataclastic flow to recrystallization-accommodated dislocation creep, as found previously in albite aggregates (Tullis and Yund, 1985, 1987). At our experimental strain rate of $2 \times 10^{-5}\ s^{-1}$, the first manifestation of crystal plastic processes is the onset of mechanical twinning at ~600°C. The first generation of free dislocations, forming dense tangles, occurs at ~700°C. The first clearly identifiable recrystallized grains are noted at ~800°C. Because the strength of samples in the recrystallization-accommodated dislocation creep regime remains high until an interconnected matrix of recrystallized grains forms (Hirth and Tullis, 1992), even at 1000°C local stresses are high enough at low sample strains to produce some cracking and grain-scale faulting. Healing and recrystallization of the crushed material produced by this faulting, in turn, hasten the development of a recrystallized matrix.

Anorthosite does not undergo a transition to steady-state dislocation creep in laboratory deformation at confining pressures <750 MPa, even at high temperatures. In this it is very similar to the behavior of quartz (e.g., Kronenberg and Tullis, 1984; Mainprice and Paterson, 1984) and albite (Tullis and Yund, 1980), although quite different from olivine (e.g., Karato et al., 1986). In the earth, dislocation creep of quartz and feldspar is clearly not limited to depths >20 km. We believe that the strong pressure dependence of dislocation creep in experimentally deformed quartz and feldspar reflects a dependence on water fugacity. Such a dependence has been determined for oxygen self-diffusion in feldspar and quartz (Farver and Yund, 1990, 1991), and is strongly indicated for dislocation creep (e.g., Tullis and Yund, 1989).

4.5. Comparison with other materials

Recently there have been detailed studies of the brittle-to-plastic transition in calcite aggregates (Fredrich et al., 1989) and in quartz aggregates (Hirth and Tullis, 1991). The transitions in these materials are more similar to one another than they are to those in feldspar aggregates, in the sense that feldspar is the only material in which there is a regime of fully distributed cracking with no associated crystal plasticity.

In calcite aggregates, the brittle-to-crystal plastic transition can be accomplished in laboratory experiments at room temperature by pressure increase alone (Heard, 1960; Fredrich et al., 1989). For a strain rate of $10^{-5}\ s^{-1}$, the latter authors observed a transition from brittle deformation (involving cracking and mechanical twinning) to semibrittle deformation (involving cracking, twinning, and dislocation glide) at a pressure of 30 MPa. The transition does not involve marked changes in the dislocation density or the density or anisotropy of cracks, but it does correspond with a decrease in average crack length and a decrease in the fraction of the total deformation energy taken up by cracking. Fischer and Paterson (1989) found that semibrittle deformation of marble was stable for strains up to at least 20%.

In quartz aggregates, the brittle-to-crystal plastic transition in laboratory experiments cannot be accomplished at room temperature by pressure increase alone; temperatures of >500°C are required (Hirth and Tullis, 1991). However, like calcite, there is no regime of distributed cracking alone, without crystal plasticity. Hirth and Tullis (1991) have found that nonporous quartzites undergo faulting to significantly higher temperatures and pressures than feldspar aggregates (500°C compared to 250°C, even at 1500 MPa), despite the fact that the first dislocation generation occurs at much lower temperatures (300°C compared to 700°C), and the onset of steady-state, crack-free dislocation creep also occurs at much lower temperatures (800°C compared to 1000°C). Thus at moderate pressure quartzites show a transition with increasing temperature from faulting, to semibrittle flow involving high dislocation densities and distributed cracks, to dislocation creep, whereas feldspar aggregates have an additional regime of cataclastic flow, involving only distributed cracking. The absence of cleavage planes in quartz may be important for this difference in behavior.

Amphiboles and pyroxenes may be quite similar to feldspars in their deformation behavior, as they also have two good cleavages and require even higher temperatures for dislocation creep (e.g., Kronenberg and Shelton, 1980). For clinopyroxenite samples deformed at 600°C and $10^{-5}\ s^{-1}$, Kirby and Kronenberg (1984) observed a transition from faulting to semibrittle flow involving penetrative microfracturing and mechanical twinning at pressures >450 MPa; they did not explore lower temperatures at these intermediate pressures where distributed cracking without mechanical twinning might be expected. In naturally deformed amphiboles, grain flattening and preferred orientations due to multiple grain-scale faulting have been observed (Allison and LaTour, 1977).

4.6. Implications for natural deformation

It is impossible at present to quantitatively extrapolate these laboratory data on the brittle-to-crystal plastic transition in anorthosite to natural conditions, because of the unknown and possibly different dependences of the processes involved on pressure and strain rate. There is good reason to believe that dislocation processes operative at 1000–1500 MPa at laboratory strain rates are similar to those characteristic of much shallower levels in the crust at natural strain rates, perhaps owing to the presence of an as-yet undetermined water fugacity term in the flow law. Thus the processes associated with the transition to dislocation creep must be studied in the laboratory at high confining pressures. Microcrack propagation rates may also depend on water fugacity (e.g., Atkinson and Meredith, 1987); however, the length and anisotropy of cracks depend directly, and therefore differently, on pressure. Hence studies at laboratory strain rates and high confining pressures may not produce the same sequence of processes as operative in the earth.

The brittle–ductile transition at some places in the continental crust occurs at approximately 300°C, ~500 MPa, and $10^{-14}\ s^{-1}$ (e.g., Sibson, 1984). This is too low a temperature to be associated with the onset of crystal plasticity in feldspars (~450°C; Voll, 1976), and is commonly interpreted to reflect the onset of crystal plasticity in quartz. However, considering that feldspars form the stress-supporting matrix phase of many 'basement' rocks, it seems possible that this may reflect the transition from faulting to cataclastic flow in feldspars. There is good evidence for distributed cataclastic flow in naturally deformed feldspar aggregates; an anorthosite deformed at low grade, in regions adjacent to the Outer Hebrides thrust, shows pervasive fracturing, with conjugate sets of multiple grain-scale faults, generally on one or more of the cleavage planes (Brown and Macaudiere, 1984). Similarly, in granitic rocks deformed at low grade, the quartz grains are relatively undeformed, whereas the feldspar matrix has deformed by distributed cracking (Evans, 1988).

If the brittle–ductile transition in the crust does correlate with the transition from faulting to cataclastic flow in feldspars, there are several implications. First, the strength in the cataclastic flow regime is temperature-dependent, indicating that on a stress vs. depth plot the high-stress peak associated with the friction–dislocation creep transition would be significantly reduced. Second, the extremely fine grain size resulting from cataclastic flow would produce ductile shear zones with a subsequent increase in temperature into the regime of recrystallization-accommodated dislocation creep (Tullis et al., 1990), because fine recrystallized grains are substantially weaker than coarse work-hardened grains. Finally, cataclastic flow is a dilatant process that may promote high permeability, as well as producing an extremely fine grain size. These two factors may promote a reaction to weak hydrous phases (e.g., Janecke and Evans, 1988) or a switch in the dominant deformation mechanism to pressure solution (Tullis and Yund, 1991). Either change would substantially decrease the rock strength.

Acknowledgments

This work was supported by USGS contract 14-08-0001-G-1363; we are extremely grateful for this support. Jafar Hadizadeh did many of the early experiments on anorthosite and participated in much of the optical analysis. We thank Greg Hirth and Terry Tullis for helpful discussions, Bill Collins for thin sections, and an anonymous reviewer for detailed comments which led to substantial improvements in the paper.

References

Allison, I.S. and LaTour, T.E. (1977). Brittle deformation of hornblende in a mylonite: A direct geometrical analogue of ductile deformation by translation gliding. *Can. J. Earth Sci.* **14**, 1953–1959.

Ashby, M.F. and Hallam, S.D. (1986). The failure of brittle solids containing small cracks under compressive stress states. *Acta Metall.* **34**, 497–510.

Atkinson, B. and Meredith, P. (1987). The theory of sub-critical crack growth with applications to minerals and rocks. In *Fracture Mechanics of Rock* (ed. B.K. Atkinson), pp. 111–166. Academic Press, Orlando.

Borg, I., Friedman, M., Handin, J., and Higgs, D.V. (1960). Experimental deformation of St. Peter sand: A study of cataclastic flow. *Geol. Soc. Am. Mem.* **79**, 133–191.

Brace, W.F. and Kohlstedt, D.L. (1980). Limits on lithospheric stress imposed by laboratory experiments. *J. Geophys. Res.* **85**, 6248–6252.

Brown, W.L. and Macaudiere, J. (1984). Microfracturing in relation to atomic structure of plagioclase from a deformed meta-anorthosite. *J. Struc. Geol.* **6**, 579–586.

Byerlee, J.D. (1968). Brittle–ductile transition in rocks. *J. Geophys. Res.* **73**, 4741–4750.

Carter, N.L. and Kirby, S.H. (1978). Transient creep and semi-brittle behavior of crystalline rocks. *Pure Appl. Geophys.* **116**, 807–839.

Chester, F.M., Friedman, M., and Logan, J.M. (1985). Foliated cataclasites. *Tectonophysics* **111**, 139–146.

Edmond, J.M. and Paterson, M.S. (1972). Volume change during the deformation of rocks at high pressures. *Int. J. Rock Mech.* **9**, 161–182.

Evans, J.P. (1988). Deformation mechanisms in granitic rocks at shallow crustal levels. *J. Struct. Geol.* **10**, 437–443.

Farver, J.R. and Yund, R.A. (1990). The effect of hydrogen, oxygen, and water fugacity on oxygen diffusion in alkali feldspar. *Geochim. Cosmochim. Acta* **54**, 2953–2964.

Farver, J.R. and Yund, R.A. (1991). Oxygen diffusion in quartz: Dependence on temperature and water fugacity. *Chem. Geol.* **90**, 55–70.

Fischer, G.J. and Paterson, M.S. (1989). Dilatancy during rock deformation at high temperatures and pressures. *J. Geophys. Res.* **94**, 17607–17617.

Fredrich, J.T., Evans, B., and Wong, T.-f. (1989). Micromechanics of the brittle to plastic transition in Carrara marble. *J. Geophys. Res.* **94**, 4129–4145.

Goldsmith, J.R. (1982). Plagioclase stability at elevated temperatures and water pressures. *Am. Mineral.* **67**, 653–675.

Hadizadeh, J. and Tullis, J. (1986). Transition from brittle faulting to cataclastic flow for anorthosite: both P and T are required. *EOS* **67**, 372–373.

Hadizadeh, J. and Tullis, J. (1992). Cataclastic flow and semi-brittle deformation of anorthosite. *J. Struct. Geol.* **14**, 57–64.

Hadizadeh, J. and Rutter, E.H. (1983). The low temperature brittle–ductile transition in a quartzite and the occurrence of cataclastic flow in nature. *Geologische Rund.* **72**, 493–509.

Heard, H.C. (1960). Transition from brittle fracture to ductile flow in Solenhofen limestone as a function of temperature, confining pressure and interstitial fluid pressure. *Geol. Soc. Am. Mem.* **79**, 193–226.

Heard, H.C. (1972). Steady-state flow in polycrystalline halite at pressure of 2 kilobars. In *Flow and Fracture of Rocks* (ed. H. C. Heard, I. Y. Borg, N. L. Carter, and C. B. Raleigh), pp. 191–210. *Geophysical Monograph 16*. American Geophysical Union, Washington, D.C.

Hirth, G. and Tullis, J. (1989). The effects of pressure and porosity on the micromechanics of the brittle-ductile transition in quartzite. *J. Geophys. Res.* **94**, 17825–17838.

Hirth, G. and Tullis, J. (1991). Mechanisms responsible for the brittle-ductile transition in experimentally deformed quartz aggregates. *EOS* **72**, 286.

Hirth, G. and Tullis, J. (1992). Dislocation creep regimes in quartz aggregates. *J. Struct. Geol.* **14**, 145–159.

Ismail, I.A.H. and Murrell, S.A.F. (1990). The effect of confining pressure on stress-drop in compressive rock fracture. *Tectonophysics* **175**, 237–42.

Janecke, S.U. and Evans, J.P. (1988). Feldspar-influenced rock rheologies. *Geology* **16**, 1064–1067.

Karato, S.-I., Paterson, M.S., and Fitz Gerald, J.D. (1986). Rheology of synthetic olivine aggregates: Influence of grain-size and water. *J. Geophys. Res.* **91**, 8151–8176.

Kirby, S.H. and Kronenberg, A.K. (1984). Deformation of clinopyroxenite: evidence for transition in flow mechanisms and semi-brittle behavior. *J. Geophys. Res.* **89**, 3177–3192.

Kranz, R.L. (1980). The effect of confining pressure and stress difference on static fatigue in granite. *J. Geophys. Res.* **85**, 1854–1866.

Kronenberg, A.K. and Shelton, G.L. (1980). Deformation microstructures in experimentally deformed diabase. *J. Struct. Geol.* **2**, 341–353.

Kronenberg, A.K. and Tullis, J. (1984). The effect of grain size, water, and pressure on the flow strengths of quartz aggregates. *J. Geophys. Res.* **89**, 4281–4297.

Kronenberg, A.K., Rossman, G.R., Yund, R.A., and Huffman, A.R. (1989). Stationary and mobile hydrogen defects in potassium feldspar. *EOS* **70**, 1406.

Mainprice, D.H. and Paterson, M.S. (1984). Experimental studies of the role of water in the plasticity of quartzites. *J. Geophys. Res.* **89**, 4257–4270.

Mandl, G., de Jong, L.N.J., and Maltha, A. (1977). Shear zones in granular material: An experimental study of their structure and mechanical genesis. *Rock Mech.* **9**, 95–144.

Marshall, D.B. and McLaren, A.C. (1977). Deformation mechanisms in experimentally deformed plagioclase feldspars. *Phys. Chem. Minerals* **1**, 351–370.

Montgomery, C.W. and Brace, W.F. (1975). Micropores in plagioclase. *Contrib. Mineral. Petrol.* **52**, 17–28.

Nemat-Nasser, S. and Horii, H. (1982). Compression-induced non-planar crack extension with application to splitting, exfoliation, and rock burst. *J. Geophys. Res.* **87**, 6805–6822.

Nur, A. and Simmons, F.G. (1970). The origin of small cracks in igneous rocks. *Int. J. Rock Mech. Min. Sci.* **7**, 307–314.

Paterson, M.S. (1978). *Experimental Rock Deformation; The Brittle Field.* Springer–Verlag, New York.

Rutter, E.H. (1986). On the nomenclature of mode of failure transitions in rocks. *Tectonophysics* **122**, 381–387.

Rutter, E.H. and Brodie, K.H. (1991). Lithosphere rheology – a note of caution. *J. Struct. Geol.* **13**, 363–367.

Schock, R.H., Heard, H.C., and Stephens, D.R. (1973). Stress–strain behavior of a granodiorite and two graywackes on compression to 20 kilobars. *J. Geophys. Res.* **78**, 5922–5941.

Seifert, K.E. (1969). Strength of Adirondack anorthosite at elevated temperatures and pressures. *Geol. Soc. Am. Bull.* **80**, 2053–2060.

Shelton, G.L., Tullis, J., and Tullis, T.E. (1981). Experimental high temperature and high pressure faults. *Geophys. Res. Lett.* **8**, 55–58.

Shimada, M. (1986). Mechanism of deformation in dry porous basalt at high pressures. *Tectonophysics* **121**, 153–173.

Shimada, M. and Cho, A. (1990). Two types of brittle fracture of silicate rocks under confining pressure and their implications in the earth's crust. *Tectonophysics* **175**, 221–235.

Shimada, M. and Yukutake, H. (1983). Fracture strength of dry silicate rocks at high confining pressures and activity of acoustic emission. *Tectonophysics* **60**, 159–172.

Sibson, R.H. (1984). Roughness at the base of the seismogenic zone: contributing factors. *J. Geophys. Res.* **89**, 5791–5799.

Stesky, R.M. (1977). Mechanisms of high temperature frictional sliding in Westerly granite. *Can. J. Earth Sci.* **15**, 361–375.

Tapponier, P. and Brace, W.F. (1976). Development of stress-induced microcracks in Westerly granite. *Int. J. Rock Mech. Min. Sci.* **13**, 103–112.

Thatcher W. and Hill, D.P. (1991). Fault orientations in extensional and conjugate strike-slip environments and their implications. *Geology* **19**, 1121–1124.

Tse, S.T. and Rice, J.R. (1986). Crustal earthquake instability in relation to the depth variation of frictional slip properties. *J. Geophys. Res.* **91**, 9452–9472.

Tullis, J. and Yund, R.A. (1977). Experimental deformation of dry Westerly granite. *J. Geophys. Res.* **82**, 5705–5717.

Tullis, J. and Yund, R.A. (1980). Hydrolytic weakening of experimentally deformed Westerly granite and Hale albite. *J. Struct. Geol.* **2**, 439–451.

Tullis, J. and Yund, R.A. (1985). Dynamic recrystallization of feldspar: A mechanism for ductile shear zone formation. *Geology* **3**, 238–241.

Tullis, J. and Yund, R.A. (1987). Transition from cataclastic flow to dislocation creep of feldspar: Mechanisms and microstructures. *Geology* **15**, 606–609.

Tullis, J. and Yund, R.A. (1989). Hydrolytic weakening of quartz aggregates: the effects of water and pressure on recovery. *Geophys. Res. Lett.* **16**, 1343–1346.

Tullis, J. and Yund, R.A. (1991). Diffusion creep in feldspar aggregates: experimental evidence. *J. Struct. Geol.* **13**, 987–1000.

Tullis, J., Dell'Angelo, L.N., and Yund, R.A. (1990). Ductile shear zones from brittle precursors in feldspathic rocks: the role of dynamic recrystallization. In *The Brittle–Ductile Transition in Rocks* (ed. A.G. Duba, W.B. Durham, J.W. Handin, and H.F. Wang), pp. 67–81. *Geophysical Monograph 56*. American Geophysical Union, Washington, D.C.

Voll, G. (1976). Recrystallization of quartz, biotite, and feldspars from Erstfeld to the Levantina nappe, Swiss Alps, and its geological implications. *Schweiz. Min. Petrol. Mitt.* **56**, 641–647.

Wong, T.-f. (1990). A note on the propagation behavior of a crack nucleated by a dislocation pileup. *J. Geophys. Res.* **95**, 8639–8646.

Wong, T.-f. (1982). Micromechanics of faulting in Westerly granite. *Int. J. Rock Mech. Min. Sci.* **19**, 49–64.

Wong, T.-f. and Biegel, R. (1985). Effects of pressure on the micromechanics of faulting in San Marcos gabbro. *J. Struct. Geol.* **7**, 737–749.

Yund, R.A., Blanpied, M.L., Tullis, T.E., and Weeks, J.D. (1990). Amorphous material in high strain experimental fault gouge. *J. Geophys. Res.* **95**, 15589–15602.

Chapter 5

Stabilization of Faulting by Cumulative Slip

Teng-fong Wong, Yaojun Gu, Takashi Yanagidani[1]*, and Yusheng Zhao*

Department of Earth & Space Sciences, State University of New York, Stony Brook, NY 11794-2100, U.S.A.

[1] *Present address: Department of Mineral Science and Technology, Kyoto University, Kyoto, Japan.*

Abstract

Repeated earthquake ruptures along seismogenic faults generate significant amounts of wear materials, resulting in thickening of the gouge zone and possibly stabilization of faulting with the accumulation of slip. To investigate the stabilization process, the frictional sliding behavior was monitored as a function of cumulative slip when a simulated gouge layer of ultrafine quartz was sheared in a conventional triaxial setup. In a typical experiment at a confining pressure between 50 and 110 MPa, the sliding behavior evolved from dynamic instability to stable sliding through several sequential stages. Supercritical oscillations marked the onset of dynamic instability. Tens and sometimes hundreds of repeated stick–slip events ensued, and the stress drop amplitude decreased monotonically with cumulated slip. As slip accumulated, a period doubling stage was observed, with the stress drop amplitude showing complex bifurcations. This was followed by self-sustained, periodic oscillations which ultimately led to stable sliding. The period of the self-sustained oscillations was observed to be inversely proportional to the load-point velocity. Using a rate- and state-dependent friction law, we infer from the self-sustained oscillation data that the characteristic slip L is on the order of 1 μm. Our data also imply that the friction parameter $[A(B - A)]^{1/2}$ is between 0.2 and 0.4 MPa. The slip displacement X_c over which repeated stick–slip instabilities occurred decreased with increasing normal stress. Experiments on three different country rocks (Westerly granite, Frederick diabase, and Blair dolomite) showed that X_c also increases with increasing rock hardness. Our data can be interpreted with a simple wear model. Comparison with seismologic data suggests that stabilization by cumulative slip in our experiments is more efficient than that induced by geologic offset. The apparent difference may be due to the

FAULT MECHANICS AND TRANSPORT PROPERTIES OF ROCKS
ISBN 0-12-243780-2

structural complexity of tectonic faults and the possibility that healing processes operative in the interseismic period are different from the evolution effect observed in room-temperature friction experiments.

1. Introduction

The recurrence of seismic activity along preexisting faults has important seismotectonic implications (e.g., Sykes and Quittmeyer, 1981; Kanamori and Allen, 1986). An earthquake rupture results in a slip discontinuity, and consequently the geologic offset accumulates with increasing number of earthquake cycles. Part of the frictional work input in an earthquake cycle is dissipated by frictional wear processes, and the accumulation of wear material results in a thickening of the gouge zone with increasing geologic offset (e.g., Robertson, 1982; Scholz, 1987).

Not much is known about the effect of cumulative slip on earthquake faulting behavior. Wesnousky (1990) recently attempted to isolate the effect of cumulative slip on earthquake 'productivity' by normalizing the number of earthquakes with respect to the fault length and the rate of slip accumulation. He studied five major strike–slip fault zones in southern California, and concluded that the seismic 'productivity' decreases with increasing cumulative slip. He noted that the stabilization effect may be related to his previous observation (Wesnousky, 1988) that the accumulation of geologic offset tends to reduce the structural complexity along a fault zone.

In a recent analysis of the stress drop characteristics of intraplate and interplate earthquakes, Scholz (1990) suggested three factors for the observation that intraplate earthquakes have relatively high static stress drops in comparison with interplate earthquakes: the geologic slip rates for intraplate faults are several orders of magnitude slower than the plate boundaries; their earthquake recurrence time is proportionally longer; and their cumulative slip is much less. The first two factors are probably related to the laboratory observation that the stress drop amplitude of a stick–slip event increases with decreasing loading velocity and increasing time of contact (Dieterich, 1972; Ohnaka, 1973a; Engelder et al., 1975; Teufel and Logan, 1978; Wong and Zhao, 1990). With regard to the third factor, the effect of cumulative slip on stick–slip instability has not been systematically studied in the laboratory.

We recently presented data on the apparent decrease of stress drop with cumulative slip in an experiment on simulated quartz gouge showing hundreds of stick–slip events (Wong and Zhao, 1990). However, the published data suffer from several limitations. An external load cell was used and its relatively low resolution renders it impossible to monitor the detailed behavior in the transition from repeated dynamic instabilities to stable sliding. A fundamental understanding of this transition is important not only in earthquake mechanics but also in nonlinear dynamics. It has been proposed that nonlinearity of the friction constitutive relation (Gu et al., 1984; Hobbs, 1990) or imperfection in the loading system (Nussbaum and Ruina, 1987; Huang and Turcotte, 1990) may induce this transition to take up a route through period doubling and chaotic oscillations. Except for a single set of period-doubling data presented by Ruina (1981), we are not aware of any observations of such nonlinear dynamical phenomena in rock friction experiments. In this study, we improved our servocontrol feedback system and designed an internal load cell with the explicit purpose of monitoring the details of the transitional behavior.

Our previous experiments were conducted on a single country rock (Westerly granite), rendering it difficult to interpret the physical mechanism for the stabilization and to generalize the experimental results to tectonic faulting. In this study, we tested Frederick diabase and Blair dolomite in addition to Westerly granite. We conducted preliminary microstructural observations, and we developed a wear model to interpret the laboratory data. Some preliminary ideas on the scaling to earthquake rupture and implications on stabilization of faulting by wear processes will also be discussed.

2. Transition from Dynamic Instability to Stable Sliding: Period Doubling and Self-Sustained Oscillations

It is not uncommon to observe a combination of episodic stick–slip events and stable sliding in a single experiment. However, a number of technical difficulties have to be overcome before one can obtain repeatable data with high resolution on the transition from dynamic instability to stable sliding. In most high-pressure rock frictional experimental configurations, the amount of slip available is limited, and hence the number of cyclic stick–slip instabilities observable in an experiment may be too few for any systematic trend to be detected. The use of fine-grained simulated gouge is useful in this regard, since in a previous study on ultrafine quartz gouge, Wong and Zhao (1990) observed that tens and sometimes hundreds of repeated stick–slip events might occur in a single frictional sliding experiment, thus allowing one to systematically investigate the effects of both cumulative slip and loading velocity on the frictional instability behavior.

As the frictional sliding behavior evolves from dynamic instability to stable sliding, the magnitude of the stress drops and the timescale of the oscillations change dramatically. To resolve the second-order changes in stress drop characteristics, it is preferable to use an internal load cell in the close vicinity of the sample so that noise due to O-ring friction along the loading piston can be minimized. The data acquisition system should cover a sufficiently broad frequency range so that dynamic instabilities with characteristic timescale on the order of 0.1 ms (Yanagidani et al., 1989) as well as quasistatic oscillations with timescale on the order of 10 s (Scholz et al., 1972) can be monitored.

2.1 Experimental procedure

2.1.1. Servocontrolled test system and data acquisition system

Experiments were performed in a servocontrolled test system in the conventional triaxial configuration. The confining pressure (monitored by a Heise gauge and a strain gauge pressure transducer) ranged from 50 MPa to 110 MPa, and it was maintained at a constant level to within 0.5 MPa. The axial load was measured with a probable error of 1% using an internal load cell. The frequency response of the signal conditioning system was up to 500 kHz.

The axial displacement was measured outside the pressure vessel with a differential transformer (LVDT) mounted between the moving piston and the fixed upper platen. This was used as the servocontrol feedback signal, and was continuously compared with the command signal from a 16-bit function generator. Except for one run (Table 1), the *axial displacement rate* of the loading ram was maintained constant at 1 $\mu m\ s^{-1}$ up from the initial loading stage to the onset of self-sustained periodic oscillations, at which

Table 1

Sample	Pressure (MPa)	Normal stress (MPa)	Period doubling	Self-sustained oscillation	Load point velocity[a] (μm s^{-1})
TY4	70	114–122	×	×	1.15
TY8	70	112–124	×	×	0.12–1.15
GU1	100	159–173	×	×	1.15
GU2	100	158–166	×	×	1.15
GU3	100	162–171		×	1.15
GU4	80	129–140		×	0.12–1.15
GU5	90	143–156	×	×	0.12–1.15
GU6[b]	70	114–123		×	0.29–5.77
GU10	110	175–185	×	×	0.12–1.15
GU22[c]	50	82–91	×	×	1.15

[a] Range of perturbation in the self-sustained oscillation stage.
[b] The load-point velocity in this experiment was maintained constant at 5.77 μm s^{-1} until the onset of self-sustained oscillations in the transition from dynamic instability to stable sliding. In all the other experiments, it was maintained constant at 1.15 μm s^{-1} until the onset of self-sustained oscillations in the transition from dynamic instability to stable sliding.
[c] Period doubling was observed in both the destabilization phase and the stabilization phase.

point the rate would be decreased sequentially by a ratio of 2:1 over a range of 1 to 2 orders of magnitude (see Figure 6a). To avoid ambiguity in the following discussion, the reference displacement rate resolved along the direction of the sawcut surface (which is higher than the axial displacement rate by a factor of 1.155) will be referred to as the *load-point velocity*. The frequency response of the signal conditioning system for the LVDT was up to 2 kHz.

The analog signals of differential stress and axial displacement were acquired using an x–y recorder and a strip-chart recorder. The digital data were acquired using two separate systems covering different frequency bands. The first system acquires the load cell, pressure transducer, and LVDT signals at a rate of up to 500 samples/s through a 14-bit A/D convertor. For studying the details of the dynamic instability process, a second system connects the load cells and LVDT to a (LeCroy 9400A) digital storage oscilloscope with a sampling rate of 10 ns/word.

2.1.2. Sample configuration: country rock, gouge, and jacket material

The country rock was Westerly granite of diameter 25.4 mm and nominal length 63.5 mm. The angle of the sawcut was 30° and the sliding surfaces were ground with a 100-grit wheel. The simulated gouge was ultrafine quartz with a median particle size of 6.4 μm supplied by the U.S. Silica Company, who also provided the gouge particle size distribution shown in Figure 1. Both the country rock and simulated gouge material were room dry. A fixed weight of quartz powder (0.38 g) was sandwiched between the sawcut surfaces to form a simulated gouge layer of nominal thickness of about 0.3 mm. The sample assembly was jacketed with five layers of polyolefin tubing (three of which were 0.3-mm thick and two of which were 0.5 mm thick). The interfaces between the loading piston and the country rock were lubricated with molykote (MoS_2).

On the basis of our observations of samples retrieved after a run, we believe that the axes of the two sawcut rock pieces remained parallel even after they had undergone as

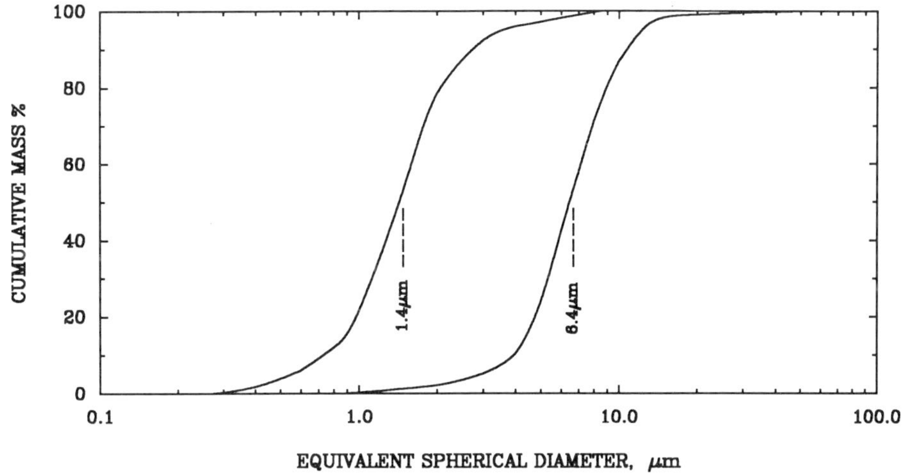

Figure 1. Particle size distribution of ultrafine quartz gouge. The coarse type (with a median particle dimension of 6.4 μm) and the finer type (with a median particle dimension of 1.4 μm) were used in the first and second part of this study, respectively.

much as 10 mm of slip displacement along the sawcut. An appropriate correction for the differential stress was made to account for area reduction as a function of cumulative slip on the frictional sliding surface.

2.2. Observations

Although the simulated gouge for this phase of our study is somewhat coarser than that used by Wong and Zhao (1990), the qualitative behavior was similar. The onset of frictional instability occurred after a relatively small amount of stable slip. The stress drop amplitude initially increased and then decreased monotonically as a function of cumulative slip, ultimately leading to stable sliding (Figures 2a and 3a). The stress drop amplitudes were somewhat smaller, and the slip distances over which cyclic stick–slip events occurred were shorter than those reported by Wong and Zhao (1990) for the same confining pressure and load-point velocity. This discrepancy can be attributed to the difference in gouge particle size.

2.2.1. Transition from stable sliding to cyclic stick–slip

The transition from stable sliding to cyclic stick–slip was marked by supercritical instability behavior. The stress drop events went through a transition from being subaudible to highly audible. In nine experiments at confining pressures ranging from 70 to 110 MPa (Table 1), the stress drop amplitude increased monotonically (Figures 2a and 3a) until it reached a peak, after which it decreased with slip (Figures 2c and 3c). In one experiment conducted at 50 MPa confining pressure, the oscillational behavior in the transition from stable sliding to cyclic stick–slip was more complicated. Starting with stable sliding, the frictional behavior evolved to subaudible periodic oscillations of increasing amplitude, which was followed by period doubling oscillations leading to cyclic stick–slip instabilities (Figure 4).

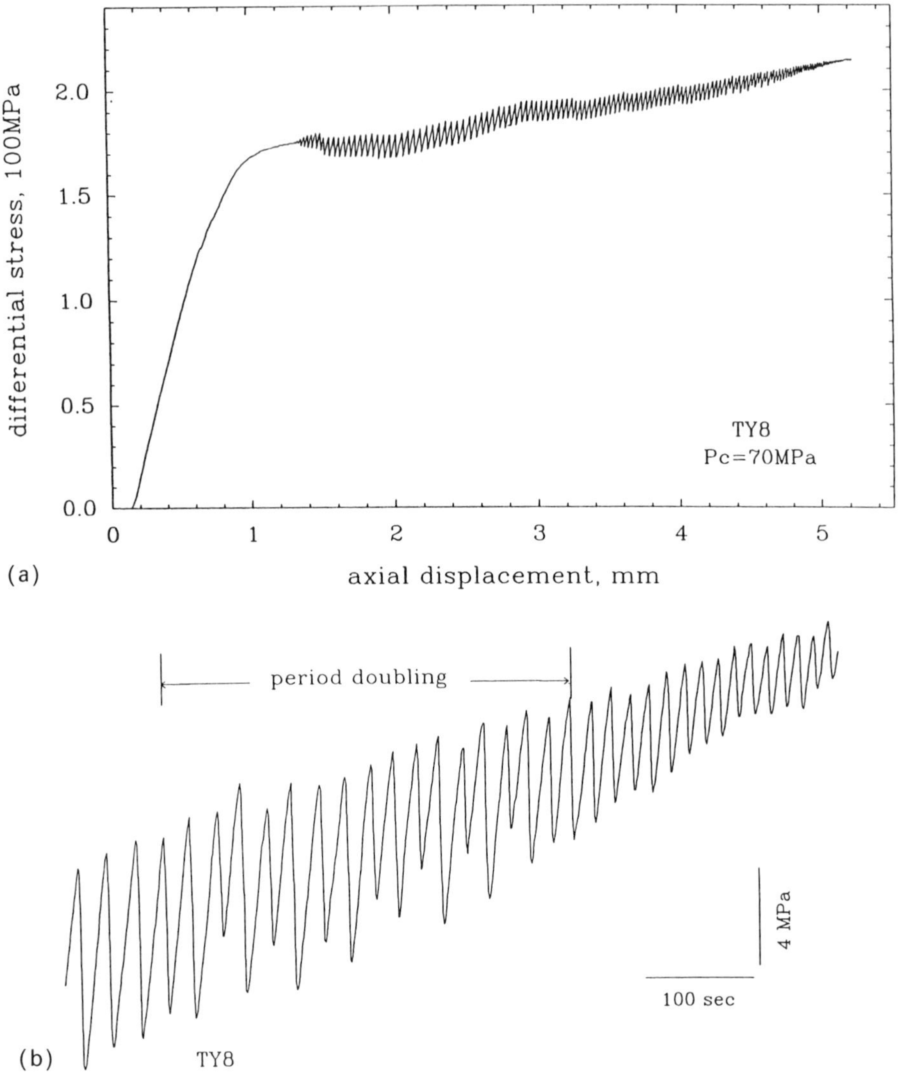

Figure 2. (a) Complete record of the differential stress versus axial displacement in experiment TY8. A summary of the experimental conditions is given in Table 1. (b) Stress drops recorded by the internal load cell for axial displacement from 4.2 mm to 5 mm. Period doubling behavior was evident.

2.2.2. Cyclic stick–slip: dynamic stress drop

An example of our high-frequency data for an audible stress drop in the cyclic stick–slip regime is shown in Figure 5. The timescale of the dynamic instability (fractions of a millisecond) is 2–3 orders of magnitude shorter than what has been reported previously by Rummel et al. (1978) and Shimamoto et al. (1980). This discrepancy may be attributed to two factors. First, the dynamic stress drop may be more rapid owing to the significantly finer gouge used in the present study. Second, the high-frequency signals from the 'ringing' of the loading frame evident in our data were not seen in the previous studies,

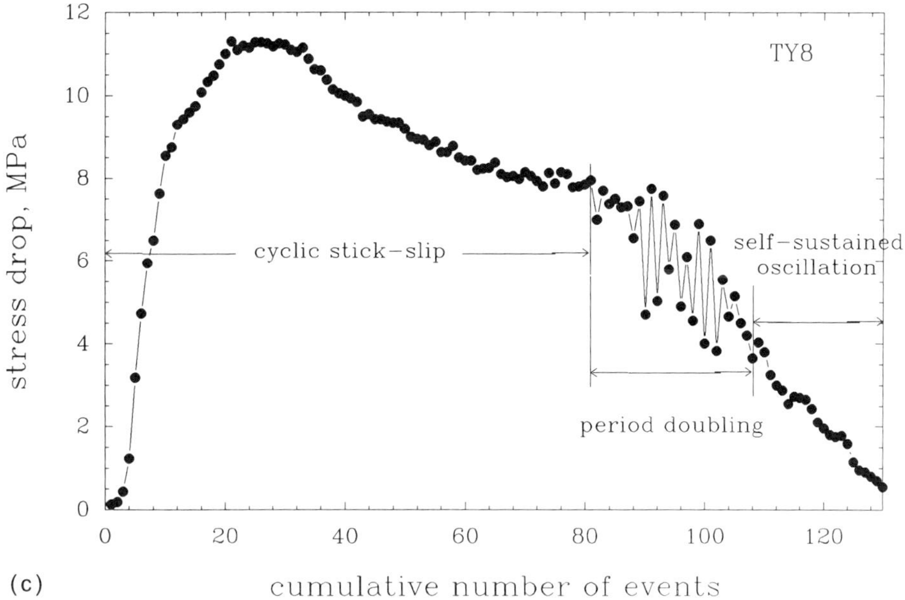

(c)

Figure 2. *Continued.* (c) Differential stress drop amplitude as a function of cumulative number of events. The three stages of cyclic stick–slip instabilities, period doubling, and self-sustained oscillations are highlighted.

implying that the previous results were filtered and hence might give an overestimate of the instability timescale.

2.2.3. Period doubling

With an internal load cell and improvement in our function generator for the servocontrol command, we were able to observe details of period doubling and self-sustained oscillations which could not be resolved using our old system (figure 4b of Wong and Zhao, 1990). In the period-doubling stage, small and large stress drop events alternate with one another (Figures 2b, 3b and 4). Usually some of the large events were audible, whereas the small ones were inaudible.

Period doubling was observed in seven out of ten experiments we conducted (Table 1). In one experiment (GU22) at the lowest confining pressure (50 MPa), period doubling was observed in the transition from stable sliding to dynamic instability and vice versa. In the other six experiments, period doubling was observed only in the transition from dynamic instability to stable sliding. The bifurcations are highlighted in a plot of the stress drop as a function of the cumulative number of event (Figures 2c and 3c). The bifurcation patterns can be very complex for one experiment (GU5 in Figure 3), and yet relatively simple for another (TY8 in Figure 2).

2.2.4. Self-sustained oscillations

In the transition from cyclic stick–slip to stable sliding, period doubling led to inaudible, quasistatic oscillations similar to those first reported by Scholz et al. (1972). Such periodic oscillations were observed in all our experiments (Table 1), and we will refer to them as 'self-sustained oscillations.' In this stage, if we maintained the load-point velocity

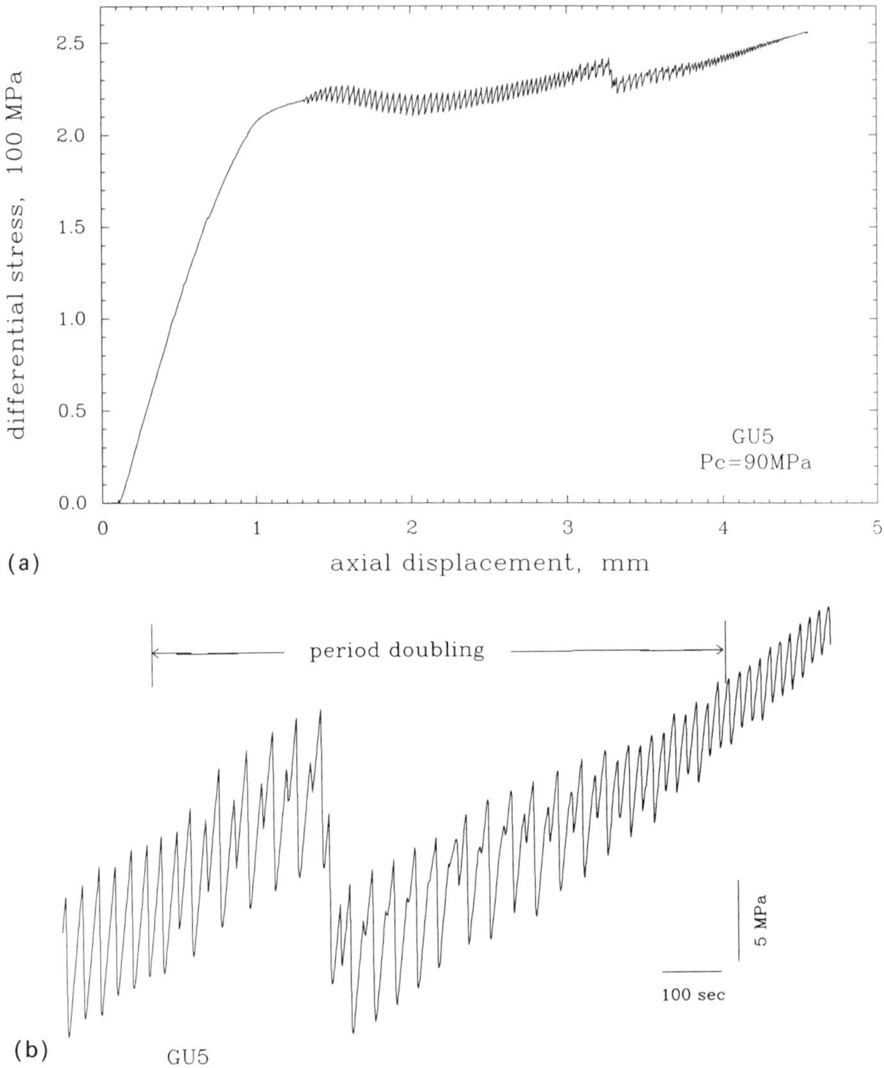

Figure 3. (a) Complete record of the differential stress versus axial displacement in experiment GU5. A summary of the experimental conditions is given in Table 1. (b) Stress drops recorded by the internal load cell for axial displacement from 2.8 mm to 4.2 mm. The period doubling behavior was more complex than that for TY8 shown in Figure 2b.

constant, then the amplitude of the self-sustained oscillations would decay with slip, ultimately leading to stable sliding. In most of our experiments, we sequentially decreased the load-point velocity by a ratio of 2:1. As shown in Figure 6a, such a velocity perturbation would induce the oscillation amplitude to instantaneously increase and then decrease gradually with further slip at the new load-point velocity.

The self-sustained oscillations differed from dynamic instability events in two respects. First, the period of the oscillation was much longer, ranging from several seconds to several minutes. Second, the recurrence time of the dynamic events had a positive correlation with the stress drop amplitude, whereas the period of the self-sustained

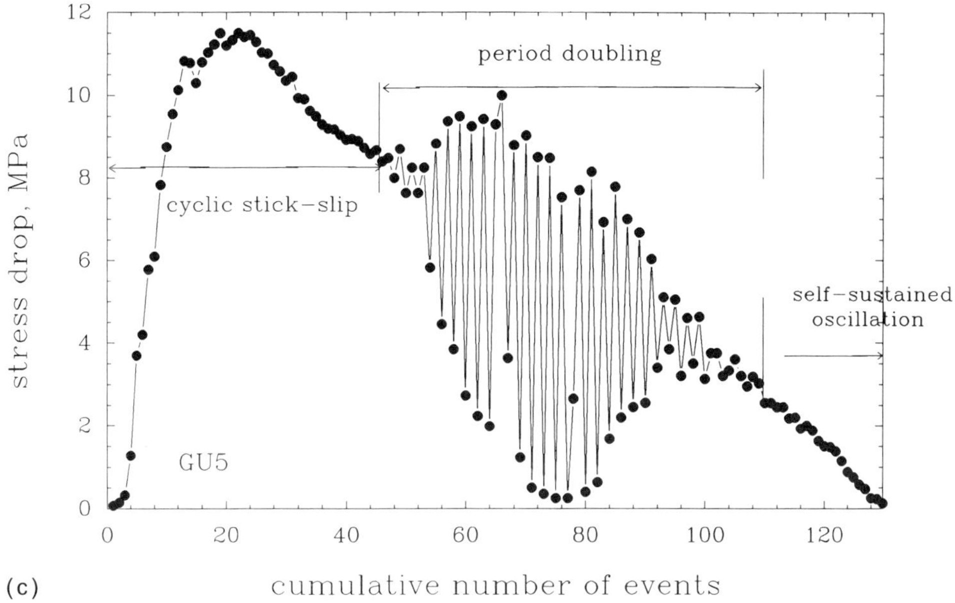

(c)

Figure 3. *Continued.* (c) Differential stress drop amplitude as a function of cumulative number of events. The three stages of cyclic stick–slip instabilities, period doubling, and self-sustained oscillations are highlighted.

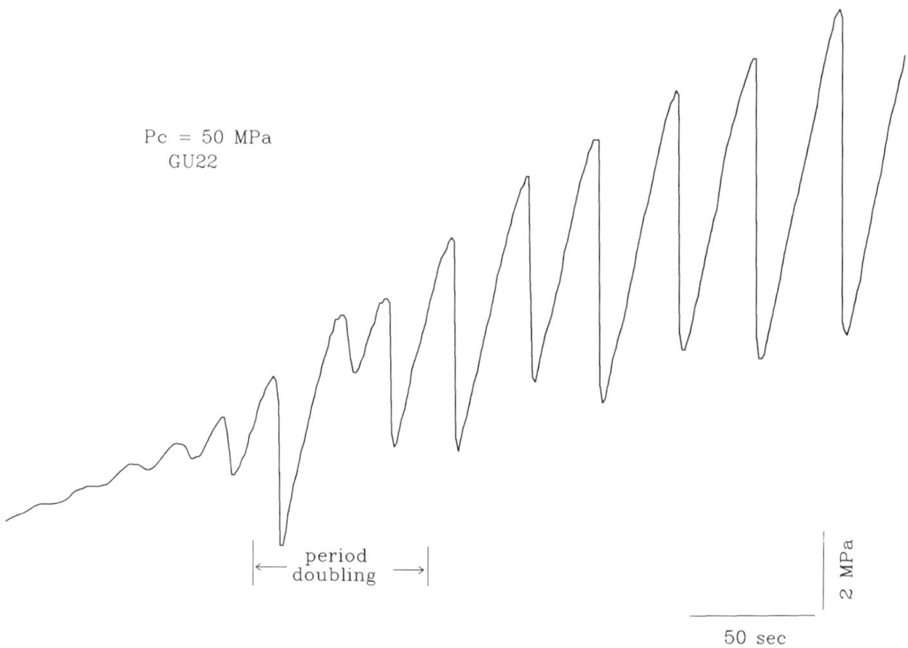

Figure 4. Supercritical oscillations recorded by the internal load cell in experiment GU22 during the transition from stable sliding to dynamic instabilities. Period doubling was observed in the initial destabilization phase. The experimental conditions are given in Table 1.

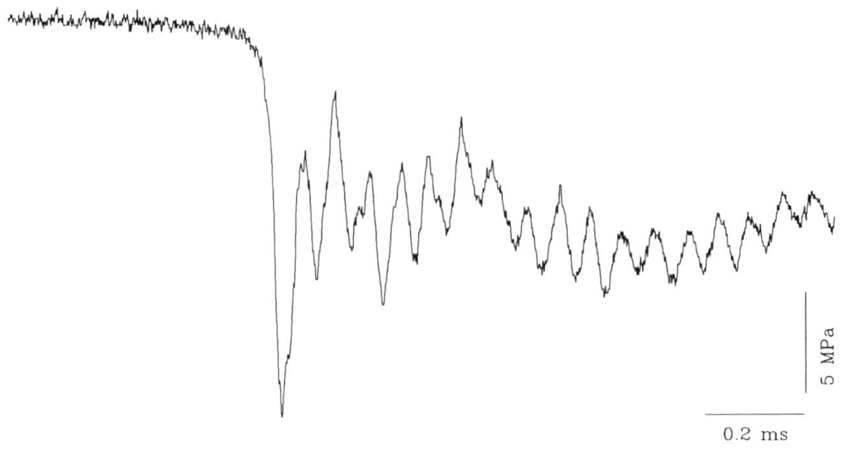

Figure 5. A dynamic instability in experiment TY4 recorded using high-frequency data acquisition system. Note the relatively short timescale (on the order of 0.1 ms) and the oscillations from 'ringing' of the loading frame.

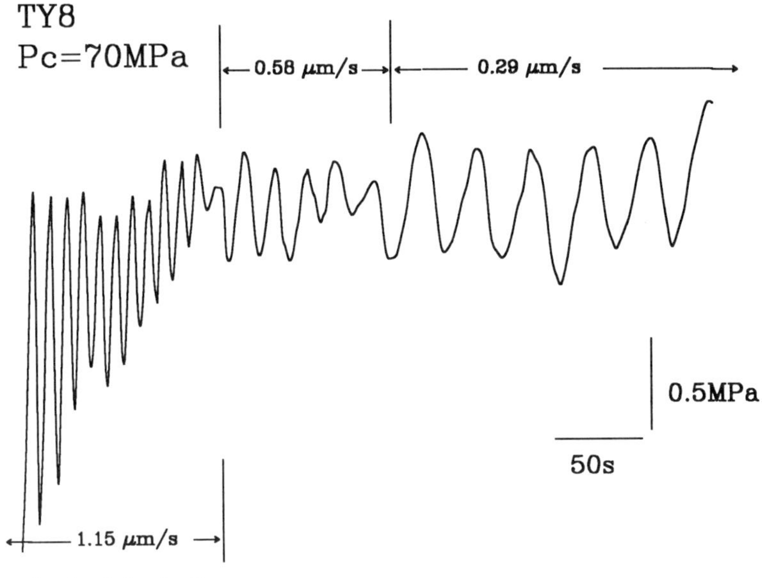

(a)

Figure 6. (a) Quasistatic, self-sustained oscillations observed in the neutral stability stage of TY8. The load-point velocity was maintained constant at 1.15 μm s^{-1} until the onset of self-sustained oscillations. The oscillation amplitude decreased from cycle to cycle at a fixed load-point velocity, and the period of the oscillations increased with decreasing load-point velocity.

oscillations did not show any systematic dependence on the stress drop. In Figure 6b, we compile the data for 37 oscillations in a single experiment (TY8) as the load-point velocity decreased from 1.15 μm s^{-1} to 0.12 μm s^{-1}. For each load-point velocity, we determined the mean and standard deviation of the oscillation periods. The oscillation

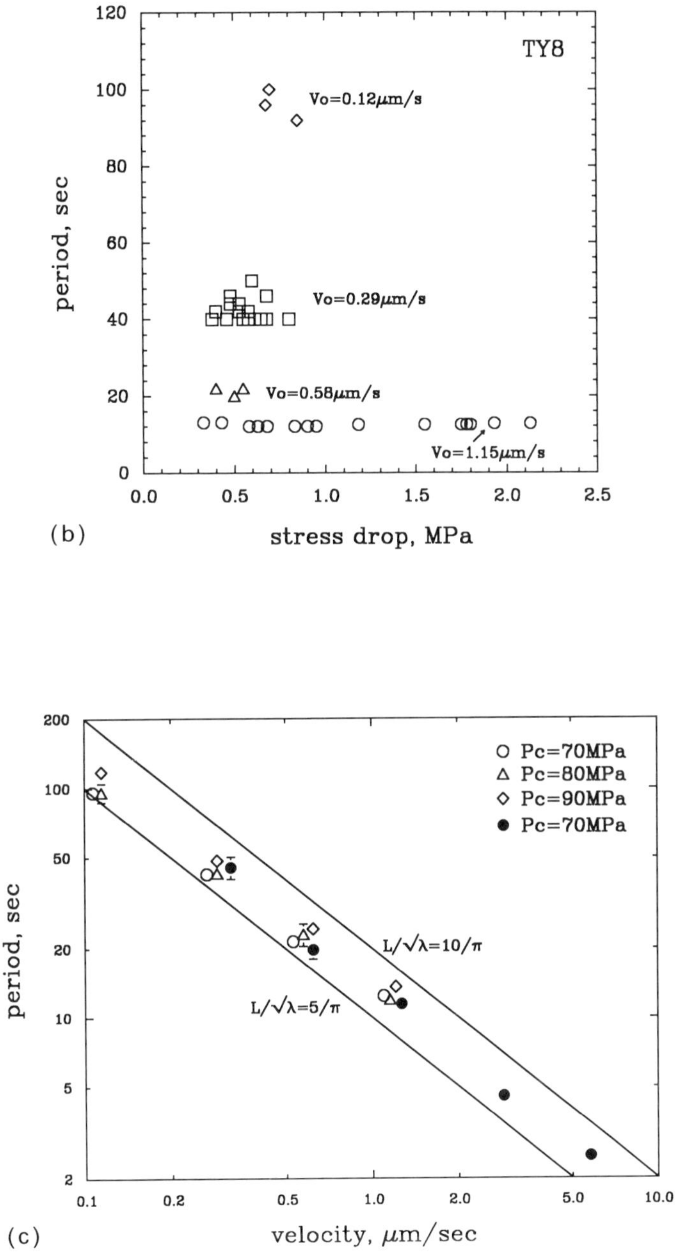

Figure 6. *Continued.* (b) Period of the self-sustained oscillations versus the stress drop amplitude for four different load-point velocities from experiment TY8. The period was basically independent of the stress drop amplitude. There was more scatter when the load point velocity was slow because the command signal for the servocontrol system was not as smooth. (c) The mean oscillation period versus the load-point velocity. The bars indicate the standard deviation, and if they are not shown, the standard deviation is smaller than the symbol itself. The two straight lines correspond to constant values of $L/\sqrt{\lambda}$ shown in micrometers.

period data are plotted as a function of the load point velocity in Figure 6c, which also includes data from three other experiments. The data follow a linear trend with a slope of -1 on a log–log plot.

2.3. Discussion

In all our experiments we observed an initial phase of supercritical oscillations in which the effect of cumulative slip was to destabilize the frictional sliding behavior. This transient destabilization was followed by prolonged phase of stabilization due to cumulative slip. Since the stiffness of the test machine remained unchanged in an experiment, the stabilization (or destabilization) had to come from a systematic change of the friction constitutive parameters, such that the critical stiffness of the frictional surface would decrease (or increase) with increasing slip. How can these experimental observations be interpreted with existing friction models? To a first approximation, our frictional sliding experiment can be analyzed as a single-degree-of-freedom spring–slider system (Shimamoto et al., 1980). A friction constitutive relation which predicts the occurrence of period doubling and self-sustained oscillations in such a system is the rate- and state-dependent formulation of Dieterich (1979) and Ruina (1981). We compare below our observations with the Dieterich–Ruina theory of rock friction.

Rice and Ruina's (1983) linear perturbation analysis has shown that at neutral stability, the period T of the self-sustained oscillations should be inversely proportional to the load point velocity V_0:

$$T = \frac{2\pi}{V_0} \frac{L}{\sqrt{\lambda}} \tag{1}$$

Here L is the characteristic slip parameter, and $\lambda = (B - A)/A$ where A and B are constitutive parameters which characterize the instantaneous and evolution response of friction to a velocity perturbation (Rice, 1983). It should be noted that in some friction literature, D_c (instead of L) is used to denote the characteristic slip, and if the constitutive parameters are normalized with respect to the normal stress σ_n so that $A = \sigma_n a$ and $B = \sigma_n b$, then $\lambda = (b - a)/a$.

Although self-sustained oscillations have been reported for sliding on granite (Scholz et al., 1972) and dolomite (Weeks and Tullis, 1985) surfaces, we are not aware of any previous observations of the systematic dependence of oscillation period on load-point velocity. In our experiments, only limited data are available during the supercritical oscillation phase (Figure 4), but the quality of the self-sustained oscillation data in the transition from dynamic instability to stable sliding (Figures 2c, 3c, and 6a) is good enough for a quantitative comparison with eq. (1). According to this relation, a log–log plot of T versus V_0 should have a slope of -1, with the intercepts corresponding to constant values of the friction parameter $L/\sqrt{\lambda}$. The two straight lines bracketing our data in Figure 6c correspond to $L/\sqrt{\lambda}$ of 1.4 μm to 2.5 μm, respectively.

Observing from experimental measurements (e.g., Dieterich, 1981; Ruina, 1981; Tullis and Weeks, 1986) that $B - A$ is comparable in magnitude to A, Rice and Tse (1986) assumed that $\lambda = 1$ in their numerical simulations. If we adopt this approximation, then our self-sustained oscillation data imply that L is between 1.4 μm and 2.5 μm. This range is in the lower end of quasistatic laboratory measurements of L. Previous studies indicate that L decreases with decreasing country rock roughness and gouge particle size (Dieterich,

1981; Biegel et al., 1989). Using crushed Westerly granite gouge, Dieterich (1981) observed that L decreased with decreasing gouge particle size to attain values on the order of 10 μm in his direct shear experiments on smooth (#600 grit) surfaces under relatively low normal stresses. It seems that L also decreases with increasing normal stress and slip. At normal stresses comparable to ours but using quartz gouge significantly coarser than ours, Marone et al. (1990) reported L values as low as 1 μm in triaxial experiments on #320 grit surfaces. In rotary shear experiments on crushed Westerly granite, Blanpied (1989) observed that L decreased by one order of magnitude after a cumulative slip of 10 mm. Since our #100 grit sliding surfaces were relatively rough, the low L values in our experiments are possibly related to the ultrafine gouge size, elevated normal stresses, and cumulative slip. However, it is difficult to quantify the relation without a firm microphysical understanding of the friction constitutive parameters.

Since the critical stiffness k_{cr} is given by $(B - A)/L$ (Rice and Ruina, 1983), we can relate $L/\sqrt{\lambda}$ to the geometric mean of $B - A$ and A:

$$[A(B - A)]^{1/2} = \frac{k_{cr} L}{\sqrt{\lambda}} \tag{2}$$

At neutral stability, the critical stiffness equals the test machine stiffness. For our triaxial test system, the stiffness k varies somewhat with pressure, ranging from 105 MPa mm^{-1} to 116 MPa mm^{-1}. Using values of $L/\sqrt{\lambda}$ inferred from the data in Figure 6c, we estimate $[A(B - A)]^{1/2} (= \sigma_n [a(b - a)]^{1/2})$ to be between 0.2 and 0.4 MPa from eq. (2). This range is comparable to what has been measured for silicate materials (e.g., Dieterich, 1981; Tullis and Weeks, 1986; Biegel et al., 1989).

To our knowledge, there has been only one previous observation of period doubling in rocks, which was by Ruina (1981), who observed a transition from self-sustained oscillations to period doubling in quartzite as the effective stiffness of his direct shear system was reduced by electronic control. Ruina's (1981) experiment was interpreted by Gu et al. (1984) in terms of the complexity of the friction constitutive law. They showed that if one incorporates two state variables in the Ruina formulation, then Hopf bifurcation occurs in the transitional regime between dynamic instability and stable sliding (Figure 7). For a given load-point velocity, the mode of oscillation is sensitive to the nondimensional stiffness ratio k/k_{cr}.

Since the cumulative slip in Ruina's (1981) experiment was relatively short and the friction constitutive parameters probably did not change much, destabilization occurred if the machine stiffness was reduced. On the other hand, if the machine stiffness remains unchanged (as in our experiments), and if the frictional response becomes more velocity weakening or if the characteristic slip distance decreases, resulting in an increase in the critical stiffness k_{cr}, then a transition from stable sliding to cyclic stick–slip instabilities involving supercritical oscillations (such as those shown in Figure 4) will occur. If the frictional response becomes less velocity weakening or if the characteristic slip distance increases, then a transition from dynamic instability to stable sliding will occur. Within the transitional regime, self-sustained oscillations, period doubling and even chaotic oscillations are possible (Figure 7).

A fixed value of k/k_{cr} was used in each of the simulations of Gu et al. (1984). A more realistic model of Ruina's (1981) and our experiments should consider the effect of varying k/k_{cr} as a function of slip. We have carried out a number of dynamic simulations using parameters identical to those used by Gu et al. (1984). We improve on their

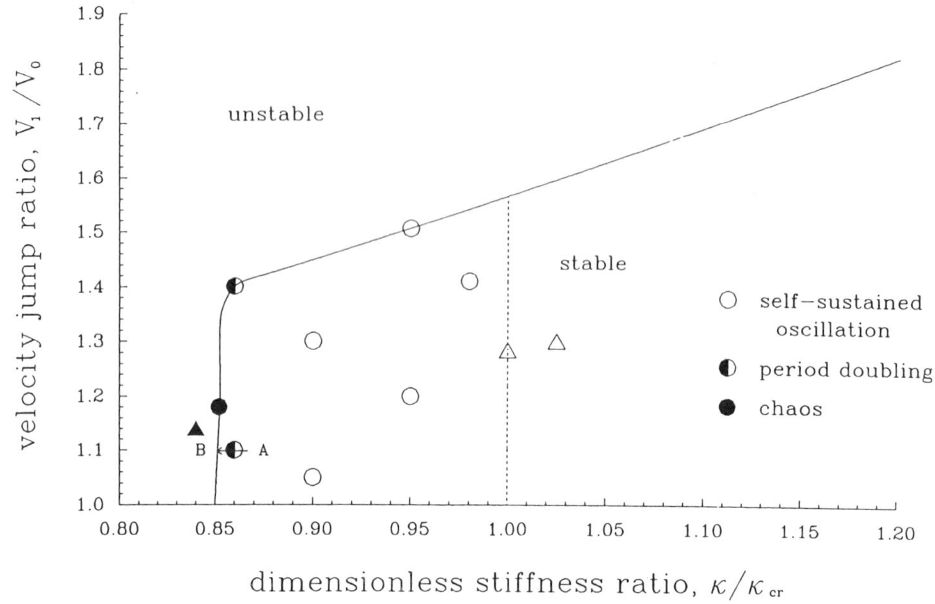

Figure 7. Stability boundary for a single-degree-of-freedom spring–slider system based on a rate- and state-dependent friction law with two state variables (after Gu et al., 1984). The frictional sliding behavior is sensitive to both the relative stiffness and the velocity perturbation (from a reference value V_0 to V_1). In the transition from dynamic instability to stable sliding, oscillation modes include period doubling, and self-sustained and chaotic oscillations.

quasistatic analysis by incorporating inertia. We also allow the normalized stiffness to vary as a function of the normalized slip. Since the detailed results will be presented in a future publication, we will only provide two sample calculations here.

Both simulations are for destabilization along the path from A to B shown in Figure 7. The nonlinear dynamical behavior is sensitive to the ratio of the change in normalized stiffness to the change in normalized slip. In both simulations, the ratios are negative, implying a decrease of k/k_{cr} as a function of slip. The first simulation shown in Figure 8a is for a ratio of -1×10^{-4}. The oscillation mode evolves from quasistatic, periodic motion to period doubling, and finally to dynamic instability. This is in qualitative agreement with the initial supercritical oscillations observed in GU22 (Figure 4). If we use a significantly higher ratio (-3×10^{-3}), then the transition follows a route with self-sustained periodic oscillations but without period doubling (Figure 8b). The latter simulation is analogous to the supercritical oscillations we observed in experiments conducted at confining pressures higher than 50 MPa. Keeping in mind that the stabilization phase corresponds to a path in the reverse direction (from B to A in Figure 7), one can see that the variation of stress drop amplitude as a function of cumulative number in the numerical simulation (Figure 8c) is in good qualitative agreement with our observations in experiments which showed period doubling in the stabilization phase (Figures 2c and 3c). In our experiments, complex bifurcations corresponding to period doubling were often observed in the stabilization phase but seldom observed in the destabilization phase. This implies that the rate by which the critical stiffness decreases at the stabilization phase is less than the rate by which the critical stiffness increases in the destabilization phase.

It should also be noted that if there is asymmetry in the loading arrangement, then period doubling may occur with a relatively simple friction constitutive relation such as the classical static/kinetic law (Nussbaum and Ruina, 1987; Huang and Turcotte, 1990). However, such a model does not predict self-sustained oscillations. Nevertheless, it would be important in the future to investigate the sensitivity of the transitional behavior to imperfections in the loading and sample configurations by changing the lubrication and jacketing in our experiments.

3. Stabilization of Faulting by Wear Processes

In quasistatic tests on simulated silicate gouge, the destabilization of frictional sliding induced by cumulative slip has been observed extensively. Dieterich (1981), Blanpied (1989), and Biegel et al. (1989) reported a transition from velocity strengthening to weakening over slip distances ranging from 1 mm to 10 mm. As we discussed before, the characteristic slip L was also observed to decrease with cumulative slip in some of these studies. An increase in velocity weakening or a decrease in L would result in an increase in the critical stiffness k_{cr}, thus destabilizing the frictional sliding behavior.

This destabilization behavior was observed in the initial supercritical stage of our experiments. However, this stage is relatively short in comparison with the second stage of stabilization. The very significant stabilization we observed is in apparent contradiction to some quasistatic sliding experiments which suggest an asymptotic approach to a 'steady-state' frictional regime with cumulative slip. If such a 'steady state' can be achieved, then the cyclic stick–slip events will be perfectly periodic with a characteristic stress drop. This type of strictly periodic stick–slip events as analogs of characteristic earthquakes was never observed in any of our experiments, implying that 'steady-state' friction was not attained in our cyclic stick–slip experiments. It should be noted that the stabilization effect is not uniquely observed in ultrafine quartz gouge, since we have also observed similar behavior in the frictional sliding of bare surfaces of Tennessee sandstone, although the initial destabilization stage in the sandstone is longer, involving more cycles of self-sustained oscillations and period doubling. It is possible that in comparison to quasistatic slip, the dynamic slip in cyclic stick–slip instabilities is more efficient in generating damage within the gouge layers and the country rock surfaces, which contributes to the fault stabilization that is observed in cyclic stick–slip experiments but is absent from some quasistatic sliding tests.

To investigate the micromechanical process responsible for the stabilization of faulting by cumulative slip, we have conducted microstructural observations on the deformed specimens. These microstructural data will be presented in a future publication. Under the scanning electron microscope, the (#100 grit) sliding surfaces of the country rock appeared to be relatively rough before an experiment, but after several millimeters of sliding the same surfaces would assume a polished appearance and be covered with fine-scale ploughing features. These observations suggest that because of their relatively high hardness, the quartz gouge particles continuously wore off materials from the country rock which were then incorporated into the gouge zone. Under the optical microscope, we have observed layers of wear materials in thin sections of samples which had slid over a relatively long distance of several millimeters. Because its particle sizes

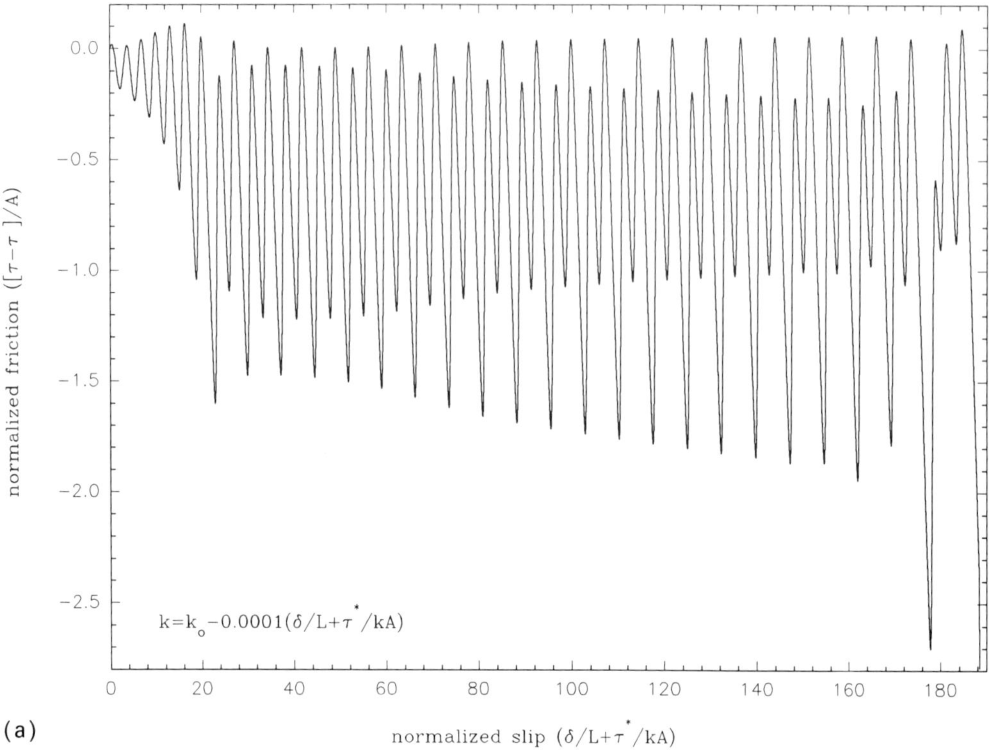

normalized friction ([τ−τ*]/A)
normalized slip (δ/L+τ*/kA)
k=k_o−0.0001(δ/L+τ*/kA)

(a)

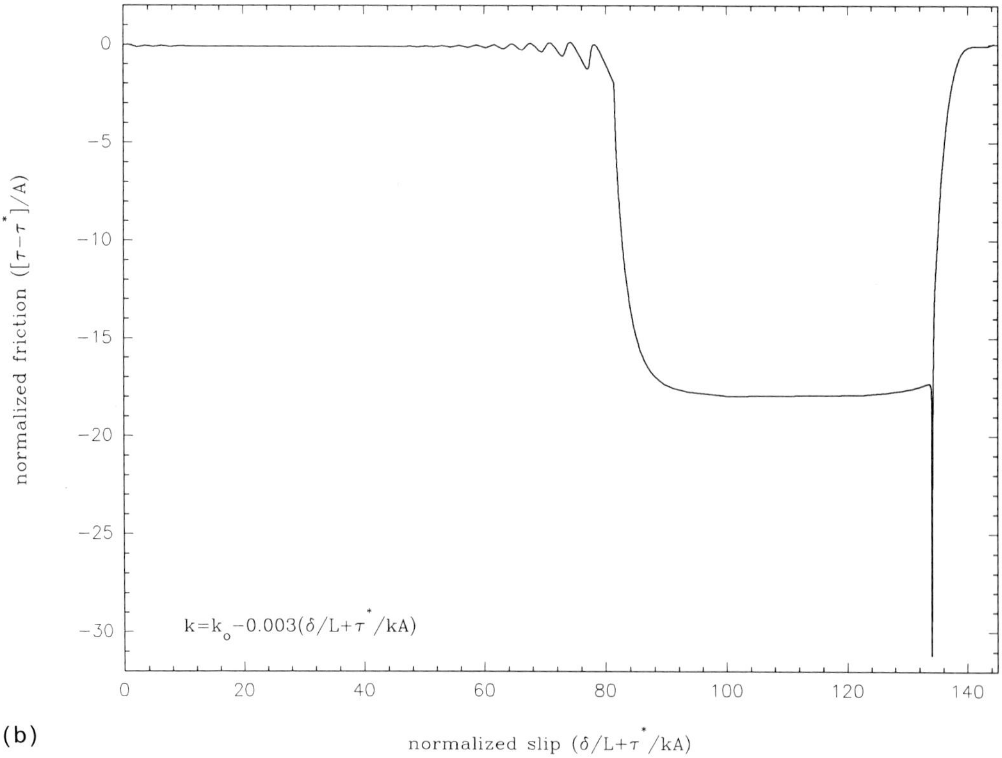

normalized friction ([τ−τ*]/A)
normalized slip (δ/L+τ*/kA)
k=k_o−0.003(δ/L+τ*/kA)

(b)

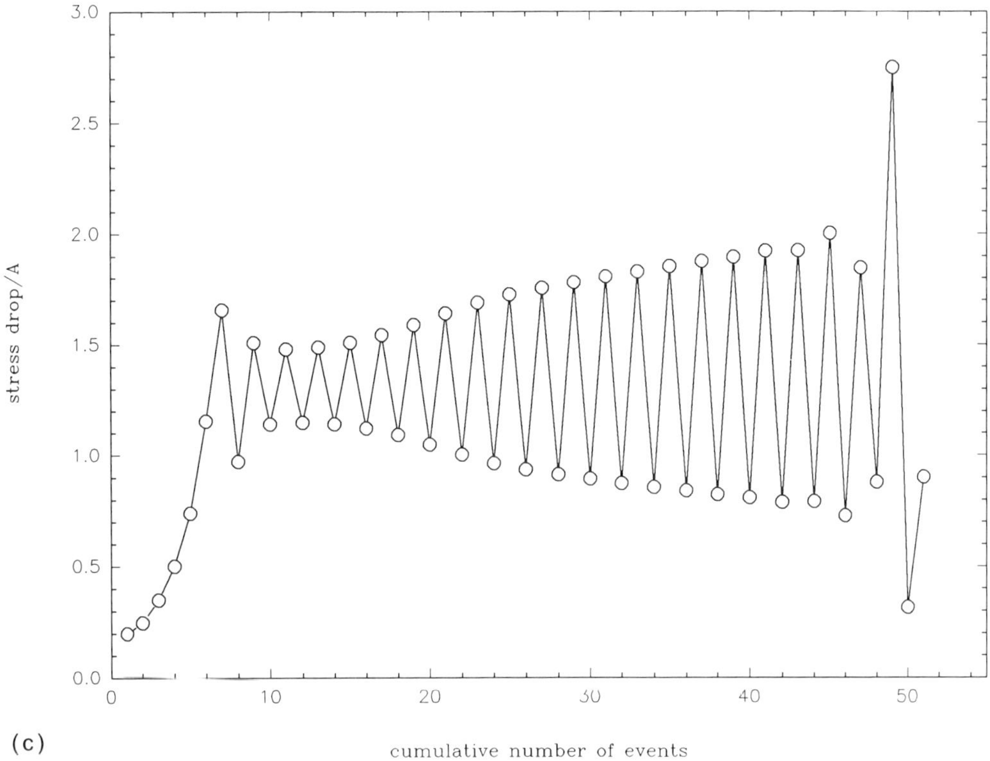

(c)

Figure 8. Numerical simulation for a single-degree-of-freedom spring–slider system based on a rate- and state-dependent friction law with two state variables. The input parameters ($B_1/A = 1$, $B_2/A = 0.84$, $L_1/L_2 = 0.048$) were identical to those used by Gu et al. (1984) for their fig. 13. We have improved on their quasistatic analysis by incorporating inertia in our simulations. The velocity perturbation was $V_1/V_0 = 1.1$ and the stiffness ratio was decreased along the path from A to B shown in Figure 7. (a) The normalized friction as a function of normalized slip as the normalized stiffness decreased linearly with normalized slip according to the relation shown. Note the transition of oscillation mode from self-sustained oscillations to period doubling. (b) The normalized friction stress versus normalized slip. Note that the ratio of the change in normalized stiffness to the change in normalized slip is increased by a factor of 30 from that in (a), and the destabilization route does not involve period doubling. (c) Normalized stress drop as a function of cumulative number of events for the simulation shown in (a). The direction of change in the simulation is opposite to that in the stabilization stage, but the simulation results are qualitatively similar to experimental data shown in Figures 2c and 3c.

were already ultrafine, the quartz gouge did not undergo appreciable comminution. We did observe extensive (Riedel shear) localization in the sheared gouge layers. To further investigate the role of the wear processes in stabilizing the faulting behavior, we carried out a comparative study on several country rocks of different bulk hardness.

3.1. Experimental procedure

We discuss here our data for Westerly granite, Frederick diabase, and Blair dolomite. The Frederick diabase samples were cored from the same block studied by Fredrich and Wong (1986), and the Blair dolomite was from Brace's (1964) block. These experiments were conducted before we installed the internal load cell, so all the stress measurements were from an external load cell. The simulated gouge and the sample configuration were identical to those of Wong and Zhao (1990), and hence they differed in two major aspects from those described in the previous section. First, the samples for this part of the study were smaller (diameter 18.4 mm and length 50.8 mm). Second, the simulated gouge was finer with a median grain size of 1.4 μm (Figure 1). All the experiments were performed at an axial displacement rate of 1 $\mu m\ s^{-1}$, corresponding to a load-point velocity of 1.15 $\mu m\ s^{-1}$.

3.2. Observations

As we discussed above, O-ring friction along the loading piston renders it difficult to resolve from external load cell data the detailed oscillational behavior in the transition from dynamic instability to stable sliding. Otherwise, the frictional sliding behavior in all the three country rocks was similar in that the onset of frictional instability was marked by supercritical oscillations and that beyond a certain point cyclic stick–slip events occurred with decreasing amplitude, leading to stable sliding.

For each run we determined the slip distance X_c (resolved along the sawcut surface) over which cyclic stick–slip instabilities were observed. This measurement is subject to the uncertainty in picking the onset of supercritical oscillation and the termination of dynamic instabilities owing to noise from O-ring friction. Since the confining pressure was fixed in an experiment, the normal stress would vary owing to the stress drops and changes in frictional strength with cumulative slip. Our data for the critical slip distance X_c as a function of the normal stress for nine experiments are compiled in Figure 9a.

Two features are worth noting. First, the critical slip distance X_c decreased with increasing normal stress for a given rock type. Second, it shows a positive correlation with the bulk hardness of the country rock. We were motivated by these data to formulate a simple wear model for the stabilization process.

3.3. A simple wear model for the stabilization process

Our model is based on two assumptions. First, the wear rate is assumed to be given by Archard's (1953) law, which is widely used in tribology (e.g., Lim and Ashby, 1987). This law was originally developed on the basis of the adhesion theory of friction, although it may also be derived from dimensional analysis (Drucker, 1976; Shaw, 1977). Archard's law predicts that the thickness of the wear product t is given by

$$t = \frac{\kappa}{3}\frac{\sigma_n X}{H} \tag{3}$$

where σ_n is the normal stress, H is the hardness, and κ is a nondimensional wear coefficient expected to be in the range of 0.1–1 (Scholz, 1987). Geological and laboratory observations summarized by Robertson (1982) and Scholz (1987) suggest that such a linear relation exists between fault zone thickness and cumulative slip.

Our second assumption is motivated by Byerlee and Summers' (1976) observation that the frictional sliding behavior in the laboratory is stabilized by an increase in gouge thickness. We postulate that the termination of stick–slip instability in our experiments was due to the accumulation of a critical thickness t_c of wear material. From eq. (3), we obtain the following relation for the critical slip distance X_c at the termination of stick–slip instability (and the onset of final stage of stable sliding),

$$\sigma_n X_c = \frac{3}{\kappa} t_c H \tag{4}$$

In other words, our model predicts that the product $\sigma_n X_c$ should be proportional to the bulk hardness H (Ohnaka, 1973b) of the country rock. Our data (Figure 9b) are suggestive of such a linear trend corresponding to $t_c/\kappa \approx 33$ μm, implying that the wear product had a thickness of less than 33 μm. We are in the process of using optical and scanning electron microscopes to characterize quantitatively the thickness of the wear products.

The wear coefficient κ is expected to be dependent on a number of factors. Figure 9c shows three different groups of data for Westerly granite. The first group is identical to that shown in Figure 9a. The second group corresponds to ultrafine quartz gouge sandwiched between Westerly granite preheated to 600°C; and the third group corresponds to coarser gouge sandwiched between unheated Westerly granite. The grain boundary strength of the preheated Westerly granite is expected to be lowered by thermal cracking (e.g., Fredrich and Wong, 1986), and consequently the wear coefficient may be higher (Scholz, 1987). The coarser quartz gouge was probably more efficient in wearing off the country rock, thus resulting in a high wear coefficient. Consequently, the critical slip distances X_c for the second and third groups were lower than those for the first group.

4. Seismological Implications

Geological observations show that wear processes operate during seismic faulting (Robertson, 1982). It is therefore of fundamental importance to understand wear mechanisms under controlled conditions, especially during dynamic instabilities. Two major implications can be drawn from our results. First, the wear process associated with frictional instability tends to reduce the structural complexity of the sliding surface, and thereby stabilize the faulting behavior. Second, in the transition from dynamic instability to stable sliding, nonlinear dynamical phenomena such as period doubling and self-sustained oscillations do occur in the laboratory under high-pressure conditions.

Although stick–slip instability is generally accepted as a physical analog of crustal earthquakes (Brace and Byerlee, 1966), a frictional sliding surface in the laboratory is not an exact replica of a seismogenic fault. To what extent do we expect the nonlinear dynamics of earthquake rupture to resemble experimental observations? How can our experimental data on the stabilization effect be scaled to crustal faulting? What are their limitations?

We have in our experimental assembly only a single degree of freedom for slip, whereas a geologic fault has geometric heterogeneity with multiple degrees of freedom. A more realistic fault model would be based on a continuum description. Using a rate- and state-dependent friction law with velocity cutoff, Horowitz and Ruina (1989) recently showed that temporal and spatial patterns of the slip in a two-dimensional continuum model can be extremely complex even for a model without any built-in complexity in geometry or heterogeneity in material properties. Accelerating aseismic slip and propagating creep waves analogous to what have been observed along some fault zones

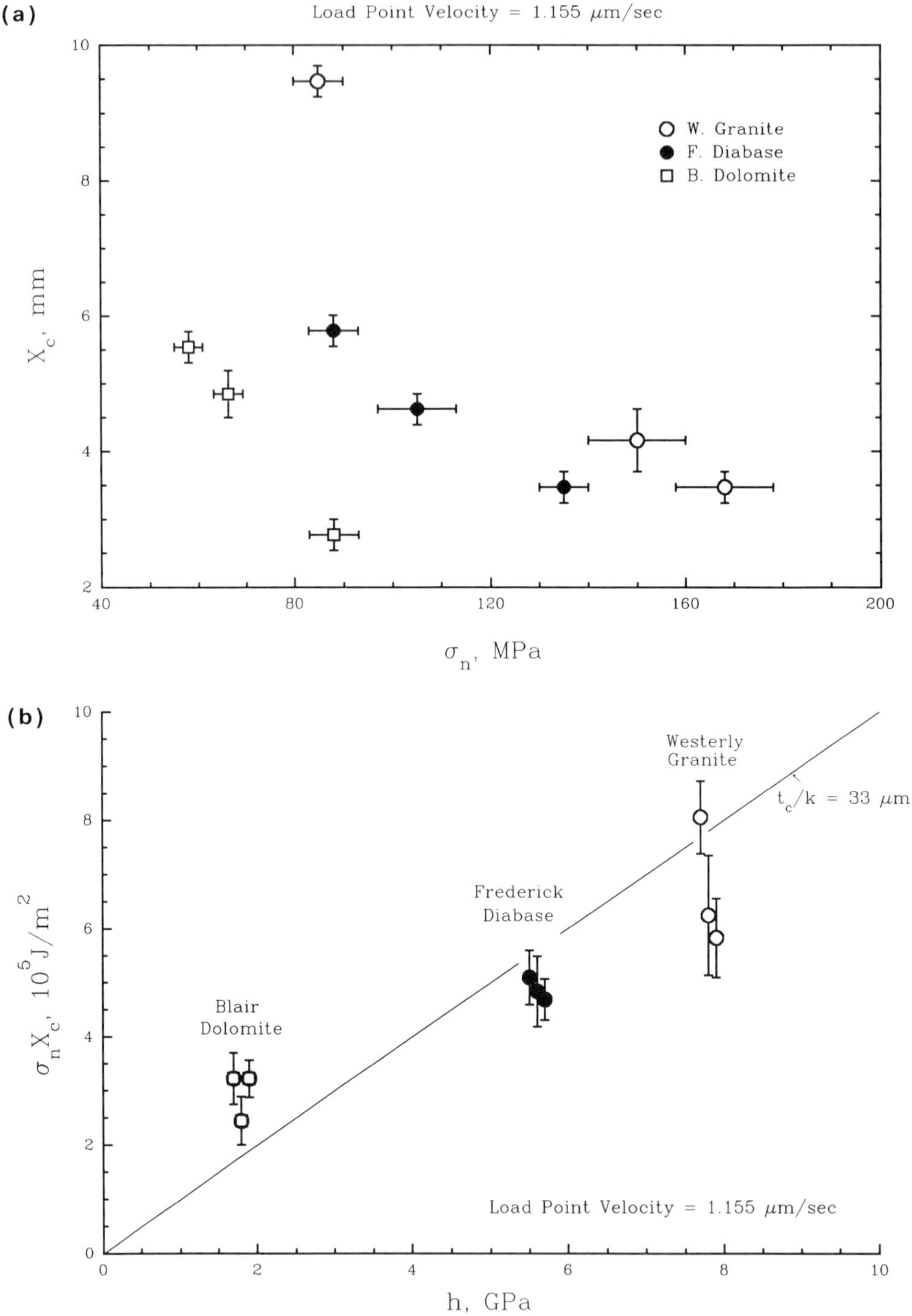

Figure 9. (a) Critical slip distance for stabilization X_c as a function of normal stress for ultrafine quartz gouge sandwiched between sawcut samples of Westerly granite, Frederick diabase, and Blair dolomite, respectively. (b) The product of the stabilization distance and normal stress as a function of the bulk hardness of the country rock. The hardness values are from Ohnaka (1973b). The straight line corresponds to $t_c/\kappa = 33\ \mu m$ in eq. (4).

(c)

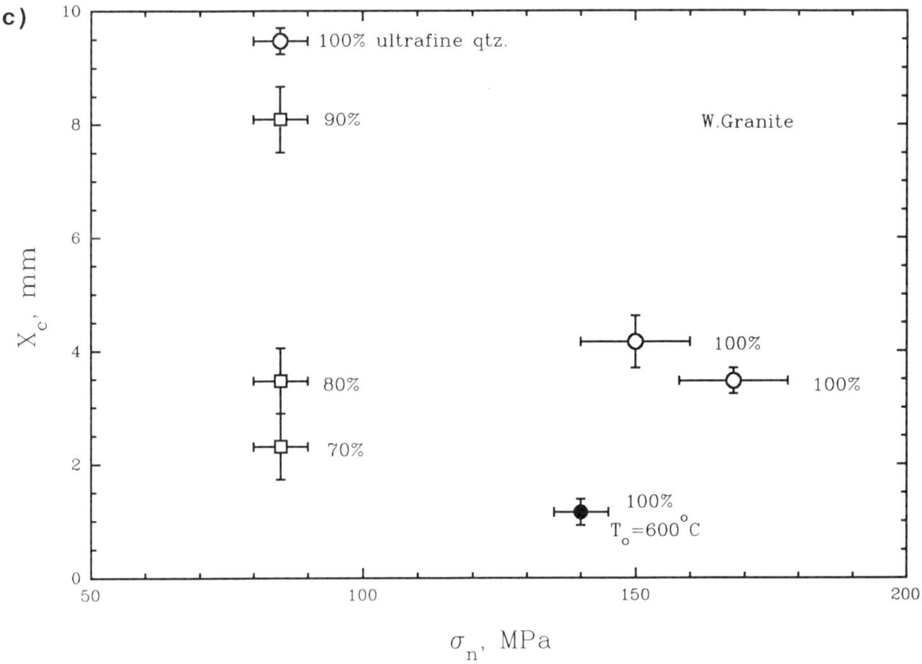

(c) Critical slip distance for stabilization X_c as a function of normal stress for three different types of experiments. The open circles are for ultrafine quartz gouge sandwiched between unheated sawcut samples of Westerly granite. The filled circle is for ultrafine quartz gouge sandwiched between sawcut samples of Westerly granite preheated at 600°C. The open squares are for a mixed gouge of ultrafine quartz (median particle size 1.4 μm) and coarse quartz (median particle size 25 μm) sandwiched between unheated sawcut samples of Westerly granite. The percentages of the ultrafine quartz in the mixtures are also shown.

can be seen in the numerical simulations. Tse and Rice (1986) and Okubo (1989) have considered continuum models with depth variation in friction and inertia respectively, but since they did not consider a wide range of frictional parameters, it is unclear what kinds of complex dynamical behavior may occur in such models.

Numerical simulations of multiple-degree-of-freedom spring–slider (e.g., Cao and Aki, 1986; Nussbaum and Ruina, 1987; Carlson and Langer, 1989; Huang and Turcotte, 1990) and cellular automaton models (e.g., Bak and Tang, 1989; Ito and Matsuzaki, 1990) also suggest that the behavior can be much more complex than our experimental results. However, Rudnicki (1988) emphasized that the conclusions of these studies have been varied and seem difficult to reconcile with one another, suggesting a strong sensitivity to choice of parameters.

A major problem for the formulation of a physical model of earthquake rupture has been the scaling of laboratory observations to tectonic faults. An approach commonly adopted for the Dieterich–Ruina type of friction law is to regard the friction constitutive parameters A and B for given rock types to be universally applicable to laboratory and crustal scale processes, and to regard the characteristic slip L as a scalable parameter, with the earthquake values 4–5 orders of magnitude higher than laboratory values (Rice, 1983). If we use a simplistic dimensional argument and assume that the critical stabilization distance X_c in eq. (4) scales with L, then corresponding to an increase of L from 1 μm (our experimental data) to 100 mm (considered to be appropriate for earthquakes), the critical slip for stabilization will increase from about 10 mm to 1 km.

In comparison, the stabilization inferred from seismological data occurs over a slip on the order of 10–100 km (Wesnousky, 1990). There are several complications not accounted for in such a simplistic model. Our data (Figure 9c) suggest several mechanisms by which the stabilization distance X_c may be shortened in the laboratory, and therefore the above extrapolation possibly represents an upper bound. That the seismological values seem to be significantly longer implies that there are crustal-scale mechanisms not duplicated in rock friction experiments. The topographic structure of our sawcut surfaces are relatively simple in comparison to the type of three-dimensional structural complexity along a fault zone mapped out by Wesnousky (1988), and hence it is expected that wear processes can reduce the structural complexity of a laboratory sliding surface in a more efficient manner. Finally, the effect of time dependence may not be fully accounted for in our friction experiments. In the interseismic phase of an earthquake cycle, a hydrothermal process such as pressure solution may be operative (Angevine et al., 1982; Sibson, 1986; Power and Tullis, 1989; Gratier and Gamond, 1990), and such a healing process lasting for tens to thousands of years is not expected to be identical to the short-term evolution effect studied in room-temperature friction experiments. Systematic studies of the phenomenon under elevated temperatures in the presence of pore water (Blanpied et al., 1991; see also Chapter 9 of this volume) should provide important physical insight on this question.

5. Conclusion

The use of ultrafine quartz as simulated gouge allows us to observe a large number of cyclic stick–slip events in a single experiment, so that the transient destabilization and subsequent stabilization of faulting by cumulative slip can be studied systematically in the laboratory. With an internal load cell and improved servocontrol system, we were able to observe the nonlinear dynamical behavior in the destabilization phase and in the transition from dynamic instability to stable sliding. The sequential occurrence of period doubling and self-sustained oscillations in a single experiment was observed for the first time under high-pressure conditions.

Our experimental observations are in qualitative agreement with what have been predicted for a single-degree-of-freedom spring–slider system based on a rate- and state-dependent friction law. We showed how experimental observations at neutral stability on the oscillation period as a function of load-point velocity can be used to infer the friction constitutive parameters. For our experiments, we infer the characteristic slip L to be several micrometers. We also estimate the parameter $[A(B-A)]^{1/2}$ to be between 0.2 MPa and 0.4 MPa.

The slip displacement over which repeated stick–slip instabilities occurred decreased with increasing normal stress and with decreasing rock hardness. Our data can be interpreted with a simple wear model. Comparison with seismological data suggests that stabilization by cumulative slip in our experiments is more efficient than that induced by geologic offset. Two possible explanations for the apparent difference are that tectonic fault zones have more complex geometry and structure, and that long-term healing processes operative in the interseismic period may not be duplicated in room-temperature friction experiments.

Acknowledgments

A pilot project using the ultrafine quartz gouge was initiated at the suggestion of Bill Brace when the senior author was associated with his laboratory. We have benefited from comments by Charlie Sammis and an anonymous reviewer. We are also grateful to Ron Biegel who provided us with tabulated data from his previous experiments. This research was supported by the National Science Foundation under grant EAR-8707532 and the U.S. Geological Survey under grant 14-08-0001-G1807.

References

Angevine, C.L., Turcotte, D.L., and Furnish, M.D. (1982). Pressure solution lithification as a mechanism for the stick–slip behavior of faults. *Tectonics* **1**, 151–160.

Archard, J.F. (1953). Contact and rubbing of flat surfaces. *J. Appl. Phys.* **24**, 981–988.

Bak, P. and Tang, C. (1989). Earthquakes as a self-organized critical phenomenon. *J. Geophys. Res.* **94**, 15635–15637.

Biegel, R.L., Sammis, C.G., and Dieterich, J.H. (1989). The frictional properties of a simulated gouge with fractal particle distribution. *J. Struct. Geol.* **11**, 827–836.

Blanpied, M.L. (1989). Friction Constitutive Behavior and Textural Evolution of Experimental Faults in Granite. Ph.D. thesis, Brown University.

Blanpied, M.L., Lockner, D.A., and Byerlee, J.D. (1991). Fault stability inferred from granite sliding experiments at hydrothermal conditions. *Geophys. Res. Let.* **18**, 609–612.

Brace, W.F. (1964). Brittle fracture of rocks. In *State of Stress in the Earth's Crust* (ed. W.J. Judd), pp. 111–178. Elsevier, New York.

Brace, W.F. and Byerlee, J.D. (1966). Stick–slip as a mechanism for earthquakes. *Science* **153**, 990–992.

Byerlee, J. and Summers, R. (1976). A note on the effect of fault gouge thickness on fault stability. *Int. J. Rock Mech. Min. Sci.* **13**, 35–36.

Cao, T. and Aki, K. (1986). Seismicity simulation with a rate and state dependent friction law. *Pure Appl. Geophys.* **124**, 487–513.

Carlson, J.M. and Langar, J.S. (1989). Properties of earthquakes generated by fault dynamics. *Phys. Rev. Let.* **62**, 2632–2635.

Dieterich, J. (1972). Time-dependent friction in rock. *J. Geophys. Res.* **77**, 3690–3697.

Dieterich, J.H. (1979). Modeling of rock friction 1. Experimental results and constitutive equations. *J. Geophys. Res.* **84**, 2161–2168.

Dieterich, J.H. (1981). Constitutive properties of faults with simulated gouge. In *Mechanical Behavior of Crustal Rocks* (ed. N.L. Carter, M. Friedman, J.M. Logan, and D.W. Stearns), pp. 102–120. American Geophysical Union Monograph No. 24.

Drucker, D.C. (1976). Elementary results of dimensional analysis for friction and wear in steady state sliding. *Wear* **40**, 129–133.

Engelder, J.T., Logan, J.M., and Handin, J. (1975). The sliding characteristics of sandstone on quartz fault-gouge. *Pure Appl. Geophys.* **113**, 68–86.

Fredrich, J.T. and Wong, T.-f. (1986). Micromechanics of thermally induced cracking in three crustal rocks. *J. Geophys. Res.* **91**, 12743–12764.

Gratier, J.P. and Gamond, J.F. (1990). Transition between seismic and aseismic deformation in the upper crust. In *Deformation Mechanisms, Rheology and Tectonics* (ed. R.J. Knipe and E.H. Rutter), pp. 461–473. Geological Society special publication No. 54.

Gu, J.-c., Rice, J.R., Ruina, A.L., and Tse, S.T. (1984). Slip motion and stability of a single degree of freedom elastic system with rate and state dependent friction. *J. Mech. Phys. Solids* **32**, 167–196.

Hobbs, B.E. (1990). Chaotic behavior of frictional shear instabilities. *Proc. 2nd Int. Conf. on Rock Burst and Seismicity in Mines*, in press.

Horowitz, F.G. and Ruina, A. (1989). Slip patterns in a spatially homogeneous fault model. *J. Geophys. Res.* **94**, 10279–10298.

Huang, J. and Turcotte, D.L. (1990) Are earthquakes an example of deterministic chaos? *Geophys. Res. Let.* **17**, 223–226.

Ito, K. and Matsuzaki, M. (1990). Earthquakes as self-organized critical phenomena. *J. Geophys. Res.* **95**, 6853–6860.

Kanamori, H. and Allen, C.R. (1986). Earthquake repeat time and average stress drop. In *Earthquake Source Mechanics* (ed. S. Das, J. Boatwright, and C.H. Scholz), pp. 227–236. American Geophysical Union Monograph No. 37.

Lim, S.C. and Ashby, M.F. (1987). Wear mechanism maps. *Acta Metall.* **35**, 1–24.

Marone, C., Raleigh, C.B., and Scholz, C.H. (1990). Frictional behavior and constitutive modeling of simulated fault gouge. *J. Geophys. Res.* **95**, 7007–7025.

Nussbaum, J. and Ruina, A. (1987). A two degree-of-freedom earthquake model with static/dynamic friction. *Pure Appl. Geophys.* **125**, 629–656.

Ohnaka, M. (1973a). Experimental studies of stick–slip and their application to the earthquake source mechanism. *J. Phys. Earth* **21**, 285–503.

Ohnaka, M. (1973b). The quantitative effect of hydrostatic confining pressure on the compressive strength of crystalline rocks. *J. Phys. Earth* **21**, 125–140.

Okubo, P. (1989). Dynamic rupture modeling with laboratory-based constitutive relations. *J. Geophys. Res.* **94**, 12321–12335.

Power, W.L. and Tullis, T.E. (1989). The relation between slickenside surfaces in fine-grained quartz and the seismic cycle. *J. Struct. Geol.* **11**, 879–893.

Rice, J.R. (1983). Constitutive relations for fault slip and earthquake instabilities. *Pure Appl. Geophys.* **121**, 443–475.

Rice, J.R. and Ruina, A.L. (1983). Stability of steady frictional slipping. *J. Appl. Mech.* **50**, 343–349.

Rice, J.R. and Tse, S.T. (1986). Dynamic motion of a single degree of freedom system following a rate and state dependent friction law. *J. Geophys. Res.* **91**, 521–530.

Robertson, E.C. (1982). Continuous formation of gouge and breccia during fault displacement. In *Issues in Rock Mechanics* (ed. R.E. Goodman and F.E. Heuze), Proc. 23rd Symp. Rock Mech. pp. 397–404.

Rudnicki, J.W. (1988). Physical models of earthquake instability and precursory processes. *Pure Appl. Geophys.* **126**, 531–554.

Ruina, A.L. (1981). Friction Laws and Instabilities: A Quasistatic Analysis of Some Dry Frictional Behavior. Ph.D. thesis, Brown University.

Rummel, F., Acheid, H.J., and John, C. (1978). Dilatancy and fracture-induced velocity changes in rock and their relation to frictional sliding. *Pure Appl. Geophys.* **116**, 743–764.

Scholz, C.H. (1987). Wear and gouge formation in brittle faulting. *Geology* **15**, 493–495.

Scholz, C.H. (1990). *The Mechanics of Earthquakes and Faulting.* Cambridge University Press.

Scholz, C.H., Molnar, P., and Johnson, T. (1972). Detailed studies of frictional sliding of granite and implications for the earthquake mechanism. *J. Geophys. Res.* **77**, 6392–6406.

Shaw, M.C. (1977). Dimensional analysis for wear systems. *Wear* **43**, 263–266.

Shimamoto, T., Handin, J., and Logan, J.M. (1980). Specimen-apparatus interaction during stick-slip in a triaxial compression machine: a decoupled two-degree-of-freedom model. *Tectonophysics* **67**, 175–205.

Sibson, R.H. (1986). Earthquakes and rock deformation in crustal fault zones. *Ann. Rev. Earth Plan. Sci.* **14**, 149–175.

Sykes, L.R. and Quittmeyer, R.C. (1981). Repeat time of great earthquakes along simple plate boundaries. In *Earthquake Prediction, An International Review* (ed. D. Simpson and P. Richards), pp. 217–247. American Geophysical Union.

Teufel, L.W. and Logan, J.M. (1978). Effect of displacement rate on the real area of contact and temperatures generated during frictional sliding of Tennessee sandstone. *Pure Appl. Geophys.* **116**, 840–865.

Tse, S.T. and Rice, J.R. (1986). Crustal earthquake instability in relation to the depth variation of frictional slip properties. *J. Geophys. Res.* **91**, 9452–9472.

Tullis, T.E. and Weeks, J.D. (1986). Constitutive behavior and stability of frictional sliding of granite. *Pure Appl. Geophys.* **124**, 383–414.

Weeks, J.D. and Tullis, T.E. (1985). Frictional sliding of dolomite: a variation in constitutive behavior. *J. Geophys. Res.* **90**, 7821–7826.

Wesnousky, S.G. (1988). Seismological and structural evolution of strike–slip faults. *Nature* **335**, 340–342.

Wesnousky, S.G. (1990). Seismicity as a function of cumulative geologic offset: some observations from southern California. *Bull. Seismol. Soc. Am.* **80**, 1374–1381.

Wong, T.-f. and Zhao, Y. (1990). Effects of load point velocity on frictional instability behavior. *Tectonophysics* **175**, 177–195.

Yanagidani, T., Wong, T.-f., and Gu, Y. (1989). Experimental observation of the dynamics of high pressure stick–slip instability. *EOS, Trans. Am. Geophys. Union* **70**, 1303.

Permeability and Fluid Flow in Rocks

Chapter 6
On the Measurement of Permeability in Anisotropic Rocks

Yves Bernabé
Chevron Oil Field Research Company, P.O. Box 446, La Habra, CA 90633-0446, U.S.A.

Abstract

The production of oil and gas in many reservoirs around the world is seriously affected by their highly heterogeneous and/or anisotropic structure. From the fluid flow standpoint, it is well accepted that heterogeneity and anisotropy are two closely related properties. Inhomogeneous materials are usually thought to appear homogeneous but anisotropic when considered at a scale much larger than the largest scale of heterogeneity. The purpose of the present chapter is to show that, even at large scales, difficulties may arise owing to the presence of impermeable boundaries. As an example, the steady-state flow configuration traditionally employed in laboratory measurements of permeability was analyzed using a 2-D finite difference scheme. The flow pattern in a two-dimensional rectangular sample was found to be significantly perturbed by the no-flow condition imposed by the impermeable jacket on the sides of the sample. The amplitude of the disturbance depended on the orientation of the sample axis with respect to the principal directions of the permeability tensor, the sample aspect ratio, and the ratio of the principal permeabilities. As a consequence, the permeability of anisotropic rocks can be significantly underestimated if traditional 1-D analysis is used. The results of the numerical simulations compared well with experimental data obtained for synthetic porous materials and an electrical analogue. To the author's knowledge, this constitutes the first experimental verification of the tensor form of Darcy's law in a highly anisotropic porous material. Furthermore, the numerical analysis used here did not depend on scale. Thus, the conclusions of this study are not limited to the laboratory scale but may be relevant to problems at larger scales. Finally, a novel method for measuring the full permeability tensor on a cubic rock sample is proposed.

FAULT MECHANICS AND TRANSPORT PROPERTIES OF ROCKS
ISBN 0-12-243780-2

1. Introduction

A quick glance at almost any outcrop should be sufficient to convince anyone of the strongly inhomogeneous character of crustal rocks. Furthermore, the vast amount of data obtained from well logs or new, emerging geophysical tools such as S-wave splitting (Crampin, 1985) suggests that most crustal rocks are also significantly anisotropic. The economical implications are important because major production problems occur in numerous oil and gas reservoirs with strongly heterogeneous and/or anisotropic structure (Lasseter et al., 1986).

The origin of anisotropy in rocks was recently discussed by Lake (1988) from the standpoint of the transport properties. One conclusion is that directionality of the pore structure (i.e., preferred orientation of microcracks or of nonspherical grains) can only produce moderate permeability anisotropy (for experimental evidence, see Rice et al., 1970). Strong permeability anisotropy is more likely to originate from fine-scale heterogeneities. Examples of highly anisotropic formations include sand–shale sequences (Haldorsen and Lake, 1984; Desbarats, 1987; Begg et al., 1989; Deutsch, 1989), eolian deposits (Lindquist, 1983; Goggin et al., 1988) and jointed or fractured rock masses (Long et al., 1982; Oda, 1985).

An important consequence is that anisotropy, like heterogeneity, is scale dependent (Dagan, 1986). In fact, scaling up permeability from the inch or foot scale (cores or well logs) to hundreds of feet (grid blocks in numerical simulations) has become a crucial task for the reservoir engineers (Haldorsen, 1986; Lake et al., 1989). Many techniques have been proposed to perform this task, more commonly known as 'determining grid block effective permeabilities.' Severe difficulties are often encountered, particularly in the case of multiphase flow (Hewett and Behrens, 1990). The fundamental equation of fluid flow in porous media, Darcy's law, is itself the result of such a scaling process, from pore to sample scale. The Stokes equations are transformed by volume averaging into a set of linear equations including Darcy's law. However, a difficulty remains – namely, to demonstrate that the coefficient of proportionality thus obtained (i.e., permeability) is a material property, that is, is independent of the boundary conditions. Similarly, to be useful, scaled-up (or effective) permeabilities should not depend on the boundary conditions. It is commonly believed that this requirement is automatically satisfied if the spatial domain considered (i.e., grid block) is much larger than the largest scale of heterogeneity contained in it, so that the material can be considered globally homogeneous (for example, see Begg et al., 1989).

The goal of the present chapter is to show that, even at scales exceeding the largest scale of heterogeneity, difficulties may arise in anisotropic materials because of the interference of impermeable boundaries with the flow. For this purpose, the steady-state flow of a single liquid phase through a rectangular, two-dimensional piece of homogeneous, anisotropic material was analyzed using a 2-D finite difference scheme (generalizing to three dimensions is in principle straightforward, although in practice the calculations become nearly intractable). This particular geometry was chosen because it corresponds to the configuration traditionally employed in laboratory measurements. Furthermore, the numerical analysis used here does not depend on scale. Therefore, the conclusions of this study are not limited to the laboratory scale but may be relevant to problems at larger scales. Many effective permeability studies make use of an identical geometry (Desbarats, 1987; White and Horne, 1987; Begg et al., 1989; Deutsch, 1989; Kasap and Lake, 1989; King, 1989; Sáez et al., 1989; Durlofsky and Chung, 1990). Also, impermeable boundaries such as sealing faults commonly divide reservoirs into sometimes quite small compartments (for example, see Smith, 1980). Special techniques are often required for interpreting well tests in the vicinity of such impermeable boundaries (Kuchuk and Kabir, 1988).

2. Background

2.1. Darcy's law

Darcy's law is most frequently formulated as follows (bold characters denote vectors and tensors, italic characters represent scalar quantities):

$$\mathbf{Q} = -\frac{k}{\eta}\nabla P \tag{1}$$

where $\mathbf{Q}$ is the macroscopic flux rate, that is the volume of fluid crossing unit area per unit time, k is the scalar permeability, η is the fluid viscosity, and P is the macroscopic fluid pressure. Equation (1) corresponds to one-dimensional flow in that $\mathbf{Q}$ and ∇P are colinear, and, consequently, only one space variable is needed. Clearly, eq. (1) only applies to isotropic porous materials since, by definition, k does not depend on the flow direction. An important experimental result is that eq. (1), the scalar (or 1-D) form of Darcy's law, has been observed to hold in isotropic media, provided that laminar flow conditions prevail, that is, at low Reynolds numbers.

In the case of anisotropic materials, a generalized form of Darcy's law has been suggested, the tensor form in which permeability is represented as a positive definite, symmetric tensor $\mathbf{k}$. The components of $\mathbf{Q}$ are then written as follows (the usual summation convention is used when indices are repeated):

$$Q_i = -\frac{k_{ij}}{\eta}\frac{\partial P}{\partial x_j} \tag{2}$$

where k_{ij} represents the components of the permeability tensor $\mathbf{k}$. Note that, in the rest of the chapter, the components of $\mathbf{k}$ in a given reference frame $\{x, y, z\}$ will be indicated by k_{xx}, k_{xy}, k_{yy}, and so forth. k_1, k_2, and k_3 will denote the principal permeabilities, with the convention, $k_1 > k_2 > k_3$.

Three main approaches have been used to derive eq. (2).

(1) A conceptual model of porous medium is assumed in which the Stokes equations are solved (Férrandon, 1948; Guin et al., 1971; Bear, 1972; Scheidegger, 1974; Neuman, 1977). In this approach, simple pore geometries, most frequently one-dimensional pores, are usually considered. Capillary models are the simplest examples but more sophisticated stochastic models may also be used. Averaging must then be performed on a so-called 'representative element volume' (or REV).

(2) Volume-averaging of the Stokes equations on an REV leads to a closure problem, the solution of which determines the permeability tensor (Slattery, 1972; Gray and O'Neill, 1976; Crapiste et al., 1986; Whitaker, 1986). In this case, the pore geometry need not be specified. However, the tensor $\mathbf{k}$ can be rigorously demonstrated to be a material property only if the pore structure is assumed spatially periodic. Because of this, the volume-averaging approach is equivalent to the third one described below (Bourgeat et al., 1988).

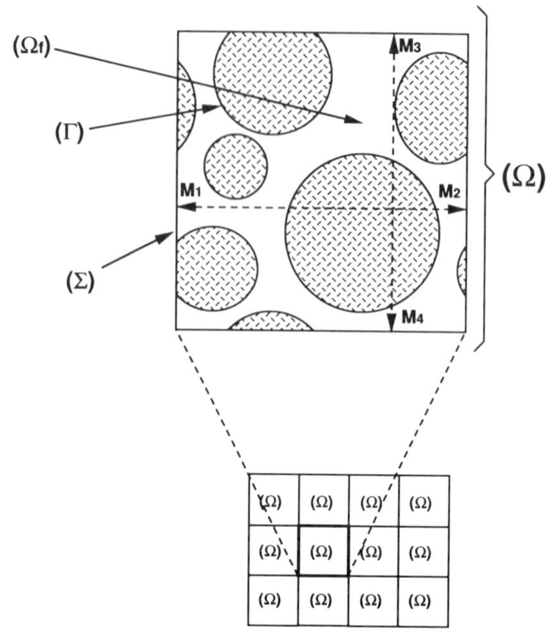

Figure 1. An idealized 2-D periodic porous medium. The unit cells are indicated by (Ω). (Ω_f) denotes the volume occupied by the fluid. (Γ) is the liquid–solid interface and (Σ) the external boundary of the cell. The periodic boundary condition on (Σ) simply imposes identical **v**'s at any given point on (Σ) and its corresponding point by a translation of one period (for example, $\mathbf{v}[M_1] = \mathbf{v}[M_2]$ and $\mathbf{v}[M_3] = \mathbf{v}[M_4]$).

(3) The third approach is based on an asymptotic method also known as homogenization (Ene and Sanchez-Palencia, 1975; Keller, 1980; Bourgeat, 1984; Mei and Auriault, 1989; Barrère et al., 1990). This approach is remarkably clear and concise although as general as any other. In order to introduce the details of the problem, a brief description largely inspired from Barrère et al. (1990) is given below.

It is first assumed that the porous medium considered contains only two scales of heterogeneity: the microscopic (or pore) scale, L_m, and the macroscopic scale, L_M. The ratio of these two scales, L_m/L_M, is much smaller than 1. Mei and Auriault (1989) recently attempted to generalize the method to an arbitrary number of scales. A second important hypothesis is that the material consists of a periodic arrangement of unit cells, as illustrated in Figure 1. Similar to the definition of an REV, the size of the unit cell, L_c, is assumed to be much larger than L_m and much smaller than L_M. The relevant scaling parameter is defined by $\varepsilon = L_c/L_M \ll 1$. The flow problem on a unit cell can be expressed in the following way:

$$\nabla p = \eta \nabla^2 \mathbf{v} \qquad \text{in } (\Omega_f) \tag{3a}$$

$$\nabla \cdot \mathbf{v} = 0 \qquad \text{in } (\Omega_f) \tag{3b}$$

$$\mathbf{v} = \mathbf{0} \qquad \text{on } (\Gamma),\ \mathbf{v} \text{ periodic on } (\Sigma) \tag{3c}$$

where **v** is the local velocity, p is the local pressure, (Ω_f) is the volume of the fluid phase in the unit cell, (Γ) is the fluid–solid interface, and (Σ) is the unit cell boundary (see

Figure 1). Equations (3a) and (3b) are the Stokes equation and the continuity equation, respectively. The boundary conditions (3c) consist of a no-flow condition on the internal surface (Γ) and a periodic condition on the external boundary. Here, 'periodic condition' simply means that $\mathbf{v}$ at any given point on (Σ) must be identical to $\mathbf{v}$ at the corresponding point by a translation of one period (see Figure 1). With ε a very small parameter, the following asymptotic expansions can be written:

$$\mathbf{v}(\mathbf{x}) = \varepsilon^2 \mathbf{v}_0(\mathbf{x}, \mathbf{x}/\varepsilon) + \varepsilon^3 \mathbf{v}_1(\mathbf{x}, \mathbf{x}/\varepsilon) + \cdots \tag{4}$$

$$p(\mathbf{x}) = p_0(\mathbf{x}) + \varepsilon p_1(\mathbf{x}, \mathbf{x}/\varepsilon) + \cdots \tag{5}$$

Thus, the macroscopic space variable, $\mathbf{x}$, and the microscopic space variable, $\mathbf{y} = \mathbf{x}/\varepsilon$, can be separated. $\mathbf{x}$ is associated with the smooth, 'regional' variations of the terms of the expansions (4) and (5), and $\mathbf{y}$ with their rapid, 'local' changes. The ultimate goal of the method is to eliminate $\mathbf{y}$. The operator $\boldsymbol{\nabla}$ is replaced by $\boldsymbol{\nabla}_x - 1/\varepsilon \boldsymbol{\nabla}_y$, where the subscripts x and y denote which space variable is being considered. Remembering that this procedure must be independent of ε approaching 0, the terms corresponding to different powers of ε can be examined separately. The lowest terms produce the following results: as written in (4) and (5), the expansion of $\mathbf{v}$ does not contain terms in ε^0 and ε^1, and p_0 does not depend on $\mathbf{y}$; $\mathbf{v}_0$ and $\mathbf{v}_1$ are related by $\boldsymbol{\nabla}_x \cdot \mathbf{v}_0 + \boldsymbol{\nabla}_y \cdot \mathbf{v}_1 = 0$ (note that, by averaging over the unit cell, this equation is transformed into the macroscopic continuity equation, $\boldsymbol{\nabla}_x \cdot \langle \mathbf{v}_0 \rangle = 0$); finally, the boundary value problem (3) becomes a local boundary value problem in which $\mathbf{x}$ is considered a parameter.

$$-\eta \nabla_y^2 \mathbf{v}_0 + \boldsymbol{\nabla}_y p_1 = -\boldsymbol{\nabla}_x p_0 \qquad \text{in } (\Omega_f) \tag{6a}$$

$$\boldsymbol{\nabla}_y \cdot \mathbf{v}_0 = 0 \qquad \text{in } (\Omega_f) \tag{6b}$$

$$\mathbf{v}_0 = \mathbf{0} \qquad \text{on } (\Gamma),\ (\mathbf{v}_0, p_1) \text{ periodic on } (\Sigma) \tag{6c}$$

The solutions of this problem are given by

$$\mathbf{v}_0(\mathbf{x}, \mathbf{y}) = \mathbf{A}(\mathbf{y}) \boldsymbol{\nabla}_x p_0(\mathbf{x}) \tag{7a}$$

$$p_1(\mathbf{x}, \mathbf{y}) = \mathbf{B}(\mathbf{y}) \boldsymbol{\nabla}_x p_0(\mathbf{x}) \tag{7b}$$

where $\mathbf{A}$ is a second-order tensor (it can be proved that $\mathbf{A}$ is symmetric) and $\mathbf{B}$ a vector. $\mathbf{A}$ and $\mathbf{B}$ are themselves solutions of a boundary value problem which no longer contains $\mathbf{x}$.

$$-\eta \nabla_y^2 A_{ij} + \frac{\partial B_j}{\partial y_i} = -\delta_{ij} \qquad \text{in } (\Omega_f) \tag{8a}$$

$$\frac{\partial A_{ij}}{\partial y_i} = 0 \qquad \text{in } (\Omega_f) \tag{8b}$$

$$A_{ij} = 0 \qquad \text{on } (\Gamma),\ (A_{ij}, B_i) \text{ periodic on } (\Sigma) \tag{8c}$$

Averaging eq. (7a) over the unit cell eliminates $\mathbf{y}$ and yields Darcy's law, $\langle \mathbf{v}_0 \rangle + \mathbf{k}/\eta \boldsymbol{\nabla} p_0 = \mathbf{0}$, where the permeability tensor $\mathbf{k}$ is equal to $-\eta \langle \mathbf{A} \rangle$. One can now see how important the periodicity hypothesis is. The 'weak' periodic boundary conditions on (Σ) have a small effect compared to that of the 'strong' no-flow conditions on (Γ). The term 'weak' refers here to the fact that neither p nor $\mathbf{v}$ is specified, while specific values of p or $\mathbf{v}$ (for example, $\mathbf{v} = \mathbf{0}$) are imposed by 'strong' boundary conditions. Thus, $\mathbf{A}$ and $\mathbf{k}$ depend principally on (Γ), or, in other words, on the pore geometry. Note, however, that the boundary value problem (8) is extremely difficult to solve in a unit cell with a realistic pore geometry. So far, solutions of similar boundary value problems are known only for very simple geometries, such as regular arrays of spheres or ellipsoids (Zick and Homsy, 1982; Meedoga et al., 1989).

One can iterate this procedure and apply it at the macroscopic level to Darcy's law and the continuity equation. Thus, effective permeabilities, independent of the 'external' boundary conditions can be calculated (Quintard and Whitaker, 1987; Mei and Auriault, 1989; Sáez et al., 1989; Durlofsky and Chung, 1990). Other attempts have been made to eliminate the effect of the boundary conditions on the effective permeability, but using less general methods (White and Horne, 1987; Kasap and Lake, 1989). Note that the requirement of independence on the 'external' boundary conditions is even more important in the macroscopic case than in the microscopic case because the macroscopic boundary value problem does not explicitly include a strong 'internal' boundary condition such as no flow at the pore-solid interface.

2.2. Experimental data and interpretation

A wide survey of the literature suggested that there was, at present, no adequate experimental verification of the tensor form of Darcy's law in highly anisotropic materials. In past experimental studies, the data gathered were often insufficient to determine fully the permeability tensor. Usually, permeability has been measured along two mutually perpendicular directions, traditionally vertical and horizontal or parallel and perpendicular to the bedding (for example, see von Ruhl and Schmid, 1957, or Greenkorn et al., 1964). In these and many similar studies, the ratio of the permeabilities in mutually perpendicular directions was found to be between 1 and 0.3, sometimes as low as 0.1. Exceptionally small ratios were observed by Auzerais et al. (1990). Johnson and Hughes (1948) systematically measured permeability as a function of the azimuth in an oriented hollow core. They observed permeability ratios of the order of 0.5. Each directional measurement was performed using a different portion of the core and it is unclear whether the changes of permeability observed are due to directionality or merely to inhomogeneity. To palliate this problem, it seems preferable in natural materials to measure the full permeability tensor on a single sample. A possible procedure is proposed in Section 4.

Johnson and Hughes' results were interpreted by Scheidegger (1954) using the so-called 'directional permeabilities' which are obtained by 'projecting' Darcy's law on the axis of the sample. However, Case (1971) demonstrated that the tensor form of Darcy's law is incompatible with such a projection. The directional permeabilities were used in the past mainly because it was believed that the so-called 'ellipse of permeability' could be constructed by plotting their square root (or the inverse of their square root) in a polar coordinate system (Scheidegger, 1954). Marcus and Evenson (1961) showed that it can successfully be done only when the measurements are performed in directions parallel to either ∇P or $\mathbf{Q}$. In these two cases, eq. (2) (or its inverse) can be treated as a scalar equation. Unfortunately, measurements verifying the requirements above can only be achieved in infinitely thin disklike or needlelike samples (Marcus and Evenson, 1961). More importantly, the very concept of directional permeability implicitly assumes that the stream lines are straight and parallel. This is certainly not true in inhomogeneous materials and, as will be shown later, impermeable boundaries may produce severe distortion of the stream lines in homogeneous, anisotropic media. One notable exception occurs, of course, when the flow direction is parallel to one of the principal permeability directions. In summary, the early works mentioned above attempted to capture a fundamentally multidimensional phenomenon within a one-dimensional framework. This approach may appear to succeed only for approximately isotropic, homogeneous materials. For instance, as an exercise, one can calculate the square roots and inverse square roots of k_{xx} and $k_{xx}^{(-1)}$ for a mildly anisotropic porous medium and plot them as a function of θ in a polar coordinate system. If reasonable (i.e., $\pm 5\%$), random errors are added, all the curves plotted will look approximately elliptic.

3. Finite-difference Simulation

3.1. Description

In conventional laboratory measurements, rock samples are shaped into right cylinders with a circular cross-section (in two-dimensions, they can simply be represented as rectangles). A steady-state flow is forced through the sample by applying constant pressures P and $P + \delta P$ on opposite faces. The sides of the plug are jacketed with impermeable material. Transient or periodic flow methods can also be employed but are not considered here, although new and interesting work can be cited in this area (for instance, Kamath et al., 1990, cleverly applied transient flow methods to quantify heterogeneity in laboratory samples).

The equation of single-phase flow in a two-dimensional anisotropic, homogeneous, porous medium can be derived from Darcy's law and the conservation of mass as shown below:

$$k_{xx}\frac{\partial^2 P}{\partial x^2} + 2k_{xy}\frac{\partial^2 P}{\partial x\,\partial y} + k_{yy}\frac{\partial^2 P}{\partial y^2} = \eta\beta_S\frac{\partial P}{\partial t} \tag{9}$$

where β_S is the effective fluid storage capacity, that is, the volume of fluid which must be injected into or released from a unit volume of rock to cause a unit pressure change, and t is the time. The boundary and initial conditions corresponding to 2-D steady-state permeability measurements are given by:

$$P\Big|_{x=0} = P_0 \qquad \forall t > 0 \tag{10a}$$

$$P\Big|_{x=L} = P_0 + \delta P \qquad \forall t > 0 \tag{10b}$$

$$k_{xy}\frac{\partial P}{\partial x}\Big|_{y=0} + k_{yy}\frac{\partial P}{\partial y}\Big|_{y=0} = k_{xy}\frac{\partial P}{\partial x}\Big|_{y=W} + k_{yy}\frac{\partial P}{\partial y}\Big|_{y=W} = 0 \tag{10c}$$

$$P\Big|_{t=0} = P_0 \qquad \forall x \tag{10d}$$

where L and W are the sample length and width, respectively, P_0 is the initial pore pressure inside the sample, and δP is the pressure difference suddenly applied across the sample at $t = 0$ and kept constant afterward. If the sample is represented by a rectangular grid with square grid blocks, an approximate finite-difference solution is obtained on this grid by replacing the space partial derivatives by explicit central finite differences and the time derivative by an explicit forward difference. The ratio,

$\tau k_{max}/\xi^2 \eta \beta_S$, where ξ is the grid block size and τ is the time step, must be taken lower than 1/6, a value which ensures computational stability in the isotropic case. The simulation is performed by imposing the appropriate initial and boundary conditions, and then allowing the system to relax, eventually reaching a steady state. In order to determine when the steady state is attained, the total flux rate in the x direction is calculated at various positions along the sample. Initially, the flux rate is very high at $x = L$ and nearly zero everywhere else. The difference gradually decreases with time until the flux rate becomes constant over the entire sample. At that time, the apparent permeability can be calculated as the ratio of the flux rate by the global pressure gradient, $\delta P/L$. In this form, the calculations are highly time consuming. But the computation time can be reduced significantly by using the same scheme with adequately chosen initial pressure fields. For instance, the equilibrium pressure field of a previous simulation can be used as the initial state for another, provided that L and W are the same for both simulations. The main advantage of this method is that it can also be used with transient or sinusoidal flow conditions.

3.2. Results

As expected, the finite-difference scheme appeared to be stable when isotropic materials were considered (i.e., $k_{xx} = k_{yy}$, $k_{xy} = 0$), even for grids as small as 10×10. However, numerical instability developed for anisotropic materials as the ratio of the principal permeabilities, k_2/k_1, and the angle of the principal directions with the sample axis, θ, were decreased. For k_2/k_1 smaller than 0.03, it was necessary to lower greatly both the time step and grid block size. For $\theta = 15^\circ$ and $k_2/k_1 = 0.01$, the grid needed so much refinement that the simulation could not be run in a reasonable amount of time.

Square samples were first considered. The angle θ was varied from 0 to 90° by increments of 15° and the permeability ratio ranged from 0.01 to 1. The apparent permeabilities normalized to the values of the xx component of the permeability tensor, $k_{xx} = k_1 \cos^2 \theta + k_2 \sin^2 \theta$, are plotted in Figure 2. In the case of a vanishing permeability ratio (i.e., $k_2 = 0$), the normalized apparent permeability is equal to $1 - \tan \theta$ for $0^\circ \leq \theta \leq 45^\circ$, 0 for $45^\circ \leq \theta < 90^\circ$, and 1 for $\theta = 90^\circ$ (see Figure 2). As expected, normalized apparent permeabilities equal to unity were obtained in two cases: for isotropic materials (i.e., $k_2/k_1 = 1$), and when the measurements were made along the principal directions (i.e., $\theta = 0$ and 90°). However, the apparent permeability was underestimated with respect to k_{xx} by up to one order of magnitude at angles near 45° and low permeability ratios. The steady-state pressure fields corresponding to these extreme cases were severely distorted (see Figure 3). Note that, for permeability ratios higher than 0.3, the discrepancy was not greater than 20%, a typical value of the experimental error in permeability measurements. Thus, the effect of the impermeable jacket can be neglected in rocks with moderate anisotropy, and the 1-D form of Darcy's law can be used with k_{xx} taken as the scalar permeability. After considering square samples, simulations were run on samples with aspect ratios other than 1. The apparent permeabilities normalized to k_1 are plotted in Figure 4. As expected from the discussion in Section 2.2., the values approached $k_{xx} = k_1 \cos^2 \theta + k_2 \sin^2 \theta$ when the aspect ratio vanished (i.e., disklike sample) and the inverse of $k_{xx}^{(-1)} = (1/k_1) \cos^2 \theta + (1/k_2) \sin^2 \theta$ when the aspect ratio became infinite (i.e., needlelike sample).

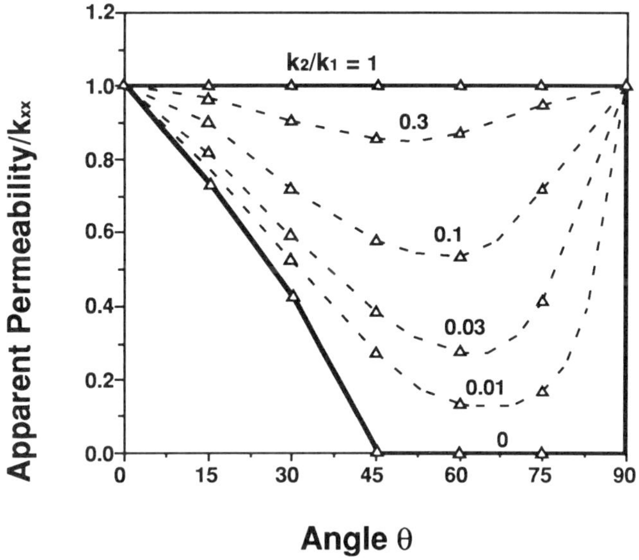

Figure 2. The apparent permeability of a square sample, normalized to k_{xx}, as a function of θ. The data points are connected by visually smoothed dashed lines corresponding to different permeability ratios as indicated above each line. The thick broken lines represent the two limiting cases: $k_2/k_1 = 1$ (i.e., isotropic), and $k_2/k_1 = 0$.

3.3. Comparison with experimental results

3.3.1. Electrical analog

Marcus and Evenson (1961) performed physical simulations of single-phase flow in a 2-D homogeneous, anisotropic material using an electrical analog consisting of sheets of conductive paper in which regularly spaced, parallel slits had been cut. This arrangement was meant to imitate a sand body with thin, discontinuous shale layers. The conductivity ratio (analogous to the permeability ratio k_2/k_1) depended on the length and spacing of the cuts. Conductivity measurements were made with different sample orientations, aspect ratios, and conductivity ratios. For comparison, finite-difference simulations were run under conditions matching those of Marcus and Evenson. The calculated apparent permeabilities normalized to k_1 are shown in Figures 5a–e as a function of k_2/k_1. To facilitate the comparison, the straight lines corresponding to $k_{xx}/k_1 = \cos^2\theta + (k_2/k_1)\sin^2\theta$ were also plotted. At first, it seems that k_{xx}/k_1 fits Marcus and Evenson's data better than the finite-difference results. But careful examination shows that, within the experimental scatter, the data points do not lie on straight lines but rather comprise curved lines which are not tangent to the k_{xx}/k_1 line at the point (1, 1) as they should be. This is probably due to uncertainty in Marcus and Evenson's values for the angle θ. As a matter of fact, they did not measure the principal permeability directions directly but inferred them from the direction of the cuts. This led to slight errors because the paper was itself mildly anisotropic and the amount and direction of anisotropy were unknown. Using only the data points nearest to the point (1, 1), that is, verifying $k_2/k_1 > 0.75$, the values of θ giving the best fit tangents were computed. The corresponding k_{xx}/k_1 lines are indicated by dashed lines in Figure 5a to e.

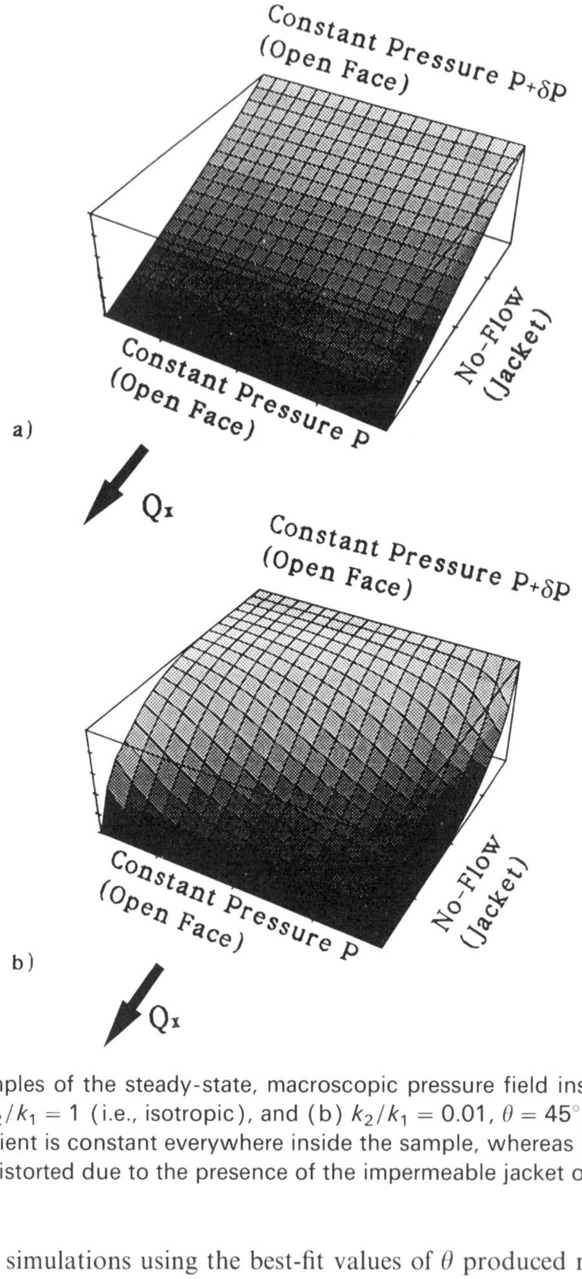

Figure 3. Examples of the steady-state, macroscopic pressure field inside the sample for two cases: (a) $k_2/k_1 = 1$ (i.e., isotropic), and (b) $k_2/k_1 = 0.01$, $\theta = 45^\circ$. Notice that in (a) the pressure gradient is constant everywhere inside the sample, whereas in (b) the pressure field is strongly distorted due to the presence of the impermeable jacket on both sides of the sample.

Finite-difference simulations using the best-fit values of θ produced results in excellent agreement with Marcus and Evenson's data. Finally, their results for various aspect ratios are indicated in Figure 4 for comparison with the finite-difference simulations. The agreement is very good in the case where $k_2/k_1 = 0.3$, while the poorer quality of the fit when $k_2/k_1 = 0.6$ may again be due to uncertainty in θ. In summary, these results appear satisfactory but not conclusive, mainly because the range of permeability ratios covered was not wide enough. Also, a conclusive experimental verification would preferably involve real fluid flow through an anisotropic porous medium rather than an electrical analog.

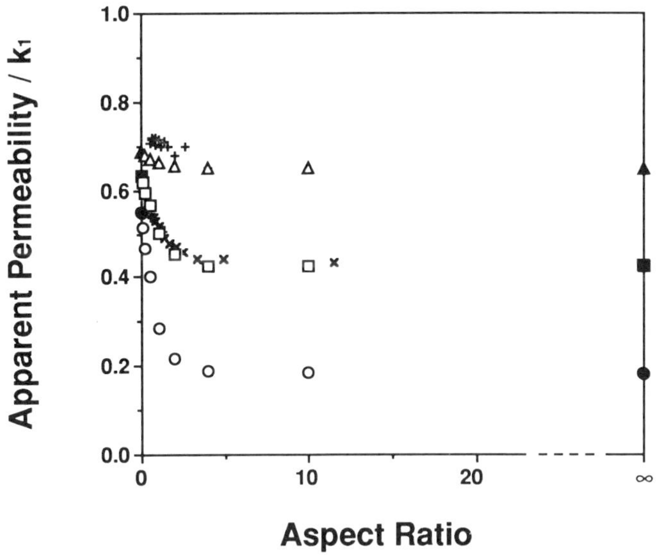

Figure 4. The apparent permeability normalized to k_1 for samples with different aspect ratios L/W. Triangles correspond to $k_2/k_1 = 0.6$, $\theta = 60^\circ$; squares to $k_2/k_1 = 0.3$, $\theta = 45^\circ$; and circles to $k_2/k_1 = 0.1$, $\theta = 45^\circ$. Experimental results by Marcus and Evenson (1961) are also plotted for comparison ($+$: $k_2/k_1 = 0.6$, $\theta = 60^\circ$; $\times$: $k_2/k_1 = 0.3$, $\theta = 45^\circ$). The solid symbols indicate the theoretical values for zero and infinite aspect ratios (i.e., for infinitely thin disklike and needlelike samples, respectively).

3.3.2. Synthetic anisotropic material

For the reasons stated above, it was decided to carry out a suite of permeability measurements on a highly anisotropic porous material. For that purpose, a well-controlled, reproducible, synthetic material seemed more appropriate than natural rocks. A large piece of synthetic porous material was prepared by stacking 42 flat porous plates of identical properties and dimensions. The plates were 15.2 cm long, 2.54 cm wide, and 1.5 mm thick. It was hoped that the free space between two plates would act as a high-permeability channel, analogous to a fracture. The stacked plates were held in place by means of a layer of hard epoxy on the sides, as shown in Figure 6.

A glass manufacturer was requested to prepare the plates from isotropic, homogeneous sintered glass. The principal permeability directions of the stack could therefore be assumed to be parallel and perpendicular to the plates, as illustrated in Figure 6. Notice that cutting, straightening, and the other processes applied to the plates are not expected to change the principal permeability directions, but rather to reinforce them. Several cubic samples, 2.54 cm in length, were cut at 0, 15°, 30°, and 45° with respect to the assumed k_1 direction. Note that, in the reference frame $\{x, y, z\}$, corresponding to the principal permeabilities k_1, k_2, and k_3, the flow equations do not depend on z and the problem can be considered two-dimensional. The apparent permeability of each sample was measured and it was found that a permeability ratio of only 0.6 had been obtained because the plates were too smooth to prevent the 'fractures' from closing under the small confining pressure needed during a permeability measurement.

In a second attempt, one or two lengths of 0.16-cm wide drawing tape were mounted on the top faces of identical sintered glass plates, alternatively at the center and along

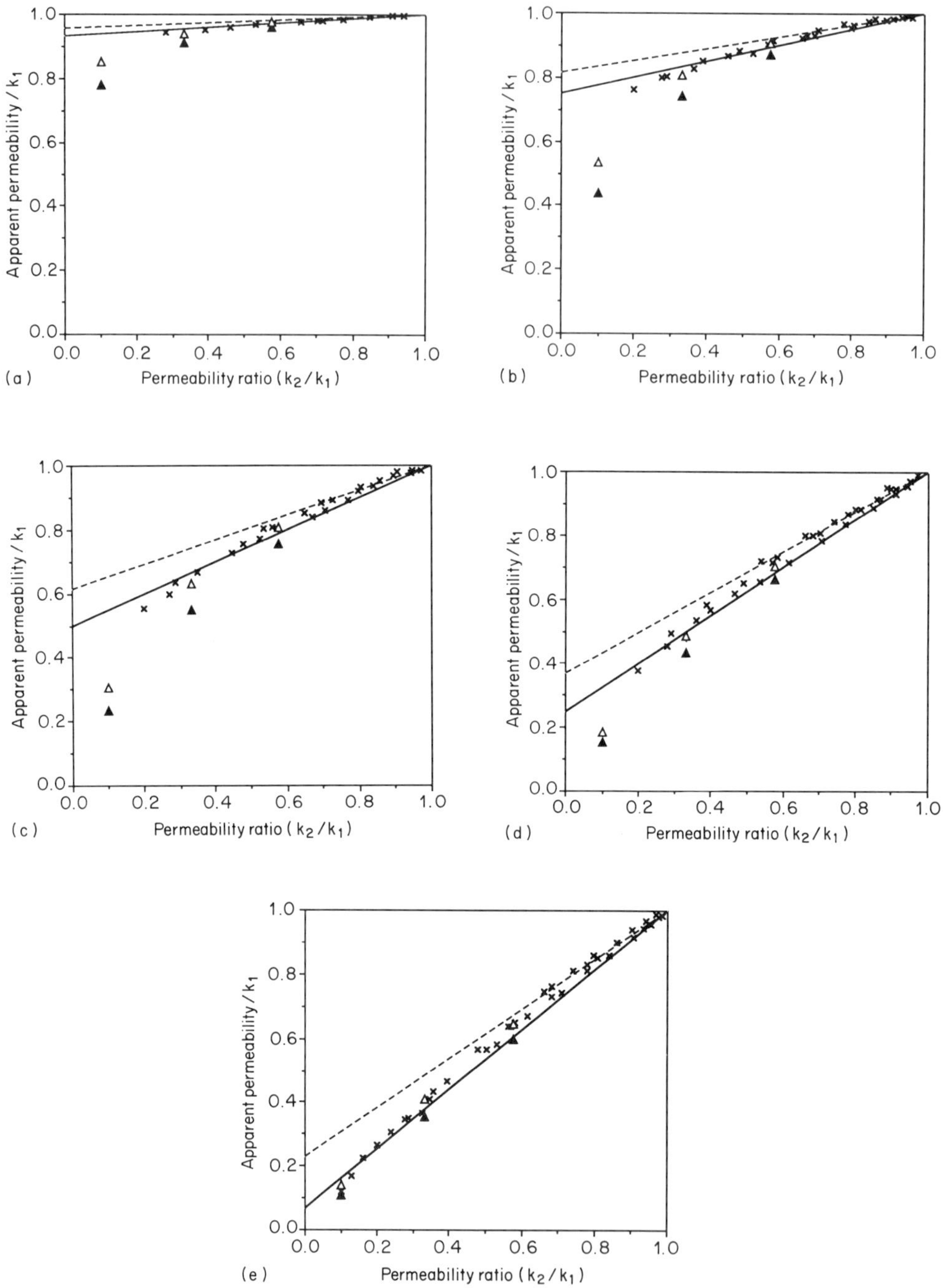

Figure 5. Comparison of Marcus and Evenson's data ($\times$'s) and the results of the present finite-difference simulations (triangles). The sample aspect ratio was 1.75 in all cases. The two straight lines represent k_{xx}/k_1 as a function of k_2/k_1 for the nominal and best-fit values of θ respectively (see explanations in the text). Solid lines and solid triangles denote the nominal values, whereas dashed lines and open triangles indicate the best-fit values: (a) 15°, 12°; (b) 30°, 25.4°; (c) 45°, 38.4°; (d) 60°, 52.6°; (e) 75°, 61.4°.

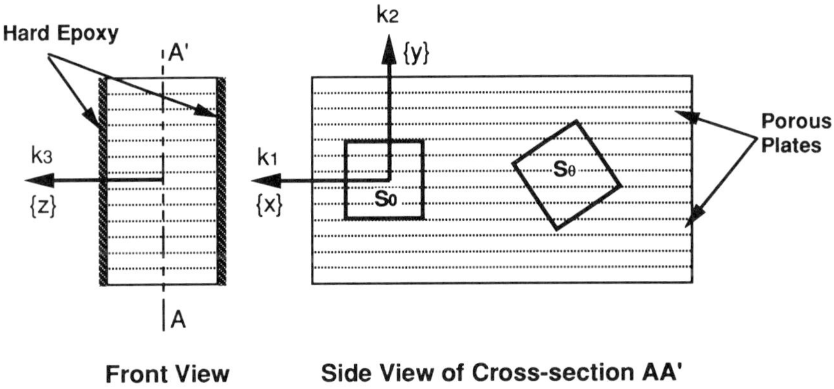

Figure 6. An illustration of the moderately anisotropic material with $k_2/k_1 = 0.6$. It was prepared by stacking sintered glass plates (see text). The principal directions of the permeability tensor are indicated by thick arrows. Examples of cubic samples cut at various angles to k_1 and k_2 are also drawn.

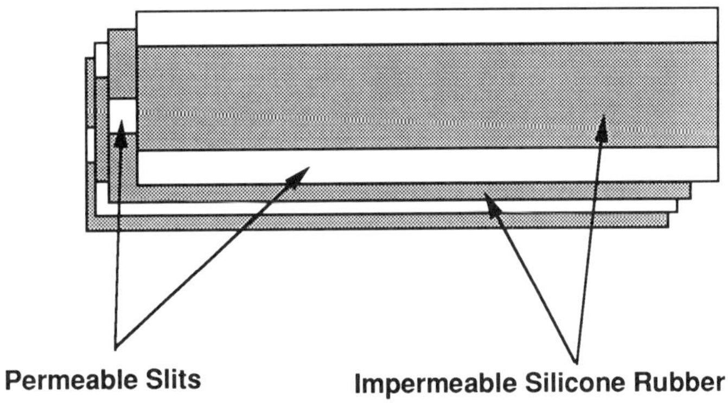

Figure 7. An illustration of the strongly anisotropic material with $k_2/k_1 = 0.02$. It was prepared by stacking sintered glass plates partially coated with impermeable silicone rubber (see text).

the sides. The plates were then coated on the same face with a thin layer of impermeable silicone rubber. After curing, the tape was removed, leaving open slits, as shown in Figure 7. Forty-two coated plates were stacked as before, producing a material with a greatly reduced permeability across the plates. Note that this material imitates periodic, discontinuous sand–shale sequences. This time, a permeability ratio of 0.02 was achieved. Each cubic sample contained from 17 to 25 plates depending on the orientation, a number large enough to ensure that the macroscopic properties did not vary significantly from sample to sample. Nevertheless, the homogeneity of the material was verified by cutting pairs of samples for each orientation. For each pair, the measured permeabilities were generally found to agree with 10%.

The experimental setting shown in Figure 8 was used for the permeability measurements. The sample was installed in a pressure vessel and subjected to a small confining pressure of about 1 MPa to prevent bypass flow between the jacket and the sample. A rough vacuum was applied to the sample for a few minutes in order to remove

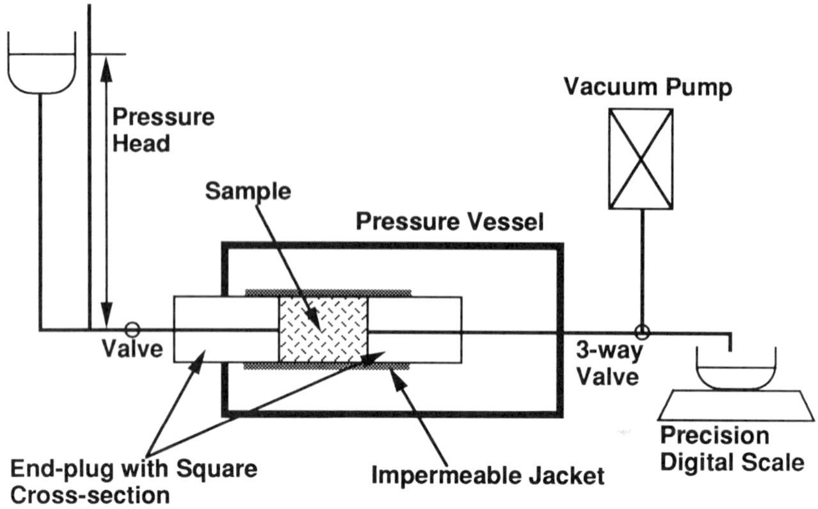

Figure 8. The experimental setup used for steady-state permeability measurements.

the air contained in it. The valves were opened and the pore fluid, a mixture of distilled water and glycerol, was let free to flow through the sample. After about 15 minutes the sample was considered saturated and the permeability measurement was performed. The pressure difference across the sample, δP, was simply taken to be the average of the two pressure heads read at the beginning and the end of each test. The tests were designed so as to minimize the changes in pressure head. The mass of fluid flowing through the sample during the duration of the test was measured using a precision digital scale. Knowing the test duration and the fluid density, the average flow rate was evaluated. Note that, since the samples were very permeable, the pressure drop in the system itself was not negligible. It was separately measured for various flow rates using the same experimental setting, but with no sample. A correction for δP, linearly related to Q, was thus determined.

For each measurement, the procedure was repeated for three different values of δP. In a plot of Q versus δP, the three data points were always observed to fall very close to a straight line passing through the origin. The 1-D form of Darcy's law could thus be used to calculate an apparent permeability. The fluid viscosity was adjusted by changing the concentration of glycerol in order to ensure maximum precision in all cases. The viscosity and density of the pore fluid were measured twice daily during the course of the study and every time the fluid was changed. With this precaution, the main source of uncertainty was reading the fluid level in the column (± 0.2 mm). Overall, the uncertainty was estimated to be on the order of 5%. In addition, the reproducibility was tested and found to be on the order of 10%. This discrepancy probably reflects variations in confining pressure, which was difficult to control at such a low level, and in initial saturation.

For most samples, the measurements were conducted while cycling the confining pressure. The apparent permeability decreased by a maximum of 15% when the confining pressure was increased from roughly 1 to 4 MPa. A slight hysteresis was generally observed during the unloading cycles. Since these effects were of the same order of magnitude as the experimental uncertainty, all the data were grouped together regardless of the confining pressure. The total scatter of the data points was 15%.

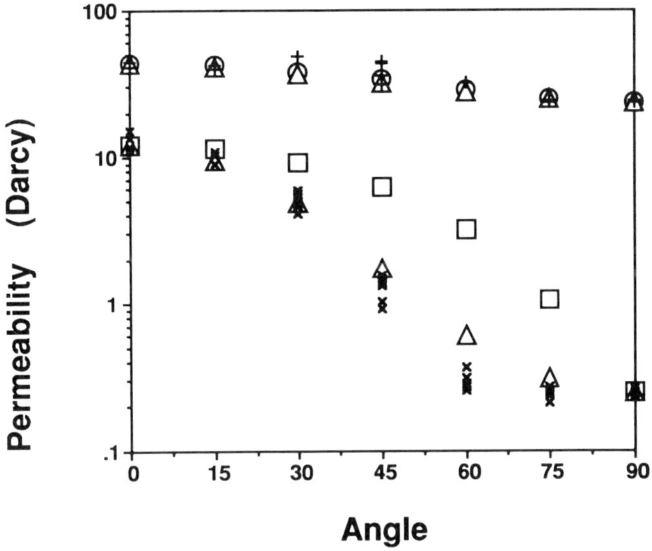

Figure 9. Comparison of the permeability measurements for the two synthetic materials ($+: k_2/k_1 = 0.6$; $\times: k_2/k_1 = 0.02$) with the apparent permeabilities from the finite-difference simulations (triangles). Part of the scatter of the experimental data points is due to variable confining pressures. The corresponding k_{xx}'s are also plotted (circles and squares, respectively).

Assuming that the principal permeabilities k_1 and k_2 for the two synthetic materials were given by the mean apparent permeabilities for $\theta = 0°$ and $90°$, numerical simulations of the experiments were run. The experimental data and the results of the simulations are compared in Figure 9. In addition, the theoretical values of the xx component of the permeability tensor, $k_{xx} = k_1 \cos^2 \theta + k_2 \sin^2 \theta$, were also plotted in Figure 9. In this way, the effect of the impermeable jacket can be visually appraised. Clearly, in the moderately anisotropic medium, the effect of the impermeable jacket was not significant. In contrast, in the highly anisotropic material, k_{xx} was up to one order of magnitude greater than the apparent permeabilities observed. This confirms that, in anisotropic media, fluid flow is fundamentally three-dimensional and a simple 1-D analysis is unlikely to be adequate.

The agreement between the experimental data and the numerical simulations is generally good. This observation demonstrates that macroscopic flow rates can be accurately predicted from Darcy's law even in media with a very low k_2/k_1, provided that the boundary conditions have been appropriately taken into account. To my knowledge, this constitutes the first experimental verification of the tensor form of Darcy's law in an anisotropic material with a permeability ratio as low as 0.02.

4. A New Experimental Procedure for Measuring the Permeability Tensor

Besides the technique of Johnson and Hughes (1948), a number of methods have been proposed to measure the full permeability tensor in rock or soil. An interesting attempt was made by Fontugne (1969) on prepared soils. He performed two flow measurements simultaneously, with the fluid outlets aligned with the assumed principal directions. He used numerical analysis to determine the ratio of the principal permeabilities from the observed flow rates. The amplitude of the principal permeabilities were measured separately. Rose (1982) designed an experimental procedure to force the stream lines in a permeability measurement to be straight lines parallel to the sample axis. However, Rose's method involves trial-and-error reshaping of the sample, which makes it rather impractical. Adams et al. (1988) inferred the directions and amplitudes of the principal permeabilities from the approximately elliptic shape of the radial encroachment of a viscous fluid in the porous medium. Young (1989) used a minipermeameter equipped with an elliptic tip seal. This technique is particularly remarkable in that it can be utilized on outcrops. However, the measurements are sometimes difficult to interpret because of the extremely complex flow geometry.

Note that none of the methods cited above, except perhaps the last, is capable of accommodating the most general three-dimensional cases. They all assume that the sample axis can be oriented parallel (or perpendicular) to one of the principal directions. In practical applications, this is not too serious a restriction, however, because in many rocks of interest at least one of the principal directions can be guessed from the bedding planes, preferred orientation of microcracks, and the like. It is interesting to remark that the commonly used concept of horizontal and vertical permeabilities implicitly assumes that the principal permeability directions in-situ are horizontal and vertical. This is not always true, however, and in practice plugs should be taken in directions parallel and perpendicular to the bedding planes when visible, rather than parallel and perpendicular to the axis of the core, as is often done.

If the principal directions cannot be guessed, a different procedure becomes necessary. As stated earlier, it is desirable to measure the full permeability tensor in a single sample. The ideal solution would be to impose periodic boundary conditions as in Quintard and Whitaker (1987), Sáez et al. (1989), Mei and Auriault (1989), and Durlofsky and Chung (1990). Unfortunately, this is very difficult to implement in practice. Alternatively, it may be possible to devise a suite of 'independent' flow measurements, each one with a different set of Neuman and Dirichlet boundary conditions which can easily be applied in the laboratory. This method may yield enough information to allow one to infer the full permeability tensor. A similar idea was applied by White and Horne (1987) in a different context.

A possible procedure is sketched in Figure 10 in the two-dimensional case (here again, generalization to three dimensions is theoretically straightforward), involving six steady-state measurements (15 in 3-D). Among them, two are conventional, longitudinal measurements (3 in 3-D), and the rest 'diagonal' measurements. Note that, in order to avoid problems at the corners of the sample, the fluid is not allowed to flow through the entire face of the sample but only through a 'window,' as seen in Figure 10. Some of the diagonal measurements will produce identical results if the material is perfectly homogeneous (specifically, $Q_3 = Q_5$ and $Q_4 = Q_6$). However, natural rocks are likely to show some degree of inhomogeneity, and redundant measurements may be useful to detect unacceptable heterogeneity levels.

To infer the permeability tensor from the flow measurements data, that is, the observed Q's and δP's, a trial-and-error method can be used. The steady-state flow equations are solved numerically for each configuration using trial values of the three components of $\mathbf{k}$, k_{xx}, k_{xy}, and k_{yy}. The calculated flow rates are compared to the

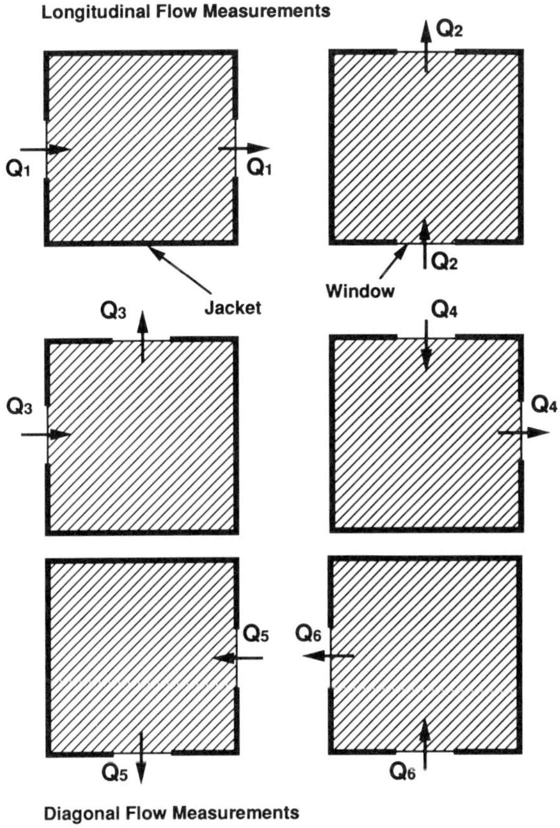

Figure 10. A two-dimensional illustration of the different flow configurations ('longitudinal' and 'diagonal') used for determining the full permeability tensor in a single cubic sample. The sample has the same orientation in all cases.

observed ones and the variance is computed, $\Sigma \mathbf{w}_i \{Q_i^{\text{calc}} - Q_i^{\text{obs}}\}^2$, where i varies from 1 to 6. The discrete weighting function $\mathbf{w}_i$ is introduced to compensate for the redundancy of some of the diagonal measurements. In this study, several functions were tried but no major change in the results was observed. A standard optimization scheme can then be used to converge toward the point of minimum variance (Fox, 1971). However, it should be remembered that, in complex situations such as this one, optimization methods may fail to find the absolute minima.

In order to test the viability of the method described above, I asked L.J. Durlofsky to run fine-grid simulations of the flow configurations shown in Figure 10 with the appropriate boundary conditions, using his well-validated finite-element code (Durlofsky and Chung, 1990). He was to provide the input data for the optimization calculations, that is, the flow rates computed by his program, but the values of **k** he used were not to be revealed. The goal was to imitate real measurement conditions and see whether the values of **k** used by Durlofsky could be retrieved with an acceptable accuracy.

In any optimization scheme, the forward flow problem has to be solved a large number of times. Too much grid refinement would therefore make the proposed method impractical. It is even felt that the application of the proposed method to the laboratory would be very limited unless the code could run on a desktop computer. Thus, a coarse grid with triangular elements was selected. A mixed finite-element scheme was used to solve the steady-state flow equations on that grid. In this kind of formulation, Darcy's law and the continuity equation are considered separately. P and $\mathbf{Q}$ are treated as independent variables. Mixed finite-element methods yield more accurate flow rates than regular finite-element schemes (Thomasset, 1981). They also allow easy implementation of relatively complex boundary conditions such as the ones in Figure 10. It was found, however, that for a given $\mathbf{k}$ the mixed finite-element code used here overestimated the flow rates slightly, probably because of the coarseness of the grid. Thus, the code was first empirically calibrated by introducing a correction factor slightly lower than unity. No further adjustments of the correction factor were allowed afterwards. It is recognized that this empirical procedure is not satisfactory, but the only alternative was to devise a more sophisticated finite-element program, capable of greater grid refinement, which would have been difficult to do in a limited amount of time.

Durlofsky's fine-grid simulations were run on six cases corresponding to different homogeneous materials (one was isotropic) with arbitrary orientations with respect to the principal directions of the permeability tensor. In general, the variance appeared to be a smooth function of k_{xx}, k_{xy}, and k_{yy}, although relative minima were encountered in two cases. It was nevertheless possible to retrieve $\mathbf{k}$ in all cases. The accuracy varied from 10% for the largest tensor components to 30% for the lowest. The error on the orientation of the principal directions was no greater than 5°. In addition, an inhomogeneous sample with an approximately fractal distribution of permeability was also tested. The permeability tensor was found within 20% of the effective $\mathbf{k}$ as determined with Durlofsky and Chung's (1990) technique. These results are considered very promising, especially since there is so much room left for improving resolution and accuracy. For instance, the grid can be significantly refined if faster and more powerful workstations are utilized. However, the theoretical problem of the uniqueness of the solution has not been tackled yet.

5. Conclusions

In this study, an experimental verification of the tensor form of Darcy's law was, for the first time, reported in a highly anisotropic material. In my opinion, previous works were not conclusive for the following reasons. Most authors considered natural rocks or soils which, in general, exhibit a spatial variability large enough to obscure the experimental results. Also, at the laboratory scale, most materials considered in past studies were not sufficiently anisotropic to permit discriminating tests to be conducted. In this work, the effect of the impermeable jacket on the apparent permeability was clearly observed because the material employed had a permeability ratio as low as 0.02. Finally, in many previous studies, the experimental data were interpreted within an inadequate one-dimensional framework.

Another important consequence of this work is that traditional 1-D analysis of laboratory measurements may lead to significant underestimate of permeability in anisotropic media even if the material is homogeneous at the sample scale. Again, this effect can be neglected in moderately anisotropic materials (as a rule of thumb, rocks with a permeability ratio k_2/k_1 higher than 0.3 should qualify as 'moderately anisotropic'). Also, it is important to realize that the absolute scale is less relevant to this problem than the 'aspect ratio.'

Lastly, the results of this study led to the elaboration of a novel method for measuring the full permeability tensor in a single cubic sample. Preliminary tests on computer-generated data are considered promising, although it is recognized that more work is required. From the theoretical standpoint, the uniqueness of the solution needs to be demonstrated. Also, for practical applications, a numerical scheme with 3-D capability will have to be incorporated. In the future, it will also be interesting to investigate the effect of anisotropy on transient and periodic flow.

Acknowledgments

I am grateful to the management of Chevron Oil Field Research Company for giving permission to publish this work; to Louis Durlofsky for his help and many enlightening discussions; and to John Hayes and Doug Fryer, who greatly helped in fabricating the synthetic materials and performing the permeability measurements. Carolyn Morrow and Mike Batzle critically reviewed this paper. Finally, I want to take advantage of this opportunity to express again my gratitude to Bill Brace for a wonderful time at M.I.T. learning from him.

References

Adams, Z.K.L., Russel, W.B., and Rebenfeld, L. (1988). Radial penetration of a viscous liquid into a planar anisotropic porous medium, *Int. J. Multiphase Flow* **14**, 203–215.

Auzerais, F.M., Ellis, D.V., Luthi, S.M., Dussan, E.B., and Pinoteau, B.J. (1990). Laboratory characterization of anisotropic rocks. SPE paper 20602 presented at the 65th. SPE Ann. Tech. Conf., New Orleans.

Barrère, J., Caltagirone, J.-P., and Gipouloux, O. (1990). Détermination numérique de la perméabilité en milieu poreux périodique tridimensionnel (in French). *C. R. Acad. Sci. Paris II* **310**, 347–352.

Bear, J. (1972). *Dynamics of Fluids in Porous Media.* Elsevier, New York.

Begg, S.H., Carter, R.R., and Dranfield, P. (1989). Assigning effective values to simulator gridblock parameters for heterogeneous reservoirs. *SPE Reservoir Eng.* **4**, 455–463.

Bourgeat, A. (1984). Homogenized behavior of two-phase flows in naturally fractured reservoirs with uniform fractures distribution. *Comp. Meth. Appl. Mech. Eng.*, **47**, 205–216.

Bourgeat, A., Quintard, M., and Whitaker, S. (1988). Éléments de comparaison entre la méthode d'homogénéisation et la méthode de prise de moyenne avec fermeture (in French). *C. R. Acad. Sci. Paris II* **306**, 463–466.

Case, C.M. (1971). Projecting of Darcy's law as related to the ellipse of direction. *Water Resources Res.* **7**, 1354–1356.

Crampin, S. (1985). Evaluation of anisotropy by shear-wave splitting. *Geophysics* **50**, 142–152.

Crapiste, G.H., Rotstein, E., and Whitaker, S. (1986). A general closure scheme for the method of volume averaging. *Chem. Eng. Sci.* **41**, 227–235.

Dagan, G. (1986). Statistical theory of groundwater flow and transport: pore to laboratory, laboratory to formation and formation to regional scale. *Water Resources Res.* **22**(9), 120S–134S.

Desbarats, A.J. (1987). Numerical estimation of effective permeability in sand-shale formations. *Water Resources Res.* **23**, 273–286.

Deutsch, C. (1989). Calculating effective absolute permeability in sandstone/shale sequences. *SPE Formation Evaluation* **4**, 343–348.

Durlofsky, L.J. and Chung, E.Y. (1990). Effective permeability of heterogeneous reservoir regions. In *2d European Conf. on Math. of Oil Recovery* (eds. D. Guérillot and O. Guillon), pp. 57–64. Arles, Éditions Technip, Paris.

Ene, H.I. and Sanchez-Palencia, E. (1975). Équations et phénomènes de surface pour l'écoulement dans un modèle de milieu poreux (in French). *J. Méc.* **14**, 73–108.

Férrandon, J. (1948). Les lois de l'écoulement de filtration (in French). *Le Génie Civil* **125**(2), 24–28.

Fontugne, D.J. (1969). Permeability measurement in anisotropic media. M.S. thesis, Syracuse University.

Fox, R.L. (1971). *Optimization Methods for Engineering Design*. Addison-Wesley, Reading, Mass.

Goggin, D.J., Chandler, M.A., Kocurek, G., and Lake, L.W. (1988). Patterns of permeability in eolian deposits: Page sandstone (Jurassic), Northeastern Arizona. *SPE Formation Evaluation*, **3**, 297–306.

Gray, W.G. and O'Neill, K.Z. (1976). On the general equations for flow in porous media and their reduction to Darcy's law. *Water Resources Res.*, **12**, 148–154.

Greenkorn, R.A., Johnson, C.R., and Shallenberger, L.K. (1964). Directional permeability of heterogeneous anisotropic porous media. *Soc. Pet. Eng. J.*, **4**, 124–132.

Guin, J.A., Kessler, D.P., and Greenkorn, R.A. (1971). Permeability tensor for anisotropic nonuniform porous media. *Chem. Eng. Sci.*, **26**, 1475–1478.

Haldorsen, H.H. (1986). Simulator parameter assignment and the problem of scale in reservoir engineering. In *Reservoir Characterization* (eds. L.W. Lake and H.B. Carroll), pp. 293–339. Academic Press, New York.

Haldorsen, H.H. and Lake, L.W. (1984). A new approach to shale management in field scale problems. *Soc. Pet. Eng. J.*, **24**, 447–457.

Hewett, T.A. and Behrens, R.A. (1990). Considerations affecting the scaling of displacements in heterogeneous permeability distributions. SPE paper 20739 presented at the 65th. Ann. Tech. Conf., New Orleans.

Johnson, W.E. and Hughes, R.V. (1948). Directional permeability measurements and their significance. *Producers Monthly*, **13**, 17–25.

Kamath, J., Boyer, R.E., and Nakagawa, F.M. (1990). Characterization of core scale heterogeneities using laboratory pressure transients. SPE paper 20575 presented at the 65th. Ann. Tech. Conf., New Orleans.

Kasap, E. and Lake, L.W. (1989). An analytical method to calculate the effective permeability tensor of a grid block and its application in an outcrop study. SPE paper 18434 presented at the SPE Symp. on Reservoir Simulation, Houston.

Keller, J.B. (1980). Darcy's law for flow in porous medium and the two space method. In *Nonlinear Partial Differential Equations in Engineering and Applied Science* (eds. R.L. Sternberg, A.J. Kalinowski and J.S. Papadakis), pp. 429–443. Dekker, New York.

King, P.R. (1989). The use of renormalization for calculating effective permeability. *Transport in Porous Media*, **4**, 37–58.

Kuchuk, F.J. and Kabir, S. (1988). Well test interpretation for reservoirs with a single linear no-flow barrier. *J. Pet. Sci. Eng.*, **1**, 195–221.

Lake, L.W. (1988). The origins of anisotropy. *J. Pet. Tech.*, **40**, 395–396.

Lake, L.W., Kasap, E., and Shook, M. (1989). Pseudofunctions, the key to practical use of reservoir description. Paper presented at the 2d Int. Conf. on North Sea Oil and Gas Reservoirs, Trondheim, Norway.

Lasseter, T.J., Waggoner, J.R., and Lake, L.W. (1986). Reservoir heterogeneities and their influence on ultimate recovery. In *Reservoir Characterization* (eds. L.W. Lake and H.B. Carroll), pp. 545–559. Academic Press, New York.

Lindquist, S.J. (1983). Nugget formation reservoir characteristics affecting production in the overthrust belt of southwestern Wyoming. *J. Pet. Tech.*, **35**, 1355–1365.

Long, J.C.S., Remer, J.S., Wilson, C.R., and Witherspoon, P.A. (1982). Porous media equivalents for networks of discontinuous fractures. *Water Resources Res.*, **18**, 645–658.

Marcus, H. and Evenson, D.E. (1961). *Directional Permeability in Anisotropic Porous Media.* Water Resources Center, Contribution no. 31, Hydrology Lab., University of California at Berkeley.

Meedoga, N.J., King, I.P., and Arulanandan, K. (1989). An expression for the permeability of anisotropic granular media. *Int. J. Numer. Anal. Methods Geomech.*, **13**, 575–598.

Mei, C.C. and Auriault, J.-L. (1989). Mechanics of heterogeneous porous media with several spatial scales. *Proc. R. Soc. London*, **A426**, 391–423.

Neuman, S.P. (1977). Theoretical derivation of Darcy's law. *Acta Mech.*, **25**(3–4), 153–170.

Oda, M. (1985). Permeability tensor for discontinuous rock masses. *Geotech.*, **35**(4), 483–495.

Quintard, M. and Whitaker, S. (1987). Écoulement monophasique en milieu poreux: effet des hétérogénéités locales (in French). *J. Méca. Théor. Appl.*, **6**(5), 691–726.

Rice, P.A., Fontugne, D.J., Latini, R.G., and Barduhn, A.J. (1970). Anisotropic permeability in porous media. *Ind. Eng. Chem.*, **62**, 23–31.

Rose, W.D. (1982). New method to measure directional permeability. *J. Pet. Tech.*, **34**, 1142–1144.

Sáez, A.E., Otero, C.J., and Rusinek, I. (1989). The effective homogeneous behavior of heterogeneous porous media. *Transport in Porous Media*, **4**, 213–238.

Scheidegger, A.E. (1954). Directional permeability of porous media to homogeneous fluids. *Geofisica Pura e Applicata*, **28**, 75–91.

Scheidegger, A.E. (1974). *The Physics of Flow through Porous Media*, 3d edn., University of Toronto Press, Toronto.

Slattery, J.C. (1972). *Momentum, Energy and Mass Transfer in Continua.* McGraw-Hill, New York.

Smith, D.A. (1980). Sealing and non-sealing faults in Louisiana Gulf Coast salt basin. *Am. Assoc. Pet. Geol. Bull.*, **64**(2), 145–172.

Thomasset, F. (1981). *Implementation of Finite Element Methods for Navier-Stokes Equations.* Springer-Verlag, New York.

von Ruhl, W. and Schmid, C. (1957). Uber das Verhaltnis dcr vertikalen zur horizontalen absoluten Permeabilitat von Sandsteinen (in German). *Geol. Jahrb.*, **74**, 447–461.

Whitaker, S. (1986). Flow in porous media I: a theoretical derivation of Darcy's law. *Transport in Porous Media*, **1**, 3–25.

White, C.D. and Horne, R.N. (1987). Computing absolute transmissibility in the presence of fine-scale heterogeneity. SPE paper 16011 presented at the SPE Symp. on Reservoir Simulation, San Antonio.

Young, G.R. (1989). Determining permeability anisotropy from a core plug using a minipermeameter. M.S. thesis, University of Texas at Austin.

Zick, A.A. and Homsy, G.M. (1982). Stokes flow through a periodic array of spheres. *J. Fluid Mech.*, **115**, 13–26.

Chapter 7
Permeability and Relative Permeability in Rocks

Stephen C. Blair and James G. Berryman
University of California, Lawrence Livermore National Laboratory, P.O. Box 808, Livermore, CA 94550, U.S.A.

Abstract

Important features of the topology of the pore space of rocks can be usefully quantified by analyzing digitized images of rock cross-sections. One approach computes statistical correlation functions using modern image-processing techniques. These correlation functions contain information about porosity, specific surface area, tortuosity, formation factor, and elastic constants, as well as the fluid permeability and relative permeability. The physical basis of this approach is discussed and examples of the results for various sandstones are presented. The analysis shows that Kozeny–Carman relations and Archie's empirical laws must be modified to account for finite percolation thresholds in order to avoid unphysical behavior in the calculated relative permeabilities.

1. Introduction

Most rocks are porous to some degree. Methods of characterizing that porosity have been studied for many years in order to understand the physical properties of rocks and to gain insight into possible scenarios for their formation. Two such properties of great practical interest are the fluid permeability and the electrical conductivity of rocks containing conducting fluids. In order for an electron, an ion, or a fluid particle to travel from one part of a rock to another distant part, it is necessary for the particle to follow a tortuous path through the rock, avoiding the solid/insulating grains that compose the rock. Thus, it may be expected on very general physical grounds that there should be a close quantitative relationship between the fluid permeability and the electrical conductivity. No such rigorous relationship has yet been found, however.

Brace (1977) and later Paterson (1983) and Walsh and Brace (1984) have shown that a type of Kozeny–Carman relation may be used to relate the fluid permeability to the electrical formation factor. Brace (1977) found that such a formula could predict the fluid permeability k of a rock within about a factor of 2 over nine orders of magnitude

FAULT MECHANICS AND TRANSPORT PROPERTIES OF ROCKS
ISBN 0-12-243780-2

in the variation of k if the hydraulic radius and formation factor of the rock was known. Bernabé, Brace, and Evans (1982) studied the dependence of permeability on porosity and pore geometry in hot-pressed calcite and found significant deviations from Kozeny–Carman in the vicinity of the percolation threshold. Walsh and Brace (1984) rederived the Kozeny–Carman relation and wrote it in terms of porosity, specific surface area, and formation factor. Berryman and Blair (1986) proposed using image-processing methods as a means of estimating the parameters in this Kozeny–Carman relation and tested this approach on porous glass and various sandstones. The results from image processing were similar to those of Brace (1977) in that the predicted and measured permeabilities generally agreed well within a factor of 2. In fact, the authors considered these results to be suspiciously good, and subsequently showed (Berryman and Blair, 1987) that the success of the method was due in part to the fortuitous cancellation of two compensating errors: (a) the Kozeny–Carman relation is an approximation based on an implicit assumption of smooth-walled pores, and (b) the image-processing method has inherent resolution limitations that force the measured specific surface area to be consistent with the assumptions in the derivation of the Kozeny–Carman relation.

If we can predict the permeability for single-phase flow through a porous material from digitized information obtained from cross-sections of rocks, then we may also be able to obtain estimates of traditional relative permeability as well when two fluid phases are present. With traditional relative permeability (Corey, 1977; Bear, 1979; Dullien, 1979), interactions between the two fluid phases are neglected and each fluid phase is treated as a single phase flowing through a porous matrix composed of the solid phase and the other fluid phase. The traditional approach to relative permeability is only an approximation to the real situation in two-phase flow, and many researchers have challenged the assumption that the interactions between the fluid phases are negligible (Raats and Klute, 1968; deGennes, 1983; Thigpen and Berryman, 1985; Whitaker, 1986; Berryman, et al., 1988; Rose, 1988, 1989; Kalaydjian, 1990; Rothman, 1990). If the interactions are significant, then the generalization of the image processing method for estimating relative permeability coefficients is not automatic. Nevertheless, the interactions must depend on the fluid–fluid interface area, which is a quantity that can be obtained from images of cross-sections.

In Section 2, we summarize the analysis of the permeability for single-phase flow and of the relative permeability for two-phase flow. We include a discussion of the cross terms that arise due to the fluid–fluid interaction. In Section 3, we present a simple but rigorous result relating the fluid–fluid interface to other measured surface areas. In Section 4, we analyze Kozeny–Carman relations for single-phase flow, for two-phase flow without fluid–fluid interaction, and for two-phase flow with such interaction. We also consider the possibility that the two fluids have mixed wettability characteristics. In Section 5, we describe methods used to obtain and quantify cross-sections of sandstones that display the two distinct fluid phases occupying the pore space of a rock. Section 6 summarizes our results for Berea sandstone. A discussion of the results to date and the possibilities for extensions in the future is presented in the final section.

2. Permeability and Relative Permeability

We will first consider a single-phase fluid and then generalize to the two-phase problem. Throughout this discussion, we assume that the porous solid frame is rigid.

2.1. Single-phase flow

Consider a porous medium of total volume Ω filled with a fluid occupying the subvolume Ω_f. The points of contact between the saturating fluid and the solid form the interior surface Γ_I. The exterior fluid surface through which the fluid may enter or exit the volume Ω is Γ_E. We assume that the applied stress tensor Π_{ij} is known on Γ_E. The applied stress takes the form

$$\Pi_{ij} = -p\,\delta_{ij} + \sigma_{ij} \tag{1}$$

where the viscosity tensor σ is related to the fluid velocity field $\mathbf{v}$ as

$$\sigma_{ij} = \mu(v_{i,j} + v_{j,i}) \tag{2}$$

for $i, j = 1, 2, 3$. In this notation, i and j index the directions in a cartesian coordinate system ($x = x_1$, $y = x_2$, $z = x_3$) and the subscript appearing after a comma refers to a partial derivative; thus, $v_{2,3} = \partial v_2/\partial z$. The fluid pressure is p and the fluid viscosity is μ. Then the energy dissipation in the fluid is given by (Landau and Lifshitz, 1959)

$$\mathscr{D} = \frac{1}{2\mu\Omega}\int_{\Omega_f} \sigma_{ij}\sigma_{ij}\,d^3x \geq 0 \tag{3}$$

where the summation convention is assumed for repeated indices in (3). If we neglect body forces (e.g., gravity) and suppose that the macroscopic applied pressure gradient arises due to the pressure difference ΔP across a distance ΔZ in the direction $\hat{z}$, then the relationship between the microscopic stresses and the macroscopic forces is easily seen to be given by (Berryman and Milton, 1985)

$$\mathscr{D} = \frac{k}{\mu}\left(\frac{\Delta P}{\Delta Z}\right)^2 \tag{4}$$

or equivalently

$$\mathscr{D} = -\mathbf{J}\cdot\hat{z}\,\frac{\Delta P}{\Delta Z} \tag{5}$$

where $\mathbf{J}$ is the volumetric flow rate per unit area and

$$\mathbf{J} = -\frac{k}{\mu}\frac{\Delta P}{\Delta Z}\,\hat{z} \tag{6}$$

Equation (6) is Darcy's law in the absence of body forces and the new constant appearing in (4) and (6) is the fluid permeability or Darcy's constant k.

In the presence of body forces, we generalize (6) to

$$\mathbf{J} = \frac{k}{\mu}\mathbf{X} \tag{7}$$

where

$$\mathbf{X} = -\nabla p + \rho\mathbf{g} \tag{8}$$

The density of the fluid is ρ and the gravitational acceleration vector is $\mathbf{g}$.

2.2. Two-phase flow

Now suppose that the pore space is completely filled with two immiscible fluids 1, 2. These fluids could be water and oil or water and air or any other pair of Newtonian fluids. Then the total energy dissipation in the two fluids will quite generally be given by the sum of the energy dissipation in the individual fluids,

$$\mathscr{D} = \frac{1}{2\mu_1\Omega}\int_{\Omega_1} \sigma_{ij}^{(1)}\sigma_{ij}^{(1)}\,d^3x + \frac{1}{2\mu_2\Omega}\int_{\Omega_2} \sigma_{ij}^{(2)}\sigma_{ij}^{(2)}\,d^3x \geq 0 \tag{9}$$

where μ_n and Ω_n are the viscosity and volume of the nth fluid.

Furthermore, by analogy with the single-phase results, we must have in general that

$$\frac{1}{2\mu_n\Omega}\int_{\Omega_n} \sigma_{ij}^{(n)}\sigma_{ij}^{(n)}\,d^3x = \mathbf{J}^{(n)}\cdot\mathbf{X}^{(n)} \tag{10}$$

where $\mathbf{J}^{(n)}$ is the volumetric flow rate of the nth fluid and $\mathbf{X}^{(n)}$ is the force on the nth fluid. However, since there is more than one force present, eq. (7) should be generalized to take into account the possibility that a force applied to one fluid may cause the other fluid to move as well. In particular, we expect the motion of one fluid to tend to drag the other along with it wherever the two fluids are in contact. Following the notation of Rothman (1990), we can write this resulting relation as

$$\mathbf{J}^{(1)} = L_{11}(\theta)\mathbf{X}^{(1)} + L_{12}(\theta)\mathbf{X}^{(2)} \tag{11}$$

$$\mathbf{J}^{(2)} = L_{21}(\theta)\mathbf{X}^{(1)} + L_{22}(\theta)\mathbf{X}^{(2)} \tag{12}$$

where $\theta = \Omega_1/(\Omega_1 + \Omega_2)$ is the saturation of (fraction of the pore space filled by) phase 1. The L's are linear coefficients that we show later are proportional to permeability and inversely proportional to viscosity.

Now define the force vector

$$f = \begin{pmatrix} \mathbf{X}^{(1)} \\ \mathbf{X}^{(2)} \end{pmatrix} \tag{13}$$

and the coefficient matrix

$$\mathscr{L} = \begin{pmatrix} L_{11} & L_{12} \\ L_{21} & L_{22} \end{pmatrix} \tag{14}$$

Then we see that the total dissipation can be written in the form

$$\mathscr{D} = f^{\mathrm{T}}\mathscr{L}f \geq 0 \tag{15}$$

Since the force vector is arbitrary, we see that the coefficient matrix must be positive definite, which implies that

$$L_{11}\;L_{22} - L_{12}L_{21} > 0 \tag{16}$$

In closely related work, Thigpen and Berryman (1985) have shown that the average fluid velocities $\mathbf{V}^{(n)}$ obey the momentum equations

$$\begin{pmatrix} C_{11} & C_{12} \\ C_{21} & C_{22} \end{pmatrix}\begin{pmatrix} \mathbf{V}^{(1)} \\ \mathbf{V}^{(2)} \end{pmatrix} = \begin{pmatrix} \phi_1 & 0 \\ 0 & \phi_2 \end{pmatrix}\begin{pmatrix} \mathbf{X}^{(1)} \\ \mathbf{X}^{(2)} \end{pmatrix} \tag{17}$$

where $\phi_1 = \theta\phi$, $\phi_2 = (1-\theta)\phi$, and the C's are drag coefficients that must satisfy the relations

$$C_{11} + C_{21} \geq 0, \qquad C_{22} + C_{12} \geq 0, \qquad C_{12} = C_{21} \leq 0 \tag{18}$$

to insure that the drag always opposes the motion and to guarantee the symmetry of the action–reaction (i.e., momentum conservation). Since the fluid velocities are related to the volumetric flow rates by

$$\begin{pmatrix} \mathbf{J}^{(1)} \\ \mathbf{J}^{(2)} \end{pmatrix} = \begin{pmatrix} \phi_1 & 0 \\ 0 & \phi_2 \end{pmatrix} \begin{pmatrix} \mathbf{V}^{(1)} \\ \mathbf{V}^{(2)} \end{pmatrix} \tag{19}$$

the permeability matrix is then directly related to the drag coefficients. Substituting (19) into (11)–(12) and comparing to (17), we find that the coefficient matrix $\mathscr{L}$ is given in terms of the C's as

$$\begin{pmatrix} L_{11} & L_{12} \\ L_{21} & L_{22} \end{pmatrix} = \frac{1}{\Delta} \begin{pmatrix} \phi_1^2 C_{22} & -\phi_1\phi_2 C_{12} \\ -\phi_1\phi_2 C_{12} & \phi_2^2 C_{11} \end{pmatrix} \tag{20}$$

where $\Delta = C_{11}C_{22} - C_{12}^2$. Thus, it follows directly from momentum conservation for linear flow of two immiscible fluids that the coefficients for the cross terms are equal:

$$L_{12} = L_{21} \tag{21}$$

This equality has been postulated by Kalaydjian (1990) and by Rothman (1990) on the basis of Onsager reciprocity (Fitts, 1962; Landau and Lifshitz, 1969). A thorough analysis of the possible pitfalls contained in such a postulate has been given by Fitts (1962). Rothman's numerical experiments have also substantially confirmed the result (21), while Kalaydjian's analytical results – obtained assuming that (21) holds – are in good agreement with those of Rothman when the fluid viscosities are equal. Furthermore, it follows directly from (18) and (20) that

$$L_{12} \leq (1 - \theta)L_{11}/\theta, \qquad L_{12} \leq \theta L_{22}/(1 - \theta) \tag{22}$$

Both Kalaydjian's analytical and experimental results are in good agreement with the combined bounds implied by (22). Rothman's numerical studies also provide experimental confirmation of (22). The details of these results will be reported elsewhere.

3. Specific Surface Area of the Fluid–Fluid Interface

There is one rigorous formula (a simple sumrule) that can be easily derived relating the various internal surface areas s_W, s_{NW}, s_T, and s_I (where the subscripts refer respectively to wetting, nonwetting, total pore, and fluid–fluid interface areas). Figure 1 shows the simplest geometrical arrangement when one of the two fluids completely wets the internal surface area of the rock. We will call this condition 'normal wettability.' Then, the nonwetting fluid introduces a fluid–fluid interface. From Figure 2 it is clear that the total surface area of the wetting fluid is now $s_W = s_T + s_I$, while the surface area of the nonwetting fluid is the same as that of the fluid–fluid interface $s_{NW} = s_I$. Thus, the sum of the two gives the general result

$$s_W + s_{NW} = s_T + 2s_I \tag{23}$$

Indeed, in the hypothetical situation where the two fluids occupy separate pore spaces, we see easily that $s_W + s_{NW} = s_T$ since then there is no fluid–fluid interface. The result

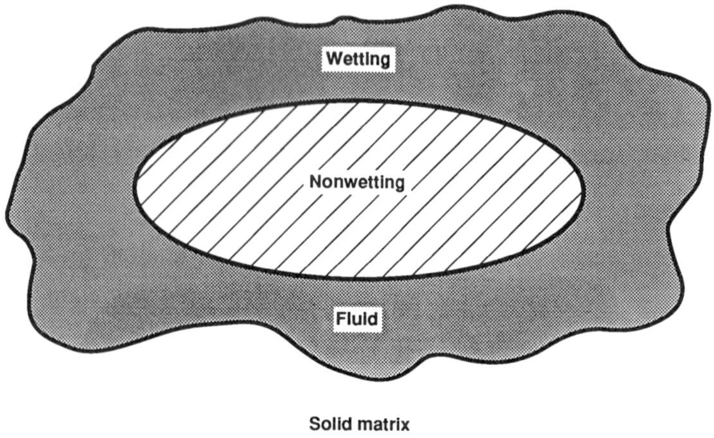

Figure 1. Geometrical arrangement of wetting and nonwetting fluids in a typical pore.

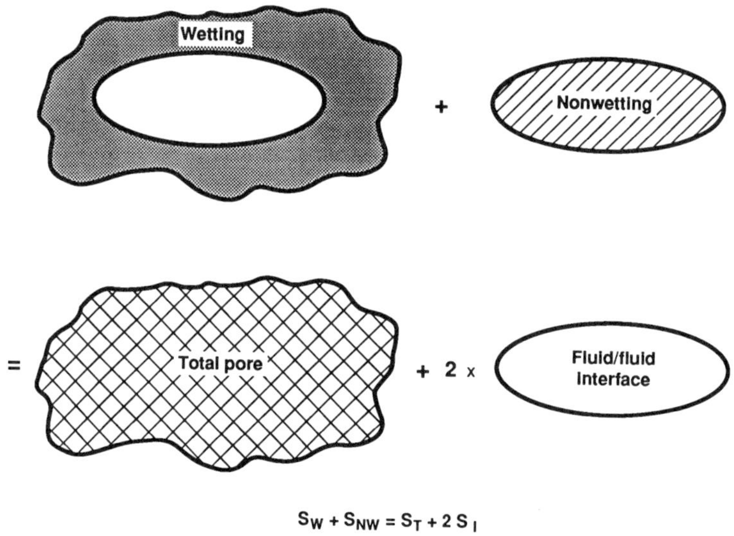

Figure 2. Illustration of the sumrule presented in eq. (23).

for segregated fluids is therefore a special case of (23) with $s_I = 0$. Furthermore, if the two fluids have 'mixed wettability' properties so that the pictures in Figures 1 and 2 are oversimplified, then (23) is still valid. For 'mixed wettability' conditions, both fluids wet different parts of the internal surface area. Quite generally the surface area of one fluid will be of the form $s_W = \omega s_T + s_I$ while the other will be of the form $s_{NW} = (1 - \omega)s_T + s_I$, where ω is some fraction $0 \leq \omega \leq 1$. Summing these two expressions again gives (23).

We stress that (23) is not an approximation, but a rigorous result when the pore space is filled with two fluid phases. If more than two fluid phases are present, then (23) must be generalized to account for all the interfaces between pairs of fluids. The utility of a rigorous result such as (23) becomes apparent when we consider that measurements of the various specific surface areas may proceed almost independently, yet we may need to know whether the results are consistent for a given rock sample. One simple check on the results that follows immediately from (23) is the condition

$$s_W + s_{NW} - s_T \geq 0 \tag{24}$$

If (24) is violated to a significant degree by the data for these three specific surface areas, then we may be sure that our experimental method has some inconsistency, bias, or error that must be eliminated. The second important practical use for (23) comes from the fact that it may be used to compute the fluid–fluid interface area per unit volume, since it implies that

$$s_I = \tfrac{1}{2}(s_W + s_{NW} - s_T) \tag{25}$$

Equation (25) will be useful whenever we need quantitative information about the interface area.

4. Kozeny–Carman Relations

Since Darcy's constant is difficult to model, many approximate methods have been developed to estimate fluid permeability. Among these techniques are the effective-medium-based methods (Koplik et al., 1984; Wong et al., 1984; Doyen, 1988), percolation theory; (Balberg, 1986; Katz and Thompson, 1986, 1987; Thompson et al., 1987), the approaches using rigorous variational bounds (Berryman and Milton, 1985; Rubinstein and Torquato, 1989), and a great variety of methods based on Kozeny–Carman empirical relations. Our experience has been that intelligent use of the Kozeny–Carman relations can give excellent results, so we will concentrate on that approach here.

4.1. Single-phase flow

The Kozeny–Carman relation for single-phase flow derived by Walsh and Brace (1984) and used extensively by Berryman and Blair (1986, 1987) is given by

$$k = \frac{\phi^2}{2s^2F} = \frac{\phi^3}{2s^2\tau^2} \tag{26}$$

where ϕ is the porosity, s is the specific surface area, F is the electrical formation factor, τ is the electrical tortuosity, and the relationship

$$\tau^2 = \phi F \tag{27}$$

has been used in the second part of eq. (26).

4.2. Two-phase flow: traditional relative permeability

When two fluid phases are present in the pore space, we will distinguish two types of wetting behavior. Normal wettability will refer to the situation where one phase tends to wet the pore surface area, while the second phase tends not to wet it as long as the other phase is present in sufficient quantity. Mixed wettability will refer to the situation where both phases tend to wet distinct parts of the pore space regardless of their relative concentrations. We will now consider both of these situations in turn, while neglecting the possibility of cross terms in the relative permeabilities.

4.2.1. Tortuosity

We first examine the effect of tortuosity on two-phase flow. If we can predict the permeability for single-phase flow through a porous material from digitized information obtained from cross-sections of rocks, we should also be able to obtain reasonable estimates of the traditional relative permeabilities. When both fluids percolate (form a continuous connected phase) throughout the porous sample, we will assume that to a first approximation the effective tortuosity for each fluid phase is the same as that for the porous solid

$$\phi F = \tau^2 = \tau_W^2 = \tau_{NW}^2 \tag{28}$$

For both phases to percolate, the saturation level must be moderate (say, in the range 20–80%, depending on the porosity) and the permeability/porosity must not be very small. Then, the assumption (28) is both warranted and necessary in the present context, since it is unlikely that the relevant electrical formation factor measurements can generally be made for both phases.

Equation (28) is clearly an approximation that must be modified when the saturation level of either phase approaches the percolation threshold. In this case, we can introduce into (28) additional factors of

$$(\phi_W - \phi_W^c)^\alpha, \qquad (\phi_{NW} - \phi_{NW}^c)^\beta \tag{29}$$

where the critical volume fractions or percolation thresholds for the wetting and nonwetting phases are given respectively by ϕ_W^c and ϕ_{NW}^c and the exponents α and β are to be determined through experiment (Bernabé et al., 1982). Typical values of these exponents have been shown (Archie, 1942; Madden, 1976; Corey, 1977; Sen et al., 1981; Sen, 1984) to be $\alpha \simeq \beta \simeq 2$ for water saturations down to about 0.15–0.20 ($\simeq \phi_W^c/\phi$). Equation (28) is then modified to

$$\tau^2 = \tau_W^2 \left(\frac{\phi_W - \phi_W^c}{\phi} \right)^\alpha = \tau_{NW}^2 \left(\frac{\phi_{NW} - \phi_{NW}^c}{\phi} \right)^\beta \tag{30}$$

In general, we expect the percolation thresholds to be small fractions $\phi_W^c \simeq \phi_{NW}^c \simeq 0.02$. With these volume fractions, we can also associate critical saturation levels $\theta_W^c = \phi_W^c/\phi$ and $\theta_{NW}^c = 1 - \phi_{NW}^c/\phi$. (Note the difference in the definitions of the ϕ's and the θ's: The ϕ's are volume fractions, so ϕ_W is the volume fraction of the wetting phase and ϕ_{NW} is the same for the nonwetting phase, and we have the sumrule $\phi_W + \phi_{NW} = \phi$, since we assume the wetting and nonwetting phases together saturate the pore volume. In contrast, θ_W^c is the critical saturation value of the wetting fluid at wetting percolation, whereas θ_{NW}^c is the critical saturation value of the *wetting* phase at *nonwetting* percolation. This definition is made to simplify the formulas later.) There is no simple relationship between the critical saturations θ_W^c and θ_{NW}^c, since the shape of the effective pore space for each phase is different. However, it is reasonable to suppose that $\theta_W^c \simeq 1 - \theta_{NW}^c$. For example, if the porosity ϕ is about 20% and critical volume fraction for percolation in the rock pores is about 2%, then the critical saturation θ for the wetting phase is on the order of 10%, while for the nonwetting phase the critical value of θ is about 90%. In the absence of other information about the percolation thresholds, we may assume they take their limiting values of $\phi_W^c = 0$ and $\phi_{NW}^c = 0$.

4.2.2. Normal wettability

Applying the Kozeny–Carman relation (26) and the approximation (28) to the separate fluid phases, we have

$$k_W = \frac{\phi_W^3}{2s_W^2\tau_W^2} \tag{31}$$

and

$$k_{NW} = \frac{\phi_{NW}^3}{2s_{NW}^2\tau_{NW}^2} \tag{32}$$

If the picture in Figure 1 is typical of the fluid arrangement in the pores, then we expect to find that $s_W = s_T + s_I$ and $s_{NW} = s_I$. The relative permeabilities are given by

$$k_{rW} = \frac{k_W}{k} = \frac{\theta^3}{(s_W/s_T)^2} = \frac{\theta^3}{(1 + s_I/s_T)^2} \tag{33}$$

and

$$k_{rNW} = \frac{k_{NW}}{k} = \frac{(1-\theta)^3}{(s_{NW}/s_T)^2} = \frac{(1-\theta)^3}{(s_I/s_T)^2} \tag{34}$$

so the tortuosity τ has completely cancelled out of these expressions.

Again, if the picture in Figure 1 is representative of the distribution of nonwetting fluid to wetting fluid, then to a reasonable approximation we should expect that

$$s_I \simeq (1-\theta)s_T \tag{35}$$

Equation (35) will be valid when the grain surface is fairly smooth, or when our *estimate* of the grain surface area is obtained using a low-resolution method such as image processing. Otherwise, we generally expect (35) to be replaced by $s_I \ll (1-\theta)s_T$, since the fluid–fluid interface will be smooth regardless of the roughness of the grain surface. So, continuing to neglect the effects of the changing tortuosity close to a percolation threshold and using (35), we should find that

$$k_{rW} \simeq \frac{\theta^3}{(2-\theta)^2} \tag{36}$$

and

$$k_{rNW} \simeq (1-\theta) \tag{37}$$

The results (36) and (37) show that we should expect a strong asymmetry between the relative permeabilities for the wetting and the nonwetting phases. These first approximations to the relative permeabilities show that k_{rW} for the wetting phase is a strongly nonlinear function of the wetting phase saturation θ. Physically, this makes sense because the wetting phase is trapped between the solid and the nonwetting fluid, so the typical open throat for the wetting phase is significantly smaller in the presence of the nonwetting phase. By contrast, the nonwetting phase sees a porous medium that is very similar to that of the solid matrix – the effective pores have just been reduced in size by the coating action of the wetting phase. Until percolation threshold effects become important, the relative permeability for this phase should be approximately proportional to $(1-\theta)$ as shown in (37).

4.2.3. *Mixed wettability*

If we can measure the various specific surface areas using either image-processing methods (as will be described later in this chapter) or some other means, then it is straightforward to determine whether the condition of normal wettability or that of mixed wettability applies. We check for mixed wettability by comparing the two specific surface areas s_{NW} and s_I. For normal wettability we have $s_{NW} = s_I$, while for mixed wettability we will find that $s_{NW} > s_I$. The interface area itself may still be computed unambiguously from the rigorous formula (25). Strictly speaking, our notation should change now, since the phase labelled NW is also a wetting phase under these conditions. However, we will keep the notation and simply be aware that the NW-phase is assumed to wet a smaller proportion of the pore wall than does the W-phase.

Now recalling that for mixed wettability $s_W = \omega s_T + s_I$ and $s_{NW} = (1 - \omega)s_T + s_I$ where ω is the relative wetting fraction, the formulas (33) and (34) must be modified so that the relative permeabilities are given by

$$k_{rW} = \frac{k_W}{k} = \frac{\theta^3}{(s_W/s_T)^2} = \frac{\theta^3}{(\omega + s_I/s_T)^2} \tag{38}$$

and

$$k_{rNW} = \frac{k_{NW}}{k} = \frac{(1-\theta)^3}{(s_{NW}/s_T)^2} = \frac{(1-\theta)^3}{(1 - \omega + s_I/s_T)^2} \tag{39}$$

Since we now have a wetting fraction ω as well as the saturation parameter θ, it becomes more difficult to do any sort of general analysis of the relative permeabilities in such complex circumstances. The image processing methods for obtaining these parameters should nevertheless provide reasonable estimates of all these parameters and therefore of the relative permeabilities as well.

4.3. Two-phase flow: relative permeability including cross terms

The theory of the relative permeability in the presence of fluid–fluid interactions has still to be developed. We will have to be satisfied here with preliminary results. Let the subscripts 1, 2 refer respectively to wetting and nonwetting fluid phases. Then a reasonable approximation to the coefficient matrix $\mathscr{L}$ that is consistent with what we know at present is

$$\begin{pmatrix} L_{11} & L_{12} \\ L_{12} & L_{22} \end{pmatrix} = \begin{pmatrix} k_{rW}k/\mu_W & \chi \\ \chi & k_{rNW}k/\mu_{NW} \end{pmatrix} \tag{40}$$

where we have used the empirically derived expression for the cross terms

$$\chi \simeq \min[(1-\theta)L_{11}/\theta \quad \theta L_{22}/(1-\theta)] \tag{41}$$

which has been found to agree well with the available data of Kalaydjian (1990) and Rothman (1990). The motivation for considering a relation of the form (41) comes from the rigorous bounds (22), based on momentum conservation. Equation (41) tends to be a good estimator of the cross terms everywhere except in the vicinity of $\theta \simeq 0.5$, where (41) overestimates the cross terms by nearly a factor of two for the real data of Kalaydjian (1990).

5. Sample Preparation and Image-processing Methodology

In this section we describe the methodology used to produce and analyze images of Berea sandstone in which the pore space was segregated into regions occupied by simulated wetting and nonwetting phases. First, a low-melting-point bismuth metal alloy was imbibed into a dry evacuated core sample of Berea sandstone, at a temperature of approximately 100°C. After imbibition, the sample was cooled to room temperature, solidifying the metal alloy. The bulk saturation with the metal alloy was determined to be 57% based on weight. The core sample was then cut into slabs in the direction perpendicular to its axis, and examination of the slabs showed that the saturation decreased along the length of the core from bottom to top.

One slab, taken from the lower section of the sample, was chosen for our analysis. This slab was vacuum-impregnated with epoxy and a standard petrographic thin section was prepared. The thin section was examined using a scanning electron microscope (SEM) in backscatter mode. In this mode the metal (nonwetting) phase appeared bright, while the epoxy (wetting) phase appeared dark and the mineral phase (grains) had intermediate brightness. Hardcopy images of the pore structure were produced as photographic prints and negatives. The negatives were then digitized using an Eikonix scanning digitizer. Once they were in digital form the image scale was determined and the images were cropped using standard techniques.

Images of the total porosity and segments of the pore space occupied by the wetting and nonwetting phases were produced as follows. The thresholding technique used to identify pore space filled by the nonwetting metal phase was a slightly modified version of the method described by Berryman and Blair (1986). Unfortunately, this technique did not work as well for the epoxy phase, especially in areas along the metal–epoxy interface. This problem was overcome by applying thresholding to sections of the image to delineate the epoxy–grain and metal–grain boundaries in each section. A mosaic was then constructed from the sections to produce an image of the total pore space. Finally, an image of the wetted (epoxy-filled) pore space was produced by subtracting the nonwetted (metal filled) pore space from the total pore space. Comparison of the digital images with the original prints was very useful in segregating the pore space.

While we are aware that this simulation produced by first imbibing a nonwetting phase and subsequently injecting a wetting phase into the pore space may not produce two-phase spatial distributions representative of simultaneous two-phase flow, we believe that the insight gained from this approach is valuable nevertheless.

6. Results

Total porosity and the amount of pore space occupied by the nonwetting and wetting phases was determined from four images with magnifications of 54× and 100×. A representative section is shown in Figure 3a. The associated image displaying the segregated pore space is shown in Figure 3b. These figures show that, as expected, the nonwetting phase filled most of the large pores. The fraction of pore space occupied by each phase was simply calculated and results are presented in Table 1. The value of total porosity was determined to be 0.23 and this agrees well with values determined by laboratory measurements. Laboratory data indicate that the bulk saturation of the rock with the nonwetting phase was approximately 0.57. However, saturation varied

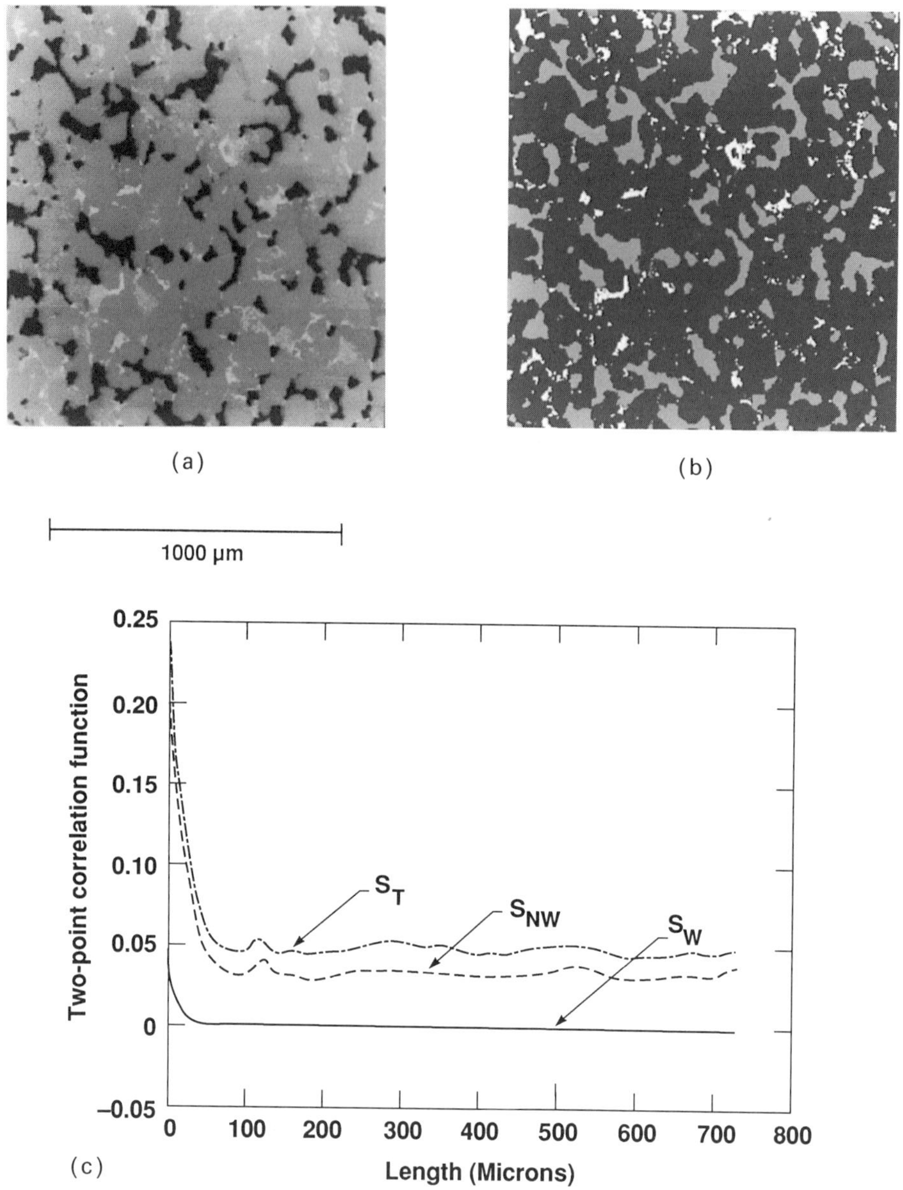

Figure 3. (a) Before image of Berea sandstone at $\times 54$ magnification. (b) Image of Berea sandstone after segmentation into solid grains (black), nonwetting (gray), and wetting (white) phases. (c) Two-point correlation functions for Berea sandstone: S_2 for total pore space, S_2^W for the wetted pore space, and S_2^{NW} for the nonwetted pore space.

widely over the length of the core and our value of 0.77 at the location of the slab provides precise data on the fine-scale distribution of the nonwetting phase.

Two-point autocorrelation functions were calculated for images of the total pore space and for pore space occupied by each phase, following the technique developed by Berryman (1985), and Berryman and Blair (1986). From these data for the total pore

Table 1. Summary of parameters for Berea sandstone measured using image processing techniques

Region	Total pore space	Wetted pore space	Nonwetted pore space
Fraction of total volume	$\phi = 0.226$	$\phi_W = 0.051$	$\phi_{NW} = 0.175$
Fraction of total pore space	1.00	$\theta = 0.23$	$1 - \theta = 0.77$
Slope of autocorrelation $S_2'(0)$ (μm^{-1})	−0.0268	−0.0220	−0.0094
Specific surface area (μm^{-1})	$s_T = 0.1072$	$s_W = 0.0881$	$s_{NW} = 0.0377$
Specific interface area (μm^{-1})	$s_I = 0.0093$	–	–
Wetting fraction ω	–	$\omega = 0.735$	$1 - \omega = 0.265$
Effective hydraulic radius (μm)	2.11	0.58	4.64
Relative permeability (dimensionless)	–	$k_{rW} = 0.018$	$k_{rNW} = 3.69$
Apparent percolation threshold	–	$\theta_W^c = 0.25$	$\theta_{NW}^c = 0.75$

space the value of the two-point correlation function S_2 at each length increment was calculated as the radial average of values for that length. S_2^{NW} and S_2^{W} were calculated in a similar way from the respective autocorrelation functions. The two-point correlation functions for the total pore space and the nonwetted and wetted pore space shown in Figure 3b are presented in Figure 3c. This figure shows that the correlation function for the nonwetting, imbibed phase closely resembles the correlation function for the total pore space. This is typical for the images we analyzed and is consistent with our previous observation that the nonwetting phase sees a porous medium that is similar to the solid matrix. Mean grain size can also be estimated from the correlation function, and our results indicate a value of 80 μm for this section.

We also analyzed four images with magnifications in the range 220–1100×. Images at this magnification were used to estimate the specific surface areas, following the work of Debye et al., (1957), who showed that the derivative of the two-point correlation function near the origin is proportional to the specific surface area, and of Berryman and Blair (1987), who related the resolution appropriate for measuring s_T to the correlation length computed for the material. A representative image is shown in Figure 4a, and the segregated pore space is shown in Figure 4b. The two-point autocorrelation functions S_2, S_2^{NW}, S_2^{W} for the total pore space and the respective sections of pore space occupied by each phase are shown in Figure 4c. This figure shows that at this scale the correlation function for the total pore space is not dominated by either pore phase. To estimate the specific surface area s_T for the total pore space, we fitted a line to the first

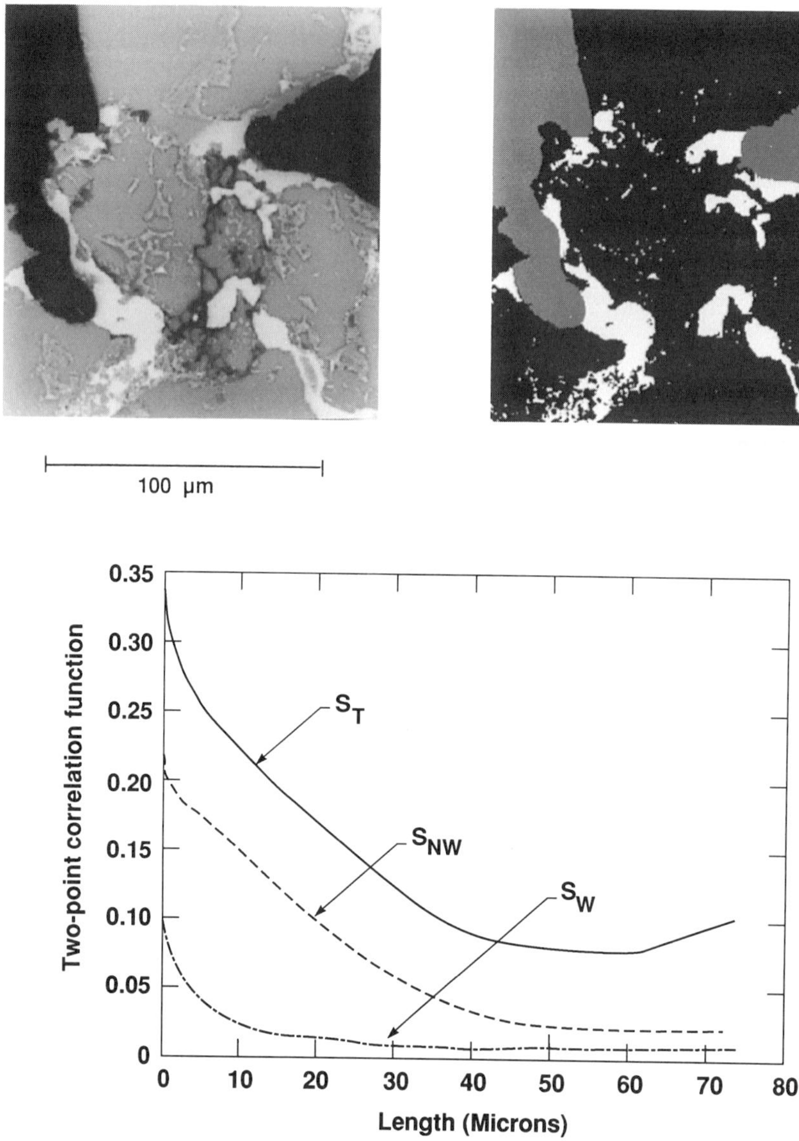

Figure 4. (a) Before image of Berea sandstone at × 540 magnification. (b) Image of Berea sandstone after segmentation into solid grains (black), nonwetting (gray), and wetting (white) phases. (c) Two-point correlation functions for Berea sandstone: S_2 for total pore space, S_2^W for the wetted pore space, and S_2^{NW} for the nonwetted pore space.

few points of the function S_2 for each image of the total pore space (see Berryman and Blair (1986) for a description of the method). The slopes of these lines were then averaged to produce a value of S_2' for the total pore space of the cross-section; this value is given in Table 1. Specific surface area (s_T) was then calculated as $s_T = -4S_2'(0)$ following Berryman and Blair (1986). Values of specific surface area for both the wetting (s_W) and nonwetting phases (s_{NW}) were calculated in a similar fashion. These measured values of specific surface area were then used to calculate specific area of the interface (s_I) following eq. (25). Results are given in Table 1. The value of s_T of 0.1072 for the total pore space agrees well with the values of 0.1109 and 0.1231 we reported previously for a different sample of Berea sandstone (see Berryman and Blair, 1986). In addition, these results also show that the pore space occupied by the wetting phase has the larger contribution to the total specific surface area.

Permeability for the cross section was estimated using the computed values of porosity and specific surface area in conjunction with eq. (26). Assuming a value of 10.0–15.0 for the formation factor, we estimate the intrinsic permeability of the rock to be 220–150 md. This is within a factor of 2 or 3 of the value of 350 md measured in the laboratory for the specimen. The estimated value of the wetting fraction was $\omega = 0.74$ and the ratio of the interface surface area to the total surface area was found to be $s_I/s_T = 0.087$, both of which are consistent with observations of the images. Thus, the wetting phase does in fact cover most of the surface area and the wetting and nonwetting phases do not actually have much interfacial contact throughout the sample.

We also estimated relative permeability for the wetting and nonwetting phases, first using (38) and (39). Estimated values of relative permeability for both phases are given in Table 1. These results show qualitatively that the wetting phase dominates the specific surface area, while the nonwetting phase dominates the permeability owing to its higher partial saturation. This also indicates that for the imaged cross section, the wetting phase is filling the dead-end pores, and that much of the specific surface area is contained in those dead-end pores.

Although the results for the wetting phase were quantitatively sensible ($k_W \simeq 0.02$), the results for the nonwetting phase were not, since the computed relative permeability for the nonwetting phase was found to be greater than unity ($k_{rNW} \simeq 3.7$). It is possible that the relative permeability found in an actual two-phase flow measurement might be greater than unity if the wetting phase acts as a lubricant for the nonwetting phase (by relaxing the no-slip boundary condition), thus permitting it to flow through the pores more easily than in the absence of the wetting phase. But the relative permeability cannot be greater than unity in our procedure because such lubrication effects are neglected. Furthermore, Berryman and Blair (1987) present a theorem stating that the permeability of a less-porous material is *always* less than or equal to that of a more-porous material when the pore space of the less-porous material is contained in the pore space of the more-porous one. Since the nonwetting phase has intruded into the rock's pore space and not created new pore space in the process, this problem satisfies the conditions of the theorem and therefore we may conclude that it is physically unacceptable to ignore the effects of percolation thresholds and the increase in tortuosity for the nonwetting phase as we have done so far.

Thus, to improve our estimates of relative permeability, we *must* incorporate eq. (30) into eqs. (31) and (32). Assuming $\alpha = \beta = 2$, eqs. (38) and (39) can be written as

$$k_{rW} = \frac{\theta^3(\theta - \theta_W^c)^2}{(s_W/s_T)^2} \tag{42}$$

and

$$k_{rNW} = \frac{(1-\theta)^3(\theta_{NW}^c - \theta)^2}{(s_{NW}/s_T)^2} \tag{43}$$

where the relative permeabilities are no longer expressed as functions of saturation and the surface areas alone, but also depend on the percolation thresholds. We do not have an independent measure of the value of the percolation threshold for the nonwetting phase θ_{NW}^c, but it is interesting to note that we can use (43) together with the theorem mentioned previously (showing that $k_{rNW} \leq 1$ is required physically) to obtain an estimate of the value of the percolation threshold. This computation shows that $(1 - \theta_{NW}^c)\phi \geq 0.05$ and that $\theta_{NW}^c \leq 0.75$. These numbers are completely consistent with the fact that the measured nonwetting saturation $1 - \theta \simeq 0.77 \geq 0.25 \simeq 1 - \theta_{NW}^c$. The computation (by construction) yields a physically consistent set of numbers for the various permeabilities and the threshold values.

Another nonrigorous relation we can use to gain some information is the approximate relation $\theta_W^c \simeq 1 - \theta_{NW}^c$ between the percolation thresholds for the wetting and the nonwetting phases. Thus, we find that $\theta_W^c \simeq 0.25$, with some (possibly significant) experimental error. The measured value of the wetting saturation was $\theta = 0.23$, so the wetting phase was imbibed at close to its percolation threshold. The value of the relative permeability obtained from (33) is already so low that any corrections from the presence of the percolation threshold value in (42) will be neglected. Clearly, the wetting phase permeability is very low in this example, and we would be pushing the image processing method well beyond its limitations to try to give a more precise estimate.

Accurate estimates of the percolation threshold values also should not be expected from this approach since the exponents α and β (assumed here to follow Archie's law) must be obtained simultaneously along with the threshold values.

Specific surface area for each phase was also used with the fractional pore volume of each phase to compute the effective hydraulic radius $r = \phi/s$, for the total and segregated pore space. These values are also shown in Table 1 and indicate that the effective hydraulic radius of the pore space occupied by the nonwetting phase is larger than that of the total pore space.

7. Discussion

The analysis in this paper has stressed two-phase flow through porous media, but it should be clear that generalization to multiphase flow is possible. For example, Oak et al. (1990) summarize the current understanding of relative permeability for three-phase flow. In the presence of water, gas, and oil, our analysis should be generalized from the 2×2 system in (11)–(12) to a 3×3 system with six independent coefficients instead of three. The result (23) for the relationship among the surface areas is also easily generalized to

$$s_{\text{water}} + s_{\text{gas}} + s_{\text{oil}} = s_{\text{pore}} + 2s_{w/g} + 2s_{w/o} + 2s_{g/o} \tag{44}$$

The rigorous result for two-phase flow (23) has been shown to be a very valuable tool for checking the consistency of computed surface areas, and (44) may also prove useful for three-phase flow analysis.

Our analysis of relative permeability has shown that Kozeny–Carman relations and Archie's empirical laws must be modified to account for a finite percolation threshold. Our continuing effort to estimate physical properties of rocks from images has therefore led us to an unexpected result. A simplified theory that ignores the percolation thresholds leads to unphysical values (>1) for the relative permeability of the nonwetting phase. Inclusion of the percolation thresholds in the calculation leads to a set of physically reasonable numbers for the various permeabilities, and threshold values themselves. This approach cannot be used to find accurate estimates of the percolation threshold values since the exponents α and β must be determined simultaneously, but with assumed values for the exponents a consistent picture of the physical parameters can be obtained.

Acknowledgments

The authors thank D.H. Rothman for helpful discussions and for sharing the results of his numerical experiments prior to publication. Deepak Agrawal provided the sample for analysis and N.G.W. Cook contributed several useful suggestions. This work was performed under the auspices of the U.S. Department of Energy by the Lawrence Livermore National Laboratory under contract No. W-7405-ENG-48 and supported specifically by the DOE Office of Basic Energy Sciences, Division of Engineering and Geosciences.

References

Archie, G.E. (1942). The electrical resistivity log as an aid in determining some reservoir characteristics. *Trans. AIME* **146**, 54–62.

Balberg, I. (1986). Excluded-volume explanation of Archie's law. *Phys. Rev. B* **33**, 3618–3620.

Bear, J. (1979). *Hydraulics of Groundwater*, pp. 206–213. McGraw-Hill, New York.

Bernabé, Y., Brace, W.F., and Evans, B. (1982). Permeability, porosity, and pore geometry of hot-pressed calcite. *Mech. Mater.* **1**, 173–183.

Berryman, J.G. (1985). Measurement of spatial correlation functions using image processing techniques. *J. Appl. Phys.* **57**, 2374–2384.

Berryman, J.G. (1987). Relationship between specific surface area and spatial correlation functions for anisotropic porous media. *J. Math. Phys.* **28**, 244–245.

Berryman, J.G. and Blair, S.C. (1986). Use of digital image analysis to estimate fluid permeability of porous materials: Application of two-point correlation functions. *J. Appl. Phys.* **60**, 1930–1938.

Berryman, J.G. and Blair, S.C. (1987). Kozeny–Carman relations and image processing methods for estimating Darcy's constant. *J. Appl. Phys.* **62**, 2221–2228.

Berryman J.G. and Milton, G.W. (1985). Normalization constraint for variational bounds on fluid permeability. *J. Chem. Phys.* **83**, 754–760.

Berryman, J.G., Thigpen, L., and Chin, R.C.Y. (1988). Bulk elastic wave propagation in partially saturated porous solids. *J. Acoust. Soc. Am.* **84**, 360–373.

Brace, W.F. (1977). Permeability from resistivity and pore shape. *J. Geophys. Res.* **82**, 3343–3349.

Corey, A.T. (1977). *Mechanics of Heterogeneous Fluids in Porous Media*, pp. 92–120, Water Resources Publications, Fort Collins, Colo.

Debye, P., Anderson, H.R., and Brumberger, H. (1957). Scattering by an inhomogeneous solid. II. The correlation function and its application. *J. Appl. Phys.* **28**, 679–683.

deGennes, P.G. (1983). Theory of slow biphasic flows in porous media. *Physico-Chem. Hydrodyn.* **4**, 175–185.

Doyen, P.M. (1988). Permeability, conductivity, and pore geometry of sandstone. *J. Geophys. Res.* **93**, 7729–7740.

Dullien, F.A.L. (1979). *Porous Media: Fluid Transport and Pore Structure*. Academic Press, New York.

Fitts, D.D. (1962). *Nonequilibrium Thermodynamics: A Phenomenological Theory of Irreversible Processes in Fluid Systems*, pp. 37–40, 152–154. McGraw-Hill, New York.

Kalaydjian, F. (1990). Origin and quantification of coupling between relative permeabilities for two-phase flows in porous media. *Transport in Porous Media* **5**, 215–229.

Katz, A.J. and Thompson, A.H. (1986). Quantitative prediction of permeability in porous rock. *Phys. Rev. B* **34**, 8179–8181.

Katz, A.J. and Thompson, A.H. (1987). Prediction of rock electrical conductivity from mercury injection measurements. *J. Geophys. Res.* **92**, 599–607.

Koplik, J., Lin, C., and Vermette, M. (1984). Conductivity and permeability from microgeometry. *J. Appl. Phys.* **56**, 3127–3131.

Landau, L.D. and Lifshitz, E.M. (1959). *Fluid Mechanics*, pp. 53–54, Pergamon Press, London.

Landau, L.D. and Lifshitz, E.M. (1969). *Statistical Physics*, pp. 374–378, 398. Addison-Wesley, Reading, Mass.

Madden, T.R. (1976). Random networks and mixing laws. *Geophysics* **41**, 1104–1125.

Oak, M.J., Baker, L.E., and Thomas, D.C. (1990). Three-phase relative permeability of Berea sandstone. *J. Petroleum Technol.* **42**, 1054–1061.

Paterson, M.S. (1983). The equivalent channel model for permeability and resistivity in fluid-saturated rock – A re-appraisal. *Mech. Mater.* **2**, 345–352.

Raats, P.A.C. and Klute, A. (1968). Transport in soils: The balance of momentum. *Soil Sci. Soc. Am. Proc.* **32**, 452–456.

Rose, W. (1988). Measuring transport coefficients necessary for the description of coupled two-phase flow of immiscible fluids in porous media. *Transport in Porous Media* **3**, 163–171.

Rose, W. (1989). Data interpretation problems to be expected in the study of coupled fluid flow in porous media. *Transport in Porous Media* **4**, 185–198.

Rothman, D.H. (1990). Macroscopic laws for immiscible two-phase flow in porous media: Results from numerical experiments. *J. Geophys. Res.* **95**, 8663–8674.

Rubinstein, J. and Torquato, S. (1989). Flow in random porous media: Mathematical formulation, variational principles, and rigorous bounds. *J. Fluid Mech.* **206**, 25–46.

Sen, P.N. (1984). Grain shape effect on dielectric and electrical properties of rocks. *Geophysics* **49**, 586–587.

Sen, P.N., Scala, C., and Cohen, M.H. (1981). A self-similar model for sedimentary rocks with application to the dielectric constant of fused glass beads. *Geophysics* **46**, 781–795.

Thigpen, L. and Berryman, J.G. (1985). Mechanics of porous elastic materials containing multiphase fluid. *Int. J. Eng. Sci.* **23**, 1203–1214.

Thompson, A.H., Katz, A.J., and Krohn, C.E. (1987). The microgeometry and transport properties of sedimentary rock. *Adv. Phys.* **36**, 625–694.

Walsh, J.B. and Brace, W.F. (1984). The effect of pressure on porosity and the transport properties of rock. *J. Geophys. Res.* **89**, 9425–9431.

Whitaker, S. (1986). Flow in porous media II: The governing equations for immiscible, two-phase flow. *Transport in Porous Media* **1**, 105–125.

Wong, P.-Z., Koplik, J., and Tomanic, J.P. (1984). Conductivity and permeability of rocks. *Phys. Rev. B* **30**, 6606–6614.

Chapter 8
The Determination of Permeability and Storage Capacity: Pore Pressure Oscillation Method

G.J. Fischer

Research School of Earth Sciences, Australian National University, Canberra 2601, Australia

Abstract

With the aim of exploring the possibility of monitoring the changes in the pore space as they take place during deformation of rocks at high pressure and temperature, we have implemented a method which utilizes the measurement of the attenuation and phase retardation of a sinusoidal pore pressure variation as it propagates through a sample under test. In this paper, the theory and some design considerations as well as detailed discussion of adequate data analysis are presented. The results of our measurement are presented in Chapter 9 of this volume.

1. Introduction

With the increasing appreciation of the role of fluids and their movement through the Earth's crust and upper mantle there is a need for an experiment in which the permeability of tight rock can readily be measured and investigations into the interaction between rock masses and fluids under controlled conditions are also possible. Provided that data are gathered on a timescale sufficiently short in comparison with the rate of changes taking place within the rock, it seems feasible to obtain both sets of information simultaneously. In the following two sections the two most popular methods of measuring permeability are briefly overviewed with the aim of discussing their applicability in terms of permeability determination as well as their usefulness for the characterization of changes taking place during fluid flow through rock masses. The remaining part of the paper deals with a laboratory technique which utilizes an oscillating pore pressure system and which is a strong candidate for experiments where high sensitivity, expediency, and a measurement on 'tight' porous bodies are needed.

FAULT MECHANICS AND TRANSPORT PROPERTIES OF ROCKS
ISBN 0-12-243780-2

1.1. Constant-pressure-difference method

When the determination of permeability of a confined or unconfined sample is needed, the method of constant pressure difference is the easiest to use. To verify its usefulness and to find its practical limitations, we consider the boundary value problem

$$\frac{\partial p^2}{\partial x^2} - \frac{\beta_s \eta}{k} \frac{\partial p}{\partial t} = 0 \tag{1}$$

with initial and boundary conditions

$$\begin{aligned} p(x, 0) &= 0 \qquad \text{for } 0 < x < L \\ p(0, t) &= p_1 \\ p(L, t) &= p_2 \end{aligned} \tag{2}$$

where

p = the pressure in the sample
x = the distance from the downstream face of the sample
t = the time elapsed from the start of the experiment
L = the length of the sample
k = the permeability of the sample
η = the dynamic viscosity of the pore fluid
β_s = the storage capacity per unit volume of the sample–fluid system. It is given by (Brace et al., 1968)

$$\beta_s = C_b + \phi C_f - (1 + \phi) C_r$$

where
C_b = the bulk compressibility of the sample
C_f = the compressibility of the pore fluid
C_r = the mean compressibility of the rock-forming minerals
ϕ = the porosity of the sample

If we are interested in the steady-state part only, i.e., in the situation where the pressure at each point is independent of time and is only a function of x, eq. (1) reduces to

$$\frac{\partial p^2}{\partial x^2} = 0 \tag{3}$$

which is satisfied by

$$p = p_1 + (p_2 - p_1)\frac{x}{L}$$

For experimental purposes, however, when the usefulness of the steady-state method is to be evaluated, it is also essential to know the behavior of the transient part of the solution in the time domain.

Since our boundary conditions are independent of time, we shall proceed as suggested by Carslaw and Jaeger (1959, p. 94), and for (1) and (2) we obtain

$$p = p_1 + (p_2 - p_2)\frac{x}{L} + \frac{2}{L}\sum_{n=1}^{\infty} \sin\frac{n\pi x}{L} e^{-Kn^2\pi^2 t/\beta_s L^2} \int_0^L (p_1 - p_2)\frac{\xi}{L} \sin\frac{n\pi\xi}{L}\, d\xi$$

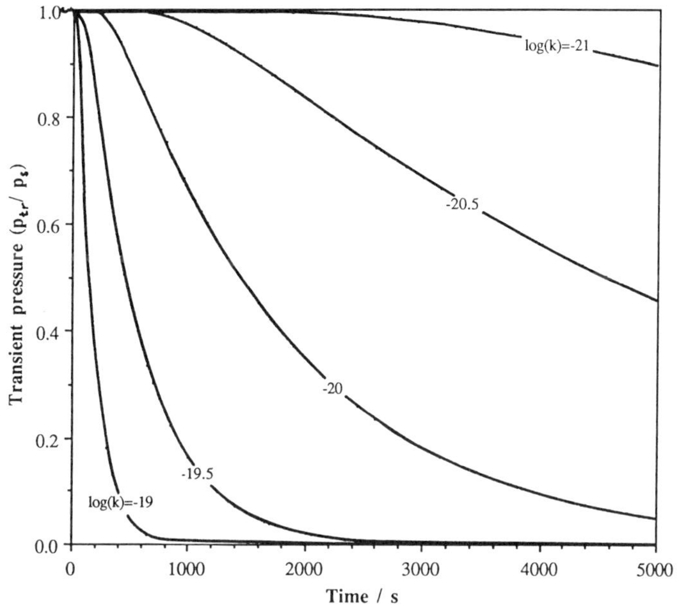

Figure 1. The decay of the transient part of the pressure history within a sample measured by the method of constant pressure difference. Numbers on curves signify log (permeability/m^2). Pressure normalized to steady-state values p_s.

where $K = k/\eta$, which upon integration results in

$$p = p_1 + (p_2 - p_1)\frac{x}{L} + \frac{2(p_2 - p_1)}{\pi} \sum_{n=1}^{\infty} \frac{(-1)^{n+1}}{n} \sin\frac{n\pi x}{L} e^{-Kn^2\pi^2 t/\beta_s L^2} \tag{4}$$

Figure 1 shows the behavior of the transient part of (4) in the time domain for various permeabilities at a constant storage capacity per unit volume. After a sufficient time has elapsed and the transient part has decayed below a significant level, distribution within the sample should be linear and Darcy's law can be used directly for the estimation of the permeability.

For the sake of illustration, assume an experiment with water (viscosity $\eta = 380\ \mu$Pa s) flowing through a 2-cm long quartz aggregate of 10 mm diameter. The storage capacity of our downstream reservoir is $10^{-15}\,\mathrm{m^3\,Pa^{-1}}$ and we want to maintain a pressure difference of $\Delta p = 10$ MPa across the specimen. It is easy to see from eq. (4) and Figure 2 that the maximum departure from the steady-state linear pressure distribution will always drift to $x = L/2$, and if this is to be less than, say, $\Delta p/100$, a waiting time of about 6400 s (107 minutes) is necessary for a sample with permeability of $10^{-20}\,\mathrm{m^2}$ and storage capacity per unit volume of $10^{-9}\,\mathrm{Pa^{-1}}$. If we monitored the pressure with a pressure transducer of resolution of $p_i = 100$ kPa, it would take time

$$t_i = \frac{\beta_d p_i \eta}{kA\,\Delta p} = 50\,\mathrm{s}$$

where

A = the cross-sectional area of the sample
β_d = storage of the downstream reservoir; it is defined as volume V_f of fluid at pressure

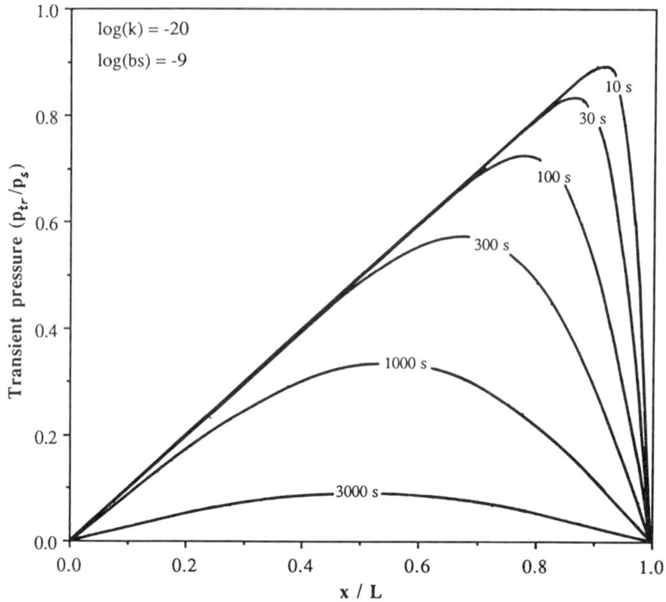

Figure 2. The departure of the pressure within a sample from the steady-state level for various times. The pressures in the upper and lower reservoirs are kept at constant levels. Pressure normalized to steady-state values p_S.

p_0, which must be injected/released from the reservoir in order to cause a unit change of pressure in that reservoir

to build up the pressure increment detectable by our transducer.

1.2. Transient step method

Because it is technically much easier to measure changes in pressure than flow rate in low-permeability experiments, Brace et al. (1968) introduced and applied the method of transient step. In this method, two reservoirs are connected through a sample. A sudden increase of pressure is induced in one of the reservoirs and the system is then allowed to equilibrate to a new pressure level, which is determined by the magnitude of the step, the storage capacity of the sample, and the storage capacities of the reservoirs. One is thus interested in the transient part of the solution, from which the sample permeability (and also storage capacity or bulk compressibility) can be derived. The equations describing such a situation are

$$\frac{\partial^2 p}{\partial x^2} - \frac{\beta_s \eta}{k}\frac{\partial p}{\partial t} = 0 \qquad \text{for } 0 < x < L \text{ and } t > 0$$

$$p = 0 \qquad \text{for } 0 < x < L \text{ and } t = 0$$

$$p = p_d(t) \qquad \text{for } x = 0 \text{ and } t > 0$$

$$p = p_u(t) \qquad \text{for } x = L \text{ and } t > 0$$

where d and u subscripts refer to the downstream and upstream reservoirs, respectively.

These pressures are defined by

$$\left(\frac{\partial p}{\partial x}\right)_{x=0} - \frac{\beta_d \eta}{kA}\frac{dp_d}{dt} = 0 \qquad \text{for } t > 0$$

$$p_d = 0 \qquad \text{at } t = 0$$

$$\left(\frac{\partial p}{\partial x}\right)_{x=L} + \frac{\beta_u \eta}{kA}\frac{dp_u}{dt} = 0 \qquad \text{for } t > 0$$

$$p_u = P_0 \qquad \text{at } t = 0$$

Brace et al. used a simplified version of this problem assuming no storage capacity of the sample and found a solution in the form

$$p = P_0\left(\frac{V_2}{V_1} + V_2\right)e^{-\alpha t} + P_f$$

where α is a function of permeability, V_1 and V_2 are the volumes of the upstream and downstream reservoirs, and P_f is the final equilibrated pressure.

This was further improved by Lin (1977) and the boundary value problem and its application has finally undergone a very thorough treatment by Hsieh et al. (1981) and Neuzil et al. (1981), who described the pressure history in both reservoirs in terms of a set of dimensionless parameters. From the comparison of the theoretical solution for these dimensionless parameters with the experimental data drawn in the same scale, one can easily graphically estimate the permeability and the storage capacity of the rock under study.

2. Sinusoidal Oscillation

2.1. General solution for linear pressure transfer into the downstream reservoir

The transient step method first used by Brace et al. (1968) and refined by Hsieh et al. (1981) and Neuzil et al. (1981) is very easy to use, but it is a grahical method relying on measurements taken over long periods of time. In cases of samples of low permeability its interpretation can easily be complicated by a noisy signal. The Fourier spectrum of the initial step contains all frequencies and so the transient step method is insensitive to any frequency dependence of the measured parameters. For this reason an application of a pressure variation of controlled frequency is desirable. It also permits computerized data processing via the fast Fourier transform (FFT) and that, as will be shown later, leads to an improvement of the precision due to its 'signal enhancement' effects. Thus another boundary value problem is presented in which the initial pressure is constant and uniformly distributed throughout the specimen; sinusoidal pressure variation is induced at time zero in the 'upstream' reservoir and linear pressure transfer takes place at the 'downstream' face connected to a 'downstream' reservoir with storage capacity

β_d. The problem is described by

$$\frac{\partial^2 p}{\partial x^2} - \frac{\beta_s \eta}{k}\frac{\partial p}{\partial t} = 0 \qquad \text{for } 0 < x < L \text{ and } t > 0 \tag{5}$$

$$p(x, 0) = 0 \qquad \text{for } 0 < x < L$$

$$p(0, t) = p_d(t) \qquad \text{for } t \geq 0$$

$$p(L, t) = p_u(t) \qquad \text{for } t \geq 0$$

p_d and p_u are functions of time and obey equations

$$\frac{\beta_d \eta}{kA}\frac{dp_d}{dt} - \left(\frac{\partial p}{\partial x}\right)_{x=0} = 0 \qquad \text{for } t > 0$$

$$p_d(0) = 0$$

$$p_u = P_A \sin(\omega t + \delta)$$

In this set of equations

p = the pressure in the sample
p_d = the pressure in the downstream reservoir
p_u = the pressure in the upstream reservoir
x = the distance from the downstream face of the sample
t = the time elapsed from the start of the experiment
P_A = the amplitude of the generated oscillation
L = the length of the sample
A = the cross-sectional area of the sample
k = the permeability of the sample
η = the dynamic viscosity of the pore fluid
β_s = the storage capacity per unit volume of the sample–fluid system. It is given by (Brace et al. (1968))

$$\beta_s = C_b + \phi C_f - (1 + \phi)C_r$$

where

C_b = the bulk compressibility of the sample
C_f = the compressibility of the pore fluid
C_r = the mean compressibility of the rock-forming minerals
ϕ = the porosity of the sample
β_d = storage of the downstream reservoir (defined as volume V_f of fluid at pressure p_0, which must be injected/released from the reservoir in order to cause a unit change of pressure in that reservoir)

The solution takes the form (see Appendix A):

$$p = p_{\text{per}} + p_{\text{tr}} = P_A \alpha \sin(\omega t + \delta + \vartheta) + 2P_A K \beta_s AL$$

$$\times \sum_{n=1}^{\infty} \frac{(\beta_s L^2 \omega \cos\delta - K\psi_n^2 \sin\delta)\left[\cos(\xi\psi_n) - \dfrac{\beta_d \psi_n}{\beta_s AL}\sin(\xi\psi_n)\right]}{(K^2\psi_n^4 + \beta_s^2 L^4 \omega^2)(\beta_d \psi_n^2 - \beta_s AL)\cos\psi_n + (\beta_s AL + 2\beta_d)\sin\psi_n]} \psi_n^2 e^{-K\psi_n^2 t/\beta_s L^2} \tag{6}$$

and for $\delta = 0$,

$$p = \alpha P_A \sin(\omega t + \vartheta) + 2P_A K\beta_s^2 A L^3 \omega$$

$$\times \sum_{n=1}^{\infty} \frac{\cos(\xi\psi_n) - \dfrac{\beta_d \psi_n}{\beta_s AL}\sin(\xi\psi_n)}{(K^2\psi_n^4 + \beta_s^2 L^4\omega^2)[(\beta_d\psi_n^2 - \beta_s AL)\cos\psi_n + (\beta_s AL + 2\beta_d)\sin\psi_n]}\psi_n^2 e^{-K\psi_n^2 t/\beta_s L^2} \tag{7}$$

α and ϑ being defined by

$$\alpha = \left| \frac{\kappa(1+i)\cosh[\kappa x(1+i)] + \dfrac{\beta_d i\omega}{KA}\sinh[\kappa x(1+i)]}{\kappa(1+i)\cosh[\kappa L(1+i)] + \dfrac{\beta_d i\omega}{KA}\sinh[\kappa L(1+i)]} \right| \tag{8}$$

$$\vartheta = \arg\left\{ \frac{\kappa(1+i)\cosh[\kappa x(1+i)] + \dfrac{\beta_d i\omega}{KA}\sinh[\kappa x(1+i)]}{\kappa(1+i)\cosh[\kappa L(1+i)] + \dfrac{\beta_d i\omega}{KA}\sinh[\kappa L(1+i)]} \right\} \tag{9}$$

2.2. Limiting cases

When the volume of the downstream reservoir becomes negligibly small, there is no flow across the lower boundary. The boundary condition for $x = 0$ becomes

$$\left(\frac{\partial p}{\partial x}\right)_{x=0} = 0 \qquad \text{for } t > 0$$

The solution of (5) is

$$p = \alpha P_A \sin(\omega t + \delta + \vartheta) + 2P_A K\pi^2 \sum_{n=1}^{\infty} \frac{(-1)^n(2n+1)^2 4\beta_s L^2\omega\cos\delta - K(2n+1)^2\pi^2\sin\delta}{K(2n+1)^4\pi^4 + 16\beta_s^2 L^4\omega^2}$$

$$\times \psi_n^2 e^{-K\psi_n^2 t/\beta_s L^2} \tag{10}$$

where

$$\alpha = \left|\frac{\cosh[\kappa x(1+i)]}{\cosh[\kappa L(1+i)]}\right|, \qquad \vartheta = \arg\left\{\frac{\cosh[\kappa x(1+i)]}{\cosh[\kappa L(1+i)]}\right\}$$

and similarly for $\delta = 0$,

$$p = \alpha P_A \sin(\omega t + \vartheta) + 8P_A K\beta_s L^2\omega\pi^2 \sum_{n=1}^{\infty} \frac{(-1)^n(2n+1)^2\cos\dfrac{(2n+1)\pi x}{2L}e^{-K(2n+1)^2\pi^2/4\beta_s L^2}}{K(2n+1)^4\pi^4 + 16\beta_s^2 L^4\omega^2} \tag{11}$$

Another special case arises when the storage of the downstream reservoir approaches infinity, i.e., pressure in that reservoir remains constant. Such a case is described by (Appendix B)

$$p = \alpha P_A \sin(\omega t + \delta + \vartheta) + 2\pi PK \sum_{n=0}^{\infty} \frac{n(-1)^n(\beta_s L^2\omega\cos\delta - Kn^2\pi^2\sin\delta)}{K^2n^4\pi^4 + \beta_s^2 L^4\omega^2}\sin(n\pi\xi)e^{-Kn^2\pi^2/4\beta_s L^2}$$

where

$$\alpha = \left|\frac{\sinh[\kappa x(1+i)]}{\sinh[\kappa L(1+i)]}\right|, \qquad \vartheta = \arg\left\{\frac{\sinh[\kappa x(1+i)]}{\sinh[\kappa L(1+i)]}\right\}$$

2.3. Steady-state solution for semi-infinite sample

In many cases (for design purposes, for example) we can assume that the upstream face will not feel the effects of the downstream end. To simplify the calculations one can proceed as if one had a semi-infinite sample. Eq. (5) will apply, with boundary and initial conditions

$$p(0,t) = P_A \sin(\omega t) \qquad \text{and} \qquad p(x,0) = 0 \tag{12}$$

which results in (Carslaw and Jaeger, 1959, p. 65)

$$p = P_A \sin\left(\omega t - x\sqrt{\omega\eta\beta_s/2k}\right)e^{-x\sqrt{\omega\eta\beta_s/2k}} \tag{13}$$

An alternative, simpler derivation is presented in Appendix C.

2.4. Design considerations

Upstream end When designing the 'upstream' end of the apparatus one is primarily concerned with the volumes involved and with the kinetics of the flow through the upper face of the sample. It is desirable to keep the volume in the upstream reservoir as small as possible, if dilatancy and permeability are to be measured during one deformation experiment, hence one must evaluate first how much fluid will actually be passing through the upstream face of the sample.

Equation (13) is an equation of wave propagation whose amplitude diminishes exponentially with the depth of penetration. It is worth noticing that the rate of decay is frequency-dependent: the lower the frequency, the further a disturbance of detectable level will propagate. The maximum pressure variation at any point x, or the range of pressure p_r, is then simply given by

$$p_r = 2P_A e^{-x\sqrt{\pi\eta\beta_s/kT}} \tag{14}$$

where T is the period of oscillation.

The rate of flow across any cross-section normal to the pressure gradient, provided that Darcy's law holds, is given by

$$Q_x = \frac{k}{\eta} A \frac{\partial p}{\partial x} \tag{15}$$

where

$$\frac{\partial p}{\partial x} = -P_A\sqrt{\frac{\omega\eta\beta_s}{2k}}\exp\left(-x\sqrt{\frac{\omega\eta\beta_s}{2k}}\right)[\cos(\omega t - x\sqrt{\omega\eta\beta_s/2k}) + \sin(\omega t - x\sqrt{\omega\eta\beta_s/2k})$$

The total volume of fluid which passes through the face of the sample in one direction

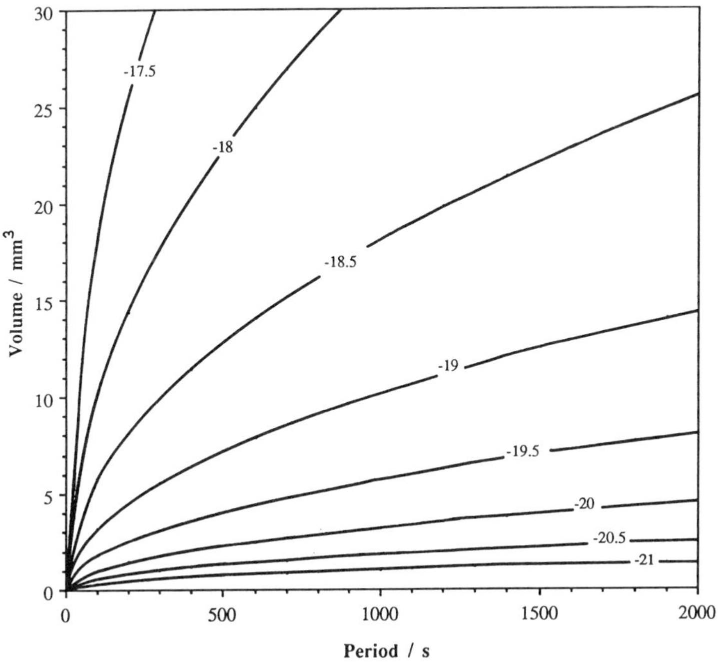

Figure 3. The volume flow across the upstream face during one half-period as a function of period plotted for various permeabilities and a storage capacity per unit volume of 10^{-9} Pa^{-1}. Amplitude of oscillation is 5 MPa.

during one half-period is thus

$$Q = \frac{k}{\eta} A \int_{-\pi/4\omega}^{3\pi/4\omega} \left(\frac{\partial p}{\partial x}\right)_{x=0} d\tau = P_A A \sqrt{\frac{2Tk\beta_s}{\eta\pi}} \tag{16}$$

(notice that the time history of the pressure gradient is out of phase with pressure by $T/8$).

The relation (16) is graphically represented by Figure 3, which illustrates the effects of the period/frequency on the volume displacements.

It is seen from eq. (13) that a maximum or minimum pressure will occur at x when

$$\omega t - x\sqrt{\frac{\omega\eta\beta_s}{2k}} = (2n+1)\frac{\pi}{2} \tag{17}$$

i.e.,

$$t_x = \frac{(2n+1)\dfrac{\pi}{2} + x\sqrt{\omega\eta\beta_s/2k}}{\omega} \tag{18}$$

with even n representing a maximum, odd n representing a minimum.

From eq. (18) the time lag τ, the apparent wave velocity v_a, and the wavelength λ can easily be derived:

$$\tau = x\sqrt{\eta\beta_s T/k\pi} \tag{19}$$

$$v_a = 2\sqrt{k\pi/\eta\beta_s T} \tag{20}$$

$$\lambda = 2\sqrt{k\pi T/\eta\beta_s} \tag{21}$$

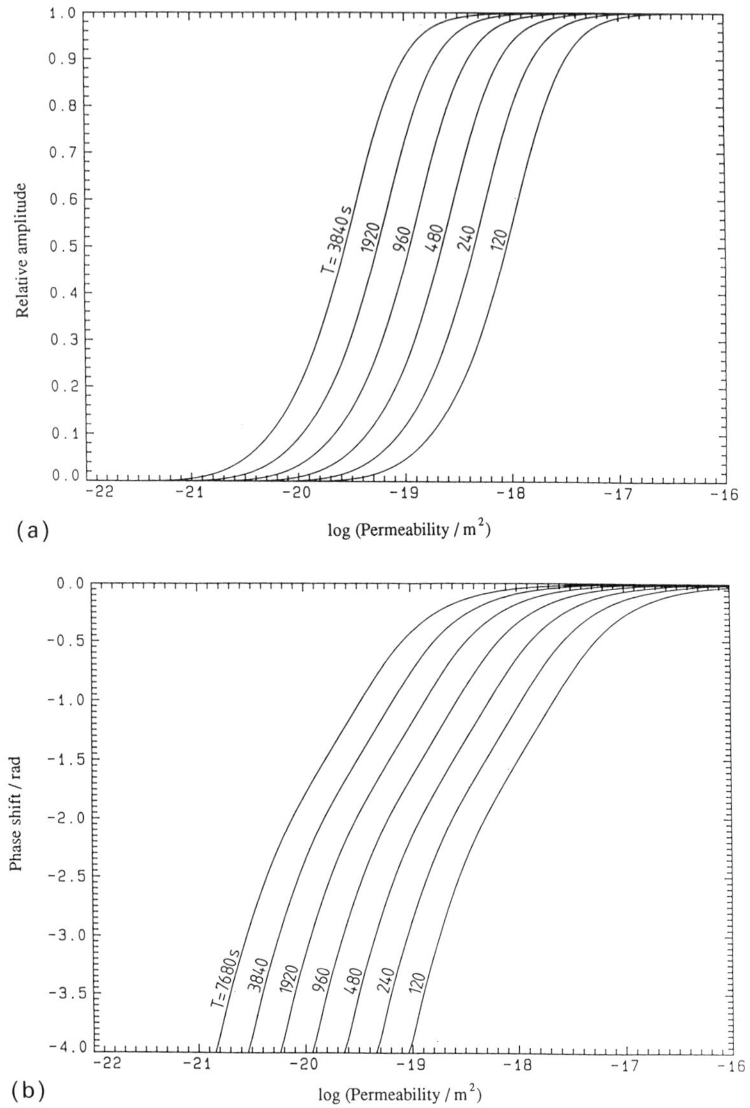

Figure 4. (a) Attenuation as a function of permeability plotted for various periods of oscillation @ $\beta_s = 10^{-9}\,\text{Pa}^{-1}$ and $\beta_d = 10^{-15}\,\text{m}^3\,\text{Pa}^{-1}$. (b) Phase shift as a function of permeability plotted for various periods of oscillation @ $\beta_s = 10^{-9}\,\text{Pa}^{-1}$ and $\beta_d = 10^{-15}\,\text{m}^3\,\text{Pa}^{-1}$. (@ = Critical temperature $T_k = 550$ K at 250 MPa pressure).

It should be pointed out that (a) the apparent wave velocity (eq. 20) is only the velocity of the wave motion and does not represent the actual velocity of the moving fluid and (b) it is proportional to the square root of the frequency of oscillation.

Downstream end While designing the 'downstream' end of the apparatus, one is primarily concerned with the sensitivity that is to be achieved. The sensitivity is determined by the characteristics of the sample, i.e., permeability, storage capacity,

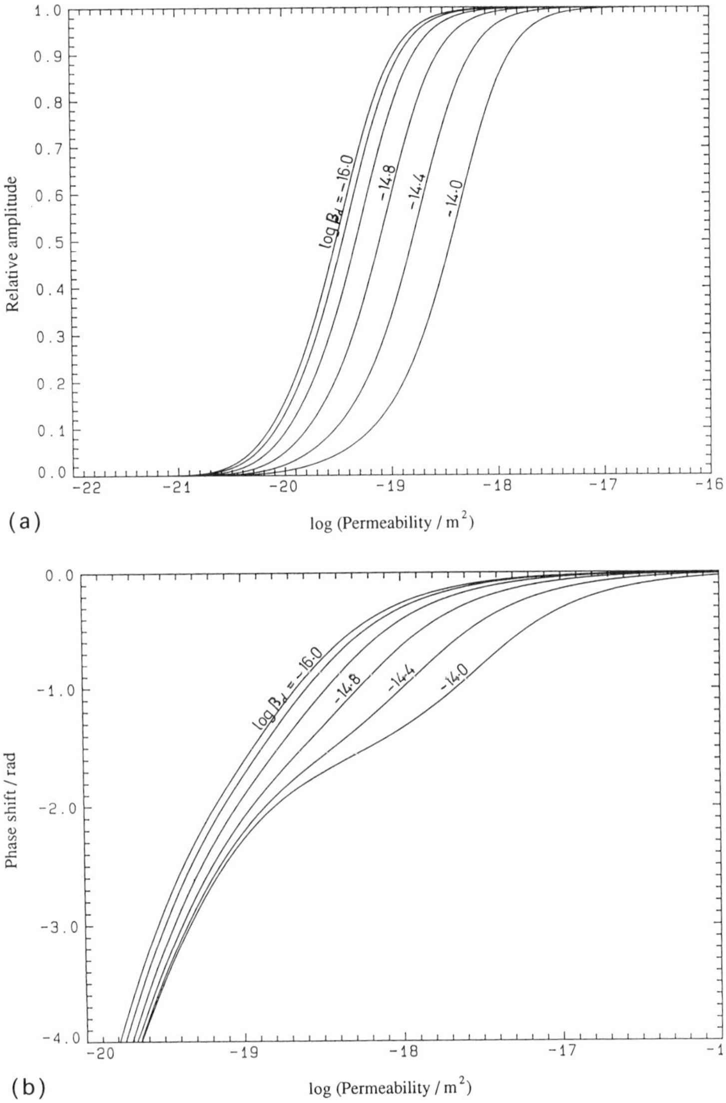

Figure 5. (a) Attenuation as a function of permeability. The influence of downstream storage capacity. $\beta_s = 10^{-9}\,\text{Pa}^{-1}$; period of oscillation $T = 1920\,\text{s}$. (b) Phase shift as a function of permeability. The influence of downstream storage capacity. $\beta_s = 10^{-9}\,\text{Pa}^{-1}$; period of oscillation $T = 1920\,\text{s}$.

length, and diameter, and by the properties of the downstream reservoir, namely its storage capacity. From Figures 4 to 9, which show the relationship between various parameters from eqs. (8) and (9), one can make a number of important observations:

1. The amplitude of the observed oscillations decays and the phase shift grows appreciably with increasing length of the sample.
2. An increase in the period of oscillation results in an increase in the propagated range of pressure and a decrease in the phase lag (but not the time lag).

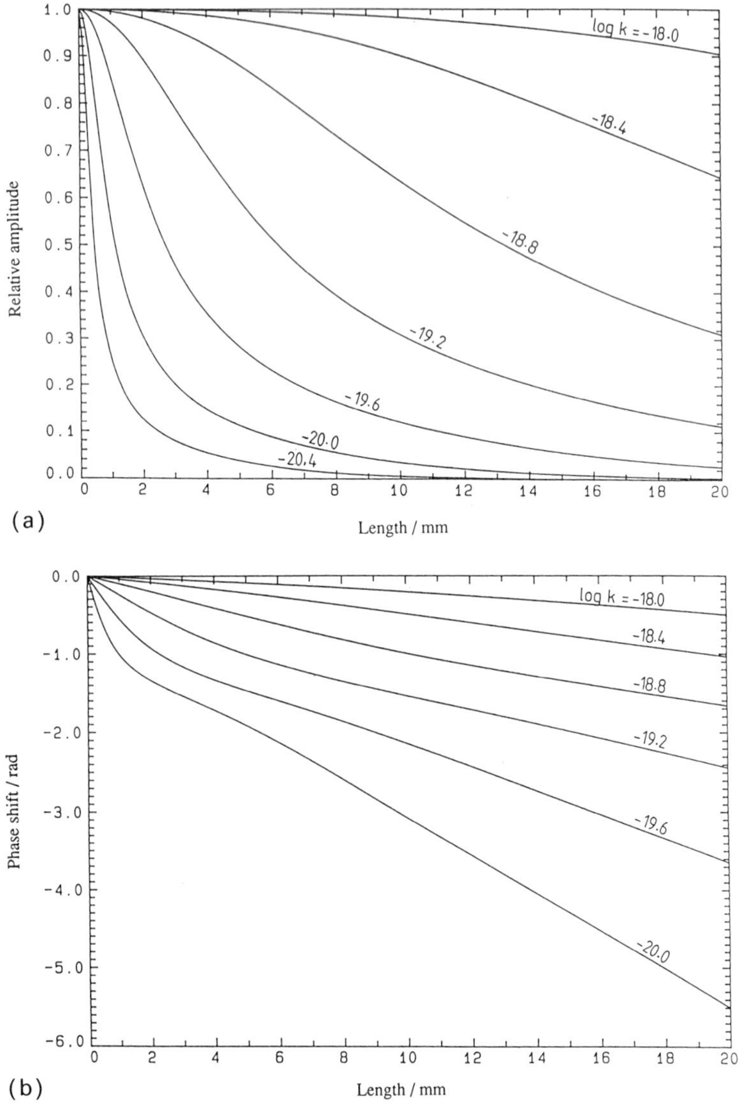

Figure 6. (a) The effect of length on attenuation at various permeabilities at a given sample storage capacity, downstream storage capacity, and oscillation period. (b) The effect of length on phase shift at various permeabilities at a given sample storage capacity, downstream storage capacity, and oscillation period.

3. The tolerance to the storage capacity of the sample of a given permeability is also enhanced by lowering the frequency. For example, if we needed an amplitude ratio of 0.01 or better, then an oscillation period of 240 s would be useless when applied to a rock characterized by $k = 10^{-19.5}\,\mathrm{m}^2$ and $\beta_s = 10^{-9.5}\,\mathrm{m}^3\,\mathrm{Pa}^{-1}$, while 2000 s would be sufficient.
4. Similarly, for every amplitude ratio and phase shift, there is a certain threshold of the downstream storage below which a rock of given parameters could not be successfully tested.

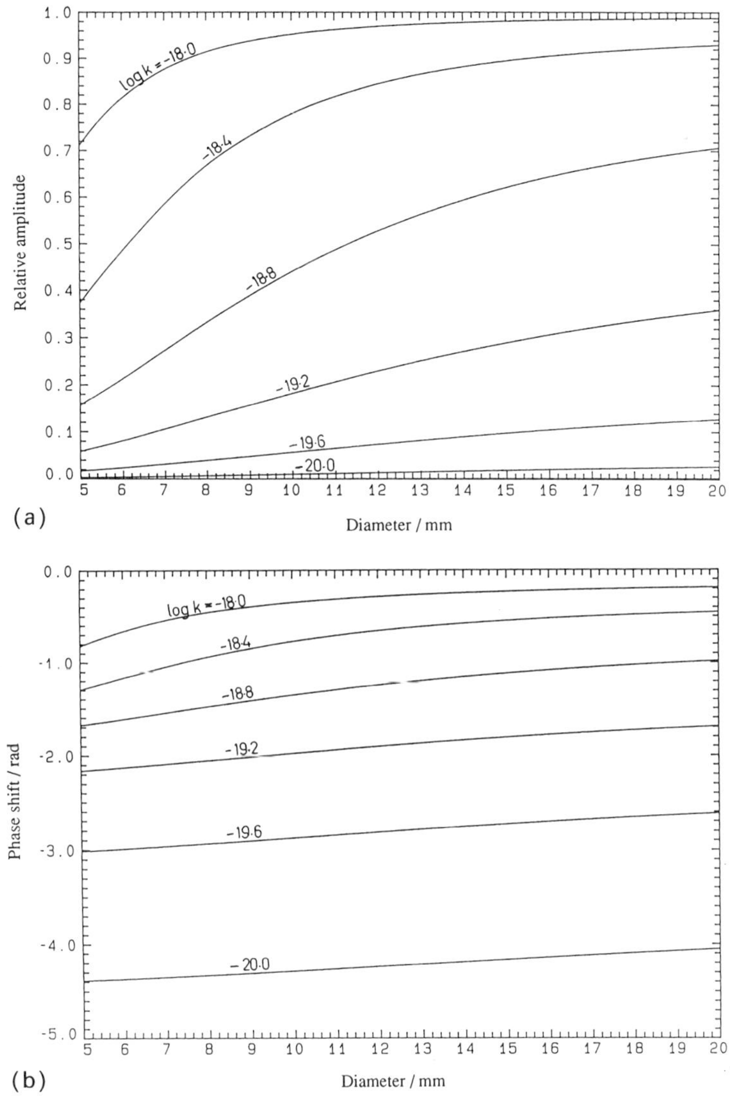

Figure 7. (a) The effect of diameter on attenuation at various permeabilities at a given sample storage capacity, downstream storage capacity, and oscillation period. (b) The effect of diameter on phase shift at various permeabilities at a given sample storage capacity, downstream storage capacity, and oscillation period.

In conclusion the following statements can be made:

(a) Permeability influences primarily the phase shift, while the sample storage capacity is mainly responsible for the amplitude damping.
(b) Shortening the length of the sample and/or increasing its diameter will increase the amplitude ratio and decrease the phase shift.
(c) Decreasing the volume of the downstream reservoir will result in a decrease of the storage capacity β_d, which will lead to larger α and therefore improve significantly the resolution of a measurement.

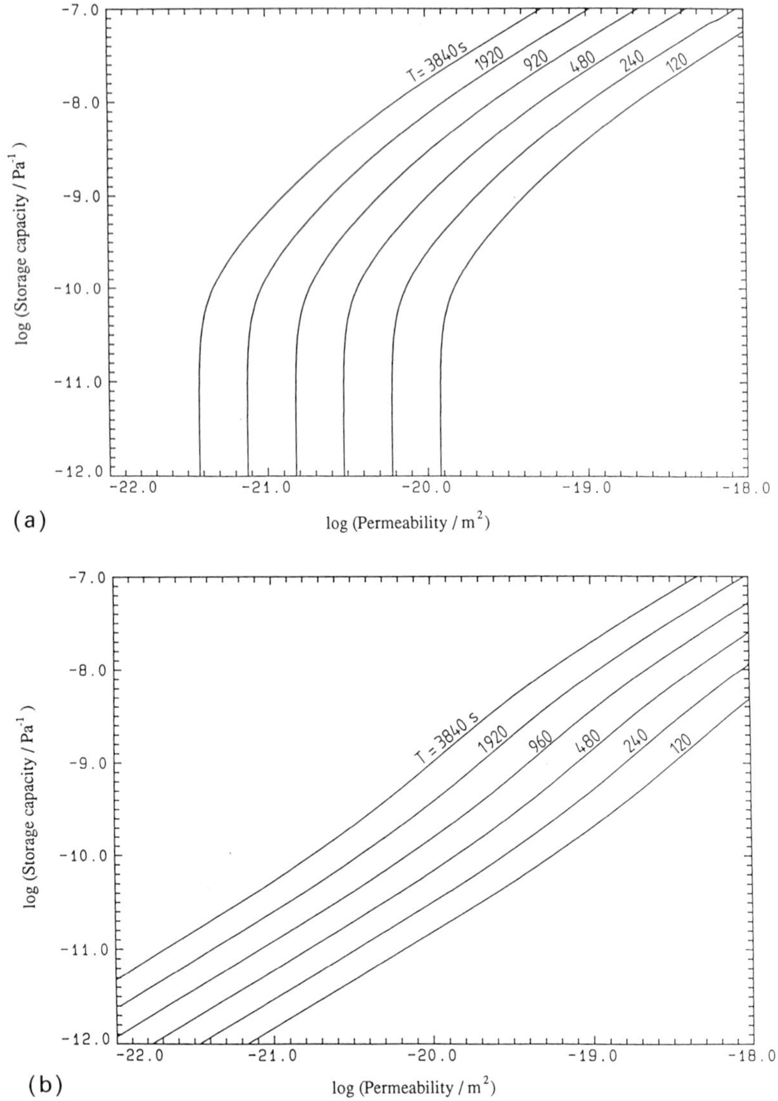

Figure 8. (a) The relation between sample storage capacity per unit volume and permeability, for a given attenuation factor. (b) The relation between sample storage capacity per unit volume and permeability, for a given phase shift.

When choosing the frequency of oscillation it should be realized that although the increase in period results in a more favorable amplitude ratio, it is always advantageous to keep the duration of the experiment as short as practicable. A decrease in phase lag does not necessarily imply a decrease in time lag.

2.5. Data processing

After a sufficient time has elapsed, the transient part of eqs. (6) and (7) will diminish below a significant level and the pressure history will settle into a steady state.

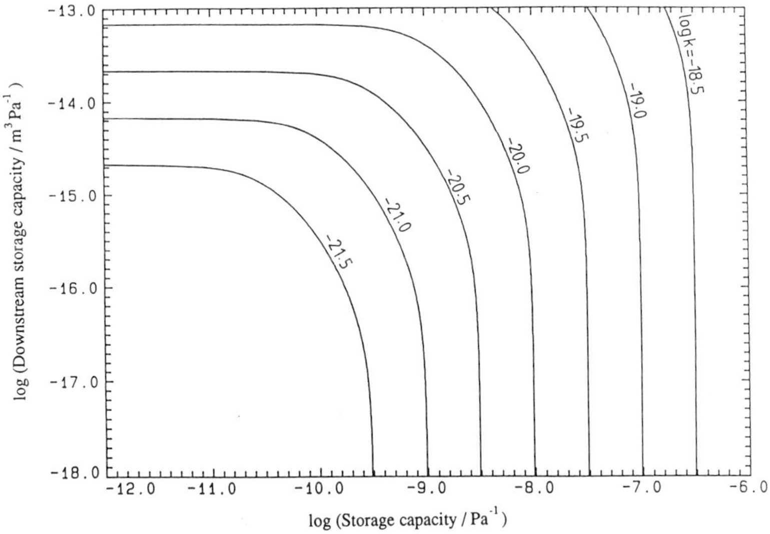

Figure 9. The relation between downstream storage capacity, sample storage capacity per unit volume, and permeability, for a given period, attenuation factor, and phase lag. Plotted at various permeabilities.

In order to obtain the values for permeability and sample storage capacity, one has to solve simultaneously eqs. (8) and (9) for k and β_s, where α and ϑ are determined experimentally.

The signal from the pressure transducers is a sum of three components: (1) true pressure signal due to the forced sinusoidal cycling, (2) true pressure signal due to temperature variations, and (3) electrical noise due to temperature effects on the electronic components, RF signal, background noise, etc. Because the signal of interest lies at one particular frequency, we can make use of the Fourier transform as a tool for elucidating the values of α and ϑ and also as a narrowband filter or a spectrum analyser which resolves the input signal into discrete frequency cells of width $\Delta f = 1/\Delta T$, where ΔT is the length of the data record. A sinusoidal input will presumably appear in only one of these cells, while the remaining signal, if it is a band-limited white noise, will be more or less uniformly distributed among all cells. If the spectral density of the white noise is D_0, then the average noise power of any single cell will be $\Delta f\, D_0$ and if ρ_s is the RMS value of the signal for the cell which also contains the sinusoidal component, then the signal-to-noise ratio (SNR) of the output is

$$(\mathrm{SNR})_{\mathrm{out}} = \frac{\rho_s^2}{\Delta f\, D_0} \tag{22}$$

At the input the average noise power is BD_0, where B is the bandwidth of the input signal, while the power of the sinusoid remains at ρ^2. Hence the input SNR is

$$(\mathrm{SNR})_{\mathrm{in}} = \frac{\rho_s^2}{BD_0} \tag{23}$$

From a comparison of (22) and (23) we see that a signal enhancement takes place during the processing operation. According to Glisson et al. (1970), we shall call that enhancement the processing gain (PG) and express it in decibels as

$$\mathrm{PG} = 10 \log\left[\frac{(\mathrm{SNR})_{\mathrm{out}}}{(\mathrm{SNR})_{\mathrm{in}}}\right] \tag{24}$$

Substitution of (22) and (23) into (24) results in

$$\mathrm{PG} = 10 \log\left(\frac{B}{\Delta f}\right) = 10 \log(BT) \tag{25}$$

This is, however, the maximum processing gain, which we may not achieve in a practical application. The reason lies in the nature of the finite discrete Fourier transform (DFT) (see, for example, Glisson and Sage (1969)), which transforms the input signal spectrum into equispaced samples at multiples of $1/T$. So the amplitude spectrum of a sinusoidal signal of the duration T and frequency f_0 is

$$|S_A(f)| = \left|\frac{\sin[\pi(f - f_0)T]}{\pi(f - f_0)T}\right| \tag{26}$$

and the finite DFT will yield

$$S_A(n/T) = \left|\frac{\sin(\pi n - \pi f_0 T)}{(\pi n - \pi f_0 T)}\right| \qquad \text{for } n = 0, 1, 2, \ldots \tag{27}$$

Thus, since f_0 is an integer multiple n_0 of $1/\Delta T$, eq. (27) is equal to zero for every n except for $n = n_0$, where it is equal to unity. When none of the DFT samples coincides with the signal frequency, that is, when

$$f_0 = \frac{2n_0 + 1}{2T}$$

then the finite DFT yields samples which are nonzero for any n. The maximum of (26) occurs at $n = n_0$ and $n = n_0 + 1$, where

$$\left|S\left(\frac{n_0}{T}\right)\right| = \left|S\left(\frac{n_0 + 1}{T}\right)\right| = \frac{2}{\pi} \tag{28}$$

Hence the actual output signal-to-noise ratio can drop by as much as almost 4 dB when the DFT sampling does not follow the signal frequency, Figure 10.

We proceed now to investigate the influence of white background noise on the determination of the phase and amplitude of the oscillation in the downstream reservoir. Figure 11 shows an amplitude spectrum in frequency domain for a fictitious experiment.

The signal is found at an integral frequency n_0 and it has a relative amplitude of 10^{-1}. At all frequencies there is a background noise present, the amplitude of which reaches 10^{-5}. Therefore, we might expect that at the frequency n_0 a noise of amplitude as much as 10^{-5} can be present. If we use $S\,e^{i\phi_s}$ for the sinusoidal signal and $N\,e^{i\phi_N}$, where ϕ_N is random, for the noise signal, the resultant signal will be

$$R\,e^{i\phi_R} = S\,e^{i\phi_S} + N\,e^{i\phi_N} \tag{29}$$

Thus R and ϕ_N are given by

$$R = S\sqrt{1 + 2\frac{N}{S}\cos(\phi_N - \phi_S) + \left(\frac{N}{S}\right)^2} \tag{30}$$

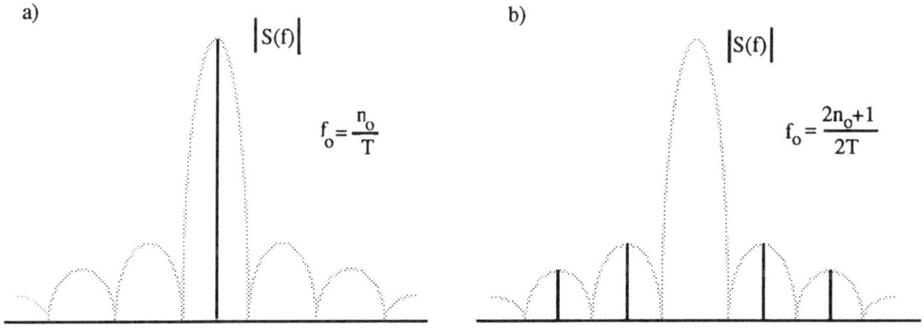

Figure 10. Discrete spectra of a time-limited sinusoid. (a) The signal frequency is an integral multiple of $1/T$. (b) The signal frequency is an odd multiple of $1/2T$. (After Glisson et al., 1970).

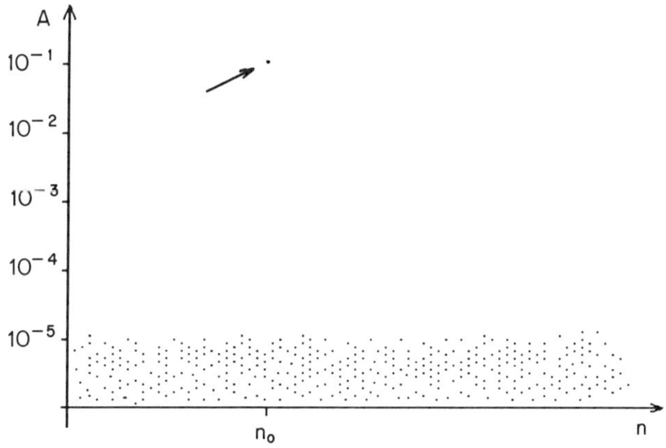

Figure 11. A sample spectrum of a fictitious experiment.

and

$$\phi_R = \arctan\left[\frac{N\sin(\phi_N - \phi_S)}{S + N\cos(\phi_N - \phi_R)}\right] + \phi_S \tag{31}$$

Since we do not know the actual SNR at the particular frequency n_0, eqs. (30) and (31) can only be used as an estimation of the worst possible case.

Other important factors which can seriously contaminate the pressure signal, but which we can be partially corrected, are linear drift, such as a steady leak from the reservoir, and linear modulation. When both of these phenomena are present, and we denote the sinusoidal signal as $s(t) = A\sin(\omega t)$, the modulating function as $m(t) = 1 + \bar{m}t$, and the drift as $d(t) = \bar{d}t$, then the resultant signal will be

$$r(t) = m(t)s(t) + d(t)$$

If we consider a data record of a length nN, where N is the number of samples per

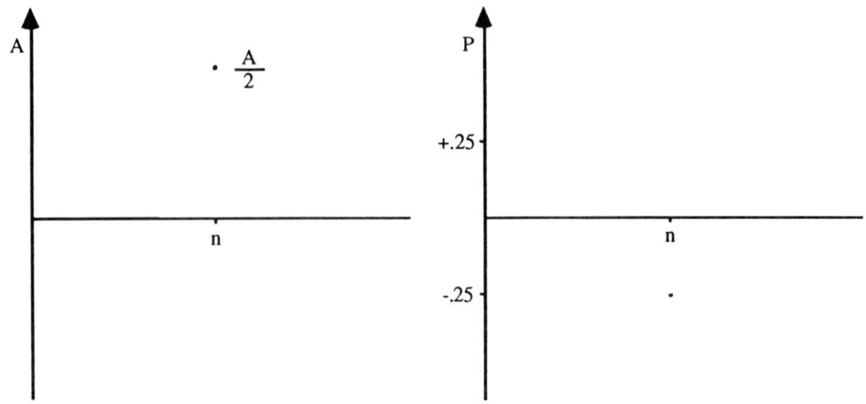

Figure 12. A discrete Fourier transform of a sinusoidal signal $s(t)$.

cycle and n is the number of cycles of a duration T_0, then

$$S(v) = \frac{1}{nN} \sum_{\tau=0}^{nN-1} A \sin\left(\frac{2\pi\tau}{N}\right) e^{-2i\pi v\tau/nN}$$

and

$$S(v) = \frac{A}{2} e^{-i\pi/2} \qquad \text{where } v = n \tag{32}$$

(Figure 12).

The modulated sine wave $S_m(t) = m(t)\,s(t)$ will be transformed into

$$S_m(v) = S(n)[M(v-n) - M(v-n)]$$

that is,

$$S_m(v) = \frac{A}{2} \frac{\bar{m}T_0 n^2}{\pi(n^2 - v^2)} e^{-i\pi} \qquad \text{for } v < n$$

$$S_m(v) = \frac{A}{2}\left(1 + \frac{\bar{m}nT_0}{2}\right) \exp\left[-i\left((\pi/2) + \frac{\bar{m}T_0}{4\pi}\right)\right] \qquad \text{for } v = n \tag{33}$$

$$S_m(v) = \frac{A}{2} \frac{\bar{m}T_0 n^2}{\pi(v^2 - n^2)} \qquad \text{for } v > n$$

(Figure 13).

Finally, the transform of $d(t)$ results in

$$D(v) = \frac{\bar{d}T_0}{2\pi}$$

$$D(v) = \frac{\bar{d}T_0 n}{2\pi v} \exp\left[i\left(\frac{\pi}{2} + \frac{\pi v}{nN}\right)\right]$$

$$= \left|\frac{\bar{d}T_0 n}{2nv}\right| \exp\left[i\left(\frac{\pi v}{nN} + \frac{\pi}{2}\operatorname{sign}(\bar{d})\right)\right] \tag{34}$$

(Figure 14).

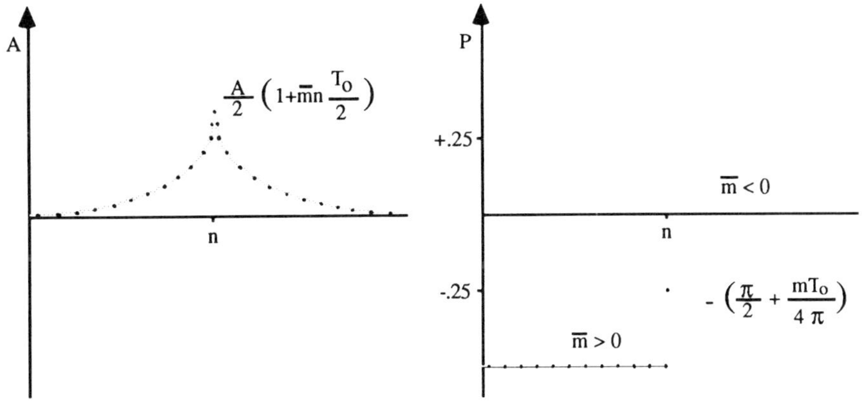

Figure 13. The DFT of a modulating function $m(t)$.

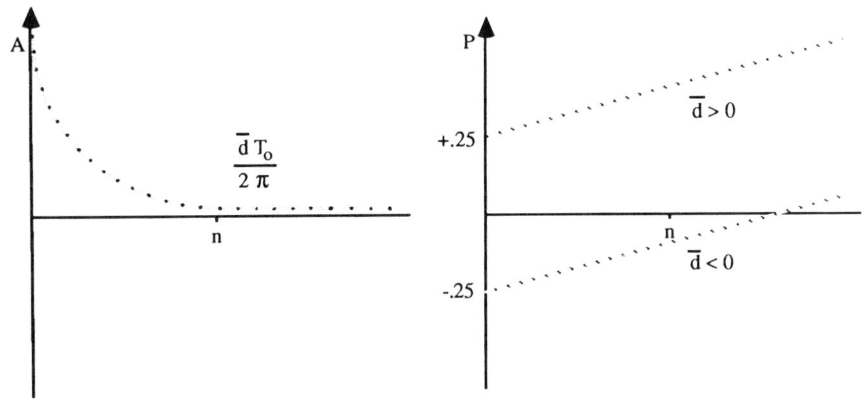

Figure 14. The DFT of a linear drift $d(t)$.

3. Summary

In comparing the methods discussed above one can suggest the following list of pros and cons for each method.

(1) *Constant pressure difference.* A relatively slow method not suitable for experiments on low-permeability materials, incapable of measurements of storage capacity. Its advantage is the simplicity of the experimental setup and data processing when high precision is not important. Its particular strength lies in dissolution–precipitation experiments where large volumes of fluid need to be passed through a sample.

(2) *Transient step method.* This method has received a lot of attention in recent years (e.g., Kamath et al., 1990) and has developed into an experimentally simple yet fast and precise method for the determination of both permeability and storage capacity. Its limitations are the insensitivity to any frequency-dependent phenomena, difficult removal of random noise from experimental data, and also difficult removal of systematic contamination of data due to, for example, a steady leak.

(3) *Sinusoidal oscillation method.* Requires a somewhat more complicated experimental

setup (drivers, etc.) than the previous methods; it is therefore more expensive to implement. It offers continuous data collection, and frequency-dependent phenomena can be studied. FFT data processing offers a powerful signal-to-noise ratio enhancement due to its filtering effect, resulting in an increased resolution/sensitivity and consequently in the possibility of measuring very low permeabilities and storage capacities. Owing to its steady-state character, systematic contaminations such as leaks and modulation can easily be recognized and often removed from the data record.

Acknowledgment

I am indebted to Ian Jackson and Robert Bitmead for numerous discussions and suggestions they have contributed throughout the course of this project, and to Mervyn Paterson for his continuous support. The reviewers, R. Kranz and Y. Bernabé, provided very stimulating and constructive criticism that improved the quality of this chapter greatly.

References

Brace, W.F., Walsh, J.B., and Frangos, W.T. (1968). Permeability of granite under high pressure. *J. Geophys. Res.* **73**, 2225–2236.

Carslaw, H.S. and Jaeger, J.C. (1959). *Conduction of Heat in Solids*, 2d edn. Oxford University Press.

Glisson, T.H. and Sage, A.P. (1969). On discrete and complex representation of real signals. *Proc. 12th Midwest Symp. Circuit Theory.*

Glisson, T.H., Black, C.I., and Sage, A.P. (1970). The digital computation of discrete spectra using the fast fourier transform. *IEEE Trans. Audio Electroacoust.* **Au-18.3**, 1127–1141.

Hsieh, P.A., Tracy, J.V., Neuzil, C.E., Bredehoft, J.D., and Silliman, S.E. (1981). A transient laboratory method for determining the hydraulic properties of 'tight' rocks – 1. Theory. *Int. J. Rock Mech. Sci. Geomech. Abstr.* **18**, 245–252.

Kamath, J. Boyer, R.E., and Nakagawa, F.M. (1990). *Characterisation of Core Scale Heterogenities Using Laboratory Pressure Transients.* Society of Petroleum Engineers publication SPE 20575.

Lin, W. (1977). *Compressible Fluid Flow Through Rocks of Variable Permeability.* Lawrence Livermore Laboratory, Livermore, Rept. UCLR-52304.

Neuzil, C.E., Cooley, C., Silliman, S.E., Bredehoft, J.D., and Hsieh, P.A. (1981). A transient laboratory method for determining the hydraulic properties of 'tight' rocks – 2. Application. *Int. J. Rock Mech. Sci. Geomech. Abstr.* **18**, 253–258.

Appendix A

From eq. (5), upon Laplace transformation, the subsidiary equation for the transformed time variable σ is

$$\frac{\partial^2 \bar{p}}{\partial x^2} - q^2 \bar{p} = 0 \tag{A1}$$

with boundary conditions

$$\left(\frac{d\bar{p}}{dx}\right)_{x=0} - \lambda\sigma\bar{p} = 0 \qquad \text{at } x = 0 \tag{A2}$$

and

$$\bar{p} = P_A \frac{\omega \cos \delta + \sigma \sin \delta}{\sigma^2 + \omega^2} \quad \text{at } x = L \tag{A3}$$

Then λ and q are given as

$$\lambda = \frac{\beta_d}{KA}, \quad q = \sqrt{\frac{\beta_s \sigma}{K}} \quad \text{where } K = \frac{k}{\eta} \tag{A4}$$

The solution of the subsidiary equation is

$$\bar{p} = P_A \frac{\omega \cos \delta + \sigma \sin \delta}{\sigma^2 + \omega^2} \frac{(q + \lambda\sigma)e^{qx} + (q - \lambda\sigma)e^{-qx}}{(q + \lambda\sigma)e^{qL} + (q - \lambda\sigma)e^{-qL}} \tag{A5}$$

which can be rewritten as

$$\bar{p} = P_A \frac{\omega \cos \delta + \sigma \sin \delta}{\sigma^2 + \omega^2} \frac{q \cosh(qx) + \lambda\sigma \sinh(qx)}{q \cosh(qL) + \lambda\sigma \sinh(qL)} \tag{A6}$$

Functions (A5) or (A6) do not appear in the tables of transforms, hence we shall perform the inverse Laplace transformation using the Bromwich integral

$$p = \frac{1}{2\pi i} \int_{c-i\infty}^{c+i\infty} e^{+\sigma t} \bar{p}(\sigma)\, d\sigma$$

to obtain

$$p = \frac{P_A}{2\pi i} \int_{c-i\infty}^{c+i\infty} \frac{\omega \cos \delta + \sigma \sin \delta}{\sigma^2 + \omega^2} \frac{q \cosh(qx) + \lambda\sigma \sinh(qx)}{q \cosh(qL) + \lambda\sigma \sinh(qL)} e^{\sigma t}\, d\sigma \tag{A7}$$

where c is a real positive constant, chosen to be sufficiently large so that all singularities of $p(\sigma)$ lie to the left of the line $(c - i\infty; c + i\infty)$. σ is now, unlike the previous treatment, considered to be complex. The integral (A7) can be evaluated by residue calculus; $p(\sigma)$ is a single-valued function of σ, therefore there are no branch cuts or branch points in the complex domain. An examination of eq. (A7) reveals that two of the poles of the integrand are located at $\sigma_{1,2} = \pm i\omega$ and that the rest of the poles σ_n lie on the negative real axis. Therefore the path of integration will follow a large semicircle Γ of radius R enclosed by the line $x = c$, where c can be any real positive number (Figure A1).

Because the integral over the semicircle Γ vanishes as R approaches infinity, the evaluation of the contour integral reduces to the evaluation of the line integral along $(c - i\infty;\ c + i\infty)$. The value of this integral, in the limit, can be determined using Cauchy's theorem as $2\pi i$ times the sum of the residues at the poles of its integrand:

$$p = 2\pi i \sum_{n=1}^{\infty} \operatorname{Res}(\sigma_n)$$

where $\operatorname{Res}(\sigma_n)$ are the residues calculated at poles σ_n of the integrand of eq. (A7).

The couple of poles $\sigma_{1,2}$ at $\pm i\omega$ indicates that the solution has a periodic component and the poles σ_n, which, as will be shown, are negative real numbers, represent the transient part of the solution which diminishes with time. Let us now evaluate σ_n. The associated singularities are determined by

$$q \cosh(qL) + \lambda\sigma \sinh(qL) = 0 \tag{A8}$$

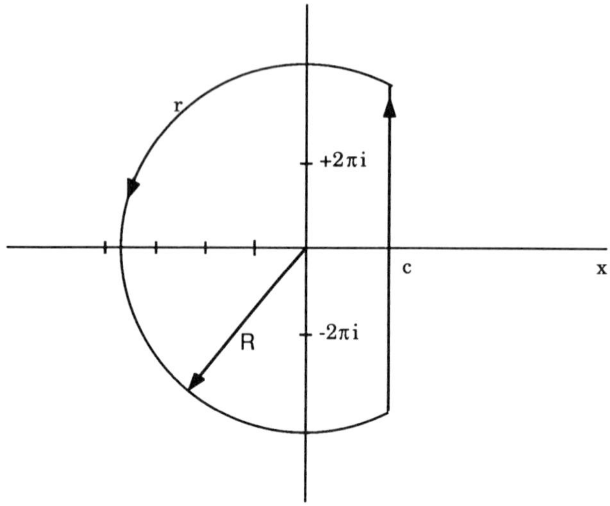

Figure A1. Integration path.

This equation can only be satisfied if qL is imaginary, i.e., $qL = i\psi$ where ψ is a real number. Therefore, using (A4)

$$\sigma_n = -\frac{K\psi_n^2}{\beta_s L^2} \tag{A9}$$

where ψ_n are the roots of the equation

$$\psi \cos \psi - \lambda \frac{K\psi^2}{\beta_s L} \sin \psi = 0 \tag{A10}$$

Substituting (A4) into (A10) gives

$$\tan \psi = \frac{\beta_s AL}{\beta_d \psi} \tag{A11}$$

where $\psi_n \to (2n+1)\pi/2$ as $n \to \infty$.

Since the integrand of (A7) can be expressed as a fraction of two polynomials, the residue at each pole can be determined by

$$\mathrm{Res}(\sigma) = \frac{f_n(\sigma)}{f'_d(\sigma)}$$

$f_n(\sigma)$ being the function of the nominator and $f'_d(\sigma)$ the derivative of the denominator, i.e.,

$$\mathrm{Res}(\sigma_n) = \frac{\omega \cos\delta + \sigma \sin\delta}{\sigma^2 + \omega^2} \frac{q\cosh(qx) + \lambda\sigma \sinh(qx)}{d/d\sigma[q\cosh(qL) + \lambda\sigma\sinh(qL)]} e^{\sigma t} \tag{A12}$$

which after substitution results in

$$\mathrm{Res}(\sigma_n) = 2K\beta_s AL\psi_n^2 \frac{(\beta_s L^2 \omega\cos\delta - K\psi_n^2 \sin\delta)\left[\cos(\xi\psi_n) - \dfrac{\beta_d \psi_n}{\beta_s AL}\sin(\xi\psi_n)\right] e^{-K\psi_n^2 t/\beta_s L}}{(K^2\psi_n^4 + \beta_s^2 L^4 \omega^2)[(\beta_d \psi_n^2 - \beta_s AL)\cos\psi_n + \psi_n(\beta_s AL + 2\beta_d)\sin\psi_n]}$$

where $\xi = x/L$ and ψ_n are the roots of $\psi = \arctan(\beta_s AL/\beta_d \psi)$. The solution for the transient part is therefore given as

$$p_{\text{tr}} = 2P_A K\beta_s AL \sum_{n=1}^{\infty} \frac{(\beta_s L^2 \omega \cos\delta - K\psi_n^2 \sin\delta)\left[\cos(\xi\psi_n) - \dfrac{\beta_d \psi_n}{\beta_s AL}\sin(\xi\psi_n)\right]}{(K^2\psi_n^4 + \beta_s^2 L^4 \omega^2)[(\beta_d \psi_n^2 - \beta_s AL)\cos\psi_n + \psi_n(\beta_s AL + 2\beta_d)\sin\psi_n]} \times \psi_n^2 e^{-K\psi_n^2 t/\beta_s L^2} \tag{A13}$$

The solution for the periodic part is found in a similar way. We write

$$\operatorname{Res}(\sigma_{1,2}) = \frac{\omega\cos\delta + \sigma\sin\delta}{d/d\sigma(\sigma_n^2 + \omega^2)} \frac{q\cosh(qx) + \lambda\sigma\sinh(qx)}{q\cosh(qL) + \lambda\sigma\sinh(qL)} e^{\sigma t} \tag{A14}$$

which results in

$$p_{\text{per}} = P_A \alpha[\sin(\omega t + \delta + \vartheta)] \tag{A15}$$

where

$$\alpha = \left| \frac{\kappa(1+i)\cosh[\kappa x(1+i)] + \dfrac{\beta_d i\omega}{KA}\sinh[\kappa x(1+i)]}{\kappa(1+i)\cosh[\kappa L(1+i)] + \dfrac{\beta_d i\omega}{KA}\sinh[\kappa L(1+i)]} \right| \tag{A16}$$

$$\vartheta = \arg\left\{ \frac{\kappa(1+i)\cosh[\kappa x(1+i)] + \dfrac{\beta_d i\omega}{KA}\sinh[\kappa x(1+i)]}{\kappa(1+i)\cosh[\kappa L(1+i)] + \dfrac{\beta_d i\omega}{KA}\sinh[\kappa L(1+i)]} \right\} \tag{A17}$$

and

$$\kappa = \sqrt{\frac{\beta_s \omega}{2K}} \tag{A18}$$

Finally, the solution of the boundary value problem is the sum of the periodic and transient parts, and therefore given as

$$p = p_{\text{per}} + p_{\text{tr}} = P_A \alpha \sin(\omega t + \delta + \vartheta) + 2P_A K\beta_s AL \sum_{n=1}^{\infty} \frac{(\beta_s L^2 \omega \cos\delta - K\psi_n^2 \sin\delta)\left[\cos(\xi\psi_n) - \dfrac{\beta_d \psi_n}{\beta_s AL}\sin(\xi\psi_n)\right]}{(K^2\psi_n^4 + \beta_s^2 L^4 \omega^2)[(\beta_d \psi_n^2 - \beta_s AL)\cos\psi_n + (\beta_s AL + 2\beta_d)\sin\psi_n]} \times \psi_n^2 e^{-K\psi_n^2 t/\beta_s L^2} \tag{A19}$$

and for $\delta = 0$

$$p = \alpha P_A \sin(\omega t + \vartheta) + 2P_A K\beta_s^2 AL^3 \omega \sum_{n=1}^{\infty} \frac{\cos(\xi\psi_n) - \dfrac{\beta_d \psi_n}{\beta_s AL}\sin(\xi\psi_n)}{(K^2\psi_n^4 + \beta_s^2 L^4 \omega^2)[(\beta_d \psi_n^2 - \beta_s AL)\cos\psi_n + (\beta_s AL + 2\beta_d)\sin\psi_n]} \times \psi_n^2 e^{-K\psi_n^2 t/\beta_s L^2} \tag{A20}$$

α and ϑ being defined by eqs. (A16) and (A17).

Appendix B

The subsidiary boundary value problem is

$$\frac{\partial^2 \bar{p}}{\partial x^2} - q^2\bar{p} = 0, \qquad \bar{p}(x, 0) = 0, \qquad \bar{p}(0, t) = 0$$

$$\bar{p}(L, t) = P_A \frac{\omega\cos\delta + \sigma\sin\delta}{\sigma^2 + \omega^2}$$

and leads to the solution

$$p = \frac{P_A}{2\pi i}\int_{c-i\infty}^{c+i\infty} \frac{\omega\cos\delta + \sigma\sin\delta}{\sigma^2 + \omega^2}\,\frac{\sinh(qx)}{\sinh(qL)}\,e^{\sigma t}\,d\sigma$$

which upon integration results in

$$p = \alpha P_A \sin(\omega t + \delta + \vartheta) + 2\pi PK \sum_{n=0}^{\infty} \frac{n(-1)^n(\beta_s L^2\omega\cos\delta - Kn^2\pi^2\sin\delta)}{K^2n^4\pi^4 + \beta_s^2 L^4\omega^2}\sin(n\pi\xi)\,e^{-Kn^2\pi^2/4\beta_s L^2}$$

Appendix C

Equation (5) is linear and homogeneous with constant coefficients, hence we may, in this case, separate the variables by writing

$$p = XT \tag{C1}$$

where X is a function of x only and T of t only. After substitution we arrive at

$$\frac{1}{X}\frac{d^2X}{dx^2} = \frac{\alpha}{T}\frac{dT}{dt} \tag{C2}$$

where $\alpha = \beta_s\eta/k$. The two sides of this equation are functions of completely independent variables; they can only be equal if each is equal to a constant, which we choose to call λ^2. The solution of the partial differential equation (5) is thus reduced to that of the two ordinary differential equations

$$\frac{d^2X}{dx^2} - X\lambda^2 = 0 \tag{C3}$$

$$\frac{dT}{dt} - \frac{T}{\alpha}\lambda^2 = 0 \tag{C4}$$

It is easy to see that e^{ax} and e^{bt} would satisfy eqs. (A23) and (A24), respectively, provided that $a = \pm\lambda$ and $b = \lambda^2/\alpha$. From (C1) the solution of (5) is

$$p = Ce^{ax+bt} \tag{C5}$$

Since $b = a^2/\alpha$, this can be rewritten as

$$p = C\exp(bt \pm x\sqrt{\alpha b}) \tag{C6}$$

Putting $b = \pm i\gamma$ (where $i = \sqrt{-1}$) we obtain

$$p = C\exp(\pm i\gamma t \pm x\sqrt{\alpha\gamma}\sqrt{\pm i}) \tag{C7}$$

or

$$p = C \exp\left(\pm x \sqrt{\frac{\gamma\eta\beta_s}{2k}} \right) \exp\left[\pm i \left(\gamma t \pm x \sqrt{\frac{\gamma\eta\beta_s}{2k}} \right) \right]$$

Equation (C7) contains a number of solutions from which other particular solutions can be derived by addition, such as

$$p = C \exp\left(-x \sqrt{\frac{\gamma\eta\beta_s}{2k}} \right) \left\{ \exp\left[i \left(\gamma t - x \sqrt{\frac{\gamma\eta\beta_s}{2k}} \right) \right] - \exp\left[-i \left(\gamma t - x \sqrt{\frac{\gamma\eta\beta_s}{2k}} \right) \right] \right\} \tag{C8}$$

which results in

$$p = C_1 \sin\left(\gamma t + x \sqrt{\frac{\gamma\eta\beta_s}{2k}} \right) e^{-x\sqrt{\gamma\eta\beta_s/2k}} \tag{C9}$$

If we further substitute ω for γ and put C_1 equal to P_A, the function which satisfies the eq. (5) and the boundary condition (12) is

$$p = P_A \sin(\omega t - x\sqrt{\omega\eta\beta_s/2k}) e^{-x\sqrt{\omega\eta\beta_s/2k}} \tag{C10}$$

Equation (C10) represents the steady-state part of the solution of the boundary value problem. When a complete solution, including the transient part, is needed, the best procedure is to use the Laplace transform, and this has been done by Carslaw and Jaeger (1959, p. 317).

Chapter 9

Measurement of Permeability and Storage Capacity in Rocks During Deformation at High Temperature and Pressure

G.J. Fischer and M.S. Paterson

Research School of Earth Sciences, Australian National University, Canberra 2601, Australia

Abstract

Both permeability and storage capacity have been determined in a marble, a limestone, and a sandstone during deformation at high pressure and temperature. The method used utilizes the attenuation and phase retardation of a sinusoidal pore pressure oscillation as it propagates through a sample under test. These parameters were measured at a confining pressure of 300 MPa, argon pore fluid pressures ranging from 20 to 270 MPa, temperatures to 873 K and strains of 0, 10% and 20%. The results for marble indicate that temperature elevated to moderate levels can enhance both the permeability and storage capacity owing to the anisotropy in thermal expansion of the constituent minerals, while the activation of plastic processes at higher temperatures causes the collapse of void spaces and reduction of permeability and storage capacity. It has been observed at all temperatures, however, that when the pore fluid pressure approaches the confining pressure, both measured parameters increase rapidly, and also that ongoing deformation can, when pore pressure remains relatively high, have a profound effect on the transport and storage properties of polycrystalline materials, pushing the measured permeability to up to two orders of magnitude above the level of virgin rock. A model analysis is presented from which information regarding the in-situ pore structure under the experimental conditions can be deduced (permeability, storage capacity, pore structure, crack aspect ratio).

1. Introduction

Many geological and environmental processes involve interactions between introduced fluid phases and preexisting solid phases in rocks. The kinetics of these processes are fundamentally dependent on the permeability of the rock, which in turn is influenced

FAULT MECHANICS AND TRANSPORT PROPERTIES OF ROCKS
ISBN 0-12-243780-2

strongly by the mechanical history. In an earlier study, we measured the changes that occur in the volume of the connected void space during deformation in the ductile field at various temperatures and at pore pressures up to 300 MPa confining pressure for a suite of rocks (Fischer and Paterson, 1989). However, the potential significance of these changes for the kinetics of fluid–rock interactions can only be fully appreciated when the permeability of the pore structure is also known (Fischer and Paterson, 1985; Fischer, 1988). In the present paper we report the results of measurements of the permeability and the storage capacity per unit volume at various stages of deformation and at various temperatures and pore pressures in samples of the same rocks as previously studied, namely, Carrara marble, Solnhofen limestone, and Gosford sandstone. Before presenting the results, however, we describe the method used for the measurements. Although the method was developed for this work, it is also of wider interest, for example for studies of heat conduction, electrical resistivity/conductivity, and diffusion, and has been already exploited in such areas (see, for example, Cowan, 1961; Cerceo and Childers, 1963; Bennion and Goss, 1971). Recently, a similar permeability study has also been published by Kranz et al. (1990).

2. The Method

The method used is based on the measurement of the attenuation and phase retardation of an oscillation of the pore fluid pressure as it propagates through the specimen. The theory is presented and discussed by Fischer (Chapter 8, this volume). In its application, a sinusoidal pressure oscillation is induced at one end of the specimen and the pressure response is monitored at the other end. We find that the response settles down eventually into another sinusoidal oscillation which is attenuated and phase-shifted relative to the applied oscillation. The magnitudes of the attenuation and phase shift depend on the following.

(1) The given experimental parameters, namely:

L = length of specimen
A = cross-sectional area of specimen
T = period of oscillation of pore fluid pressure
η = dynamic viscosity of pore fluid
B_d = storage capacity of the downstream reservoir (storage capacity is defined, at a given fluid density, as the increase in volume of pore fluid stored per unit increase in pore fluid pressure)

These parameters are known or are readily determined in calibration tests.

(2) The unknown properties of the sample, namely:

k = permeability
β_c = storage capacity per unit volume

It is the purpose of the experiment to determine these quantities.

The experiment is described mathematically by the continuity equation

$$\frac{\partial^2 p}{\partial x^2} - \frac{\beta_c \eta}{k}\frac{\partial p}{\partial t} = 0, \qquad \text{for } 0 < x < L \text{ and } t > 0 \tag{1}$$

where x is the distance along the sample measured from the downstream face, p is the fluid pressure at position x, and t is the time elapsed from the beginning of the experiment. The pressure is initially zero everywhere and the boundary conditions consist of (a) a prescribed sinusoidal pressure p_u of amplitude P and period T applied to the upstream face, and (b) a linear pressure transfer at the downstream face to a reservoir of known storage capacity in which the pressure is the dependent quantity p_d. The solution of (1) under these conditions consists of the sum of a transient term and a steady-state term (Fischer, Chapter 8), the value of the steady-state term at the downstream face being of the form

$$p_d = \alpha P \sin(\omega t - \delta) \tag{2}$$

where α is the attenuation factor, δ is the phase lag, and $\omega = 2\pi/T$.

Permeability and storage capacity per unit volume are then implicitly described by the functions

$$\alpha = \left| \frac{\kappa(1+i)\cosh[\kappa x(1+i)] + \dfrac{B_d i\omega}{KA}\sinh[\kappa x(1+i)]}{\kappa(1+i)\cosh[\kappa L(1+i)] + \dfrac{B_d i\omega}{KA}\sinh[\kappa L(1+i)]} \right|$$

and

$$\delta = \arg\left\{ \frac{\kappa(1+i)\cosh[\kappa x(1+i)] + \dfrac{B_d i\omega}{KA}\sinh[\kappa x(1+i)]}{\kappa(1+i)\cosh[\kappa L(1+i)] + \dfrac{B_d i\omega}{KA}\sinh[\kappa L(1+i)]} \right\}$$

where

$$\kappa = \sqrt{\frac{\beta_c \pi \eta}{Tk}}$$

Alternatively, when explicit expressions for k and β_c are preferred, one can make use of two graphically determined parameters ψ and γ and proceed as follows. At $x = 0$ we have

$$\alpha = |R(\gamma, \psi)| = \left| \frac{1}{\cosh[\psi(1+i)] + \gamma\psi(1+i)\sinh[\psi(1+i)]} \right| \tag{3}$$

and

$$\delta = \arg[R(\gamma, \psi)] = \arg\left[\frac{1}{\cosh[\psi(1+i)] + \gamma\psi(1+i)\sinh[\psi(1+i)]} \right] \tag{4}$$

where

$$\gamma = \frac{B_d}{AL\beta_c} \tag{5}$$

$$\psi = L\left(\frac{\pi\eta\beta_c}{Tk}\right)^{1/2} \tag{6}$$

Thus, if δ is to be expressed in cycles,

$$\alpha = \sqrt{\{\mathrm{Re}[R(\gamma, \psi)]\}^2 + \{\mathrm{Im}[R(\gamma, \psi)]\}^2} \tag{7}$$

and

$$\delta = \frac{1}{2\pi} \arctan \frac{\mathrm{Im}[R(\gamma, \psi)]}{\mathrm{Re}[R(\gamma, \psi)]} + \frac{n}{2}, \qquad n \text{ an integer} \tag{8}$$

where if

$$\begin{aligned} X_D &= \cosh \psi \cos \psi + \gamma\psi(\sinh \psi \cos \psi - \cosh \psi \sin \psi) \\ Y_D &= \sinh \psi \sin \psi + \gamma\psi(\cosh \psi \sin \psi + \sinh \psi \cos \psi) \\ A &= X_D^2 + Y_D^2 \end{aligned}$$

then

$$\mathrm{Re}[R(\gamma, \psi)] = \frac{X_D}{A}, \qquad \mathrm{Im}[R(\gamma, \psi)] = -\frac{Y_D}{A}$$

From the measurements of α and δ, the values of the dimensionless parameters ψ and γ can be evaluated (Figure 1), and thence, from (3) and (4), we obtain

$$\beta_c = \frac{B_d}{LA\gamma} \tag{9}$$

$$k = \frac{\pi\eta\beta_c}{T}\left(\frac{L}{\psi}\right)^2 = \frac{\pi\eta B_d L}{TA\psi^2\gamma} \tag{10}$$

Before discussing the apparatus and the values of the experimental parameters, it is useful to note from (9) that

$$\gamma = \frac{\text{storage capacity of downstream reservoir}}{\text{storage capacity of specimen}}$$

and also that, according to Brace et al. (1966), the storage capacity per unit volume of the specimen is related to its porosity φ and elastic parameters by

$$\beta_c = \varphi(\beta_f - \beta_m) + (\beta_b - \beta_m) \tag{11}$$

where β_f is the compressibility of the pore fluid, β_b is the bulk compressibility of the specimen, and β_m is the mean compressibility of the component minerals. The storage capacity of the downstream reservoir can be expressed as

$$B_d = V_r(\beta_f + \beta_r) \tag{12}$$

where V_r is the volume of the reservoir and $\beta_r = (1/V_r)(dV_r/dp)$. The values of β_c and β_d may thus vary appreciably with pressure and temperature.

3. The Apparatus

The deformation apparatus is basically that described by Paterson (1970), with internal furnace and internal load cell. Argon is used as both confining pressure and pore fluid. A modification permits pore fluid reservoirs to be connected to both ends of the specimen (Figures 2 and 3). The upstream reservoir consists of the same volumometer as used in

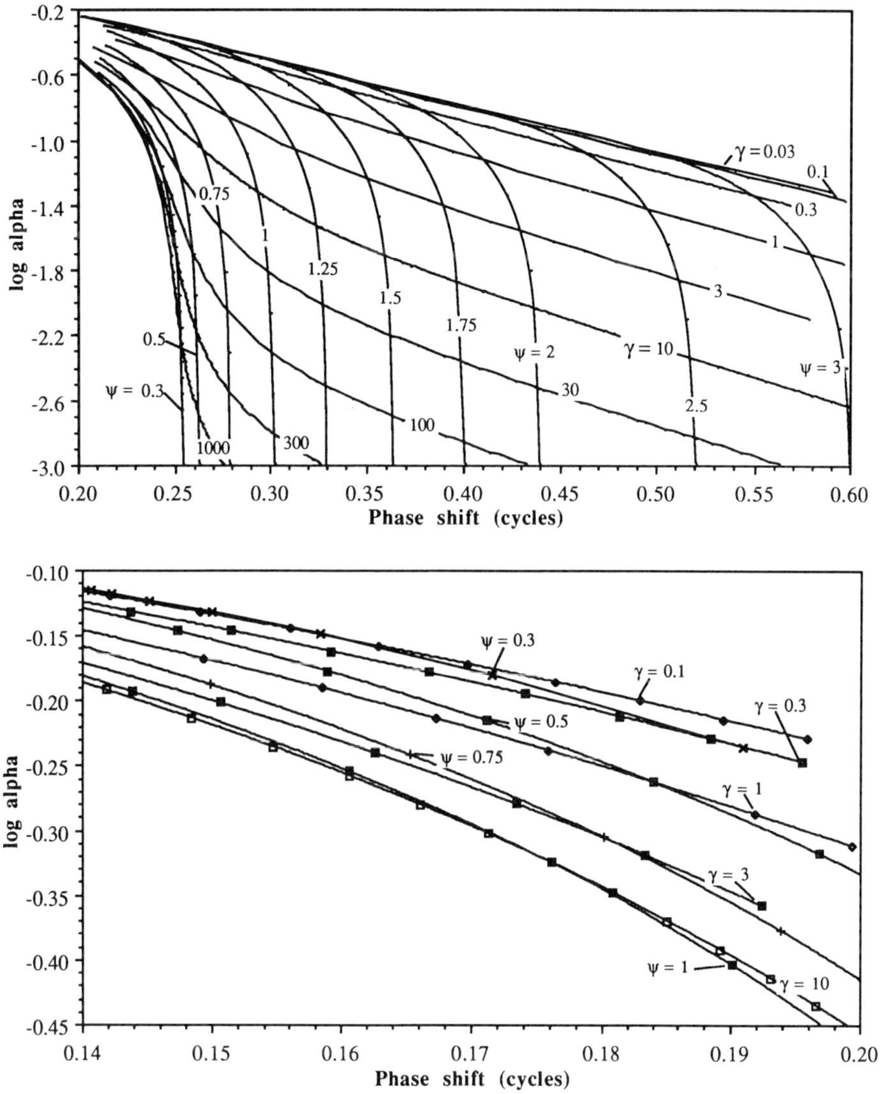

Figure 1. Relationship of measured attenuation factor α and phase lag δ to dimensionless parameters ψ and γ used for graphical determination of permeability and of storage capacity per unit volume.

Fischer and Paterson (1989) except that, for reasons of expediency associated with generating the sinusoidal pressure oscillation, a hydraulic motor has been used in place of the electric motor. The downstream reservoir consists only of a pressure gauge and connecting tubing, in which wire inserts are placed to minimize the volume. Extending the lower limit of measurable permeability as far as possible depends both on minimizing this volume and on maximizing the resolution of the pressure transducers.

Each pressure transducer consists of a Minalpha resistance wire (similar to Manganin) in a DC-excited Wheatstone bridge circuit. In order to match the high, 14-bit resolution of the forcing signal, the bridge signal is processed with a monolithic instrumentation

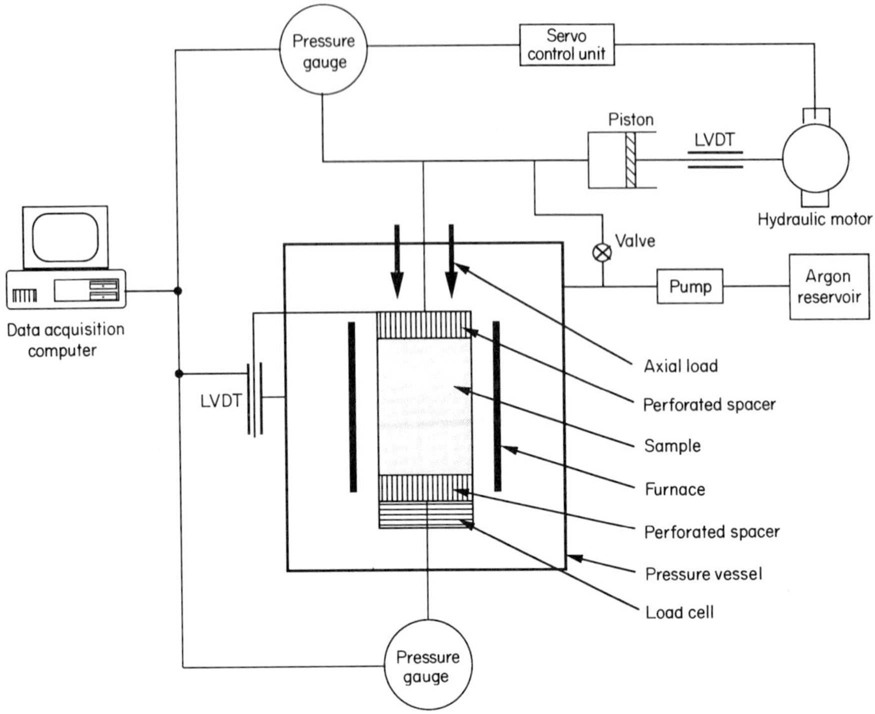

Figure 2. Schematic layout of the experimental set-up.

amplifier (Analog Devices type AD624C) providing −110 dB common mode rejection, followed by a hybrid isolation amplifier (Burr Brown type ISO 100 CP) which provides hum loop isolation. This procedure permits a resolution of 0.001 MPa relative to a base level of up to 500 MPa. The amplified transducer signal incorporates 0–500 MPa offset so that only the sinusoidal component is applied to the 14-bit analog-to-digital converter. In the case of the upstream reservoir, this signal is also used as feedback to the servocontroller.

The control of the mean pore fluid pressure and the magnitude of the forced pressure oscillation in the upstream reservoir is effected by a servosystem consisting of a series of amplifiers, the motor driving the piston of the volumometer, and the pressure transducer that provides the feedback (Figure 4). The mean pore pressure is controlled by means of a potentiometer setting and the pressure oscillation by a command signal from a signal generator. In our case, the command signal is of sinusoidal waveform but any other practical boundary condition to equation (1) can be simulated, including a constant pore pressure difference between upstream and downstream reservoirs at any base pressure level up to 400 MPa or more. The confining pressure is maintained constant to within approximately 0.5 MPa by a simple on–off controller.

The generation of the driving signal and the data acquisition are implemented with an IBM Personal Computer, fitted with a Data Translation analog I/O card (model DT2801). To relieve the computer of the intensive task of calculating the sine function during the course of an experiment, that function is predetermined and tabulated to 14-bit amplitude resolution, with 1024 samples per cycle. The computer clocks the digitized function into a digital-to-analog converter in order to constitute the sinusoidal

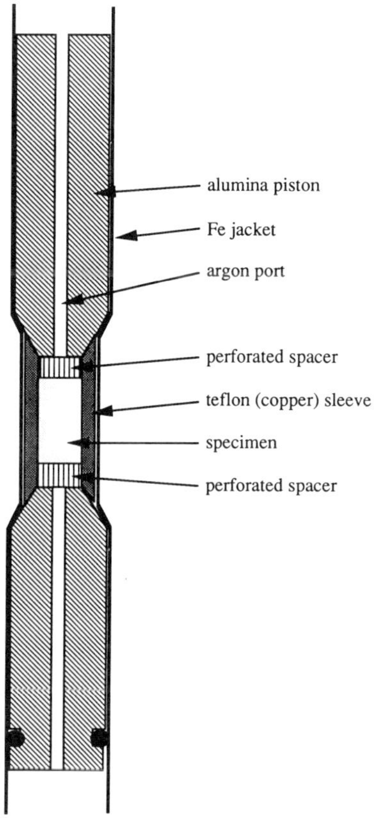

Figure 3. Details of specimen assembly.

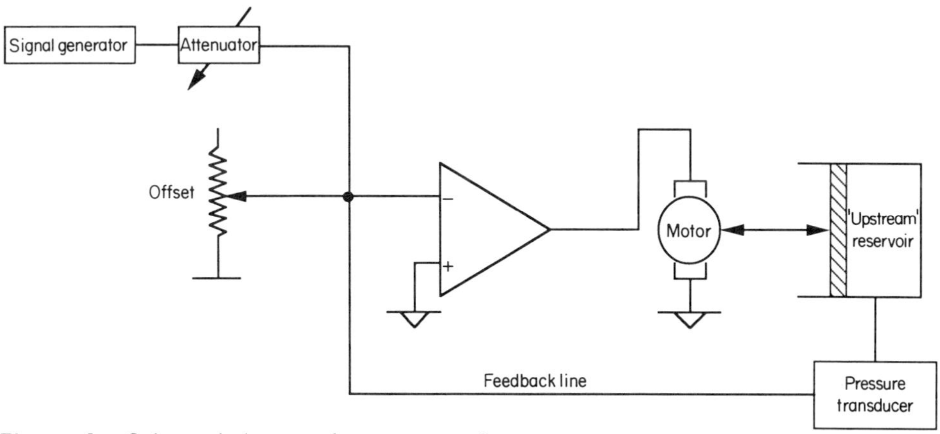

Figure 4. Schematic layout of servosystem for pore pressure.

command signal. Confining pressure, axial load, axial strain, upstream reservoir volume and temperature signals from the various transducers are processed using monolithic instrumentation amplifiers (Analog Devices AD524C) prior to analog-to-digital conversion by the DT2801 card and data logging within the computer.

A Fast Fourier Transform program for discrete data is used to process the raw pore pressure data in order to obtain the values of the attenuation factor α and the phase lag δ. These values are then used to obtain the permeability k and the storage capacity per unit volume β_c of the specimen, as set out under 'Method.'

4. Experimental Parameters and Limits of Measurement

Specimens 10 mm in diameter and 5–20 mm long have been used. With the hydraulic motor fitted, the limits on period of pore pressure oscillation were primarily determined by computational constraints and were between 10 and 32 000 seconds. Values of compressibility and viscosity of argon which have been used are given in Table 1.

The storage capacity B_d of the downstream reservoir can be determined at 300 MPa confining pressure in a calibration experiment at room temperature in which the volumometer upstream is used to measure the total storage capacity of the system, first with the downstream reservoir blanked off by a thin carbide disk placed below the hollow dummy specimen, and second with the disk removed; the difference gives the values of B_d.

In the actual implementation of this simple calibration, however, a mistake has been made, which was only discovered after the completion of our test and when a recalibration was no longer possible. In this paper therefore the values of B_d have been obtained by the multiplication of the volume of our downstream reservoir (850 mm^3) by the compressibility of argon. The small volume changes of the actual reservoir due to pressure cycling have thus been neglected. Figure 5 shows the values of storage capacity in the downstream reservoir actually used. Note that only room-temperature values were used, since most of the volume of our downstream reservoir is sufficiently far from the hot region to be practically independent of the experimental temperature.

The minimum detectable permeability can be estimated as follows. Assuming that the pore pressure oscillation range is limited to 20 MPa, that is, $P = 10$ MPa, so as to be kept fairly small relative to the pore pressure itself, the pressure measurement sensitivity of 0.001 MPa permits a minimum value of the attenuation factor α of 0.0001 to be

Table 1. Viscosity and compressibility of argon. For each temperature and pressure, the first number is the compressibility in units of GPa^{-1} and the second is the viscosity in units of 10^{-6} Pa s

	Pressure (MPa)						
Temperature (K)	20	50	100	150	200	250	300
293	47.8	11.6	3.73	2.07	1.40	1.05	0.84
	30.1	46.5	68.6	87.3	104	121	136
473	47.0	15.6	5.83	3.21	2.11	1.55	1.19
	36.2	43.1	55.9	68.7	81.3	93.6	106
673	47.0	16.8	6.93	3.97	2.64	1.94	1.50
	44.5	48.4	56.0	64.5	73.4	82.5	91.6
873	47.5	17.4	7.51	4.42	3.00	2.21	1.72
	52.4	55.0	60.2	66.2	72.7	79.6	86.7

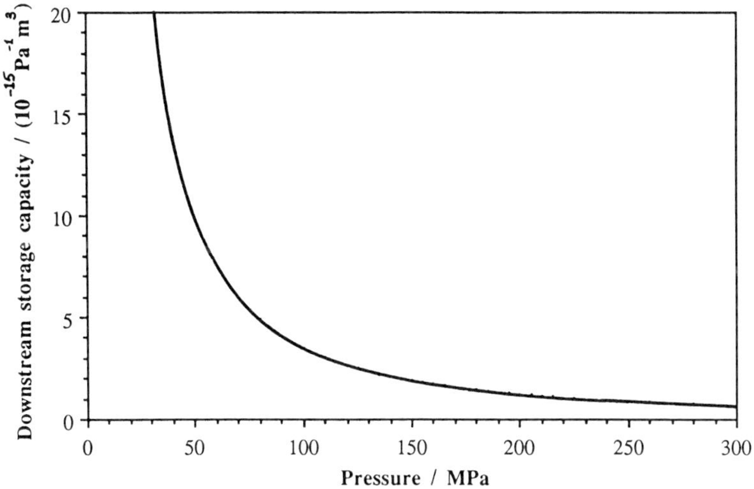

Figure 5. Downstream storage capacity as a function of pressure.

detected. The phase lag should be limited to less than one cycle if ambiguity concerning the number of complete cycles is to be minimized. Thus, from an extrapolation of Figure 1, the values of ψ and γ for detectability are limited such that approximately $\psi^2\gamma < 7500$. Then, for a specimen 10 mm in diameter and 5 mm long, an oscillation period of 4000 s, and typical values of η and B_d (Table 1) of 150 μPa s and 10^{-15} m^3 Pa^{-1}, the minimum detectable permeability k would be 10^{-24} m^{-2}. Such a value would, of course, be only of order of magnitude significance, and if at least $\pm 10\%$ accuracy were to be achieved, the lower measurable limit would be around 10^{-23} m^{-2}. Similarly, the maximum measurable permeability can be derived. From Figure 1, the minimum value of $\psi^2\gamma$ for reliable graphical determination is around 0.01. For a specimen 10 mm in diameter and 20 mm long, an oscillation period of 50 s and the values of η and B_d of 75 μPa s and 10^{-14} m^3 Pa^{-1}, respectively (for higher permeability the measurement can be conducted at lower pressure, i.e., with more compressible and less viscous fluid), the maximum detectable permeability would be 10^{-15} m^2, or the maximum measurable with $\pm 10\%$ accuracy would be around 10^{-16} m^2.

Although the above graphical method offers advantages in terms of speed and convenience, it is not very reliable for work with very permeable materials. Here, a numerical inversion gives more satisfactory results. In practice, the parameters that can still be detected are constrained by the limits of 0.995 for the attenuation factor α and about 10^{-3} cycle for the phase lag δ, the latter being determined by the Fourier transform procedure in conjunction with the signal-to-noise ratio enhancement. Thus, if the maximum value of sample storage capacity is limited by 10^{-13} m^3 Pa^{-1}, for a 20 mm $\times$ 10 mm sample, the maximum permeability that can be measured with a 10% accuracy is around 10^{-16} m^2.

5. Specimen Material and Jacketing

The rocks studied are listed in Table 2 and are described in more detail by Edmond and Paterson (1972) and Fischer and Paterson (1989). These specimens were cored out

Table 2. Materials studied (from Fischer and Paterson, 1989). The porosities in the table have been obtained from calibration runs using a mild steel specimen. For some conditions where no calibration data are available, a simple interpolation of a quadratic fit was performed. Interpolated values are marked by italics

	Initial density ($kg\ m^{-3}$)	Initial porosity	Pore fluid pressure (MPa)	Porosity at 300 MPa confining pressure			
				Room temp.	473 K	673 K	873 K
Carrara marble	2690	0.011	50	0.006	0.007	0.005	*0.004*
			150	0.008	0.007	0.004	0.003
			250	0.011	0.014	0.009	*0.005*
Solnhofen limestone	2560	0.062	50	0.056	0.058	0.027	0.014
			150	0.058	0.067	*0.030*	0.015
			250	0.060	0.081	0.030	0.011
Gosford sandstone	2450	0.152	50	0.120	*0.103*	0.072	0.062
			150	0.140	0.123	0.086	0.071
			250	0.145	*0.130*	0.096	0.081

of the same slabs as used by Fischer and Paterson (1989) for dilatancy studies. They were dried in an oven at 383 K for at least 48 hours prior to an experiment. The predeformation porosity listed in Table 2 is the interconnected porosity deduced from the work of Fischer and Paterson (1989) to be present after establishing a given temperature and pore pressure at 300 MPa confining pressure.

In permeability measurement, particular care must be taken to seal the cylindrical surface of the specimen within the jacket (Figure 3). We originally intended to rely on an annealed copper sleeve inserted between the specimen and the main iron jacket, since the usual tests carried out on drawn glass and roughly machined aluminum dummy specimens indicated a satisfactory seal. However, it was then found that with a dummy specimen cored out of a glass plate, an appreciable communication between the upstream and downstream reservoirs could not be avoided at room temperature even at the highest effective confining pressures. A Teflon sleeve of about 2 mm wall thickness was, however, found to provide an adequate seal and was used in tests up to 473 K. At 673 K and above, a copper gasket was found to be satisfactory for a cored glass specimen and was therefore used for the rock specimens on the assumption that it would be also satisfactory for the rock, although the nature of the surface cracking may be different from that in glass since it may involve opening of grain boundaries deeper into the specimen; in a more refined study this aspect may require further testing by grinding of the cylindrical surfaces of the specimens.

6. Testing Procedure

The permeability and storage capacity at various pore pressures was determined on both undeformed and deformed specimens, using argon as pore fluid. In the case of an undeformed specimen, it was first subjected to the confining pressure, then heated and, finally, the pore fluid was introduced. In order to explore reproducibility and hysteresis, two sets of measurements were made on each specimen during a run. The first set was made during constant-pore-pressure pauses between increments of pore pressure, usually from 50 MPa up to 250 MPa, and the second during pauses between decrements of pore pressure. Between the two sets, the pore pressure was always increased to 270–280 MPa so that all measurements in the second set were made along a path of decreasing pore pressure.

In the case of specimens undergoing deformation, the initial procedure was similar. After equilibrating the pore pressure at its minimum level (ca. 50 MPa), deformation at a strain rate of $10^{-4}\,\mathrm{s}^{-1}$ was initiated. Permeability measurements at various pore pressures were made at strain intervals of 10%. During the measurement, deformation was continued at a strain rate that gave rise to zero dilatancy, after which the strain rate of $10^{-4}\,\mathrm{s}^{-1}$ was resumed. Dilatancy during straining was monitored between the permeability measurements, as in Fischer and Paterson (1989).

The amplitude and period of pore pressure oscillation were adjusted according to the conditions of each run with the aim of keeping the time of data acquisition to a minimum, while also keeping the amplitude as small as possible so as to ensure approximate constancy of pore fluid and rock properties during the cycling. Thus amplitudes P of between 2 MPa at 50 MPa pore pressure and 10 MPa for 250 MPa pore pressure were used, while the periods varied from rock to rock for particular strains, as shown in Tables 3–5. Since, as will be discussed later, subsequent measurements revealed some dependence of measured permeability on the oscillation amplitude and period, the choice of various amplitudes and periods, as just described, has probably inadvertently introduced some systematic, second-order variations into the permeability results.

In carrying out an experiment it was generally found that transient effects had decayed to a negligible level within the first cycle and so the data from the second cycle onwards were taken for processing. Normally, data for six cycles were analyzed, using the Fourier transform procedure of Glisson et al. (1970) as applied by Jackson and Paterson (1987). An example is given in Figure 6 (this example has also been chosen to illustrate that a drift in mean pressure due to a steady leak into or out of the downstream reservoir can be corrected for; changes in specimen properties themselves during the measurement interval are less easily identified and allowed for but they will generally result in a distortion of the sinusoidal shape of the waveform). The raw data are shown in Figure 6a, the amplitude and frequency spectra in the time domain for cycles 3–8, inclusive, in Figures 6b and 6c, and the same spectra after drift removal in Figures 6d and 6e. The sought values of amplitude ratio (attenuation factor) α and phase lag ϑ ($= -2\pi\delta$) occur at integer frequency 6 (indicated by arrow) and are, respectively 0.073 and -1.842 radians, which correspond to $\alpha = 0.073$ and $\delta = 0.293$ cycles from equations (7) and (8) and hence, from (9), (10) and Table 1, give $k = 2 \times 10^{-20}\,\mathrm{m}^{-2}$ and $\beta_c = 3.55 \times 10^{-10}\,\mathrm{Pa}^{-1}$ for permeability and storage capacity per unit volume, respectively.

The Fourier transform processing has the advantage of enhancing the signal-to-noise ratio. In the example just given, this ratio for the attenuation factor, as seen in Figure 6d, is more than three decades. While this case was selected as a favorable one, the signal-to-noise ratio for the attenuation factor was always at least 500:1. The uniform scatter of points in Figure 6e simply indicates that, apart from the phase lag signal point at integer frequency 6, the phases of the noise are random (white noise). The experiments have not been repeated to check reproducibility but the consistent trends observed in the results that follow indicate useful reproducibility for a first study.

Table 3. Experimental data and results of analysis for Carrara marble

Pore pressure (MPa)	Period (s)	α	δ (cycles)	$\log(k/\text{m}^{-2})$	$\log(\beta_c/\text{Pa}^{-1})$	φ	x (nm)	$\frac{\beta_c}{\varphi\beta_f}$	a/c
293 K, 0% strain									
50	2000	0.0023	0.297	−21.54	−10.14	0.6	10	1.04	(34)
80	2000	0.0035	0.291	−21.57	−10.36	0.7	8	1.19	(76)
100	2000	0.0042	0.281	−21.59	−10.54	0.7	8	1.10	(28)
160	2000	0.0060	0.278	−21.61	−10.72	0.8	7	1.25	(37)
200	2000	0.0070	0.280	−21.60	−10.76	0.9	6	1.39	(42)
250	2000	0.0534	0.247	−20.78	−10.73	1.1	13	1.62	50
270	1000	0.1171	0.236	−20.15	−10.52	1.2	25	2.65	120
250	2000	0.0692	0.245	−20.67	−10.62	1.1	15	2.09	88
225	2000	0.0235	0.258	−21.11	−10.72	1.0	10	1.56	(52)
200	2000	0.0177	0.258	−21.20	−10.84	0.9	10	1.15	(17)
180	2000	0.0165	0.259	−21.20	−10.78	0.9	10	1.16	(20)
150	2000	0.0145	0.262	−21.19	−10.61	0.8	11	1.47	(74)
120	2000	0.0124	0.288	−21.18	−10.08	0.7	13	4.16	685
95	2000	0.0104	0.294	−21.17	−9.95	0.7	13	3.95	909
70	2000	0.0081	0.286	−21.14	−9.94	0.6	16	2.92	959
50	2000	0.0060	0.283	−21.11	−9.87	0.6	16	1.93	822
293 K, 10% strain									
50	1000	0.0092	0.270	−20.71	−9.89	0.9	17	1.23	(203)
90	1000	0.0166	0.291	−20.72	−9.73				
150	500	0.0155	0.388	−20.57	−9.47				
200	500	0.0608	0.346	−20.06	−9.17				
250	500	0.1187	0.301	−19.85	−9.20				
270	500	0.2061	0.254	−19.64	−9.30				
250	500	0.1781	0.268	−19.68	−9.21				
200	500	0.0822	0.322	−19.94	−9.15				
150	500	0.0381	0.337	−20.20	−9.30				
110	500	0.0296	0.283	−20.25	−9.66				
50	1000	0.0286	0.256	−20.21	−9.69				

293 K, 20% strain									
50	250	0.0258	0.291	−19.28	−9.54	1.8	45	1.38	(333)
100	250	0.0525	0.287	−19.29	−9.70				
145	120	0.0867	0.284	−18.88	−9.68				
172	60	0.1419	0.268	−18.41	−9.63				
200	60	0.2585	0.240	−18.16	−9.53				
250	60	0.6959	0.145	−17.60	−9.25				
260	60	0.8966	0.081	−17.29	−9.32				
250	60	0.8076	0.114	−17.43	−9.18				
220	60	0.5382	0.180	−17.78	−9.35				
150	60	0.1445	0.263	−18.35	−9.59				
115	120	0.1390	0.258	−18.59	−9.53				
75	250	0.1776	0.231	−18.64	−9.69				
50	500	0.1909	0.226	−18.71	−9.48				
473 K, 0% strain									
50	1000	0.0093	0.276	−20.53	−9.63	0.7	27	2.13	1256
95	1000	0.0175	0.272	−20.54	−9.79	0.6	31	4.30	1476
150	500	0.0521	0.253	−19.95	−9.93	0.7	53	5.18	951
200	500	0.0868	0.248	−19.84	−9.87	1.0	42	6.41	810
250	500	0.3263	0.205	−19.31	−9.61	1.4	55	11.37	1139
250	500	0.1739	0.234	−19.60	−9.72	1.4	40	8.83	859
225	500	0.1243	0.237	−19.72	−10.07	1.2	40	3.93	372
150	1000	0.0289	0.282	−20.51	−9.69	0.7	28	9.00	1819
120	1000	0.0179	0.300	−20.62	−9.60	0.6	29	9.46	2678
60	2000	0.0044	0.327	−21.22	−9.60	0.7	12	2.94	1675
473 K, 10% strain									
50	2000	0.0066	0.277	−21.07	−9.77	0.6	17	1.80	890
100	1000	0.0066	0.341	−21.04	−9.64				
150	1000	0.0174	0.320	−20.80	−9.59				
200	1000	0.0214	0.355	−20.79	−9.47				
245	1000	0.0404	0.366	−20.55	−9.22				
265	500	0.0403	0.328	−20.32	−9.48				
242	1000	0.0632	0.301	−20.40	−9.38				
223	1000	0.0461	0.319	−20.50	−9.37				

Table 3. *Continued*

Pore pressure (MPa)	Period (s)	α	δ (cycles)	$\log(k/\mathrm{m}^{-2})$	$\log(\beta_c/\mathrm{Pa}^{-1})$	φ	x (nm)	$\frac{\beta_c}{\varphi\beta_f}$	a/c
200	100	0.4460	0.205	−18.00	−9.07				
250	50	0.4287	0.223	−17.71	−8.93				
170	1000	0.0188	0.341	−20.80	−9.50				
122	1000	0.0174	0.322	−20.71	−9.45				
100	1000	0.0118	0.319	−20.80	−9.52				
50	2000	0.0112	0.275	−20.83	−9.56				
473 K, 20% strain									
60	250	0.1208	0.245	−18.59	−9.36	1.4	127	2.55	1342
100	100	0.1009	0.269	−18.49	−9.34				
150	100	0.1178	0.311	−18.53	−9.08				
200	100	0.4460	0.205	−18.00	−9.07				
250	50	0.4287	0.223	−17.71	−8.93				
255	100	0.8533	0.101	−17.46	−8.86				
202	100	0.7222	0.137	−17.66	−8.98				
175	100	0.4498	0.205	−17.94	−8.96				
127	100	0.3566	0.223	−17.96	−8.88				
60	100	0.1511	0.250	−18.08	−9.01				
673 K, 0% strain									
50	4000	0.0030	0.291	−21.67	−9.83	0.5	10	1.76	828
100	2000	0.0031	0.325	−21.64	−9.89	0.4	13	4.61	1625
200	1000	0.0034	0.382	−21.57	−9.98	0.6	10	6.58	961
250	1000	0.0065	0.347	−21.38	−9.99	0.9	8	5.86	613
250	1000	0.0080	0.344	−21.29	−9.91	0.9	9	7.04	762
180	1000	0.0031	0.396	−21.54	−9.90	0.5	12	8.30	1455
120	2000	0.0037	0.343	−21.64	−9.82	0.4	13	6.95	2093
40	4000	0.0024	0.296	−21.67	−9.77	0.5	10	1.55	(779)

				673 K, 10% strain					
50	4000	0.0019	0.268	−21.94	−10.38	0.6	6	0.41*	(−639)
120	2000	0.0029	0.301	−21.86	−10.21				
180	1000	0.0038	0.353	−21.58	−10.00				
250	1000	0.0033	0.513	−21.58	−9.60				
250	1000	0.0069	0.438	−21.35	−9.55				
200	1000	0.0030	0.437	−21.64	−9.79				
150	2000	0.0053	0.296	−21.69	−10.12				
100	2000	0.0032	0.305	−21.73	−10.03				
50	4000	0.0024	0.268	−21.87	−10.29				
				673 K, 20% strain					
52	1000	0.0036	0.376	−20.62	−9.72	0.7	25	1.72	746
95	1000	0.0140	0.352	−20.28	−9.53				
150	500	0.0101	0.506	−20.14	−9.29				
205	500	0.0327	0.425	−19.84	−9.23				
255	500	0.0580	0.455	−19.54	−8.86				
270	500	0.0631	0.469	−19.45	−8.73				
255	500	0.0642	0.455	−19.47	−8.78				
205	500	0.0458	0.403	−19.71	−9.15				
150	500	0.0234	0.456	−19.83	−9.07				
92	1000	0.0248	0.340	−20.02	−9.31				
50	1000	0.0105	0.313	−20.15	−9.56				

Table 4. Experimental data and results of analysis for Solnhofen limestone

Pore pressure (MPa)	Period (s)	α	δ (cycles)	$\log(k/m^{-2})$	$\log(\beta_c/Pa^{-1})$	φ	x (nm)	$\frac{\beta_c}{\varphi\beta_f}$	a/c
293 K, 0% strain									
50	1000	0.0459	0.271	−19.76	−9.20	5.6	8	0.97	(−26)
80	1000	0.0714	0.263	−19.78	−9.41	5.7	8	1.30	(122)
100	500	0.0449	0.273	−19.79	−9.68	5.7	8	0.98	(−5)
125	1000	0.1067	0.244	−19.80	−9.90	5.8	8	0.81	(−38)
150	500	0.0614	0.263	−19.80	−9.92	5.8	8	0.99	(0)
200	250	0.0388	0.282	−19.78	−10.05	5.9	8	1.09	(10)
225	250	0.0429	0.289	−19.77	−9.98	6.0	8	1.43	(40)
250	250	0.0469	0.297	−19.76	−9.93	6.0	8	1.88	71
270	250	0.0532	0.312	−19.70	−9.78	6.0	8	2.91	139
250	250	0.0540	0.300	−19.69	−9.83	6.0	8	2.36	110
225	250	0.0499	0.291	−19.70	−9.88	6.0	8	1.80	74
190	250	0.0434	0.298	−19.72	−9.80	5.9	8	1.79	90
150	500	0.0756	0.263	−19.70	−9.77	5.8	8	1.40	(63)
130	500	0.0664	0.267	−19.71	−9.70	5.8	8	1.36	(69)
95	1000	0.1041	0.255	−19.70	−9.42	5.7	9	1.64	199
50	1000	0.0610	0.267	−19.63	−9.10	5.6	10	1.22	(193)
293 K, 10% strain									
40	1000	0.0276	0.287	−19.94	−9.09	5.0	7	0.95	(−57)
100	500	0.0358	0.301	−19.96	−9.48				
150	500	0.0614	0.289	−19.87	−9.56				
200	500	0.0729	0.293	−19.88	−9.60				
250	500	0.1275	0.316	−19.63	−9.17				
245	500	0.1471	0.301	−19.57	−9.17				
200	500	0.1089	0.286	−19.69	−9.42				
147	500	0.0592	0.313	−19.86	−9.37				
88	1000	0.0699	0.277	−19.92	−9.27				
50	1000	0.0411	0.289	−19.89	−9.03				

				293 K, 20% strain					
40	250	0.0161	0.050	−19.05	−8.27	6.1	17	5.17	5383
100	250	0.0528	0.447	−19.07	−8.71				
155	100	0.0697	0.423	−18.73	−8.92				
200	100	0.0862	0.456	−18.50	−8.63				
230	100	0.0726	0.516	−18.23	−8.22				
250	50	0.2130	0.336	−17.92	−8.69				
265	50	0.4051	0.248	−17.64	−8.70				
250	50	0.3686	0.260	−17.70	−8.71				
225	50	0.1935	0.337	−18.02	−8.78				
210	50	0.1673	0.360	−18.02	−8.70				
150	100	0.1921	0.331	−18.23	−8.62				
100	100	0.2184	0.267	−18.21	−8.90				
80	100	0.1255	0.325	−18.33	−8.69				
50	100	0.0644	0.406	−18.33	−8.30				
				473 K, 0% strain					
50	1000	0.0646	0.272	−19.51	−8.85	5.8	11	1.55	(611)
95	1000	0.1320	0.247	−19.49	−9.19	6.2	10	1.66	294
140	1000	0.1899	0.246	−19.47	−9.07	6.6	10	3.59	653
150	500	0.1100	0.262	−19.45	−9.33	6.7	10	2.15	263
200	500	0.1789	0.245	−19.34	−9.38	7.3	10	2.71	257
250	500	0.2481	0.245	−19.23	−9.16	8.1	10	5.54	499
270	500	0.2337	0.259	−19.26	−9.07	8.4	10	7.37	627
240	500	0.1645	0.276	−19.40	−9.13	7.9	9	5.72	546
185	500	0.0941	0.294	−19.57	−9.19	7.1	8	3.84	479
130	1000	0.1518	0.250	−19.55	−9.18	6.5	9	2.59	446
85	1000	0.1041	0.258	−19.54	−9.08	6.1	10	1.84	442
40	1000	0.0503	0.246	−19.54	−9.77	5.7	10	0.14*	(−1268)
				473 K, 10% strain					
40	1000	0.0378	0.311	−19.72	−8.61	5.6	9	2.09	1618
80	1000	0.0773	0.293	−19.71	−8.78				
123	500	0.0475	0.370	−19.75	−8.83				
180	500	0.0705	0.362	−19.71	−8.88				

Table 4. *Continued*

Pore pressure (MPa)	Period (s)	α	δ (cycles)	$\log(k/m^{-2})$	$\log(\beta_c/Pa^{-1})$	φ	x (nm)	$\frac{\beta_c}{\varphi\beta_f}$	a/c
245	500	0.0922	0.356	−19.68	−8.94				
255	500	0.0967	0.357	−19.66	−8.93				
260	500	0.1783	0.234	−19.53	−9.83				
225	500	0.1413	0.241	−19.59	−9.84				
200	500	0.0903	0.347	−19.64	−8.89				
150	500	0.0742	0.345	−19.64	−8.83				
100	500	0.0492	0.366	−19.65	−8.69				
50	1000	0.0538	0.317	−19.65	−8.51				
473 K, 20% strain									
50	500	0.2178	0.246	−18.51	−8.59	5.0	39	3.27	2521
85	250	0.2010	0.273	−18.44	−8.63				
142	100	0.1367	0.325	−18.39	−8.82				
200	100	0.1780	0.333	−18.30	−8.72				
260	100	0.2182	0.336	−18.10	−8.50				
260	100	0.2176	0.342	−18.03	−8.40				
210	100	0.3234	0.263	−18.05	−8.72				
158	100	0.1911	0.319	−18.21	−8.62				
91	100	0.1159	0.343	−18.26	−8.54				
46	100	0.0602	0.352	−18.30	−8.49				
673 K, 0% strain									
60	4000	0.0084	0.356	−21.27	−8.96	2.8	3	2.92	1681
100	4000	0.0192	0.312	−21.17	−9.13	2.9	3	3.66	1197
125	4000	0.0262	0.286	−21.15	−9.35	3.0	3	2.91	633
150	2000	0.0139	0.343	−21.18	−9.32	3.0	3	4.00	776
200	1000	0.0173	0.321	−20.91	−9.53	3.0	4	3.71	467
250	1000	0.0816	0.255	−20.33	−9.63	3.0	8	4.03	382
250	1000	0.0363	0.289	−20.68	−9.57	3.0	5	4.62	457

200	2000	0.0499	0.271	−20.77	−9.50	3.0	5	3.98	512
160	4000	0.0596	0.266	−20.90	−9.34	3.0	4	4.21	758
140	4000	0.0542	0.266	−20.88	−9.32	3.0	4	3.61	742
90	4000	0.0322	0.279	−20.91	−9.13	2.9	4	3.19	1139
50	4000	0.0111	0.315	−21.09	−8.98	2.7	4	2.31	1424
673 K, 10% strain									
78	4000	0.0117	0.342	−21.34	−9.02	2.8	3	3.52	1575
112	4000	0.0229	0.314	−21.23	−9.10				
180	2000	0.0193	0.313	−21.21	−9.48				
250	2000	0.0419	0.325	−20.98	−9.21				
250	2000	0.0507	0.298	−20.91	−9.31				
200	2000	0.0305	0.306	−21.05	−9.37				
150	2000	0.0132	0.381	−21.25	−9.15				
100	4000	0.0181	0.350	−21.26	−8.91				
50	4000	0.0068	0.366	−21.37	−8.91				
673 K, 20% strain									
50	2000	0.0047	0.473	−20.97	−8.85	2.9	4	2.90	2068
100	2000	0.0171	0.392	−20.80	−8.96				
145	2000	0.0279	0.379	−20.76	−9.00				
200	500	0.0347	0.371	−20.20	−9.13				
250	500	0.1175	0.291	−19.77	−9.09				
250	500	0.1213	0.300	−19.74	−8.99				
220	500	0.0547	0.352	−20.04	−9.05				
200	1000	0.0714	0.322	−20.21	−9.02				
160	2000	0.0571	0.324	−20.53	−9.00				
115	2000	0.0162	0.428	−20.85	−8.91				
70	4000	0.0145	0.386	−21.03	−8.86				

Table 5. Experimental data and results of analysis for Gosford sandstone

Pore pressure (MPa)	Period (s)	α	δ (cycles)	$\log(k/m^{-2})$	$\log(\beta_c/Pa^{-1})$	φ	x (nm)	$\frac{\beta_c}{\varphi\beta_f}$	a/c
293 K, 0% strain									
20	500	0.0075	0.532	−19.43	−8.14	11.1	6	1.36	(671)
50	250	0.0689	0.279	−18.84	−8.98	12.0	11	0.75	(−113)
80	100	0.0957	0.276	−18.51	−9.15	12.8	15	1.05	(12)
100	100	0.1469	0.247	−18.43	−9.43	13.2	16	0.75	(−35)
140	100	0.2861	0.212	−18.26	−9.73	13.9	19	0.59*	(−35)
170	50	0.2180	0.229	−18.15	−9.76	14.2	21	0.71	(−19)
200	50	0.2516	0.234	−18.11	−9.50	14.4	21	1.58	(32)
225	50	0.3266	0.221	−18.01	−9.43	14.5	24	2.10	53
250	50	0.3656	0.224	−17.94	−9.23	14.5	26	3.90	120
270	50	0.6101	0.171	−17.59	−9.01	14.4	39	7.14	232
250	50	0.4980	0.191	−17.77	−9.22	14.5	32	3.99	124
220	50	0.4004	0.209	−17.88	−9.28	14.5	28	2.93	94
180	50	0.3378	0.210	−17.95	−9.55	14.3	26	1.24	(16)
140	100	0.4535	0.183	−18.02	−9.51	13.9	24	0.98	(0)
100	100	0.2396	0.230	−18.21	−9.25	13.2	21	1.13	(20)
70	100	0.1226	0.270	−18.34	−8.95	12.5	19	1.37	(96)
50	250	0.1630	0.249	−18.46	−8.81	12.0	17	1.11	(50)
293 K, 10% strain									
50	240	0.0300	0.333	−19.25	−8.97	9.1	9	1.01	(6)
100	100	0.0630	0.312	−18.87	−9.22				
150	100	0.2225	0.239	−18.46	−9.39				
200	50	0.0858	0.359	−18.58	−9.16				
250	50	0.1111	0.390	−18.40	−8.87				
260	50	0.3407	0.257	−17.93	−8.84				
220	50	0.1161	0.375	−18.38	−8.87				
170	100	0.2591	0.256	−18.38	−9.02				
118	240	0.1622	0.274	−18.88	−9.05				
50	240	0.0397	0.354	−19.10	−8.71				

				293 K, 20% strain					
50	100	0.2040	0.254	−17.82	−8.38	9.6	45	3.73	1246
100	50	0.3816	0.213	−17.52	−8.71				
150	25	0.5585	0.174	−17.14	−8.86				
205	25	0.9210	0.070	−16.67	−8.95				
250	25	0.9920	0.023	−16.18	−8.93				
230	25	0.9842	0.032	−16.31	−8.91				
205	25	0.9561	0.053	−16.51	−8.84				
170	25	0.9277	0.067	−16.61	−8.88				
150	50	0.8443	0.100	−17.04	−8.72				
125	50	0.5924	0.175	−17.28	−8.48				
80	100	0.6967	0.142	−17.32	−8.37				
50	100	0.4073	0.207	−17.47	−8.22				
				873 K, 0% strain					
50	240	0.0186	0.311	−19.14	−9.18	6.2	15	0.62	(−242)
100	240	0.1055	0.256	−18.71	−9.25	6.6	23	1.14	(39)
150	100	0.0705	0.287	−18.68	−9.30	7.1	23	1.58	(95)
200	100	0.1445	0.261	−18.49	−9.30	7.6	26	2.22	135
250	100	0.1768	0.263	−18.47	−9.23	8.1	25	3.26	185
260	100	0.2340	0.246	−18.35	−9.21	8.2	28	3.58	200
250	100	0.2975	0.227	−18.24	−9.20	8.1	33	3.50	204
220	100	0.1588	0.265	−18.47	−9.22	7.8	26	2.93	188
170	100	0.1119	0.269	−18.53	−9.29	7.3	26	1.89	124
120	240	0.1153	0.270	−18.73	−9.05	6.8	22	2.20	263
60	240	0.0282	0.310	−19.03	−9.08	6.3	17	0.94	(−29)
				873 K, 10% strain					
50	100	0.0054	0.417	−19.42	−9.10	5.0	14	0.92	(−48)
100	100	0.0257	0.343	−19.10	−9.08				
150	100	0.0727	0.298	−18.85	−9.07				
200	100	0.1234	0.278	−18.74	−9.10				
250	50	0.1651	0.288	−18.36	−8.91				
225	100	0.1773	0.275	−18.60	−8.92				
180	100	0.1564	0.263	−18.59	−9.04				

Table 5. *Continued*

Pore pressure (MPa)	Period (s)	α	δ (cycles)	$\log(k/m^{-2})$	$\log(\beta_c/Pa^{-1})$	φ	x (nm)	$\frac{\beta_c}{\varphi\beta_f}$	a/c
150	100	0.0557	0.315	−18.96	−9.08				
100	100	0.0322	0.340	−19.00	−8.99				
50	100	0.0078	0.394	−19.28	−9.02				
873 K, 20% strain									
50	25	0.1420	0.239	−17.77	−8.74	9.6	48	1.10	(65)
100	25	0.3508	0.201	−17.66	−8.81				
150	25	0.6703	0.138	−17.46	−8.84				
200	25	0.8301	0.099	−17.35	−8.77				
250	25	0.9454	0.058	−17.10	−8.52				
250	25	0.9289	0.066	−17.17	−8.55				
200	25	0.9021	0.076	−17.18	−8.61				
150	25	0.8944	0.077	−17.11	−8.79				
115	25	0.4935	0.174	−17.55	−8.86				
100	25	0.3776	0.197	−17.62	−8.74				
50	25	0.1821	0.229	−17.66	−8.78				

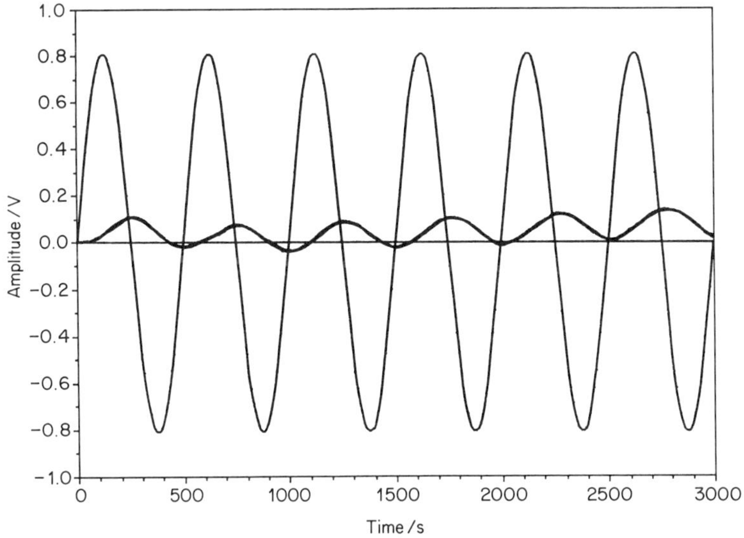

Figure 6a. The experimental data from a measurement on Solnhofen limestone at 300 MPa confining pressure, 200 MPa pore pressure, strain 10% and period 500 s. The last four cycles are considered to be unaffected by the transient and are processed.

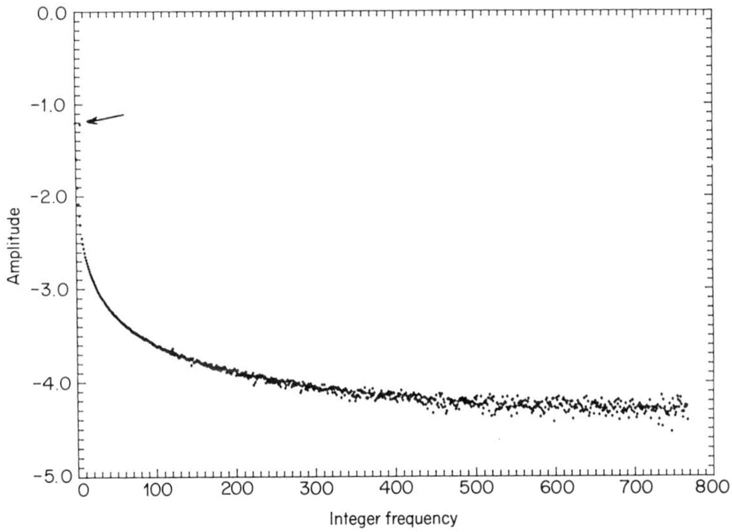

Figure 6b. The amplitude spectrum for the data of Figure 6a.

7. Results

From the measurements of attenuation factor and phase lag, the values of permeability k and storage capacity per unit volume β_c shown in Figures 7, 8, and 9 have been obtained for Carrara marble, Solnhofen limestone, and Gosford sandstone, respectively, at various temperatures, pore pressures and stages of deformation. In all cases the

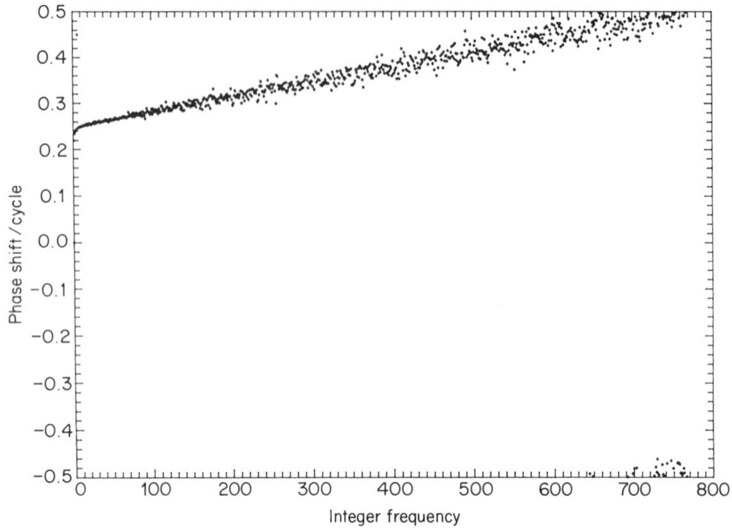

Figure 6c. The phase spectrum for the data of Figure 6a.

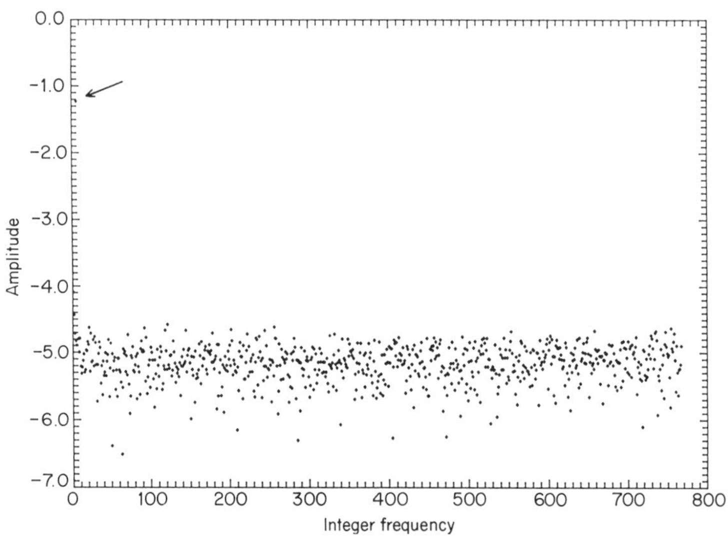

Figure 6d. The amplitude spectrum for the data of Figure 6a after drift removal.

confining pressure was 300 MPa and the deformation was carried out at the minimum pore pressure listed, that is, at 50 MPa or thereabouts. The results for zero strain and 10% strain were obtained on the same specimen for each rock, but a separate specimen was used for the 20% strain cases, with the exception that for the Gosford sandstone at 873 K separate specimens were also used for the zero strain and 10% strain cases.

In view of the considerable hysteresis observed in k and β_c during cycling of the pore pressure (Figures 7–9), some further measurements were made on Solnhofen limestone at room temperature, 200 MPa confining pressure, and 150 MPa pore pressure, in order

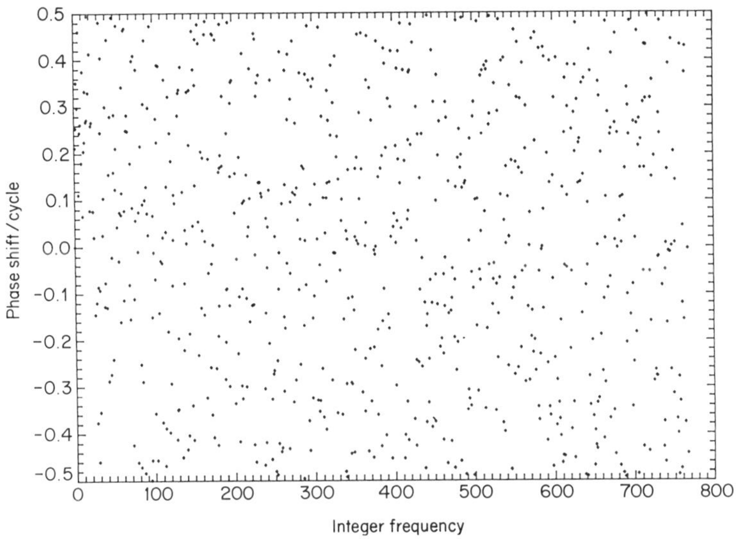

Figure 6e. The phase spectrum for the data of Figure 6a after drift removal.

to explore the influence of amplitude and frequency of the pore pressure oscillations used in the determination of k and β_c. The results are given in Figure 10. The effects are larger than had been anticipated and indicate that variations in the amplitudes and frequencies used in the measurements for Tables 3–5 and Figures 7–9 may have introduced minor systematic variations in k and β_c apart from the effects of the strain and the pore pressure themselves. Thus the effect of deformation may be slightly underrepresented, especially in the case of Gosford sandstone, and the effect of pore pressure variation may be slightly overrepresented. However, the distortions are not expected to exceed about ± 0.2 in $\log k$ and $\log \beta_c$ and in most cases should be less.

The actual values of the porosity φ under all of the above conditions were not measured, because the range of the pore fluid volumometer used was inadequate for the full range of pore pressures. Additional argon had to be added to or released from the pore fluid system at an intermediate stage in order to keep the volumometer within range. However, the values of φ measured in the preceding dilatancy study (Fischer and Paterson, 1989) can be expected to apply fairly well where the conditions are comparable, because the specimens were taken from the same blocks. These values of φ are listed in Tables 3–5. Values of φ during pore pressure cycling after deformation are not available.

Extensive microstructural studies have also not been carried out, in part because of difficulties encountered in specimen preparation that would have required more time to resolve than was available and, in part, because of doubts that such observations would have adequate resolution and would accurately reflect the situation in situ because of additional effects introduced by the subsequent temperature and pressure reduction. However, general guidance to likely microstructures is available in the extensive observations of Wong and collaborators (Fredrich and Wong, 1986; Fredrich et al., 1989; Wong et al., 1989; Zhang et al., 1990).

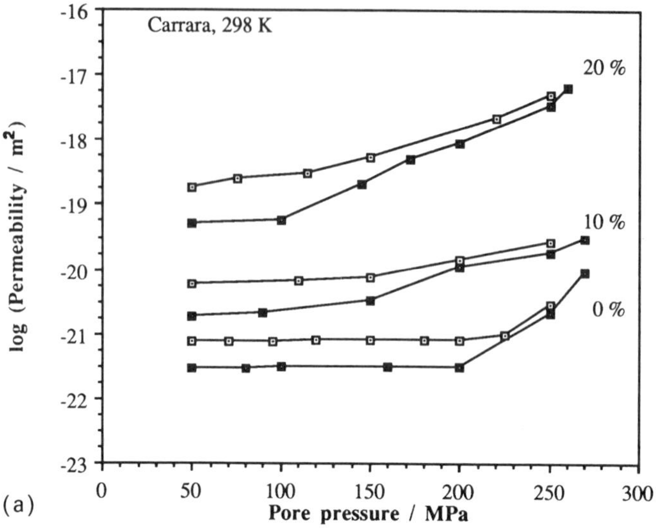

(a)

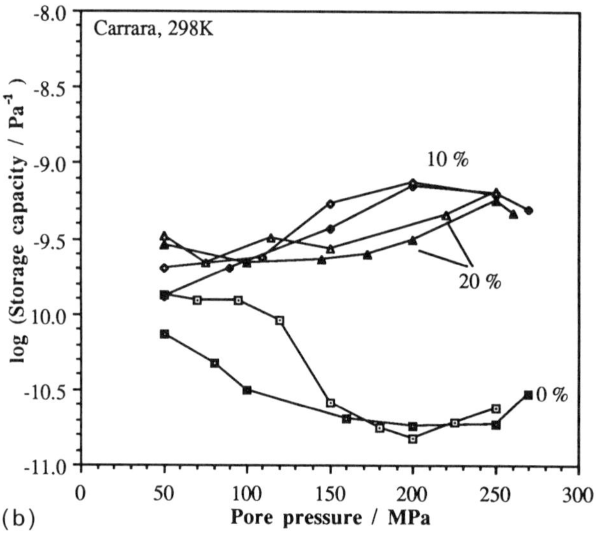

(b)

Figure 7a–f. Values of permeability k and storage capacity per unit volume β_c for Carrara marble as a function of temperature, deformation and pore pressure, all at 300 MPa confining pressure. Pore fluid is argon.

8. Discussion of Results

The above results show the following interesting trends at constant confining pressure.

(1) The permeability k generally increases with increasing pore pressure in a given case, more or less parallel to the accompanying increase in porosity (Fischer and Paterson, 1989). There is also, for a given pore pressure, a trend for k to increase with increase in porosity between the different rocks as well as between different states of deformation of each rock (cf. porosity trends in Fischer and Paterson, 1989, figures 6,

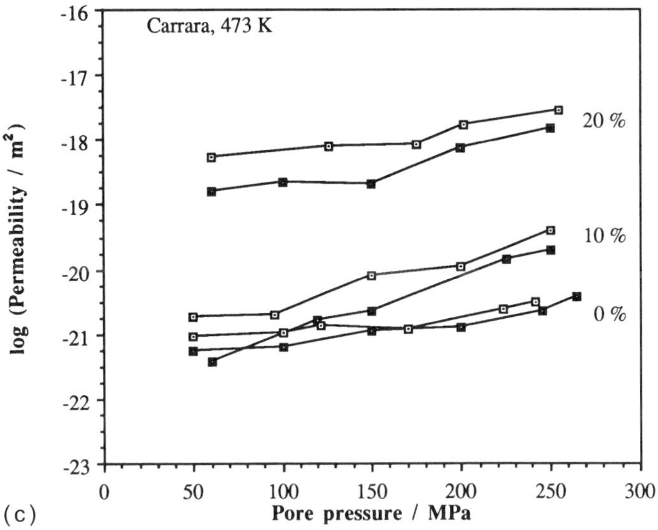

(c)

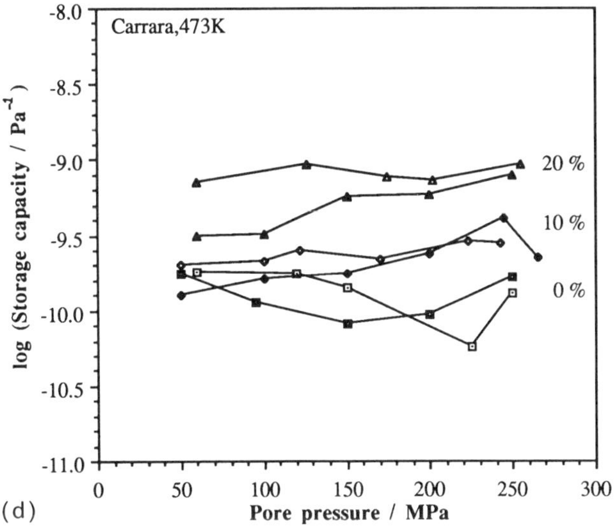

(d)

Figure 7. *Continued.*

8 and 10). However, the correlation with porosity is not a strict one, since for undeformed Carrara marble at room temperature no change in k is obtained from 50 to 200 MPa pore pressure (Figure 7a) whereas the previous observations show some increase in porosity over this range (Fischer and Paterson, 1989, table 1).

(2) The storage capacity per unit volume β_c often first decreases to a minimum and then increases again with increasing pore pressure under given conditions. This behavior probably reflects a competition between two opposing tendencies as the pore pressure is increased. Thus, on one hand, β_c decreases with decreasing compressibility of the argon and, on the other, β_c increases with increasing porosity and, possibly, also with increasing volumetric compliance of the pore space (cf. equation (12)). For a given pore

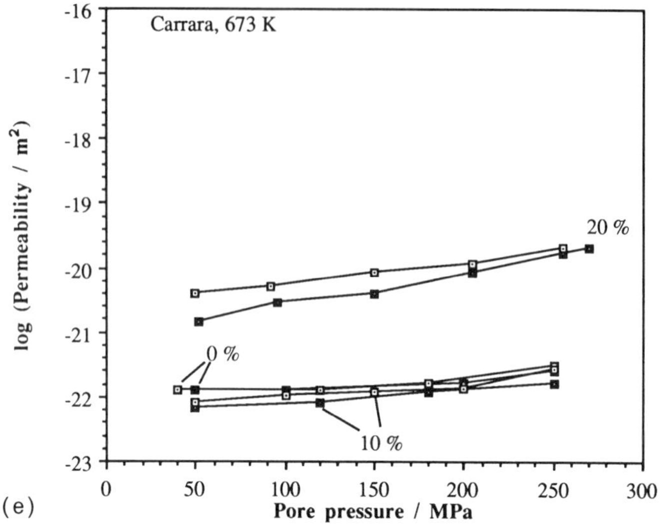

(e)

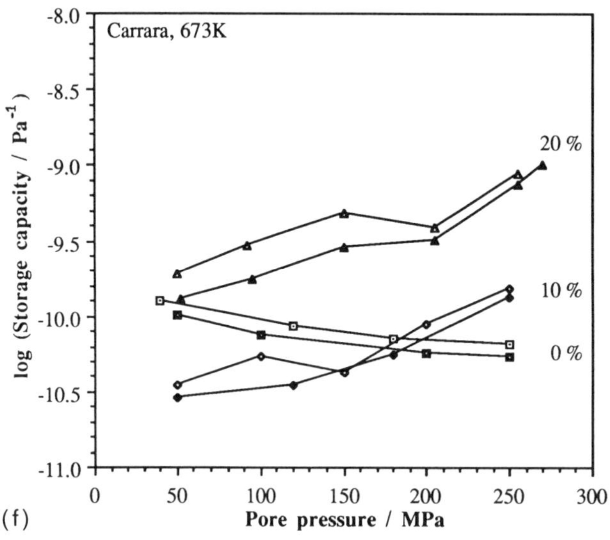

(f)

Figure 7. *Continued.*

pressure, the variations in β_c between different rocks and different deformation states of each rock seem to reflect primarily the differences in porosity. However, it is to be noted that the trends in k and β_c with deformation are not always in the same direction, especially between 0 and 10% strain for Solnhofen limestone and Gosford sandstone, and so, again, any correlation with porosity is not a strict one in detail.

(3) The general effect of increase in temperature is a decrease in both k and β_c, accompanying a decrease in porosity due presumably to increased thermal relaxation effects and a reduction of dilatancy during deformation (Fischer and Paterson, 1989). An exception to the general trend with temperature is seen in the case of undeformed Carrara marble, for which k and β_c increase from room temperature to 473 K. This

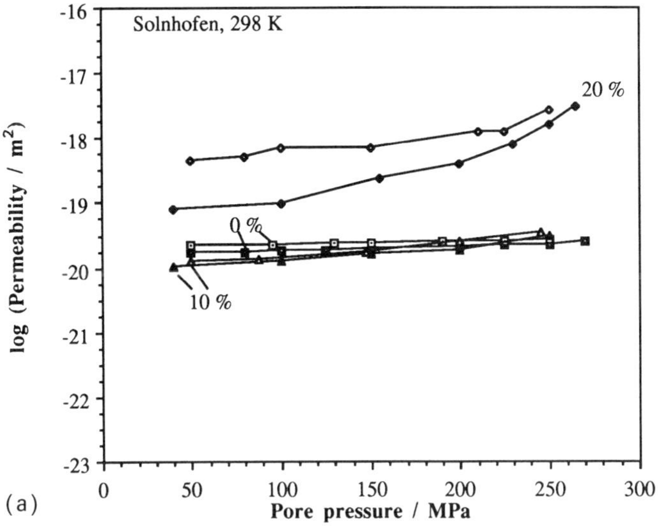

(a)

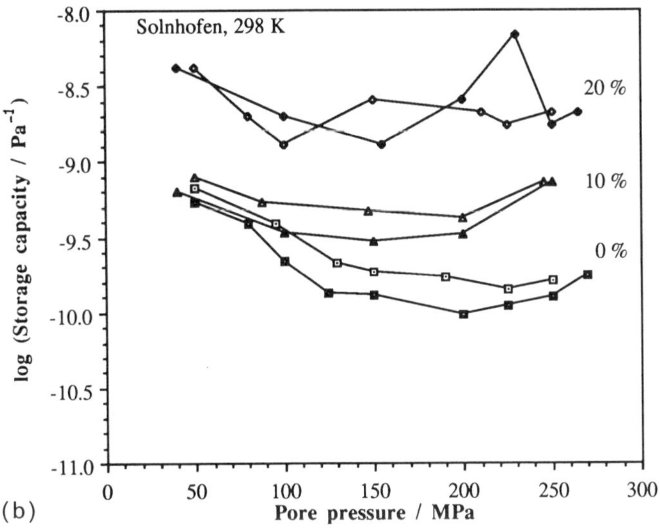

(b)

Figure 8a–f. Values of permeability k and storage capacity per unit volume β_c for Solnhofen limestone as a function of temperature, deformation and pore pressure, all at 300 MPa confining pressure. Pore fluid is argon.

effect can be attributed to microcracking associated with anisotropy in thermal expansion of the grains, which is counteracted by thermal relaxation effects at still higher temperatures, when k and β_c decrease again. The absence of such an effect in Solnhofen limestone may be related to its much smaller grain size or to masking by effects associated with its greater initial porosity.

(4) For Solnhofen limestone the range of variation in k and β_c, both with deformation and with cycling of pore pressure at a given deformation, is less than for Carrara marble. This reduced responsiveness may be related to a reduction in the amount of intragranular microcracking at the much smaller grain size.

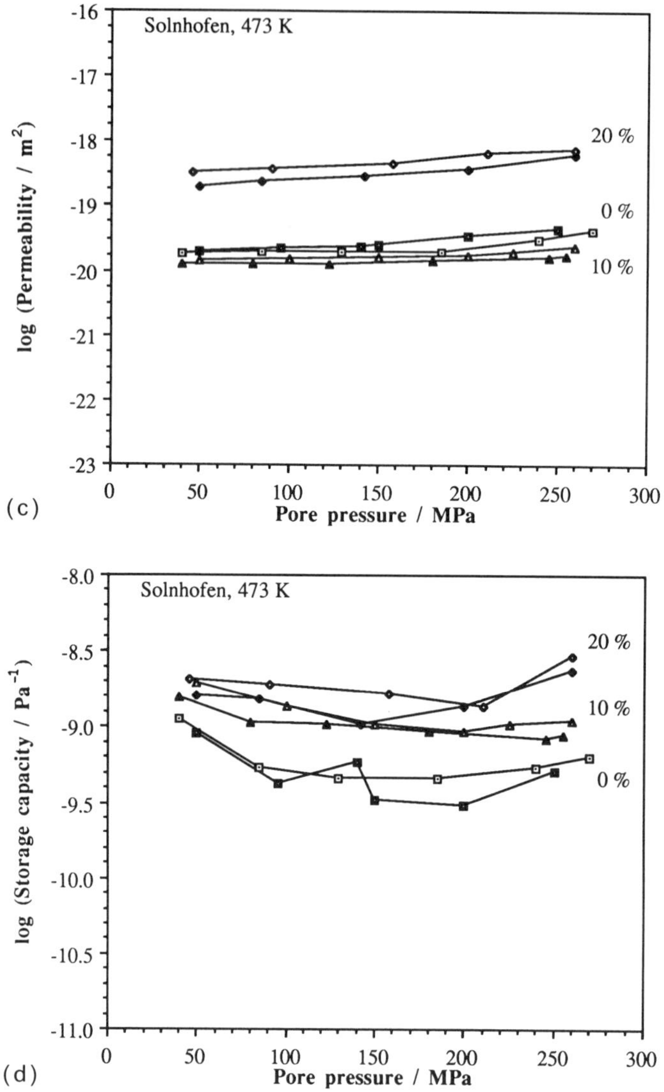

Figure 8. *Continued.*

(5) Gosford sandstone gives values of k and β_c that are generally higher than those for the calcite rocks, and the dependence on temperature is somewhat less.

(6) Measurements made in a sequence of increasing pore pressures generally give lower values of k and β_c than those made in a decreasing sequence. Thus there is a hysteresis in the porosity and/or other changes that accompany change in pore pressure. The cases of Solnhofen limestone at elevated temperature are exceptional in that the hysteresis is of reversed sense. The hysteresis in k and β_c may be related to their dependence, at given mean pore pressure, on the amplitude and frequency of the small pore pressure oscillation used in its measurement (Figure 10). Both effects can probably be related to

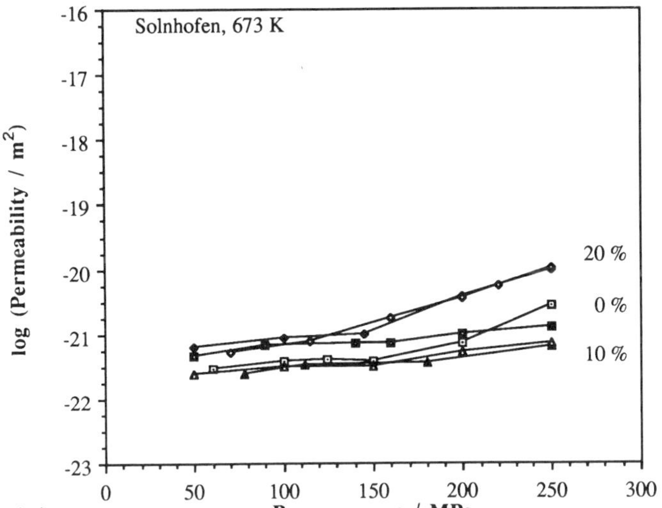

(e)

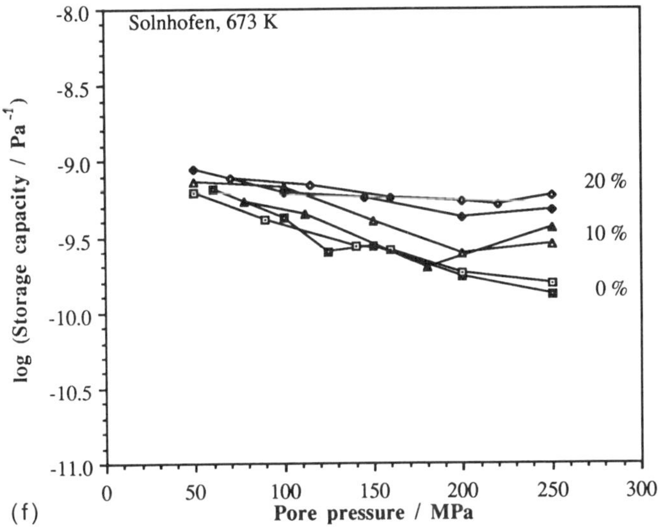

(f)

Figure 8. *Continued.*

non-elastic changes in pore structure that occur during change in pore pressure. These changes will be time-dependent because of the thermally activated processes that also give rise to the temperature dependence of k and β_c. The reversed sense of the hysteresis in Solnhofen limestone at the higher temperatures may also involve thermal recovery effects. However, any nonlinear behavior distorting the sinusoidal form of p_d is evidently confined to the first cycle or two of the small oscillations and so is obscured by the normal transient term; that is, the first cycle or two at a given mean pore pressure presumably 'conditions' the rock into the quasistationary state that is measured in later cycles, but which is different from the quasistationary state reached after returning from an excursion to markedly different mean pore pressures.

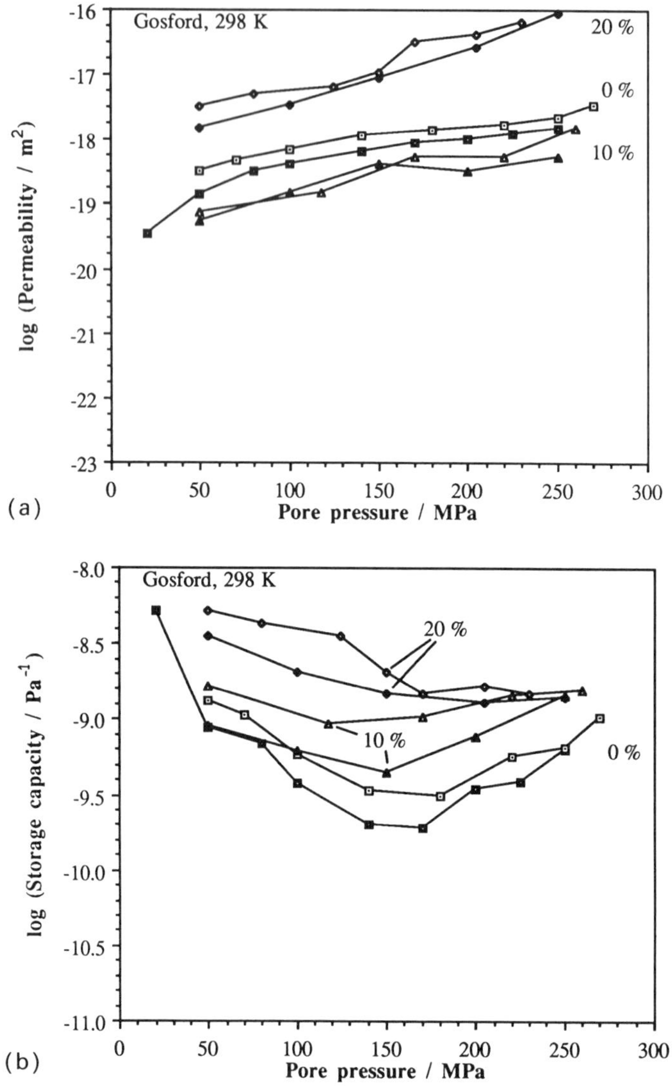

Figure 9a–d. Values of permeability k and storage capacity per unit volume β_c for Gosford sandstone as a function of temperature, deformation and pore pressure, all at 300 MPa confining pressure. Pore fluid is argon.

9. A Model Analysis

We now analyze the above results in terms of models relating the porosity φ, permeability k, and storage capacity per unit volume β_c to the pore structure of the rock, as characterized by parameters describing the size, shape and connectivity of the pores. At this stage, only mean values of these parameters will be discussed, no detailed analysis of pore parameter distributions being attempted.

For interpretation of the permeability, the equivalent channel or Kozeny–Carman

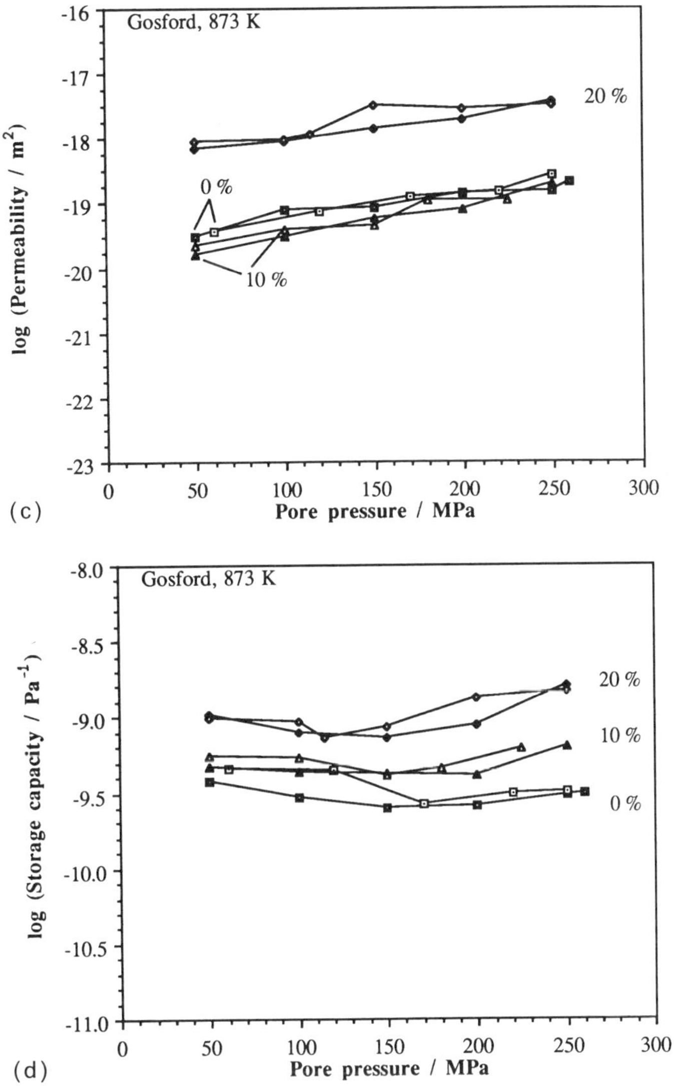

Figure 9. *Continued.*

model is assumed, as described by Paterson (1983) (see also Bear, 1972, Ch. 5; Scheidegger, 1974, Ch. 6; Dullien, 1979; Walsh and Brace, 1984). This model treats the pore space as if it consists of pipelike channels that are fully characterized, in respect of fluid flowing through them, by a length L_c and a cross-sectional parameter R called the hydraulic radius and defined as the ratio of the pore volume to the solid–fluid interfacial area (in effect, the cross-sectional area-to-perimeter ratio). The tortuosity of the channels is represented by $\tau = (L_c/L)^2$, where L is the length of the specimen in the direction of macroscopic flow. In terms of these three quantities, the permeability can be expressed as

$$k = C_1 \varphi \frac{R^2}{\tau} \tag{13a}$$

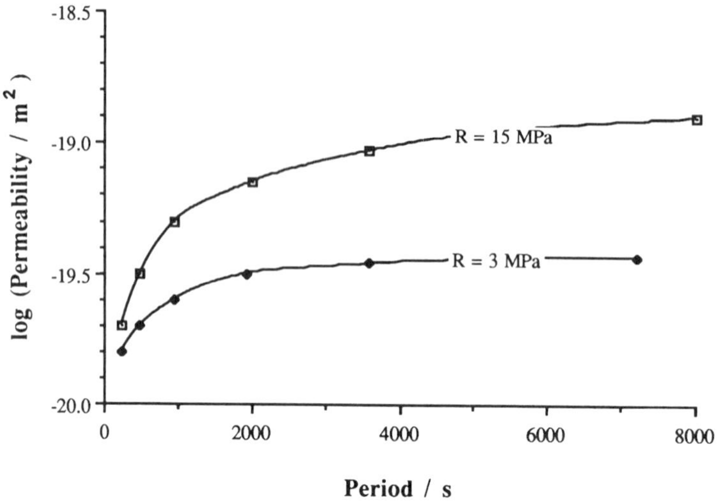

Figure 10. The dependence of the permeability of Solnhofen limestone on the frequency and amplitude of the pore pressure oscillation at room temperature. Confining pressure is 200 MPa, pore pressure is 150 MPa. R (range) $= 2P$.

where C_1 is a numerical factor depending somewhat on the actual pore shape. Hydrodynamical considerations indicate values for C_1 of 1/2 for a circular cross-section, 3/5 for a triangular cross-section, and 1/3 for a slot (Wyllie and Spengler, 1952). In the absence of direct information, the value of the tortuosity τ is commonly obtained from the formation factor associated with electrical resistivity in the case of a conducting pore fluid, using the expression $\tau = F\varphi$ (Paterson, 1983). Combining this expression with Archie's empirical law $F = \varphi^{-m}$ (Archie, 1942) and substituting in (13a) gives

$$k = C_1 \varphi^m R^2 \tag{13b}$$

Brace et al. (1965) conclude that m is close to 2 for a wide range of rocks with porosities varying from 0.001 to 0.1, while at higher porosities in granular media Archie (1942) found values of about 1.3. In anticipation that we shall be mainly concerned with microcrack porosity, we shall assume $m = 2$ and $C_1 = 1/3$ for a first analysis. Since for a slot-shaped cross-section the hydraulic radius is essentially equal to one-half of the minimum cross-sectional dimension x, we thus obtain from (13b) the approximate expression

$$x \approx \frac{3.5k^{\frac{1}{2}}}{\varphi} \tag{14}$$

The same expression also roughly represents the case of channels of equant-shaped cross-section in granular media of porosities of 0.2–0.4, and so it will be used comprehensively in the following analysis.

For interpreting the storage capacity per unit volume β_c, we use the theory of elasticity of porous bodies, as set out by Walsh (1965) and O'Connell and Budiansky (1974). From Walsh (1965, equations (6) and (11)) we can write, for small porosities,

$$\beta_b = \beta_m \left(1 + C_2' \frac{a^3}{v}\right) \tag{15}$$

where a is the maximum half-dimension of the pore, v is the fractional volume of the region containing the pore (equal to $1/N$, where N is the number density of pores), and C_2' is a constant having the values $(3/2)(1-\nu)/(1-2\nu)$, $(16/9)(1-\nu^2)/(1-2\nu)$ and $(4\pi/3)(1-\nu^2)/(1-2\nu)$ (where ν is the Poisson ratio) for spherical, penny-shaped and cylindrical (plane strain) pores, respectively. Then, using the relation $\varphi = (4\pi/3)Na^2c$ for spheroidal pores of radius a and half-height c, or a similar expression for cylindrical pores, we can rewrite (15) as

$$\beta_b = \beta_m\left(1 + C_2\varphi\frac{a}{c}\right) \tag{16}$$

where C_2 is a constant, the value of which varies from about 0.4 for spherical pores and $\nu = 0.1$, to about 2.3 for elongate pores and $\nu = 0.3$. The constant C_2 can be put equal to unity as a typical value, with an error of no more than about a factor of 2. Then substituting (16) into (11) we obtain

$$\frac{a}{c} \approx 1 + \frac{\beta_f}{\beta_m}\left(\frac{\beta_c}{\varphi\beta_f} - 1\right) \tag{17}$$

The following points may be noted.

(1) The effect of unconnected porosity can be taken into account by increasing the value of β_m appropriately in (17).

(2) Since the compressibility of the rock/pore fluid system cannot be smaller than the compressibility of the pore fluid alone, it follows that $\beta_c/(\varphi\beta_f)$ is never less than unity. This corresponds to $a/c \geq 1$ in eq. (17).

(3) When experimental uncertainties in β_c and φ are taken into account, there will be a lower limit to the value of a/c that can be reliably determined. From (9) and (12), neglecting β_r relative to β_f, we have $\beta_c/(\varphi\beta_f)$ equal to $V_r/(LA\gamma\varphi)$. If we allow factors of uncertainty of $1 + u_i$ for each of the quantities V_r, φ and γ ($i = 1, 2, 3$, respectively), the minimum value of $\beta_c/\varphi\beta_f - 1$ that can be relied upon is $(1 + u_1)(1 + u_2)(1 + u_3)$. We estimate that in the present study the values of x_i are $|u_1| < 0.1$ and $|u_{2,3}| < 0.2$, leading to a minimum reliable value of $\beta_c/(\varphi\beta_f)$ of 1.58, say 1.6, and a lower limit of 0.63, say 0.6. Therefore values of a/c for the cases $\beta_c/(\varphi\beta_f) < 1.6$ are regarded as unreliable, as indicated by brackets in Tables 3–5. Where $\beta_c/(\varphi\beta_f) < 0.6$ (indicated by an asterisk), it must be presumed that there is further error in the data that is so far unaccounted for.

The results of the calculation of x and a/c, using eqs. 14 and 17, are included in Tables 3–5 and, for the undeformed cases, plotted in Figure 11. The mineral compressibilities β_m used were obtained from Carmichael (1984) using the Hill average and the values of (dK/dp) and (dK/dT) listed there. The values of $\beta_c/(\varphi\beta_f)$ are also listed in the tables and used to determine acceptability of a/c values, as mentioned above. At the present levels of accuracy, medium ($\sim 10^2$) and large ($\sim 10^3$) aspect ratios should be approximately resolvable at the higher pore pressures (>150 MPa, say), but only large aspect ratios at the lower pore pressures; aspect ratios less than $\sim 10^2$ are not resolved (note that the use of a less-compressible pore fluid would improve the resolution).

For all three undeformed rocks at room temperature, where resolution permits, it would appear from the a/c values in Tables 3–5 that the porosity is initially of a crack-like type with a medium aspect ratio, and that this aspect ratio tends to increase somewhat with increasing pore pressure. If the pore openings also either increase or

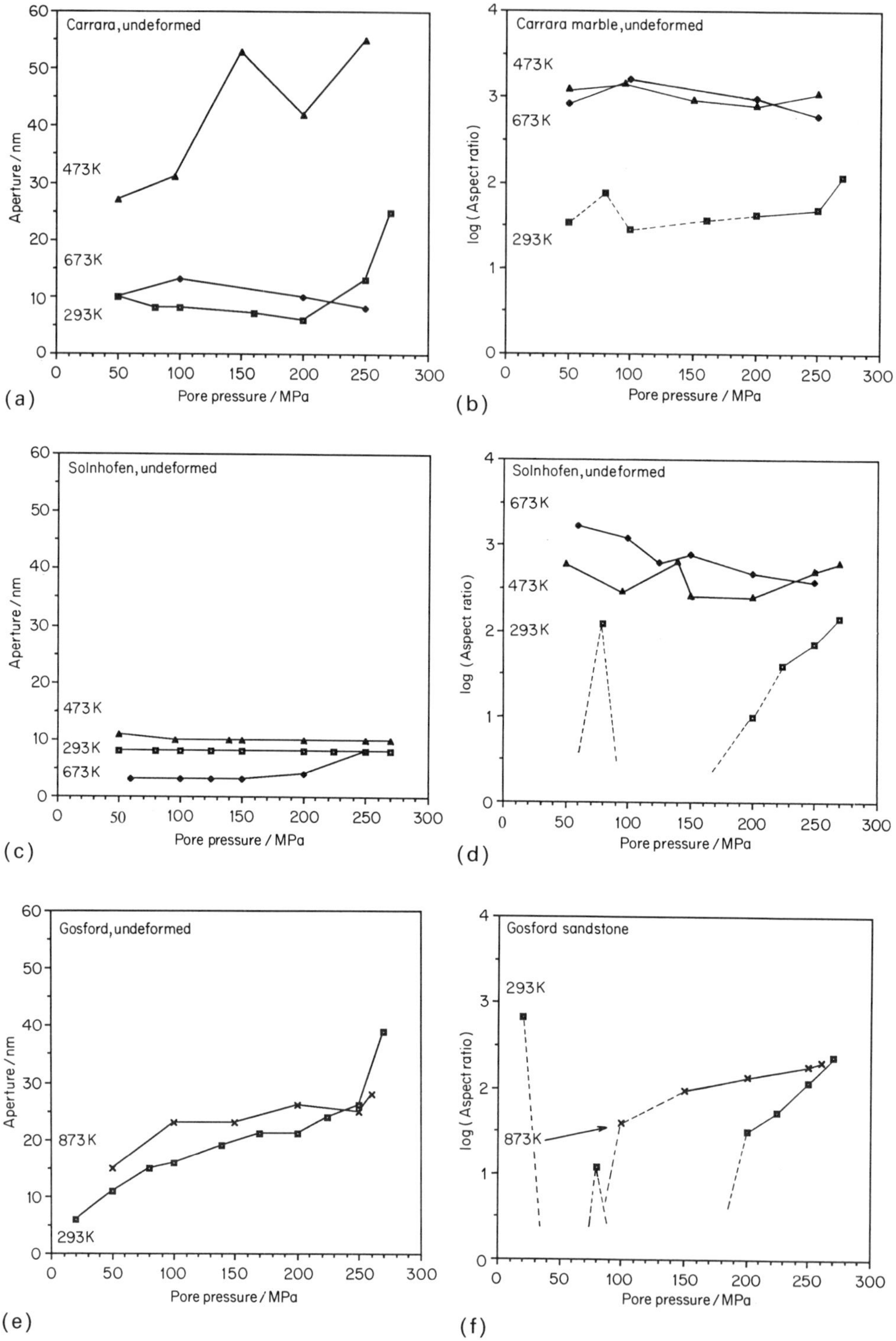

Figure 11a–f. Values of pore apertures x and aspect ratios a/c as a function of temperature and pore pressure, all at 300 MPa confining pressure.

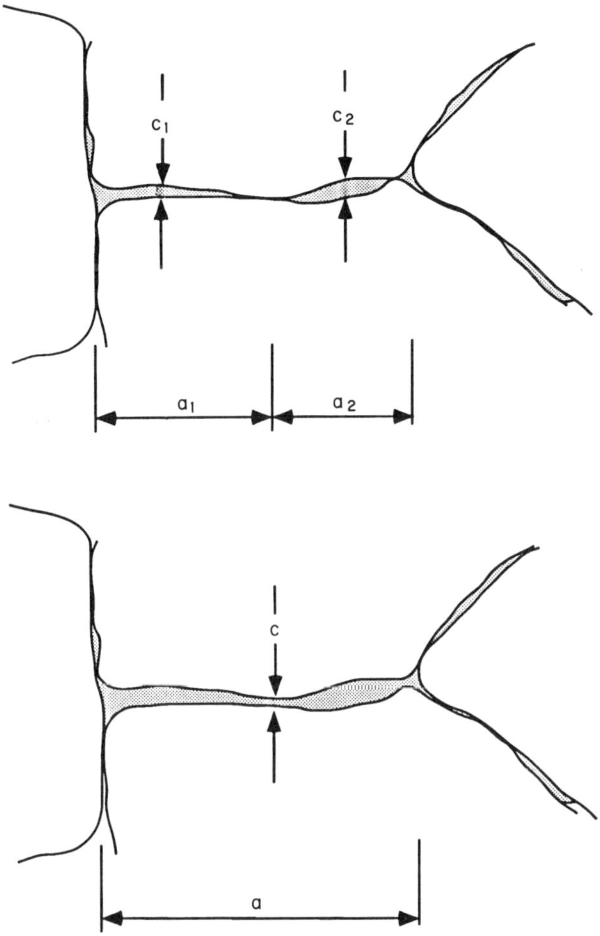

Figure 12. An illustration of how a marked increase in aspect ratio can result from a small increase in crack opening.

remain approximately constant, then the increase in a/c with increasing pore pressure may reflect an increase in length of the cracks, possibly due to the effect illustrated in Figure 12. In the case of Carrara marble, heating the undeformed rock to 473 K leads to a notable increase in aspect ratios, suggesting propagation of the cracks under thermal stresses, as remarked earlier. This effect is less marked in the Solnhofen limestone and is not evident in the Gosford sandstone (although only a relatively high temperature was investigated for the latter). Deformation also tends to lead to large aspect ratios, again probably associated with crack propagation.

The values of the model pore apertures x are, in most cases, of the order of 10 nm (Tables 3–5), in contrast to values mostly in the 40–120 nm range for Westerly granite (Wong et al., 1989). If the viscosity of argon flowing through such small apertures were higher than the microscopic values assumed, the values of x would be correspondingly higher, but it is not expected that such an effect would change the order of magnitude of the deduced apertures. However, it is possible that these low values are not representative of the actual mean pore apertures because of the inadequacy of the equivalent channel model to take into account the effect of substantial variations in pore dimensions. Insofar as the pore structure might consist of more or less distinct cavities linked by necks of smaller dimensions ('neck and bulge' structure), the aperture deduced for the 'equivalent channel' would be biased towards the neck dimensions rather than the mean aperture because of the markedly nonlinear relationship between aperture and flow resistance. A similar effect would arise if a significant proportion of the pores were connected on only one side ('dead-end' pores). That a pore structure of some such type exists is also indicated by the calculation of a model channel spacing under the assumption of uniform channel cross-section. Taking this cross-sectional area to be approximately πac, then the mean channel spacing l would be $l = (\pi ac/\varphi)^{1/2} = c\{\pi(a/c)/\varphi\}^{1/2}$. If we put $x = c$ then the values of the mean channel spacing are predicted to be much smaller than the grain size; for example, for each of the undeformed rocks, l is then of the order of 1 μm or less. However, such a close spacing of pores or cracks is not seen in thin sections of such rocks (for example, Fredrich et al., 1989). Since the aspect ratio determination is based on linear elastic behavior, which should yield mean values, it would appear that the above unrealistic spacing values result from the model aperture values being smaller than the mean pore apertures, perhaps by up to an order of magnitude or so. Thus we conclude that the pore structure involves a wide range of apertures, such as might be represented by a neck-and-bulge model, or a high proportion of dead-end pores.

The undeformed Carrara marble, when heated to 473 K, shows a substantial increase in model pore apertures without proportional increase in porosity, and the value of l, as defined in the previous paragraph, is now of the order of 10 μm, only one order of magnitude less than the grain size. This situation suggests that now the variation in pore aperture or the proportion of dead-end pores is much less and that the pore structure approximates more closely to a continuously connected crack network, which also applies after deformation. At 673 K, the situation is partially reversed again, possibly as a result of some thermal recovery. Such effects are much less marked in the other rocks.

The small size of the deduced pore apertures could have important consequences in geological situations involving fluid transport in such 'tight' rocks. The submicron-sized apertures will be strongly affected by even very small amounts of dissolution in or deposition from the pore fluid and so will be particularly prone to either growth or blocking during fluid flow.

10. Conclusion

By means of the procedure of oscillating the argon pore pressure at one end of a specimen and measuring the attenuation and phase shift of the oscillation during its propagation through the specimen, values of the permeability and storage capacity have been obtained for specimens of marble, limestone, and sandstone at various pore pressures, temperatures, and stages of deformation in experiments at 300 MPa confining pressure. Analysis of these results in terms of models for permeability and pore elasticity indicate mean pore apertures that are initially of the order of 10–20 nm and aspect ratios of the order of 100 or less. After heating or deformation, the apertures and aspect ratios both increase, but higher temperatures can lead to reductions in apertures again (especially in Solnhofen limestone), suggesting thermal relaxation. Extending such studies to active pore fluids such as water should provide a powerful tool in fluid–rock interaction studies.

Acknowledgments

We are indebted to John Phillip, who first suggested use of the oscillatory technique for our study and we are also grateful to Herbert Niesler and Norm Schram for invaluable assistance at various stages in the design and implementation of the experiment and to Graeme Horwood for maintenance of the high-pressure apparatus. Bob Krantz and Yves Bernabé improved the manuscript appreciably by their review comments and suggestions.

References

Archie, G.E. (1942). The electrical resistivity log as an aid in determining some reservoir characteristics. *Trans. AIME*, **146**,54–67.

Bear, J. (1972). *Dynamics of Fluids in Porous Media.* American Elsevier, New York.

Bennion, D.W. and Goss, M.J. (1971). A sinusoidal pressure response method for determining the properties of a porous medium and its in-situ fluid. *Soc. Pet. Eng. J.*, paper SPE 3541.

Brace, W.F., Orange, A.S., and Madden, T.M. (1966) The effect of pressure on the electrical resistivity of water-saturated crystalline rock. *J. Geophys. Res.* **70**, 5669–5678.

Carmichael, R.S. (ed.) (1984). *Handbook of Physical Properties of Rocks*, vol. 3. CRC Press, Boca Raton.

Cerceo, M. and Childers, H.M. (1963). Thermal diffusivity by electron bombardment heating. *J. Appl. Phys.* **34**, 1445–1449.

Cowan, R.D. (1961). Proposed method of measuring thermal diffusivity at high temperatures. *J. Appl. Phys.* **32**, 1363–1370.

Dullien, F.A. (1979). *Porous Media: Fluid Transport and Pore Structure.* Academic Press, New York.

Edmond, J.M. and Paterson, M.S. (1972). Volume changes during the deformation of rocks at high pressures. *Int. J. Rock Mech. Min. Sci.* **9**, 161–182.

Fischer, G.J. (1988). Permeability and storage capacity during deformation at elevated temperatures (abstract). *EOS, Trans. Am. Geophys. Union* **69**, 484.

Fischer, G.J. and Paterson, M.S. (1985). Dilatancy and permeability in rocks during deformation at high temperature and pressure (abstract). *EOS, Trans. Am Geophys. Union* **66**, 1065.

Fischer, G.J. and Paterson, M.S. (1989). Dilatancy during rock deformation at high temperatures and pressures. *J. Geophys. Res.* **94**, 17607–17617.

Fredrich, J.T. and Wong, T-f. (1986). Micromechanics of thermally induced cracking in three crustal rocks. *J. Geophys. Res.* **91**, 12743–12764.

Fredrich, J.T., Evans, B., and Wong, T-f. (1989). Micromechanics of the brittle to plastic transition in Carrara marble. *J. Geophys. Res.* **94**, 4129–4145.

Glisson, T.H., Black, C.I., and Sage, A.P. (1970). The digital computation of discrete spectra using the fast Fourier transform. *IEEE Trans. Audio Electroacoustics* **Au-18**, 3.

Jackson, I. and Paterson, M.S. (1987). Shear modulus and internal friction of calcite rocks at seismic frequencies: pressure, frequency and grain size dependence. *Physics of the Earth and Planetary Interiors* **45**, 349–367.

Krantz, R.L., Saltzman, J.S., and Blacic, J.D. (1990). Hydraulic diffusivity measurements on laboratory rock samples using an oscillating pore pressure method. *Int. J. Rock Mech Min. Sci.* **27**, 345–352.

O'Connell, R.J. and Budiansky, B. (1974). Seismic velocities in dry and saturated cracked solids. *J. Geophys. Res.* **79**, 5412–5426.

Paterson, M.S. (1970). A high-pressure, high-temperature apparatus for rock deformation. *Int. J. Rock Mech Min. Sci.* **7**, 517–526.

Paterson, M.S. (1983) The equivalent channel model for permeability and resistivity in fluid-saturated rock – a reappraisal. *Mech. Mater.* **2**, 345–352.

Scheidegger, A.E. (1974). *The Physics of Flow through Porous Media*, 3rd edn., pp. 135–144. University of Toronto Press.

Walsh, J.B. (1965). The effect of cracks on the compressibility of rock. *J. Geophys. Res.* **70**, 399–411.

Walsh, J.B. and Brace, W.B. (1984). The effect of pressure on porosity and the transport properties of rock. *J. Geophys. Res.* **89**, B11, 9425–9431.

Wong, T-f., Fredrich, J.T., and Gwanmesia, G.D. (1989). Crack aperture statistics and pore space fractal geometry of Westerly granite and Rutland quartzite: Implications for an elastic contact model of rock compressibility. *J. Geophys. Res.* **94**, B8, 10267–10278.

Wyllie, M.R.S. and Spangler, M.B. (1952). Application of electrical resistivity measurements to problem of fluid flow in porous media. *Bull. Am. Assoc. Petr. Geol.* **36**, 359.

Zhang, J., Wong, T-f., and Davis, D.M. (1990). Micromechanics of pressure-induced grain crushing in porous rocks. *J. Geophys. Res.* **95**, B1, 341–352.

Chapter 10

Growth of Grain Contacts in Halite by Solution-transfer: Implications for Diagenesis, Lithification, and Strength Recovery

Stephen H. Hickman[1] and Brian Evans[2]

[1] *U.S. Geological Survey, 345 Middlefield Road, Menlo Park, CA 94025, U.S.A.*

[2] *Department of Earth, Atmospheric, and Planetary Sciences, Massachusetts Institute of Technology, Cambridge, MA 02139, U.S.A.*

Abstract

Lithification of a sediment to form a rock may involve cementation, diagenetic reactions, or compaction under load. In these experiments, convex halite lenses were pressed against flat halite plates at 50°C in a specially designed microscope stage. A saturated brine surrounded the samples, which were observed during the experiment in transmitted and reflected light. No time-dependent convergence was observed between the two crystals, even at mean normal stresses of up to 14 MPa at the contact. In all experiments, however, the contact (or neck) between the two crystals grew with time as material dissolved from the surrounding lens surfaces, diffused through the pore fluid, and precipitated at the neck. Neck growth rates did not appear to correlate with the applied load, but did systematically increase as the misorientation between the two crystals decreased. Our analysis of the shapes of fluid inclusions formed along the grain boundary within the neck suggests that the grain boundary energy is about 1.8 times greater than the fluid–solid interfacial energy. Neck growth appears to be driven by the reduction of interfacial energy rather than by mechanical loads. Assuming that the interfacial energy is isotropic, and incorporating some geometric simplifications, two models of neck growth were formulated. The rate-controlling steps in the models were either precipitation or diffusion in the pore fluid. The data fit either model equally well. Both models predict that neck growth rate will be rapid at first but will decrease with time, as was observed. Neck growth will lead to an increase in real area of contact between grains in a granular aggregate even without the introduction of supersaturated solutions and may be important in the induration of sediments and the strengthening of fault gouge.

FAULT MECHANICS AND TRANSPORT PROPERTIES OF ROCKS
ISBN 0-12-243780-2

1. Introduction

Porosity and pore geometry strongly affect many rock properties, including permeability, elastic moduli, shear strength, electrical resistivity, seismic attenuation, and seismic anisotropy (Walsh, 1965; Brace, 1977; Budianski and O'Connell, 1976; Walder and Nur, 1984). In the presence of aqueous fluids, porosity in the Earth can be modified by a number of solution-transport processes, including pressure solution, crack healing, and cementation. In sediments, such processes may be the dominant agents of diagenesis and lithification. Knowledge of the kinetics of these processes, and their dependence upon temperature, stress, fluid chemistry, and pore geometry is essential in predicting the temporal and spatial evolution of physical properties in the Earth and in understanding the recovery of fault strength between earthquakes (e.g., Angevine et al., 1982).

Lithification of a sediment to form a rock may involve cementation, diagenetic reactions, or compaction under load (Maxwell, 1964; Dapples, 1971). Each process, in itself, may be rather complex and can take place concurrently with, or exclusive of, the others. Transfer of minerals by solution and precipitation may fill the interstices of the rock, form an interlocking texture, establish new grain boundaries, and indurate, or harden, the sediment (e.g., Pittman, 1972). Such cementation may occur in several generations and contain several different minerals (Meyers, 1974).

The source of cement is sometimes problematic. Cement minerals may be introduced into the rock from external sources by advection through the pore fluid. Alternatively, cement may arise from an internal source, such as a stylolyte at which pressure solution is occurring (Bathurst, 1958; Trurnit, 1968; Friedman, 1975; Tada and Siever, 1989). If the amount of material being precipitated as cement and overgrowths exceeds the amount of material dissolved at sources within the formation, the rock is a net importer of material (Houseknecht, 1987). If the reverse is true, the rock is a net exporter of minerals (Engelder, 1984). A formation that is a net importer may later become a net exporter, leading to decementation. Replacement of one cement by another is also possible (Blatt et al., 1980, ch. 9).

During lithification, close-fitting contact zones between grains, which we call necks, are established. The relationship between cementation, compaction, and neck growth is shown in Figure 1. In an open system, where minerals may be introduced through the pore fluids, cement may form a junction, or neck, between two abutting grains. A grain boundary may be formed between the two grains, providing that the cement is the same phase as the grains and that the energy of the grain boundary is less than twice that of the solid–liquid interfaces (Smith, 1948). Cementation (Figure 1a) leads to a decrease in pore volume and an increase in rock density, even though there is no relative motion between two arbitrary points fixed within the grains (i.e., there is no convergence between the two grains). The rate of cementation depends on the rates of diffusion, advection, and precipitation, and on the ratio of the interfacial and grain boundary energies.

If no new minerals are introduced and if there is no compaction, the total porosity – and also the density – must remain constant. Even so, a neck region may be formed by cementation (Figure 1b). For the case shown, this cement is locally derived from sources along the fluid–solid interface. The shape of the pore also changes with time, because dissolution occurs at some points along the pore–solid interface, while precipitation occurs at the neck region. The exact shape of the pore is determined by relative interfacial

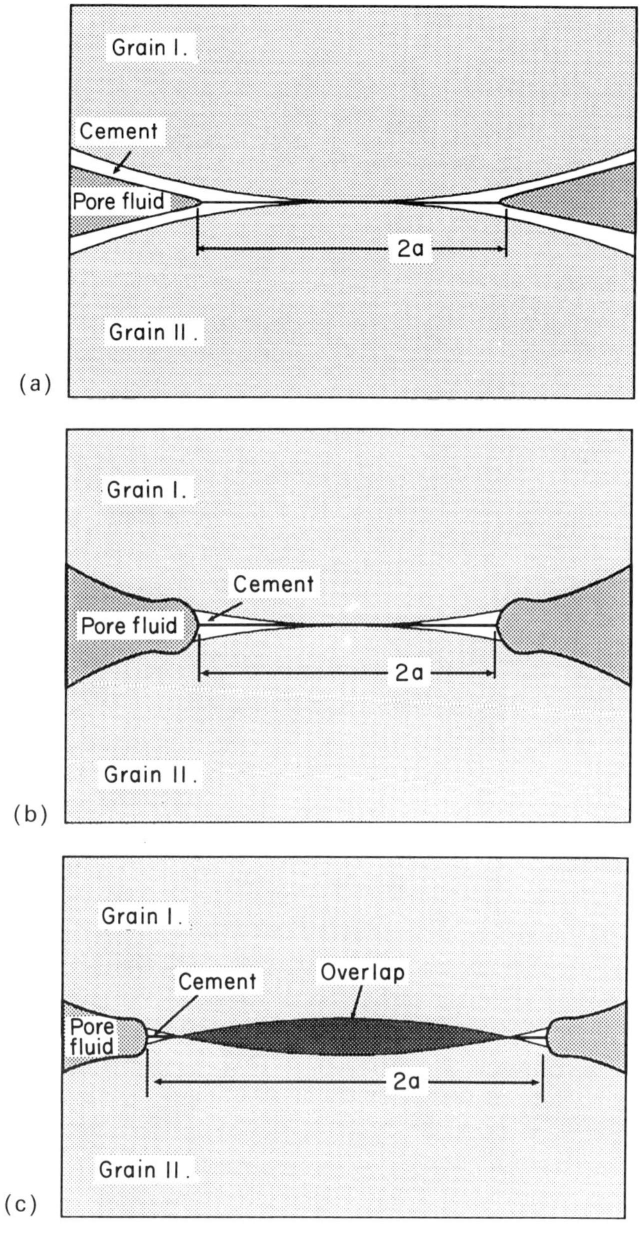

Figure 1. The relationship between cementation, pressure solution, neck growth, and compaction. (a) In a chemically open system, minerals may be introduced through the pore fluids and precipitate as cement to form a junction, or neck, of diameter $2a$, between two abutting grains. In this case, there is no relative motion between arbitrary points in the grains and, hence, no compaction. (b) For a chemically closed system with no compaction the total porosity is constant, but a neck region may still be formed by cementation. In this case the cement must be locally derived from the fluid-solid interface. (c) If compaction occurs, the material in the contact zone between the two crystals must deform enough to remove overlap material from the neck region (shown here in black). A neck region is also formed in this case, but two arbitrary points in abutting grains converge, resulting in compaction of the aggregate.

energies and the competition between the rates of dissolution, precipitation, and diffusion of the solute in the fluid.

If compaction occurs, the material in the zone between the two crystals must deform enough to remove overlap material from the neck region (Figure 1c). Deformation at the neck may occur by pressure solution, plastic flow, diffusional flow, or fracture. Material arising from the overlap or source region may be precipitated within the rock as cement or as an overgrowth, or removed entirely from the formation. As in the preceding two cases, a neck region is also formed, but the distance between two arbitrary points in abutting grains constantly decreases. This relative convergence of points within different grains results in the overall compaction of the aggregate.

As part of an investigation of pressure solution and solution-transfer processes, we observed the morphology at the contact between convex lenses of halite pressed against flat lenses (disks) of halite or fused silica under load in a heated microscope stage. The rate of convergence between the two lenses was monitored continuously using reflected light interferometry, as discussed in detail in a previous paper (Hickman and Evans, Evans, 1991). Convergence (i.e., compaction) was observed only when a halite lens and a fused-silica disk were pressed together in brine; in that case, the convergence rate depended upon the applied load and the size of the contact spot between the two sample halves. When a convex lens of halite was pressed against a halite disk, the material in the contact zone deformed by plastic flow during the first few moments of loading. No further compaction was observed after this initial yielding, even when the samples were immersed in a saturated brine. Although compaction was not observed, neck growth did occur when two halite lenses were pressed together in brine and a grain boundary was formed between the two crystals.

In this chapter, we present additional data on the rates of neck growth when convex halite lenses are pressed against halite disks in brine (henceforth referred to as ball-and-plate experiments) and discuss the mechanisms and kinetics of this process. Two observations suggest that neck growth in these experiments was driven by a reduction in the total solid–liquid interfacial energy near the contact zone. First, neck growth and cementation continued even though no creep convergence was observed. Second, the microstructure of the interface was similar to structures formed during crack healing. Although these experiments involved neck formation and cementation only at a single contact, the results also give insight into solution-transfer processes in general. Such processes are probably important in the lithification of other common single-phase sediments and will have a profound influence on the physical properties of sediments and fault gouges.

2. Experimental Technique

In this study, a convex halite lens and a flat halite disk were observed while under load in a heated microscope stage (Figure 2). To begin the experiment, the halite samples, a spring, and any necessary fused-silica spacers were immersed in saturated brine inside a Teflon chamber. The chamber was surrounded on all sides by circulating water from a constant-temperature bath. However, the brine within the Teflon sample chamber remained stagnant throughout the experiment. Special care was taken to insure that the temperature within the sample chamber was spatially uniform and constant in time (within 0.1°C), that the brine was completely saturated, and that the springs were

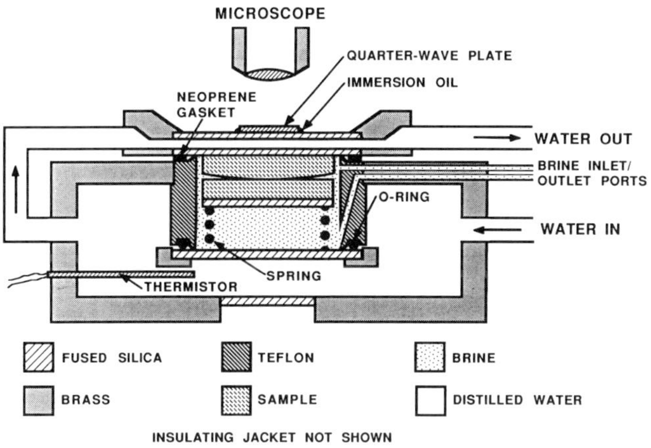

Figure 2. Spring-actuated loading frame used for in-situ observation of pressure solution and cementation in halite. Brine was injected into the chamber through the inlet and outlet ports, but remained stagnant during the experiment. Circulating water from a constant-temperature bath maintained the temperature of the brine chamber within $\pm 0.1\,^{\circ}C$. Loads were adjusted by inserting spacers beneath the spring and thereby changing the spring compression. Viewing the contact region between the two lenses in monochromatic (sodium) light allowed us to monitor the separation between the two lenses using reflected-light interferometry, providing a sensitive measure of the convergence between the lens and the disk (see Hickman and Evans, 1991). In these experiments, although an instantaneous elastic–plastic convergence occurred upon application of load, no creep compaction (i.e., time-dependent convergence) was observed. The quarter-wave plate helped to eliminate back reflections from the interface between the air and the fused silica window.

creep- and corrosion-resistant. Sample dimensions were measured using a microscope equipped with a dial indicator. Loads were determined from the spring calibration curves, the measured sample-chamber depth, and the sample heights; adjustments in load were made using fused-silica spacers. All dimensions used in calculating loads were measured before the experiment and corrected for thermal expansion.

In most cases, samples were weighed before and after each experiment. These measurements invariably indicated a slight gain in sample weight (0.3–0.5%). Unfortunately, as the samples were removed from the chamber a small amount of hot brine always remained on the lenses, despite efforts to remove it with compressed air. The brine would then evaporate, leaving behind a visible residue of crystallized salt. We believe the observed weight gain resulted from this precipitation. Further details of the device, sample preparation, and measurement technique are given by Hickman (1989) and Hickman and Evans (1991).

Observations of the morphology of the contact region and measurements of convergence were made using time-lapse photography in transmitted white light or reflected sodium light (Figure 3). The contact area was measured directly from the photomicrographs, using a digitizing table. When the sample was illuminated by reflected light, rays reflected from the disk–brine interface interfere with rays from the brine–lens interface. The interference produces a pattern around the central contact, commonly called Newton's rings. The distance between the two lenses, h, can be determined at

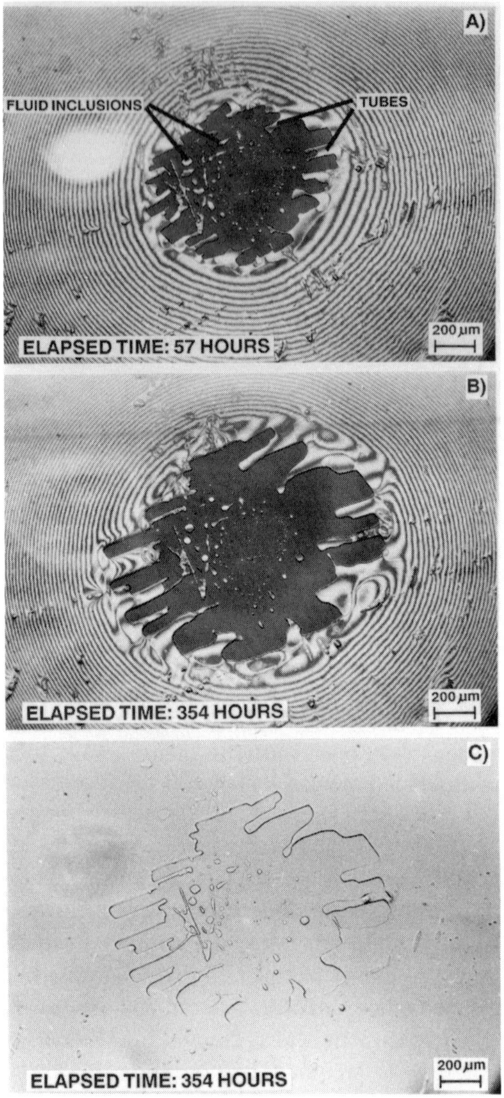

Figure 3. (a) and (b) Reflected sodium light interferograms of a convex halite lens pressed against a flat halite lens (sample PSH 15) in brine for the times shown. The dark lobate, or dendritic, region in the center of each photo is the contact spot formed between the two lenses and contains fluid-filled tubes and residual fluid inclusions. The dark concentric bands are interference minima and are partially disrupted by imperfections in the lens surface and through localized mass transfer near the contact spot (moat formation; see text). As these lenses were immersed in saturated brine (refractive index = 1.376), the change in lens separation between adjacent interference minima is 0.214 μm. (c) Transmitted light photomicrograph of this sample taken at the same time as (b).

any point from:

$$h(r) = \frac{m\lambda}{2n_D} \tag{1}$$

where m is the order of the interference minima, λ is the wavelength of sodium light in a vacuum, and n_D is the refractive index of sodium light in saturated brine ($n_D = 1.376$ at 50.2°C; Hickman, 1989). The amount of convergence that occurred during each experiment was determined by monitoring the change in either area or radius of these interference minima (Hickman and Evans, 1991). No convergence was observed in any of the experiments discussed in this study.

The exact method by which samples were prepared was important. In preliminary experiments we observed that standard abrasive polishing techniques produced thick layers of surface damage in the halite (Hickman and Evans, 1991, fig. 1). These surfaces roughened when they were exposed to brine and the Newton's rings were disrupted. In addition, the rate of neck-growth between two abrasively polished lenses was slower than that between two unpolished cleavage chips or between samples prepared using a combination of chemical polishing and high-temperature heat treatment.

In all the experiments reported here, the samples were prepared by cleaving a blank from a single NaCl crystal with a razor blade, dissolving this blank to the appropriate size and curvature using porous polymer laps wetted with distilled water, polishing with 0.1 μm alumina grit in isobutyl alcohol, and baking in air at 650°C for 14 hours (Hickman and Evans, 1991). This procedure produced lenses with a minimum of surface damage.

3. Results

3.1. Neck growth and grain boundary formation

Seven experiments were done with convex halite lenses pressed against halite disks at approximately 50°C under loads of 0.8–4.4 N (Table 1). In all experiments, the radius of curvature of the convex lens was determined using interferograms taken at the beginning of each experiment (see Hickman and Evans, 1991). In each experiment conducted at low loads (PSH 15, PSH, 32, and PSH 33), the lens curvature was uniform and the radius of curvature was determined from the first ten interference minima. In the high-load experiments, however, the free surfaces of the lenses just outside of the contact spot exhibited significant deflection when the load was applied, resulting in a localized increase in the radius of curvature of 10–38%. The radii of curvature reported for these experiments (Table 1) and used in the following analysis were computed as the average curvature over the region that was eventually engulfed by the growing contact spot. Thus, these radii differ from the global radii (i.e., determined from the first ten interference minima, but avoiding the deformed region near the contact spot) reported for the same experiments by Hickman and Evans (1991).

In all the experiments, the contact spot between the lens and the disk broadened with time as the neck grew, producing a monotonic decrease in mean normal stress with time (see below). The lobate or fingered morphology of the contact zone (the dark central region in Figure 3) is similar to structures resulting from crack healing in halite and other minerals (Roedder, 1984; Smith and Evans, 1984). The curvature of the lenses near the contact zone changed with time through localized dissolution of halite near the neck region, but because the fringes farther from the contact region were stationary, it is clear that there was no relative motion between the two lenses (Hickman and Evans, 1991).

The structure and energy of the grain boundary formed depends on the relative

Table 1. Neck growth experiments

Experiment	Temperature[a] (°C)	Local radius of curvature[b] (cm)	Load (N)	Grain-boundary misorientation (±1°)	Fabrication[c]	Experiment duration (h)	Weight change (%)	Peripheral moat formation
PSH 14	50.10 ± 0.06	9.21	3.74 ± 0.13	5.5° twist	Dry	98.6	NR[d]	Pronounced
PSH 15	50.10 ± 0.06	9.62	0.837 ± 0.029	10.6° twist	Wet	353.8	NR	Pronounced
PSH 16	50.17 ± 0.06	10.55	4.12 ± 0.13	7.1° twist	Wet	185.6	+0.3	Pronounced
PSH 19	50.19 ± 0.08	9.15	4.44 ± 0.11	39.3° twist	Wet	68.1	+0.5	None
PSH 20	50.21 ± 0.06	10.87	4.22 ± 0.11	41.3° twist	Wet	166.4	+0.3	None
PSH 32	50.20 ± 0.06	8.21	0.993 ± 0.025	25° symmetric tilt + 7.2° twist	Wet	70.7	+0.3	Slight
PSH 33	50.21 ± 0.08	8.06	0.868 ± 0.024	25° symmetric tilt + 5 ± 2° twist	Wet	433.4	+0.3	Slight
HHP 1	90 ± 1	Flat	522[e]	12.0° twist	Wet	16.1	NR	–

[a] Except for HHP 1, where temperatures were recorded only with a thermometer, mean values shown were recorded by precision thermometer in fluid return line and uncertainties are maximum excursions recorded by thermistor shown in Figure 2.

[b] For the convex lens. Radii for the high-load experiments determined from the initial curvature under load as averaged over the region that was eventually engulfed by the growing neck. Radii for all other experiments determined from the first ten interference minima (see text).

[c] Wet: Brine added directly to contact between the two lenses before microscope stage lid closed and sample chamber filled with brine (see text). Dry: Lid to microscope stage closed and brine injected into the sample chamber without adding brine directly to contact.

[d] Not recorded.

[e] Load of 522 N results in mean normal stress of 5.2 ± 0.5 MPa across approximately 1 cm^2 contact area.

orientation of the two crystals and on the orientation of the grain boundary plane. In these experiments, a variety of grain boundary orientations were used (Table 1). In the pure twist-boundary configuration, both the planoconvex lens and the disk were polished with their flat surfaces parallel to (001). Thus, once pressed together, the ⟨001⟩ axes in both crystals were parallel to one another and perpendicular to the grain boundary. The angle between ⟨100⟩ in the disk and ⟨100⟩ in the convex lens (i.e., the twist angle) ranged from 6° to 41° in these experiments and was determined by reference to marks scribed into the perimeter of each lens. In the mixed tilt/twist-boundary configuration, a special polishing jig was employed to insure that the flat surfaces of both the disk and the planoconvex lens were at $12\frac{1}{2}°$ to (001) and cozonal with ⟨100⟩. The samples were then assembled so that (001) in the two crystals were out of parallel by 25° and bisected by the grain boundary. The resulting 25° symmetric tilt boundary was superimposed on a small (5–7°) twist component, which was measured using the method described previously. The determination of misorientation angles in these experiments was approximate, and tilt components of a degree or two might also be present in the nominally pure-twist boundaries.

Fluid inclusions were often formed at the interface between the two crystals as the neck region grew. Because the contact area was a small fraction of the total area of the crystals, it was difficult to recover samples without fracturing this boundary. To produce a more robust interface, we performed an additional experiment, HHP1 (see Table 1). In this experiment, we used two nominally flat cleavage chips which were polished using the same procedure as for the halite lenses. These chips were pressed together at 90°C under a much higher load, 522 N. Although we were not able to observe the boundary during loading, the interface which formed was much larger in area and could be recovered intact.

When observed in plan view in the optical microscope, the fluid inclusions formed during experiment HHP1 (Figure 4a) appeared to be similar to those formed during the experiments employing polished lenses (cf. Figure 3). A thin section made perpendicular to the boundary in HHP1 revealed that the inclusions were thin, tabular pores with negative crystal faces (Figure 4b). The presence of discrete fluid inclusions along the grain boundary precludes the existence of a continuous free fluid film. Substantial plastic flow must have occurred during this experiment, as is shown by the variable extinction of the crystals on either side of the boundary (Figure 4c).

The shape of the inclusions gives valuable information on the relative energies of the grain boundary and the fluid–solid interface. Measurements of the cross-sectional dimensions of 18 inclusions are given in Table 2. The dimensions of a few inclusions were measured in plan view to confirm the visual impression that these were equidimensional (i.e., $L_1 \cong L_2$ in Figure 4d). In addition, because the top and bottom of the inclusions were flat, we were able to obtain independent measures of the thickness, H, of two of these inclusions by viewing them in plan view when illuminated in reflected light by a variable-wavelength monochrometer. For these determinations, we measured the change in wavelength of the illumination required to offset the interference minima produced at a given location by one order number. By using these two combinations of λ and interference minima order numbers, i.e., $\lambda_1 m$ and $\lambda_2(m+1)$, we eliminated m from eq. (1) and determined H directly. The thickness of a third inclusion formed during experiment PSH15 was measured by focusing on the top and bottom surface of the pore with a calibrated focusing knob after the sample was split apart. Estimated uncertainties in L_1 and L_2 (see Table 2) result primarily from irregularities in inclusion

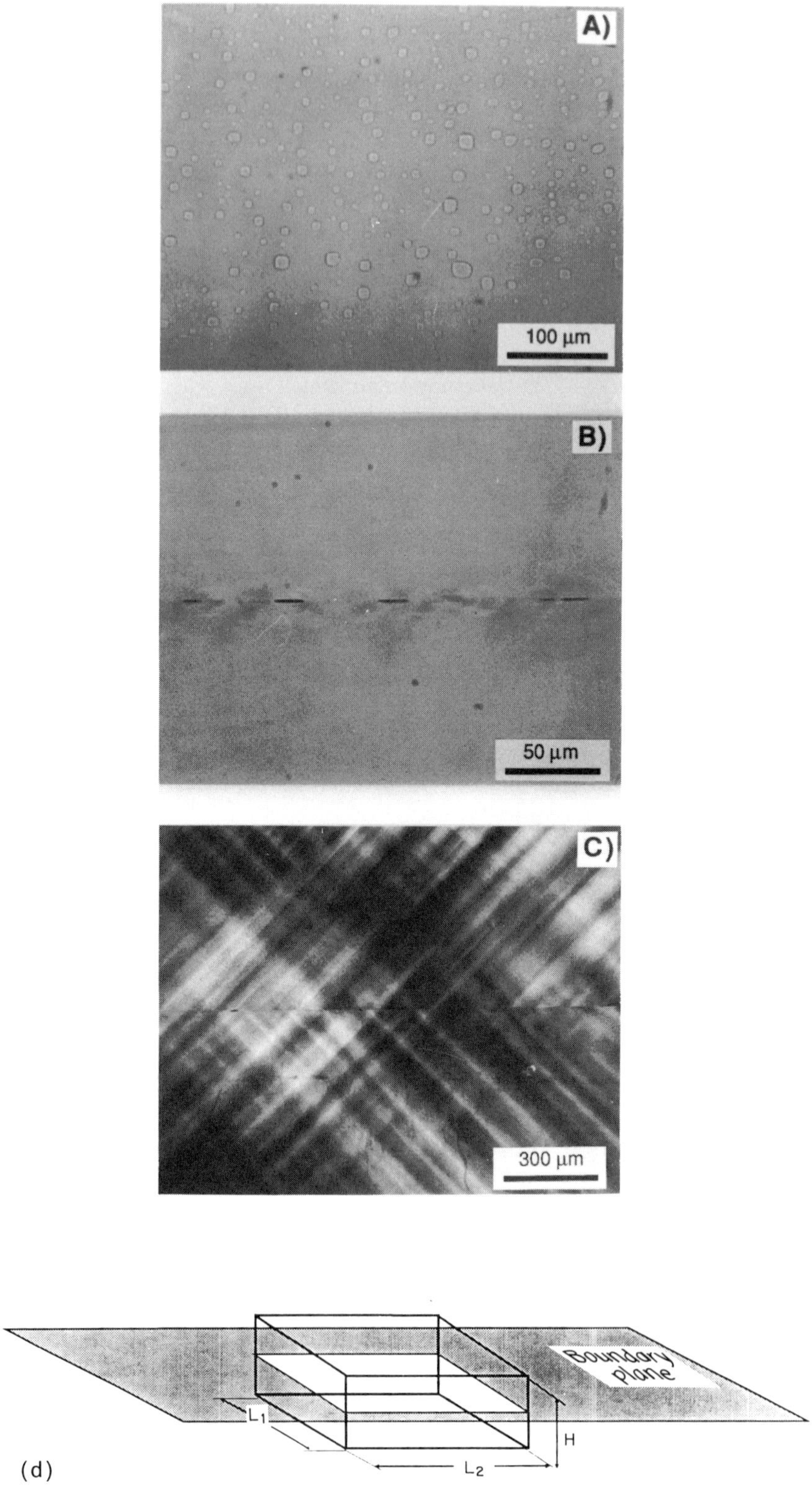

(d)

shape. The relative uncertainties in inclusion thickness as measured in transmitted light are on the order of 20–30%, due largely to inherent resolution limitations of the optical microscope.

3.2. Contact stresses and compaction

The stresses in the region of contact between the two crystals are determined by the total applied load and the geometry of the crystals (see Johnson, 1987, for a review). In our experiments the load was applied by a compressed spring. Because the amount of convergence between the two crystals was negligible, the total compression, and hence the applied load, was roughly constant during the experiment. The contact area, however, did change. Immediately after loading, the two crystals deformed to establish a flattened neck region. The contact area continued to increase due to neck growth, leading to a monotonic decrease in the mean normal stress with time.

The mean normal stress was estimated from the ratio of the applied load to the instantaneous area of contact between the crystals and is shown as a function of time in Figure 5. Mean normal stresses ranged from 1 MPa to 14 MPa, depending on the magnitude of the applied load and the elapsed time. The experiments may be separated into two groups; those with high mean normal stresses (loads of 3.74 N and above: PSH 14, 16, 19, and 20); and those with low mean normal stresses (loads below 1 N: PSH 15, 32, and 33). The mean stresses along the grain boundaries in both classes decreased rapidly as the neck diameter increased.

It is worth noting that neck growth in these experiments is analogous to high-temperature sintering in that both involve material transport driven by interfacial tension. At temperatures high enough to allow Coble or Nabarro–Herring creep in powdered metals and ceramics, interfacial tensions can provide sufficient driving force for compaction-induced densification and pore removal during sintering, even without the application of an external load (e.g., Herring, 1951). Although the combinations of interfacial tensions and applied loads in the halite ball-and-plate experiments were insufficient to cause convergence of the crystals at a measurable rate, the rapid dissolution of salt into the brine did allow for solution-transfer through the pore and precipitation of halite at the neck. This material flux altered the geometry of the pore space in the region adjacent to the contact zone and cemented the two crystals together.

Figure 4. (a) Transmitted light photomicrograph of sample HHP 1, showing residual fluid inclusions formed along the grain boundary. The grain boundary is parallel to the field of view. Note that the inclusions are roughly square in plan view. (b) Transmitted light photomicrograph of sample HHP 1, but with the optic axis of the microscope lying in the plane of the grain boundary. Note that the fluid inclusions formed along the grain boundary are quite thin. The tabular form (i.e., low aspect ratios) of these inclusions indicates that the ratio of grain boundary energy to solid–liquid interfacial energy is relatively high (see text). (c) Same view as presented in (b) but with crossed Nichols and at a lower magnification. The variable extinction on either side of the grain boundary is indicative of significant plastic deformation. (d) Schematic of fluid inclusions observed along halite grain boundaries in these experiments. The boundary plane is shaded.

Table 2. Wetting characteristics of grain-boundary fluid inclusions in halite

Experiment	Number of inclusions	Inclusion dimensions (μm)				Inclusion aspect ratio, α_i	Γ_b/Γ_{sl}
		L_1	L_2	H	How measured (see notes)		
HHP 1	15	5.6–21.2	–	0.55–0.95	*a*	0.070 ± 0.018	1.86 ± 0.04
HHP 1	2	28.5	27.5	1.12	*b*	0.040 ± 0.011	1.92 ± 0.02
		26.5	37.5	1.15	*b*	0.036 ± 0.010	1.93 ± 0.02
PSH 15	1	32	29	3	*c*	0.098 ± 0.042	1.80 ± 0.08

[a] All dimensions measured in transmitted light with optic axis of microscope in plane of grain boundary. Estimated uncertainties in L_1 and H are $\pm 15\%$ and ± 0.2 μm, respectively.

[b] All dimensions measured with optic axis of microscope perpendicular to grain boundary. L_1 and L_2 measured in transmitted light; H measured in reflected light using monochrometer (see text). Estimated uncertainty in L_1 and L_2 is $\pm 15\%$; uncertainty in H is ± 0.14 μm.

[c] Inclusion dimensions measured after sample split apart along grain boundary. L_1 and L_2 measured using scanning electron microscope; H determined from measurements on mating sample halves using calibrated focusing knob on optical microscope. Uncertainties are about ± 3 μm in L_1 and L_2 and ± 1 μm in H.

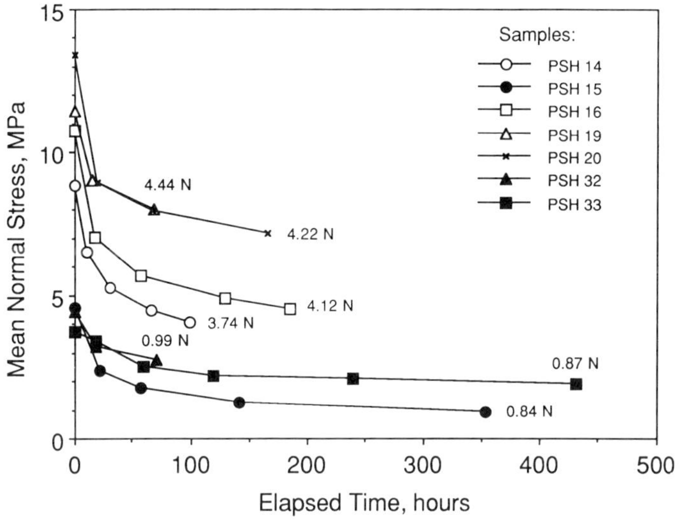

Figure 5. Mean contact normal stress, calculated as the applied load divided by the measured contact spot area between the halite lenses, as a function of elapsed time for the ball-and-plate experiments from this study. The applied loads were held constant during each experiment and ranged from 0.84 to 0.99 N and 3.7 to 4.4 N.

3.3. Kinetics of neck growth

Data for neck diameter as a function of time for the seven ball-and-plate experiments are presented in Figure 6. In all experiments, neck growth was initially rapid but decelerated with time. Neither the initial nor the final growth rate correlate with the magnitude of the applied load. For example, compare PSH 14 and 20, both high-load experiments, with PSH 15 and 33, two low-load experiments. More systematic is the relationship between the neck growth rate and the crystallographic misorientation between the two halite crystals. Experiments where the misorientation was relatively small ($<11^\circ$) tended to have higher growth rates than those experiments where the lattice misorientation of the two crystals was relatively large.

4. Discussion

4.1. Boundary structure in halite

Grain boundary structure is one important element in determining the rate of ionic transport during deformation by pressure solution. High-diffusivity fluid films have often been postulated to occur along mineral boundaries when aqueous pore fluids are present (Weyl, 1959; Rutter, 1976, 1983; Robin, 1978; Lehner and Bataille, 1985; Tada et al., 1987; Spiers and Schutjens, 1990).

For a fluid film to be stable thermodynamically, the energy of a grain boundary must be greater than twice that of the solid–liquid interface, otherwise wetting will occur (Smith, 1948). In the case of two solid–liquid interfaces intersected by a grain boundary, the dihedral angle formed by the solid–liquid interfaces at their junction with the grain

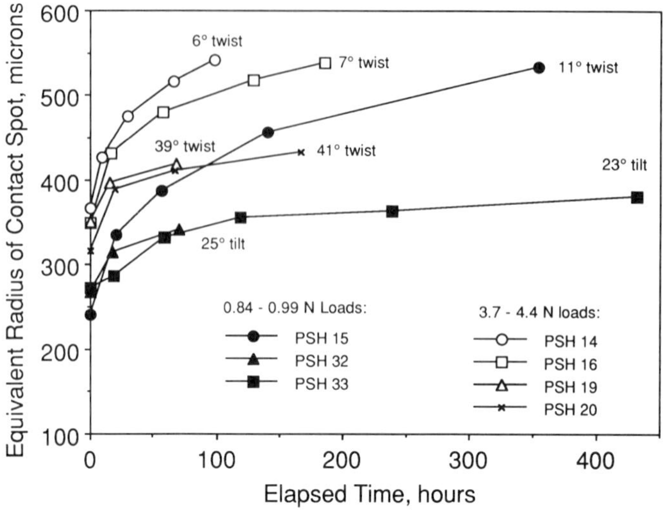

Figure 6. Equivalent radii of contact spots formed in these experiments as a function of elapsed time. For each data point, the area of the dendritic contact formed between the two halite lenses was measured digitally and the equivalent radius computed as the radius of a circle of an equivalent area. Notice that the radii increase rapidly at the beginning of each experiment and then more slowly as time progresses. Importantly, neck growth rates at comparable elapsed times are greatest for lenses with the smallest relative misorientations (i.e., 6–11° twist angles).

boundary is determined by a vector force balance of interfacial tensions (Blakely, 1973). When the dihedral angle is zero, wetting occurs, and a fluid film is stable. Thus, the dihedral angle strongly influences the distribution of fluids within the solid. Provided that transport of the solid component is possible, the pore geometry will adjust to provide vector equilibrium of the interface tensions.

Because many physical properties are critically dependent on pore geometry, the dihedral angle has been explored in several petrologic systems: granites with melt (Jurewicz and Watson, 1985), basaltic rocks with melt (Bulau et al., 1979; Vaughan et al., 1982; Watson, 1982; Cooper and Kohlstedt, 1982; Jurewicz and Watson, 1984; 1985), quartzites with melt and water (Watson and Brenan, 1987; Laporte and Watson, 1991; Lee et al., 1991), and limestones with water (Hay and Evans, 1988). In all these experiments, which were done under a wide variety of pressures, temperatures, and fluid chemistries, the dihedral angles observed were 40° or larger, indicating that the fluids were nonwetting.

Although these wetting-angle experiments suggest that fluid films with the same thermodynamic properties as water are not present along grain boundaries, the experiments give no information on other changes in structure, mobility, or diffusivity of grain boundaries which might result from the presence of water or other fluids. For example, there might be chemisorbed species segregated near the boundary. Furthermore, nonequilibrium structures have been proposed to arise during deformation (Lehner and Bataille, 1985; Spiers and Schutjens, 1990) or during migration of boundaries (Urai, 1985; Urai et al., 1986), although we saw no evidence to support such structures in these experiments.

In a previous paper (Hickman and Evans, 1991), we argued that the inclusion of fluid-filled pores along the growing boundary shows that fluid films do not exist in the halite–halite contact regions. No intermediate nonequilibrium structures were observed during the establishment of a neck during the ball-and-plate experiments reported here. However, because of the small neck diameters and the torques caused by unloading of the samples, it was not possible to remove the samples without fracturing the boundary between them. The interface formed during experiment HHP1, however, was strong enough to allow observations of the inclusions and the boundary, both in plan view and in cross-section. Within the resolution of the optical microscope, there was no evidence for a fluid film along the boundary.

The shape of the inclusions along the grain boundary is determined by the minimization of interfacial energy and gives information about the ratio of grain boundary energy to the solid–liquid interfacial energy. The strong tendency for fluid inclusions to develop planar negative faces is probably due to cusps in the dependence of interfacial energy on orientation (Blakely, 1973). The surface with lowest energy is probably $\{100\}$. Inclusions along the boundaries have much different aspect ratios from those within the halite crystals: stationary inclusions within halite crystals tend to be cubes (e.g. Anthony and Cline, 1971), while those lying along the grain boundaries in our experiments are tabular, i.e., $L_1 \cong L_2 \gg H$.

The interfacial energy of an inclusion not lying along a grain boundary is given by the product of the specific interfacial energy, Γ_{sl}, and the total surface area of the inclusion $(2L_1L_2 + 2L_1H + 2L_2H)$. For inclusions within the crystal, the shape with minimum interfacial energy is a sphere when surface energies are isotropic or a cube when one plane of a crystal with cubic symmetry has a particularly low energy, as in halite.

For an inclusion of dimensions, $L_1 = L_2 \equiv L$ and H, lying on the boundary with specific grain boundary energy of Γ_b (see Figure 4d), the total energy of the inclusion and an arbitrary area of boundary surrounding the inclusion, A_b, is given by

$$E_i = \Gamma_b(A_b - L^2) + \Gamma_{sl}(2L^2 + 4LH) \tag{2}$$

For a given volume $V = L^2H$, the inclusion will assume a shape for which the total interfacial energy is minimized. Computing $\partial E_i/\partial L$, setting the derivative equal to zero, simplifying, and defining the aspect ratio $\alpha_i = H/L$ gives:

$$\frac{\Gamma_b}{\Gamma_{sl}} = 2(1 - \alpha_i) \tag{3}$$

When $\Gamma_b/\Gamma_{sl} = 2$, $\alpha_i \to 0$, and the fluid wets the boundary. When $\Gamma_b = 0$, as in the interior of the crystal, $\alpha_i \to 1$, and the inclusion is cubic. For the 18 inclusions we examined (Table 2), the ratio Γ_b/Γ_{sl} ranges from 1.80 to 1.93.

Although the boundary energy in halite for twist angles of 11–12° (i.e., samples HHP1 and PSH15) is about 0.9 that of the critical energy for wetting, there was no evidence for a fluid film in any of the ball-and-plate experiments or in experiment HHP1, nor was there evidence for nonequilibrium boundary structures. Boundaries with inclusions were formed immediately upon loading in all the ball-and-plate experiments regardless of the angle of misorientation. The fact that boundaries with inclusions are formed even when misorientations are larger than 12° suggests that, at least for the angles measured here, the grain boundary energy is only weakly dependent on angle for large twist angles. Furthermore, boundaries were formed even though considerable plastic flow occurred along the interface in the ball-and-plate geometry (Hickman, 1989) and throughout the crystals in HHP1 (Figure 4c). Thus, the presence of dislocations in these experiments did not seem to produce a nonequilibrium structure of the island-channel type envisioned by Lehner and Bataille (1985) or Spiers and Schutjens (1990).

4.2. Kinetics of the growth of grain contacts

Three experimental observations suggest that the growth of the interface in the halite ball-and-plate experiments was governed more by interfacial tensions than by stress or deformation in the contact zone. First, growth of the grain boundary in these experiments occurred even when there was no measurable creep deformation (Hickman and Evans, 1991). Second, the changes in surface curvature of the pore near the interface (Figures 3a and b), the dendritic morphology of the interface itself, and the fluid inclusions formed along the interface were reminiscent of crack healing structures observed in halite (S.H. Hickman, unpublished experiments) and other minerals (Hollister and Crawford, 1981; Roedder, 1984; Smith and Evans, 1984; Hickman and Evans, 1987; Brantley et al., 1990). Third, as discussed in more detail below, the rate of neck growth seems to be affected more by the angle of misorientation between the two crystals than by the magnitude of the applied load (Figure 6).

For the case of a material with isotropic interfacial energy, the transport of material will be governed by the Gibbs–Thompson relation:

$$\Delta\mu = \kappa\Gamma_{sl}V_m \tag{4}$$

where $\Delta\mu$ is the change in specific free energy (chemical potential) in transferring material between an interface with a curvature κ and a planar interface, and V_m is the molar volume (e.g., Blakely, 1973). Thus, interfaces with positive curvature (hills, bumps) have higher free energy per mole of diffusing species than those with negative curvature (valleys, holes). The resulting net transfer of material tends to fill in negatively curved surfaces and remove positively curved interfaces.

The interfacial tension of the solid–liquid interfaces is resisted by the tension of the grain boundary, which depends directly on the specific grain boundary energy. Grain boundary structure and energy are determined by several factors, including the relative misorientation of the two crystals, the orientation of the boundary plane itself, and chemical impurity effects. But, to first order, high-angle boundaries ($>15°$) tend to have higher energy than boundaries with relatively low misorientation angles ($<10°$). The effects of the grain boundary tension can be seen in our growth data (Figure 6). The growth rates were lowest in those samples where boundary misorientation, and therefore the boundary tension, were largest.

Detailed observation of the geometry of the pore near the neck suggests that the anisotropy of solid–liquid interfacial energy also affects the neck growth process in a substantial way. In the pure twist-boundary experiments, the broadening in the spacing of the interference fringes near the neck shows that a flat region roughly parallel to (001) in the two lens is developing. Square fingers of crystalline material grow outward into the pore space, and the neck region departs substantially from a simple circular plan (Figure 3b). These changes and observations of the cross-sectional shape of the fluid inclusions formed in HHP1 suggest that the dihedral angles at the junction of the grain boundary with the pore are, on average, very blunt. It is probable, however, that the pore surface must be affected in the region very near the boundary, perhaps on a scale which cannot be observed optically. Regardless, it is quite clear that anisotropy of interfacial energy has a strong influence on the geometry which is not accounted for by eq. 4.

4.3. Models for neck growth driven by reduction in interfacial tension

The results of the experiments may be summarized briefly as follows. The lack of measurable convergence between the two crystals suggests that the pore geometry was modified solely by transport of material within the pore fluid, with no transport along the grain boundary or through the lattice. In most of these experiments (see Table 1), the modification of Newton's rings in the pore indicates that halite was being dissolved in the moat region near the neck. The lack of correlation of normal load with neck growth rate, the similarity of neck morphology to crack-healing structures, and the lack of creep convergence imply that reduction of interfacial energy caused the transport.

Coblenz et al. (1980) review the analytical models for predicting sintering rates driven by interfacial tension in materials with isotropic interfacial energies. The geometry of the pore close to the neck is particularly important, since the gradients in surface curvature drive the material transport. Pore shape will be affected both by the location of the source (i.e., whether the solid transported to the neck is derived near the neck region, or whether the solids are derived more-or-less uniformly from the entire pore space) and by the balance of interfacial tensions at the junction of the pore with the grain boundary (Coblenz et al., 1980). In particular, this balance of interfacial tensions exerts a strong influence on the local surface curvature and, through eq. 4, on the rate of material transport to the neck region.

Models for surface curvature at the neck which assume isotropic solid–liquid interfacial energy (e.g., Coblenz et al., 1980) predict that the radius of curvature will approach its minimum value when the angle between the two solid–liquid surfaces in contact with the grain boundary (the dihedralangle) approaches 180° (e.g., Figures 1b and 1c). In this case, $\Gamma_b/\Gamma_{sl} \ll 2$ and the grain-boundary tension exerts a negligible influence on the local surface curvature. Conversely, as $\Gamma_b/\Gamma_{sl} \rightarrow 2$, then the dihedral angle will approach 0° and the radius of curvature near the neck will be much larger (e.g., Figure 1a). Assuming isotropy of interfacial energy, P.-Y.F. Robin (personal communication, 1991) estimates that the radius of curvature will be 10 times larger for a dihedral angle corresponding to $\Gamma_b/\Gamma_{sl} = 1.8$ than when the pore surface is unaffected by the grain boundary. However, the neck geometry in our experiments does not appear to follow these systematics. Paradoxically, even though the grain boundary energy is rather large, the anisotropy of solid–liquid interfacial energy is so great that the actual dihedral angle is apparently quite blunt, at least on the optical scale. Because of this complexity, the true influence of dihedral angle on local surface curvature in our experiments is unknown and has not been explicitly incorporated in the following analysis.

Assuming that interface tension was the sole driving force for neck growth in the halite experiments, three elements are essential to adapt the initial-stage sintering models to the halite experiments: the geometry of the pore space must be specified so that curvatures can be calculated; the source and sink regions must be identified; and the rate-limiting step in the sequence of dissolution, transport, and precipitation must be known. We consider two models, one in which the flux of material is limited by the rate of precipitation and a second in which the rate-limiting step is the diffusive transport of the solute. In both models the growth of the neck region is driven by chemical potential gradients due to variations in curvature. Neither model takes account of

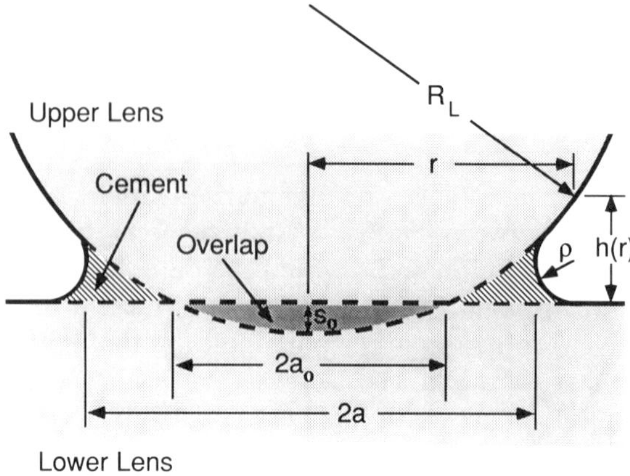

Figure 7. Schematic of the ball-and-plate sample geometry employed in these experiments, where a spherical lens (of radius R_L) and a flat lens (plate) are pressed together, illustrating the geometrical assumptions employed in the model for neck growth rate-limited by precipitation at the neck region. In this model, material is assumed to dissolve far from the contact and precipitate as cement at the highly curved contact region (neck) between the two grains. The local curvature ρ at the perimeter of the neck is given by the radius of a circle tangent to the upper and lower lenses at that point. In all of the experiments discussed in this paper, an instantaneous convergence s_0 occurred between the two lenses when the load was initially applied, forming an initial contact spot of radius a_0. Following this initial compaction, no time-dependent convergence was observed. In the actual experiments, R_L was much larger than pictured here.

anisotropy of interfacial energy, except that the dihedral angle is artificially set equal to 180°.

In the first model we assume that precipitation rates are slow compared with those of electrolyte diffusion and dissolution (Appendix A). To simplify the geometry of the neck region, suppose that the curvature in the sink region is that of a small circle tangent to both the sphere and the plate (Figure 7). The flux of precipitating halite is assumed to be linearly related to the chemical potential difference between the highly curved neck region and the more gently curved lens. The grain boundary energy is postulated to have negligible influence on the curvature at the pore interface. Under these restrictions, the rate of growth of the neck is given by

$$\frac{da}{dt} \approx \frac{4k_{-}\Gamma_{sl}V_m^2 R_L}{RT(a^2 - a_0^2)} \tag{5}$$

where a is the radius of the neck, a_0 is the radius of the neck at time $t = 0$, k_{-} is a precipitation rate constant, Γ_{sl} is the specific surface energy of the solid–liquid interface, V_m is the molar volume of the solid, R_L is the radius of the spherical lens and RT is the Boltzman term. Then, the radius of the neck at any time, $a(t)$, is given by:

$$a^3 - 3aa_0^2 + 2a_0^3 \approx \frac{12k_{-}\Gamma_{sl}V_m^2 R_L t}{RT} \tag{6}$$

The neck growth rate, da/dt, is plotted as a function of $R_L/(a^2 - a_0^2)$ in Figure 8. Because the neck growth rate was rapid during the initial part of the experiment, i.e., large values

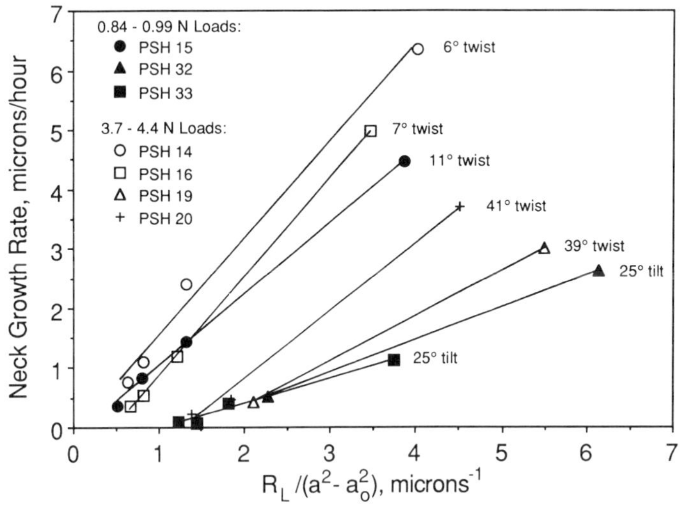

Figure 8. Test of the precipitation rate-limited model for neck growth, where we have plotted the instantaneous neck growth rate against $R_L/(a^2 - a_0^2)$ for the seven ball-and-plate experiments discussed in this chapter. Note that neck growth rates are much greater for low-angle ($\leq 11°$) twist boundaries than for high-angle twist or tilt boundaries. Thus, the failure of these data to fall on a straight line as predicted by the model may be due, at least in part, to the dependence of grain boundary energy on boundary misorientation. Comparison of the low-angle twist boundary results (PSH 14, PSH 15, and PSH 16) also suggests that the magnitude of the applied load has little or no impact upon neck growth rates in these experiments.

of $R_L/(a^2 - a_0^2)$, the data points cluster near the origin of the plot. Consequently, it is difficult to check the fit of the data to the model rigorously. Although the effect of dihedral angle was not incorporated in the model, neck growth rates were clearly slower for experiments with greater grain-boundary misorientations and, presumably, greater grain-boundary energies.

In the second model, the rate-limiting step is assumed to be the diffusion of the dissolved components of the solid through the liquid (Appendix B). If electrolyte transport is slow, but dissolution and precipitation are fast, then the entire drop in chemical potential occurs along the transport path. The maximum driving force for neck growth is the difference between the chemical potential of halite at the source region and that at the sink, divided by a characteristic path length. We assume that the source region is confined to a moat around the neck between the two grains (Figure 9). For simplicity, the moat region is taken to be approximately circular in cross section. The diffusion of the halite through the brine is presumed to be governed by Fick's laws.

The neck growth rate predicted on the basis of the second model is similar in form to that predicted by the first:

$$\frac{da}{dt} \approx \frac{4\pi D C_0 \Gamma_{sl} V_m R_L a}{RT(a^2 - a_0^2)^2} \tag{7}$$

where D is the diffusion coefficient of the solute through the solvent and C_0 is the concentration of the solute in solution in equilibrium with a planar interface. The neck

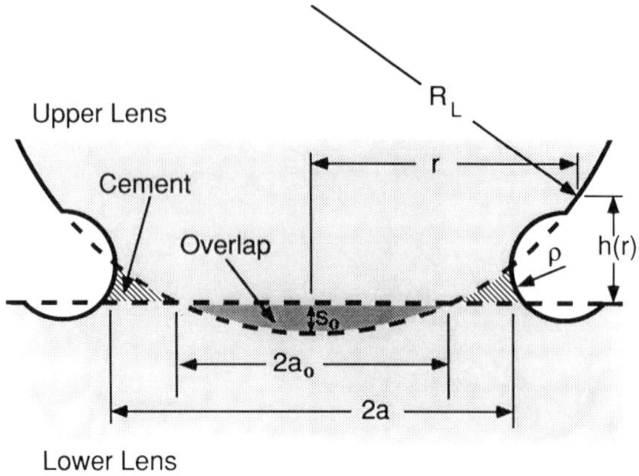

Figure 9. Schematic of the ball-and-plate sample geometry employed in these experiments, illustrating the geometrical assumptions employed in the diffusion rate-limited model for neck growth. In this model, the cement binding the two lenses together is presumed to be derived entirely from the lens surfaces immediately adjacent to the neck, undercutting these surfaces and producing a moat. As in the previous model (Figure 7), mass transfer and cementation in this model are driven by gradients in surface curvature, but the local curvature ρ at the neck increases more rapidly with increasing neck size in the diffusion-limited model. Note, however, that in the actual experiments the radius of the lens was much greater than the neck radius or the moat radius.

radius is given as a function of time by:

$$[a^2 - 2a_0^2]^2 - a_0^4\left[1 - 4\ln\frac{a}{a_0}\right] = \frac{16\pi D C_0 \Gamma_{sl} V_m R_L t}{RT} \tag{8}$$

A plot of neck growth rate versus the parameter $R_L a/(a^2 - a_0^2)^2$ is shown in Figure 10. Here again, the data cluster about the origin and do not provide a rigorous test of the model but the effect of boundary misorientation on neck growth is still evident.

Both models predict rapid neck growth when the neck radius is small, owing to large negative curvature and narrow pore dimensions near the neck. Each also predicts that growth rates will be influenced by a kinetic term which is thermally activated – the first model will be influenced by the precipitation rate constant, the second by the solute diffusion coefficient. The first model does not explain the moat region formed near the neck in some of these experiments, whereas the second model is consistent with moat formation. Both models are inadequate because they do not allow for the known anisotropy of solid–liquid interfacial energy, because the geometry of the transport path is quite simplified, and because the grain boundary energy is assumed to have no impact on local surface curvature. To include anisotropy of interfacial energy and local perturbations to surface curvature at the neck would require the inclusion of interfacial torques as well as interfacial tension; such treatment is well beyond the scope of this study.

Neither sintering model includes the possible effects of stress upon the dissolution rate, precipitation rate, or the interfacial energy. Considerable effort has been expended to observe directly the effect of stress and strain on dissolution through experiments and theory (for example, see Sprunt and Nur, 1977; Bosworth, 1981; Reuschlé et al., 1988, and works cited therein). We were not able to observe any obvious effects of loading

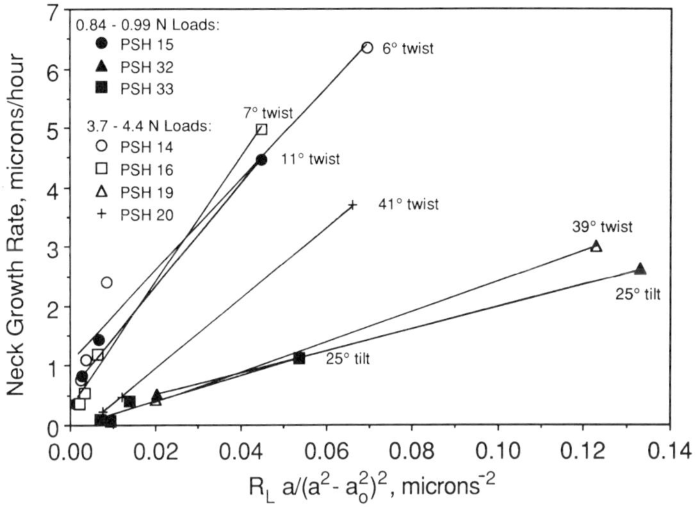

Figure 10. Test of the diffusion rate-limited model for neck growth, where we have plotted the instantaneous neck growth rate against $R_L a/(a^2 - a_0^2)^2$ for the seven ball-and-plate experiments discussed in this paper. As in Figure 8, note that neck growth rates are much greater for low-angle ($\leq 11°$) twist boundaries than for high-angle twist or tilt boundaries, suggesting that grain boundary energy (which was not incorporated in either model) may be quite important in controlling the neck growth rates. As before, comparison of the low-angle twist boundary results (PSH 14, PSH 15, and PSH 16) suggests that the magnitude of the applied load has little or no impact upon neck growth rates.

on the neck growth process during these experiments. However, the rapid decrease in mean normal stress (Figure 5) and increase in contact spot size (Figure 6) at the beginning of these experiments may have masked effects due to variations in stress or initial plastic strain near the contact region.

5. Applications

5.1. Diagenesis and lithification

Cementation and diagenesis may be quite complicated processes. In some cases, it is clear that the cementing material must have originated outside the formation; for example, when calcite cements a quartzose matrix. But if the cement and the matrix are the same phase, both the source and sink of the cement may be in the same rock. Such was the case in these experiments, in which the transport of material and change in pore shape were driven by interfacial tension. Thus, cementation, or formation of necks, may occur in a closed system without porosity reduction. Further, even when an external source of cement is present, interfacial tension will continue to influence the pore shape. For example, even in an oversaturated solution, the precipitation rate should be greater in regions of negative curvature than in those with positive curvature.

In rocks where the cement and the matrix are the same phase, the cementing material may be identified by cathodoluminescence (e.g., Sprunt and Nur, 1979; Houseknecht, 1984). Using this technique, one may distinguish between regions of a grain which have slight variations in trace impurities. The impurities are usually thought to be incorporated into the host mineral, often quartz, during precipitation from the fluid. Impurity variations indicate slight changes in pore fluid chemistry and can be used to identify material deposited at different times. It is important to realize that, while changes in the cathodoluminescence of newly crystallized quartz may identify the chemistry of the pore fluid present during precipitation, they do not identify the location of the cement source, i.e., whether the dissolved cement arose from a local source or was imported into the rock from a distance.

5.2. Strength recovery along faults

Solution transfer processes may also be important (Angevine et al., 1982; Means, 1987; Power and Tullis, 1989) in the evolution of fault rocks. During the interseismic period of the earthquake loading cycle, some form of healing may restore fault strength (Scholz et al., 1986). Field observations of faulted rocks seem to indicate that induration commonly occurs. Strength recovery through solution-transfer processes has been clearly demonstrated through laboratory measurements of frictional strength of quartz gouge before and after healing in water at high temperature and pressures (Frederich and Evans, 1992).

Lithification resulting from either neck growth or pressure solution might be a mechanism to establish cemented bonds between gouge particles (Angevine et al., 1982) and increase the strength (or equivalently, the coefficient of friction) along faults during quiescent periods. The mechanism for neck growth envisioned here differs in driving force from pressure solution: the former is driven by a reduction in interfacial area, while the latter is driven by mechanical forces. Both might strengthen the gouge material by establishing increased contact area between gouge particles, as would cementation owing to the influx of dissolved material from external sources. Each process would be expected to predominate under certain conditions of loading, temperature, fluid flow, and mineralogy. Only neck growth, however, could lead to induration and strengthening of rocks even without porosity reduction. Furthermore, neck growth should lead both to an increase in real area of contact between grains and an increase in the strength of those contacts, by replacing a relatively weak adhesive contact between grains with a (presumably stronger) semicoherent grain boundary.

Detailed experimental observations of the effect of induration on the strength of naturally formed rocks have yet to be made, but Chester and Logan (1986) tested some rocks with varying damage states from the Punchbowl Fault Zone, CA. The undeformed sandstone country rock was approximately 30% stronger than the damaged rock near the fault zone. Although the systematics are not clear, neck growth should lead to a strengthening of fault gouge with time. In laboratory experiments, the strengths of rock joints deforming by frictional processes are also thought to depend on the real area of contact (for reviews, see Dieterich, 1979; Dieterich and Conrad, 1984; and Scholz, 1990). Since neck radius in our experiments grows approximately as $t^{1/4}$ to $t^{1/3}$ (see Figure 6), an increase in real area of contact due to diffusive mass transfer should cause the strength of an individual fracture surface or intergranular contact to increase roughly as $t^{1/2}$ or $t^{2/3}$. Although we cannot predict the exact rates at which fractures or intergranular contacts will strengthen with time, our data clearly suggest that the strength should increase quite quickly at first and then increase more slowly as time progresses. This interpretation is in qualitative agreement with the experimental observations of Frederich and Evans (1992), who showed that most of the strength recovery in simulated quartz gouges healed in water at 636°C and pore pressures of 75 MPa for times of up to 2 hours occurs in the first hour of healing.

6. Conclusions

These experiments show that a neck region containing a grain boundary can be established between two halite crystals while under load and with brine present. The morphology and energy of the boundary are markedly different from that expected if a simple fluid layer were present between the crystals. Our analysis of inclusion shape suggests that the grain boundary energy is about 1.8 times greater than that of the fluid–solid interfacial energy, or about 0.9 times the critical energy for wetting.

The growth of the junction or neck between two abutting grains occurs even when the fluids are stagnant and does not require the introduction of solute from an external source. In the halite–halite experiments, the rate of growth of the junction between the two abutting grains does not seem to depend on the applied load, but does depend on the misorientation angle between the two grains. As the grain boundary between the two abutting grains grows, the curvature of the crystals in the pore region near the neck changes in most of our experiments. Thus, much of the solution transport in the pore appears to be confined to a region near the growing neck.

Assuming that the growth process is driven solely by the reduction of interfacial energy, and incorporating some geometric simplifications, two models were formulated to describe the kinetics of neck growth. In one model, the rate-limiting step in transport of material is assumed to be precipitation of halite from the pore fluid; in the second model, the transport of solute in the fluid is assumed to be slower than either dissolution or precipitation of the solid. Both models fail to describe some aspects of the experiments, and the data fit either model equally well. Both also predict that the neck growth rate will be very rapid initially but will decrease with time, as observed. For a curved lens pressed against a flat plate, these models predict that the growth rate will be proportional to the radius of curvature of the lens, to the interfacial energy of the solid–liquid interface, and to a thermally activated term which describes the rate-limiting step: either a precipitation rate constant or a diffusion coefficient for electrolyte transport.

Acknowledgment

P.-Y.F. Robin kindly supplied a calculation of the effect of dihedral angle on the curvature at the neck for an isotropic solid, as well as a detailed review of the manuscript which greatly aided our revisions. We also thank E.H. Rutter, M. Blanpied, and R. Fournier for thoughtful reviews. Funding for this work was provided by NSF (EAR # 8904061) and USGS # 14-08-0001-G1806.

References

Angevine, C.L., Turcotte, D.L., and Furnish, M.D. (1982). Pressure solution lithification as a mechanism for the stick-slip behavior of faults. *Tectonics* **1**, 151–160.

Anthony, T.R. and Cline, H.E. (1971). Thermal migration of liquid droplets through solids. *J. Appl. Phys.* **42**, 3380–3387.

Bathurst, R.C.G. (1958). Diagenetic fabrics in some British Dinantian limestones. *Liverpool Manchester Geol. J.* **2**, 11–36.

Blakeley, J.M. (1973). *Introduction to the Properties of Crystal Surfaces*. Pergamon Press, Oxford.

Blatt, H., Middleton, G., and Murray, R. (1980). *Origin of Sedimentary Rocks*. Prentice-Hall, Englewood Cliffs, N.J.

Bosworth, W. (1981). Strain-induced preferential dissolution of halite. *Tectonophysics* **78**, 509–525.

Brace, W.F. (1977). Permeability from resistivity and pore shape. *J. Geophys. Res.* **82**, 3343–3349.

Brantley, S.L., Evans, B., Hickman, S.H., and Crerar, D.A. (1990) Healing of microcracks in quartz: Implication for fluid flow. *Geology* **18**, 136–139.

Budiansky, B. and O'Connell, R.J. (1976) Elastic moduli of a cracked solid. *Int. J. Solids Struct.* **12**, 81–97.

Bulau, J.R., Waff, H.S., and Tyburozy, J.A. (1979). Mechanical and thermodynamic constraints on fluid distribution in partial melts. *J. Geophys. Res.* **84**, 6102–6108.

Chester, F.M. and Logan, J.M. (1986). Implications for mechanical properties of brittle faults from observations of the Punchbowl fault zone, California. *Pure Appl. Geophys.* **124**, 79–106.

Coblenz, W.S., Dynys, J.M., Cannon, R.M., and Coble, R.L. (1980). Initial stage solid state sintering models: A critical analysis and assessment. In *Sintering Processes, Materials Science Research*, vol. 13 (ed. G.C. Kuczynski). Plenum Press, New York.

Cooper, R.F. and Kohlstedt, D.L. (1982). Interfacial energies in the olivine-basalt system. In *High Pressure Research in Geophysics, Advances in Earth and Planetary Sciences*, vol. 12 (ed. S. Akimoto and J.H. Manghnani), pp. 217–228. Center for Academic Publications, Tokyo.

Dapples, E.C. (1971). Some concepts of cementation and lithification of sandstones. *Am. Assoc. Petrol. Geol. Bull.* **56**, 3–25.

Dieterich, J.H. (1979). Modeling of rock friction I: Experimental results and constitutive equation. *J. Geophys. Res.* **84**, 2161–2168.

Dieterich, J.H. and Conrad, G. (1984). Effect of humidity on time- and velocity-dependent friction in rocks. *J. Geophys. Res.* **89**, 4196–4202.

Engelder, T. (1984). The role of pore water circulation during the deformation of foreland fold and thrust belts. *J. Geophys. Res.* **89**, 4319–4326.

Frederich, J. T. and Evans, B. (1992). Strength recovery along simulated faults by solution transfer processes. In *Proc. 33rd u.v. Symposium on Rock Mechanics* (ed. W. Wawersik), A. A. Balkema, Rotterdam, in press.

Friedman, G.M. (1975). The making and unmaking of limestones or the downs and ups of porosity: Address of the retiring president. Society of economic paleontologists and mineralogists. *J. Sediment. Petrol.* **45**, 379–398.

Hay, R.S. and Evans, B. (1988). Intergranular distribution of pore fluid and the nature of high-angle grain boundaries in limestone and marble. *J. Geophys. Res.* **93**, 8959–8974.

Herring, C. (1951). Surface tension as a motivation for sintering. In *Physics of Powder Metallurgy* (ed. W.E. Kingston), pp. 143–179. McGraw-Hill Book Co., New York, NY.

Hickman, S.H. (1989). Experimental studies of pressure solution and crack healing in halite and calcite. Ph.D. thesis, Massachusetts Institute of Technology, Cambridge, Mass.

Hickman, S.H. and Evans, B. (1987). Diffusional crack-healing in calcite: The influence of crack geometry upon healing rate. *Phys. Chem. Min.* **15**, 91–102.

Hickman, S.H. and Evans, B. (1991). Experimental pressure solution in halite: the effect of grain/interphase boundary structure. *J. Geol. Soc. Lond.* **148**, 549–560.

Hollister, L.S. and Crawford, M.L. (ed.) (1981). *Short Course in Fluid Inclusions: Applications to Petrology*. Mineralogical Association of Canada, Calgary.

Houseknecht, D.W. (1984). Influence of grain size and temperature on intergranular pressure solution, quartz cementation, and porosity in a quartzose sandstone. *J. Sediment. Petrol.* **54**, 348–361.

Houseknecht, D.W. (1987). Intergranular pressure solution in four quartzose sandstones. *J. Sediment. Petrol.* **58**, 228–246.

Johnson, K.L. (1987). *Contact Mechanics*. Cambridge University Press.

Jurewicz, S.R. and Watson, E.B. (1984). Distribution of partial melt in a felsic system: The importance of surface energy. *Contrib. Mineral. Petrol.* **85**, 25–29.

Jurewicz, S.R. and Watson, E.B. (1985). The distribution of partial melt in a granitic system: The application of liquid phase sintering theory. *Geochem. Cosmochim. Acta* **49**, 1109–1121.

Laporte, D. and Watson, E.B. (1991). Direct observation of near equilibrium pore geometry in synthetic crustal lithologies. *J. Geol.* **99**, 873–878.

Lee, V.W., Mackwell, S.J., and Brantley, S.L. (1991). The effect of fluid chemistry on wetting textures in novaculite. *J. Geophys. Res.* 96, 10023–10377.

Lehner, F.K. and Bataille, J. (1985). Nonequilibrium thermodynamics of pressure solution. *Pure Appl. Geophys.* **122**, 53–85.

Maxwell, J.C. (1964). Influence of depth, temperature, and geologic age on porosity of quartzose sandstones. *Am. Assoc. Petrol. Geol. Bull.* **48**, 697–709.

Means, W.D. (1987). A newly recognized type of slickenside striation. *J. Struct. Geol.* **9**, 585–590.

Meyers, W.J. (1974). Carbonate cement stratigraphy of the Lake Valley Formation (Mississippian) Sacramento Mountains, New Mexico. *Sediment. Petrol.* **44**, 837–861.

Pittman, E.D. (1972). Diagenesis of quartz in sandstones as revealed by scanning electron microscopy. *J. Sediment. Petrol.* **42**, 507–519.

Power, W.L. and Tullis, T.E. (1989). The relationship between slickenside surfaces in fine-grained quartz and the seismic cycle. *J. Struct. Geol.* **11**, 879–893.

Reuschlé, T., Trotignon, L., and Gueguen, Y. (1988). Pore shape evolution by solution transfer: thermodynamics and mechanics. *Geophys. J.* **95**, 535–547.

Robin, P.-Y.F. (1978). Pressure solution at grain-to-grain contacts. *Geochim. Cosmochim. Acta* **42**, 1383–1389.

Roedder, E. (1984). *Fluid Inclusions, Reviews in Mineralogy, 12*, pp. 70–77. Mineralogical Society of America, Washington, D.C.

Rutter, E.H. (1976). The kinetics of rock deformation by pressure solution. *Phil. Trans. R. Soc. Lond.* **A283**, 203–219.

Rutter, E.H. (1983). Pressure solution in nature, theory, and experiment. *J. Geol. Soc. Lond.* **140**, 725–740.

Scholz, C.H. (1990). *The Mechanics of Earthquakes and Faulting*. Cambridge University Press.

Scholz, C.H., Aviles, C.A., and Wesnousky, S.G. (1986). Scaling differences between large interplate and intraplate earthquakes. *Bull. Seismol. Soc. Am.* **76**, 65–70.

Sibley, D.F. and Blatt, H. (1976). Intergranular pressure solution and cementation of the Tuscarora orthoquartzite. *J. Sediment. Petrol.* **46**, 881–896.

Sippel, R.F. (1968). Sandstone petrology, evidence from luminescence petrography. *J. Sediment. Petrol.* **38**, 530–554.

Smith, C.S. (1948). Grains, phases, and interfaces: An interpretation of microstructure. *Trans. Am. Inst. Min. Metall. Petrol. Eng.* **175**, 15–51.

Smith, D.L. and Evans, B. (1984). Diffusional crack healing in quartz. *J. Geophys. Res.* **89**, 4125–4136.

Spiers, C.J. and Schutjens, P.M.T.M. (1990). Densification of crystalline aggregates by fluid phase diffusional creep. In *Deformation Processes in Minerals, Ceramics, and Rocks* (ed. D. Barber and P. Meredith), pp. 334–353, Unwin Hyman, London.

Sprunt, E.S. and Nur, A. (1977). Experimental study of the effects of stress on solution rate. *J. Geophys. Res.* **82**, 3013–22.

Sprunt, E.S. and Nur, A. (1979). Microcracking and healing in granites: new evidence from cathodoluminescence. *Science* **205**, 495–497.

Tada, R. and Siever, R. (1989). Pressure solution during diagenesis. *Annu. Rev. Earth Planet. Sci.* **17**, 89–118.

Tada, R., Maliva, R., and Siever, R. (1987). A new mechanism for pressure solution in porous quartzose sandstone. *Geochim. et Cosmochim. Acta* **51**, 2295–2301.

Trurnit, P. (1968). Pressure solution phenomena in detrital rocks. *Sediment. Geol.* **2**, 89–114.

Urai, J.L. (1985). Water-enhanced dynamic recrystallization and solution transfer in experimentally deformed carnallite. *Tectonophysics* **120**, 285–317.

Urai, J.L., Spiers, C.J., Zwart, H.J., and Lister, G.S. (1986). Weakening of rock salt by water during long-term creep. *Nature* **324**, 554–557.

Vaughan, P.J., Kohlstedt, D.L., and Waff, H.S. (1982). Distribution of the glass phase in hot-pressed, olivine-basalt aggregates: An electron microscopy study. *Contrib. Mineral. Petrol.* **81**, 253–61.

Walder, J. and Nur, A. (1984). Porosity reduction and crustal pore pressure development. *J. Geophys. Res.* **89**, 11539–11548.

Walsh, J.B. (1965). The effects of cracks on the compressibility of rock. *J. Geophys. Res.* **70**, 381–389.

Watson, E.B. (1982). Melt infiltration and magma evolution. *Geology* **10**, 236–240.

Watson, E.B. and Brenan, J.M. (1987). Fluids in the lithosphere, 1. Experimentally-determined wetting characteristics of CO_2–H_2O fluids and their implications for fluid transport, host-rock physical properties, and fluid inclusion formation. *Earth Planet. Sci. Lett.* **85**, 497–515.

Weyl, P.K. (1959). Pressure solution and the force of crystallization – a phenomenological theory. *J. Geophys. Res.* **64**, 2001–2025.

Appendix A: Precipitation Rate-Limited Growth

When precipitation is much slower than electrolyte diffusion and dissolution, neck growth rates in the solution-transport regime will be limited by the rate at which new cement precipitates at grain-to-grain contacts. The following model for precipitation rate-limited neck growth is obtained through a simple modification of Coblenz et al.'s (1980) model for initial-stage sintering rate-limited by evaporation–condensation. Our model differs from that of Coblenz et al. primarily in that it treats transport of the solid constituent through solution as opposed to vapor-phase transport, and that it allows for an initial component of elastic–plastic convergence prior to the onset of neck growth. Although derived to match the ball-and-plate geometry employed in our halite experiments, both models presented in these appendices may easily be modified to treat the case of two spherical surfaces in contact.

For the geometry used in these experiments (Figure 7), the separation between the two lenses, h, at a distance, r, from the center of contact, but outside of the contact spot, is given by

$$h(r) = (R_L^2 - a_0^2)^{1/2} - (R_L^2 - r^2)^{1/2} \tag{A1}$$

where R_L is the radius of curvature of the convex lens and a_0 is the radius of the contact spot formed through elastic–plastic deformation immediately upon the application of load at time $t = 0$ (see Hickman and Evans, 1991). Since $r \ll R_L$ in these experiments, we expand both terms in eq. (A1) using the binomial theorem and, discarding terms higher than second order, obtain

$$h(r) \approx \frac{r^2 - a_0^2}{2R_L} \tag{A2}$$

An approximate expression for surface curvature at the neck, ρ, as a function of neck radius a is then derived by assuming that ρ is equal to the radius of a circle tangent to the upper and lower lenses at that point, i.e.,

$$\rho \approx \frac{h(r = a)}{2} \approx \frac{a^2 - a_0^2}{4R_L} \tag{A3}$$

Note that, although a increases monotonically with time in the halite–halite experiments discussed in this paper, no time-dependent convergence was observed; consequently, a_0 is fixed for each experiment.

The change in chemical potential $\Delta\mu$ in transfering one mole of atoms from the relatively flat pore wall to the highly curved neck is (e.g., Blakely, 1973)

$$\Delta\mu \equiv \mu_{\text{neck}} - \mu_{\text{pore}} = (\kappa_{\text{neck}} - \kappa_{\text{pore}})\Gamma_{sl} V_m \tag{A4}$$

where Γ_{sl} is the specific solid–liquid interfacial energy, V_m is the molar volume, and the surface curvature, κ, is given by

$$\kappa = \frac{1}{\rho_1} + \frac{1}{\rho_2}$$

and ρ_1 and ρ_2 are the principal radii of curvature of the surface in question (defined positive if the surface is convex). Since, in these experiments, $R_L \gg a \gg \rho$, we set $\kappa_{\text{pore}} = 0$ and $\kappa_{\text{neck}} = -1/\rho$ to obtain from eq. (A4):

$$\Delta\mu \approx \frac{-\Gamma_{sl} V_m}{\rho} \tag{A5}$$

Combining eq. (A5) with the definition of the activity,

$$\mu = \mu_0 + RT \ln \gamma C \tag{A6}$$

where RT is the Boltzman term, γ is the activity coefficient, and C is the concentration (mole fraction) of the solute in solution. Assuming γ to be a constant, we obtain the expression

$$\frac{C_0}{C_\rho} \approx \exp\left(\frac{\Gamma_{sl} V_m}{\rho RT}\right) \tag{A7}$$

where C_0 and C_ρ represent solute concentrations in equilibrium with a planar interface and the curved neck, respectively.

If the rate of neck growth is limited by precipitation at the neck, then the precipitation rate j can be expressed as

$$j = k_-\left(\frac{C_0 - C_\rho}{C_\rho}\right) \tag{A8}$$

where k_- is a precipitation rate constant and the quantity in brackets represents the fractional supersaturation in the solution immediately adjacent to the neck. For $\Gamma_{sl} V_m/\rho PT \ll 1$, eq. (A7) may be expanded using a Taylor series and, neglecting terms higher than first order, combined with eq. (A8) to yield

$$j \approx k_-\left[\frac{\Gamma_{sl} V_m}{\rho RT}\right] \tag{A9}$$

where j is in mol $\text{cm}^{-2}\,\text{s}^{-1}$. Taking the precipitation rate as uniform over the entire neck region (of height 2ρ), the neck growth rate, da/dt, is simply:

$$\frac{da}{dt} = jV_m \approx \frac{4k_-\Gamma_{sl} V_m^2 R_L}{RT(a^2 - a_0^2)} \tag{A10}$$

Equation (A10) is separable and, subject to the initial condition $a(t = 0) = a_0$, can be integrated to give:

$$a^3 - 3aa_0^2 + 2a_0^3 \approx \frac{12k_-\Gamma_{sl} V_m^2 R_L t}{RT} \tag{A11}$$

Note that eq. (A11) predicts that neck radius a should be proportional to $t^{1/3}$ as $a_0 \to 0$.

Appendix B: Diffusion Rate-Limited Growth

When electrolyte diffusion is much slower than dissolution and precipitation, neck growth rates will be limited by the rate at which material can diffuse through the pore fluid. To treat the case of growth rate-limited by diffusion, we assume that all cement incorporated into the neck is derived from surfaces immediately adjacent to the neck and that the contribution of sources far from the neck is negligible. This localized dissolution of material should lead to undercutting of the grain surfaces around the neck (i.e., moat formation) in a manner similar to that predicted for initial-stage sintering via surface diffusion. This process of moat formation will cause ρ to increase more rapidly with increasing neck radius than predicted by eq. (A3) (see discussion of undercutting in Coblenz et al., 1980).

A rough expression for the radius of curvature at the neck in the presence of undercutting is obtained by assuming that the moat is contained within the intersection of a circular torus and the upper and lower lenses (Figure 9). By requiring local conservation of volume and equating the volume of this moat with the volume of cement between $r = a_0$ and $r = a$, we obtain

$$\rho \approx \frac{a^2 - a_0^2}{\sqrt{8\pi R_L a}} \tag{B1}$$

where, since $\rho \ll a \ll R_L$, we have approximated the moat volume as the volume of a circular torus with minor and major radii ρ and a, respectively.

The flux of dissolved material into the neck region is given by Fick's first law:

$$\mathbf{J} = -\frac{DC_0 \nabla\mu}{V_m RT} \tag{B2}$$

where D is the diffusion coefficient of the solute moving through the solution. An approximate expression for the magnitude of the chemical potential gradient $|\nabla\mu|$ is obtained by dividing eq. (A5) by a diffusional path length l. Assuming that material diffuses from the outer perimeter of the moat to the neck (see Figure 9) we set $l \approx 2\rho$ to obtain

$$|\nabla\mu| \approx -\frac{\Gamma_{sl} V_m}{2\rho^2} \tag{B3}$$

Combining eqs. (B1), (B2) and (B3) gives

$$|\mathbf{J}| \approx \frac{4\pi D C_0 \Gamma_{sl} R_L a}{RT[a^2 - a_0^2]^2} \tag{B4}$$

Assuming a uniform flux into the neck region, the neck growth rate, da/dt, is then

$$\frac{da}{dt} = |\mathbf{J}| V_m \approx \frac{4\pi D C_0 \Gamma_{sl} R_L V_m a}{RT[a^2 - a_0^2]^2} \tag{B5}$$

Equation (B5) is separable and, subject to the initial condition $a(t = 0) = a_0$, can be integrated to give

$$[a^2 - 2a_0^2]^2 - a_0^4\left[1 - 4\ln\frac{a}{a_0}\right] \approx \frac{16\pi D C_0 \Gamma_{sl} V_m R_L t}{RT} \tag{B6}$$

Chapter 11

The Influence of H_2O and CO_2 on Melt Migration in Two Silicate Liquid–Olivine Systems

G.N. Riley Jr.[1] *and D.L. Kohlstedt*[2]

[1]*Dept. of Materials Science and Engineering, Bard Hall, Cornell University, Ithaca, NY 14853, U.S.A.*

**Present Address: American Superconductor Corporation, 149 Grove St., Watertown, MA 02172, U.S.A.*

[2]*Department of Geology and Geophysics, Pillsbury Hall, University of Minnesota, Minneapolis, MN 55455, U.S.A.*

Abstract

Experiments were conducted on a synthetic melt plus olivine and a natural melt plus olivine system to determine the influence of H_2O and CO_2 on the kinetics of melt migration via porous flow. For each experiment, a melt migration couple was formed between a disk of polycrystalline San Carlos olivine and a disk of fully dense glass, which was saturated with either H_2O or CO_2 and was in chemical equilibrium with San Carlos olivine. For comparison, 'volatile-free' glasses were also used. The melt migration experiments were conducted at a temperature of 1300°C and a confining pressure of 300 MPa for times of 6 and 7 hours for couples involving the synthetic and the natural glasses, respectively. Melt migrated into the dunite during the course of the experiment, in order to decrease the interfacial free energy of the system. Melt migration profiles indicate that the amount of volatiles dissolved in the liquids investigated here has only a modest influence on (i) the rate of melt infiltration and (ii) the volume of melt that a rock can absorb.

1. Introduction

Recently, the kinetics of melt migration through a dunitic matrix was quantified for a 'volatile-free' silicate liquid plus olivine material (Riley et al., 1990; Riley and Kohlstedt, 1991). However, volatiles (e.g., H_2O or CO_2) dissolved in silicate melts significantly

FAULT MECHANICS AND TRANSPORT PROPERTIES OF ROCKS
ISBN 0-12-243780-2

influence such physical properties as density and viscosity (Scarfe, 1986; Mysen, 1976, 1977). Hence volatiles are expected to influence the kinetics of magma transport via porous flow. Because significant amounts of H_2O and CO_2 are thought to exist in regions of the mantle (e.g., Michael, 1988; Tingle and Green, 1987), this investigation examines the effect of dissolved volatiles on melt migration in two systems composed of silicate liquid plus olivine.

In the present study, volatiles were dissolved in a synthetic and a natural silicate melt. To characterize melt infiltration into a dunitic matrix, melt migration couples were formed between these melts and fine-grained aggregates of olivine. In one series of experiments, H_2O and CO_2 were dissolved in a synthetic potassium silicate melt. In the other series of experiments, H_2O was dissolved in a natural MORB liquid. In both cases, 'volatiles-free' melts were used for reference. Melt migration profiles were determined so that the influence of the volatiles dissolved in the liquid on the kinetics of melt migration in upper-mantle type systems could be assessed.

2. Experimental Techniques

As illustrated schematically in Figure 1, melt migration couples were formed between disks of silicate glass and disks of nearly fully dense, polycrystalline olivine. A disk of polycrystalline San Carlos olivine ($(Mg_{0.9}Fe_{0.1})_2SiO_4$) was used as a sink in each of the melt infiltration experiments. To synthesize a cylinder of polycrystalline olivine, San Carlos crystals were hand-selected on the basis of the absence of inclusions and then ground in a tungsten carbide shatterbox to yield a fine powder. Subsequent classification via sedimentation of this material in distilled water yielded a sub-10-μm olivine powder. A texturally equilibrated dunite was formed by hot-pressing this powder in a nickel jacket at a temperature of about 1300°C and a confining pressure of 300 MPa for a time of about 6 hours.

A synthetic glass from the K_2O–MgO–FeO–SiO_2 system and a naturally occurring glass of mid-oceanic ridge basalt composition were used in the melt infiltration experiments performed here. The composition of the synthetic glass was determined from the MgO–SiO_2–K_2O ternary phase diagram such that the liquid was in chemical

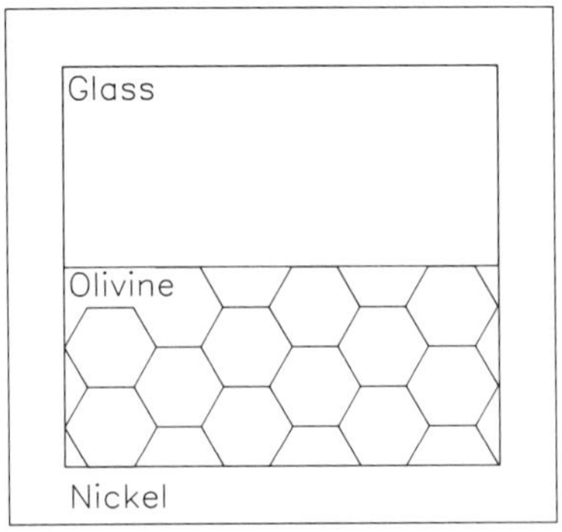

Figure 1. A schematic diagram of the geometry used for the melt infiltration experiments in which a source – a disk of glass – was placed on top of a sink – a disk of fine-grained olivine. The melt migration couple is encapsulated in nickel.

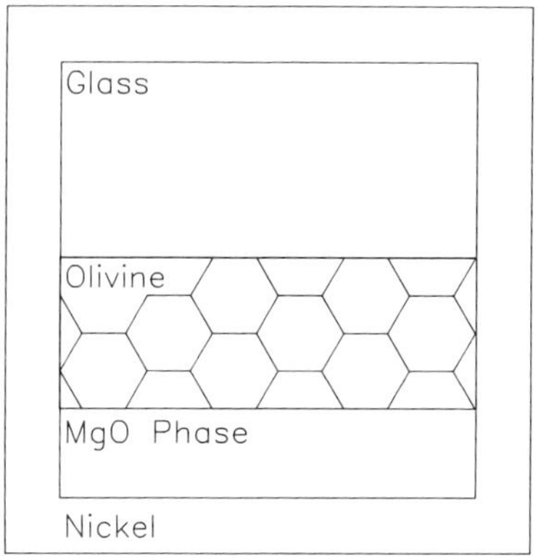

Figure 2. A schematic diagram of the assembly used to form the glasses that were in chemical equilibrium with both the San Carlos olivine and the volatile, in which the glass cylinder (top), the olivine disk (middle), and the volatiles-bearing magnesium-oxide phase (bottom) are shown encapsulated in nickel.

equilibrium with forsterite at a temperature slightly less than 1300°C. FeO was substituted for MgO in the liquid composition determined from the ternary phase diagram such that the final ratio of FeO to MgO matched published results for chemical equilibrium between a silicate liquid and San Carlos olivine (Roeder and Emslie, 1970). To form the batch of synthetic glass, stoichiometric ratios of reagent-grade component powders were first mixed together. Metal oxide and potassium carbonate powders were the sources of the cations in the synthetic glass. A small amount (about 1 wt%) of ammonium nitrate was added to the mixture to act as a fining agent during the formation of the glass. The powder mixture was then melted in a platinum crucible in air at 1400°C for 2 hours. To form batches of mid-oceanic ridge basalt, MORB powders were melted under a flowing CO/CO_2 gas mixture in a nickel crucible with a small amount of olivine powder at a temperature equal to that of the ensuing melt infiltration experiment.

To form a silicate liquid which is in chemical equilibrium with both the San Carlos olivine and the volatile, a disk of polycrystalline olivine was placed between a cylinder of glass and a disk of the volatile-bearing MgO phase, as illustrated in Figure 2. The MgO phases used to equilibrate the melt with H_2O and CO_2 were brucite ($Mg(OH)_2$) and magnesite ($MgCO_3$), respectively. For the H_2O-bearing glasses, the ratio of the weight of glass to the weight of brucite was about 15 in the anneal of the synthetic glass and about 18 in the anneal of the natural glass. For the CO_2-bearing synthetic glass, the ratio of the weight of glass to the weight of magnesite was about 20. Further, a synthetic glass which was 'free' of volatiles was formed by using periclase (MgO) in the place of brucite or magnesite during the anneal. The weight of periclase was not determined prior to the run, but the volume of periclase used in this anneal was about equal to the volumes of the volatile-bearing phases employed in the other anneals. Each assembly was encapsulated in a nickel jacket and hot-pressed for 2 hours at 300 MPa and 1300°C, the pressure and temperature of the ensuing melt infiltration experiment.

Melt migration couples were created by first grinding and sectioning the dunite and the glass cylinders to form disks about 7 mm in diameter and 2 mm thick. A pair of disks was then encapsulated in a nickel jacket to form a melt migration couple. The sample geometry, illustrated in Figure 1, ensures that the discs are in intimate contact with each other. The oxygen fugacity within the melt migration couple is set by the nickel jacket at the Ni–NiO solid-state buffer, which is about 10^{-7} atm at 1300°C, a value well within the stability field of San Carlos olivine (Nitsan, 1974). The experiments were performed in a gas-medium pressure vessel. Each couple was (1) brought slowly to the experimental conditions (approximately 1.25 hours to 1300°C and 300 MPa); (2) held at these conditions for 6 or 7 hours for the experiments employing the synthetic or the natural glasses, respectively; and (3) quenched to 1000°C at roughly 100°C min^{-1} before being cooled slowly to ambient conditions (approximately 45 minutes). For steps (1) and (3), the pressure and the temperature were increased together and decreased together, respectively. For step (2), the temperature changed by less than 2°C and the pressure varied by less than 10 MPa. The variation in temperature along the length of the couple was maintained at less than 1°C for each experiment by independently controlling the power to each half of a two-zone furnace.

The melt migration profiles – the change in melt fraction with distance across the couple – was determined using an X-ray mapping technique developed by Riley (1990). In the case of the synthetic glass, the maps were formed using potassium $K\alpha$ X-rays. In the case of the MORB, they were generated using calcium $K\alpha$ X-rays.

3. Results and Discussion

3.1. Volatile solubility

Three types of analyses were performed to detect the presence of voltatiles dissolved in the silicate glasses. (1) After the glasses had been chemically equilibrated with olivine and the volatiles-bearing MgO phase (Figure 2), they were inspected optically to determine whether vesicles were present. (2) The volatile species dissolved in the silicate glasses were characterized with Fourier transform infrared (FTIR) spectroscopy. (3) The amount of volatiles dissolved in the glasses was calculated by difference, based on electron microprobe analyses; the weight percentage of volatiles dissolved in each 'volatile-rich' glass was presumed to be equal to the difference between the total weight percentage of the component oxides for the 'volatile-free' glass and that of the 'volatile-rich' glass.

Large (up to 2.5 mm) vesicles were located at the top of the sample charges (Figure 2) containing the glasses which were annealed in the presence of brucite and magnesite; vesicles were not observed elsewhere in these sample charges. No vesicles were observed in the synthetic glass which was annealed in the presence of periclase.

The presence of vesicles at the top of the sample charge after the glass had been annealed with the olivine and the MgO phase (Figure 2) suggests that the melt is saturated with the volatile. Further, because vesicles were only seen at the top of the charge, significant exsolution of the volatiles could not have occurred during the quench at the end of the chemical equilibration anneal. Presumably, upon decomposition of the brucite (at about 350°C) to form periclase and H_2O and the magnesite (at about 900°C) to form periclase and CO_2, vesicles are formed containing the respective volatile phase (H_2O or CO_2). Subsequent to the formation of the vesicles, the volatile slowly dissolves into the amorphous phase and, to a limited degree, into the olivine until the amorphous phase and the olivine become saturated with the volatile. The volatile-filled

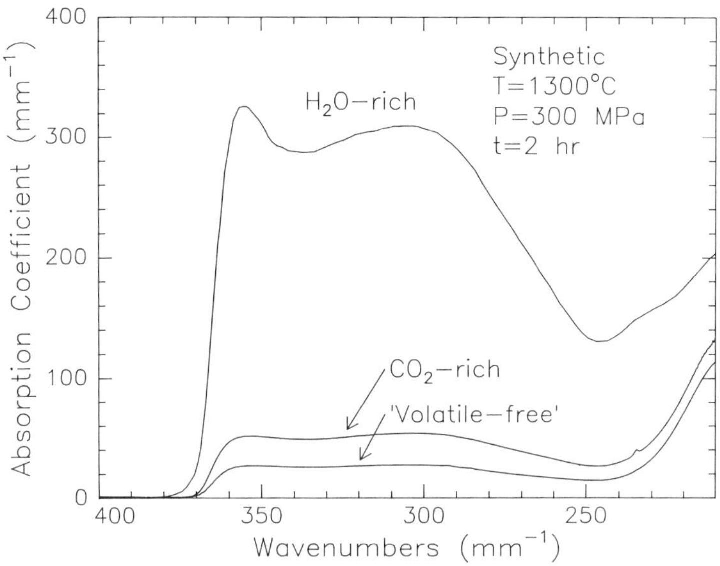

Figure 3. FTIR spectra for the synthetic glasses.

vesicles rise to the top of the sample charge driven by buoyancy forces because of their relatively low density. Note that the solubilities of H_2O and CO_2 are relatively low in olivine (Mackwell and Kohlstedt, 1990; Tingle and Green, 1988) and relatively high in silicate liquids/glasses (Mysen, 1977; McMillan et al., 1986; Paterson, 1982).

Transmission infrared spectra were acquired using an IBM series 40 FTIR. Each of the spectra was acquired at room temperature after purging the sample chamber with dried air. Figure 3 shows the FTIR spectra for the three synthetic glasses. The spectrum acquired for the 'volatile-free' glass has a broad double-peak located between 350 and 250 mm^{-1}. The spectrum acquired for the CO_2-rich glass has a more intense, broad double peak in the same region as the spectra for the 'volatile-free' glass and a small peak centered at 235 mm^{-1}. The spectrum acquired for the H_2O-bearing glass has a very intense double peak in the same region in the spectra as for the other two synthetic glasses. Figure 4 shows the FTIR spectra for the two natural glasses. The spectrum acquired for the 'volatile-free' glass has a peak located at approximately 350 mm^{-1}. The spectrum acquired for the H_2O-rich glass has a very intense peak located in the same region as that for the 'volatile-free' glass.

The type of volatile dissolved in the glass may be identified by correlating the positions of the absorption peaks in the FTIR spectra (Figures 3 and 4) with the positions of the absorption bands for volatile species in silicate glasses. In Figure 3, maxima are identified at about 305 and 355 mm^{-1} in the spectra from the 'volatile-free' and the H_2O-rich glasses and at about 235, 305, and 355 mm^{-1} in the spectra from the CO_2-rich glass. In Figure 4, absorption peaks occur at 355 mm^{-1} for the 'volatile-free' and the H_2O-rich glasses. Peak positions of 305 and 355 mm^{-1} correspond to absorption bands caused by OH stretching (Bartholomew et al., 1980) and a peak position of 235 mm^{-1} corresponds to the absorption band caused by molecular CO_2 (Mysen, 1976). For the synthetic glasses, the amount of hydroxyl in the H_2O-rich glass is about 20 times that in the CO_2-rich glass and 40 times that in the 'volatile-free' glass. For the natural

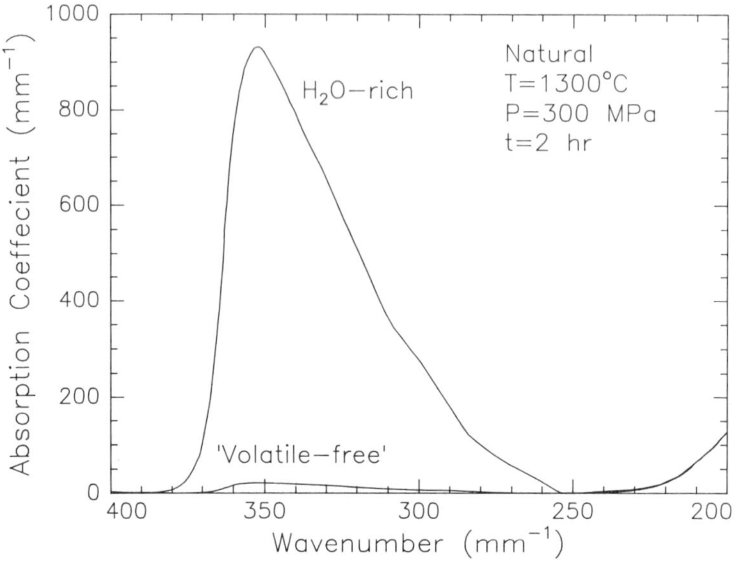

Figure 4. FTIR spectra for the natural glasses.

Table 1. Compositions of the synthetic glasses (wt%)

Sample	'Volatiles-free'	H_2O-rich	CO_2-rich
SiO_2	65.17	64.60	64.26
MgO	10.70	11.19	11.14
FeO	7.85	7.75	6.78
K_20	16.52	16.62	17.07
Al_2O_3	0.06	0.09	0.08
NiO	0.02	0.03	0.03
Total	100.32	100.08	99.36

glasses, the amount of hydroxyl in the H_2O-rich glass is about 100 times that in the 'volatile-free' glass.

Average compositions of the respective glasses within 50 μm of the boundary with the dunite were calculated from quantitative electron microprobe analyses of the synthetic glasses after the melt infiltration experiments (Table 1). The total amount of the component oxides increases from 99.36 wt% for the H_2O-rich glass to 100.32 wt% for the 'volatile-free' glass. On the basis of quantitative analyses of the natural glasses prior to the melt infiltration experiments (Table 2) the total amount of the component oxides increases from 99.35 wt% for H_2O-rich glass to 100.66 wt% for the 'volatile-free' glass.

Quantitative electron microprobe analyses reveal a systematic variation in the total weight percentage of the component oxides in the glasses in which different amounts of volatiles are dissolved. Because the same standards were used to determine the compositions of the volatile-rich and the 'volatile-free' glasses, the difference between the total weight percentage of the component oxides in the volatile-rich glass and that in the 'volatile-free' glass should be approximately equal to the weight percentage of

Table 2. Compositions of the natural glasses (wt%)

Sample	'Volatiles-free'	H_2O-rich
SiO_2	50.74	50.96
MgO	12.75	9.97
FeO	8.77	8.82
CaO	8.854	9.51
K_2O	0.05	0.06
Na_2O	2.72	2.68
Al_2O_3	14.67	15.67
TiO_2	1.49	1.55
NiO	0.23	0.20
Cr_2O_3	0.03	0.04
MnO	0.16	0.15
Total	100.45	99.61

Table 3. Effective dihedral angle (degrees)

Sample	Θ_{eff} ($\pm 1.5°$)
Synthetic	36.2
Synthetic + CO_2	36.6
Synthetic + H_2O	35.7
Natural	38.3
Natural + H_2O	38.0

the volatiles dissolved in the volatile-rich glass. On the basis of this analysis, about 0.24 and 0.96 wt% of volatiles were dissolved in the CO_2 and H_2O-rich synthetic glasses, respectively, and about 0.84 wt% of volatiles was dissolved in the H_2O-rich natural glass.

3.2. Dihedral angle

Values for the effective dihedral angle for each of the silicate melt–olivine systems, determined using a method outlined by Stickles and Hucke (1964), are listed in Table 3. Interfacial angles – the angle between two grains at the junction between the melt pocket and the grain boundary – were measured on SEM micrographs of the two-phase region in each of the couples. Because the surfacc section recorded by the SEM micrograph is a random cut through a three-dimensional glass (melt) network, a range of interfacial angles is expected, even for two-phase systems in which the dihedral angle is single-valued. Because previous work has shown that the median of the interfacial angle distribution is the best measure of an 'effective' dihedral angle, Θ_{eff}, (Stickles and Hucke, 1964; Jurewicz and Jurewicz, 1986), this value is quoted in Table 3.

Table 3 indicates that the effective dihedral angle does not change significantly with the addition of either H_2O or CO_2 to the synthetic melt or with the addition of H_2O

to the natural melt. Although the effective dihedral angles for the H_2O-rich melts are slightly lower than those for the 'volatile-free' melts, the magnitude of the difference is within the uncertainty of the measurement. These results imply that the variation of the effective dihedral angle with volatiles content is small for changes in the volatiles content from ~0.1 wt% to ~1 wt% for these two silicate melt compositions. Because less than ~1 wt% of volatiles was dissolved in the melts investigated in the present study and none of the melts were truly free of volatiles (Riley and Kohlstedt, 1991), the small variation in the magnitude of the effective dihedral angle between the volatile-rich and the 'volatile-free' systems is not altogether unexpected. However, the difference between the effective dihedral angle of the 'volatile-free' synthetic melt material and that of the 'volatile-free' natural melt material is statistically significant.

3.3. Melt migration profiles

3.3.1. Natural melts

Melt migration profiles for the experiments in which the glasses are 'free' of volatiles are presented in Figure 5. The uncertainty in the distance of any given point from the boundary between the two disks is equal to the physical width of the row of X-ray maps from which the melt fraction was determined. The uncertainty in melt fraction is equal to the standard error of the distribution of melt fractions for a given position across the couple. The error bars shown in Figure 5 are representative of those for all experiments performed in this study. For clarity, the error bars are omitted from subsequent melt migration profiles.

The melt migration profiles for the experiments involving the natural glasses are nearly flat, as shown in Figure 6. For both the H_2O-rich and the 'volatile-free' couples, the melt fraction varies by less than 20% across the dunite. However, more melt infiltrated

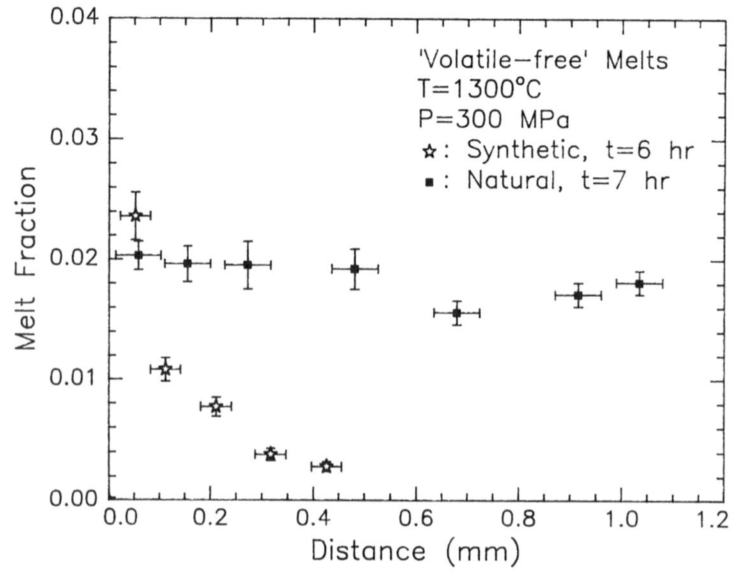

Figure 5. Melt migration profiles for experiments involving glasses in which no volatiles were added.

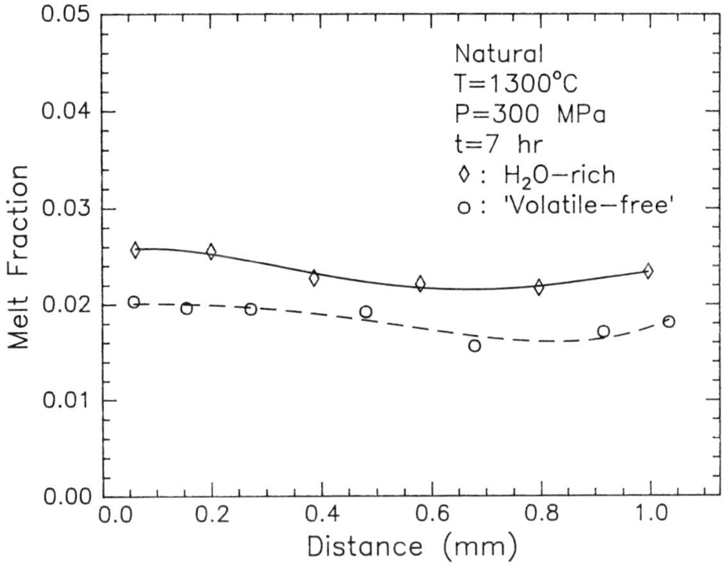

Figure 6. The melt migration profiles for the experiments involving the natural glasses. The curves are for identification purposes only.

into the sink disk during the experiment involving the H_2O-rich melt than during the one involving the 'volatile-free' melt. The melt fraction at the boundary between the source and the sink (the saturation melt fraction, ϕ_0) is about 0.026 for the H_2O-rich couple and about 0.020 for the 'volatile-free'couple.

Recently, thermodynamic models have been proposed which predict that the maximum amount of liquid that a polycrystalline material will absorb (i.e., the equilibrium liquid or saturation melt fraction) increases as the dihedral angle decreases. For a simplified two-dimensional geometry, the saturation melt fraction increases from 0 for a dihedral angle of 60° to about 0.09 for a dihedral angle of 0 (Jurewicz and Watson, 1985; Riley, 1990). Although the uncertainties in the values for the effective dihedral angles are larger than their differences, the effective dihedral angle for the H_2O-rich material is lower than that for the 'volatile-free' material, qualitatively consistent with these models.

It should be noted in the case of the 'volatile-free' melt migration couple, for example, that the melt fraction increases from about 0.015 near the middle of the sink to about 0.018 far from the boundary between the source and the sink. This slight increase in melt fraction from the central region to the far region of the sink disk in the melt migration profiles for the experiments involving the natural glasses (Figure 6) occurs because melt transport is rapid along the region of contact between the nickel capsule and the olivine aggregate. Thus melt was available to infiltrate slowly into the disk via porous flow from the sides and the rear face of the melt-sink disk, as well as dirctly from the source disk. Therefore, the regions near the edges of the melt-sink disk became enriched in melt relative to the interior regions near the center of the aggregate.

3.3.2. Synthetic melts

The melt migration profiles for the experiments involving the synthetic glass are presented in Figure 7. The amounts of melt (glass) in the sink disks in the CO_2-rich couple and the 'volatiles-free' couple are statistically indistinguishable from one another, while more

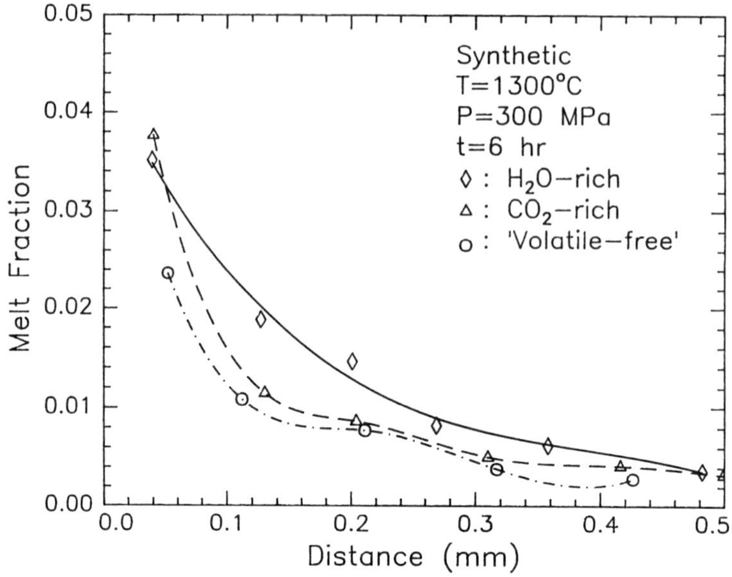

Figure 7. The melt migration profiles for the experiments involving the synthetic glasses. The curves are for identification purposes only.

melt is present in the sink disk in the H_2O-rich couple. In contrast to the behavior found for the experiments involving the natural melts, the 'dunitic' aggregates did not saturate with melt during the experiments involving the synthetic glasses.

For all three melt migration couples involving the synthetic melt, the melt fraction at the boundary between the source and the sink (ϕ_0) is about 0.045; the average effective dihedral angle is about 36°. For the natural melt materials, Θ_{eff} is about 38° and ϕ_0 is about 0.025. The decrease in saturation melt fraction with increasing dihedral angle observed in this comparison of samples with synthetic and natural melts is again at least qualitatively consistent with the models which predict the saturation melt fraction as a function of the dihedral angle.

One systematic source of uncertainty in the melt migration profiles is a natural consequence of the technique employed to determine the profiles. Because each X-ray map has a finite width, it samples a gradient in the melt fraction; this effect is particularly pronounced near the boundary between the source and the sink. Hence, extrapolation of the trend of the melt fraction data to the boundary between the source and the sink may result in significant error in the value for ϕ_0. For samples involving the synthetic melt (Figure 7) the melt migration profile for the 'volatile-free' couple is almost identical to that for the CO_2-rich couple except for the value of the melt fraction at the point closest to the boundary between the source and the sink. Because (1) large gradients in the melt fraction generally exist in regions close to the boundary between the source and the sink and (2) the melt migration profiles for the 'volatile-free' and the CO_2-rich couples are statistically indistinguishable from each other except for the value of the melt fraction nearest the boundary, the actual changes in the melt fraction with distance into the sink disks are probably about the same in these couples.

3.4. Rate of melt migration

The rate of melt infiltration can be (a) calculated from compaction theory (e.g., McKenzie, 1984) as modified to describe melt migration driven by capillary forces and resisted by matrix deformation (Riley et al., 1990) and (b) estimated from the experimentally determined melt migration profiles. In the first case, one of the scaling parameters, w_0, provides a value for the rate of melt migration at the boundary between the source and the sink:

$$w_0 = \frac{(1 - \phi_0)W_0}{\phi_0} \sim \frac{\gamma_{sl}}{d}\left(\frac{K_0}{3\eta^*\mu}\right)^{1/2} \tag{1}$$

where W_0 is the scaling velocity of the matrix, γ_{sl} is the solid–liquid interfacial free energy, d is the grain size, K_0 is the reference permeability of the partial melt, η^* is the effective viscosity of the matrix, and μ is the viscosity of the melt (Riley and Kohlstedt, 1991). Phenomenologically, the reference permeability can be expressed as

$$K_0 = \frac{d^2\phi_0^n}{b} \tag{2}$$

where n is the melt fraction exponent and b is a constant that depends on the geometry of the interconnected network of melt (e.g., McKenzie, 1984).

If the composition of the melt does not markedly affect the viscosity of the matrix, then, on the basis of eq. (1), the difference between the rate of migration of the natural melt and that of the synthetic melt can be ascribed to the respective values for (i) the viscosity of the melt, (ii) the equilibrium melt fraction, and (iii) the solid–liquid interfacial free energy. The viscosities of the 'volatile-free' synthetic and natural melts are about 23 and 9 Pa s, respectively (Bottinga and Weill, 1972). The equilibrium melt fractions are about 0.045 and 0.02 for the 'volatile-free' synthetic and natural melts, respectively. The difference between the solid–liquid interfacial free energies of the respective melts is probably not large in view of the relatively small difference between the effective dihedral angle for these melts. If these values for the materials parameters are used in eq. (1) and if $n = 1$ (Riley and Kohlstedt, 1991), then the migration rate of the 'volatile-free' natural melt should only be about 15% larger than that of the 'volatile-free' synthetic melt. This comparison can be made more quantitative by combining eqs. (1) and (2) and inserting $b = 3000$ and $\gamma_{sl} = 0.5\ \mathrm{J\ m^{-2}}$ (Cooper and Kohlstedt, 1982); an estimated value of $\eta^* = 9 \times 10^{10}$ Pa s was obtained for matrix deformation via Coble creep from the melt migration results of Riley and Kohlstedt (1991) by extrapolating from 1255°C to 1300°C and using a grain size of 10 μm. With these values, $w_0 = 7.8 \times 10^{-10}\ \mathrm{m\ s^{-1}}$ for the 'volatile-free' synthetic melt and $w_0 = 8.2 \times 10^{-10}\ \mathrm{m\ s^{-1}}$ for the 'volatile-free' natural melt.

In the second case, the characteristic rate of melt migration, w_{exp}, is defined as

$$w_{exp} = \frac{x_{exp}}{t} \tag{3}$$

where x_{exp} is a characteristic distance of melt infiltration and t is the duration of the experiment. In samples employing a disk of melt as the source and a disk of fine-grained olivine as the sink, the melt migration profiles can be fitted well by an exponential of the form

$$\phi = \phi_0 \exp(-x/x_{exp}) \tag{4}$$

(Riley, 1990). For the 'volatile-free' synthetic melt, this procedure yields a melt migration rate of $4 \times 10^{-9}\,\mathrm{m\,s^{-1}}$. For the 'volatile-free' natural melt, the melt migration rate is $>2 \times 10^{-8}\,\mathrm{m\,s^{-1}}$; in these samples, only a limit can be obtained because the source disks are almost saturated with melt.

The complete reason for the discrepancy by a factor of 5 to greater than 25 between the rates of melt migration calculated from compaction theory and those estimated from the experimental data (Figure 5) is not clear. In part the discrepancy arises because the rate of melt migration obtained from the scaling velocity in compaction theory is only a lower bound; however, simulations reveal that the velocity of the melt increases by only about a factor of 2 with increasing distance into the sink (Riley and Kohlstedt, 1991). It is also possible that the difference between the respective rates is due to the effect of melt composition on matrix viscosity; for example, if some of the cations from the melt segregate to the grain boundaries, then the rate of grain boundary diffusion and hence of Coble creep will likely be sensitive to the ionic composition of the melt. Results of simulations based on compaction theory must be compared to experimentally determined melt migration profiles to determine the matrix viscosity for each melt composition. This work is in progress.

For both the synthetic and the natural melts, the rate of melt infiltration is enhanced to only a limited degree by the dissolution of H_2O into the melt (Figures 6 and 7). If the melt migration profiles for the synthetic melt are fitted to eq. (4), the melt migration rate is about $4 \times 10^{-9}\,\mathrm{m\,s^{-1}}$ for the 'volatile-free' and CO_2-rich melts, while it is about $5 \times 10^{-9}\,\mathrm{m\,s^{-1}}$ for the H_2O-rich melt. Likewise, for the natural (MORB) melts, because the melt migration profiles are similar in form for the 'volatiles-free' and H_2O-rich melts but significantly more melt is present in the latter case, the flux of melt into the sink, ϕw, and thus the rate of melt migration, must also be enhanced by the addition of H_2O to the natural (MORB) melt.

4. Conclusions

(1) An increase in the amount of H_2O dissolved in a synthetic potassium-rich silicate melt from ~0.01 wt% to ~1 wt% results in an increase of only about 20% in the rate of melt migration and does not markedly change the saturation melt fraction.

(2) Dissolution of ~0.1 wt% of CO_2 in a synthetic potassium-rich silicate melt does not measurably change the rate of melt migration or the saturation melt fraction relative to that measured for a 'volatile-free' melt.

(3) An increase in the amount of H_2O dissolved in a natural MORB melt from ~0.01 wt% to ~1 wt% increases the rate of melt migration by only a limited degree and increases the melt fraction at which the dunite saturates by about 25%.

(4) The limited influence of the amount of H_2O dissolved in the melt on the kinetics of melt migration suggests that the dunite may be saturated with H_2O in the 'volatile-free' runs.

(5) The MORB migrates into the dunite much more rapidly than does the synthetic melt. The difference between the viscosities of the respective melts appears to be the only factor that contributes significantly to the disparity between the respective migration rates, unless the viscosity of the matrix is influenced by the composition of the melt.

Acknowledgments

The authors thank Dr. Ken Chyung at Corning, Inc. for supplying the synthetic glass and John Hunt of the Materials Science Center of Cornell University for assisting with the electron microprobe analyses. We also benefited from discussions with Frank Richter. The support of the National Science Foundation through grant EAR-8916438 is gratefully acknowledged.

References

Bartholomew, R.F., Butler, B.L., Hoover, H.L., and Wu, C.K. (1980). Infrared spectra of a water-containing glass. *J. Am. Ceram. Soc.* **63**, 481–485.

Bottinga, Y. and Weill, D.F. (1972). The viscosity of magmatic silicate liquids: a model for calculation. *Am. J. Sci.* **272**, 438–475.

Cooper, R.F. and Kohlstedt, D.L. (1982). Interfacial energies in the olivine-basalt system. In *High Pressure Research of Geophysics* (eds. S. Akimoto and M.H. Maughnani), vol. 12, pp. 217–228. Center for Academic Publications, Tokyo.

Jurewicz, S.R. and Jurewicz, A.J.G. (1986). Distribution of apparent angles on random sections with emphasis on dihedral angle measurements. *J. Geol. Res.* **91**, 9277–9282.

Jurewicz, S.R. and Watson, E.B. (1985). The distribution of partial melt in a granitic system: the application of liquid phase sintering theory. *Geol. Cosmol. Acta* **49**, 1109–1121.

Mackwell, S.J. and Kohlstedt, D.L. (1990). Diffusion of hydrogen in olivine: implications for water in the mantle. *J. Geol. Res.* **95**, 5079–5088.

McKenzie, D. (1984). The generation and compaction of partially molten rock. *J. Petrol.* **25**, 713–765.

McMillan, P.M., Peraudeau, G., Holloway, J., and Coutures, J.P. (1986). Water solubility in a calcium aluminosilicate melt. *Contrib. Mineral. Petrol.* **94**, 178–182.

Michael, P.J. (1988). The concentration, behavior, and storage of H_2O in the suboceanic upper mantle: implication for mantle metasomatism. *Geol. Cosmol. Acta* **52**, 555–566.

Mysen, B. (1976). The role of volatiles in silicate melts: solubility of carbon dioxide and water in feldspar, pyroxene and feldspathoid melts to 30 kb and 1625°C. *Am. J. Sci.* **276**, 969–996.

Mysen, B. (1977). The solubility of H_2O and CO_2 under predicted magma genesis conditions and some petrological and geophysical implications. *Rev. Geol. Space Phys.* **15**, 351–361.

Nitsan, U. (1974). Stability field of olivine with respect to oxidation and reduction. *J. Geol. Res.* **79**, 706–711.

Paterson, M.S. (1982). The determination of hydroxyl by infrared absorption in quartz, silicate glasses and similar materials. *Bull. Mineral.* **105**, 20–29.

Riley, G.N. Jr. (1990). Liquid distribution and transport in silicate liquid–olivine materials. Ph.D. thesis, Cornell University.

Riley, G.N. Jr. and Kohlstedt, D.L. (1991). Kinetics of melt migration in upper mantle-type rocks. *Earth Planet. Sci. Lett.*, **105**, 500–521.

Riley, G.N. Jr, Kohlstedt, D.L., and Richter, F.M. (1990). Melt migration in silicate liquid–olivine system: an experimental test of compaction theory. *Geophys. Res. Lett.* **17**, 2101–2104.

Roeder, P.L. and Emslie, R.F. (1970). Olivine–liquid equilibrium. *Contrib. Mineral. Petrol.* **29**, 275–289.

Scarfe, C.F. (1986). Viscosity and density of silicate melts. In *Short Course in Silicate Melts* (ed. C.M. Scarfe), pp. 57–92. Mineralogical Association of Canada.

Stickles, C.A. and Hucke, E.E. (1964). Measurement of dihedral angles. *Trans. Metall. Soc. AIME.* **230**, 795–801.

Tingle, T.N. and Green, H.W. II (1987). Carbon solubility in olivine: implications for upper mantle evolution. *Geology* **15**, 324–326.

Tingle, T.N. and Green, H.W. (1988). Experiments and observations bearing on solubility and diffusivity of carbon in olivine. *J. Geol. Res.* **93**, 15289–15304.

Fracture Characterization and Physical Properties of Rock

Chapter 12

Fluid-driven Cyclic Propagation of a Joint in the Ithaca Siltstone, Appalachian Basin, New York

Alfred Lacazette[1] and Terry Engelder

Department of Geosciences, The Pennsylvania State University, University Park, PA 16802, U.S.A.

[1] Present address: Texaco Inc., Exploration and Production Technology Dept., 3901 Briarpark, Houston, TX 77042, U.S.A.

Abstract

Crack-seal veins, millifractures, and joints with rhythmic c-type plume patterns are common examples of cyclic crack propagation in rocks. Although in some cases cyclic propagation could result from periodic external forcing by far-field stress changes or fluid-pressure pulsation, the regularity and rhythmic nature of several types of fractures suggest that cyclic propagation also arises from dynamic instability of the fracture–fluid–rock system. A 40–90-m-long cross-fold joint that propagated within a single bed of the Devonian Ithaca Siltstone near Watkins Glen, New York has a plumose surface morphology with multiple arrest lines indicating that cracking occurred in increments rather than in one smooth rupture. The crack increments increase in overall length in the propagation direction over the final 28-m portion of the exposed end of the study joint with the largest increments increasing in length from 0.6 m to 1.0 m. At least three conceptual models based on linear elastic fracture mechanics and fluid flow along joints can be imagined to explain incremental crack growth under conditions of constant stress and pore pressure: the compressibility-limited propagation model; the flow-limited propagation model; and the infiltration-limited propagation model. This surface morphology of the study joint provides constraints on the propagation process so that the growth of the joint may be analyzed in terms of these three models. Based on quantitative evaluation of the cracking process, compressibility-limited propagation is favored and the driving fluid is identified as a gas rather than a brine. The gas is identified as a natural gas on the basis of geological constraints.

FAULT MECHANICS AND TRANSPORT PROPERTIES OF ROCKS
ISBN 0-12-243780-2

1. Introduction

The morphology of joints and veins in rocks suggests that crack propagation sometimes occurs in spurts during either reactivation of open ruptures or repeated cracking through a narrow zone which subsequently fills with vein material. For example, crack-seal veins consist of the repeated overgrowth of cements in new crack traces which followed old (Ramsay, 1980). These overgrowths are seeded in crystallographic continuity on the surface of cracks which themselves may have propagated through previous overgrowth. The crossing pattern of millifractures in Mesozoic limestones of the Italian Apennines (Geiser and Sansone, 1981) and in carbonates of the Central Appalachians also indicate repeated rupture. Millifractures are seen as hairline cracks often associated with the development of disjunctive cleavage in limestones throughout foreland fold–thrust belts. Crack-seal veins and millifractures may differ only by the fact that millifractures do not propagate through or follow existing veins. A prominent feature on the surface of joints within Devonian shales and siltstones of the Appalachian Plateau is a regular pattern of fan-shaped plumose markings (rhythmic c-type plumes)* indicating propagation and arrest through many cycles either horizontally within a single bed (Bahat and Engelder, 1984), or vertically between beds (Helgeson and Aydin, 1988, 1989). Hence, cyclic propagation is a feature that crack-seal veins, millifractures, and joints with rhythmic c-type plumes all have in common. The stress which drives these cracks appears to develop by an increase of the crack fluid pressure which momentarily exceeds the total stress acting inward from the surrounding rock by an amount that is sufficient to overcome the resistance of the rock to fracture propagation. This form of fluid-induced crack growth is a type of natural hydraulic fracturing (Secor, 1965; Beach, 1977; Engelder, 1985, Engelder and Lacazette, 1990).

The extremely uniform orientation of many joint and vein sets and the regularity of episodic fracturing implied by crack-seal veins and joints with rhythmic c-type plumes is striking. From these observations we conclude that it is unlikely that such regular, cyclic propagation was caused by repeated reactivation under varying conditions of tectonic stress or catastrophic injection of pulses of high-pressure fluid from an external source. Our notion is that cyclic crack propagation represents a momentary disequilibrium in the local crack–fluid–rock system followed quickly by a return to stable conditions. If so, crack interactions or other mechanisms can cause a crack to become unstable under conditions of constant remote stress and matrix pore pressure.

This chapter presents detailed observations of incremental growth of a vertical joint contained in a single bed of Devonian siltstone in the Appalachian plateau near Ithaca New York, U.S.A. To explain these observations we consider three conceptual models for this type of natural hydraulic fracturing and conclude that the compressibility of the fluid within the joint was responsible for cyclic crack propagation and that the joint was driven by a gas, which geologic evidence suggests was natural gas.

2. Incremental Crack Growth in the Ithaca Siltstone

Joints in the Ithaca siltstone at Watkins Glen, New York formed by in-situ natural hydraulic fracturing (Engelder and Lacazette, 1990). The origin flaws were 1–3 cm fossils

* c-Type plumes belong to a class of joint surface morphology with curved or irregular plume axes (Bahat and Engelder, 1984).

Figure 1. Photograph of joint initiation point in the Ithaca Siltstone at Watkins Glen, New York.

and flute casts (Bahat and Engelder, 1984; Engelder and Lacazette, 1990). Calculations using linear elastic fracture mechanics and poroelasticity theory (assuming 3 km of overburden) show that the vertical stress and pore pressure were about 80 and 65 MPa, respectively, at the time of joint formation (Engelder and Lacazette, 1990).

Rhythmic c-type plume patterns indicate that joints propagated in discrete increments (Bahat and Engelder, 1984). These patterns are well-developed on the surfaces of cross-fold joints in exposures of essentially flat-lying Devonian siltstones of the Ithaca Formation at Watkins Glen, New York (Bahat and Engelder, 1984; Engelder, 1986; Engelder, 1987). Joints in siltstones of the Ithaca Formation are vertical and are restricted to individual beds that are always less than 1 m thick. Siltstone beds are separated from each other by laminas of shale or thick shale beds. Composite joints cutting both siltstones and shales have vertical dimensions up to at least 10 m (Helgeson and Aydin, 1989). The initiation point of the fractures are found by tracing fan-like patterns of delicate barbs or hackles back to a central origin (Figure 1, photograph of initiation point). The hackles sometimes end abruptly at the perimeter of the fan, indicating that the crack arrested. Successive propagation increments are characterized by a smooth area beyond the previous arrest line which gives way to progressively more intense hackle so that another fan is developed beyond the point of arrest of the previous event (Figure 2, photograph of propagation cycle). The point of arrest of the crack front is again marked by the abrupt cessation of hackle. Although some joints in the siltstones show rhythmic c-type plumes, others show s-type plumes (Bahat and Engelder, 1984) which are characterized by a continuous plume without detectable arrest lines. Individual joints are highly planar and maintain a very consistent orientation.

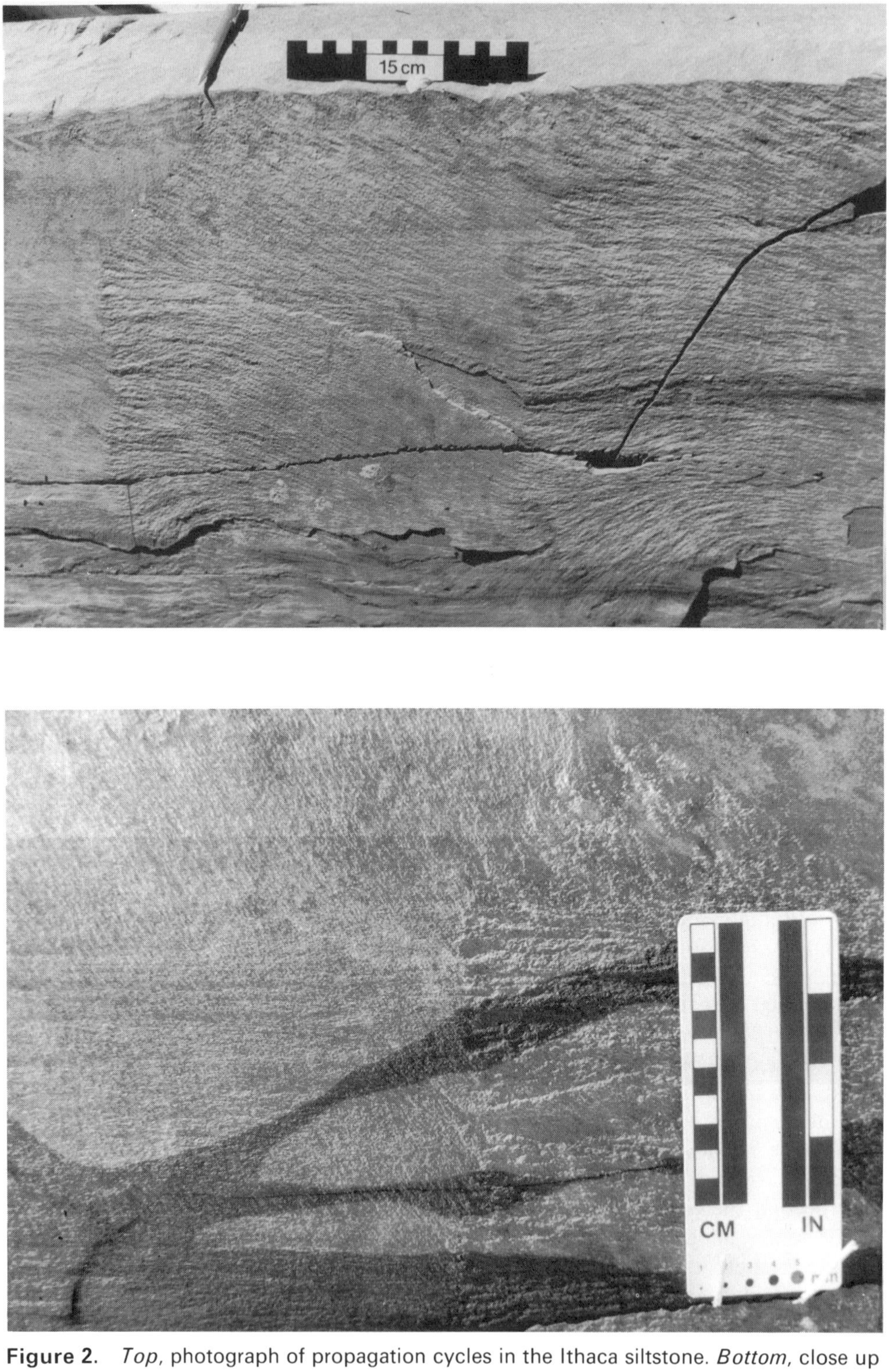

Figure 2. *Top*, photograph of propagation cycles in the Ithaca siltstone. *Bottom*, close up of arrest line showing the sharp boundary between coarse and fine hackle.

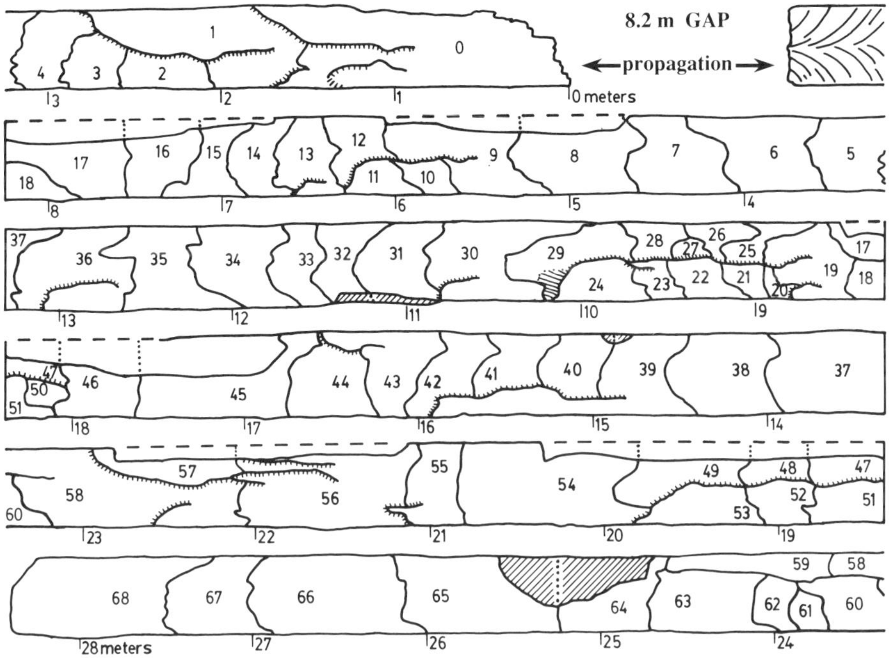

Figure 3. Map drawn from photographs of the surface of a vertical joint contained in a siltstone bed that is isolated in shale at Watkins Glen, New York. The fracture propagated in discrete increments which are numbered in order of development as determined by morphological criteria. Arrest lines as shown mark the position of the fracture front at the termination of each fracturing cycle. The study joint initiated within the 8.2-m gap and propagated outwards. A 3.6-m segment of the study joint surface to the northwest (right) of the gap contains an S-type plume (Bahat and Engelder, 1984). Tick marks show where side-cracks rejoin the main fracture plane. Diagonal lines indicate local destruction of surface morphology by weathering. Dashed lines and dotted lines respectively show the inferred location of the top of the bed and the arrest line where parts of the bed are missing or weathered. Successive map segments match up exactly where they terminate in vertical lines whereas open segment ends indicate slight overlap between segments. See text for further discussion.

To further understand the mechanism of incremental crack growth, the surface of one well-exposed cross-fold joint (hereafter referred to as 'the study joint') in the Ithaca Formation was mapped in detail from photographs. A map of the joint surface is given in Figure 3. The increments are numbered in order of their propagation sequence as determined by observations of the surface morphology. A map of the exposed part of the upper surface of the bed that contains the study joint is given in Figure 4. The study joint was confined to a 0.41-m-thick bed and contained 68 separate propagation increments. It terminated in solid rock. During road construction, one wall of the study joint was ripped away so that the terminal arrest line is both preserved and exposed (Figure 3). The arrest lines of each increment and other surface features of the study joint were marked with chalk. This surface was then photographed with scales so that the area of each fracture increment could be determined from the photographs with a digital planimeter in the laboratory. Replicate determinations of increment areas on

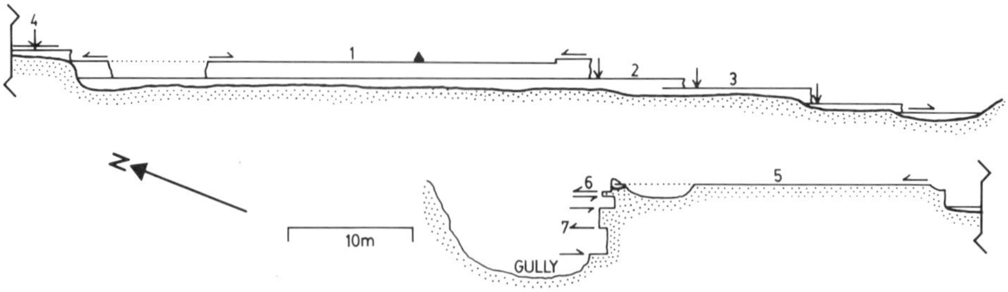

Figure 4. Map of the exposed parts of the upper surface of the siltstone bed containing the study joint shown in Figure 3. Arrows show the location of initiation points, joints propagated outwards from the initiation points. Half-arrows show the propagation direction of joints where the initiation point is not contained in the exposed surface. Joint 1 is that shown in Figure 3. Only joint 1 shows arrest lines. All other joints in the bed show S-type plumes (Bahat and Engelder, 1984). Solid triangle shows the location of segment 45 where the study joint undergoes a slight change of orientation. Joint 2 is the location of the measurements of joint surface morphology of Brown and Scholz (1985). Note that joint 3 terminates just north of its initiation point indicating asymmetrical propagation. Joints 1, 3, and 5 show terminal arrest lines. '4' denotes the intersection point of two approximately coplanar joints with opposite propagation directions.

different photographs with different scales shows that the areas thus determined are repeatable to $\pm 0.015\ m^2$. We find that the mean bed thickness from direct measurements in the field is 0.443 ± 0.046 m. The average bed thickness determined using the cumulative areal measurements and the measured joint length is 0.41 m, which is within the range of the independently measured bed thickness. The area of each fracture increment was converted to a normalized increment length by dividing increment areas by the average bed thickness (0.41 m). Figure 5 gives the increment area plotted against the increment number as given in Figure 3. Figure 3 was prepared from smaller-scale photographs than those used for the area determinations given in Figure 5. The delicate surface morphology of the joints at this outcrop is rapidly being degraded by weathering. Portions of the surface layer either had peeled off or were in the process of peeling when this study was undertaken and the studied joint face may soon be rendered unusable for mapping.

The overall orientation of the study joint is 338/86, although it changes slightly at increment 45 (Figure 3). This increment has a relatively rough and broken-up surface unlike the smooth surface of the rest of the study joint. Southeast of increment 45 the study joint strikes counterclockwise to and dips more shallowly by 1° or less than the remainder of the face, which lies to the northwest of this increment.

Although the surface of the study joint is quite planar and smooth overall, *side-cracks* propagate away from the main joint surface (Figures 3 and 6). Side-cracks are always clockwise to the main fracture face and deviate from its orientation by up to roughly 1°. Where side-cracks leave the main fracture, the hackle is continuous and the surface is smooth, planar, and unbroken. A distinct change of orientation cannot be detected at the point of departure from the main fracture. In general, side-cracks terminated within solid rock after propagating for a short distance and the septum of rock that separated the main fracture from the side-crack (if one ever existed) has broken away so that the entire surface of the side-crack is exposed and the connecting fracture is at

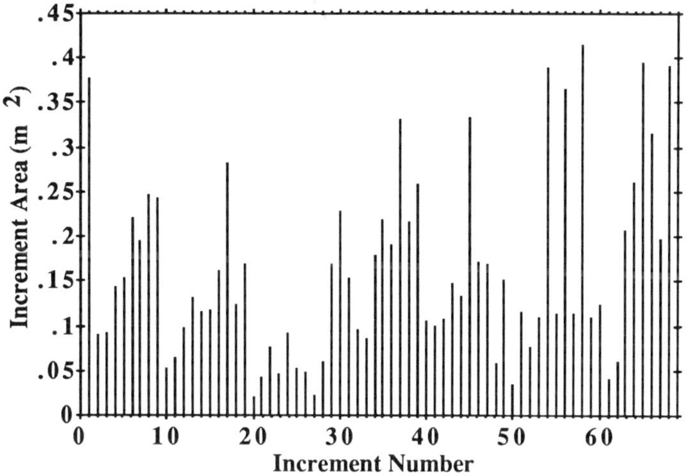

Figure 5. Surface area of increments shown in Figure 4 as determined by digital planimetry of photographs in the laboratory. Repeat determinations on different photographs show that the areas are reproducible to $\pm 0.015\ m^2$.

a high counterclockwise angle to the main fracture face. In one or two cases large side-cracks propagated far enough so that up to 1 cm of rock separates the side-crack tip from the main fracture and the entire surface of the side crack is not exposed (these cracks can be traced on the top surface of the bed). At a few locations the side-cracks smoothly rejoin the main fracture surface so that no distinct break is present, although the surfaces of reconnection are more steeply inclined to the main fracture face than the remainder of the side-crack. These areas are indicated by long tic marks with or without associated arrest lines (see increments 29 and 0, Figure 3). Side-cracks always resulted in a tongue which ran ahead of the main face along either the top or the bottom of the bed prior to arrest. Although the arrest lines of fracturing events that did not result in side-cracks have an overall tendency towards being convex in the propagation direction (see increments 4, 5, 8, 30–34, 66–68), they also show a tendency to propagate farther along the surfaces of the bed (see increments 6, 7, 17).

On the basis of the surface morphology of the study joint and other field observations we draw six major conclusions regarding incremental propagation:

(1) *Crack velocity increased during each propagation event.* The fan pattern accompanying incremental growth is characterized by a subtle increase in intensity of the barbs or hackles up to the point of their abrupt termination. Based on the fractography of ceramics, this increase in intensity of the barbs or hackles suggests an increase in crack velocity (Kulander et al., 1979; Kulander and Dean, 1985). Although it is argued below that the joints in the Ithaca Siltstone propagated in a quasi-static manner, the crack velocity must have increased until halting abruptly.

(2) *Incremental crack growth is quasistable.* If a crack becomes unstable it can accelerate up to about 0.38 times the shear-wave velocity of the material (Broek, 1987). Crack branching may occur after the fracture has reached or exceeded 0.13–0.19 times the shear wave velocity (Broek, 1987). During incremental growth of joints in the Ithaca Formation there is no evidence for this type of high-velocity crack branching. We reject

Figure 6. Photographs of the side-crack of increments 56 and 57. *Top*, entire side-crack. *Bottom*, departure point of the side-crack in increment 56.

the hypothesis that the side-cracks represent branching associated with high-velocity fracture propagation for several reasons. (i) The side-cracks themselves are quite planar and show no tendency to curve or fork (Figure 6). (ii) Side-cracks are sometimes contained entirely within an increment so that they must have developed prior to the peak velocity of the fracture as indicated by the maximum intensity of hackle development adjacent to the arrest line (see increments 0, 36, 58; Figure 3) yet side-cracks do not develop near the arrest lines of these same increments. (iii) No increase of hackle development is observed at the point of deviation of side-cracks, rather the hackle intensity continues to smoothly increase in intensity toward the arrest lines of both the side-crack and the main fracture (Figure 6). Therefore a local velocity increase cannot have lead to branching. (iv) Branches typically subtend an angle of about 15° relative to the main fracture (Broek, 1987), which is more than an order of magnitude greater than the observed angle subtended by side cracks.

Although evidence indicates that the study joint was a relatively low-velocity fracture, propagation of each increment ends abruptly without the large number of very fine cracks in complex en echelon patterns as is found at the termination of stable cracks in granite (e.g., Segall and Pollard, 1983). This may also indicate a sudden termination after acceleration. The lack of branches only implies that velocities sufficient to generate branches were not achieved. The crack could have become unstable and propagated under conditions of K_I close or equal to K_{IC}.

(3) *The initial propagation increment was large.* Although 13 or 14 other joints are contained within the adjacent exposed portion of the same bed (Figure 4) only the study joint shows a rhythmic c-type plume. Within the outcrop as a whole, joint faces showing rhythmic c-type plumes are much less common than joints showing s-type plumes. The morphology of the study joint shown in Figure 3 suggests that the apparent scarcity of rhythmic c-type plumes may result from large initial propagation increments relative to the parts of the joints showing cyclic increments in the majority of cases. Hence, one is more likely to observe the initial increment on the relatively small preserved and/or exposed portions of the joints. The propagation direction of the study joint was clearly outwards from the missing segment (increment 0) and the portion to the northwest of the missing segment does not show arrest lines. Also, increment 0 is much larger than any of the succeeding increments. These observations suggest that the initial propagation event occurred as a single continuous rupture between 13.5 and 20 m in total length (assuming symmetrical propagation about the initiation point) and that the arrest line that divides increment 0 from increment 1 was the first arrest line to form on the exposed portion of the study joint. Because most of the exposed portions of other joints within this bed contain initiation points and are much shorter than the study joint, only the large initial increments may be presently observable.

(4) *The study joint propagated under conditions of significant differential stress.* The combination of extreme planarity and close spacing of the joints at this outcrop indicate that these fractures propagated under conditions of significant differential stress so that the crack parallel far-field compressive stresses were large relative to the crack normal stress (Olson and Pollard, 1989). The tendency of these fractures to develop planar side-cracks may be indicative of propagation at or near the limit of pure tensile failure (see Jaeger and Cook, 1979 for a discussion of hybrid tensile–shear failure). Alternatively, temporary perturbations of the stress field resulting from minor interactions with an adjacent joint, material inhomogeneities, or other causes could have produced minor deviations of the crack path. The constancy of orientation of the entire joint, the total

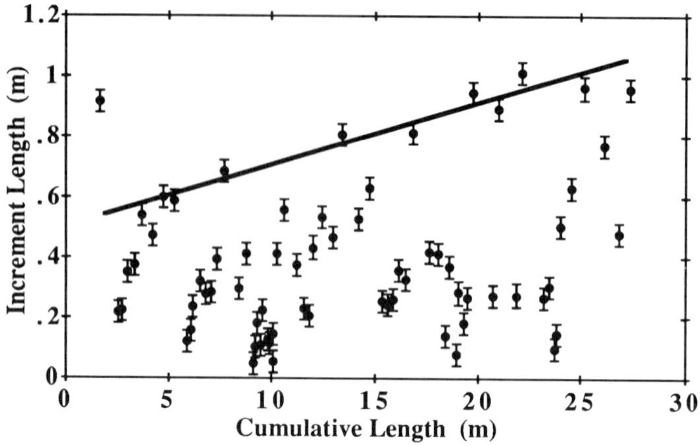

Figure 7. Normalized length of each fracture increment, with error bars, vs. cumulative length of the study joint surface shown in Figure 4. Normalized lengths were computed by measuring the area of each fracture increment on photographs and dividing by the bed thickness (0.41 m). Note the tendency for the largest increments to increase in size with increasing joint length.

absence of side-cracks from many increments, the small number and size of side-cracks relative to the entire joint, and the tendency of some side-cracks to smoothly rejoin the main fracture face all indicate that the great majority of the study joint formed as a pure mode I fracture.

(5) *The maximum size of the increments increased as the study joint grew.* Figure 7 shows the normalized length of fracture increments vs. the cumulative length. Normalized lengths were calculated from the area data of Figure 5 by dividing by the bed thickness (0.41 m). (Note that no attempt was made to include the extra fracture surface area represented by side-cracks in the computation of the normalized increment lengths when compiling the data for Figures 5 and 7.) Figure 5 shows that the length of successive fracture increments increases and decreases in a cyclical pattern. However, comparison with Figure 3 shows that small increments occur after side-crack-forming events. For example, increments 2–4 and increments 40–42 formed after side-cracking events and are especially small. Post-side-crack increments may be smaller because after side-crack formation the crack front is configured as a corner crack rather than as an edge crack and the geometric factor (see eq. (4)) is larger (Tada et al., 1973; Murakami, 1987), which favors failure at a lower driving stress. Increments which do not immediately follow side-cracking events have surface areas between 0.10 and 0.43 m^2, whereas the surface areas of increments that follow side-cracking events are generally less than 0.10 m^2 and always less than 0.15 m^2 in size. Increments that followed side-cracking events must therefore be considered anomalous because they represent a variable and much smaller degree of instability than events that began at fairly even through-the-bed arrest lines. One prominent feature of these data is the presence of a uniformly sloping cut-off at large increment lengths above which no data are present (Figure 7). This cut-off is composed of the largest increments which began at uniform, straight arrest lines. These show the most regular behavior and, hence, are most indicative of changing fracture behavior. Increment 1, which is an apparent exception to this rule, is discussed in a later section.

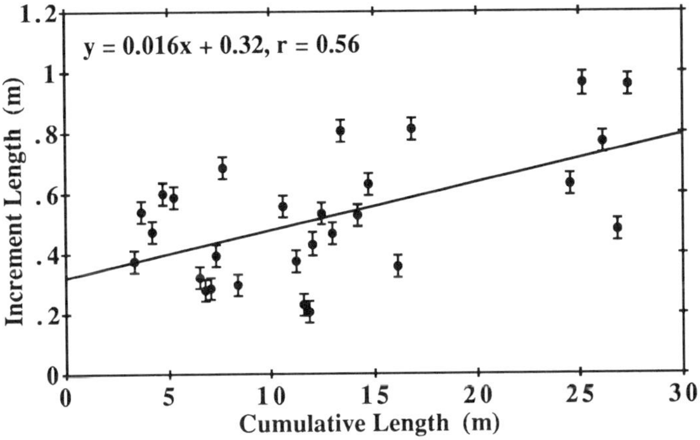

Figure 8. Linear best fit to normalized length vs. cumulative normalized length data for increments that did not follow side-cracking events.

Figure 8 shows the same data as Figure 7 but with the data points of post-side-crack increments removed and a linear best fit to the data. Although the data show substantial scatter, the correlation coefficient of 0.56 indicates that a significant amount of the variation in increment length is correlated with increasing joint length. The overall increment size is thus seen to increase with increasing joint length as well as the maximum increment size.

(6) *The total length of the study joint is less than 90 m.* Figure 4 shows a map of the bed containing the study joint and its location within the bed. Although the study joint is concealed to the north of the missing section, all but two of the joints that are exposed in the gully at the north end of the bed have propagation directions opposite to that of the north section of the study joint and one of them (joint 6) is not continuous to the south. The remaining one (joint 7) is approximately coplanar with joint 1 but cannot be confidently correlated with any points in the southern portion of the bed because of the length of the covered section and because of the tendency of these joints to undergo slight changes in orientation along their lengths. It is therefore possible that joint 7 is the northern section of joint 1.

The bed that contains the joints thins slightly from the location of joint 1 towards the north and appears to pinch out or undergo a lateral change in lithology in the gully. In either case, neither the bed nor the joints are present on the north side of the gully, so that joint 1 had a maximum possible length of about 90 m.

3. Conceptual Models for Incremental Crack Growth

This section describes a set of three conceptual models to define the ways in which natural hydraulic fractures may be driven. Quantitative application of the models leads to predictions of increment size and to the stress cycling that occurs during propagation. The models do not treat the problem of initiation of an incremental propagation event. Sammis and Julian (1987) show that an isolated flat crack in compression that is loaded by internal fluid pressure cannot propagate unstably. If cyclic propagation is to occur by any of the mechanisms described below, then some mechanism must allow excess fluid pressure to build up within the joint. Any such mechanism must release the joint so that an unstable increment may develop with $K_I \geq K_{IC}$, and so that the fracture accelerates as it propagates.

Although interactions between cracks within a set may be important for releasing a restrained crack tip, our models assume no crack interaction during the propagation event. Other types of cycling mechanism include far-field stress variation and/or pulses of fluid injected into either the fracture or the pore space of the rock mass from an external reservoir. The following models are end-member idealizations and real fractures may combine elements of all of them. All of these models allow for propagation at either a single crack tip or simultaneously at both crack tips.

To drive the crack, fluid pressure inside the joint must counterbalance both the total earth stress acting perpendicular to the crack and fracture toughness effects (eq. 8). On the rapid timescale of joint propagation we assume that changes in driving stress are solely a function of internal crack pressure. Therefore, the four major parameters controlling driving stress are: (1) increases in joint volume during joint propagation; (2) flow from pore space in the host rock to the joint; (3) compressibility of the fluid within the joint; (4) flow within the joint including flow from the main body of the joint into the newly created portion of the joint tip. The first parameter acts to decrease driving stress whereas the next two act to maintain driving stress. The fourth controls both the timescale of the stress drop and the stress distribution within the joint. By applying these four parameters in various combinations, one can imagine three qualitative models for incremental joint propagation: the infiltration-limited model; the compressibility-limited model; and the flow-limited model (Figure 9).

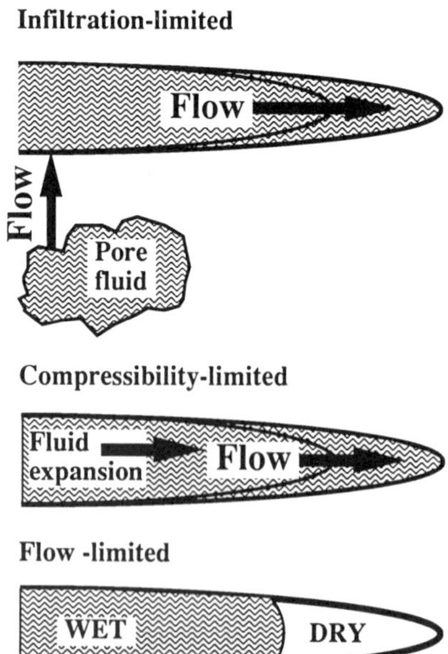

Figure 9. Schematic illustration of the three conceptual models for the hydraulic control of crack propagation. The velocity vs. stress intensity regions appropriate to each model are shown in Figure 12. See text for discussion.

Infiltration-limited propagation

Infiltration-limited propagation is described elsewhere (i.e., Segall and Pollard, 1983). The infiltration-limited model assumes that the driving fluid is incompressible. The rate of infiltration of fluid into the joint from adjacent pore space in the host rock maintains fluid pressure to counterbalance total stress and fracture toughness. Throughout propagation K_I remains at a level where crack growth can occur and joint propagation is stable.

Using fracture mechanics, Secor (1965, 1969) and Engelder and Lacazette (1990) predict that the internal pressure required to extend a joint decreases as the joint grows. The pressure decline occurs because of decreasing influence of fracture toughness with increased joint size (eqs. 4 and 5). Furthermore, infiltration resulting from growth of a joint set may lead to decreased remote pore pressure and thus to a decline in the poroelastic component of the crack perpendicular stress (eq. 5). If the rock volume around the joint is not instantaneously recharged, the increase of the joint volume due to growth must decrease the fluid pressure within a joint. Hence one consequence of joint growth is the development of a fluid pressure gradient between the joint and the pore space of the rock mass. Pore fluid spontaneously drains into the joint. The internal pressure may thereby be maintained at a level sufficient to cause joint propagation. Joint growth is arrested either when the rock mass pore pressure drops below the level required to cause crack propagation or when the far-field earth stress changes. A pore pressure decrease could be caused either by drainage of pore fluid into the joints of the growing joint set or by drainage of the rock mass through the joint system into another region of the rock mass.

Steady-state, stable crack growth occurs if the rate of fluid flow into the joint is sufficient to maintain internal pressure so that the joint slowly grows, where $K_I \leq K_{IC}$ and the fluid pressure is equal throughout the joint. In this case, the rate of fluid flow into the joint is equal to the rate of increase of the joint volume due to growth.

Incremental crack growth may occur if circumstances allow the crack to become unstable where $K_I > K_{IC}$. The size of the increment produced by unstable propagation is limited by the pore volume of the rock surrounding the crack that is hydraulically connected to the fracture on the timescale of fracture propagation. The size of the hydraulically connected region will be controlled by a combination of porosity and permeability of the wall rock, and crack velocity. Arrest occurs once the increase of fracture volume decreases the fluid pressure so that $K_I < K_{IC}$. Recharge of the adjacent host rock and joint follows arrest. During this period of infiltration the fluid pressure within the joint increases.

The infiltration-limited model predicts that during incremental growth the crack velocity will gradually decline as the pressure within the joint declines. The porosity and permeability of the Ithaca Siltstone are so small (Table 1) that only very small increments could form by this mechanism. This model is therefore discarded as a mechanism of cyclic cracking and incremental growth in this case. However, infiltration-limited propagation is considered to cause continuous slow crack propagation during the development of the first increment of the study joint.

Compressibility-limited propagation

Expansion of the fluid within the joint can maintain the crack driving stress even though propagation causes an increase in crack volume. For our compressibility-limited model, propagation distance during a rupture event is a function of the compressibility of the

Table 1. Physical properties of the Ithaca Siltstone

Property	Value	Orientation[a]
Bulk density	$2620\ kg\ m^{-3}$	
Young's modulus	56 GPa	BN
Young's modulus	73 GPa	BP
Intrinsic compressibility	$1.3 \times 10^{-11}\ Pa^{-1}$	BP
Bulk compressibility	$4.8 \times 10^{-11}\ Pa^{-1}$	BP
Biot's constant (α)	0.7	BP
Poisson's Ratio (ν)[b]	0.17	BN
Permeability	<0.01 md	BP
Porosity	<2%	
K_{IC} (c_C = 82 mm)[c]	$2.66\ MPa\ m^{1/2}$	BN
K_{IC} (c_C = 11 mm)	$1.74\ MPa\ m^{1/2}$	BN
K_{IC} (c_C = 2 mm)	$1.26\ MPa\ m^{1/2}$	BN
K_{IC} (c_C = 80 mm)	$2.02\ MPa\ m^{1/2}$	BP
K_{IC} (c_C = 11 mm)	$1.31\ MPa\ m^{1/2}$	BP

Data from A. Lacazette; I. Meglis; Scott (1989); and Evans et al. (1989).
[a] BP, bedding parallel; BN, bedding normal.
[b] Measurement of ν taken from a silty mudstone of the West River Formation.
[c] c_C, Critical crack length at the onset of rapid crack propagation.

fluid within the joint. In this model, rates of fluid flow into the newly created portion of the joint are equal to or greater than the speed of joint propagation. Thus fluid pressure changes at an equal rate within all portions of the joint at all times during the fracturing process. This model also assumes that infiltration into the joint from pores in the host rock is insignificant during propagation. Arrest occurs because the increase in joint volume due to crack growth will eventually reduce the fluid pressure within the entire joint until the fluid drive is counterbalanced by fracture toughness and the crack perpendicular compressive stress.

If this model applies, then the ratio of new joint volume to old joint volume remains constant, so that each successive increment is slightly larger than the preceding one. Such a ratio applies because the size of each increment is controlled by the compressibility of the fluid within the joint. For example, a joint that is filled with gas should show larger increments than one that is filled with water. If the fluid is a gas-saturated brine, then the increment size will depend on the solubility of the gas in the brine and on the relative compressibilities of the two fluids. The compressibility of a two-phase fluid composed of gas and gas-saturated brine is between that of a gas and that of a brine and depends on the relative proportions of the two fluids. Osif (1988) presents extensive new data on the compressibility of mixed gas–brine systems.

Flow-limited propagation

Infiltration-limited and compressibility-limited models assume that the crack tip propagates slowly enough to maintain contact with fluid inside the joint. For these two models, the rate of fluid flow from the preexisting part of the joint into the newly created portion of the joint is faster than the speed of joint propagation. If the crack tip is restrained and is suddenly released, the crack, at least momentarily, becomes unstable. In this case, the crack may outrun the fluid within the crack (Sammis and Julian, 1987). In the most extreme case of a flow-limited model the rate of fluid flow into the propagating joint tip is infinitely slow relative to the rate of joint propagation. A case similar to this is described by Secor and Pollard (1975). Cracking under conditions where flow to the tip is limited is roughly equivalent to inserting a flat-jack into a slot and pressurizing it until rupture occurs. Velocity of the crack tip increases because K_I increases with crack length. The crack may arrest for one or both of two reasons: (1) The dry crack tip may outrun the zone in which the stresses generated by the pressurized portion of the crack overcome the remote crack normal compressive stresses. (2) Enough fluid may drain into the crack tip so that the driving stress is relieved. This mechanism necessarily leads to cycling, although the time between cycles could be short if fracturing resumes immediately upon equilibration of fluid pressure. This model for incremental crack growth is rejected as indicated by discussion in the next section.

4. Discussion: Mechanism of Cyclic Crack Propagation in the Ithaca Siltstone

4.1. Causes of instability and crack front acceleration

This section considers possible mechanisms to explain crack front acceleration during formation of an increment. Infiltration-limited propagation is not compatible with cyclic crack propagation in the Ithaca Siltstone because crack velocity increased during each cycle and because of the low permeability of the siltstone (Table 1). The boundary between increments 0 and 1 or possibly between increments 1 and 2 marks the onset of cyclic propagation of the study joint after an initial period of steady-state infiltration-limited propagation. Concerning the flow-limited model, Sammis and Julian (1987) show that, if an isolated flat crack with a dry tip is subjected to a far-field compression, it will not propagate unstably when its internal fluid pressure increases. Finally, the relationship between K_I and fluid pressure for the case of a crack with uniform fluid pressure is given by eq. (4). Inspection of eq. (4) shows that an increase of crack fluid pressure is necessary to increase K_I. Acceleration of the fracture is, therefore, prohibited for all of the models because there is a decline of fluid pressure during propagation. However, morphological evidence suggests that the fracture velocity increased during each cycle. If so, then before propagation began, the fluid pressure must have risen above the critical fluid pressure required to propagate the crack. Hence, the crack tip must have been restrained in some way so that the buildup of fluid pressure could occur without being dissipated by subcritical extension of the study joint. We can imagine several possible sources for this restraint.

The tendency for the study joint front to be convex in the propagation direction shows that bedding-plane tractions influenced propagation. It is possible that bedding tractions may have helped restrain the tip after propagation ceased. However, bedding tractions should have been less effective at restraining the stationary fracture tip than the moving tip because creep along the bedding planes while the fracture front was stationary should have allowed relaxation of the bedding tractions. Since bedding plane tractions did not prohibit dynamic propagation, this is an unlikely mechanism for stationary crack-tip restraint.

Other possible restraining mechanisms have their origin in the geometry of the crack tip. Small-scale irregularities of the arrest lines could have served to restrain the tip somewhat. Arrest lines have a coarser texture than the remainder of the crack and some are slightly curved. The roughness resulting from the presence of numerous fracture tongues may have effectively produced a larger radius of curvature of the crack tip. This effect could have inhibited reinitiation by decreasing the stress concentration and thereby increasing the pressure required to reinitiate cracking. A short, bent tip would also restrain crack propagation. But rough calculations using stress intensity factors for cracks with an infinitesimally short bent tip (Murakami, 1987) and $\sigma(\theta)_{\max}$ theory (Erdogan and Sih, 1963; Ingraffea, 1987) show that for a tip deviation of 1–3° out of the crack plane the propagation pressure is increased by 0.01 to 0.6 Pa for joints with c between 10 and 100 m. Thus, bent tips provide only a small contribution to the overpressure, although the calculation did not take into account the crack parallel compressive loads. If a joint turns out of parallel with the principal compressive stress, the pressure required to propagate the crack increases because of the resolved compression across the tip.

A scenario for incremental propagation can be envisioned if subcritical propagation returned the crack to its optimal orientation while the fluid pressure built. Once the optimal orientation was achieved, the crack tip was released and the crack could accelerate because K_I was slightly above K_{IC}. As above, crack arrest occurred when the fluid pressure drop inside the crack caused the stress intensity to decline to K_{IC}. Formation of the arrest line then began the process anew. It is possible that the development of the complex boundary between increments 0 and 1 (Figure 3) could have initiated this process, which then became self-perpetuating. However, such a tip-restraint model seems unlikely because of the small amount of overpressure that the arrest lines can provide and because the complex fracture progression seen in Figure 3 does not seem to favor such a delicate balance.

Another explanation for the velocity increase is that the fracture toughness of the rock declined during a propagation cycle. Some materials show decreased fracture toughness with increasing fracture velocity (Broek, 1987). Such rate-dependent effects may have amplified the effects of crack tip restraint to produce the velocity increase. These dynamic fracture toughness changes could have led to acceleration under a steadily declining load. This hypothesis is impossible to test for the Ithaca Siltstone owing to a lack of relevant data.

In the next subsection we argue that a pressure drop of 1–5 MPa occurred during a propagation event. None of the above mechanisms explain this much overpressure. Another possibility is that interactions between the study joint and adjacent fractures provided a trigger for cyclic propagation. Propagation of a nearby joint could have resulted in a pressure drop within that joint and thereby decreased the total horizontal stress on the walls of the study joint. A drop in total horizontal stress may have released the study joint tip and thus initiated a propagation increment. Cyclic propagation may have begun when adjacent joints began to interfere. The large size of increment 1 may result from the transition from steady-state infiltration-limited propagation to cyclic propagation. Weak crack-tip interactions with an adjacent joint may have produced the complex geometry of the arrest lines between increments 0 and 1 and between increments 1 and 2.

4.2. Pressure-drop during increment formation

The pressure in excess of σ_h (the minimum horizontal total stress) that is required to drive a joint is the result of fracture toughness (eqs. 4 and 5). Longer fractures require a lower excess pressure if the fracture is modeled as an isolated penny-shaped crack or infinite flat tunnel crack in a homogeneous medium. These models assume that all parts of the fracture 'feel' all other parts of the fracture through the elastic medium. This may not be an appropriate model for the study joint. The study joint was confined to a single bed that was quite thin relative to the length of the study joint. Tractions along the bounding bedding planes may have resulted in mechanical isolation of the crack tip so that joint behavior was independent of joint size. If this was the case, only a portion of the fracture (the 'effective length') was mechanically coupled to the crack tip. Size-independent behavior of large fractures has been observed in some cases of failure of large engineering structures such as boilers and pipelines (Broek, 1987). Also, size-independent fracture behavior is assumed in the widely used PKN model for artificial hydraulic fractures that are confined to a single bed (Perkins and Kern, 1961; Nordgren, 1972; see Ben-Naceur, 1989 for a recent review of artificial hydrofrac models). We postulate that the study joint exhibited such size-independent behavior.

Size-independent behavior implies that the internal joint pressure required to drive the fracture must have been essentially constant once the joint reached a certain size. A maximum value for the pressure in excess of σ_h required to drive a joint within the bed that contained the study joint may be obtained by assuming that the effective length was equal to the bed thickness and that a penny-shaped geometry is appropriate for the crack tip (eq. 4). This calculation gives an excess pressure of 5.2 MPa. The propagation-induced pressure drop of a joint within this bed could not have exceeded the excess pressure in any case, so that joint interactions could have provided no more than about 5 MPa of overpressure for incremental propagation of the study joint. During cyclic propagation of the study joint the value of K_I must have remained close to K_{IC} because the fracture velocity never rose sufficiently high for branches to form (see Appendix Figure 12; preceding discussion), so that the pressure drop within the study joint during incremental propagation must therefore have been small. We suggest that 'small' was on the order of 5 MPa.

4.3. Hydraulic drive mechanism

Sammis and Julian (1987) provide an equation applicable for evaluating the stress intensity for our flow-limited model at the tip of an isolated flat tunnel crack in compression with internal fluid pressure and dry tips. Solving this for the fluid pressure within the wet part of the crack after the formation of a dry increment gives

$$P = \frac{K_{IC} + \sigma_h[\pi(c + \Delta c)]^{1/2}}{2[(c + \Delta c)/\pi]^{1/2} \sin^{-1}[c/(c + \Delta c)]} \tag{1}$$

for a vertical joint, where c is the crack half-length, Δc is the length of the dry tips, P is the crack fluid pressure, and σ_h is the crack perpendicular compressive stress.

Equations (1) and (5) were solved for the fluid pressure in excess of σ_h for values of c between 1 and 1000 m and for $\Delta c = 0.6$ m, which is the normalized length of the initial large propagation increments on the study joint face. Comparison of the results shows that the propagation pressure for a joint with a dry tip is 35–19 times greater than the propagation pressure of a uniformly pressurized fracture *after* the formation of the increment. At fracture initiation the fluid pressure must have been much higher. This calculation underestimates the required initial joint pressure for two reasons. The calculation was performed for the pressure after formation of the increment and the calculation was performed for an infinite tunnel crack that was larger than the likely effective length of the study joint. It is unlikely that any possible crack-restraint mechanisms could have provided enough restraint for the fluid pressure to build to levels sufficient for flow-limited fracturing to occur or that a significant length of dry crack could exist during propagation. We discard the flow-limited fracturing model as a possible drive mechanism and, hence, favor the compressibility-limited propagation model by elimination. The sections below show that the compressibility-limited propagation model leads to reasonable predictions of the cycling behavior and allows identication of the driving fluid within the study joint.

4.4. Identification of the joint-forming fluid

Specific predictions about the size increments associated with a given fracture length are obtained from the compressibility-limited model. One way to quantitatively evaluate our model is with the PKN model for artificial hydrofracs (Perkins and Kern, 1961; Nordgren, 1972). The PKN model is appropriate for a long, vertical fracture confined to a single bed with the fracture length in bedding much greater than the bed thickness (Figure 10). In the PKN model, the fracture pressure varies slowly along the length of the fracture, so that the aperture of any cross-section is controlled by the pressure within the fracture at that point. The geometry of a given section of the fracture is then computed by idealizing it as a two-dimensional pressurized elliptical hole (Nordgren, 1972). The area of an ellipse is given by

$$\text{Area} = \pi ab \tag{2}$$

where a is the semimajor axis of the ellipse (in this case half the bed height), and b is the semiminor axis of the ellipse (in this case half the fracture aperture at the bed center).

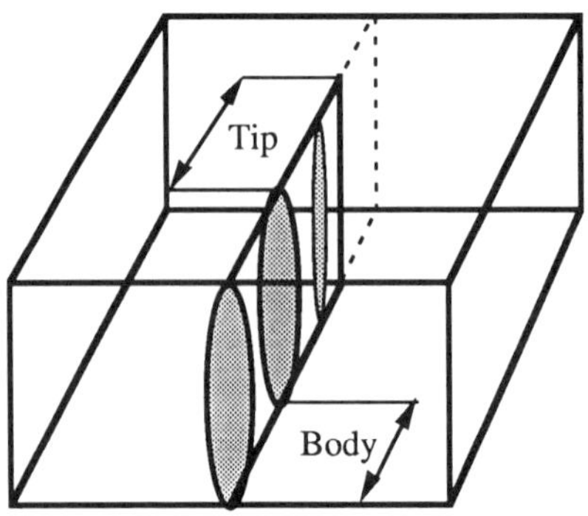

Figure 10. Geometry of the PKN fracture model as applied here. The body has a constant cross-sectional area because the fluid pressure throughout the fracture is uniform while the tip has a uniform taper due to toughness effects and tip restraint.

b is given by (Nordgren, 1972; Pollard and Segall, 1987)

$$b = \frac{2Pa(1 - \nu^2)}{E} \tag{3}$$

where P in this case is the amount by which the crack pressure exceeds σ_h. The driving stress, P, equals the driving stress, σ, as computed from eq. (4) in the Appendix. The cross-sectional area is thus constant for the case in which the pressure is uniform along the fracture, and the fracture volume is therefore equal to the fracture length times the cross-sectional area. Equations (2) and (3) were evaluated for this case using the data in Table 1.

The PKN model for artificial hydrofracs is developed in terms of the injection and flow rates of fluid within the crack from the wellbore so that the crack volume is dependent on the pressure distribution. In the compressibility-limited propagation model, the pressure is uniform throughout the fracture, so that the volume of the fracture tip region is controlled by bedding tractions and fracture toughness. To obtain an estimate of a newly formed tip segment, we assume that the segment has a uniform taper and a triangular cross-section parallel to bedding so that the volume of an increment at the crack tip is half the volume of an equivalent length of the main portion of the fracture body. Propagation leads to formation of a new increment of similar length and expansion of the old increment to a cross-section equivalent to that of the rest of the fracture. The total increase of the crack volume during propagation is therefore equal to the cross-sectional area of the main fracture body times the increment length. We assume that contraction of the main fracture body due to the propagation-induced pressure drop is negligible. This is a valid approximation because inward movement of the joint walls near the tip would result in arrest.

Any geologic fluid, either liquid, gas, or mixtures of the two, may drive a joint. In sedimentary rocks, brine and natural gas are typical pore fluids, which we will use to evaluate the compressibility-limited propagation model. We discuss natural gas in terms of pure methane and brine in terms of pure water in the arguments that follow. The depth of burial of the Ithaca Siltstone was about 3 km at the time of joint formation (Engelder and Lacazette, 1990) so that the temperature was about 90–105°C given a geotherm of 25–30°C km^{-1}. Under the pressure and temperature conditions assumed for joint propagation at Watkins Glen, the adiabatic compressibility of methane is about 1.5% MPa^{-1} (Figure 11; Angus et al., 1976), while the compressibility of water is about 0.04% MPa^{-1} throughout the pressure–temperature region of interest (Sato et al., 1988). If the driving fluid was methane and the final fracture length was 90 m, then the final 1 m long increments could have been driven by a pressure drop of 0.74 MPa, which is within the acceptable limits defined above. If the driving fluid was water, then a 27 MPa pressure drop during propagation is required to drive the final increments, which is 520% higher than the maximum estimate of 5.2 MPa that is derived above and is between 30% and 50% of the total horizontal stress, depending on the exact overburden thickness at the time of propagation.

The size of the study joint during propagation of the initial increments is not well-constrained because asymmetrical propagation is observed on joints in this outcrop. The study joint may have experienced unequal growth on either end. It is possible to estimate the minimum initial size of the study joint at the onset of cycling by noting that the study joint was at least 13 m long when cycling began (Figure 3). Choosing methane as the crack fluid leads to a pressure drop of 3.1 MPa during increment

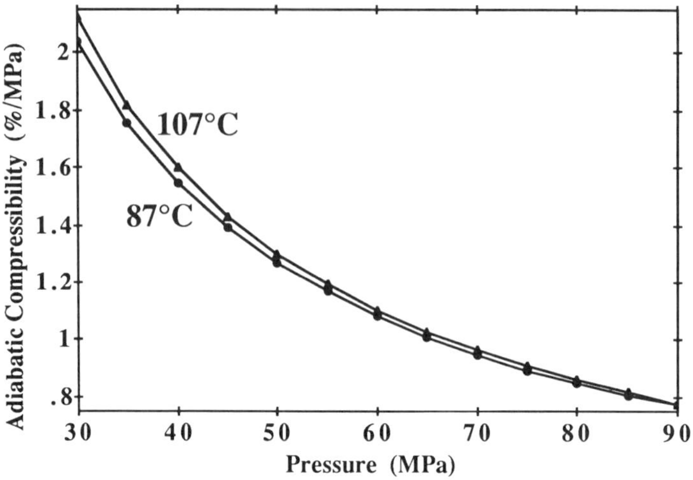

Figure 11. Adiabatic compressibility of methane as a function of pressure at two different temperatures, as calculated from the data of Angus et al. (1976).

formation, while water requires a 115 MPa pressure drop. Water is clearly too incompressible to have served as the crack fluid.

Although water is unacceptable as a driving fluid for the study joint, the data are not adequate to distinguish between geologically common gases or to resolve whether or not the fluid was actually a two-phase mixture with a small liquid fraction. It is unlikely that a liquid fraction was present, because the joints show no evidence of fill or mineralization of any kind. A natural gas with a substantial methane component is the most likely driving fluid for several reasons: (1) the Ithaca Siltstone beds that contain these joints are enclosed in organic-rich black shales and the beds themselves are black to dark gray, presumably from organic matter; (2) the temperature of the rocks at the time of joint formation was appropriate for methanogenesis; (3) natural gas is presently produced from adjacent formations in the region.

We explain the increase in size of the increments with increasing fracture length by postulating that joint growth led to a larger joint volume so that progressively larger increments were driven as the study joint grew. In the pressure–temperature region of interest the compressibility of methane changes over 100% as a function of pressure (Figure 11) whereas the compressibility of water changes by about 15% as a function of pressure (Sato et al., 1988). Formation of the joint set would have resulted in concomitant drainage of the pore fluid into the crack so that the pore pressure behind the study joint would have declined as the joint set grew. Drainage would then have led to a decrease in the total earth stress on the study joint wall and a decrease in the propagation pressure of the study joint (eq. 8). The tendency toward larger increments with increasing joint length is favored by both the change in compressibility of the driving fluid and the increase in joint size. A natural-gas drive provides a more effective mechanism than a water drive for increasing increment size as a function of joint length because of the larger compressibility change of methane in response to decreasing pressure. However, we cannot quantitatively assess this effect without better knowledge of fluid composition, stress state, fluid-pressure regime, and joint size during joint growth.

4.5. Work of joint formation

The three models discussed above are developed in terms of the hydraulic controls on the driving stress of the joint. The work of joint propagation may be understood by considering the joint and rock as a fixed-grips system. During a propagation event the remote boundaries of the rock mass do not move, but the driving stress relaxes. The driving stress for joint propagation is supplied by the fluid pressure within the joint because the crack normal stress in the earth is compressive. When a joint propagates, the work of fracture is supplied by the PV work done by the fluid within the joint and by the release of stored elastic strain energy in the rock mass. A hydraulic criterion is therefore sufficient to describe the stress of cracking but is insufficient to describe the energy balance. Both work and stress conditions must be met for a valid model of fracture behavior (Broek, 1987).

The amount of work available from expansion of methane increases with declining pressure under the conditions at which the study joint formed, and may provide another way to understand the increase in increment size. The isenthalpic Joule–Thomson coefficient (μ) is equal to $(dT/dP)_H$ for a fluid. For methane in the temperature and pressure range at which the joint propagated, μ changes sign with increasing pressure, so that at higher pressure the temperature increases on expansion while at lower pressure the temperature decreases on expansion. Less work can be extracted from methane at higher pressures because of this effect. Decreasing pressure during joint propagation would therefore have led to an increase in the amount of work that methane could do on the joint walls by expansion. If methane was the driving fluid, then increasing increment size is favored from the standpoint of the energy balance of fracturing. In contrast, μ is negative for water throughout the P–T region of interest (so that heating occurs on expansion) and increases only slightly with declining pressure (Grigull et al., 1968). Again, this aspect cannot be evaluated quantitatively without better knowledge of the conditions of joint formation and fluid composition.

5. Conclusions

The three conceptual hydraulic models provide a useful framework in which to consider the hydraulic controls of joint propagation: infiltration of host-rock pore fluid into the joint, flow of the fluid within the joint, and the compressibility of the driving fluid.

The study joint in the Ithaca Siltstone initiated by slow, steady-state infiltration-driven propagation. The change to incremental propagation probably resulted from crack interactions.

The compressibility of the driving fluid appears to be the major control on increment size. Joints in the Ithaca Siltstone at Watkins Glen were driven by natural gas.

Acknowledgments

We thank Mike Gross and Paul Scott for help with field work. Art Rose and Steve Mackwell provided helpful reviews of early versions of this manuscript. Two anonymous reviewers provided helpful comments. This work was supported by an unrestricted grant to Terry Engelder from Texaco and by Gas Research Institute contract 5088-260-1746.

References

Anderson, O.L. and Grew, P.C. (1977). Stress corrosion theory of crack propagation with applications to geophysics. *Rev. Geophys. Space Phys*, **15**, 77–104.

Angus, S., Armstrong, B., and de Reuck, K.M. (1976). *International thermodynamic tables of the fluid state – 5*: Methane. International Union of Pure and Applied Chemistry Chemical Data Series. Pergamon Press, New York.

Atkinson, B.K. (1984). Subcritical crack growth in geological materials. *J. Geophys. Res.* **89**, 4077–4114.

Atkinson, B.K. (1987). Introduction to fracture mechanics and its geophysical applications. In *Fracture Mechanics of Rock* (ed. B.K. Atkinson), pp. 1–26. Academic Press, London.

Atkinson, B.K. and Meredith, P.G. (1987a). The theory of subcritical crack growth with applications to minerals and rocks. In *Fracture Mechanics of Rock* (ed. B.K. Atkinson), pp. 111–166. Academic Press, London.

Atkinson, B.K. and Meredith, P.G. (1987b). Experimental fracture mechanics data for rocks and minerals. In *Fracture Mechanics of Rock* (ed. B.K. Atkinson), pp. 477–525. Academic Press, London.

Bahat, D. and Engelder, T. (1984). Surface morphology on cross-fold joints of the Appalachian Plateau, New York and Pennsylvania. *Tectonophysics* **104**, 299–313.

Beach, A. (1977). Vein arrays, hydraulic fractures, and pressure solution structures in a deformed flysch sequence, S.W. England. *Tectonophysics* **40**, 201–225.

Ben-Naceur, K. (1989). Modeling of hydraulic fractures. In *Reservoir Stimulation* (ed. M.J. Economides and K.G. Nolte), pp. 3-1–3-31. Prentice-Hall, Englewood Cliffs, N.J.

Broek, D. (1987). *Elementary Engineering Fracture Mechanics.* Martinus Nijhoff, Dordrecht.

Brown, S.R. and Scholz, C.H. (1985). Broad bandwidth study of the topography of natural rock surfaces. *J. Geophys. Res.* **90**, 12575–12582.

Costin, L.S. (1987). Time-dependent deformation and failure. In *Fracture Mechanics of Rock* (ed. B.K. Atkinson), pp. 167–216. Academic Press, London.

Engelder, T. (1985). Loading paths to joint propagation during a tectonic cycle: An example from the Appalachian Plateau, U.S.A. *J. Struct. Geol.* **7**, 459–476.

Engelder, T. (1986). The use of joint patterns for understanding the Alleghanian orogeny in the Upper Devonian Appalachian Basin, Finger Lakes district, New York. In *New York State Geological Association 58th Annual Meeting Field Trip Guidebook*, pp. 129–144.

Engelder, T. (1987). Joints and shear fractures in rock. In *Fracture Mechanics of Rock* (ed. B.K. Atkinson), pp. 27–70. Academic Press, London.

Engelder, T. and Lacazette, A. (1990). Natural hydraulic fracturing. In *Rock Joints* (ed. N. Barton and O. Stephansson), pp. 35–43. A.A. Balkema, Brookfield.

Erdogan, F. and Sih, G.C. (1963). On the crack extension in plates under plane loading and transverse shear. *ASME J. Basic Eng.* **85** (Dec.), 519–527.

Evans, K., Oertel, G., and Engelder, T. (1989). Appalachian stress study 2: Analysis of Devonian shale core: some implications for the nature of contemporary stress variations and Alleghanian deformation in Devonian rocks. *J. Geophys. Res.* **94**, 7155–7170.

Freiman, S.W. (1984). Effects of chemical environment on slow crack growth in glasses and ceramics. *J. Geophys. Res.* **89**, 4072–4076.

Fyfe, W.S., Price, N.J., and Thompson, A.B. (1978). *Fluids in the Earth's Crust.* Elsevier Scientific, New York.

Geiser, P.E. and Sansone, S. (1981). Joints, microfractures, and the formation of solution cleavage in limestone. *Geology* **9**, 280–285.

Gretener, P.E. (1981). Pore pressure: Fundamentals, general ramifications, and implications for structural geology. *American Association of Petroleum Geologists Educational Course Note Series 4.*

Grigull, U., Bach, J., and Reimann, M. (1968). Die eigenshaften von Wasser und Wasserdampf nach 'The 1968 IFC Formulation.' *Wärme und Soffübertragung* **1**, 202–213.

Helgeson, D. and Aydin, A. (1988). Composite joints in layered sedimentary rocks: implications for fracture permeability. Geological Society of America Abstracts with Programs 20, A318.

Helgeson, D. and Aydin, A. (1989). Use of surface features for interpretation of the propagation, interaction, and intersection of joints. Geological Society of America Abstracts with Programs 21, A64.

Ingraffea, A.R. (1987). Theory of crack initiation and propagation in rock. In *Fracture Mechanics of Rock* (ed. B.K. Atkinson), pp. 71–110. Academic Press, London.

Jaeger, J.C. and Cook, N.G.W. (1979). *Fundamentals of Rock Mechanics*, 3d edn. Chapman and Hall, New York.

Kulander, B.R., Barton, C.C., and Dean, S.L. (1979). *The Application of Fractography to Core and Outcrop Fracture Investigations*. U.S. Dept. of Energy, Morgantown, National Technical Information Service # METC/SP-79/3.

Kulander, B.R. and Dean, S.L. (1985). Hackle plume geometry and joint propagation dynamics. In *Fundamentals of Rock Joints*. (ed. O. Stephansson), pp. 85–94. Centek Publishers, Sweden.

Lawn, B. (1983). Physics of fracture. *J. Am. Ceramic Soc.* **66**, 83–91.

Lawn, B.R. and Wilshaw, T.R. (1975). *Fracture of Brittle Solids*. Cambridge University Press.

Michalske, T.A. (1983). The stress corrosion limit: Its measurements and implications. In *Fracture Mechanics of Ceramics*, vol. 5 (ed. R.C. Bradt, A.G. Evans, D.P.H. Hasselman, and F.F. Lange), pp. 277–289. Plenum Press, New York.

Murakami, Y. (ed.) (1987). *Stress Intensity Factors Handbook* (two volumes). Pregamon Press, Oxford.

Nordgren, R.P. (1972). Propagation of vertical hydraulic fractures. *Soc. Petrol. Eng.* **12**(4), 306–314.

Nur, A. and Byerlee, J.D. (1971). An exact effective stress law for elastic deformation of rocks with fluids. *J. Geophys. Res.* **76**, 6414–6419.

Osif, T.L. (1988). The effects of salt, gas, temperature and pressure on the compressibility of water. *Soc. Petrol. Eng. Reservoir Eng.* **3**, 175–181.

Olson, J. and Pollard, D.D. (1989). Inferring paleostresses from natural fracture patterns: A new method. *Geology* **17**, 345–348.

Perkins, T.K. and Kern, L.R. (1961). Widths of hydraulic fractures. *J. Petroleum Technol.* **13**, 937–949.

Pollard, D.D. and Segall, P. (1987). Theoretical displacements and stresses near fractures in rock with application to faults, joints, veins, dikes, and solution surfaces. In *Fracture Mechanics of Rock* (ed. B.K. Atkinson), pp. 277–349. Academic Press, London.

Ramsay, J.G. (1980). The crack-seal mechanism of rock deformation. *Nature* **284**, 135–139.

Sammis, C.G. and Julian, B.R. (1987). Fracture instabilities accompanying dike intrusion. *J. Geophys. Res.* **92**, 2597–2605.

Sato, H., Uematsu, M., Watanabe, K., Saul, A., and Wagner, W. (1988). New international skeleton tables for the thermodynamic properties of ordinary water substance. *J. Phys. Chem. Ref. Data* **17**, 1439–1540.

Scott, P.A. (1989). Analysis of the correlation between fracture toughness and surface roughness on joints in the Ithaca Siltstone, Watkins Glen, New York. M.S. thesis, The Pennsylvania State University.

Secor, D.T. (1965). Role of fluid pressure in jointing. *Am. J. Sci.* **253**, 633–646.

Secor, D.T. (1969). Mechanics of natural extension fracturing at depth in the earth's crust. *Geological Survey of Canada Paper 68-52*, pp. 3–48.

Secor, D.T. and Pollard, D.D. (1975). On the stability of open hydraulic fractures in the earth's crust. *Geophys. Res. Lett.* **2**, 510–513.

Segall, P. and Pollard, D.D. (1983). Joint formation in granitic rock of the Sierra Nevada: *Geological Society of America Bulletin 94*, pp. 563–575.

Tada, H., Paris, P., and Irwin, G. (1973). *The Stress Analysis of Cracks Handbook*. The Del Research Corporation, Hellertown, Pennsylvania.

Appendix: Fundamental Concepts of Fracture Mechanics and Joint Propagation

A.1. Joint initiation and propagation

Many textbooks and review articles provide an introduction to fracture mechanics (Lawn and Wilshaw, 1975; Broek, 1987; Atkinson, 1987). Chemical environment influences the velocity of crack propagation (Anderson and Grew, 1976; Atkinson, 1984; Atkinson and Meredith, 1987a). Stresses and displacements around ellipsoidal fractures are considered in detail by Pollard and Segall (1987). Fracture mechanics data for rocks and minerals are reviewed and summarized by Atkinson and Meredith (1987b). The discussion below considers only the case of pure mode I (tensile) fracture, which is commonly referred to as 'cracking.' In this paper the term 'joint' is used to describe a natural mode I rock fracture (i.e., a crack). The discussion here is framed in terms of linear elastic fracture mechanics, which assumes a linearly elastic–perfectly brittle material. Although these are excellent approximations for cracks in single crystals and for microscopic cracks in polycrystalline materials (Lawn, 1983; Atkinson, 1987), nonlinear effects are possibly significant in cases of natural rock fracture where the crack is large relative to the grain size (Ingraffea, 1987; Scott 1989; Scott et al., this volume, Chapter 16). Nevertheless, linear elastic fracture mechanics provides useful first-order approximations of the fracture behavior of rock.

In the general case of tensile fracture, the crack tip stress intensity is described by

$$K_I = -Y\sigma c^{1/2} \tag{4}$$

where K_I (i.e., K-one) is the stress intensity factor at the crack tip, Y is a geometric factor that accounts for the crack shape and loading condition, c is the crack half-length, and σ is the driving stress (compression considered positive). In this paper the driving stress is the superposition of the crack fluid pressure and the crack perpendicular earth stress. For a penny-shaped (circular disk) crack, $Y = 1.13$, while for an infinite flat tunnel crack, $Y = 1.77$ (Broek, 1987). K_I will increase with an increase in either crack length or driving stress. In order for K_I to remain constant during crack propagation the driving stress must fall. Equation (4) contains no information about the material in which the crack is contained. The critical stress intensity or mode I fracture toughness (K_{IC}) is a material property that expresses the stress intensity at which a crack propagates unstably in a given material.

If the driving stress is supplied by fluid pressure within the fracture, this fluid pressure must counterbalance both the rock strength (eq. 4) and the compressive stress acting on the joint wall. The classic model for fluid-driven joint initiation and propagation was developed by Secor (1965, 1969) who postulates that fractures initiate from randomly oriented Griffith flaws which are loaded by the in-situ pore fluid. Secor (1965, 1969) assumes that the only stress that the fracture pressure must overcome is σ_3, the least principal far-field compressive stress. Price (in Fyfe et al., 1978) and more recently Gretener (1981) have correctly pointed out that, for the propagation of vertical cracks, this analysis neglects the role of pore pressure in increasing the total stress. Fyfe et al. (1978) recognize that Secor's (1965, 1969) mechanism is viable for vertical cracks if the general law of effective stress (Nur and Byerlee, 1971) is used in place of the simple law. Engelder and Lacazette (1990) clarify this concept and develop equations to describe joint initiation and propagation based on linear elastic fracture mechanics and the equations of poroelasticity. The key difference is that across vertical cracks there is a poroelastic term that must be added to the fracture pressure equation so that

$$P_f = \frac{K_{IC}}{Yc^{1/2}} + \sigma_3 + \alpha P_p \tag{5}$$

where P_f is the fluid pressure within the crack or flaw that is required to cause propagation, α is Biot's poroelastic constant, and P_p is the in-situ pore pressure of the rock mass. It is important to note that in this paper the sign conventions differ from those of Engelder and Lacazette (1990). In eqs. (4–8) both pressure and compressive stress have positive signs. Although this convention differs from that which is normally employed in the engineering literature, it is more convenient for the calculation of the fluid pressure required to propagate the crack.

Secor's (1965, 1969) original model and Engelder and Lacazette's (1990) modification postulate that at joint initiation, P_f is equal to P_p. It is implicit in the model that P_p exceeds σ_3 so that a condition of effective tension exists at fracture initiation. P_p must also exceed the least principal total stress. The average stress within the solid skeleton of the rock (the total stress) has two components: remotely applied load (σ_3) and porelastically derived stress that results from P_p (which is given by αP_p). During cracking, part of P_p must counterbalance the applied loads while the remainder must counterbalance its own contribution to the average skeletal stress. This contribution arises because the solid skeleton of the rock is not free to expand in response to P_p. A flaw can propagate under these circumstances because it is much larger than an individual grain or pore and because αP_p is >0 and $<P_p$ for consolidated materials such as rocks. The wall of the flaw thus feels the full effects of P_p acting outward, while only part of P_p acts inward through the solid skeleton of the rock. Cracking occurs when P_p exceeds the sum of the far-field stress and its own contribution to the average skeletal stress. The amount by which P_p must exceed the sum of the inward-acting stresses is a function of the size and geometry of the flaw and the fracture toughness of the rock. This process of natural hydraulic fracturing may not be viable at the scale of individual grains and pores, where a complex stress state may exist. In the Ithaca Siltstone the origin flaws of joints are roughly 1–3 cm in size (Engelder and Lacazette, 1990).

Engelder and Lacazette (1990) define a tectonically relaxed basin as one where σ_2 and σ_3 are equal, horizontal, and a function of the overburden load so that

$$\sigma_2 = \sigma_3 = \frac{v}{1 - v}\sigma_{ve} \tag{6}$$

where σ_{ve} is the effective vertical stress supplied by the overburden load and v is Poisson's ratio. σ_{ve} is given by

$$\sigma_{ve} = \sigma_v - \alpha P_p \tag{7}$$

σ_v is equal to the weight of the saturated overburden. In a tectonically relaxed basin, eq. (5) becomes

$$P_f = \frac{K_{IC}}{Yc^{1/2}} + \frac{v}{(1 - v)}\sigma_v + \frac{1 - 2v}{1 - v}\alpha P_p \tag{8}$$

A.2. Crack velocity

Crack velocity in a linear elastic rock is a function of K_I (Anderson and Grew, 1977). Slow, stable crack propagation occurs at lower values of K_I. When $K_I = K_{IC}$, the critical stress intensity, the fracture is unstable and the velocity undergoes large increases with small increases of K_I. This behavior is shown schematically in Figure 12. The term subcritical crack propagation describes fracturing when $K_I < K_{IC}$. Subcritical crack growth mechanisms in rocks and minerals are reviewed by Costin (1987) and Atkinson and Meredith (1987a). These mechanisms include stress corrosion, dissolution, diffusion, ion-exchange, and microplasticity, although stress corrosion is the dominant mechanism by which subcritical cracking occurs in geologic materials. All of these mechanisms are affected by the chemical environment of the crack; however, microplasticity and defect diffusion may also occur in a vacuum.

In a chemically mediated process, such as stress corrosion, the controlling process is the rate of chemical reactions at the crack tip. In the subcritical regime, dramatic changes in crack velocity accompany changes in the mechanism of cracking (Figure 12) thus producing the three regions of the velocity curve in Figure 12 (Atkinson, 1984). At very slow crack velocities the crack behavior is controlled by the rate of reactions at the crack tip (Region I). As the crack-tip velocity increases it reaches the diffusion rate of reactive species to the crack tip (Region II). Velocity in Region II is unaffected by changes in K_I if diffusion-controlled rupture operates. In Region III, K_I approaches K_{IC} and the crack velocity increases to the point where the crack tip outruns the effects

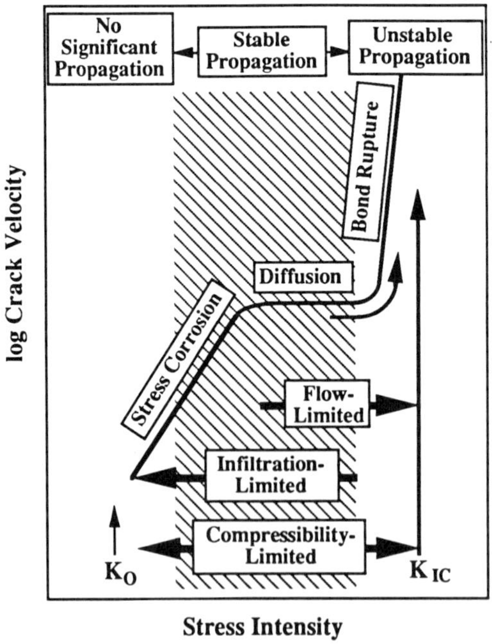

Figure 12. Schematic stress intensity vs. crack velocity diagram (after Atkinson, 1984). Diagram also shows the velocity fields appropriate to the three hydraulic models presented in Figure 9.

of the chemical environment so that crack tip behavior is controlled by bond rupture alone. K_{IC} is therefore essentially independent of chemical environment (Freiman, 1984). Although it has not been demonstrated in geologic materials that there is a limiting value of K_I below which no crack growth occurs (Atkinson, 1987), the existence of such a limit has been proven for glasses (Michalske, 1983). The subcritical crack growth limit is termed K_0 (Figure 12).

Three states in the subcritical regime pertain to the rate of joint propagation: no significant crack propagation; stable crack propagation; and unstable crack propagation (Figure 12). Although crack propagation does occur in the field of no significant propagation, it contributes very little to the overall growth of a crack increment. Stable crack propagation occurs if the change in crack length and the crack driving stress are balanced in such a manner that K_I does not increase abruptly. If K_I somehow increases abruptly then an unstable situation can develop where at least momentarily the crack-tip velocity increases greatly. Such a situation develops if the driving stress is maintained during crack propagation. For the development of cross-fold joints in the Ithaca Siltstone, K_I may have varied between the stress corrosion crack growth limit (K_0) and some value approaching K_{IC} (2.66 MPa m$^{1/2}$) as measured by Scott et al. (this volume, Chapter 14; Scott, 1989). Stable crack propagation is possible for a large range of K_I.

Chapter 13
The Influence of Hydrostatic and Uniaxial Stress on Remanent Magnetization

Randolph J. Martin
New England Research, Inc., 76 Olcott Drive, White River Junction, VT 05001, U.S.A.

Abstract

A laboratory study was carried out to examine the relationship between stress, remanent magnetization and the inclination of the remanent vector with respect to the direction of the greatest principal stress. Experiments were performed in hydrostatic compression and in uniaxial compression on specimens of Ralston diabase with an initial remanent intensity of (TRM) 190 A m^{-1}. For the hydrostatic compression experiments, a demagnetization of 12% was observed at 200 MPa regardless of the orientation of the magnetic vector with respect to the core axis. The demagnetization was nonlinear and a permanent decrease was observed at the end of a 200 MPa pressure cycle. Eight rocks were studied in uniaxial compression. The inclination of the remanent vector with respect to the axis of the rock sample varied between 5° and 75°. A demagnetization of 30 ± 2% was observed for all specimens at 200 MPa. The demagnetization was nonlinear and a permanent demagnetization was observed at the termination of the initial stress cycle. The permanent decrease in remanent intensity was independent of the orientation of the remanent vector with respect to the greatest principal stress direction and proportional to the applied stress. Subsequent cycles to the same peak load produced no additional permanent remanent demagnetization.

1. Introduction

Numerous observations of magnetic field changes accompanying tectonic and explosive events have been documented. For example, Smith and Johnston (1976) reported a 1.8 nT anomaly prior to the 1975 Thanksgiving day earthquake near Hollister, California. More recently a 1.0 nT change was estimated for the October, 1989, Loma Prieta

FAULT MECHANICS AND TRANSPORT PROPERTIES OF ROCKS
ISBN 0-12-243780-2

earthquake (Mueller and Johnston, 1990). These observations were recorded at a single station or inferred using differencing techniques between a number of stations before and after the event. The distribution of the magnetic field change throughout the epicentral region could not be determined. Continuous magnetic field measurements at two stations separated by 6 km provided a direct correlation between the temporal variation in the magnetic field, fluid flow in the epicentral region, and earthquake activity (Yamazaki and Rikitake, 1970) during the Matsushiro earthquake swarm in 1966. A more detailed study of magnetic changes due to a nuclear event was reported by Hasbrouck and Allen (1972). In this case, detailed magnetic surveys were conducted over a fault zone 1.6 km from ground zero before and after the detonation of CANNIKAN. Changes as large as 30 nT were detected along the fault trace, with lesser values laterally outward from the fault zone.

For each observation, the magnetic change can be attributed to a perturbation within the rock mass. Fluid flow or a change in the state of stress have been proposed as mechanisms for the magnetic variations. Mitzutani and Ishido (1976) proposed that the 7.0 nT magnetic field change associated with the Matsushiro swarm was attributable to fluid flow into the epicentral region. Fluid flow can produce an electrical potential by means of electrokinetic phenomena. The resulting current flow, in turn, will generate a magnetic field. Ishido and Mitzutani (1981) expanded the theoretical basis; Martin et al. (1982) experimentally corroborated the electrokinetic effect in the laboratory. Stress-related magnetic changes have been postulated for other precursory magnetic changes (Rikitake, 1968; Johnston, 1978; Rikitake et al., 1980).

Many investigators have observed changes in rock magnetization due to stress (Nagata, 1966; Ohnaka and Kinoshita, 1968; Nagata and Carleton, 1969; Ohnaka, 1969; Martin and Wyss, 1975; Jelenska, 1975; Kapicka, 1975; Kean et al., 1976; Revol et al., 1977; Martin et al., 1978; Martin, 1980; Pike et al., 1981; Lanham and Fuller, 1988; Martin and Noel, 1988). Both remanent and induced magnetization decrease with nonhydrostatic stress. In most instances, the induced magnetization is recoverable (Nagata, 1966; Ohnaka and Kinoshita, 1968; Martin, 1980). When a rock with a remanent moment is initially loaded in compression, it also demagnetizes. However, upon unloading, the remanent magnetization does not return to its prestress value, but exhibits a permanent demagnetization. The decrease at the termination of the initial stress cycle is a function of the maximum stress that the specimen experiences, the magnetic grain size, and the composition of the magnetic mineralogy. For subsequent loading cycles over the same stress path, to the same peak stress, the change in remanent magnetization is more or less reproducible and additional permanent demagnetization is minimal (Ohnaka, 1969; Bezuglaya and Skovorodkin, 1972; Kapicka, 1975; Kean et al., 1976; Martin et al., 1978; Martin, 1980; Martin and Noel; 1988).

One important result of the previous studies is that magnetic changes, in the absence of fluid flow, are stress dominated. This is particularly important for tectonomagnetic applications. While many geophysically relevant rock properties, such as seismic wave velocity, electrical resistivity, and permeability are strongly dependent on crack porosity (e.g., Brace and Orange, 1968; Spetzler and Martin, 1974; Zoback and Byerlee, 1976), changes in both remanent and induced magnetization have been shown to be independent of the development of microcrack porosity (Martin, 1980). This suggests that, at least for well-documented areas, magnetic field variations associated tectonic events can be analyzed to obtain local variations in in-situ stress. While it has been demonstrated that piezomagnetism is a stress-dominated property, the relative contribution of loading history on remanent magnetization has not been clearly established. Previous studies have shown that hydrostatic stress produces smaller decreases in remanent magnetization than those observed due to deviatoric stress (Martin and Noel, 1988). Furthermore, it has been observed that the magnetic vector for both remanent magnetization and induced magnetization rotates with respect to the greatest principal stress axis, with the application of differential stress (Powell, 1960; Hall and Neale, 1960; Domen, 1962; Ohnaka, 1969; Martin et al., 1978; Pike et al., 1981; Lanham and Fuller, 1988). One particular problem that has not been addressed in these studies is how the inclination of the magnetic vector (with respect to the greatest principal stress direction) affects the piezomagnetic response. Several previous studies have reported that the demagnetization of the component of the magnetic vector parallel to the greatest principal stress direction was systematically greater than the component normal to it (Ohnaka, 1969; Martin et al., 1978; Martin, 1980; Martin and Noel, 1988).

In the earth it is very unlikely that the magnetic field will be aligned with the greatest principal stress direction or the remanent vector. Consequently, it is of interest to consider situations where the principal stress directions are not coincident with the magnetic vector, or the axis of the dipole field. In order to approach this problem realistically, the geometric relationship between stress and magnetization must be considered. For induced magnetization, the relationship between the inducing field and the small deviatoric stress (less than 100 MPa) can be represented with a third-rank tensor. Given an induced magnetization, $\mathbf{J}_i$, the relationship between stress σ_{jk} and the change in magnetization $\Delta\mathbf{J}_i$ can be written as

$$\Delta\mathbf{J}_i = d_{ijk}\sigma_{jk} \tag{1}$$

where d_{ijk} is the piezomagnetic tensor.

Kern (1961) showed theoretically that deviatoric stress should result in a decrease in susceptibility, and that confining pressure should produce no change in susceptibility. In most discussions on induced magnetization, susceptibility is reported rather than magnetization. Noting that $\mathbf{J}_i = k_{ij}\mathbf{H}_j$, that most measurements are carried out in a constant field $\mathbf{H}$, and that the changes in $\mathbf{J}$ are directly proportional to k, it is clear that the material parameter k actually changes with stress. Since confining pressure produces only dilatational strains (in a hydrostatic state of stress), the geometry, as well as the orientation of the magnetic carriers, should remain constant during pressurization. If it is assumed that piezomagnetism can be represented with a third-rank tensor, Nye (1964) demonstrated, using symmetry arguments, that hydrostatic pressure should produce no change in magnetization. The application of shear stress introduces magnetic anisotropy in the rock, which results in a decrease in magnetism. These effects have been experimentally verified by Martin (1980) and Ramsey (1982).

With respect to remanent magnetization the situation is much more complex. The decrease in magnetization with the application of stress is nonlinear and history-dependent. A unique functional relationship between stress and remanent magnetization must be developed for each rock. The permanent decrease in magnetization due to initial loading has been shown to depend on the magnitude of the applied stress. Consequently, the permanent stress-induced demagnetization may be utilized as a paleostress indicator in some instances. The paleostress, which will be inferred from the decrease in remanent intensity, is the largest deviatoric stress that the rock has experienced since its TRM was acquired. The underlying assumption in such an analysis is that no secondary chemical and thermal processes have altered the remanent magnetism in the region under consideration.

In light of the potential of using changes in remanent intensity as in-situ stress indicators, it will be useful to understand the relationship between stress and magnetization. Since it is ultimately necessary to use these results to interpret field data, the relationship between the direction of the greatest principal stress, the inclination of the magnetic vector with respect to the applied stress, and the loading history were studied for a characteristic rock type.

2. Experimental procedure

The effect on stress-induced remanent demagnetization and the inclination of the remanent moment with respect to greatest principal stress direction was studied. The experiments were conducted on the Ralston diabase with a fixed remanent intensity (TRM) for inclinations of the magnetic vector with respect to the core axis. Ralston diabase is a low-porosity ($\phi = 0.3\%$) intrusive rock from Golden, Colorado. The mean grain diameter for the matrix is 0.7 mm; the titanomagnetite grain diameters vary between 20 and 300 μm. The rock has a low-field volume susceptibility of 1.69; a coercivity of 240 A m^{-1}, a saturation magnetization of 7.68 k A m^{-1}, and a Curie temperature of 535°C.

The measurements were performed on ground right-circular specimens 1.79 cm in diameter and 4.13 cm long. A TRM was induced in each sample with a nominal intensity of 190 A m^{-1} at inclinations of the magnetic vector with respect to the core axis ranging from 5° to 75°. A TRM was induced by cooling the specimen through the Curie temperature in an 80 A m^{-1} field. After measuring the remanent magnetization, the specimens were subjected to either hydrostatic or uniaxial stress loading conditions to a level of 200 MPa. Several loading cycles were conducted on each sample.

The deformation tests were conducted in a beryllium–copper pressure vessel as shown in Figure 1 (after Martin et al., 1978). All the components of the pressure vessel were fabricated of beryllium–copper hardened to $R_c = 39$. Beryllium-copper was used because it has a small remanent magnetization (10^{-4} A m^{-1}) and a low magnetic susceptibility. The pressure vessel was encased in a two-layer mu-metal shield which attenuated the external field to several hundred nanoteslas. The pressure vessel was bolted into a hydraulic press fitted with a beryllium–copper end plate. The conventional mild-steel end plate was replaced with a beryllium–copper end plate because nonreproducible variations in the background field were observed with the steel plate.

A schematic diagram of the sample assembly and fluxgate magnetometer sensors is also shown in Figure 1. Beryllium–copper spacers were positioned at each end of the sample, and all three pieces were jacketed in 0.13-mm thick copper. The sealed sample was then clamped to the base plug, inserted in the pressure vessel, and a confining pressure of up to 200 MPa was imposed. A differential stress was applied to the sample by advancing the piston of the hydraulic cylinder, driven with an air hydraulic booster. The force on the sample was measured to an accuracy of 1% with an external load cell.

The change in each component of the remanent magnetic vector of the sample was measured with unpotted flux gate magnetometer heads supported with a nylon holder. The magnetometers were Schonstedt DM-2220 digital magnetometers with a resolution of 1 nT over the range of 0–2000 nT. Two pairs of magnetometers were positioned as

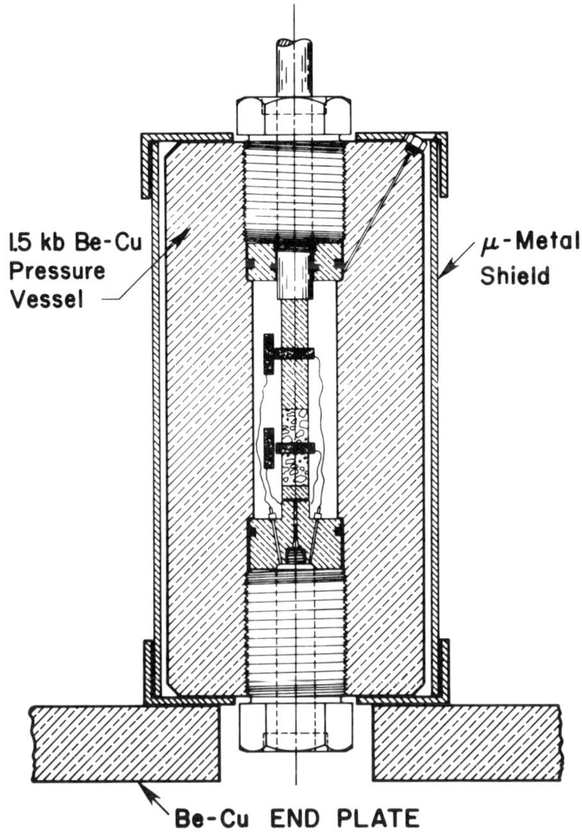

Figure 1. Scale drawing of pressure vessel, fluxgate magnetometer array, and sample assembly utilized for studying the influence of stress and pressure on remanent magnetization. (After Martin et al., 1978.)

shown in Figure 1. One pair measured the axial and radial component of magnetization of the rock sample. The second set of magnetometers was positioned around a beryllium–copper cylinder to monitor changes in the axial and radial component of the magnetic field inside the vessel during a loading cycle. The axial magnetometers were positioned as far away from each other as possible so that the magnetometers would not interact and the remanent magnetization of the sample would not bias the background field measurement in the beryllium–copper vessel. The change in remanent magnetization of the rock as a function of stress could be simply obtained with the use of these four magnetometers. This was accomplished by subtracting the axial and radial magnetometer readings for the rock sample from axial and radial readings obtained in proximity to the beryllium–copper cylinder.

3. Experimental results

For the hydrostatically pressurized specimens, the change in remanent intensity should be the same irrespective of the orientation of the magnetic vector, if the effect of sample

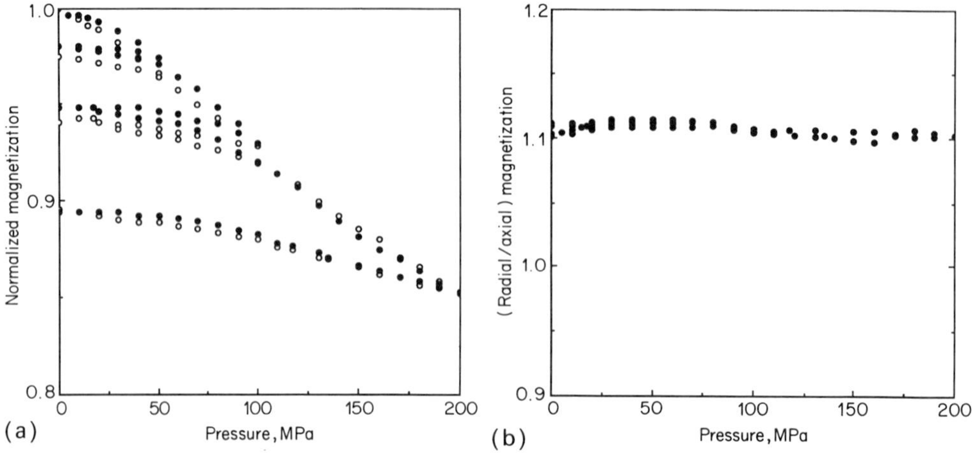

Figure 2. (a) Axial (open symbols) and radial (solid symbols) components of the remanent magnetization, normalized to their initial values, are plotted as a function of pressure for a diabase specimen. (b) The ratio of the radial to axial component of the magnetic magnetization as a function of pressure. The inclination of the magnetic vector with respect to the sample axis was 48°. Three pressurization cycles to successively higher stresses were performed.

geometry is negligible. For this case, as the orientation magnetic vector is varied with respect to the fixed coordinate system of the magnetometer array, the components of the magnetic vector should transform according to:

$$\mathbf{J}_A = \mathbf{J}_{(p=200)} \cos\theta$$
$$\mathbf{J}_R = \mathbf{J}_{(p=200)} \sin\theta \tag{2}$$

where $\mathbf{J}_A$ and $\mathbf{J}_R$ are the components of the magnetic vector measured in the axial and radial directions with respect to the magnetization reference system and $\mathbf{J}_{(p=200)}$ is the remanent intensity at a pressure of 200 MPa.

A typical result for these experiments is shown in Figure 2; the initial inclination of the remanent vector was 48°. The change in each component of remanent magnetization normalized to its prestress value is plotted as a function of confining pressure. The specimen was pressurized to a peak load of 200 MPa in three successively larger cycles: 0–50 MPa, 0–100 MPa, and 0–200 MPa. As the pressure increased, the TRM monotonically decreased. The change was not linear; the demagnetization was most pronounced at low pressures. Upon unloading, the magnetization was not reversible; each component showed a permanent decrease at the termination of a cycle. For example, at the end of the 50 MPa cycle the decrease was 2.5%, whereas at the termination of the 200 MPa cycle the decrease was 12.5%. Furthermore, the decrease in remanent intensity as a function of pressure was the same for all specimens regardless of the inclination of the magnetic vector.

Rotation of the magnetic vector is evident when the ratio of the radial to the axial component of the magnetic vector varies with pressure. In Figure 2, the ratio of the radial to the axial component of magnetization is shown as a function of confining pressure. The results indicate that the ratio remained unchanged during the pressurization sequence. Thus, in spite of the fact that the intensity of magnetization decreased

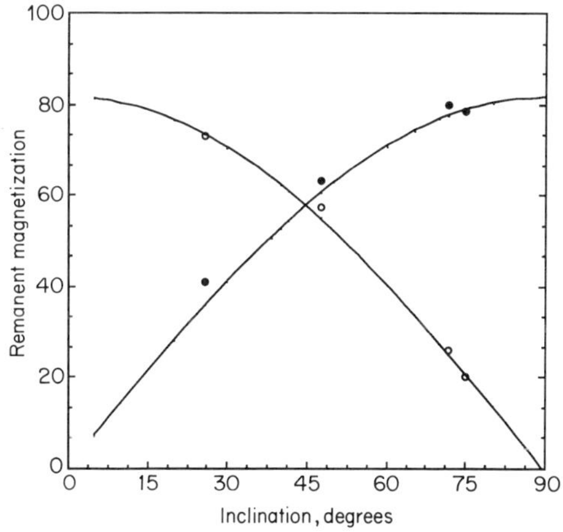

Figure 3. Change in the components of the remanent magnetization for diabase specimens at a pressure of 200 MPa are shown as a function of the inclination of the magnetic vector with respect to the sample axis. Open symbols indicate the axial component and solid symbols indicate the radial component. The solid lines are the transformation for each component using eq. 2.

as much as 13%, no permanent change in the inclination of the magnetic vector was observed due to pressurization.

Four hydrostatic experiments were carried out with initial inclinations of the magnetic vector ranging from 26° to 75°. All the specimens exhibited the same demagnetization at a peak pressure of 200 MPa. The components of magnetization at the peak stress of 200 MPa transformed according to eq. 2. The change in each component of magnetization due to a pressurization of 200 MPa is shown in Figure 3. For reference, the relationship given by eq. 2 is also shown.

The results of these experiments indicate that hydrostatic pressure causes the remanent magnetization simply to decrease in intensity, but does not affect the orientation of the vector. Consequently, the sample geometry and the orientation of the magnetic vector with respect to the core axis do not affect the observed change in magnetization due to hydrostatic pressure.

A similar suite of cyclic experiments were carried out in uniaxial compression. Specimens of diabase with remanent vectors inclined to the greatest principal stress axis were stressed to 200 MPa. This stress level was selected because previous experiments have shown that the greatest decrease in remanent intensity occurs at deviatoric stresses less than 200 MPa (Kinoshita, 1968; Ohnaka, 1969; Martin et al., 1978; Pike et al., 1981; Martin and Noel, 1988). A small confining pressure, 2.5 MPa, was applied to maintain alignment at the sample assembly. A typical result from these experiments is shown in Figure 4. The axial and radial components of the remanent vector, normalized to their initial value are plotted as a function of differential stress. In addition, the values of the components of magnetization measured with the fluxgate magnetometer assembly are presented as a function of differential stress. The values have not been corrected for the separation between the sample and the magnetometer. Consequently, these are reported as relative remanent magnetization without units. For each sample, the stress was monotonically increased to 200 MPa and then reduced to zero. The results were

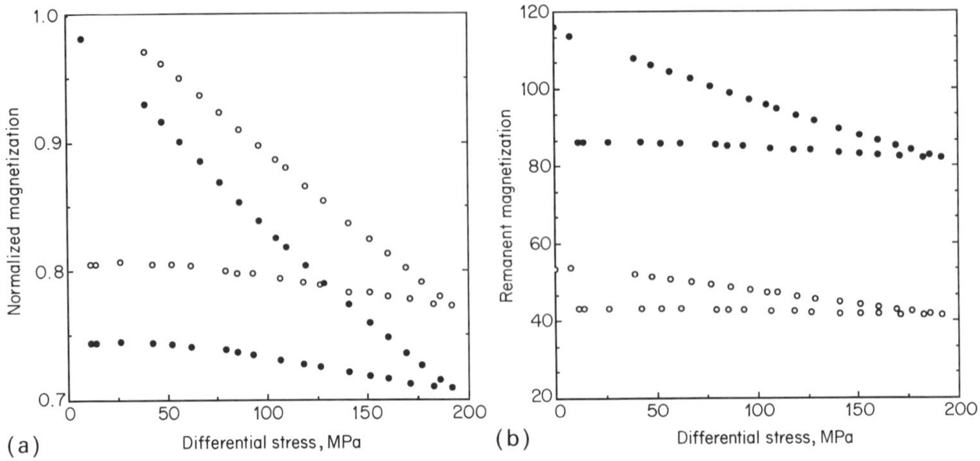

Figure 4. Axial (open symbols) and radial (solid symbols) components of the remanent magnetization (a) normalized to their initial values or (b) without normalization are plotted as a function of differential stress for a uniaxial stress experiment on a diabase specimen. The inclination of the magnetic vector with respect to the sample axis was 65°.

very consistent. Each component decreased with increasing stress. Note that the radial component of the remanent vector decreased more than the axial component with increasing stress. The initial inclination of the remanent vector was 65°. Upon unloading, the magnetization did not recover significantly and a permanent demagnetization at the termination of the cycle was observed. The single exception to this characteristic occurred on a sample with a magnetic vector inclined 4.9° to the loading axis. For this sample, the axial component decreased with load and the radial component exhibited a slight increase.

An examination of all the data showed that the decrease in remanent intensity at a differential stress of 200 MPa was the same for all specimens regardless of the orientation of the magnetic vector with respect to the loading axis. These data are presented in Figure 5; the change in remanent intensity is plotted as a function of the inclination of the magnetic vector with respect to the greatest principal stress direction. All the specimens exhibited a demagnetization of approximately 30 ± 2%.

In order to quantify this finding, the results for eight uniaxial compression tests with varying inclinations of the magnetic vector with respect to the principal stress direction were considered. The observed change in each component of magnetization at a stress of 200 MPa was plotted as a function of inclination. These results are shown in Figure 5. Since the change in remanent intensity was independent of the inclination of the magnetic vector, the observations should transform according to

$$\begin{aligned} \mathbf{J}_A &= \mathbf{J}_{(\sigma=200)} \cos\theta \\ \mathbf{J}_R &= \mathbf{J}_{(\sigma=200)} \sin\theta \end{aligned} \tag{3}$$

where $\mathbf{J}_A$ and $\mathbf{J}_R$ are components of the remanent magnetization measured with the fixed magnetometer array, $\mathbf{J}_{(\sigma=200)}$ is the total remanent intensity at a stress of 200 MPa, and θ is the inclination of the magnetic vector with respect to the specimen axis. Using

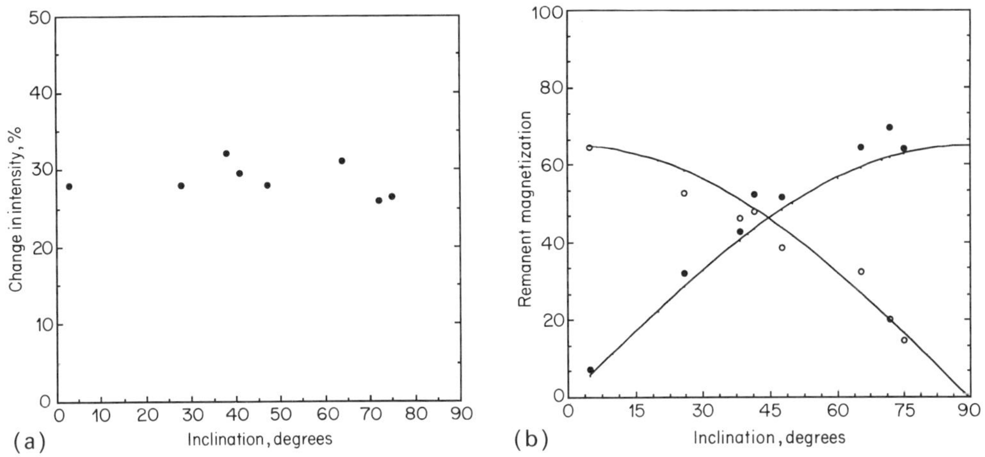

Figure 5. (a) Change in TRM for diabase samples loaded in uniaxial compression to 200 MPa plotted as a function of the inclination of the remanent magnetization to the stress axis. (b) Change in the axial (open symbols) and radial (solid symbols) components of the remanent magnetization for diabase specimens at a differential stress of 200 MPa plotted as a function of the inclination of the magnetic vector to the stress axis. The solid lines are the transformation for each component using eq. 3.

a value of 64 relative magnetic units for $\mathbf{J}_{(\sigma=200)}$, the solid lines in Figure 5 were generated using eqs. 3. The data are in good agreement with the idealized vector transformation.

The orientation of the magnetic vector initially changed with increasing stress for the diabase samples tested. The vector rotated toward the stress direction; that is, the inclination decreased with increasing stress. Upon unloading, there was typically no further change in the orientation of the vector. Results of a representative experiment are shown in Figure 6 (the same experiment shown in Figure 4). Several cycles were carried out on a specimen with an initial inclination of 65°. As the stress increased, the vector rotated toward the principal stress axis by 2.7° at a peak stress of 200 MPa. The orientation of the vector did not recover during unloading. Subsequent cyclic loading to the same peak stress resulted in a small demagnetization as the stress was increased to 200 MPa with no permanent demagnetization at the termination of the cycle. The orientation of the remanent vector remained nearly constant during the second cycle.

Several multicycle experiments were also performed. The results for an experiment consisting of two stress cycles to the same peak stress of 200 MPa are also shown in Figure 6. The remanent magnetization normalized to its initial prestress value is plotted as a function of differential stress. A demagnetization of approximately 25% occurred during the first loading cycle. During the second loading cycle, the changes in magnetization were small. The magnetization at the peak stress of 200 MPa was very nearly the same as that observed during initial loading. As the specimen was unloaded, the magnetization exhibited only a slight hysteresis with several percent additional permanent demagnetization present at the end of the cycle. This observation is in substantial agreement with the results reported previously by Kapicka (1975), Martin et al. (1978), and Martin (1980). Furthermore, an examination of the data in Figure 6 showing the ratio of the radial to axial component of magnetization over the same stress

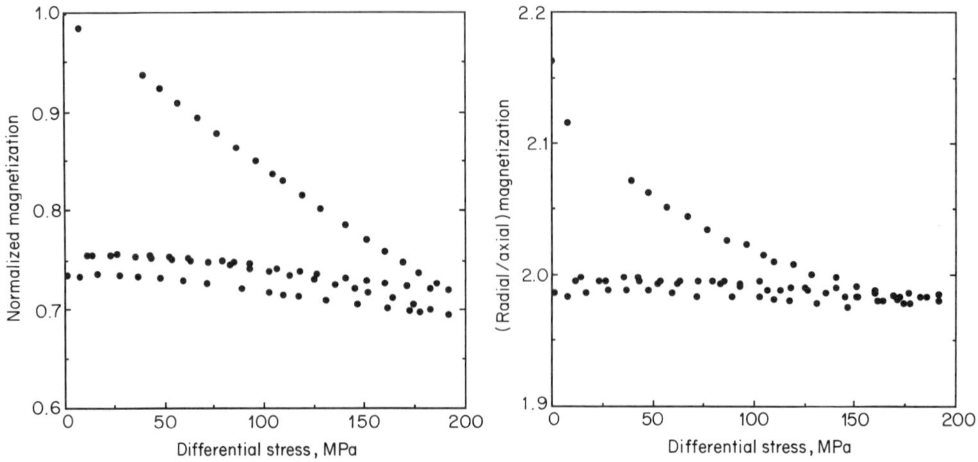

Figure 6. (a) Thermoremanent intensity (TRM), normalized to its initial value, is plotted as a function of differential stress for two loading cycles of a diabase specimen to a peak stress of 200 MPa. (b) The ratio of the radial to the axial component of magnetization is shown as a function of differential stress. The inclination of the remanent magnetization with respect to the sample axis was 65°. These data are from the same experiment shown in Figure 4.

range indicates that there was only a slight rotation of the magnetic vector during the second stress cycle.

Finally, several mixed loading regime experiments were conducted. That is, after an initial hydrostatic pressurization cycle to 200 MPa, several deformational cycles to a peak stress of 200 MPa were performed. During the pressure cycle, each component of the remanent vector demagnetized to approximately 88% of its initial value. The relative change in each component during pressurization was the same. However, when the loading regime was changed to uniaxial stress, the relative change in each component differed (Figure 7) and an increase in demagnetization was observed. Upon unloading, both components showed a pronounced hysteresis. At the termination of the cycle, an additional permanent demagnetization was noted. A third cycle was then carried out to the same peak stress. The loading–unloading curve for each component was nearly identical with the preceding cycle, with only a small hysteresis. No additional permanent demagnetization was detected at the end of the cycle.

For specimens that were initially cyclically stressed to 200 MPa and then subjected to hydrostatic pressurization, small changes in magnetization were observed. The stress-induced variations were less than 2% of the initial remanence and were reversible.

4. Discussion

The suite of experiments on Ralston diabase show that the effect of differential stress on remanent magnetization is significantly greater than that of hydrostatic compression. The remanent intensity simply decreased in magnitude due to pressurization, whereas nonhydrostatic stress initially resulted in a small rotation in the magnetic vector, as well as a decrease in magnetic intensity. For both loading regimes, permanent remanent demagnetization was observed at the termination of the initial stress or pressurization

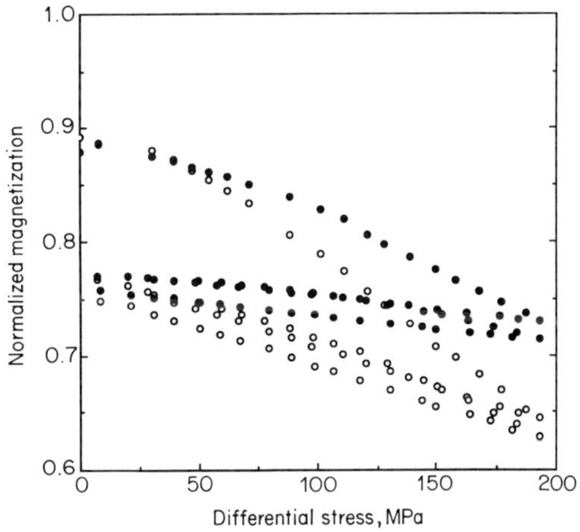

Figure 7. Axial (open symbols) and radial (solid symbols) components of the remanent magnetization normalized to their initial values are plotted as a function of differential stress for the second and third loading cycles on a diabase specimen. The first cycle was carried out in hydrostatic compression. The inclination of the magnetic magnetization with respect to the sample axis was 72°.

cycle. Furthermore, as the peak stress was augmented for sequential cycles, the magnetization at the peak stress decreased and the permanent demagnetization at the termination of the cycle increased. For subsequent stress cycles over the same path to the same peak load, the change in magnetization was not as great, the unloading was accomplished with minimal hysteresis, and no permanent demagnetization was apparent at the end of the cycle.

The results obtained in this study are in substantial agreement with those reported previously by Ohnaka (1969), Martin (1980), and Martin and Noel (1988). In each instance, pressure or uniaxial stress produced a demagnetization which increased monotonically with load and resulted in a permanent demagnetization when the stress was removed. Ohnaka (1969) and Martin et al. (1978) observed a rotation of the remanent moment during the application of stress and a permanent change in the inclination at the termination of the experiment.

Pike et al. (1981) report a rotation of the magnetic vector for a granodiorite specimen cyclically pressurized to 150 MPa. The study presented here on Ralston diabase showed no rotation of the vector, either under pressure or at the completion of a cycle. The experiments were carried out on a granodiorite specimen with a remanent intensity an order of magnitude lower than the diabase in this study.

Perhaps the most interesting result of this series of experiments is that, at least for two well-defined loading regimes and for the single rock studied, the decrease in remanent intensity was independent of the inclination of the magnetic vector with respect to the loading axis (or core axis). The effect of pressure was to simply reduce the remanent intensity, but the demagnetization was neither linear with pressurization nor reversible with depressurization. This suggests that previous studies that reported the greatest stress-induced remanent demagnetization parallel to the stress axis may not have considered a sufficient range of inclinations of the remanent vector or that the effect observed here is not universally applicable to all rocks containing multidomain and pseudo-single-domain titanomagnetitic minerals.

The nonlinearity during loading and the pronounced hysteresis during unloading prevent the remanent demagnetization from being described with a tensor. Furthermore, the change in remanent magnetization depends on loading cycle as well as the loading regime; this means that each particular rock must be individually characterized. No general set of equations can be developed that will universally describe the observed behavior.

On the domain level, the observed effect of pressure and stress can be explained in terms of domain nucleation and domain wall movement (Boyd et al., 1984; Halgedahl and Fuller, 1983). They observed that the application of stress or increase of the magnetic field produced domain nucleation and distortion of existing domains. Changing the stress or the magnetic field perturbs the free energy of the magnetic grain. The fact that initial loading, with either pressure or uniaxial stress, produces a permanent change in remanence can be explained by the fact that a portion of the TRM is metastable. The remanence induced at a high temperature in the presence of a magnetic field is the equilibrium condition for the TRM. The experiments conducted in this study were at room temperature and greater than atmospheric pressure. Consequently, the application of stress establishes a new equilibrium state within the titanomagnetite grains. The domains readjust and nucleate in order to reach an equilibrium configuration. Hydrostatic pressure causes only dilatation of the magnetic mineralogy. Consequently, the transition to the new energy state is accomplished without a rotation of the magnetic moment. For the nonhydrostatic case, shear stress causes a distortion of the magnetic carrier. Small rotations are observed in conjunction with the demagnetization. For these samples, the distortional component of the strain energy produced greater initial demagnetization than the dilatational component of the strain energy.

The applicability of stress-induced changes of remanent magnetization as a paleostress indicator requires some familiarity with the previous stress history. Stress changes within fault zones experience dynamic loading during slip and more or less static loading between events. If changes in remanent magnetization are to be used to infer stress changes, then the effect of loading rate on the demagnetization becomes important. Martin and Noel (1992) carried out a joint laboratory and field study that demonstrated that the stress-induced remanent demagnetization was independent of loading rate, at least for strain rates up to $10^2\ s^{-1}$. The field portion of the experiment was carried our during the high-explosive test ONE TON at the Nevada Test Site. A series of rock cores with a known TRM were fielded in the vicinity of the working point to test the applicability of using decreases in remanent magnetization to measure peak stress. In principle, the technique was extremely simple. Studies such as reported here have shown that during initial loading of rock samples the remanent magnetization decreases. Upon unloading, a permanent demagnetization is observed. Furthermore, the magnitude of the decrease in remanent intensity was proportional to the peak stress experienced by the rock. On the basis of this observation, a series of rock cores with known magnetic properties were grouted along the tuff–grout interface of the access drift to ONE TON at selected distances from the working point in order to test the effect as a passive stress sensor.

After the test the samples were retrieved; nearly all the samples were recovered intact. The permanent demagnetization of each sample was computed and the magnitude of the change in remanent intensity was then correlated with the stress required to produce

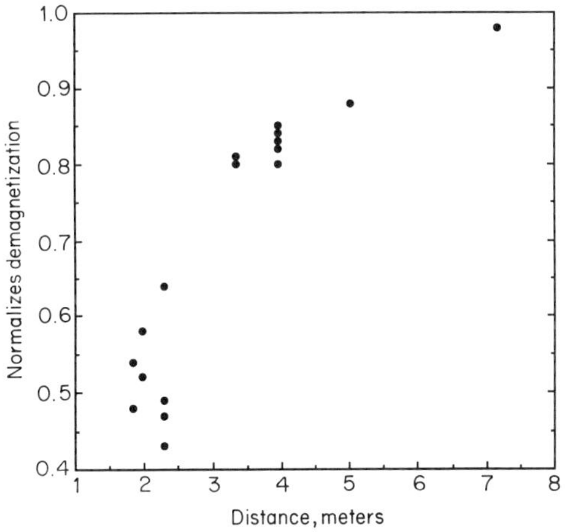

Figure 8. The postshot/preshot remanent magnetization is shown as a function of distance from the shot for specimens of Rapidan gabbro recovered after the detonation of the ONE TON experiment.

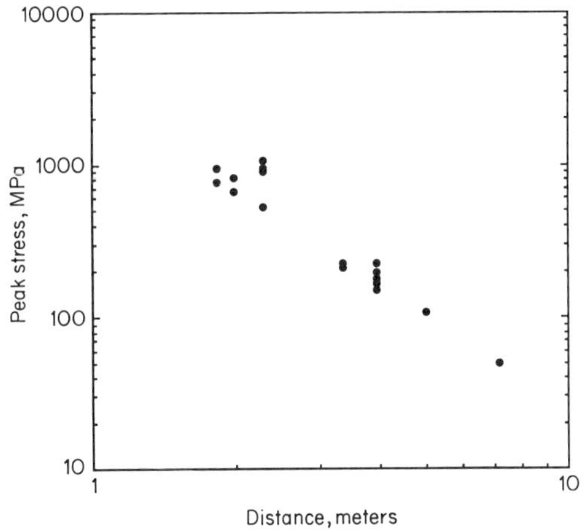

Figure 9. The inferred peak stress determined from the demagnetization of the Rapidan gabbro specimens is plotted as a function of distance from the working point on ONE TON.

the same amount of demagnetization in laboratory samples of the same rock type (Figure 8). In this way, the peak stress developed by the change was obtained as a function of range (Figure 9). The measured peak stresses were on the order of 1000 MPa within 2 m of the device and 50 MPa at a distance of 9 m from the source. These measurements are in very good agreement with other gages fielded on the shot and with empirical equations relating peak stress to range (distance from the working point) for explosive loading. The results of this study indicate that stress-induced remanent demagnetization is independent of loading rate; that is, static calibration tests are applicable to dynamic loading conditions.

Under controlled loading conditions, for simple stress histories, field and laboratory results indicate that stress-induced remanent demagnetization can be related to the peak load experienced by the rock. In the earth, direct application is much more difficult; multiple loading cycles may have occurred and mechanisms other than stress may have produced changes in the remanent magnetization. In spite of the difficulty, it may be possible to use stress-induced remanent demagnetization as a paleostress indicator.

5. Conclusions

A suite of experiments were carried out on the Ralston diabase to examine the effect of pressure and stress on remanent magnetization. For samples with a TRM, initial loading produced a pronounced decrease in magnetization. As the specimen was unloaded, very little recovery was observed. This resulted in a permanent demagnetization at the termination of the cycle. Differential stress produced a larger demagnetization than hydrostatic pressure. For example, similar specimens with the same initial remanent intensities were loaded to 200 MPa in both pressure and uniaxial stress. Demagnetizations of approximately 12% were observed during pressurization, while the change in magnetization approached 30% for a differential stress of 200 MPa. If the specimen was reloaded over the same path to the same stress, the change in magnetization was much smaller than for the initial cycle, and only a small additional increment of demagnetization was observed at the end of the cycle. If the peak stress was augmented, once the peak stress from the previous cycle was exceeded, the stress sensitivity increased noticeably. Upon unloading, there was a pronounced hysteresis and additional permanent demagnetization at zero stress.

The most interesting result of this study was that the decrease in remanent intensity at a fixed differential stress was independent of the orientation of the magnetic vector with respect to the applied stress. The observation that the decrease in remanent intensity for hydrostatic pressures up to 200 MPa is independent of the orientation of the remanent vector with respect to the core axes was anticipated. However, a similar result at deviatoric stress was not; this requires further investigation. For specimens loaded in hydrostatic compression, no rotation of the magnetic vector was evident and the rock simply demagnetized in direct proportion to the applied pressure. It is important to note that even during the first pressurization cycle, when the amount of demagnetization was large, no rotation of the vector occurred.

For uniaxial compression, however, small rotations of the magnetic moment, on the order of 3 degrees, were observed during the initial loading. During the unloading, changes in orientation were less than 0.25°. For subsequent cycles conducted in uniaxial compression to the same peak stress, the orientation of the vector remained stable. Only when the peak stress was increased above the peak for the previous cycle was additional rotation observed. Once again, as the specimen was unloaded, the inclination did not recover.

If the mode of deformation changed from hydrostatic to uniaxial stress, a greater stress sensitivity was observed for those cycles conducted under nonhydrostatic stress than for those carried out hydrostatically. Furthermore, for the first cycle after the transition, a permanent demagnetization was noted when the stress was removed. Additional cycles to the same peak stress became reproducible with no increase in the permanent demagnetization. If, on the other hand, the mode of deformation changed from uniaxial compression to hydrostatic, the observed stress sensitivity decreased and no permanent demagnetization occurred.

Acknowledgment

This research was sponsored by the Air Force Office of Scientific Research (AFSC), under Contract F49620-85-C-0135. The United States Government is authorized to reproduce and distribute reprints for governmental purposes notwithstanding any copyright notation hereon. I would like to thank James S. Noel for his assistance and suggestions throughout this study.

References

Bezuglaya, L.S. and Skovorodkin, Y.P. (1972). Thermoremanent magnetization of rocks in uniaxial compression at temperatures of 20–200°C. *Izu. Acad. Sci. USSR Phys. Solid Earth.* **6**, 385.

Brace, W.F. and Orange, A.S. (1968). Electrical resistivity changes in saturated rocks during fracture and frictional sliding. *J. Geophys. Res.* **73**, 1433–1445.

Boyd, J.R., Fuller, M., and Halgedahl, S. (1984). Domain wall nucleation as a controlling factor in the behavior of fine magnetic particles in rocks. *Geophys. Res. Lett.* **11** (3), 193–196.

Domen, H. (1962). Piezo-remanent magnetism in rock and its field evidence. *J. Geomag. Geoelect.* **13**, 66–72.

Halgedahl, S. and Fuller, M. (1983). The dependence of magnetic domain structure upon magnetization state with emphasis upon nucleation as a mechanism for pseudo-single domain behavior. *J. Geophys. Res.* **86**, 6505–6522.

Hall, J.M. and Neale, R.N. (1960). Stress effects on thermo-remanent magnetization, *Nature* **188**, 805–806.

Hasbrouck, W.P. and Allen, J.H. (1972). Quasi-static magnetic field changes associated with the Cannikin nuclear explosion. *Bull. Seismol. Soc. Am.* **62**, 1479–1487.

Ishido, T. and Mitzutani, H. (1981). Experimental and theoretical basis of electrokinetic phenomena in rock-water systems and its applications to geophysics. *J. Geophys. Res.* **86**, 1763–1775.

Jelenska, M. (1975). Stress dependence on the magnetization and magnetic properties of igneous rocks. *Pure Appl. Geophys.* **113**, 635–649.

Johnston, M.J.S. (1978). Local magnetic field observation and stress changes near a slip discontinuity on San Andreas fault. *J. Geomag. Geoelect.* **30**, 511–522.

Kapicka, A. (1975). The change of isothermal saturation remanent magnetization of basalts as a result of cyclic elastic deformation. *J. Geophys.* **41**, 633–639.

Kean, W.R., Day, R., Fuller, M., and Schmidt, V.A. (1976). The effect of uniaxial compression on the initial susceptibility of rocks as a function of grain size and composition of their constituent titanomagnetites. *J. Geophys. Res.* **81**, 861–872.

Kern, J.W. (1961). The effect of stress on susceptibility and magnetization of a partially magnetized multidomain system. *J. Geophys. Res.* **66**, 3807–3816.

Kinoshita, H. (1968). Studies on piezo-magnetization (III)-PRM and related phenomena. *J. Geomag. Geoelect.* **20**, 155–167.

Lanham, M. and Fuller, M. (1988). Weak field control of remanent magnetization changes produced by uniaxial stress cycling. *Geophys. Res. Lett.* **15**(5), 511–513.

Martin, R.J. III (1980). Is piezomagnetism influenced by microcracks during cyclic loading? *J. Geomag. Geoelect.* **32**, 741–755.

Martin, R.J. III and Noel, J.S. (1988). The influence of stress path on thermoremanent magnetization. *Geophys. Res. Lett.* **15**(5), 507–510.

Martin, R.J. III and Wyss, M. (1975). Magnetism of rocks and volumetric strain in uniaxial failure tests. *Pure Appl. Geophys.* **113**, 169–182.

Martin, R.J. III, Habermann, R.E., and Wyss, M. (1978). The effect of stress cycling and inelastic volumetric strain on remanent magnetization. *J. Geophys. Res.* **83** (B7), 3485–3496.

Martin, R.J. III, Haupt, R. W., and Greenfield, R.J. (1982). The effect of fluid flow in the magnetic field in low porosity crystalline rock. *Geophys. Res. Lett.* **9**, 1301–1304.

Martin, R.J. III, and Noel, J.S. (1992). Dynamic stress measurements using piezomagnetism. *Proc. 33rd U.S. Symposium on Rock Mechanics,* (ed. W. Wawersik), A. A. Balkema, Rotterdam, in press.

Mitzutani, H. and Ishido, T. (1976). A new interpretation of magnetic field variation associated with the Matsushiro earthquakes. *J. Geomag. Geoelect.* **28**, 179–186.

Mueller, R.J. and Johnston, M.J.S. (1990). Seismomagnetic effect generated by the October 18, 1989, ML 7.1 Loma Prieta, California, earthquake. *Geophys. Res. Lett.* **8**, 1231.

Nagata, T. (1966). Magnetic susceptibility of compressed rocks. *J. Geomag. Geoelect.* **18**, 73–80.

Nagata, T. (1969). Tectonomagnetism. *Bull. Int. Assoc. Geomagn. Aeron.* **27**, 12–43.

Nagata, T. and Carleton, B.J. (1969). Notes of piezo-remanent magnetization of igneous rocks, III. *J. Geomag. Geoelect.* **21**, 623–645.

Nye, J.F. (1964). *Physical Properties of Crystals.* Oxford University Press.

Ohnaka, M. (1969). Stability of remanent magnetization of rocks under compression – Its relation to grain size of rock forming ferromagnetic minerals. *J. Geomag. Geoelect.* **21**, 495–505.

Ohnaka, M. and Kinoshita, H. (1968). Effects of uniaxial compression on remanent magnetization. *J. Geomag. Geoelect.* **20**, 93–99.

Pike, S.J., Henyey, T.L., Revol, J., and Fuller, M.D. (1981). High-pressure apparatus for use with a cryogenic magnetometer. *J. Geomag. Geoelect.* **33**, 449–466.

Powell, D.W. (1960). Stress dependent magnetization in some quartz-dolerites. *Nature* **187**, 225.

Ramsey, R.M. (1982). An empirical tensor representation for the stress-dependence of magnetic susceptibility. MS thesis, Pennsylvania State University.

Revol, J.R., Day, R., and Fuller, M.D. (1977). Magnetic behavior of magnetite and rocks stressed to failure – relation to earthquake prediction. *Earth Planet. Sci. Lett.* **37**, 296–306.

Rikitake, T. (1968). Geomagnetism and earthquake prediction. *Tectonophysics* **6**, 59–68.

Rikitake, T., Honkura, Y., Tanaka, H., Ohshiman, N., Sasai, Y., Ishikawa, Y., Koyama, S., Kawamura, M., and Ohchi, K. (1980). Changes in the geomagnetic field associated with earthquakes in the Izu Peninsula, Japan. *J. Geomag. Geoelect.* **32**, 721–739.

Smith, B.E. and Johnston, M.J.S. (1976). A tectonomagnetic effect observed before a magnitude 5.2 earthquake near Hollister, California. *J. Geophys. Res.* **81**, 3556–3560.

Spetzler, H. and Martin, R.J. III (1974). Correlation of strain and velocity during dilatancy. *Nature* **252**, 30–31.

Yamazaki, Y. and Rikitake, T. (1970). Local anomalous change in the geomagnetic field at Matsushiro. *Bull. Earthquake Res.* **48**, 637–643.

Zoback, M.D. and Byerlee, J.D. (1976). The effect of high pressure deformation on permeability of Ottawa Sand. *Am. Assoc. Petrol. Geol. Bull.* **60**, 1531–1542.

Chapter 14

The Correlation between Fracture-toughness Anisotropy and Crack-surface Morphology of Siltstones in the Ithaca Formation, Appalachian Basin

Paul A. Scott[1], *Terry Engelder*[2], *and John J. Mecholsky Jr.*[3]

[1] *BP Research, Houston, TX 77024, U.S.A.*

[2] *Department of Geosciences, The Pennsylvania State University, University Park, PA 16802, U.S.A.*

[3] *Department of Materials Science and Engineering, University of Florida, Gainesville, FL 32611, U.S.A.*

Abstract

This paper describes the effect of bedding-plane anisotropy on the fracture toughness, K_{Ic}, and the relationship between fracture toughness and fracture surface topography of the Ithaca Formation, a Devonian-aged siltstone of the Appalachian Plateau from Watkins Glen, New York. The Ithaca siltstone is the focus of a number of studies to decipher the tectonics of the Appalachian Plateau. Our fracture toughness measurements used the chevron-notched short bar and the indentation/strength test methods. The chevron-notched short-bar test was conducted using three different sample diameters (109, 98, and 12.7 mm). The K_{Ic} results in MPa $\sqrt{m}$ are: 2.67 ± 0.07 and 1.93 ± 0.18 at 109 mm, 2.41 ± 0.13 and 1.82 ± 0.06 at 98 mm, and 1.74 ± 0.11 and 1.35 ± 0.12 at 12.7 mm, for the bedding-perpendicular and bedding-parallel fracture orientations, respectively. The indentation/strength test was conducted for cracks in the bedding-perpendicular orientation only, for which an average toughness of 1.26 ± 0.06 MPa $\sqrt{m}$ was observed. These results suggest that K_{Ic} is not a material constant, and therefore little value is placed on fracture toughness measurements using a single sample size. Although the fracture surfaces appear macroscopically to be similar, the surface topographies were quantitatively distinct in their fractal dimension D. The magnitude of D specifies the degree of correlation between points on the fracture surface, and increases as the fracture surface topography becomes more jagged.

FAULT MECHANICS AND TRANSPORT PROPERTIES OF ROCKS
ISBN 0-12-243780-2

The average fractal dimensions of the bedding-perpendicular and bedding-parallel fractures are 1.37 and 1.32, respectively. This study found a good correlation between fracture toughness and fractal dimension, with toughness increasing for an increasing fractal dimension.

1. Introduction

Evidence suggests that the difference in fracture toughness between brittle materials such as rock correlates with the amount of inelastic deformation (mainly microcracking) at the crack-tip (e.g., Friedman et al., 1972; Delaney et al., 1986). Because the roughness of the fracture surface is an indication of the degree of microcracking, there should be a direct correlation between fracture toughness and fracture surface roughness. One purpose of this study is to further verify the correlation between fracture toughness and surface roughness. Fracture toughness was measured using three sample configurations, including the chevron-notched short bar and indentation/strength techniques (Barker, 1979; Chantikul et al., 1981). Fracture surface topography was quantified using fractal analysis (Mandelbrot, 1977, 1983; Mandelbrot et al., 1984; Brown and Scholz, 1985; Mecholsky et al., 1986, 1988).

There are many different tests and specimen designs for measuring the fracture toughness of rock, and values reported in previous investigations show wide differences for a single rock type (Atkinson, 1984), despite the fact that K_{Ic} is presumed to be a material constant. A major factor contributing to these discrepancies is the zone of inelastic deformation generated by high stresses at the crack-tip. More energy is absorbed during inelastic processes such as microcracking than during elastic processes, so the fracture toughness of a material will increase with the volume of the crack-tip inelastic zone. This is confirmed by the strong correlation between sample size and fracture toughness (e.g., Barker, 1984). A second objective of this study is, therefore, a careful look at the effect of sample size on fracture toughness.

To decrease any ambiguity in our correlation of fracture toughness and surface roughness, we chose to work with one rock, characterized by a fabric anisotropy which we suspected would control the differences in the measured fracture toughness for mutually perpendicular cracks. In sedimentary rocks, a depositional fabric causes a bedding-plane anisotropy which tends to favor easy splitting parallel to bedding. As a consequence, a third objective of this study was to document the fracture toughness anisotropy in the siltstones of the Ithaca Formation, Appalachian Plateau. This objective was motivated by the wish to understand conditions which favor the propagation of large vertical joints throughout the Ithaca siltstone despite the presence of relatively weak bedding planes (see Lacazette and Engelder, this volume, Chapter 12).

2. The Ithaca Siltstone

We chose to study the fracture toughness of two siltstone beds within the Devonian Ithaca Formation of the Genesee Group, which outcrops on the Appalachian Plateau (Figure 1). These beds are exposed along NY Route 414 in Watkins Glen, New York (Engelder et al., 1987). The outcrop consists of 26 m of interbedded siltstones and shales, with siltstone beds ranging in thickness from 5 to 50 cm and shales no thicker than

Figure 1. Outcrop of the Ithaca Formation exposed along NY Route 414 in Watkins Glen, New York. The outcrop consists of 26 m of interbedded siltstones and shales. The samples for this study were taken from the two siltstone beds indicated by arrows, which have thicknesses of 12 cm and 29 cm. A meter stick is set horizontally between the beds for scale.

15 cm. However, composite beds of either siltstones or shales are as thick as 10 m. The primary anisotropy within this outcrop is bedding, which has a regional dip of approximately 1° to the south. The outcrop is cut by two systematic cross-fold joint sets with a strong correlation between joint orientation and lithology. Cross-fold joints in the shales strike 345°, whereas those in the thicker siltstones strike 335° (Bahat and Engelder, 1984). The samples for this study were taken from two dark gray, relatively homogeneous siltstone beds with thicknesses of 12 cm and 29 cm. Our test specimens come from a siltstone bed approximately 3 m above the siltstone bed described in detail by Lacazette and Engelder (Chapter 13).

Thin sections were cut parallel and perpendicular to bedding. The mineralogy of the siltstone consists of 45% quartz, 15% chlorite, 15% heavy minerals, 2% muscovite, 1% biotite, traces of plagioclase, and 22% clay mineral matrix. The majority of the grains are subrounded to subangular and have sizes on the order of 15–100 μm. Aspect ratios of the mineral grains range between 1:3 and 1:5. A planar fabric parallel to bedding (Figures 2a and 2b) is associated with the preferred orientation of the long axes of the micas, and oblate quartz grains developed during deposition, and subsequent bedding compaction and volume loss (Engelder and Oertel, 1985). Despite this planar anisotropy, the interior portion of each siltstone bed does not readily part parallel to bedding.

The siltstone has one obvious fabric anisotropy (bedding) and possibly a second, which appears as a minor pencil cleavage within interlayered shale beds (Engelder and Geiser, 1979). This tectonic fabric anisotropy developed parallel to regional fold axes of the Appalachian Plateau during the Alleghanian Orogeny (Geiser and Engelder,

(a)

(b)

Figure 2. (a) Photomicrograph of a thin-section of the Ithaca Formation siltstone cut parallel to bedding. The white grains are quartz with sizes on the order of 15–100 μm. The quartz grains in this orientation are relatively equidimensional. (b) Photomicrograph of a thin section of the Ithaca Formation siltstone cut perpendicular to bedding. A planar fabric parallel to bedding is due to the preferred orientation of the long axes of the oblate quartz grains developed during deposition, and subsequent bedding compaction and volume loss. The aspect ratios of the mineral grains range between 1:3 and 1:5.

1983). Because of these two fabric anisotropies, fracture toughness was measured for three mutually perpendicular planes in each suite of specimen sizes (Figure 3), except in the case of the indentation/strength technique. The orientations of the toughness measurements are specified in relation to the plane of the crack and the direction of crack propagation as (1) parallel to bedding in the direction of the cross-fold joint (335°); (2) perpendicular to bedding in the direction of the cross-fold joint (335°); and (3) perpendicular to bedding and in the direction of strike of local fold axes (65°). All fracture toughness tests were conducted at room temperature (72–79°F, 22–26°C) with a relative humidity between 24% and 29%.

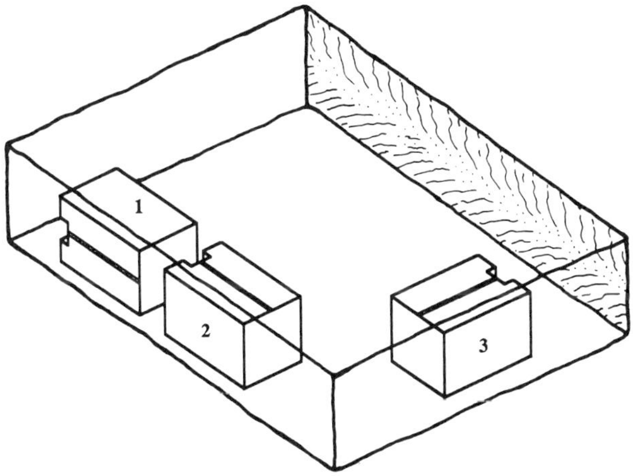

Figure 3. Schematic of a siltstone block taken from the outcrop in Watkins Glen, New York. The cross-fold joint face displays a well-developed plumose surface morphology. Fracture toughness was measured for three mutually perpendicular planes. The orientations of the toughness measurements are specified in relation to the plane of fracture and the direction of crack propagation: (1) parallel to bedding in the direction of the cross-fold joint, (2) perpendicular to bedding in the direction of the cross-fold joint, and (3) perpendicular to bedding in the direction of strike.

3. Test Methods

For fracture toughness measurements we used two different test methods: the short-bar method and the indentation/strength method. In brief, the ISRM short-bar method is an adaptation of a fracture toughness test originally designed by Barker (1977, 1979) to yield accurate toughness measurements of metal samples smaller than the requirements for a purely elastic analysis. A 'plasticity' factor is calculated to correct for nonlinearity such as microcracking. Fracture toughness tests were conducted using the short-bar method for three different sample sizes. A second method, the indentation/strength method, was employed to compare toughness values with the short-bar method. The indentation/strength method was conceived by Chantikul et al. (1981) for use on ceramic materials.

3.1. The elastic–plastic analysis for grip loading

Our first test method used the chevron-notched short-bar specimen. A complete description and LEFM analysis of this test is given by Barker (1977, 1979), Ingraffea et al. (1984), and Ouchterlony (1988). The short bar is a rectangular specimen with a notch machined along its mid-plane so that a V-shaped ligament remains (Figures 3 and 4). Plane strain fracture toughness, K_{Ic}, is determined by loading the sample at a constant load rate until the specimen fails. Tensile loads were generated by a screw-driven Instron (Canton, Mass. U.S.A.) loading machine, and applied to the samples by two steel grips which fit into a groove machined into one end of the specimen. Bending and torsional stresses which may develop due to misalignments in the linkage system were minimized by using universal joints.

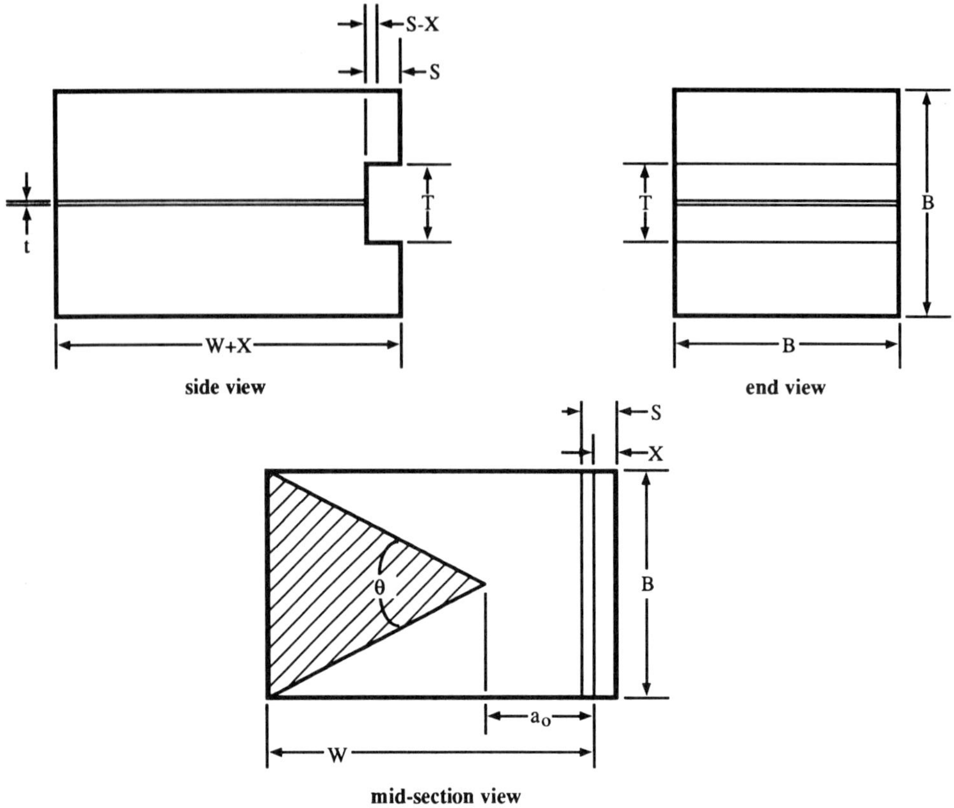

Symbol	Geometry parameter	Value
B	Diameter	B
W	Length	$1.45B$
a_0	Distance to chevron tip from load line (initial crack length)	$0.481B$
S	Grip groove depth	$0.15B$
X	Depth of load line	$0.10B$
T	Grip groove width	$0.35B$
t	Slot thickness	$\leq 0.03B$
θ	Subtended chevron angle	54.6°

Figure 4. Specimen dimensions for the large-diameter chevron-notched short-bar samples used in grip loading.

Upon initial loading, a naturally sharp crack is generated at the tip of the notch. The specimen is designed so that crack propagation is initially stable despite an ever-increasing load. When the crack length reaches a critical value a_c, the stress intensity factor K_I is equal to K_{Ic}. This occurs at or near the peak load if microcracking and plasticity effects are negligible. Thereafter the crack advances under decreasing load. Experiments and theory suggest that the critical crack length is constant for a given specimen geometry and loading configuration (Barker, 1977). An apparent fracture toughness, K_{app}, is determined by recording the peak load and using the expression

(ASTM E399)

$$K_{\text{app}} = \frac{AF_{\max}}{BW^{0.5}} \tag{1}$$

where $F_{\max}$ is the peak load, B is the specimen diameter, and W is the specimen length (Figure 4). Although the short-bar samples are rectangular in shape, the short dimension of the sample is called the diameter. The parameter A is a constant equal to 25.11 that relates the applied load and specimen geometry to the resulting crack-tip stress-intensity factor in a chevron-notched short-bar test (ASTM E399).

The ideal fracture toughness measurement is taken at the point where the stress-intensity factor as a function of crack length is a minimum. For the short-bar configuration this occurs when the crack has advanced a distance of $0.34B$ from the chevron tip, at which point the stiffness of the specimen is about half of its original value. Under ideal circumstances (i.e., for a linearly elastic material) this coincides with the maximum load. Under nonideal circumstances it may occur before or after the maximum load (Ouchterlony, 1988).

A correction for nonlinearity is made by applying a plasticity factor, determined by measuring the crack-mouth opening across the load. This correction is based on a load–displacement record as shown in Figure 5. The crack-mouth opening displacement was measured using an Instron extensometer with a maximum extension range of 0.0127 cm and a maximum nonlinearity of 0.25%. The extensometer was calibrated using an Instron High Magnification Calibrator with an accuracy of $\pm 3.8 \times 10^{-5}$ cm before each set of tests to obtain a record of the absolute displacement. At least two unloading–rcloading cycles performed during each test serve as unloading compliance curves. The plasticity factor is calculated by

$$p = \frac{x_u}{x_l} \tag{2}$$

where x_u and x_l are equal to the distances shown in Figure 5. The plasticity factor has a value between 0 and 1. If $p = 0$, this indicates that the specimen has undergone perfectly elastic growth. A value of $p = 1$ indicates that no crack growth took place and that x_l is due entirely to microcracking or plasticity.

K_{app} is also corrected for size imperfections and small deviations from the required specimen proportions due to imprecise cutting and machining. The size correction factor is given by (Ouchterlony, 1988)

$$C_{\text{size}} = \left(1 - 0.6\frac{\Delta W}{B} + 1.4\frac{\Delta a_0}{B} - 0.01\,\Delta\theta\right) \tag{3}$$

where ΔW, Δa_0, and $\Delta\theta$ denote the differences between the measured specimen dimensions and the specified values, i.e., $\Delta W/B = W/B - 1.45$ (Figure 4). The size correction factor C_{size} should have a value close to 1.00. These correction factors are applied to the apparent fracture toughness K_{app} to give the plane strain fracture toughness K_{Ic} in the following manner:

$$K_{Ic} = \left(\frac{1+p}{1-p}\right)^{1/2} C_{\text{size}} K_{\text{app}} \tag{4}$$

In some instances the unloading compliance curves did not span the fracture toughness

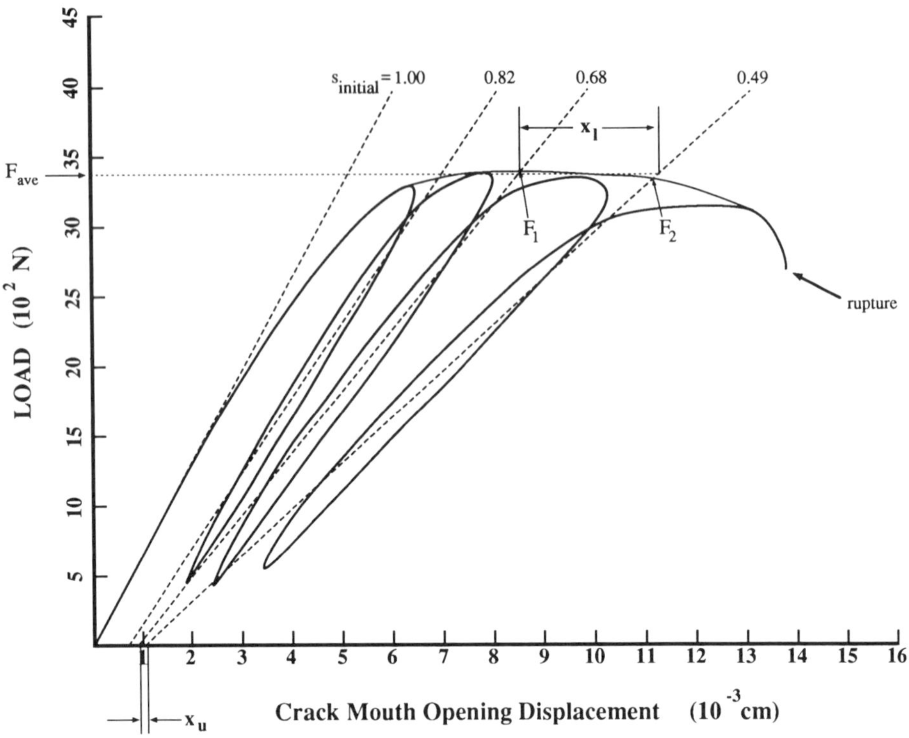

Figure 5. Typical load–displacement record for a short-bar fracture toughness test (sample 4 is shown). At least two unloading–reloading cycles are performed during each test to serve as unloading compliance curves. The dashed unloading lines are determined in the manner suggested by the ISRM (1988). The current specimen stiffness for each unloading cycle is calculated by $s = \tan$ (angle between the unloading line and the CMOD axis), and is given as a ratio of the original specimen stiffness at the top of each unloading line. A smooth approximation of the load-versus-CMOD curve is drawn between the unloading lines, and F_1 and F_2 are the intersections with the unloading lines. F_{ave} is equal to $(F_1 + F_2)/2$. The evaluation point for K_{Ic} is the maximum load on the CMOD curve. The two unloading lines whose slopes most closely span the K_{Ic} evaluation point are used to determine x_u and x_l; x_l is the horizontal distance between the unloading lines for a load equal to F_{ave}, and x_u is the horizontal distance between the unloading lines along the CMOD axis.

evaluation point and other corrections were made to the apparent fracture toughness K_{app}. These corrections involve constructions and are given by Ouchterlony (1988).

Seventeen short-bar specimens (ten oriented for bedding-perpendicular crack propagation and seven oriented for bedding-parallel crack propagation) were machined to the dimensions shown in Figure 4, with all dimensions scaled to the diameter of the block. The exact sample dimensions are given in Table 1. Two diameters were used: 109 mm and 98 mm.

3.2. The elastic–plastic analysis for flatjack loading

The chevron-notched short-bar specimen configuration was also conducted using a Terra Tek (Salt Lake City, Utah) Fractometer testing machine. However, flatjack loading

Table 1. Sample dimensions for the large-diameter chevron-notched short-bar samples used in grip loading. In the bedding-perpendicular orientation, fracture toughness was determined for crack propagation in two directions: parallel to the strike of the cross-fold joint (CF), and parallel to the strike of bedding (ST)

Sample no.	*B*, diameter (mm)	*W*, length (mm)	a_0, initial crack length (mm)	Chevron angle (deg)	Ligament area (10^{-2} m^2)
		Bedding-perpendicular (CF)			
A5	111.0	148.0	45.0	54.6	0.57
A6	109.0	147.0	42.0	54.6	0.57
A7	109.0	148.5	53.0	54.6	0.52
A8	107.5	149.0	39.5	53.5	0.59
B12	97.0	143.0	45.0	53.0	0.48
B13	98.5	144.0	47.5	54.6	0.48
B14	98.0	145.0	45.5	53.0	0.49
		Bedding-perpendicular (ST)			
B15	98.0	144.0	46.0	53.5	0.48
B16	98.0	144.0	48.5	54.6	0.47
B17	97.0	142.5	44.0	52.0	0.48
		Bedding-parallel			
A1	109.5	147.0	43.5	54.6	0.57
A2	109.0	145.5	50.0	54.6	0.52
A3	107.0	148.0	39.5	54.0	0.58
A4	109.5	145.5	40.5	54.6	0.57
B9	100.0	144.0	42,0	54.6	0.51
B10	99.0	147.5	50.0	54.6	0.48
B11	98.5	143.0	40.0	54.6	0.51

has four major differences relative to grip loading: (1) the sample dimensions are drastically smaller; (2) the geometrical ratios are slightly different; (3) the dimensions are set specifically and calibrated for the test machine; and (4) a grip groove is not necessary (Figure 6).

The Fractometer uniformly loads the specimen through the use of a flatjack consisting of an ultrathin, inflatable stainless steel bladder which is pressurized with mercury. This eliminates the use of external grips (therefore no grip grooves) and alignment problems. The bladder expands due to pressure supplied by a motor-driven pump, subjecting the sample to a tensile load. As in the case of grip loading, during the initial loading a crack is generated at the tip of the chevron notch and propages with increasing load until the critical crack length is reached. Thereafter, the pressure decreases with increasing crack length. Measurement of peak pressure is electronically converted to an apparent fracture toughness K_{app} and digitally displayed by the Fractometer. Similar to grip loading, load–displacement records with unloading–reloading cycles were produced in order to correct for nonlinearity. A displacement gage, designed by Terra Tek for use with the Fractometer, was placed around the sample to measure the crack-mouth opening displacement. A plasticity factor was determined and applied to the apparent fracture toughness in the same manner as explained above.

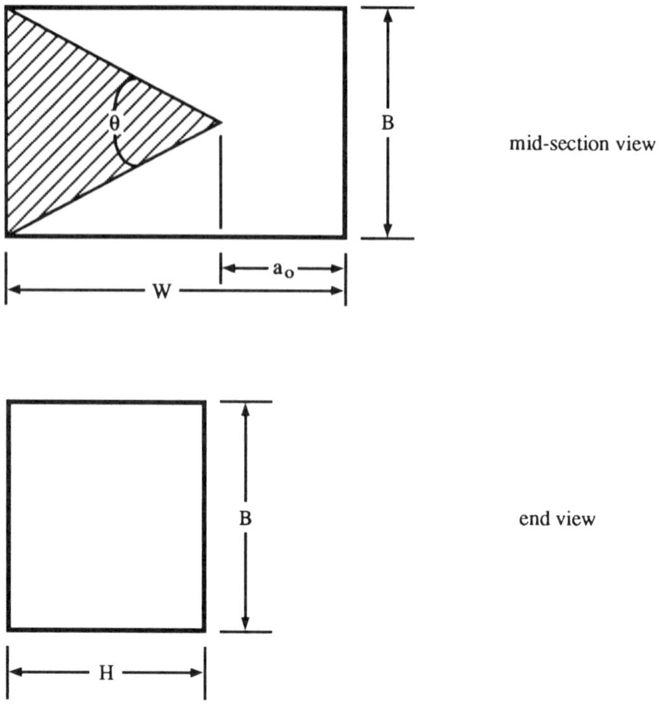

Symbol	Geometry parameter	Value
B	Diameter	12.70 mm
W	Length	19.05 mm
H	Height	11.05 mm
a_0	Distance to chevron tip (initial crack length)	6.76 mm
θ	Subtended chevron angle	54.6°

Figure 6. Specimen dimensions for the small-diameter chevron-notched short-bar samples used in flatjack loading. These samples do not have a grip groove.

The size correction factor for these tests is different from grip loading (Fractometer I Systems Owner's Manual):

$$C_{\text{size}} = C_W C_a C_\theta \tag{5a}$$

$$C_W = 1 - 0.5\frac{\Delta W}{B} \tag{5b}$$

$$C_a = 1 + \frac{\Delta a_0}{B} \tag{5c}$$

$$C_\theta = 1 - 0.01\ \Delta\theta \tag{5d}$$

where W is the specimen length, and a_0 is the distance to the chevron tip (Figure 6). A corrected value of K_{Ic} is obtained by multiplying the apparent fracture toughness by C_{size}.

Twenty-eight samples were machined to the dimensions given in Figure 6. Seventeen tests were conducted to measure the fracture toughness of bedding-perpendicular crack

Table 2. Sample dimensions for the small-diameter chevron-notched short-bar samples used in flatjack loading. In the bedding-perpendicular orientation, fracture toughness was determined for crack propagation in two directions: parallel to the strike of the cross-fold joint (CF), and parallel to the strike of bedding (ST)

Sample no.	B, diameter (mm)	W, length (mm)	H, height (mm)	a_0, initial crack length (mm)	Chevron angle (deg)
		Bedding-perpendicular (CF)			
4	12.70	18.99	10.94	6.73	54.6
5	12.70	19.06	10.89	6.81	54.6
6	12.64	19.05	11.07	6.71	54.6
7	12.71	19.05	11.06	6.71	54.6
8	12.70	19.07	11.06	6.76	54.6
9	12.70	19.06	10.95	6.71	54.6
10	12.61	19.00	11.07	6.73	54.6
11	12.67	19.04	10.81	6.71	54.6
12	12.70	19.06	10.74	6.76	54.6
13	12.69	19.03	11.05	6.86	54.6
		Bedding-perpendicular (ST)			
14	12.71	19.05	10.74	6.65	54.6
15	12.70	18.99	11.05	6.71	54.6
16	12.72	19.05	10.57	6.76	54.6
17	12.70	19.00	11.08	6.68	54.6
18	12.72	19.03	11.03	6.76	54.6
19	12.69	18.83	11.06	6.83	54.6
20	12.73	19.03	11.01	6.71	54.6
		Bedding-parallel			
21	12.70	19.04	11.06	6.73	54.6
22	12.71	19.07	11.09	6.73	54.6
23	12.68	18.88	11.06	6.76	54.6
24	12.72	19.05	10.97	6.71	54.6
25	12.70	19.03	11.08	6.86	54.6
26	12.74	19.05	11.06	6.76	54.6
27	12.72	19.05	11.08	6.86	54.6
28	12.71	19.06	10.85	6.73	54.6
29	12.72	19.10	11.04	6.83	54.6
30	12.73	19.03	11.07	6.76	54.6
31	12.72	19.03	11.05	6.91	54.6

propagation, and eleven tests were conducted to measure the toughness of bedding-parallel crack propagation. The measured sample dimensions are listed in Table 2.

3.3. The indentation/strength technique

The indentation/strength technique for fracture toughness evaluation is based on the assumption that the radial crack system generated from a Vickers indenter can be used as the dominant flaw in a flexural bend specimen. An effective K_{Ic} is determined by

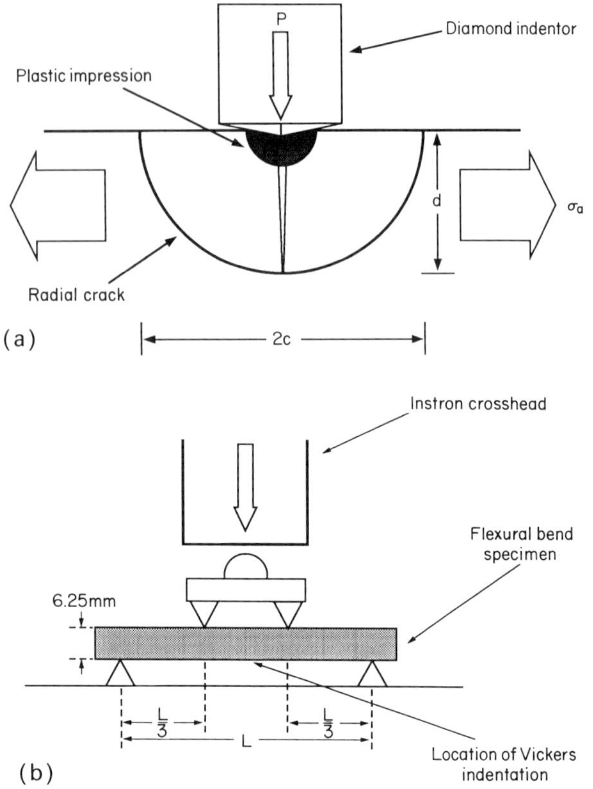

Figure 7. (a) A schematic of the Vickers-induced radial crack system, depicting the peak load P of the indentor, and the depth d and length $2c$ of the radial crack. Two driving forces act on the radial crack: (1) the applied tensile load σ_a and (2) the residual contact stresses from the impression. (b) Specimen loading configuration for symmetrical four-point bending. The Vickers-induced radial crack system is located in the tensile field on the underside of the flexural bend specimen. The beams were loaded using an Instron testing machine at a crosshead speed of 0.127 cm min^{-1} until rupture occurred. The span L of the four-point bend apparatus was 50.8 mm, and the specimens had approximate heights and widths of 6.25 mm and 12.60 mm, respectively.

measuring the strength of the beam under tensile loading, using an appropriate fracture mechanics analysis for cracks (Chantikul et al., 1981). The stability of the Vickers-induced radial crack depends on two driving forces: (1) the residual contact stresses induced by a 'plastic' impression produced when the specimen was indented; and (2) the applied tensile load generated during bending.

The stress-intensity factor owing to the residual stress component scales to the indent load P and has the form (Lawn and Fuller, 1975)

$$K_r = \frac{\chi_r P}{c^{1.5}} \tag{6}$$

where χ_r is a constant, and c is the crack dimension (Figure 7a). This was derived by explicit models of radial crack evolution from a sharp indenter for an isotropic, homogeneous material (Marshall and Lawn, 1979; Marshall et al., 1979; Lawn et al.,

1980). Detailed studies on the accommodation of the 'plastic' impression by the surrounding elastic material show that χ_r depends on the square root of the ratio of Young's modulus to hardness (E/H) (Lawn et al., 1980),

$$\chi_r = \xi_V^R \left(\frac{E}{H}\right)^{0.5} \tag{7}$$

where ξ_V^R is a material-independent constant for Vickers-produced radial cracks (Anstis et al., 1981). The stress-intensity factor for the radial crack system due to the tensile loading has the standard form (Lawn and Wilshaw, 1975)

$$K_a = \sigma_a(\pi\Omega c)^{0.5} \tag{8}$$

where σ_a is the applied tensile stress, and Ω is a crack-geometry factor. Conventional strength theory states that in the absence of any other crack-driving forces, failure will occur spontaneously from the starting flaw at some critical stress level (Chantikul et al., 1981).

Considering the contributions of the two driving forces on the crack system, we obtain a net stress-intensity factor of

$$K = K_r + K_a = \xi_V^R \left(\frac{E}{H}\right)^{0.5} \frac{P}{c^{1.5}} + \sigma_a(\pi\Omega c)^{0.5} \tag{9}$$

The applied stress as a function of crack size can be solved from eq. (9) to give

$$\sigma_a = \left(\frac{K}{(\pi\Omega c)^{0.5}}\right)\left(1 - \frac{\xi_V^R (E/H)^{0.5} P}{K c^{1.5}}\right) \tag{10}$$

The crack remains in a state of equilibrium as long as stable crack growth persists. However, at some critical applied stress the stress-intensity factor K reaches K_c and failure occurs. This is given by the maximum in the $\sigma_a(c)$ function; thus, at $d\sigma_a/dc = 0$, eq. (10) gives the maximum crack dimension and maximum applied stress:

$$c_m = \left[\frac{4\xi_V^R (E/H)^{0.5} P}{K_c}\right]^{2/3} \tag{11a}$$

$$\sigma_m = \frac{3K_c}{4(\pi\Omega c)^{0.5}} \tag{11b}$$

An expression for the toughness can be derived from eqs. (11), eliminating the crack dimension,

$$K_c = \eta_V^R \left(\frac{E}{H}\right)^{1/8} (\sigma_m P^{1/3})^{3/4} \tag{12}$$

where $\eta_V^R = [(256/27)(\pi\Omega)^{3/2}\xi_V^R]^{1/4}$ is another geometrical constant whose value has been calibrated to 0.59 ± 0.12 by averaging over the toughness results for designated 'reference' ceramics (Chantikul et al., 1981). From eq. (12) we can see that if $\log \sigma_m$ is plotted versus $\log P$, the data should have a slope of $-1/3$. Many ceramic materials obey this relationship (Cook et al., 1982).

To evaluate fracture toughness using eq. (12), flexural bend beams with approximate height and width dimensions of 6.25 mm and 12.60 mm, respectively, were machined from the Ithaca siltstone. The actual sample dimensions are given in Table 3. The presence of severe machining damage has been demonstrated to have a significant

Table 3. Sample dimensions for the flexural bend beams used in the indentation/strength test method

	Sample no.	Width (mm)	Height (mm)
Bedding-perpendicular	1	12.56	6.31
	2	12.59	6.29
	3	12.58	6.33
	4	12.60	6.29
	5	12.60	6.27
	6	12.61	6.30
	7	12.59	6.26
	8	12.61	6.29
	9	12.61	6.19
	11	12.57	6.28
	12	12.56	6.20
	14	12.62	6.26
	15	12.58	6.30
	16	12.56	6.26
	18	12.58	6.29
	19	12.61	6.29
	20	12.59	6.27

influence on strength properties of ceramics (Cook et al., 1981), so the surfaces were polished to a 3 μm finish. A single Vickers indentation was made near the center of the prospective tensile face of each test piece, taking care that the indentation was aligned with one set of pyramidal edges parallel to the long axis of the beam.

Indentation loads P ranged from 196 to 343 N and were limited at the lower end by the size of the Vickers-induced flaw. Below 196 N, the indent was not the dominant flaw, and failure occurred at another point. The beams were then loaded using an Instron testing machine and a conventional four-point bend loading apparatus with a span of 50.8 mm (Figure 7b). Each specimen was loaded at a crosshead speed of 0.127 cm min^{-1} until rupture occurred. Beam elasticity formulas were used to calculate the maximum applied tensile stress σ_m at the radial crack, and an effective fracture toughness was determined from P and σ_m using eq. (12).

4. Fracture Toughness Test Results

4.1. Grip loading

Table 4 gives the results for grip loading and is separated into groups of fracture orientation and specimen diameter. Toughness measurements were made for two bedding-perpendicular orientations and one bedding-parallel orientation (Figure 3). Because there was little difference between the two bedding-perpendicular orientations, we lumped them into a single group. There was a significant difference between the behavior of the bedding-parallel and bedding-perpendicular orientations.

An average Young's modulus E was calculated for each orientation using an equation

Table 4. Test results for the large-diameter chevron-notched short-bar samples used in grip loading

	Sample no.	F_{max} (N)	a_c (mm)	K_{app} (MPa $\sqrt{m}$)	C_{size}	Plasticity factor	K_{Ic} (MPa $\sqrt{m}$)	G_{Ic} (J m^{-2})	W_f (J)	R (J m^{-2})
Bedding-perpendicular	A5	3839	82.74	2.26	0.97	0.17	2.58	85.3	0.41	71.9
	A6	4270	79.06	2.57	0.93	0.17	2.64	89.3	0.51	89.6
	A7	3625	90.06	2.17	1.06	0.16	2.70	93.4	0.40	77.5
	A8	4492	76.05	2.72	0.89	0.15	2.75	96.9	0.51	86.7
	B12	2980	77.98	2.04	0.98	0.16	2.35	70.7	0.41	87.2
	B13	2945	80.99	1.98	1.00	0.22	2.46	77.5	0.35	74.4
	B14	2887	78.82	1.94	0.98	0.19	2.17	60.3	0.39	80.9
	B15	2949	79.32	1.99	0.98	0.24	2.50	80.0	0.35	71.8
	B16	2958	81.82	2.00	1.01	0.18	2.42	75.0	0.36	77.5
	B17	3203	76.98	2.20	0.98	0.19	2.57	84.6	0.35	73.1
Bedding-parallel	A1	2615	80.73	1.56	0.95	0.10	1.64	45.4	–	–
	A2	2838	87.06	1.71	1.04	0.14	1.95	64.1	0.25	47.9
	A3	3612	75.88	2.20	0.89	0.09	2.13	76.5	0.33	56.3
	A4	3403	77.73	2.05	0.92	0.05	1.98	66.1	0.36	63.3
	B9	2513	76.00	1.66	0.92	0.21	1.90	60.9	0.22	43.0
	B10	2260	83.66	1.49	1.01	0.24	1.82	55.9	0.28	60.0
	B11	2349	73.49	1.58	0.90	0.21	1.75	51.6	0.29	57.1

from the ISRM-suggested method (Ouchterlony, 1988),

$$E = \frac{84.5\, s_{\text{initial}}}{B} \tag{13}$$

where s_{initial} is the initial tangent slope of the load–displacement record, and B is the specimen diameter. The Young's modulus is anisotropic. For strains perpendicular to bedding, $E = 55.6 \pm 15.5$ GPa, and for strains parallel to bedding, $E = 73.2 \pm 10.6$ GPa.

The strain energy release rate, G_{Ic}, was calculated for each sample from the relation

$$G_{Ic} = \frac{(1 - \nu^2)K_{Ic}^2}{E} \tag{14}$$

where ν is the Poisson's ratio. The Poisson's ratio was not measured in this investigation, so a reasonable value of 0.25 was assumed. The calculated G_{Ic} is relatively insensitive to deviations in the Poisson's ratio.

The specific work to fracture R was determined by calculating the total work to fracture W_f divided by the projected fracture area (equivalent to the ligament area). An expression for W_f is given by the integral of the load over the crack-mouth opening displacement (CMOD) (Ouchterlony, 1988):

$$W_f = \int F\, d(\text{CMOD}) \tag{15}$$

The integral was computed numerically by integrating a cubic spline fit using Romberg integration.

4.2. Flatjack loading

Table 5 gives the fracture toughness measurements using the flatjack loading apparatus. A single specimen diameter of 12.70 mm was used and again the data are separated into two fracture orientations. The strain energy release rate G_{Ic} was computed for each sample using the Young's modulus from the previous test and eq. (14). Unlike for grip loading, the specific work to fracture could not be determined because the crack-mouth displacement gage was not calibrated to measure absolute opening.

4.3. Indentation/strength tests

The results for the indentation/strength tests are given in Table 6. This test was conducted on four or five beams at each of the four different indent loads, and crack propagation was in the bedding-perpendicular direction. In Figure 8, the fracture stress σ_m is plotted against the indent load P, in log–log space. In accordance with eq. (12) the data has a slope of $-1/3$, which suggests that the siltstone obeys the theory of the indentation/strength method. K_{Ic} was calculated from eq. (12) using a measured E/H ratio of 25.0, and G_{Ic} was determined as before. The average fracture toughness values for all the different methods, sample sizes, and fracture orientations are given in Table 7.

Table 5. Test results for the small-diameter chevron-notched short-bar samples used in flatjack loading. Six of the samples tested do not have load–displacement records, so plasticity factors and corrected K_{Ic} and G_{Ic} values are not available

Sample no.	a_c (mm)	K_{app} (MPa $\sqrt{m}$)	C_{size}	Plasticity factor	K_{Ic} (MPa $\sqrt{m}$)	G_{Ic} (J m^{-2})
			Bedding-perpendicular			
4	11.05	1.43	0.999	–	–	–
5	11.13	1.46	1.004	–	–	–
6	11.01	1.39	0.996	0.19	1.68	36.1
7	11.03	1.49	0.996	0.24	1.90	46.2
8	11.08	1.37	0.999	0.16	1.61	33.2
9	11.03	1.54	0.996	–	–	–
10	11.02	1.57	0.998	0.13	1.79	41.0
11	11.02	1.35	0.996	0.10	1.49	28.4
12	11.08	1.43	1.000	0.21	1.77	40.1
13	11.17	1.77	1.008	0.09	1.95	48.7
14	10.97	1.41	0.992	0.16	1.64	34.4
15	11.03	1.37	0.998	0.22	1.71	37.5
16	11.08	1.41	1.000	0.24	1.80	41.5
17	11.00	1.53	0.996	0.12	1.72	37.9
18	11.08	1.61	1.001	0.06	1.71	37.5
19	11.14	1.47	1.014	0.19	1.81	42.0
20	11.04	1.43	0.997	0.20	1.75	39.2
			Bedding-parallel			
21	11.05	1.07	0.998	–	–	–
22	11.05	1.10	0.997	0.10	1.21	24.7
23	11.07	1.07	1.006	0.16	1.27	27.2
24	11.03	1.23	0.996	0.10	1.35	30.7
25	11.18	1.11	1.009	0.17	1.33	29.8
26	11.09	1.10	1.001	–	–	–
27	11.18	1.00	1.008	0.31	1.39	32.6
28	11.05	1.11	0.997	–	–	–
29	11.15	1.03	1.004	0.26	1.34	30.7
30	11.09	1.09	1.001	0.16	1.28	27.6
31	11.23	1.19	1.013	0.29	1.62	44.3

5. Surface Roughness

Surface profiles were generated using a digital surface profilometer at the Lamont–Doherty Geological Observatory. The instrument is capable of recording wavelengths from 1 cm to 10 μm and amplitudes from 2 mm to 0.1 μm. Power spectra (equivalent to amplitude spectra) for the surface profiles were computed using fast Fourier transform calculations (Brown and Scholz, 1985). The power spectra I of a surface profile varies with frequency:

$$I(\omega) \propto \omega^s \tag{16}$$

where $\omega = 2\pi/\lambda$ is the angular frequency, $s = -(5\text{-}2D)$, and D is the fractal dimension (Mandelbrot, 1983; Berry and Lewis, 1980). Using this relationship, an estimate of the

Table 6. Test results for the flexural bend beams used in the indentation/strength test method. f is the flaw size, P is the indent load, and σ_m is the maximum applied stress. The flaw size was determined for nine of the samples, and calculated from the measured depth and length of the Vickers-induced radial crack

	Sample no.	f (mm)	P (N)	σ_m (MPa)	K_{Ic} (MPa$\sqrt{m}$)	G_{Ic} (J m^{-2})
Bedding-perpendicular	1	–	245.3	25.8	1.26	20.3
	2	1.64	245.3	23.2	1.17	17.5
	3	2.24	245.3	24.3	1.21	18.8
	4	–	245.3	27.7	1.33	22.7
	5	–	245.3	28.4	1.36	23.7
	6	2.00	196.2	30.8	1.36	23.7
	7	1.89	196.2	26.6	1.22	19.1
	8	–	196.2	30.4	1.35	23.3
	9	–	196.2	26.7	1.23	19.4
	11	2.74	294.3	21.5	1.15	16.9
	12	1.72	294.3	23.9	1.25	20.0
	14	–	294.3	23.8	1.25	20.0
	15	–	294.3	25.0	1.29	21.3
	16	2.55	343.4	23.5	1.28	21.0
	18	1.50	343.4	23.2	1.27	20.7
	19	3.00	343.4	22.2	1.23	19.4
	20	–	343.4	22.4	1.24	19.7

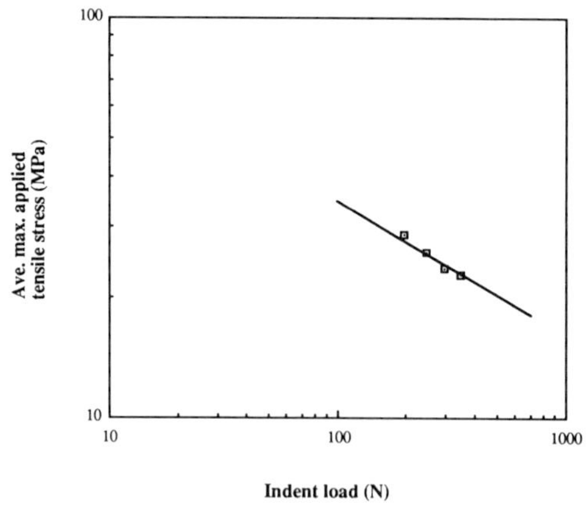

Figure 8. Plot of the average maximum applied tensile stress versus indent load in log–log space against a line with slope of $-1/3$. In accordance with eq. (12), the data fits the $-1/3$ slope fairly well, suggesting that the siltstone obeys the indentation/strength theory.

fractal dimension of a surface profile is determined from the slope of a log–log plot of power spectrum versus spatial frequency.

Surface profiles were measured for bedding-parallel and bedding-perpendicular orientations. Ten profiles, 1.4 inches (35 mm) in length at 0.1 inch (2.5 mm) intervals, were taken from each surface, parallel to the direction of crack propagation. Ensemble

Table 7. Average fracture toughness results for all methods

Test method	Fracture orientation	Specimen diameter (mm)	Critical crack length, a_c (mm)	Average K_{Ic} (MPa $\sqrt{m}$)	Average G_{Ic} (J m^{-2})
Grip loading	Bedding-perpendicular	109	82 ± 6.3	2.67 ± 0.06	91.2 ± 4.4
Grip loading	Bedding-perpendicular	98	79.3 ± 2.1	2.41 ± 0.13	74.7 ± 7.7
Grip loading	Bedding-parallel	109	80.4 ± 5.1	1.93 ± 0.18	63.0 ± 11.2
Grip loading	Bedding-parallel	98	77.7 ± 4.9	1.82 ± 0.06	56.1 ± 3.8
Flatjack loading	Bedding perpendicular	12.7	11.05 ± 0.08	1.74 ± 0.11	38.8 ± 5.0
Flatjack loading	Bedding parallel	12.7	11.11 ± 0.08	1.35 ± 0.12	31.0 ± 5.6
Indentation/strength	Bedding perpendicular	–	2.14 ± 0.49	1.26 ± 0.06	20.4 ± 1.9

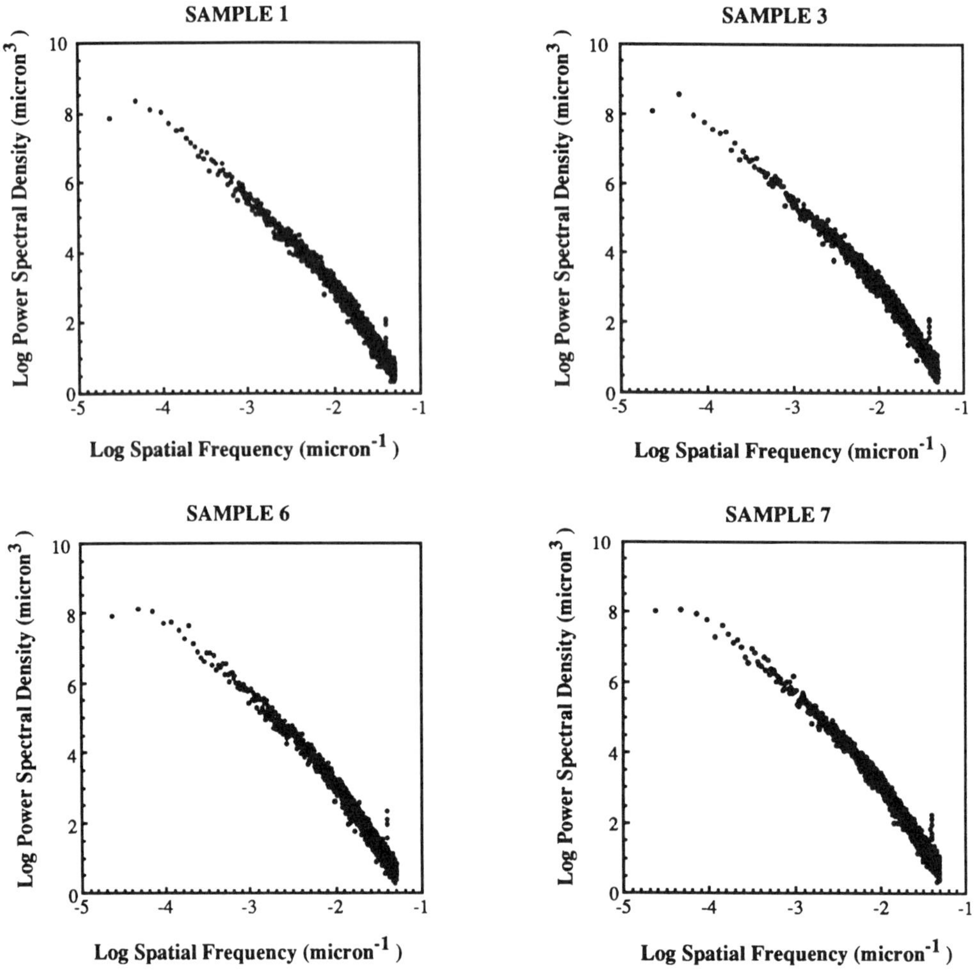

Figure 9. Log power spectral density versus log spatial frequency from the surface profiles of four samples. Samples 1 and 3 are bedding-parallel fracture surfaces and samples 6 and 7 are bedding-perpendicular fracture surfaces. The fractal dimension D is calculated for each surface from the slopes of the log–log plots using the relationship $s = -(5\text{-}2D)$, where s is the slope. The slope varies for a given surface, and tends to decrease from high to low frequency. This suggests that the fracture surface is not adequately described by a single fractal dimension.

averages were computed by stacking the individual spectra for each surface, in order to reduce false variations in power between adjacent frequencies.

The log–log plots of the power spectra for each surface are shown in Figure 9, and the corresponding fractal dimensions for two frequency bands are given in Table 8. The slopes used to calculate D were computed by a linear least-squares fit in log–log space. Examination of the power spectra plots in Figure 9 and their fractal dimensions in Table 8 shows that the slope varies for a given surface, and tends to decrease from high to low frequency, similar to the investigation of Brown and Scholz (1985). The observed variation in spectral slope with frequency suggests that a single fractal dimension does

Table 8. Fractal dimension D for arbitrary frequency bands from laboratory-generated fractures of this study and natural fractures from Brown and Scholz (1985)

Fracture surface	Sample no.	D for frequency band 10^{-3}–10^{-2} μm^{-1}	D for frequency band 10^{-2}–10^{-1} μm^{-1}	K_{Ic} (MPa $\sqrt{m}$)
Bedding-parallel (this study)	1	1.30	0.74	1.64
Bedding-parallel (this study)	3	1.34	0.74	2.13
Bedding-perpendicular (this study)	6	1.37	0.74	2.64
Bedding-perpendicular (this study)	7	1.37	0.77	2.70
Bedding-parallel (Brown and Scholz, 1985)	–	1.30	–	–
Bedding perpendicular (Brown and Scholz, 1985)	–	1.32	–	–
Bedding perpendicular (Brown and Scholz, 1985)	–	1.34	–	–

not adequately describe the entire surface. Therefore, a quantitative comparison of fracture surfaces requires an explicit statement of the variation of the fractal dimension with spatial frequency for similar frequency bands (Brown and Scholz, 1985). Two frequency ranges were selected in this study: 10^{-3}–10^{-2} μm^{-1} and 10^{-2}–10^{-1} μm^{-1}.

The fractal dimension must lie in the range $1 < D < 2$ in order to produce reliable results. Spectral slopes falling outside this range were observed at spatial frequencies of 10^{-2}–10^{-1}, corresponding to the grain size of the siltstone ($\sim$10–100 μm). A similar result was described by Brown and Scholz (1985), and they attributed this result to the profilometer traversing smooth crystal cleavage surfaces which at this scale may not be fractal.

6. Discussion

6.1. Fracture toughness versus surface morphology

The bedding-plane anisotropy was found to have an effect on the measured fracture toughness. The fracture toughness of the Ithaca siltstone is approximately 35% higher for cracks propagated perpendicular to bedding than for cracks parallel to bedding (Table 7). Previous investigations have found similar results (Hoagland et al., 1973; Schmidt, 1976, 1977; Costin, 1981) and have interpreted this to be the result of a 'bedding plane weakness.' This qualitative statement, however, does not properly describe how crack propagation is affected by the orientation of the bedding-plane anisotropy. The authors' working hypothesis is that since more energy is absorbed during inelastic processes, the toughness of the bedding-perpendicular cracks is greater owing to a larger amount of microcracking.

The calculated fractal dimensions for the laboratory-produced cracks in the Ithaca siltstone are given in Table 8 along with Brown and Scholz's (1985) results for natural joint and bedding planes in the same rock. Two observations can be made: (1) the bedding-perpendicular joints and cracks have a larger fractal dimension than the bedding-parallel planes and cracks within each study, respectively; and (2) fracture toughness increases steadily as the fractal dimension increases (Figure 10), a relation similar to that exhibited by ceramic materials (Mecholsky and Passoja, 1985; Mecholsky et al., 1988). The toughness–roughness relationship of the Ithaca siltstone and the larger fracture toughness for the bedding-perpendicular crack orientation is explained in the following manner.

Microscopic examination of the crack surfaces showed that crack propagation is predominantly intergranular with some transgranular fracture (Figures 11a and 11b). This is attributed to the quartz grains having a larger Young's modulus than the clay matrix material, making them behave as hard-particle inclusions in the relatively soft clay matrix. Elastic constants of clay minerals are not available, but a comparison of quartzites and shales is sufficient (Clark, 1966).

Owing to the oblate shape of the quartz grains, the relative irregularity of the crack path (surface roughness) will depend on the crack orientation. For the case of a bedding-perpendicular crack, crack deflection and microcracking will be large as the crack is forced to propagate around the long dimension of the flattened quartz grains. This will produce a crack-surface topography which is highly irregular. Conversely, for crack propagation parallel to bedding, the crack is now deflected around the short

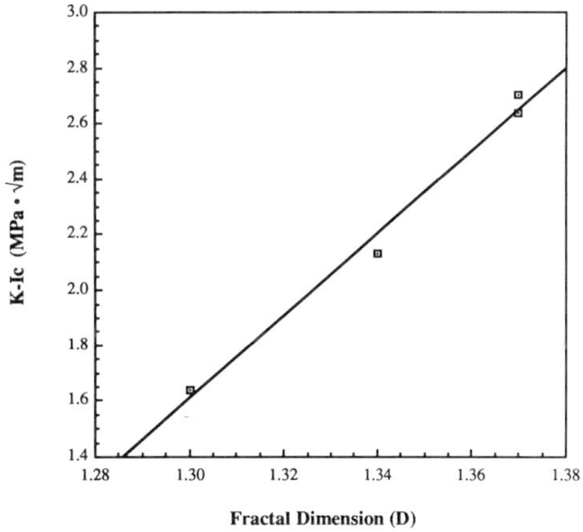

Figure 10. Fracture toughness versus fractal dimension for the Ithaca Formation siltstone. The solid line is a linear least-squares fit with a slope of 14.9 and a correlation coefficient R of 0.99.

dimension of the oblate quartz grains, which produces a relatively smooth surface topography in comparison to the bedding-perpendicular crack (Figure 12).

An irregular crack path due to crack deflection (Faber and Evans, 1983) and microcracking requires more energy for crack propagation, and will increase a material's fracture toughness. Also, the roughness of the surface topography is most likely a direct measure of the irregularity of the crack path. Therefore, we would expect the bedding-perpendicular cracks to have a larger fracture toughness and rougher fracture surface topography (larger fractal dimension) than the bedding-parallel cracks, owing to crack deflection and a greater proportion of microcracking. This relationship is demonstrated in Figure 10.

6.2. Fracture toughness measurements

The measured fracture toughness values are consistent and reproducible for a specific test method and sample size (crack length) within a single crack orientation. However, there are large toughness variations when comparing the results of one test method or sample size against another.

The crack lengths used in the fracture toughness evaluation were calculated for each specimen (Tables 4, 5, and 6). Theoretically, for the chevron-notch tests, the crack front is approximately $0.34B$ from the chevron tip when the crack reaches the critical length a_c. For the indentation/strength tests, the depth d and length $2c$ of the Vickers-induced radial crack were measured under a microscope after the experiment was run (Figure 7a). The flaw size f, equivalent to the critical crack length in the previous methods, is given by

$$f = \sqrt{dc} \tag{17}$$

The ASTM crack-length criterion requires that the specimen dimensions and crack length be at least 25 times larger than the radius of the inelastic zone, or greater than $2.5(K_{Ic}/\sigma_t)^2$. Published values for the tensile strength σ_t of siltstones are inherently low

(a)

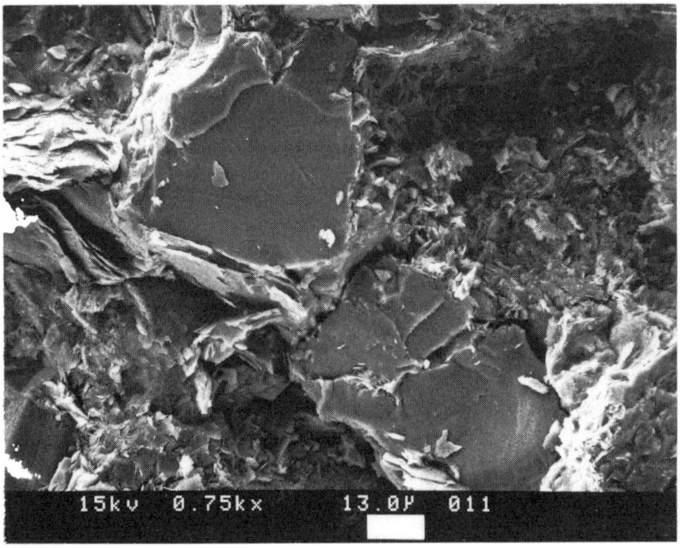

(b)

Figure 11. (a) Scanning electron microscope photograph of a bedding-perpendicular fracture surface in the Ithaca Formation siltstone. Crack propagation is predominantly intergranular (i.e., through the clay mineral matrix and around quartz grains) as indicated by the arrows pointing out holes where plucking of grains has occurred. However, there is some transgranular fracture, indicated by the smooth surfaces of the quartz grains. (b) Scanning electron microscope photograph of the boxed area in Figure 11a. The two quartz grains in the center of the photograph show fracture markings typical of transgranular fracture. However, the grains are surrounded by holes (upper right, upper left, and lower left) where the crack propagated around quartz grains.

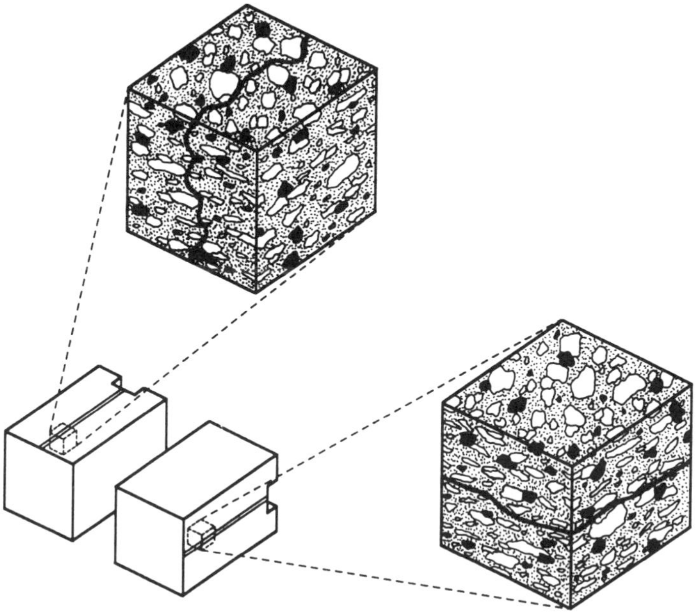

Figure 12. Schematic of the nature of crack propagation for bedding-perpendicular and bedding-parallel fractures. The white grains represent quartz, the black grains are heavy minerals, and the dotted material is the clay mineral matrix. The magnified cubes have side dimensions equivalent to 300 μm. Microscopic examination of the fracture surfaces showed that the crack propagates predominantly through the clay mineral matrix and around quartz grains. For the case of a bedding-perpendicular fracture, crack deflection and microcracking are large as the crack is forced to propagate around the long dimension of the flattened quartz grains. Conversely, for crack propagation parallel to bedding the crack is now deflected around the short dimension of the oblate quartz grains. Consequently, it is expected that the bedding-perpendicular fracture surfaces will have a more jagged topography than the bedding-parallel surfaces.

owing to the presence of natural flaws, and these values give microcrack zones with artificially large radii. This problem is avoided by assuming that the radius of the microcrack zone is no greater than two times the grain size of the siltstone, or 0.2 mm. This assumption is based on observations of microcracks on the crack surface. Therefore, the ASTM criterion requires the crack length to be greater than 5.0 mm. The crack lengths of indentation/strength tests (Table 6) are slightly smaller than required by the ASTM criterion and suggest that the measurements may be invalid. All other test methods used crack lengths larger than 5.0 mm.

The measured fracture toughnesses for the bedding-parallel and bedding-perpendicular orientations are plotted against the critical crack lengths in Figure 13. These results show that fracture toughness generally increases with crack length for the Ithaca siltstone. Similar results were discovered in previous investigations and in each case the fracture toughness was found to level off for long cracks (Ingraffea and Schmidt, 1978; Schmidt and Lutz, 1979; Weisinger et al., 1980; Ingraffea et al., 1984; Peck, et al., 1985; Kobayashi et al., 1986). The data in Figure 13 are also interpreted to gradually level off for large crack lengths.

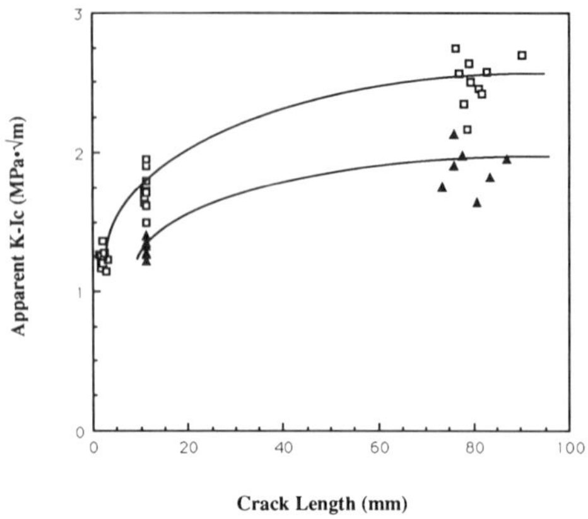

Figure 13. Apparent fracture toughness versus crack length for the two principal crack geometries. The curves are hand-drawn interpretations of the leveling off of data for long crack lengths. Triangles, bedding parallel; squares, bedding-perpendicular.

There are at least two explanations for the crack-length effect on fracture toughness. Costin (1981) proposed that the increase in toughness with crack length is due to the development of the crack-tip inelastic zone. He hypothesized that the inelastic zone is initially very small, and increases in size until it reaches a steady-state value. As the inelastic zone increases, it absorbs more energy, and therefore the toughness increases in connection with the growth of the crack. A second explanation is that the specimen dimensions and crack length of the smaller samples are too small in comparison to the inelastic zone. Consequently, the LEFM assumptions of the Westergaard and Irwin (Irwin, 1958) stress analyses are invalidated and the toughness values are incorrect. However, this study finds that even for samples satisfying the ASTM criterion, toughness increases with increasing crack length. Therefore, a third alternative is that K_{Ic} is not a material constant, and that some materials can show an increase in fracture toughness with increasing crack length. This behavior is called R-curve behavior (Evans and Faber, 1984; Shannon and Munz, 1984; Cook, et al., 1987).

Recommendations are made for future investigations concerned with measuring the fracture toughness of rock. Both test methods yield consistent results for a single sample size, as the low deviations in this study attest. However, little confidence is placed in fracture toughness measurements using only one sample size. In order to make best use of the labor-intensive samples, it is recommended that emphasis be placed on conducting tests over a relatively wide range of sample sizes (i.e., crack lengths), with just a few tests at each size, rather than conducting a large number of tests at each sample size over a small sample size range. Considering the maximum crack length (83 mm) reached in this study and those in previous studies (Ingraffea and Schmidt, 1978; Schmidt and Lutz, 1979; Weisinger et al., 1980; Ingraffea et al., 1984; Peck et al., 1985; Kobayashi et al., 1986), tests should be conducted for at least three sample sizes with a range of crack lengths spanning 30–100 mm. It is expected that the fracture toughness values over this range of crack length will begin to level off and reach an upper limit, and therefore describe the complete fracture toughness–crack length relationship. If a standardized fracture toughness value is required, then the upper limit of crack length where K_{Ic} levels off is best suited. However, from a practical view, a constant, large crack size value of K_{Ic} may be too simplified, since some fractures will propagate from small cracks.

6.3. Effect of anisotropy in K_{Ic} on initial joint propagation

The anisotropy in fracture toughness within the Ithaca siltstones favors propagation of joints in the horizontal plane provided the rock is subject to a state of lithostatic stress (i.e., horizontal and vertical stresses are equal). Indeed, if the simplest linear elastic fracture mechanics equation is applied assuming initial penny-shaped flaws of 1 cm in diameter and assuming a lithostatic state of stress, then initiation of vertical joints (using $K_{Ic} = 2.67 \pm 0.06$ MPa $\sqrt{m}$ vs. 1.93 ± 0.18 MPa $\sqrt{m}$) requires an internal fluid pressure within the flaw of about 7 MPa larger than required for horizontal joints (Engelder and Lacazette, 1990). With this anisotropy in mind, the presence of vertical joints in the Ithaca siltstone suggests that the total least horizontal stress was less than the total vertical stress. The fracture toughness anisotropy in the Ithaca siltstone means that the propagation of vertical joints is favored as long as the total vertical stress is more than 7 MPa above the total least horizontal stress. Vertical joint initiation from flaws 3 cm in diameter requires a total stress difference on the order of 3 MPa (Engelder and Lacazette, 1990).

7. Conclusions

Fracture toughness tests were conducted on the Ithaca Formation, a siltstone, to investigate the effects of planar anisotropy on fracture toughness and crack-surface topography. The chevron-notched short-bar test was conducted using three different diameters (109, 98, and 12.7 mm). The average fracture toughness results are: 2.67 ± 0.06 MPa $\sqrt{m}$ and 1.93 ± 0.18 MPa $\sqrt{m}$ at 109 mm, 2.41 ± 0.13 MPa $\sqrt{m}$ and 1.82 ± 0.06 MPa $\sqrt{m}$ at 98 mm, and 1.74 ± 0.11 MPa $\sqrt{m}$ and 1.35 ± 0.12 MPa $\sqrt{m}$ at 12.7 mm, for the bedding-perpendicular and bedding-parallel orientations, respectively. The indentation/strength tests were conducted for the bedding-perpendicular orientation only, and produced an average K_{Ic} value of 1.26 ± 0.06 MPa $\sqrt{m}$.

Planar anisotropies have a pronounced effect on the fracture toughness of rock. For each sample size, fracture toughness is 35% higher for bedding-perpendicular crack propagation as opposed to bedding-parallel crack propagation. This contrast is proposed to be due to crack deflection around quartz grains and a greater proportion of microcracking in the bedding-perpendicular fracture, which invariably creates a more jagged fracture surface. The fact that fractal dimensions of the fracture surfaces increase with fracture toughness supports this idea.

Both test methods produced consistent fracture toughness values for a single sample size. However, the toughness was found to increase with sample size (i.e., crack length), and therefore little confidence is placed on a fracture toughness study utilizing a single sample size. It is recommended that future investigators conduct toughness tests for at least three different sample sizes with a relatively large range in crack length ($\sim$30–100 mm). Our data can define the effect of crack length on fracture toughness, and this fracture toughness–crack length relationship is important when comparing results between different materials and/or studies.

Acknowledgments

The authors thank R. Queeney for his advice on fracture mechanics, D. Heitzenrater for his assistance in the indentation/strength testing and scanning electron microscopy, M. Gross for his assistance in the work of sample collection, and finally, G. Boitnott, T. Koczynski (Lamont–Doherty Geological Observatory), and T. Mackin for their help with the fractal analysis. R. Cornelius reviewed an early version of this manuscript. Aspects of this work were supported by GRI contract 5088-260-1746 to T.E.

References

Anstis, G.R., Chantikul, P., Lawn, B.R., and Marshall, D.B. (1981). A critical evaluation of indentation techniques for measuring fracture toughness: I, direct crack measurements. *J. Am. Ceram. Soc.* **64**, 533–538.

ASTM E399-83 (1985). Standard test method for plane-strain fracture toughness of metallic materials. In *Annual Book of ASTM Standards*, vol. 3.01, pp. 547–582. American Society for Testing and Materials, Philadelphia.

Atkinson, B.K. (1984). Subcritical crack growth in geological materials. *J. Geophys. Res.* **89**, 4077–4114.

Bahat, D. and Engelder, T. (1984). Surface morphology on cross-fold joints of the Appalachian Plateau, New York and Pennsylvania. *Tectonophysics* **104**, 299–313.

Barker, L.M. (1977). A simplified method for measuring plane-strain fracture toughness. *Eng. Fract. Mech.* **9**, 361–369.

Barker, L.M. (1979). Theory for determining K_{Ic} from small, non-LEFM specimens, supported by experiments on aluminum. *Int. J. Fract. Mech.* **15**, 515–536.

Barker, L.M. (1984). Specimen size effects in short-rod fracture toughness measurements. In *Chevron-Notched Specimens: Testing and Stress Analysis*, ASTM STP 855, pp. 117–133. American Society for Testing and Materials, Philadelphia.

Berry. M.V. and Lewis, Z.V. (1980). On the Weierstress–Mandelbrot fractal function. *Proc. R. Soc. London, Sec. A* **370**, 459–484.

Brown, S.R. and Scholz, C.H. (1985). Broad bandwidth study of the topography of natural rock surfaces. *J. Geophys. Res.* **90**, 12575–12582.

Chantikul, P., Anstis, G.R., Lawn, B.R., and Marshall, D.B. (1981). A critical evaluation of indentation techniques for measuring fracture toughness: II, strength method. *J. Am. Ceram. Soc.* **64**, 539–543.

Clark, S.P. (ed.) (1966). *Handbook of Physical Constants* GSA Memoir 97. Geological Society of America, New York.

Cook, R.F., Fairbanks, C.J., Lawn, B.R., and Mai, Y.W. (1987). Crack resistance by interfacial bridging: Its role in determining strength characteristics. *J. Mater. Res.* **2**, 345–356.

Cook, R.F., Lawn, B.R., Dabbs, T.P., and Chantikul, P. (1981). Effects of machining damage on the strength of a glass-ceramic. *J. Am. Ceram. Soc.* **64**, C121–C122.

Cook, R.F., Lawn, B.R., and Anstis, G.R. (1982). Fatigue analysis of brittle materials using indentation flaws: II. *J. Mater. Sci.* **17**, 1108–1116.

Costin, L.S. (1981). Static and dynamic fracture behavior of oil shale. In *Fracture Mechanics for Ceramics, Rocks, and Concrete*, ASTM STP 745, pp. 169–184. American Society for Testing and Materials, Philadelphia.

Delaney, P.T., Pollard, D.D., Ziony, J.I., and McKee, E.H. (1986). Field relations between dikes and joints: Emplacement processes and paleostress analysis. *J. Geophys. Res.* **91**, 4920–4938.

Engelder, T. and Geiser, P. (1979). The relationship between pencil cleavage and lateral shortening within the Devonian section of the Appalachian Plateau, New York. *Geology* **7**, 460–464.

Engelder, T. and Lacazette, A. (1990). Natural hydraulic fracturing. In *Rock Joints* (ed. N. Barton, and O. Stephansson), pp. 35–44. A.A. Balkema, Rotterdam.

Engelder, T. and Oertel, G. (1985). Correlation between abnormal pore pressure and tectonic jointing in the Devonian Catskill Delta. *Geology* **13**, 863–866.

Engelder, T., Geiser, P., and Bahat, D. (1987). Alleghanian deformation within shales and siltstones of the Upper Devonian Appalachian Basin, Finger Lakes District, New York. In *Geological Society of America Centennial Field Guide – Northeastern Section*, pp. 113–118. Geological Society of America, New York.

Evans, A.G. and Faber, K.T. (1984). Crack-growth resistance of microcracking brittle materials. *J. Am. Ceram. Soc.* **67**, 255–260.

Faber, K.T. and Evans, A.G. (1983). Crack deflection processes – I. Theory. *Acta Metall.* **31**, 565–576.

Fractometer I Systems 4201, 4201-C, and 4202 Owner's Manual. (1984). Terra tek Systems, Salt Lake City, Utah.

Friedman, M., Handin, H., and Alani, G. (1972). Fracture-surface energy of rocks. *Int. J. Rock Mech. Min. Science* **9**, 757–766.

Geiser, P. and Engelder, T. (1983). The distribution of layer parallel shortening fabrics in the Appalachian foreland of New York and Pennsylvania: Evidence for two non-coaxial phases of the Alleghanian orogeny. In *Contributions to the Tectonics and Geophysics of Mountain Chains*, GSA Memoir 158, p. 161–175. Geological Society of America, New York.

Hoagland, R.G., Hahn, R.G., and Rosenfield, A.R. (1973). Influence of microstructure on fracture propagation in rock. *Rock Mech.* **5**, 77–106.

Ingraffea, A.R., and Schmidt, R.A. (1978). Experimental verification of a fracture mechanics model for tensile strength prediction on Indiana limestone. In *Proc. 19th Symp. on Rock Mechanics*, pp. 247–253. Stateline, Nebr.

Ingraffea, A.R., Gunsallus, L.K., Beech, J.F., and Nelson, P.P. (1984). A short-rod based system for fracture toughness testing of rock. In *Chevron-notched Specimens, Testing and Stress Analysis*, ASTM STP 855, pp. 152–166. American Society for Testing and Materials, Philadelphia.

Irwin, G.R. (1958). Fracture. In *Handbuch der Physik* (ed. S. Flugge), vol. VI, pp. 55–1590. Springer-Verlag, Berlin.

Kobayashi, R., Matsuki, K., and Otsuka, N. (1986). Size effect in the fracture toughness of Ogino Tuff. *Int. J. Rock Mech. Min. Sci.* **23**, 13–18.

Lawn, B.R., Evans, A.G., and Marshall, D.B. (1980). Elastic/plastic indentation damage in ceramics: the median/radial crack system. *J. Am. Ceram. Soc.* **63**, 574–581.

Lawn, B.R. and Fuller, E.R. (1975). Equilibrium penny-like cracks in indentation fracture. *J. Mater. Sci.* **10**, 2016–2024.

Lawn, B.R. and Wilshaw, T.R. (1975). *Fracture of Brittle Solids.* Cambridge University Press.

Mandelbrot, B.B. (1977). *Fractals.* W.H. Freeman, San Francisco.

Mandelbrot, B.B. (1983). *The Fractal Geometry of Nature.* W.H. Freeman, San Francisco.

Mandelbrot, B.B., Passoja, D.E., and Paullay, A.J. (1984). Fractal character of fracture surfaces of metals. *Nature* **308**, 721–722.

Marshall, D.B. and Lawn, B.R. (1979). Residual stress effects in sharp-contact cracking: I. *J. Mater. Sci.* **14**, 2001–2012.

Marshall, D.B., Lawn, B.R., and Chantikul, P. (1979). Residual stress effects in sharp-contact cracking: II. *J. Mater. Sci.* **14**, 2225–2235.

Mecholsky, J.J. and Passoja, D.E. (1985). Fractals and brittle fracture. In *Fractal Aspects of Materials*, pp. 117–119. Materials Research Society, Pittsburgh.

Mecholsky, J.J., Mackin, T.J., and Passoja, D.E. (1986). Crack propagation in brittle materials as a fractal process. In *Fractal Aspects of Materials*, pp. 5–7. Materials Research Society, Pittsburgh.

Mecholsky, J.J., Mackin, T.J., and Passoja, D.E. (1988). Self-similar crack propagation in brittle materials. *Adv. Ceram.* **22**, 127–134.

Ouchterlony, F. (co-ordinator) (1988). International Society for Rock Mechanics Commission on Testing Methods. Suggested methods for determining the fracture toughness of rock. *Int. J. Rock Mech. Min. Sci. & Geochem. Abstr.* **25**, 71–96.

Peck, L., Nolen-Hoeksema, R.C., Barton, C.C., and Gordon, R.B. (1985). Measurement of the resistance of imperfectly elastic rock to the propagation of tensile cracks. *J. Geophys. Res.* **90**, 7827–7836.

Schmidt, R.A. (1976). Fracture toughness testing of limestone. *Exp. Mech.* **16**, 161–167.

Schmidt, R.A. (1977). Fracture mechanics of oil shale – unconfined fracture toughness, stress corrosion cracking, and tension test results. In *Proc. 18th Symp. on Rock Mechanics*, pp. 2A2-1 to 2A2-6. Keystone, Colo.

Schmidt, R.A. and Lutz, T.J. (1979). K_{Ic} and J_{Ic} of Westerly granite – effect of thickness and in-plane dimensions. In *Fracture Mechanics Applied to Brittle Materials*, ASTM STP 678, pp. 166–182. American Society for Testing and Materials, Philadelphia.

Shannon, J.L. and Munz, D.G. (1984). Specimen size and geometry effects on fracture toughness of aluminum oxide measured with short-rod and short-bar chevron-notched specimens. In *Chevron-Notched Specimens: Testing and Stress Analysis*, ASTM STP 855, pp. 270–280. American Society for Testing and Materials, Philadelphia.

Weisinger, R., Costin, L.S., and Lutz, T.J. (1980). K_{Ic} and *J*-resistance-curve measurements on Nevada tuff. *Exp. Mech.* **20**, 68–72.

Westergaard, H.M. (1939). Bearing pressures and cracks. *Trans. ASME, J. Appl. Mech.* **6**, 49–53.

Chapter 15

CT Imaging of Electrical Resistivity Measurements: Nonuniform Water Saturation Can Be a Problem

Eve S. Sprunt

Mobil Research and Development Corporation, Dallas Research Laboratory, 13777 Midway Road, Dallas, Texas 75244, U.S.A.

Abstract

In-situ hydrocarbon saturation is calculated using electrical resistivity data obtained from well logs. The equations used to calculate saturations contain adjustable parameters that are usually obtained from laboratory measurements on reservoir core samples at various partial hydrocarbon saturations. For these laboratory measurements to be useful in interpreting field data, it is important that the partial hydrocarbon saturation be uniformly distributed along the length of the core sample.

The saturation distribution within core samples was monitored using an X-ray CT scanner during desaturation with a porous plate technique. The scans showed that a saturation front formed at low capillary pressure. The time and capillary pressure required for the front to pass through the sample varied for the two samples studied and depended on whether the samples were in a water-wet or oil-wet state. In one case the saturation front became effectively stationary. If the capillary pressure was reduced to zero while the front was traversing the sample, the front stopped but did not dissipate within a period of days. This last observation is particularly important in that it indicates that sealing a sample with a nonuniform hydrocarbon saturation and allowing it to 'equilibrate' will not necessarily bring the sample to uniform saturation.

1. Introduction

Bill Brace was a leader in the effort to improve our understanding of the physical properties of rock, including electrical resistivity. Although he generally worked with low-porosity, low-permeability crystalline rocks, he investigated the validity of Archie's

FAULT MECHANICS AND TRANSPORT PROPERTIES OF ROCKS
ISBN 0-12-243780-2

empirical relationship (Archie, 1942), which is the basis for electrical log interpretation in the oil industry. Two forms of the Archie relationships are given by eqs. (1) and (2):

$$S_w^n = \left(\frac{a}{\phi^m}\right) \cdot \left(\frac{R_w}{R_t}\right) \tag{1}$$

and

$$I_R = \frac{R_t}{R_0} \tag{2}$$

where

S_w = water saturation (fraction)
n = saturation exponent
R_t = resistivity of the sample in ohm-m at any given saturation
R_0 = resistivity of the fully brine-saturated sample in ohm-m
R_w = resistivity of the brine in ohm-m
I_R = resistivity index
m = cementation exponent
ϕ = porosity as a fraction
a = porosity as a fraction
a = a constant, assumed here to be equal to 1

At a time when computers were not readily available for complex simulations, Brace used simple two- and three-dimensional networks of electrical resistors (Greenberg and Brace, 1969) to model brine-saturated rocks. He found that his analog models, more or less independently of the details of the network geometry, mimicked the porosity–resistivity relationship described by Archie for fully brine-saturated rocks. He also showed (Brace and Orange, 1968) that Archie's relationship for fully-saturated rocks applied to rocks with very low porosity (1–0.1%). Such rocks were beyond the scope of Archie's original work, because they are not of interest to the petroleum industry as hydrocarbon reservoirs.

In the oil industry, Archie's law, or variations of it (Waxman and Smits, 1968; Givens, 1987), is the basis for quantitative determination of hydrocarbon saturation from electrical logs. The cementation exponent m in Archie's law is the slope of the log–log plot of R_0/R_w versus porosity. The cementation exponent may be determined from well logs run through a water-saturated zone from the same geologic entity as the hydrocarbon zone which is being evaluated (Pickett, 1966), determined from measurements on core samples, or given an assumed value. The saturation exponent n in Archie's law is the slope of the log–log plot of resistivity index versus water saturation. The saturation exponent may either be determined from measurements on core samples or given an assumed value.

The accuracy of laboratory determinations of m and n was the subject of an industry-wide study conducted by the Society of Core Analysts (SCA), a chapter-at-large of the Society of Professional Well Log Analysts (Sprunt et al., 1988, 1990). The wide range of m and n values obtained for matched samples prompted development of guidelines for different aspects of the resistivity measurements (Lerner et al., 1990; Maerefat et al., 1990; A.E. Worthington et al., 1990; P.F. Worthington et al., 1990). These guidelines are a step toward improved accuracy of laboratory electrical measurements, but several aspects of the laboratory procedures are still controversial. Particularly problematic is the selection of a method for desaturating the core samples for the determination of n.

In laboratory measurements to determine n, the electrical resistivity of the sample must be measured at two or more saturations. An important assumption in these measurements is that the fluids saturating the core sample are uniformly distributed. Failure of this assumption can result in significant errors in the determination of n. Lyle and Mills (1989) showed that the correct value of n can be determined for ideal cores only when measurements are made on uniformly-saturated samples, and that the log–plot of resistivity index versus water saturation will exhibit curvature if the core is not uniformly saturated.

The requirement that the desaturation method produce a uniformly-saturated core is the most important consideration, but there are others. Some rocks do not obey a simple Archie relationship as they desaturate. Curvature of data on the log–log resistivity index–water saturation crossplot has been attributed to various pore geometries (Diederix, 1982; Dicker and Bemelmans, 1984; Swanson, 1985; Givens, 1987; P.F. Worthington et al., 1989). To determine whether a single value of n is appropriate for all saturations of interest, it is important to desaturate the core samples to water saturations equivalent to those found in hydrocarbon reservoirs. To obtain low water saturations, centrifuge and napkin desaturation are sometimes used. These two techniques rely on redistribution of fluids when the core is allowed to rest undisturbed for a period of time ranging from hours to days. However, Baldwin and Yamanashi (1989) used NMR images to show that fluid in centrifuged samples does not redistribute as quickly as previously assumed.

Another desirable attribute of a desaturation method is that it can rapidly alter the saturation. The centrifuge is fast; but, as shown by Baldwin and Yamanashi, it does not yield a uniformly saturated core sample. De Waal et al. (1989) developed a technique in which oil is injected at a constant rate at one end of the sample, and brine is expelled at the other end through a semipermeable membrane, which is permeable to brine but not to oil. CT scans of some of the 3-cm long samples showed that under some conditions an oil bank formed at the outflow end of the sample. The standard alternative to the de Waal et al. technique is desaturation through a semipermeable membrane at stepwise increasing constant pressure. The disadvantage of this method is that many weeks may be required to reach equilibrium at each pressure.

An alternative to waiting for equilibrium at each pressure during desaturation through a semipermeable membrane is to obtain multiple measurements at each pressure without waiting for equilibrium. Since this technique produced essentially colinear points in many water-wet test samples, it was assumed that uniform saturation existed along the length of the core sample even when the sample had not attained the equilibrium saturation for that capillary pressure. The CT-scan monitored measurements described in this paper show that this is not a good assumption for all samples.

2. Methods of Determining Core Saturation Distribution

Maerefat et al. (1990) compared different methods which can be used to determine the saturation distribution in core samples, including X-ray absorption (Oak and Ehrlich, 1985), X-ray computerized tomography (CT scanning) (Vinegar, 1986), nuclear magnetic resonance imaging (Vinegar, 1986), and microwave absorption (Brouse and Davis, 1981).

Multiple electrodes (Lewis et al., 1988) can also be used to determine whether the saturation is nonuniform but cannot determine the saturation.

In this study the average saturation was determined volumetrically, and the saturation distribution was determined by CT scanning. An aluminum pressure vessel was used instead of a stainless-steel pressure vessel so that sufficient X-rays could be transmitted through the experimental assembly. To increase the X-ray contrast between the oil and brine, a 50/50 mix of hexadecane–iododecane was used instead of light mineral oil. The iodine in the iododecane increased the X-ray attenuation even though the oil mixture, with a density of 0.944 $g\,cm^{-3}$, was less dense than the brine, with a density of 1.102 $g\,cm^{-3}$. At 25°C the viscosity of the doped oil was 3.1 centipoise.

Baseline CT scans were obtained before the desaturation procedure commenced. The average CT number was determined for each CT slice:

$$\text{CT number} = \left(\frac{\mu_{\text{sample}} - \mu_{\text{water}}}{\mu_{\text{water}}}\right) \cdot K \tag{3}$$

where μ_{sample} is the linear attenuation coefficient of the sample, μ_{water} is the linear attenuation coefficient of water, and K is a constant. The linear attenuation coefficient is a quantitative measure of X-ray attenuation of the material and is a function of the density and atomic number of the material and the energy of the X-rays. When the X-ray attenuation of the material in a CT scan slice increases because the iododecane–oil mixture displaces brine, the average CT number of the slice increases. The CT numbers were converted to approximate water saturations using baseline scans at 100% brine saturation and 100% oil saturation.

3. Sample Materials

The CT-scan monitored electrical resistivity measurements were performed on two carbonate grainstone samples from an oil reservoir (Sprunt et al., 1991). The wettability of adjacent samples was determined from the Amott–Harvey wettability index and from steady-state relative permeability tests (Anderson, 1986a). The samples were tested first in a 'fresh' mixed wettability condition and then in a cleaned, water-wet condition. The term mixed wettability is assumed (Anderson, 1986b) to mean that the oil-wet surfaces form continuous paths through the larger pores.

The 'fresh' samples were prepared for both the electrical resistivity measurements and the wettability tests by flushing with light oil followed by liquid propane. After the first resistivity test, the samples were cleaned with toluene by the distillation – extraction method (API RP40, 1960). The extraction of the samples with toluene altered the wettability to preferentially water-wet.

Both samples were examples of excellent reservoir-quality rock. Sample 1 was an oolitic grainstone with ambient porosity of 26% and air permeability of 1600 md. Sample 2 was a peloidal grainstone with ambient porosity of 29% and air permeability of 1200 md. The test samples were approximately 38 mm in diameter and 76 mm long.

This paper also includes electrical measurements which were not monitored with the CT scanner on Berea Sandstone, Bedford Limestone, and a synthetic material, porous alundum. The Berea Sandstone and Bedford Limestone were both obtained from surface quarries and were not from hydrocarbon reservoirs. Those samples were all water-wet and were 25 mm in diameter and 76 mm long.

4. Experimental Procedures

The CT-scan monitored experiments were performed on samples which were mounted in a rubber sleeve with a semipermeable membrane (a porcelain porous plate) at one end. The samples were vacuum-saturated with synthetic brine before they were inserted into the rubber sleeve. The sample–porcelain plate–rubber sleeve assembly was then inserted into the aluminum pressure vessel. The sample configuration is shown in Figure 1.

To insure complete brine saturation at the commencement of the electrical measurements, the samples were flushed with synthetic brine under a back-pressure of 250 psig (1.7 MPa). Upon completion of the flushing procedure, the back-pressure was removed, and a sample of the brine that had passed through the sample was collected. The resistivity of that brine sample was measured to determine R_w.

Next, a pipette was connected to the outlet end of the sample, and the system was allowed to equilibrate to insure that no residual pressure gradients remained from the flushing procedure. The confining pressure was then increased to 3300 psig (22.7 MPa), and the sample was again allowed to equilibrate. The amount of fluid expelled from the sample, as a result of porosity reduction due to pore compression when the confining pressure was applied, was measured in the pipette. The pipette was then re-zeroed. The fraction of the fluid expelled that was attributable to compression of the porous plate was not determined. It was assumed that for the high-porosity rocks in this study the

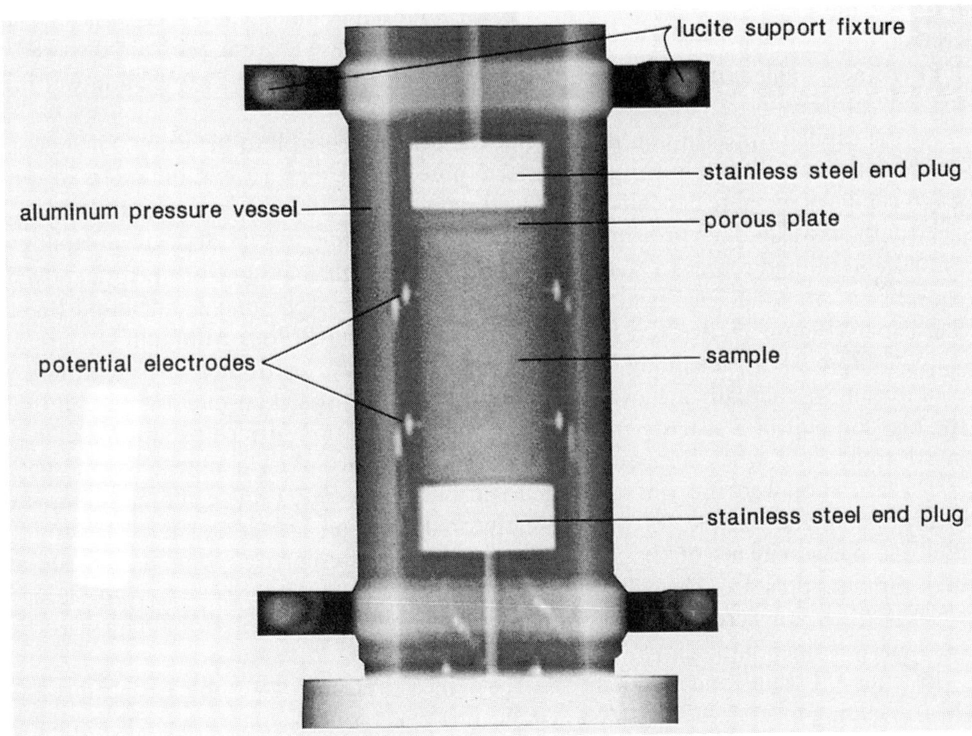

Figure 1. Digital projection X-ray from Sprunt et al. (1991) of the core sample inside the aluminum pressure vessel showing the location of the potential electrodes and the porous plate.

amount of fluid expelled from the porous plate was small in comparison to the total fluid expelled from the rock.

The sample was then ready for the first resistance measurement. The rubber sleeve into which the sample had been inserted contained two silver potential electrodes spaced 38 mm apart (Figure 1). The silver-plated stainless steel current electrodes were at both ends of the sample. The current electrode at the outflow end of the sample was separated from the sample by the porous plate. The resistance of the samples was measured at 1000 Hz using both two- and four-electrode configurations. Both two- and four-electrode configurations are commonly used in the petroleum industry (Sprunt et al., 1988, 1990).

The first resistivity measurements were obtained on the fully brine-saturated sample. Subsequent measurements were made at different partial brine saturations. The brine saturation of the sample was changed by introducing a second, immiscible liquid phase, the 50/50 mixture of hexadecane and iododecane.

In single-fluid systems the pressure in the pore spaces is commonly referred to as the pore pressure. When two immiscible fluids are present in a porous rock, a capillary pressure exists. Capillary pressure is defined in a fine-bore tube as the pressure drop across the curved liquid interface (Amyx et al., 1960),

$$P_c = \frac{(2\sigma \cos \theta)}{r} \tag{4}$$

where P_c is the capillary pressure, r is the radius of the tube, σ is the interfacial tension between the two fluids and determines which fluid will preferentially wet the solid, and θ is the contact angle which is formed when immiscible fluids contact a surface. The contact angle is measured through the denser liquid phase and ranges from 0 to 180°. If the contact angle is greater than 90°, the denser phase is considered to be non-wetting for that surface.

As the capillary pressure on the oil which is being injected into the sample increases, the oil enters smaller pore throats. The size pore throat that a given oil enters at any given capillary pressure depends on the wettability of the surface of the pore throat.

Usually in a capillary pressure measurement the sample is kept at a given capillary pressure until fluid ceases to be expelled. The sample is then assumed to be at capillary equilibrium. Previous electrical resistivity tests on water-wet samples including Berea Sandstone (Figure 2), Bedford Limestone (Figure 3), and Porous Alundum (Figure 4) were interpreted by assuming that fluid was uniformly distributed before capillary equilibrium was actually achieved. That assumption was based on the observation that the log–log plot of resistivity index versus water saturation formed an approximately linear trend even when nonequilibrium points were used. The nonlinear points near 100% saturation were thought to be due to problems associated with obtaining complete saturation of the sample. In the CT-monitored tests on the carbonate grainstones, multiple measurements of the average saturation and resistivity index were made at each capillary pressure. The CT-monitored tests indicate that nonlinear points obtained when the samples were not at capillary pressure equilibrium were probably caused by nonuniform water saturations.

Baseline CT scans and electrical resistivity measurements were obtained before the desaturation process was initiated. To begin the desaturation process, the capillary pressure was raised to 1 psig (6.9 kPa). Another set of CT scans was then obtained at the baseline positions. The average CT number of each CT scan slice was computed, and a CT number profile along the length of the sample was constructed to determine

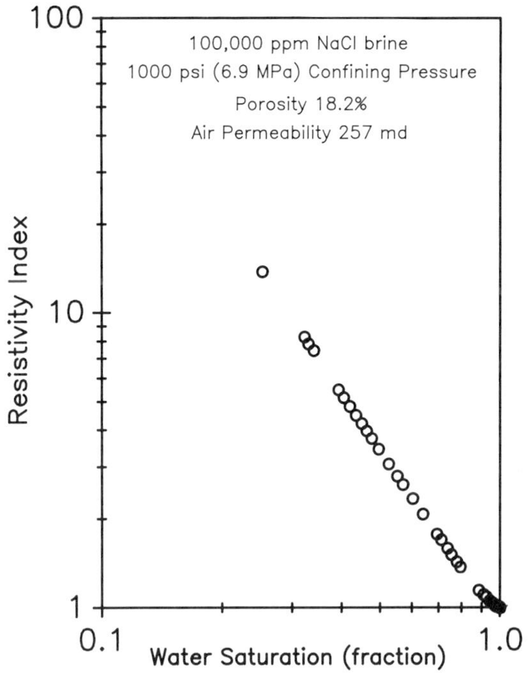

Figure 2. Resistivity index versus average water saturation crossplot for Berea Sandstone.

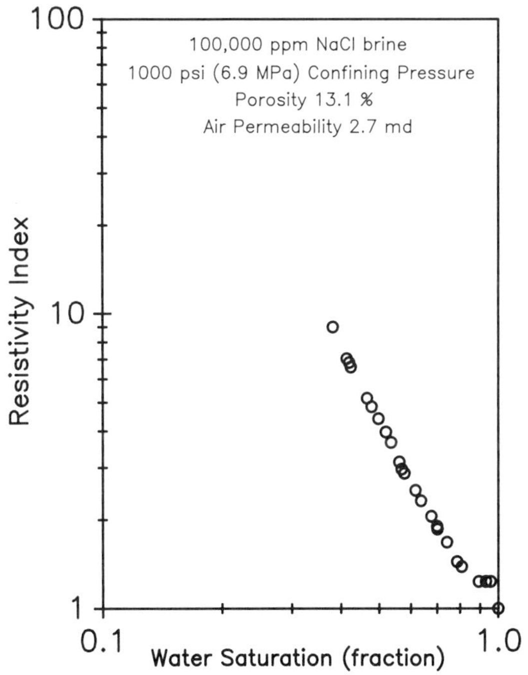

Figure 3. Resistivity index versus average water saturation for Bedford Limestone.

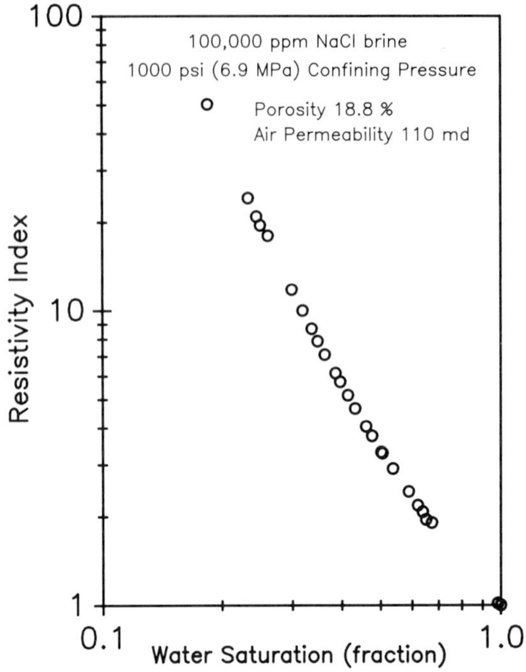

Figure 4. Resistivity index versus average water saturation crossplot for porous alundum.

whether a nonuniform fluid distribution was present. The CT scans were repeated to determine whether the fluid distribution was rapidly changing. Between series of CT scans, measurements were made of the sample resistance and corresponding volume of brine expelled from the sample. If the saturation was changing rapidly, the electrical measurements and CT scans were repeated at short time intervals. Otherwise, the electrical measurements and CT scans were repeated two or three times a day. The capillary pressure was increased when negligible brine was being expelled from the sample and when the CT number profile along the core was not changing significantly with time.

At the end of the tests on the fresh samples, distillation–extraction with toluene (API RP40, 1960) was used to cross-check the final volumetric water saturation. The toluene extraction was followed by methanol extraction to remove salt. The toluene extraction altered the wettability of the samples from mixed wettability to water-wet. The cleaned samples were resaturated with brine, and the CT-scan monitored electrical resistivity measurements were repeated.

5. Moving Front

A saturation front was observed in both carbonate grainstone samples in both the fresh and cleaned conditions. Uniform fluid distribution was not achieved until the front passed through the sample, which in these samples generally took less than 49 hours. However, in one of the two samples in the fresh condition, the front became stationary and had scarcely moved after 2 weeks. For the two samples in fresh condition, uniform

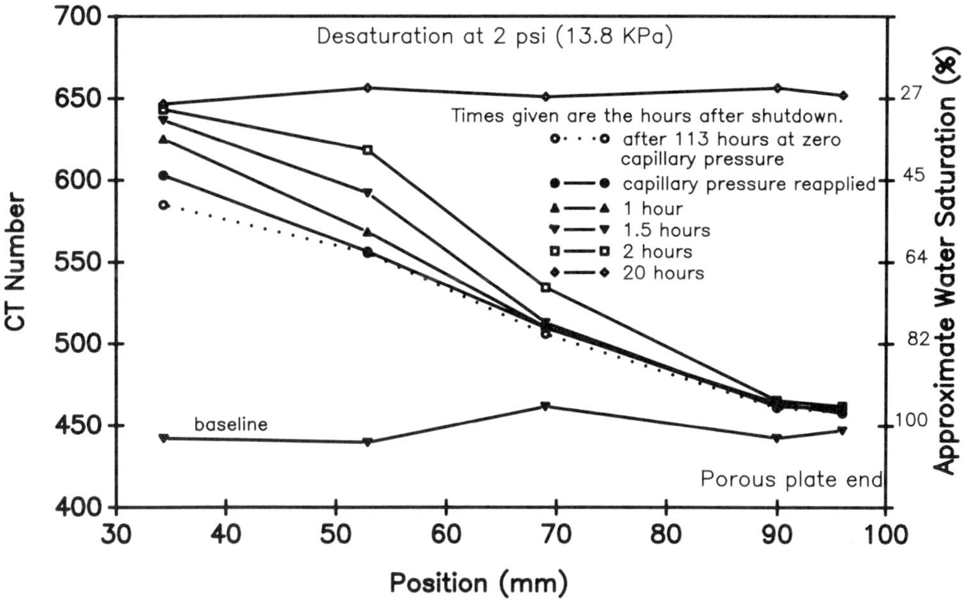

Figure 5. Crossplot from Sprunt et al. (1991) of CT number versus position along the length of core sample 1 in the fresh state, showing the progression of the oil front through the sample. Owing to experimental difficulties, the capillary pressure was reduced to zero after 1.5 hours at 2 psig (13.8 kPa) – 'shut in.' The dotted line connecting the open circles represents the saturation profile just before the capillary pressure of 2 psig (13.8 kPa) was reapplied. After 20 hours, the front passed completely through the sample. The baseline refers to the fully brine-saturated core.

fluid distribution was not obtained between 100% water saturation and around 30–40% water saturation.

In sample 1 in the fresh state (Figure 5), a moving front formed at 2 psig (13.8 kPa) oil–brine capillary pressure. Figure 6 shows an X-ray image of the sample with the front near the middle of the sample. After the front passed completely through the core sample, the average water saturation was below 30%. Little change in saturation occurred at 2 psig (13.8 kPa) after the front's passage.

When the tests were repeated on sample 1 in a cleaned, water-wet state, a moving front again formed at 2 psig (13.8 kPa) oil–brine capillary pressure. This front passed through sample 1 in about 5 hours, reducing the average water saturation to around 75% (Figure 7). In contrast with the fresh condition test, considerable fluid continued to be expelled from the core at 2 psig (13.8 kPa) by capillary desaturation after the front had passed. After 24.5 hours at 2 psig (13.8 kPa) oil–brine capillary pressure, the average water saturation was around 40%.

In sample 2, a moving front formed at 1.5 psig (10.3 kPa) oil–brine capillary pressure when the sample was in the fresh state (Figure 8), but did not form until the oil–brine capillary pressure was increased to 4 psig (27.6 kPa) after the sample was cleaned (Figure 9). In the cleaned state, a considerable amount of fluid was expelled from sample 2 at 4 psig (27.6 kPa) oil–brine capillary pressure after the front had passed through the core sample. This is similar to the observation for sample 1 in the cleaned state at 2 psig (13.8 kPa) oil–brine capillary pressure.

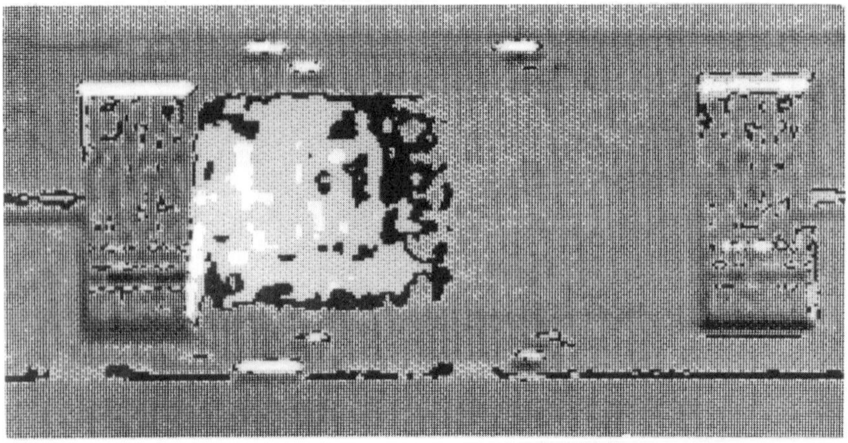

Figure 6. This image was generated by digitally subtracting two X-ray projection images. The baseline image (Figure 1), obtained before the desaturation started, was subtracted from the image obtained after 2 hours (Figure 5) of desaturation at 2 psig (13.8 kPa). 'Ghosts' of the pressure vessel and end plugs remain because the two images could not be perfectly aligned. The oil front is entering from the left. The leading edge of the front is black, with increasing oil saturation behind the front shown in lighter shades of gray.

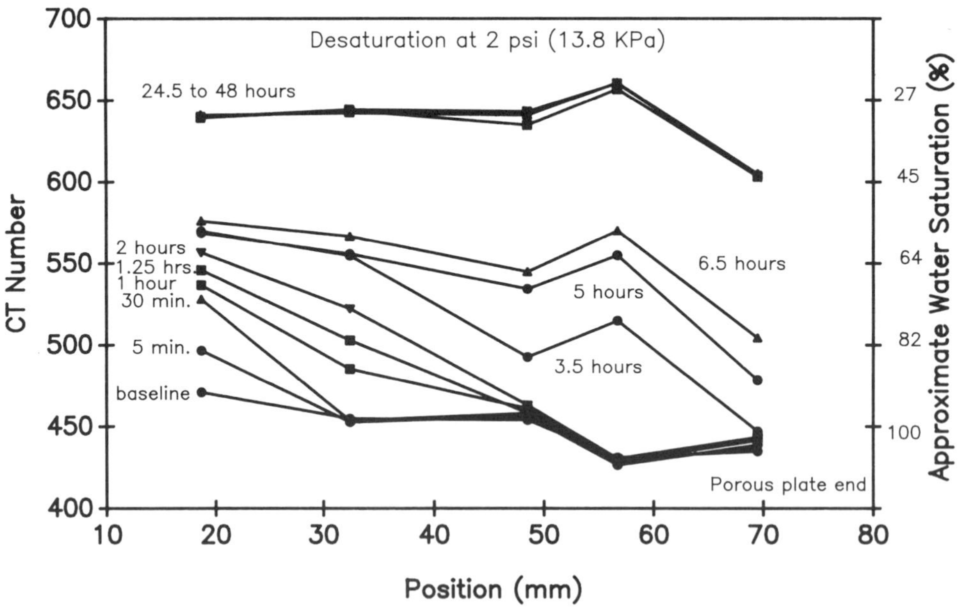

Figure 7. Crossplot from Sprunt et al. (1991) showing the fluid distribution along the length of sample 1 in the cleaned water-wet condition. The average water saturation in the core sample after 94 hours at 1 psig (6.9 kPa) was 97%. When the capillary pressure was increased to 2 psig (13.8 kPa), a front formed; but when the edge of the front reached the 48.5-mm point, the edge of the front became harder to define. The inflection at 57.4 mm is probably due to locally high porosity within the core sample.

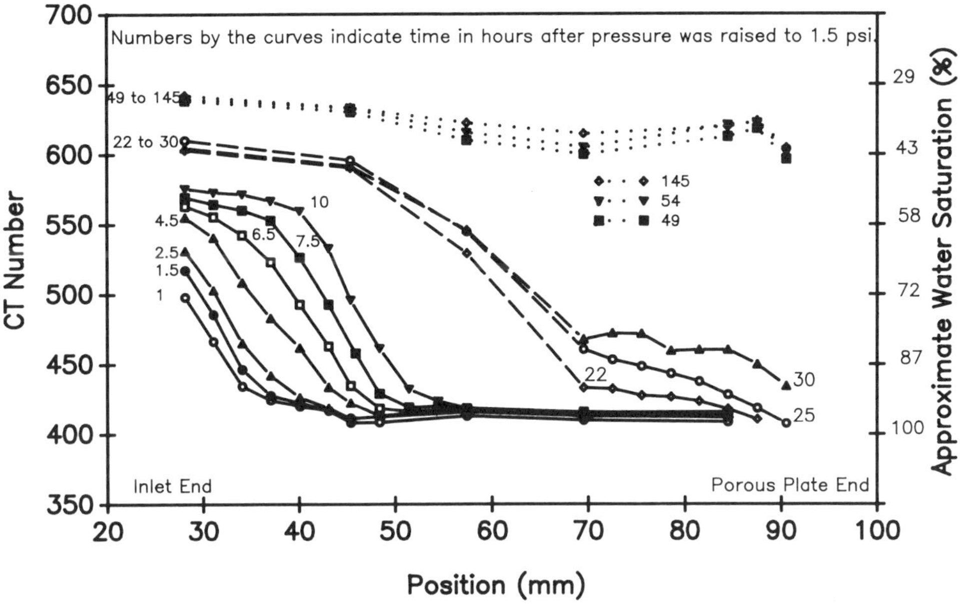

Figure 8. Crossplot from Sprunt et al. (1991) showing the fluid distribution along the length of sample 2 in the fresh mixed-wettability state at 1.5 psig (10.3 kPa).

The passage of a front manifests itself as nonlinearities in the plot of log resistivity index versus log water saturation. In this crossplot, the water saturation is the volumetrically determined, average water saturation for the core sample. The crossplot for sample 2 in a fresh state (Figure 10) illustrates these nonlinearities for the four-electrode measurement. At water saturations greater than around 85% (region A in Figure 10), the resistivity changes very little with change in water saturation. CT scans at average water saturations greater than 85% show that the front has not yet altered the water saturation between the two potential electrodes. At average saturations at which CT scans showed the front in the region between the two potential electrodes (region B in Figure 10), the resistivity at first changes more rapidly than would be expected if the volumetrically measured change in water saturation were distributed over the entire rock. Then as the front moves further along the core, the saturation between the potential electrodes changes less than the average change in saturation of the entire core sample. The log–log relationship between resistivity index and water saturation became linear (region C, Figure 10) at average saturations at which the CT scans indicated uniform water saturation along the length of the core sample. In Figure 10, only the 100% saturation point and the low-saturation data (region C in Figure 10) should be used to determine the least-squares fit line whose slope is the saturation exponent.

De Waal et al. (1989) observed pistonlike displacement in their constant-injection-rate experiments on a homogeneous clastic rock with porosity of 24.4% and permeability of 2400 md. They injected kerosene at a rate of one pore volume per day, i.e., about $0.15\ cm^3\ h^{-1}$. The fluid distribution became fairly uniformly distributed only at brine saturations less than 40%. De Waal et al. attributed the kerosene bank to the high injection rate and the large permeability contrast between the sample and the

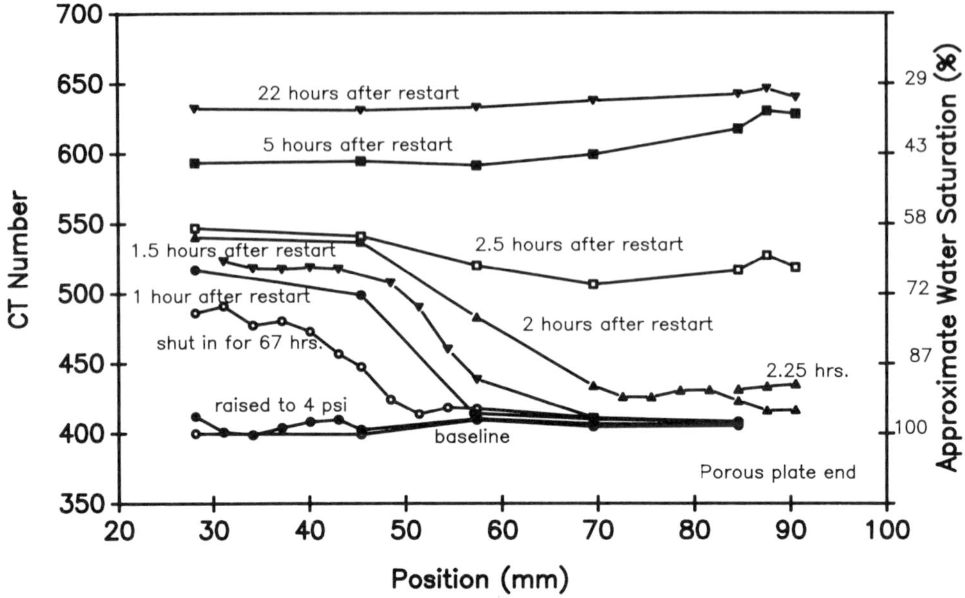

Figure 9. Crossplot from Sprunt et al. (1991) showing the fluid distribution along the length of sample 2 in the cleaned water-wet state at 4 psig (27.6 kPa). The profile marked 'raised to 4 psi' (27.6 kPa) was obtained just after the pressure was raised to 4 psig (27.6 kPa) from 2 psig (13.8 kPa). The average saturation of the sample after 23 hours at 2 psig (13.8 kPa) was 98%. After 28.5 hours at 4 psig (27.6 kPa), the capillary pressure was reduced to zero for 67 hours (shut in). The times on the other profiles are the number of hours after a capillary pressure of 4 psig (27.6 kPa) was reapplied. The front passed through the core in 2.5 hours after shut in, but capillary desaturation continued.

semipermeable membrane. In our experiments the capillary pressure, not the fluid injection rate, was controlled. An increase in oil–brine capillary pressure from 1 psig (6.9 kPa) to 1.5 psig (10.3 kPa) caused the flow rate in sample 2 in the fresh condition to increase from 0.003 $cm^3 h^{-1}$ to 0.3 $cm^3 h^{-1}$, but a front was present at both pressures. Also, the front did not form by banking behind the porous plate, which served as the semipermeable membrane in our experiments.

6. Fluid Redistribution

A front formed in sample 2 in the cleaned water-wet condition at 4 psig (27.6 kPa) but progressed through the core slowly. After 28.5 hours at 4 psig (27.6 kPa) oil–brine capillary pressure, the capillary pressure was reduced to zero. The water expulsion rate prior to the reduction in capillary pressure was 0.26 $cm^3 h^{-1}$ and was increasing with time. The sample was CT scanned immediately before and after the capillary pressure was reduced. No differences in these scans were observed. After 67 hours at zero capillary pressure, sample 2 was rescanned before an oil–brine capillary pressure of 4 psig (27.6 kPa) was reimposed. The CT number profile was the same before and after the 'shut in' (Figure 11). When the capillary pressure was reapplied, the fluid expulsion rate increased to about 2 $cm^3 h^{-1}$, and the front rapidly moved through the sample. The

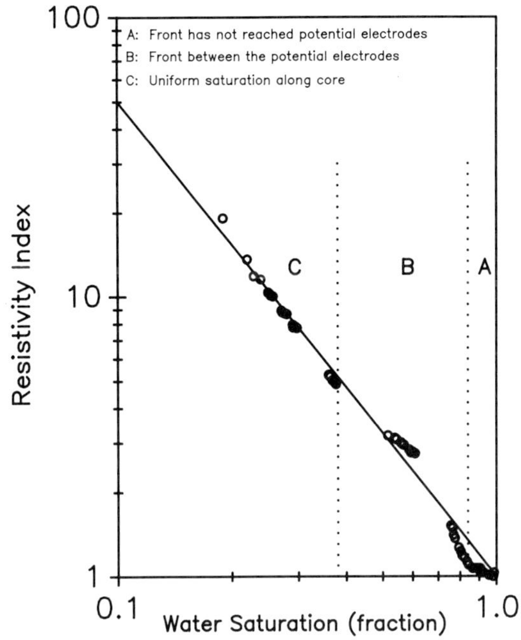

Figure 10. Resistivity index versus water saturation plot from Sprunt et al. (1991) for sample 2 in a fresh condition showing how nonuniform water saturation affects the resistivity measurements.

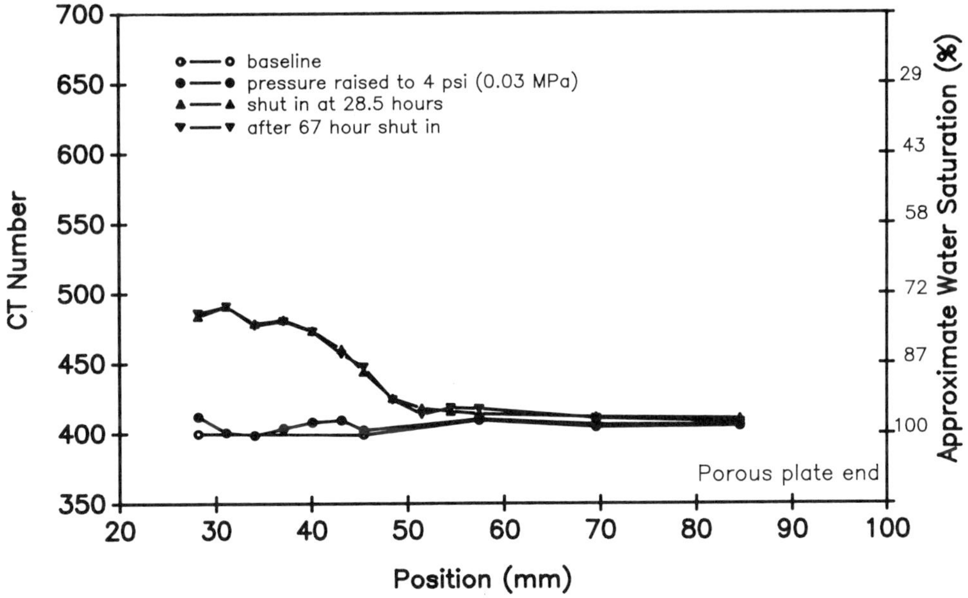

Figure 11. Crossplot from Sprunt et al. (1991) showing the saturation profile along the length of sample 2 in the cleaned water-wet condition prior to, and immediately after, shut in at 28.5 hours. The scans taken immediately before and after the reduction in capillary pressure are indistinguishable.

rapid increase in expulsion rate may have occurred because the front passed a low-permeability streak in the core. This experiment showed that without an external driving force the fluid in this sample does not redistribute.

The test using sample 1 in the fresh, mixed-wettability condition also showed that fluid does not rapidly redistribute without an external driving force. When sample 1 was being tested in the fresh condition, experimental problems developed after 1.5 hours at 2 psig (13.8 kPa) oil–brine capillary pressure. The experiment was interrupted (shut in), and the oil–brine capillary pressure was reduced to zero for a period of 113 hours. After this time, the saturation front was still sharply defined (Figure 5). The water saturation at the oil injection end of the sample was around 50%, while at the porous plate the rock was almost completely brine-saturated. The observation in this experiment is the same as that for the water-wet sample: the fluid did not redistribute without an external driving force.

7. Conclusions

The resistivity index–water saturation crossplot provides clues about the possible existence of a front if resistance measurements are obtained several times a day. In a four-electrode system, constant resistance with changing water saturation (region A, Figure 10) is a strong indication of a front, because the saturation between the potential electrodes has not changed, although the average saturation of the sample has. A secondary indication of a front is a 'hump,' which occurs because the resistance in a four-electrode system first increases faster than expected for the change in average water saturation (region B, Figure 10). The back side of the hump forms when the saturation between the potential electrodes changes less than the average change in saturation. A 'hump' followed by a linear trend should not be attributed to rock properties until the existence of nonuniform water saturation is ruled out.

Nonlinearity in the crossplot is not necessarily due to nonuniform saturation. Mismeasurement of the sample porosity will produce a monotonically increasing or decreasing slope on the resistivity index–water saturation crossplot. Figure 12 illustrates how undermeasurement of the porosity produces a downward-curving trend of the data points on the resistivity index–water saturation crossplot. In contrast, if the porosity measurement is erroneously high, the data will form an upward-curving trend. The nonlinearity due to passage of a saturation front through the sample is not the same as that due to errors in porosity measurement. The flat zone (region A, Figure 10) will not be present if the curvature is due to a porosity measurement error.

A conductive rock matrix due to immobile conductive water or conductive minerals (e.g., Givens, 1987; Diederix, 1982) causes a downward curvature of the data on the resistivity index–water saturation crossplot. The character of the nonlinearities due to a conductive rock matrix is different from that due to a front passing through the core sample.

In these CT-scan monitored experiments, measurements on fresh samples at water saturations over 40% were affected by the passage of a front with a large difference in water saturation across the front. Since the fluid was not uniformly distributed, the electrical measurements at average water saturations between 100% and 40% could not be used to determine the saturation exponent. Only the 100% water saturation point and data obtained after the front has passed through the sample (region C, Figure 10) could be used to determine the saturation exponent.

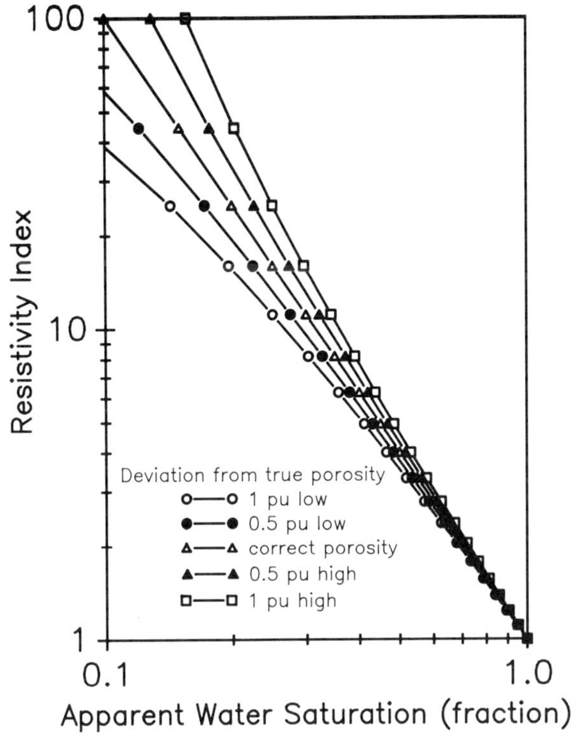

Figure 12. Effect of errors in porosity measurement on the resistivity index–water saturation crossplot; assumed $n = 2.0$, porosity 15%. When the porosity measurement is too high, the data curve upward; when the porosity measurement is too low, the data curve downward. When the correct porosity is used, the data points fall on a straight line.

The CT scans showed that fluid does not rapidly redistribute when the oil–brine capillary pressure is reduced to zero. Thus, the average water saturation points between 100% and 40% water saturation cannot be obtained by removing the capillary pressure and allowing the sample to equilibrate.

The failure of the fluid to redistribute also means that methods of desaturation such as centrifuge desaturation and napkin desaturation will not work. Baldwin and Yamanashi (1989) have already used NMR imaging to show that the nonuniform fluid distribution along the length of centrifuged samples is not eliminated when the samples are allowed to rest.

Since fluid does not rapidly redistribute, CT scanning may be used to check saturation distribution at the conclusion of the resistivity measurements. CT scanning the sample at the completion of the resistivity measurements and comparing the CT number profile with that obtained on the sample when it is clean and dry is useful in verifying uniform fluid distribution.

CT scanning is an expensive method for monitoring fluid saturation distribution. A lower-cost method of monitoring saturation distribution along the length of a core sample is X-ray absorption (Oak and Ehrlich, 1985). Multiple electrodes (Lewis et al., 1988) can be used if the monitoring method is not required to provide an independent measure of the saturation.

Nonuniform water distributions have been shown by Lyle and Mills (1989) to cause nonlinearities in the log–log crossplot of resistivity index versus water saturation. The CT-scan monitored experiments show the way in which a saturation front affects the crossplot. Given the experimental difficulties involved in obtaining uniform water saturation along the length of a core sample, fluid saturation distribution should be considered in evaluating nonlinearities in resistivity index–water saturation crossplots.

Acknowledgments

I thank Mike Davis for performing the electrical resistivity tests and Ernest Muegge for CT scanning the samples. Mary Coles provided assistance in calibrating the CT scanner and processing and interpreting the images. Discussions of this work with Kanti Desai, Sam Collins, Ben Marek, Bob Maute, Rod Schulze, Mickey Lyle, and Craig Rutley were helpful in preparing this paper. I thank Connie Terry for typing the manuscript. I thank the management of Mobil Research and Development Corporation for permission to publish this paper.

References

Amyx, J.W., Bass, D.M., and Whiting, R.L. (1960). Properties of porous media containing multiple fluid saturations. In *Petroleum Reservoir Engineering, Physical Properties*, pp. 133–210. McGraw-Hill, New York.

Anderson, W.G. (1986a). Wettability literature survey – Part 2: Wettability measurement. *J. Petrol. Technol.* **38**(11), 1246–1262.

Anderson, W.G. (1986b). Wettability literature survey – Part 1: Rock/oil/brine interations and the effects of core handling on wettability. *J. Petrol. Technol.* **38**(10), 1125–1144.

API (American Petroleum Institute) (1960). *Recommended Practice 40, Recommended Practice for Core-Analysis Procedure* (August 1960). First edition.

Archie, G.E. (1942). The electrical resistivity log as an aid in determining some reservoir characteristics. *Trans. AIME, Petrol. Branch* **146**, 54–62.

Baldwin, B.A. and Yamanashi, W.S. (1989). Persistence of nonuniform brine saturation distribution in SCA electrical resistivity core plugs after desaturation by centrifuging. *The Log Analyst* **30**, 45–48.

Brace, W.F. and Orange, A.S. (1968). Further studies of the effects of pressure on electrical resistivity of rocks. *J. Geophys. Res.* **73**(16), 5407–5420.

Brouse, D.F. and Davis, L.A. (1981). Determination of oil saturation distribution in field cores by microwave attenuation. SPE 56th Annual Fall Technical Conference, San Antonio, Texas, SPE paper 10110.

de Waal, J.A., Smits, R.M.M., de Graaf, J.D., and Schipper, B.A. (1989). Measurement and evaluation of resistivity index curves. *Transactions of the Society of Professional Well Log Analysts Symposium, Denver, Colorado.*

Dicker, A.I.M. and Bemelmans, W.A. (1984). Models for simulating electrical resistance of porous media. SPWLA 25th Annual Logging Symposium, Paper HH.

Diederix, K.M. (1982). Anomalous relationships between resistivity index and water saturations in the Rotliegend sand (The Netherlands). SPWLA 23d Annual Logging Symposium, Paper X.

Givens, W.W. (1987). A conductive rock matrix model (CRMM) for the analysis of low-contrast resistivity formulations. *The Log Analyst* **28**, 138–151.

Greenberg, R.J. and Brace, W.F. (1969). Archie's law for rocks modeled by simple networks. *J. Geophys. Res.* **74**(8), 2099–2102.

Lerner, D.B., Dacy, J.M., Raible, C.J., Rathmell, J.J., Swanson, G., and Walls, J.D. (1990). Guidelines for sample preparation and porosity measurement of electrical resistivity samples. *The Log Analyst* **31**(2), 57–63.

Lewis, M.G., Sharma, M.M., Dunlap, H.F., and Dorfman, M.H. (1988). Techniques for measuring the electrical properties of sandstone cores. 63d Annual Technical Conference and exhibition of the Society of Petroleum Engineers, Houston, Texas, SPE paper number 18178.

Lyle, W.D. and Mills, W.R. (1989). Effect of nonuniform core saturation on laboratory determination of the Archie saturation exponent. *SPE Formation Evaluation*, **4**, 49–52.

Maerefat, N.L., Baldwin, B.A., Chaves, A.A., LaTorraca, G.A., and Swanson, B.F. (1990). Guidelines for saturating/desaturation core plugs during electrical resistivity measurements. *The Log Analyst* **31**(2), 68–75.

Oak, M.J. and Ehrlich, R. (1985). A new X-ray absorption for measurement of three-phase relative permeability. SPE 60th Annual Technical Conference and Exhibition, Las Vegas, Nevada, SPE paper 14420.

Pickett, G.R. (1966). A review of current techniques for determination of water saturation from logs. *J. Petrol. Technol.* **18**(11), 1425–1433.

Sprunt, E.S., Hensel, W.M., Jr., York, C.E., and Honarpour, M.M. (1988). Compilation of electrical resistivity measurements as performed by twenty-five laboratories. *The Log Analyst* **29**(11), 13–39.

Sprunt, E.S., Maute, R.E., and Rackers, C.L. (1990). An interpretation of the SCA electrical resistivity study. *The Log Analyst* **31**(2), 76–88.

Sprunt, E.S., Desai, K.P., Coles, M.E., Davis, R.M., and Muegge, E.L. (1991). CT scan monitored electrical resistivity measurements show problems achieving homogeneous saturation. *SPE Formation Evaluation* **6**(2), 134–140.

Swanson, B.F. (1985). Microporosity in reservoir rocks: its measurement and influence on electrical resistivity. *The Log Analyst* **26**(6), 42–52.

Vinegar, H.J. (1986). X-ray CT and NMR imaging of rocks. *J. Petrol. Technol.* **38**, 257–259.

Waxman, M.H. and Smits, L.J.M. (1968). Electrical conductivities in oil bearing shaly sands. *J. Petrol. Technol.* **20**, 213–225.

Worthington, A.E., Hedges, J.H., and Pallatt, N. (1990). Guidelines for preparation of brine and determination of brine resistivity for use in electrical measurements. *The Log Analyst* **31**(1), 20–28.

Worthington, P.F., Pallatt, N., and Toussaint-Jackson, J.E. (1989). Influence of microporosity on the evaluation of hydrocarbon saturation. *SPE Formation Evaluation*, **4**(2), 203–209.

Worthington, P.F., Evans, R.J., Klein, J.D., Walls, J.D., and White, G. (1990). Guidelines for the mechanics of electrical resistivity measurement on rock samples. *The Log Analyst* **31**(2), 64–67.

Chapter 16

Fracture Detection and Characterization from Hydrophone Vertical Seismic Profiling Data

M.N. Toksöz, C.H. Cheng, and R.D. Cicerone

Earth Resources Laboratory, Department of Earth, Atmospheric, and Planetary Sciences, Massachusetts Institute of Technology, Cambridge, MA 02139, U.S.A.

Abstract

Fractures intersecting a borehole can be detected and characterized by tube waves generated by a fluid pulse injected into the borehole when a seismic wave impinges on and compresses the fracture. A vertical seismic profiling (VSP) survey, using a hydrophone array, provides the ideal data for tube wave recording and analysis. In this chapter the theory of tube wave generation in a borehole by a seismic wave is reviewed. A technique is developed to calculate the dip of a fracture from a single-offset hydrophone VSP survey from P-, SV-, and tube-wave amplitudes. In addition, the contribution to tube wave generation from particle motion parallel to the fracture plane is evaluated. On the basis of an analysis of the fluid flow regime in the fracture, it is shown that this type of particle motion causes fluid to be injected into the borehole which is several orders of magnitude less than that caused by compression of the fracture. Therefore, it is concluded that parallel-plane particle motion is negligible in terms of generating tube waves. A technique is then developed whereby the amplitude ratio of the SV-generated tube wave to the P-generated tube wave can be used to determine the dip of the fracture plane.

The complete characterization of a fracture (dip and strike of fracture plane and hydraulic conductivity) can be achieved if hydrophone VSP data are acquired from several surface source locations. This method is used to estimate fracture properties from a VSP experiment conducted at Kent Cliffs, New York and Hamilton, Massachusetts. The results are compared to fracture orientation estimates from borehole televiewer measurements and core data.

FAULT MECHANICS AND TRANSPORT PROPERTIES OF ROCKS
ISBN 0-12-243780-2

1. Introduction

The detection and characterization of naturally-occurring and artificial fractures are important considerations in many areas, including enhanced recovery projects in oil and gas development, site characterization for hazardous waste disposal, the evaluation of geothermal reservoirs, and groundwater flow studies. This has created interest in developing methods to identify fractures and to quantify their properties. The vertical seismic profiling (VSP) technique (Hardage, 1983; Balch and Lee, 1984; Toksöz and Stewart, 1984) has became a valuable diagnostic tool in evaluating formation properties in the vicinity of a borehole, including fracture identification. There has been much recent work using VSP data to identify fractures and quantify their properties (Beydoun et al., 1984, 1985; Stewart et al., 1984; Turpening, 1984; Hardin, 1986; Hardin et al., 1987; Cicerone et al., 1988; Lee, 1990).

Fractures can be detected from the tube waves in hydrophone VSP data. Huang and Hunter (1981a,b) have shown that tube waves are generated at fractures which intersect the borehole. The widely accepted mechanism for tube-wave generation (Beydoun et al., 1985) is that P-waves, incident upon a fluid-filled fracture, compress the fracture, which in turn causes a fluid pulse to be injected into the borehole. This pressure pulse causes tube waves to be generated.

Tube waves are low-frequency Stoneley waves which propagate along the borehole as guided waves (Cheng and Toksöz, 1981, 1984; Toksöz et al., 1984). The tube wave has a phase velocity which is lower than both the formation shear-wave velocity (in uncased boreholes) and the fluid velocity. The amplitude of the tube wave decays exponentially away from the borehole wall in both the fluid and the formation. Tube waves are slightly dispersive, with a velocity of about 0.9 times the fluid velocity at low frequencies and increasing slowly with increasing frequency (i.e., for last formations), and they do not undergo geometrical spreading. Tube waves can exist in all formations.

This study reviews mechanisms of tube-wave generation in a borehole intersecting open fractures by incident P- and SV-waves. Beydoun et al. (1985) develop a quantitative model of tube-wave generation for P-waves. The mechanism of tube-wave generation is compression of the fracture caused by the particle motion of a normally incident P-wave. This compression causes a finite volume of fluid to be injected into the borehole. The fluid volume is then related to the acoustic potential of the tube wave. The theory is summarized in Appendix A.

In the case of shear (SV) waves, the particle motion can have a significant component parallel to the fracture plane. A theory is developed to determine whether such particle motion contributes to tube-wave generation. This analysis of parallel plane particle motion is motivated by the observation of tube waves which are generated by SV-waves in a VSP experiment conducted at Kent Cliffs, New York (Figures 1–4). It is shown that for an SV-wave normally incident on a fracture plane this mechanism has a negligible contribution to tube-wave generation. Since both P- and SV-waves are observed to generate tube waves, and particle motion parallel to the fracture plane does not contribute to tube-wave generation, the SV-generated tube waves must be caused by the normal component of the particle motion. In the case of vertical incidence, SV-generated tube waves would therefore indicate that the fracture is dipping.

The amplitudes of both the P- and SV-generated tube waves can be used to estimate the dip of the fracture. If the incident wavenumber vector impinges on the fracture at a nonnormal angle, then both the P- and SV-wave will have a component of motion

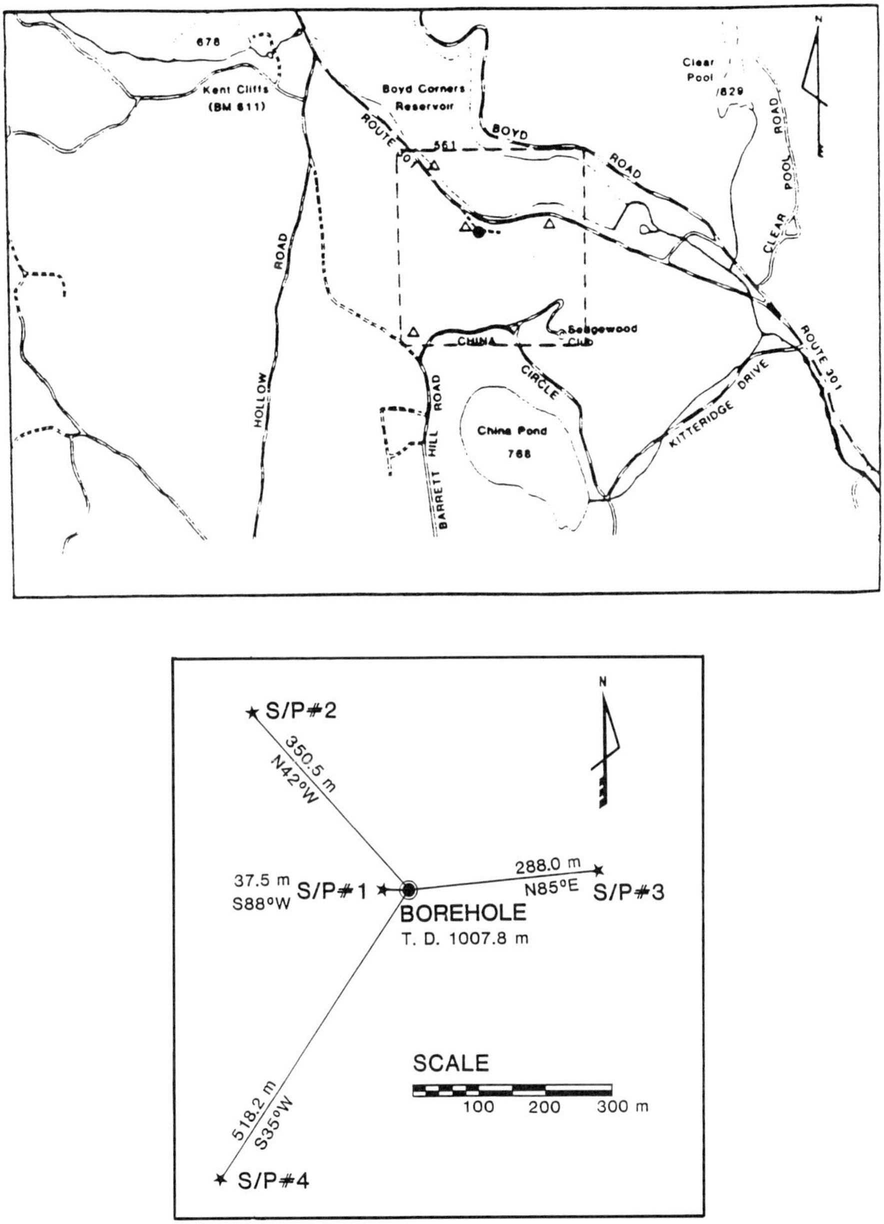

Figure 1. Location and geometry of the Kent Cliffs, New York VSP experiment.

normal to the fracture plane, and both will generate tube waves. Based on this, a method is proposed which uses P- and SV-generated tube waves to estimate the dip of the fracture plane. The advantage of this method is that it can give estimates of the dip of fractures based on information from a single-offset VSP recording. The method is applied to data recorded in a hydrophone VSP experiment recorded at Kent Cliffs, New York. The hydraulic conductivity of the fracture is determined from the frequency dependence of the amplitude ratios of tube waves to P- or SV-waves.

A more comprehensive method of fracture characterization, including the determination of the dip and strike of fracture plane and the hydraulic conductivity, requires data from two or more seismic sources at the surface. Hardin (1986) and Hardin et al. (1987) used the tube waves generated by P-waves from three or more sources located at different azimuths around the source to obtain the fracture orientations in Hamilton, Massachusetts and Mirror Lake, New Hampshire. This approach works especially well when the source distributions at the surface provide good angular coverage.

In this chapter we first discuss the mechanisms of tube-wave generation by SV-waves. Second, we describe a method for determining the dip of the fracture plane using tube waves generated by P- and SV-waves from a single surface seismic source. Third, we describe the frequency dependence of tube waves relative to incident P-waves to obtain the hydraulic conductivity. In the final section we use formal inversion techniques to obtain fracture dip, strike, and hydraulic conductivity from field experiments in Kent Cliffs, New York and Hamilton, Massachusetts.

2. Fluid Flow in a Fracture due to Parallel-Plane Motion of Fracture Waves

Field observations (Figures 2, 3, and 4) show that tube waves are generated by incident P- and SV-waves. The tube wave generation by P-waves can be explained by the fracture compressions due to the impinging wave and the resulting fluid volume injected into the borehole (Beydoun et al., 1985). The SV-waves, however, act on the fracture in two

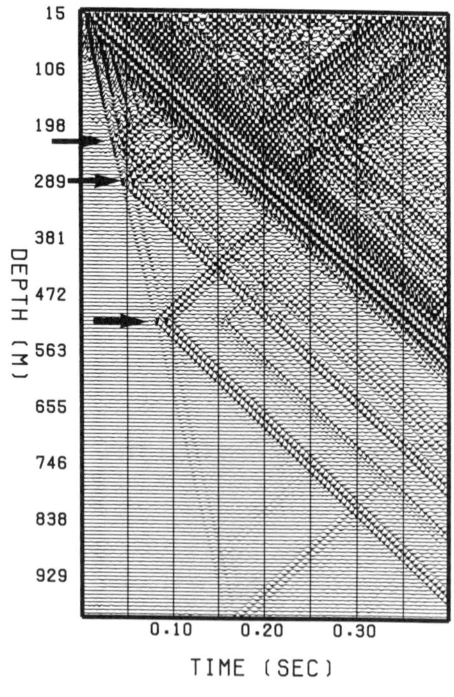

Figure 2. VSP section for shot point 1, with arrows indicating location of fractures.

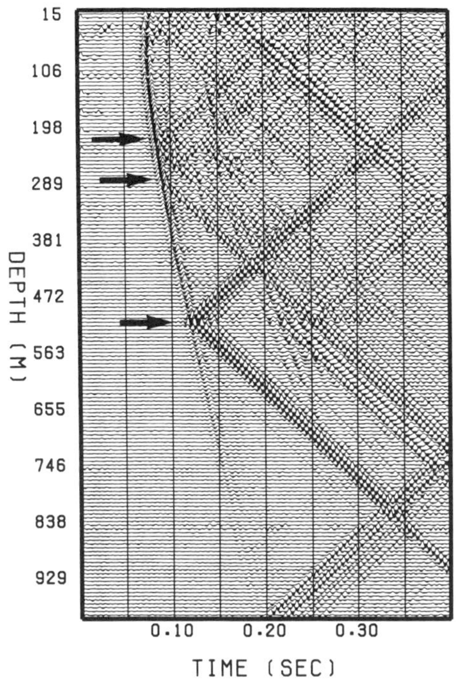

Figure 3. VSP section for shot point 2, with arrows indicating location of fractures.

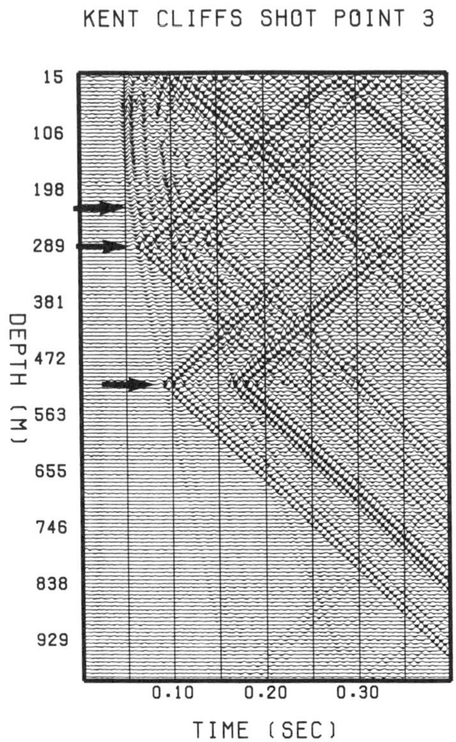

Figure 4. VSP section for shot point 3, with arrows indicating location of fractures.

ways: (1) the shear motion of the fracture walls parallel to the fracture plane, and (2) the normal component of the motion. In this section we analyze the first mechanism.

In order to examine the contribution to tube-wave generation due to fluid flow caused by particle motion parallel to the fracture plane, it is necessary to determine the velocity profile of the fluid flow in the fracture. From this, the two-dimensional volume can be obtained by integrating over the width of the fracture and over an appropriate time interval. The extreme case is considered: a vertically incident SV-wave impinging upon a horizontal, parallel plane fracture. The SV-wave displacement is assumed to have a periodic time dependence of the form $\zeta_0 \cos(\omega t + \varepsilon)$, where ζ_0 is the maximum displacement of the fracture wall.

Fluid flow is governed by the Navier–Stokes equation:

$$\rho_f\left(\frac{\partial \mathbf{v}}{\partial t} + \mathbf{v}\cdot\nabla\mathbf{v}\right) = -\nabla p + \mu\nabla^2\mathbf{v} + \frac{\mu}{3}\nabla\nabla\cdot\mathbf{v} + \rho_f g\hat{\mathbf{z}} \tag{1}$$

where $\mathbf{v}$ is the fluid velocity, ρ_f is the fluid density, p is the pressure in the fracture, μ is the viscosity, g is the gravitational acceleration, and $\hat{\mathbf{z}}$ is the unit vector in the z-direction. For a thin horizontal fracture with plane, parallel walls, we can make the following assumptions: (1) the flow regime in the fracture is purely radial; (2) nonlinear flow effects are negligible; (3) gravitational effects are negligible; (4) pressure effects are negligible (i.e., the fluid motion is entirely due to the harmonic excitation at the fracture walls); and (5) the fluid velocity varies only in the z-direction (i.e., $\mathbf{v} = v(z)\hat{\mathbf{r}}$). Therefore, the Navier–Stokes equation reduces to the simpler 1-D diffusion equation

$$\frac{\partial v}{\partial t} = \nu\frac{\partial^2 v}{\partial z^2} \tag{2}$$

where $\nu = \mu/\rho_f$ is the kinematic viscosity. The 1-D diffusion equation contains a first derivative in t and a second derivative in z; therefore, one initial condition and two boundary conditions are required for a solution. The initial condition is

$$v = 0 \text{ at } t = 0$$

and the boundary conditions are

$$v = -\omega\zeta_0 \sin(\omega t + \varepsilon) \qquad \text{at } z = -L_0/2 \text{ and } z = L_0/2$$

where L_0 is the fracture aperture. The boundary conditions were derived by taking the time derivative of the displacement at the fracture wall, with the assumption that both walls of the fracture oscillate in phase.

The solution to this problem is given by Carslaw and Jaeger (1959, p. 105):

$$v(z,t) = -\omega\zeta_0\left\{A\sin(\omega t + \phi) + 4\pi\nu\sum_{n=0}^{\infty}\frac{(-1)^n(2n+1)[4l^2\omega\cos\varepsilon - \nu(2n+1)^2\pi^2\sin\varepsilon]}{16l^4\omega^2 + \nu^2\pi^4(2n+1)^4}\right.$$
$$\left.\times\, e^{-\nu(2n+1)^2\pi^2 t/4l^2}\cos\left[\frac{(2n+1)\pi z}{2l}\right]\right\} \tag{3}$$

where

$$A = \left|\frac{\cosh[kz(1+i)]}{\cosh[kl(1+i)]}\right| = \left\{\frac{\cosh(2kz) + \cos(2kz)}{\cosh(2kl) + \cos(2kl)}\right\}^{1/2} \tag{4}$$

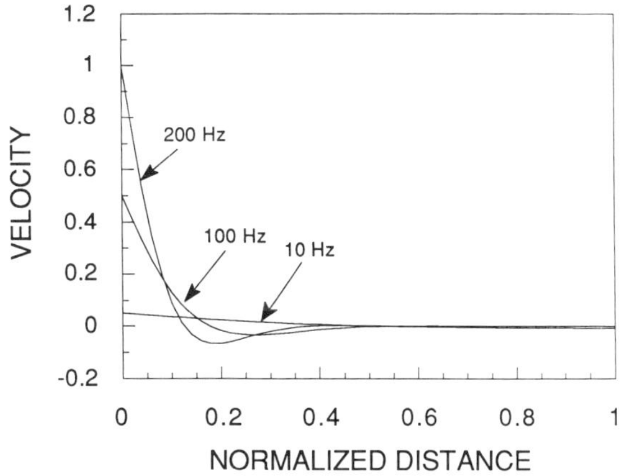

Figure 5. Normalized fluid flow velocity profile in the fracture due to parallel-plane harmonic particle motion for frequencies of 10 Hz, 100 Hz, and 200 Hz, plotted vs. the normalized distance from the fracture wall. All velocity values are normalized by the maximum value at 200 Hz. The distance from the fracture wall is normalized by half the fracture aperture, so that a distance of 0 corresponds to the fracture wall and 1 corresponds to the center of the fracture.

$$\phi = \arg\left\{\frac{\cosh[kz(1+i)]}{\cosh[kl(1+i)]}\right\} \tag{5}$$

$$k = \left(\frac{\omega}{2v}\right)^{1/2} \tag{6}$$

and $l = L_0/2$. The first term of the solution represents the periodic steady-state solution and the second term represents the transient part of the solution. The quantities A and ϕ are the amplitude and phase of the steady-state velocity field at the point z, and they are functions of the two dimensionless quantities z/l and kl.

An example of a velocity profile from this solution is shown in Figure 5. The fluid flow profile is characterized by the development of a boundary layer at the two fracture walls. The thickness of the boundary layer is given by (Schlichting, 1960)

$$\delta = \left(\frac{2v}{\omega}\right)^{1/2} \tag{7}$$

Analogous to the derivation of Beydoun et al. (1985), the 2-D volume is determined by integrating over the fracture width L_0 and over the time interval $T/2$

$$\begin{aligned} V_{2D} &= \int_{-l}^{l}\int_{0}^{T/2} v(z,t)\,dt\,dz \qquad (8)\\ &= -2\zeta_0 \int_{-l}^{l} A\cos(\varepsilon+\phi)\,dz \times \frac{64\omega\zeta_0 l^3}{\pi^2}\sum_{n=0}^{\infty}\frac{4l^2\omega\cos\varepsilon - v(2n+1)^2\pi^2\sin\varepsilon}{(2n+1)^2[16l^4\omega^2 + v^2\pi^4(2n+1)^4]}\\ &\quad \times [e^{-v(2n+1)^2\pi^2T/8l^2} - 1] \end{aligned}$$

The integral for the 2-D volume must be evaluated numerically, because A and ϕ are complex functions of z, and the integral with respect to z cannot be determined analytically.

A simpler, asymptotic formula can be derived for the 2-D volume. The problem is reformulated so that the upper boundary of the fracture is at $z = 0$. It is assumed that the fracture wall displacement ζ_0 is much smaller than the fracture width L_0. Since the kinematic viscosity for water is very small, the boundary-layer thickness will also be small relative to the fracture width. Therefore, for practical purposes, the fracture can be considered as an infinite half-space with respect to the fluid flow regime near the boundary.

Again, assuming a harmonic displacement of the form $\zeta_0 \cos(\omega t + \varepsilon)$ at the boundary, with a corresponding velocity of the form $-\omega\zeta_0 \sin(\omega t + \varepsilon)$ at the boundary, the solution for the fluid flow in the fracture is (Schlichting, 1960)

$$v(z, t) = -\omega\zeta_0 e^{-kz} \sin(\omega t + \varepsilon - kz) \tag{9}$$

The 2-D volume can be calculated analytically,

$$V_{2D} = \int_0^\infty \int_0^{T/2} v(z, t)\, dt\, dz = \frac{\zeta_0}{k} \tag{10}$$

Since the same fluid flow regime will exist at the other fracture wall, the total 2-D volume (V_{2D}^T) is obtained by simply doubling the result for the infinite half-space, therefore,

$$V_{2D} = \frac{2\zeta_0}{k} \tag{11}$$

A comparison of the exact and asymptotic solutions is shown in Figure 6, which illustrates that there is good agreement between the two solutions for typical seismic frequencies.

Now we compare the volume of fluid produced by the horizontal (i.e., shear) motion of the fracture wall with that produced by the vertical (i.e., normal) motion of the wall of the same amplitude. The problem of flow due to normal motion of the fracture wall produced by P-waves was solved by Beydoun et al. (1985) and is briefly described in Appendix A for easy reference. Using the same horizontal fracture, the same fluid parameters, and the same wave amplitudes (S_0), we calculated the fluid volumes for P- and SV-waves using equations (A-3) and (10), respectively.

The results of 2-D volume calculations for both the compression of the fracture and parallel motion of the fracture wall are shown in Figure 7. For seismic frequencies and fracture apertures of interest, the 2-D volume of fluid generated by compression of the fracture is several orders of magnitude greater than that generated by a harmonic displacement parallel to the fracture plane. Therefore, displacement parallel to the fracture plane does not significantly contribute to tube-wave generation in a borehole. In addition, the 2-D volume due to parallel motion depends only on the frequency of the excitation and the viscosity of the fluid in the fracture, and not on the fracture width, as does the 2-D volume generated by compression. It can be concluded that tube-wave generation in a borehole due to the presence of fractures is mainly due to compression of the fractures by the passage of seismic waves.

3. Fracture DIP from a Single-offset VSP

The tube wave amplitude due to the incident wave can be calculated by determining the volume of the fluid injected into the borehole, as described in Appendix A. On the

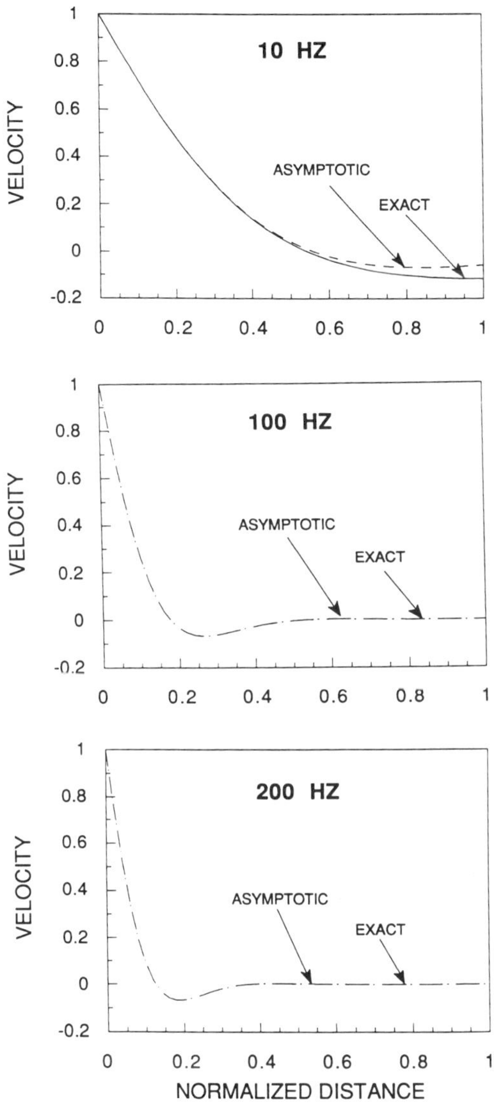

Figure 6. Comparison of the exact (dashed curve) and asymptotic (dotted curve) solutions for the fluid velocity in the fracture for frequencies of 10 Hz, 100 Hz, and 200 Hz. Velocity values for each frequency are normalized by the maximum value at that particular frequency. The normalized distance is the same as that in Figure 5.

basis of the discussions in the previous section, the fluid injection results from fracture compression in the cases of both P- and SV-waves. Given the amplitudes of tube waves generated by P- and SV-waves from a given source impinging on the same fracture, a simple method to estimate the dip of the fracture plane can be derived. This method requires tube wave amplitude data from a single offset only. It is based simply on geometric considerations.

To demonstrate intuitively how the method works, the two extreme cases are considered: a plane horizontal fracture and a plane vertical fracture. For the horizontal

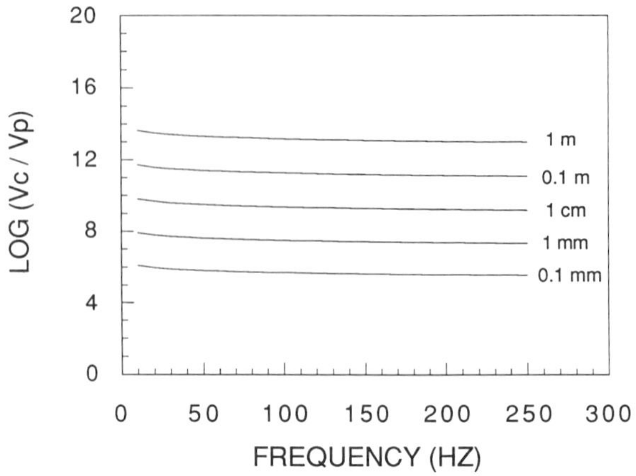

Figure 7. Logarithm of ratio of the volume due to compression to the volume due to parallel-plane harmonic excitation vs. frequency for different values of fracture aperture.

fracture, a vertically incident P-wave will have all of its particle motion normal to the fracture plane, so that all of the particle motion will compress the fracture and generate a tube wave. Conversely, a vertically incident SV-wave will have all of its particle motion parallel to the fracture plane. Since this mechanism is insignificant to tube-wave development, no tube wave will be generated. In the case of the vertical fracture, the situation is reversed. A vertically incident P-wave will have all of its particle motion parallel to the fracture plane, so that no tube wave will be generated. A vertically incident SV-wave will have its particle motion normal to the fracture, and a tube wave will be generated. If the fracture is at an angle of 45°, and the amplitudes of the incident P- and SV-waves are the same, then the tube waves generated by the P- and SV-waves should be equal, because the projection of the particle motion normal to the fracture is the same in both cases.

On the basis of this discussion, the ratio of the amplitude of the SV-wave-generated tube wave u_{Ts} (normalized by the incident SV-wave amplitude u_S) to the P-wave-generated tube wave u_{Tp}(normalized by the incident P-wave amplitude u_P) can be expressed as

$$\frac{u_{Ts}/u_S}{u_{Tp}/u_P} = \tan\theta \tag{12}$$

where θ is the angle between the vertical and the fracture normal, which is equal to the dip of the fracture. The derivation of this formula assumes that the incident waves propagate vertically, an assumption which can be removed for an incident wave propagating in a direction which is not vertical. If the incident wave propagates at an angle φ with respect to the vertical, then the tube wave amplitude ratio can be expressed as

$$\frac{u_{Ts}/u_S}{u_{Tp}/u_P} = \tan(\theta + \varphi) \tag{13}$$

where φ is measured in the updip direction of the fracture (downdip is taken as negative). This equation is valid only if the offset is in a direction perpendicular to the strike of the fracture plane. If the azimuth of the offset is in a nonperpendicular direction to the

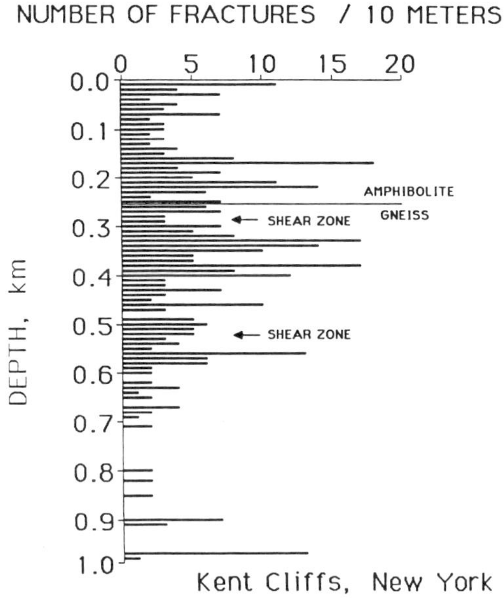

Figure 8. Number of fractures vs. depth in the well (from borehole televiewer).

strike of the fracture plane, then the calculated dip will actually be an apparent dip which will be less than the true dip.

In order to test this model of fracture dip, the technique was applied to hydrophone VSP data from an experiment conducted in the Kent Cliffs, New York area. The Kent Cliffs VSP consists of hydrophone data recorded in a 1000-meter depth borehole at a spacing of 3.28 m (10 ft) and for four offsets; one zero-offset and three offsets of varying distance from the borehole along varying azimuths (Figure 1). A geologic cross-section of the borehole, based on cuttings collected at 6.56-m (20-ft) intervals over the entire depth of the well, shows predominantly metamorphic rocks, namely amphibolite, phyllonite to a depth of about 250 m, and granitic gneiss below that. Borehole televiewer logging shows extensive fracturing throughout the entire depth of the well (Figure 8). In addition, cores were recovered from eight intervals. These cores show extensive fracturing. Three large fractures were also inferred to be present at depths of 232, 286, and 512 m, based on tube waves observed on the VSP sections (Figures 2–4). These three fractures can also be seen on the caliper log (Figure 9).

An attempt was made to apply the technique for calculating the dip to the fracture at 512 m depth. This fracture shows strong tube waves generated by both the incident P-wave and the incident SV-wave. The zero-offset section was used in order to avoid any problems associated with using an offset that may not be in a perpendicular direction to the strike of the fracture.

In order to get a relatively accurate estimate of the amplitudes of the P-, SV-, and tube waves, the VSP data were f–k filtered in order to isolate the P- and SV-waves from the tube waves. A procedure was used whereby nine traces were stacked together to enhance the signals. The stacked traces were then Fourier-transformed into the frequency domain. The ratios of the SV-generated tube wave to the P-generated tube wave were obtained by dividing the respective spectra, so that ratios were available over the usable bandwidth of the VSP data (up to about 250 Hz). The amplitude spectra for

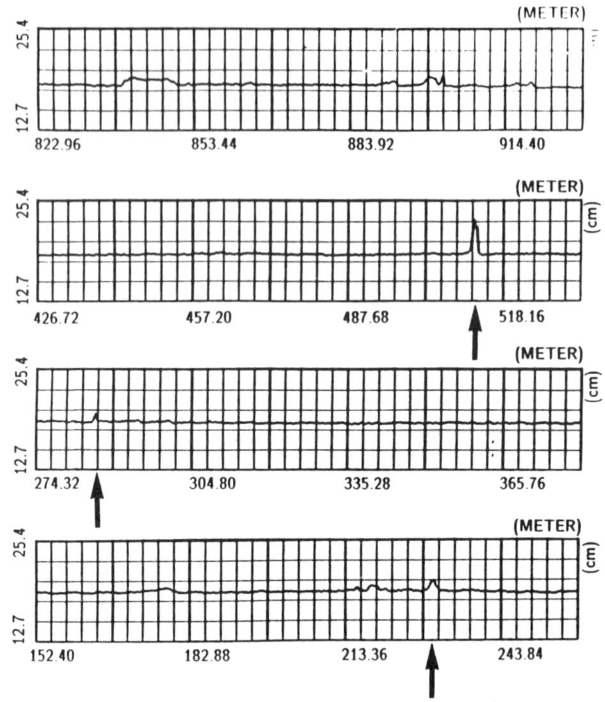

Figure 9. Caliper log for the Kent Cliffs well, with arrows indicating location of fractures.

the P-wave and tube waves are shown in Figure 10. In addition, the spatial sample rate was adequate to eliminate spatial aliasing in the bandwidth of the data. The velocities of the P- and SV-waves were obtained from full-waveform acoustic logs.

The dip of the fracture plane was determined by taking the average value of the dip angle calculations from the ratios at different frequencies. The value for the dip of the fracture calculated in this manner was about 55°, as compared to values of about 75° estimated from the borehole televiewer (Figure 11). While the data agree in a qualitative sense (i.e., that the fracture is steeply dipping), the discrepancy in the values may be attributable to the fact that the fracture geometry is more complicated than that of a simple planar feature. Hardin et al. (1987) attributed discrepancies in their calculated fracture orientations (from inversion of multioffset data) to more complex fracture geometries. Another possibility is that the borehole televiewer data for the large fracture shown in Figure 11c may be affected by the spalling of material from the fracture edges during the drilling. The thinner fracture immediately below the large fracture at Figure 11c has a dip of about 50°. Thus, we consider the dip of 55° obtained from tube wave data to be a reliable value, perhaps representing the average dip of a fracture plane without the effects of borehole spallation.

4. Estimation of Hydraulic Conductivity

There is an ambiguity in the literature regarding the term 'fracture permeability.' One definition relates to the intrinsic properties of the medium and is often referred to as

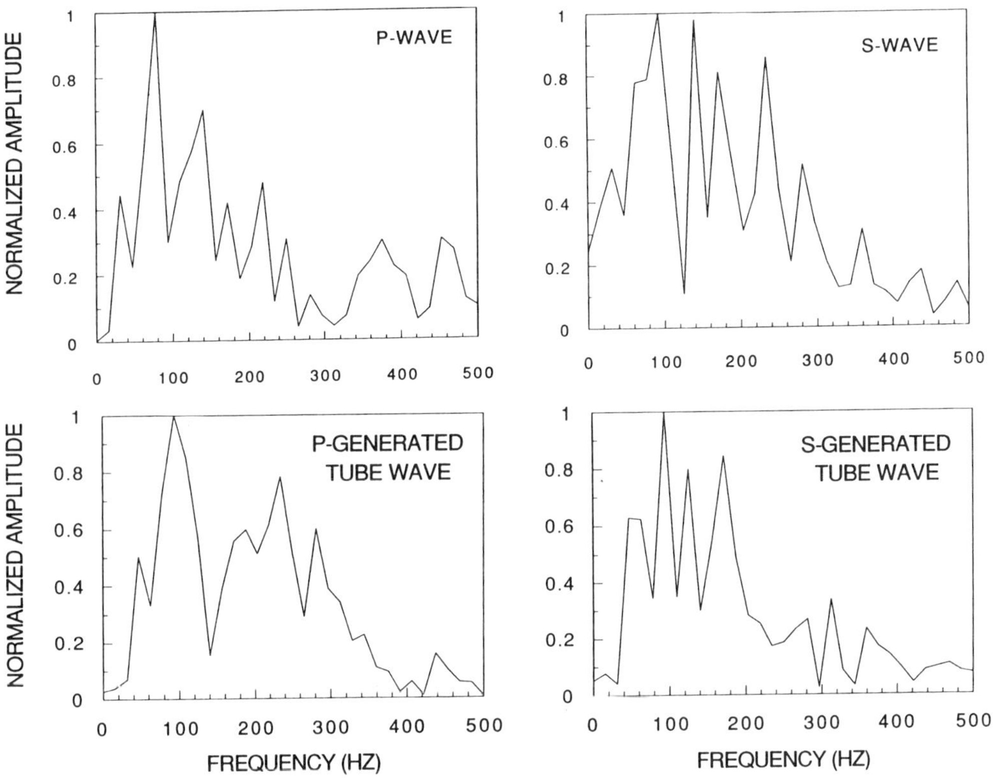

Figure 10. Amplitude spectra for P-, S-, and tube waves (both P-generated and S-generated) for the Kent Cliffs VSP data. The spectra were obtained from stacked traces in the vicinity of the fracture at 512 m depth, and each spectrum is normalized by the maximum value.

the intrinsic permeability κ, defined as (Golf-Racht, 1982, p. 157)

$$\kappa = \frac{L_0^2}{12} \tag{14}$$

where L_0 is the fracture width or fracture aperture. Another definition of permeability is related to the ability of the medium to transmit fluid, and hence is a function of the properties of both the medium and the fluid. This second definition of permeability is often also referred to as the hydraulic conductivity K and is defined by the relation (Freeze and Cherry, 1979, p. 27)

$$K = \frac{\kappa \rho_f g}{\mu} \tag{15}$$

where κ is the intrinsic permeability. In this report, the term 'hydraulic conductivity' is used, and the common unit of hydraulic conductivity is the darcy ($1 \text{ d} = 10^{-5} \text{ m s}^{-1}$).

To determine hydraulic conductivity, the tube wave amplitude is normalized by the pressure amplitude of the incident P-wave at the same receiver. The P-wave pressure response due to a displacement in the formation can be expressed, using the low-frequency

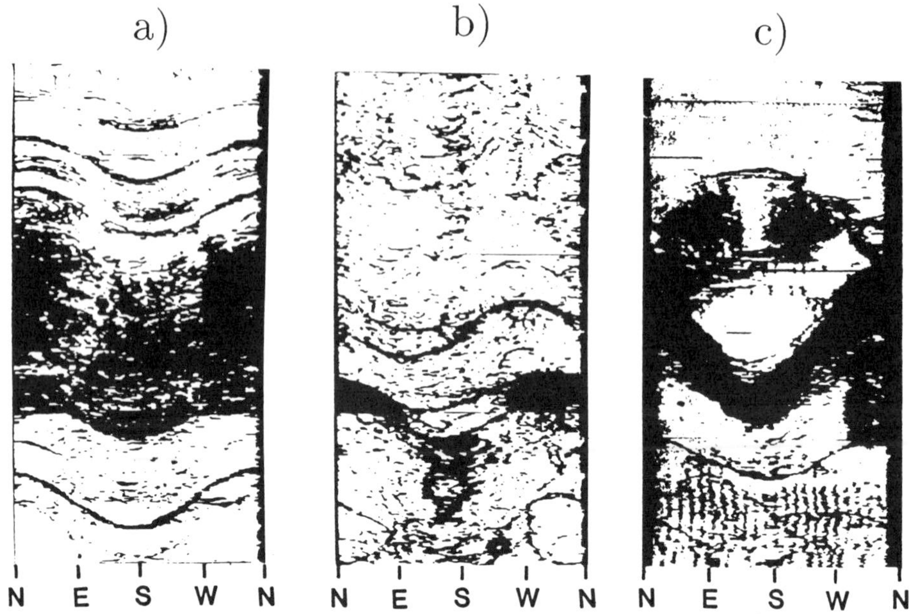

Figure 11. Borehole televiewer images for the fractures: (a) fracture at 232 m depth, strike EW, dip 50° S; (b) fracture at 286 m depth, strike EW, dip 60° SSE; (c) fracture at 512 m depth, strike EW, dip 75° S.

approximation, as (White, 1983; Beydoun et al., 1985)

$$p^{\alpha} = \zeta_0 \frac{\omega \rho_f c^2 \alpha [1 - 2(\beta \cos\theta/\alpha)^2] \cos\theta}{\beta^2 [1 - (c(\cos\theta)/\alpha)^2] \cos\varphi} \tag{16}$$

where α and β are the P- and SV-wave velocities, respectively, c is the tube-wave phase velocity, φ is the angle between the P-wave propagation vector and the fracture normal, and θ is the angle between the P-wave propagation vector and the borehole axis. The normalized tube wave pressure amplitude is then given by (Beydoun et al., 1985)

$$\frac{p^t}{p^{\alpha}} = C(L_0) \frac{\omega \beta^2 \cos\varphi [1 - (c(\cos\theta)/\alpha)^2] I_0(nR)}{\zeta_0 (\cos\theta) c^2 \alpha [1 - 2(\beta(\cos\theta)/\alpha)^2]}, \tag{17}$$

where $C(L_0)$ is given in Appendix A. The ratio of tube wave amplitude to P-wave amplitude can be determined from a single trace.

Hydraulic conductivity estimates for the three fractures in the Kent Cliffs VSP have been determined by comparing actual tube wave amplitude to P-wave amplitude ratios from the VSP data with those predicted by the theory. The data are plotted against the theoretical values for fracture dips of 50°, 60°, and 75°, which are dip estimates of the three fractures determined from the borehole televiewer data. For the fracture at 232 m depth, the data indicate that the hydraulic conductivity is in the range 0.1–0.5 darcy (Figure 12). In the case of the fracture at 286 m depth, the hydraulic conductivity lies in the range 0.5–2 darcy (Figure 13), and the data are more scattered. Finally, for the fracture at 512 m depth, the data show a large amount of scatter, and the hydraulic conductivity ranges from 1 to 10 darcy (Figure 14).

In general, there is a discrepancy between the predicted and actual amplitude ratios

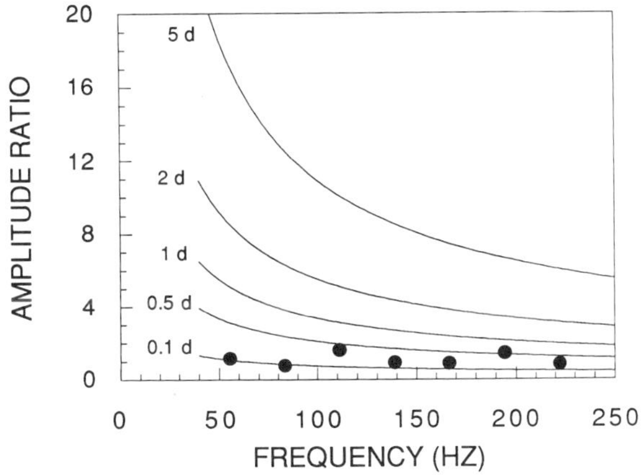

Figure 12. Theoretical tube/P amplitude ratios as a function of frequency and hydraulic conductivity and actual data for the fracture at 232 m depth.

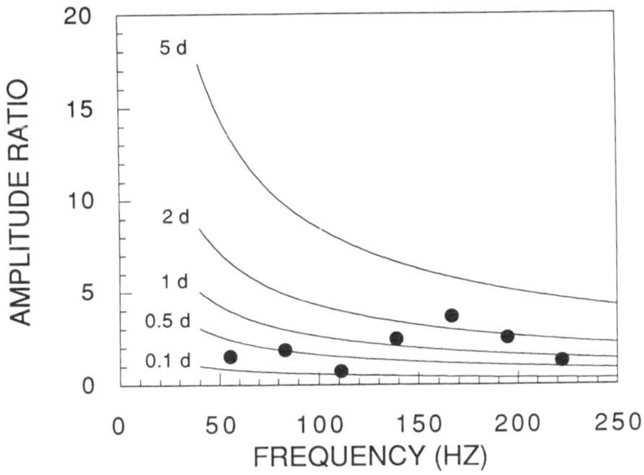

Figure 13. Theoretical tube/P amplitude ratios as a function of frequency and hydraulic conductivity and actual data for the fracture at 286 m depth.

at the low frequencies, i.e., the model predicts higher ratios at the lower frequencies than are observed in the data. A possible explanation for this is that fluid in the fracture is assumed to be inviscid in the model calculations. Burns (1987) has studied the effects of fluid viscosity on guided-wave propagation in boreholes. At frequencies above 1 kHz, these effects are negligible. However, at lower frequencies, attenuation due to viscous drag may be important, and it would act to reduce the observed amplitude ratios.

5. Inversion of Field Data for Fracture Orientation

In addition to the modeling techniques previously described for estimating dip and hydraulic conductivity, a nonlinear inversion developed by Hardin (1986) was applied to the Kent Cliffs VSP for the three fractures observed. The details of this inversion

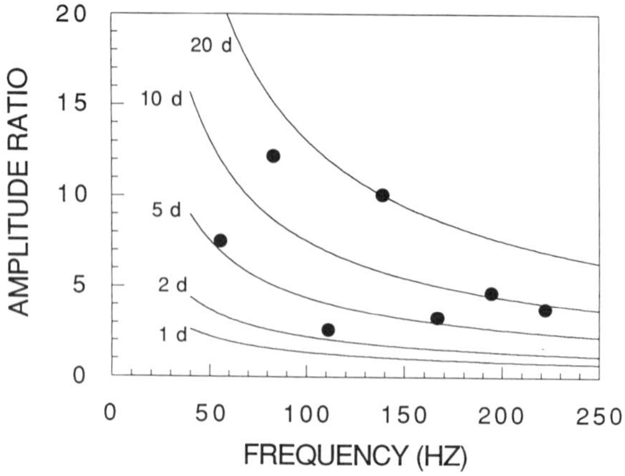

Figure 14. Theoretical tube/P amplitude ratios as a function of frequency and hydraulic conductivity and actual data for the fracture at 512 m depth.

scheme, as well as those of a similar linear inversion, are summarized in Appendix B. Tube-wave to P-wave spectral ratios from the hydrophone data were obtained for three of the offsets (shotpoints 1, 2, and 3).

Filter tests performed on the data indicate that the seismic energy is contained in the bandwidth 50–250 Hz. The source signature was recorded with a monitor geophone whose frequency response was relatively low compared to the seismic data. Therefore, the frequency spectrum of the source signal has a lower frequency content than the seismic data, and weighting based on the frequency content of the source signal would incorrectly bias the inversion (i.e., the low frequencies would be overweighted at the expense of the higher frequencies). Therefore, the weighting matrix **W** is taken to be the identity matrix **I** (i.e., each frequency point is weighted equally).

The results of the inversion are summarized in Table 1. The borehole televiewer shows that the fractures all have strike approximately east–west, and dips of 50° for the shallowest fracture to 75° for the deepest fracture, all approximately to the south. The inversion gives values of the strike as approximately north–south for the fracture at 232 m depth, and northwest–southeast for the other two fractures. In addition, the estimate of the dip for the fracture at 232 m depth is about 7° south, which is not in agreement with the borehole televiewer. The quality of VSP data was poor in this case since there were interfering sets of tube waves (Figure 3) and the tube wave amplitudes were very small in one case (Figure 4), indicating that shot point 3 was along the strike of the fracture. This assures us that the strike of the fracture is correct. The dip estimates for the other two fractures are in good agreement with the values obtained from the borehole televiewer. For the fracture at 286 m depth, the dip estimate from the inversion is 52°, compared with 60° from the borehole televiewer. The fracture at 512 m depth had a dip estimated from inversion of 69°, compared to 75° from the borehole televiewer.

The estimated values of strike obtained from the inversion do not agree as well with the borehole televiewer data. The discrepancies may be due to one or more of three causes. The first is the fracture orientation may change away from the well. The televiewer results show the fracture as it intersects the well, while the VSP results represent the average value of fracture plane for a radius of several meters around the well. The second source of discrepancy may be the limited azimuthal coverage of the three shot points

Table 1. Summary and comparisons of modeling and inversion results of attitudes and conductivities of three major fractures observed at the Kent Cliffs well, based on televiewer and VSP data

(a) Borehole televiewer and forward modeling

Fracture	Depth (m)	Strike[a]	Dip	Conductivity (d)
1	232	90°	50° S	0.1– 0.5
2	286	85°	60° S	0.5– 2.0
3	512	85°	75° S	1.0–10.0

(b) Nonlinear inversion

Fracture	Depth (m)	Strike[a]	Dip	Conductivity (d)
1	232	101°	7° S	0.2
2	286	143°	52° SW	0.6
3	512	119°	69° S	6.3

[a] Strike values are given as a bearing measured clockwise from north.

used in the inversion. This may cause the inversion itself to be relatively insensitive to the strike, owing to the fact that large changes in strike result in correspondingly small changes in the angle between the fracture normal and the direction of propagation of the incident waves. The limited azimuthal coverage may make it difficult to resolve these small changes; that is, the inverse problem may be ill-posed with respect to the strike for limited azimuthal coverage. A third and important factor is the strong anisotropy of the metamorphic rocks (phyllites) that the top 600 m of the well is in (Lee, 1990). In such a medium the seismic rays will deviate from the sagittal (i.e., source–receiver) plane and the incidence angle of the P-wave on the fracture will be different from the theoretical values based on our isotropic model. This conclusion becomes credible on the basis of results obtained by Hardin (1986) for fracture orientation estimates obtained from hydrophone data of the Britton VSP experiment, Hamilton, Massachusetts, where there is no significant anisotropy.

The geometry of the VSP experiment in Hamilton, Massachusetts is shown schematically in Figure 15. Three offsets were recorded with adequate azimuthal coverage, and the inversion was applied to the two fractures visible at 210 m and 290 m depth (Figure 16). The sources were small explosions (200 g) in holes drilled through the weathered layer into the underlying granite. The quality of the data shown in Figure 16 is excellent. The results of the inversion for fracture parameters are summarized in Table 2. These results show remarkably good agreement between the strike and dip estimates obtained from borehole televiewer data and those computed by the inversion.

The nonlinear inversion also allows for the calculation of the fracture aperture, because data are available for multiple frequency points (see Appendix B). The hydraulic conductivity can then be determined from the fracture aperture. For each of the three fractures, the values of the fracture aperture and hydraulic conductivity are also summarized in Table 1. The permeabilities calculated are approximately 0.2 darcy for the fracture at 232 m depth, 0.6 darcy for the fracture at 286 m depth, and 6.3 darcy for the fracture at 512 m depth. These results are in agreement with the ranges of hydraulic

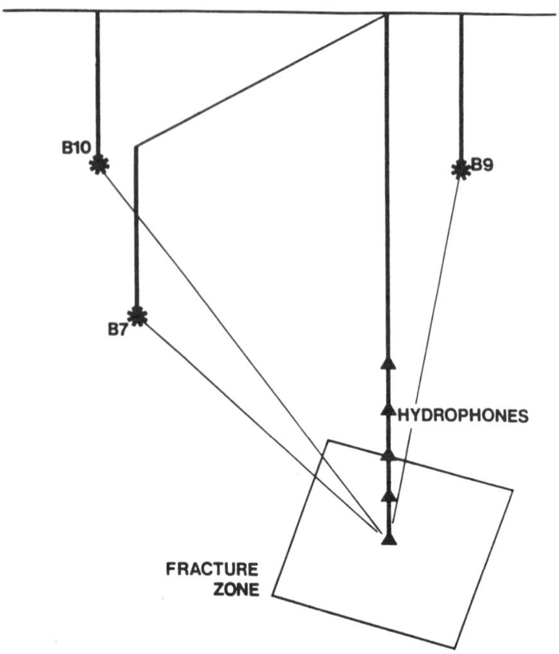

Figure 15. Geometry of the Britton VSP experiment, Hamilton, Mass.: shot point B7, located 180 m east of the well at a depth of 123 m; shot point B9, located 12 m north of the well at a depth of 62 m; shot point B10, located 146 m south of the well at a depth of 58 m.

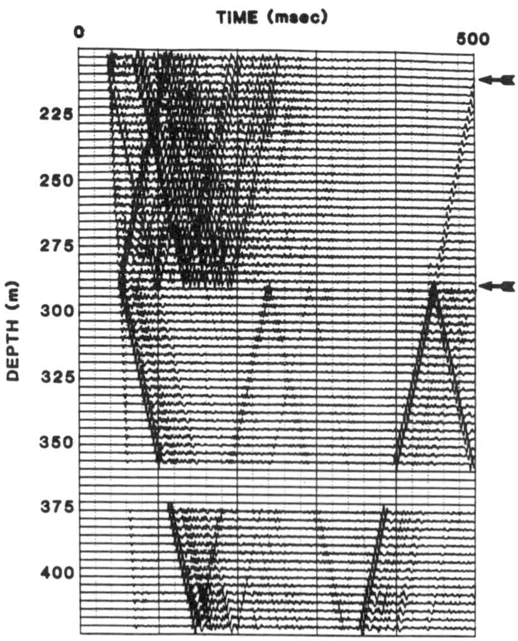

Figure 16. Hydrophone VSP data from shot point B10 of the Britton VSP experiment, showing fractures at 210 m and 290 m depth.

Table 2. Comparison of inversion results with borehole televiewer estimates of fracture strike and dip for the Britton data (from Hardin, 1986)

(a) Borehole televiewer

Fracture	Depth (m)	Strike[a]	Dip
1	210	140°	40° SW
2	290	10°	30° W

(b) Nonlinear inversion

Fracture	Depth (m)	Strike[a]	Dip	Conductivity (d)
1	210	145°	30° SW	2.0
2	290	5°	20° W	5.1

[a] Strike values are given as a bearing measured clockwise from north.

conductivity estimated from the forward modeling. These inversion estimates are probably better because they are the result of a nonlinear least-squares procedure for all three offsets, as opposed to the single-offset estimates from the forward modeling.

6. Conclusions

The model of tube-wave generation in a borehole due to interaction of P-waves with a fluid-filled fracture is reviewed. Based on observations of tube waves caused by incident SV-waves, the effect of parallel-plane harmonic motion of the fracture wall is evaluated in terms of its influence in generating tube waves. It is shown that the volume of fluid injected into the borehole is several orders of magnitude less than the volume injected by compression of the fracture. Hence, it is concluded that compression of the fracture is the dominant mechanism of tube-wave generation, thereby implying that the SV-generated tube waves are caused by the component of the SV-wave particle normal to the fracture plane.

A simple method of estimating the dip of a fracture plane, based on the amplitudes of the P- and SV-generated tube waves, is proposed. The method is tested using the VSP data from the Kent Cliffs well in New York. The estimated value for the dip was in qualitative agreement with the dip of microfractures observed in cores taken in the same interval and that from borehole televiewer data. Since the analysis was performed for only one fracture, further testing is required before the validity of the method can be established.

Hydraulic conductivity estimates for the fractures are determined from theoretical calculations based on tube to P-wave amplitude ratios for the three fractures, and compared to the actual data. The data show a large degree of scatter, and allow only ranges of hydraulic conductivity to be estimated. In addition, a strong dependence of hydraulic conductivity on the dip of the fracture plane is evident. Estimates of the fracture width based on hydraulic conductivity values range in the order of a few micrometers, which is significantly less than the observed fracture width in the borehole.

The relatively small effective fracture width may be due to the fact that real fractures have rough surfaces, resulting in a greatly increased fracture stiffness. Also, a rough fracture reduces the effective fluid flow because of tortuosity. Neither the fracture stiffness nor tortuosity are incorporated into the model, and these may be important in real fractures in the earth. Further research should focus on predicting these effects.

Nonlinear inversions for fracture aperture and fracture orientation have also been conducted. The results for fracture aperture (and hence hydraulic conductivity) are consistent with the ranges of hydraulic conductivity determined from forward modeling. The results for fracture orientation did a fairly good job of estimating fracture dip for the two deeper fractures (at 286 m and 512 m depth) at Kent Cliffs, but the values obtained for fracture strike were not consistent with the borehole televiewer data. This may be the result of the limited azimuthal coverage of the three offsets used in the inversion, which may lead to a relative insensitivity of the inversion to the strike. In addition, the anisotropy of metamorphic rocks, in which the upper half of the well penetrates, affects the seismic ray paths and the inversion results based on the isotropic model. The shallow fracture (at 232 m depth) had results which were not consistent with the borehole televiewer data. This may be the result of the relatively poor quality of the VSP tube wave data because of interference. The dip and strike estimates at the well in Hamilton, Massachusetts agreed very well with televiewer data. In this case the data quality was excellent and the medium was isotropic.

Acknowledgments

We thank Jung Mo Lee, Wafik Beydoun, and Xiaoming Tang for many helpful comments and discussions. We also thank two anonymous reviewers for their suggestions. This research has been supported by the Department of Energy (Grant No. DE-FG02-86ER13636) and the Reservoir Delineation Consortium at the Earth Resources Laboratory, Massachusetts Institute of Technology.

References

Balch, A.H. and Lee, M.W. (eds.) (1984). *Vertical Seismic Profiling: Technique, Applications, and Case Histories.* International Human Resources Development Corporation, Boston.

Beydoun, W.B., Cheng, C.H., and Toksöz, M.N. (1984). Detection of subsurface fractures and permeable zones by the analysis of tube waves. In *Vertical Seismic Profiling: Advanced Concepts* (eds. M.N. Toksöz and R.R. Stewart). Geophysical Press, London.

Beydoun, W.B., Cheng, C.H., and Toksöz, M.N. (1985). Detection of open fractures with vertical seismic profiling. *J. Geophys. Res.* **90**, 4557–4566.

Bower, D.R. (1983). Bedrock fracture parameters from the interpretation of well tides. *J. Geophys. Res.* **88**, 5025–5035.

Brown, K.M. (1972). Computer oriented methods for fitting tabular data in the linear and nonlinear sense. Department of Computer, Information, and Control Sciences, TR No. 72–13, University of Minnesota.

Burns, D.R. (1987). Viscous fluid effects on guided wave propagation in a borehole. *J. Acoust. Soc. Am.* **83**, 463–469.

Carslaw, H.S. and Jaeger, J.C. (1959). *Conduction of Heat in Solids.* Oxford University Press, London.

Cheng, C.H. and Toksöz, M.N. (1981). Elastic wave propagation in a fluid-filled borehole and synthetic acoustic logs. *Geophysics* **46**, 1042–1053.

Cheng, C.H. and Toksöz, M.N. (1984). Generation, propagation and analysis of tube waves in a borehole. In *Vertical Seismic Profiling: Advanced Concepts* (eds. M.N. Toksöz and R.R. Stewart). Geophysical Press, London.

Cicerone, R.D., Lee, J.M., Toksöz, M.N., and Alexander, S.S. (1988). Fracture detection and characterization from hydrophone VSP data, Kent Cliffs, New York. Reservoir Delineation – Vertical Seismic Profiling Consortium, Annual Report.

Freeze, R.A. and Cherry, J.A. (1979). *Groundwater.* Prentice-Hall, Englewood Cliffs, N.J.

Golf-Racht, T.D. van (1982). *Fundamentals of Fractured Reservoir Engineering.* Elsevier, Amsterdam.

Hardage, B.A. (1983). *Vertical Seismic Profiling: Principles.* Geophysical Press, London.

Hardin, E.L. (1986). Fracture characterization from attenuation and generation of tube waves. M.S. thesis, Massachusetts Institute of Technology.

Hardin, E.L., Cheng, C.H., Paillet, F.L., and Mendelson, J.D. (1987). Fracture characterization by means of attenuation and generation of tube waves at fractured crystalline rock at Mirror Lake, New Hampshire. *J. Geophys. Res.* **92**, 7989–8006.

Huang, C.F. and Hunter, J.A. (1981a). The correlation of 'tube wave' events with open fractures in fluid-filled boreholes: Current research, part A. *Geol. Surv. Pap. Geol. Surv. Can.*, **81-1A**, 361–376.

Huang, C.F. and Hunter, J.A. (1981b). A seismic 'tube wave' method for in-situ estimation of fracture permeability in boreholes. Presented at the 51st Annual International Meeting, Society of Exploration Geophysicists, Los Angeles, California.

Lee, J.M. (1990). In situ seismic anisotropy and its relationship to crack and rock fabrics. Ph.D. thesis, The Pennsylvania State University.

Marquardt, D.W. (1963). An algorithm for least squares estimation of nonlinear parameters. *SIAM J. Appl. Math.* **11**, 431–441.

Schlichting, H. (1960). *Boundary Layer Theory*, 4th edn. McGraw-Hill, New York.

Stewart, R.R., Turpening, R.M., and Toksöz, M.N. (1984). Subsurface fracture zone delineation. In *Vertical Seismic Profiling: Advanced Concepts* (eds. M.N. Toksöz and R.R. Stewart). Geophysical Press, London.

Toksöz, M.N. and Stewart, R.R. (eds.) (1984). *Vertical Seismic Profiling: Advanced Concepts.* Geophysical Press, London.

Toksöz, M.N., Cheng, C.H., and Willis, M.E. (1984). Seismic waves in a borehole, In *Vertical Seismic Profiling: Advanced Concepts* (eds. M.N. Toksöz and R.R. Stewart). Geophysical Press, London.

Turpening, R. (1984). Differential vertical seismic profiling: Fracture volume analysis. In *Vertical Seismic Profiling: Technique, Applications, and Case Histories* (eds. A.H. Balch and M.W. Lee). International Human Resources Development Corporation, Boston.

White, J.E. (1983). *Underground Sound.* Elsevier North-Holland, New York.

Appendix A. Tube-wave Generation Model

Beydoun et al. (1985) have presented a model for the generation of tube waves in a fluid-filled borehole due to the presence of fractures. In this model, fractures are idealized as parallel-plane, fluid-saturated, and imbedded in an isotropic, elastic medium. The borehole–fracture system is assumed to be in hydrostatic equilibrium, so that there are no pressure gradients which may induce fluid flow between the fracture and the borehole. A plane P-wave with wavenumber unit vector **k** impinges on the fracture with unit vector **n**. The fracture width $L(t)$ is assumed to oscillate about an equilibrium width L_0

at the same frequency as the incident P-wave, i.e.,

$$L(t) = L_0 - \zeta_0 \cos(\omega t) \tag{A1}$$

where ζ_0 is the amplitude of the fracture closure. It is assumed that this condition applies everywhere on the fracture. To calculate the volume of fluid injected into the borehole, the following assumptions are made:

1. The fracture closure ζ_0 is much smaller than the fracture width L_0.
2. The flow regime in the fracture is a linear, one-dimensional laminar flow (i.e., Stokes' law is satisfied).
3. The fluid compressibility is small.
3. The fluid injected into the borehole does not significantly perturb the borehole pressure P_0 at the fracture location, and the pressure in the fracture is essentially at P_0.
5. The wavelength of the incident P-wave is large compared to the fracture width and borehole radius.
6. The intrinsic fracture permeability κ is constant.
7. The amplitude of the fracture closure ζ_0 is related to the incident P-wave displacement u by

$$\zeta_0 = u \cos \varphi$$

where φ is the angle between the wave vector **k** and unit normal to the fracture plane **n**

$$\cos \varphi = \frac{\mathbf{n} \cdot \mathbf{k}}{|\mathbf{n}||\mathbf{k}|}$$

For the two-dimensional case of linear, laminar, incompressible fluid flow in response to the fracture closure, the flow is governed by Darcy's law:

$$q(s, t) = \frac{-\kappa L(t)}{\mu} \frac{\partial p(s, t)}{\partial s} \tag{A2}$$

where μ is the viscosity of the fluid. The fluid flow rate in the presence of a pressure gradient $\partial p(s, t)/\partial s$ is therefore related to the fracture aperture $L(t)$. The additional elevation gradient term $\rho_f g\, \partial z/\partial s$ is neglected. Bower (1983) has shown that errors due to the elevation gradient effects are small for fractures of dip $< 60°$, owing to the symmetry of the idealized fracture system with respect to the borehole axis at the point of intersection.

Because the fracture movement is periodic in T and symmetric in $T/2$, the volume of fluid injected into the borehole can be calculated from the fracture closure in a time interval of $T/2$ (i.e., from $t = 0$ to $t = T/2$):

$$V_{2D}(\kappa) = -\zeta_0 \left(\frac{\kappa}{\pi \gamma \mu} \right)^{1/2} F(\omega, \zeta_0/L_0) \tag{A3}$$

where

$$F(\omega, \zeta_0/L_0) \simeq 2\omega \int_0^{T/2} \left(\frac{T}{2} - t \right)^{1/2} \sin(\omega t)\, dt, \qquad \zeta_0 \ll L_0 \tag{A4}$$

The effective length of the fracture is defined as

$$d(\kappa) = \left(\frac{2\kappa T}{\mu\gamma}\right)^{1/2} \tag{A5}$$

where γ is the fluid compressibility. This effective length is the radial distance from the borehole at which the pressure in the fracture drops to about 10% of the value at the borehole wall, over an interval of $T/2$.

The two-dimensional result for the volume of injected fluid can be related to the axisymmetric three-dimensional geometry by comparing the 2-D and 3-D equations for steady state flow in response to a constant pressure at the effective length d:

$$q_{3D} = 2\pi R\chi q_{2D} \tag{A6}$$

where χ is a purely geometrical factor given by

$$\chi(L_0) = \frac{d(L_0)}{R\ln\left[\dfrac{R + d(L_0)}{R}\right]} \tag{A7}$$

and R is the borehole radius. Therefore, the 2-D and 3-D fluid volumes injected into the borehole during a half-cycle of fracture movement are related by the expression

$$V_{3D} = 2\pi R\chi V_{2D} \tag{A8}$$

For a propagating tube wave, most of the energy is trapped in the fluid. Therefore, the amplitude of a tube wave excited by a dilatational periodic source in the fluid can be approximated by equating the source strength to the dilatation associated with the coherent tube waves. Beydoun et al. (1985) have developed a relationship between fluid flow and tube wave pressure amplitude (neglecting phase response, body waves related to the solid, and the associated energy of the normal modes in the solid), which predicts the existence of both upgoing and downgoing tube waves of equal amplitudes.

The amplitude of the tube wave generated by the fluid pulse is determined by the tube wave dilatation in the fluid Δ_f^T. The transfer of energy from the compressional body wave to the normal mode is taken to be real-valued, and is assumed to be linear with respect to frequency. The integral of volume strain Δ_f^T over a half-period of the wave is equated with the volume of fluid injected into the borehole from the fracture over the same time period, i.e.,

$$-V(L_0) = 2\pi c \int_0^{T/2} \int_0^R \Delta_f^T(r, z = 0, t) r\, dr\, dt \tag{A9}$$

where c is the tube-wave phase velocity. The volumetric strain and amplitude of the tube wave are determined using the displacement potential for tube waves Φ_f^T, which is taken to be of the form

$$\Phi_f^T(r, z, t) = CI_0(nr)\sin(\omega t - kz) \tag{A10}$$

where $I_0(nr)$ is a modified Bessel function of order 0, $k = \omega/c$ is the vertical wave number, n is the horizontal wavenumber given by

$$n = k\left(1 - \frac{c^2}{\alpha_f^2}\right)^{1/2} \tag{A11}$$

and r is the radial distance from the axis of the borehole. Then the tube wave dilatation in the fluid can be expressed as

$$\Delta_f^T = \nabla^2 \Phi_f^T \tag{A12}$$

Integrating equation (9) twice, an expression for the volume is obtained, given by

$$-V(L_0) = 4\pi RC(L_0)\left(2 - \frac{c^2}{\alpha_f^2}\right)I_1(nR)\frac{k}{n} \tag{A13}$$

An expression for $C(L_0)$ can be obtained from this equation by substituting and rearranging, leading to the expression

$$C(L_0) = \zeta_0 \chi(L_0) L_0 \left(\frac{1}{12\pi\gamma\mu}\right)^{1/2} F(\omega, \zeta_0/L_0)\frac{(1 - c^2/\alpha_f^2)}{2(2 - c^2/\alpha_f^2)I_1(nR)} \tag{A14}$$

The parameter $C(L_0)$ depends on the properties of the fracture and serves to couple fracture movement to tube wave pressure amplitude, since pressure amplitude in the fluid is given by

$$p^t = \rho_f \omega^2 C(L_0) I_0(nr) \tag{A15}$$

Appendix B. Inversion

A technique to invert hydrophone VSP data for fracture orientation has been developed by Hardin (1986). Since the ratio of tube wave to P-wave amplitude can be expressed as a function of fracture parameters, it is a relatively straightforward procedure to set up an inverse problem to solve for these parameters. Two types of inversion are described: linear and nonlinear.

The linear inversion problem can be described for fixed frequency and fixed survey geometry. The tube wave to P-wave spectral amplitude ratio can be expressed in terms of the unknown fracture parameters as a nonlinear function given by

$$\frac{p^t}{p^\alpha} = f(L_0, \psi, \sigma) \tag{B1}$$

where L_0, ψ, and σ are the fracture aperture, strike, and dip, respectively.

Borehole pressure is discretely sampled in the time domain, windowed to isolate the P-wave and tube wave signals, and transformed into the frequency domain via the FFT, so that spectral values for both the P-wave and tube wave are available at discrete frequency points. The spectral ratios are then obtained by simply dividing the respective spectra.

In order to utilize the geometrical dependence of the spectral amplitude ratio, multiple VSP offsets are needed (i.e., at least three). For three or more offsets, a linear least-squares inversion can be used to obtain unit-vector direction cosines for the fracture normal, leaving an underdetermined, transmissivity-dependent factor common to all offsets. The problem is formulated as

$$\begin{bmatrix} (p^t/p^\alpha)_1 \\ (p^t/p^\alpha)_2 \\ \vdots \end{bmatrix} = \begin{bmatrix} p_{1,x} & p_{1,y} & p_{1,z} \\ p_{2,x} & p_{2,y} & p_{2,z} \\ \vdots & \vdots & \vdots \end{bmatrix} \begin{bmatrix} n_x \\ n_y \\ n_z \end{bmatrix} B(L_0) \tag{B2}$$

where $B(L_0)$ is given by

$$B(L_0) = C(L_0)\frac{\omega\beta^2[1-(c\cos\theta/\alpha)^2]I_0(nR)}{\zeta_0\cos\theta c^2\alpha[1-2(\beta\cos\theta/\alpha)^2]} \tag{B3}$$

and $C(L_0)$ is given by eq. (A13) in Appendix A. The parameters $p_{i,x}$, $p_{i,y}$, and $p_{i,z}$ are components of the wavenumber vector $\mathbf{p}$ corresponding to the ith shot point. Similarly, n_x, n_y, and n_z are the components of the unit fracture normal $\mathbf{n}$.

Where source frequency content and data quality permit acquisition of spectral ratios over a frequency band, a different inversion approach is potentially useful. This inversion is nonlinear and consists of finding a solution vector $\mathbf{x} = (L_0, \psi, \sigma)$ which results in a best possible fit of the model to the processed data over the source band. For a single offset, the inversion is

$$\lambda(\psi, \sigma)\mathbf{A}(L_0, \omega_i)\mathbf{x} = \mathbf{Gx} = \mathbf{d} \tag{B4}$$

where the vector $\mathbf{A}$ is given by

$$\mathbf{A}(L_0, \omega_i) = \frac{A_1(\omega_i)\cdot L_0^2}{\ln[A_2(\omega_i) + A_3(\omega_i)L_0]} \tag{B5}$$

and $\mathbf{d}$ is a vector of spectral ratios determined for all offsets and frequencies from the data. Nonlinear least-squares inversion is used to minimize the residual quantity

$$SSQ = |\mathbf{f}|^2 = |\mathbf{W}[\mathbf{Gx} - \mathbf{d}]|^2 \tag{B6}$$

where $\mathbf{W}$ is a diagonal matrix of weighting factors by which additional frequency is incorporated. The norm SSQ is taken with respect to all offsets and applicable frequencies. Each diagonal element is the ratio of the normalized source amplitude to the standard deviation of the tube wave to P-wave amplitude ratio determined from many traces at a particular frequency.

The parameter φ cannot independently specify the orientation of a plane in space. The required additional information is derived from the frequency dependence of the aperture L_0, and to a larger extent by multiple VSP offsets. Multiple offsets can be explicitly represented in the inversion scheme.

The Levenberg–Marquardt inversion (Marquardt, 1963), a damped least-squares procedure, is used to find the minimizing vector $\mathbf{x}$. Modifications by Brown (1972) improve convergence by adjusting the scalar damping parameter at each iteration. The problem is linearized by considering the local behavior of the residual norm about an a priori parameter vector. A sequence of approximations to the minimizing solution is generated by

$$\mathbf{x}_{n+1} = \mathbf{x}_n - [\alpha_n D_n + J_n^T J_n]^{-1} J_n^T \mathbf{f} \tag{B7}$$

where J_n is the numerical Jacobian describing the behavior of the components of the vector $\mathbf{f}(\mathbf{x}, \omega)$ with respect to the components of $\mathbf{x}$ in the vicinity of $\mathbf{x}_n$. The matrix D_n is a diagonal matrix consisting of the diagonal of $J_n^T J_n$, and α_n is the damping parameter. The Jacobian is estimated using the second-order central difference approximation to partial derivatives,

$$\frac{\partial}{\partial x_j}\mathbf{f}(\mathbf{x}_n, \omega_i) \simeq \frac{1}{2h_j}[\mathbf{f}(\mathbf{x}_n + h_j s_j, \omega_i) - \mathbf{f}(\mathbf{x}_n - h_j s_j, \omega_i)] \tag{B8}$$

where

$$h_j = \max(|x_j|, 0.1) \cdot \varepsilon^{1/2} \tag{B9}$$

$\mathbf{s}_j$ is a unit vector in the direction of the solution component x_j, h_j is the step size along this direction used to estimate the partial derivative, and ε is the precision of the floating point machine number unity. The many evaluations of $\mathbf{f}(\mathbf{x}, \omega)$ required are made to the next higher precision.

Implications of Rock Mechanics on Crustal Tectonics

Chapter 17

Role of Elastic Stiffness and Fault Damping during Thrust-sheet Emplacement in a Foreland Belt

E.G. Bombolakis

Department of Geology & Geophysics, Boston College, Chestnut Hill, Boston, MA 02617, U.S.A

Abstract

Field evidence from ancient and modern foreland-type thrust belts indicates that several styles of fault slip are involved in the emplacement of external thrust sheets; for example, typical earthquake faulting in the range of 1 m s^{-1} and fault creep in the range of 10^{-5}–10^{-6} m s^{-1}. Elastic stiffness and velocity-dependent fault damping are two of the fundamental factors that determine whether slip is stable or unstable. The elastic stiffnesses of sedimentary packages depend not only on elastic constants and thicknesses of the strata, but also on the elastic strain gradient with depth. Methods are developed to enable elastic stiffness and velocity-dependent fault damping parameters to be estimated with field data from thrust belts. Quantitative application of these methods is illustrated with the Hogsback thrust sheet in the Kemmerer region of the Wyoming Salient. The data indicate that some of the imbrication of the Hogsback ramp anticline probably had occurred seismically. Data analysis also indicates that the 12 km or more net slip of the Hogsback sheet accumulated principally by recurring slip events with specific rates at least 10^3 times greater than average geologic slip rates of 10^{-9}–10^{-10} m s^{-1}.

1. Introduction

Mechanical analyses of ancient thrust belts are made almost exclusively in terms of quasistatic concepts. But this approach is deficient in two important respects. First, active foreland-type belts, such as the Himalayan foreland belt, the Shotori belt in eastern Iran, the El Asnam thrust terrain of the Tell Atlas in Algeria, the western Transverse Ranges of California, and the Coalinga region of the fold-and-thrust belt of the southern Coast Ranges bordering the San Joaquin Valley, indicate that seismic

FAULT MECHANICS AND TRANSPORT PROPERTIES OF ROCKS
ISBN 0-12-243780-2

deformation is an important part of their development (Bombolakis, 1989a; Bombolakis, 1992). Their deformation includes seismic slip along imbricate faults and low-angle thrusts, and seismically induced 'instantaneous' increments of folding, as well as the fault creep and slow-motion folding usually envisaged for thrust belts. Secondly, a basic assumption in quasistatic analyses is that the equations of mechanical equilibrium are applicable when deformation proceeds slowly. This assumption is not strictly valid for fault slip involving inertial forces, even when the motion is slow in response to damping along the slip surface, as illustrated in many mechanical engineering studies of slip (Timoshenko et al., 1974).

The rate and amount of slip in mechanical engineering problems frequently are analyzed in terms of inertial forces, elastic stiffnesses, and damping resulting from constant and velocity-dependent dissipative forces along a slip surface. A similar approach is followed here with respect to the emplacement of an external thrust sheet in a foreland belt. Rock deformation studies in general, and rock friction studies in particular, show that the elastic stiffness is a critical factor determining whether slip is stable or unstable (Paterson, 1978; Scholz, 1989). Damping also is a critical factor because it can stabilize the motion (Timoshenko et al., 1974). In the special case of sliding friction of rock on rock in the laboratory, the dynamic shear stress is found to be velocity-dependent (Dieterich, 1981; Ruina, 1983; Scholz, 1989). But because dissipative forces responsible for velocity-dependent damping may be very complex, the practice in mechanical engineering is to characterize velocity-dependent damping by means of an 'equivalent viscous damping' force for which the dissipated energy equals the energy dissipated by the actual damping forces (Timoshenko et al., 1974, pp. 64, 81–88).

The common measure of damping employed in seismology is quality factor Q, or its inverse Q^{-1} (Johnston and Toksöz, 1981). Q is a ratio of stored elastic energy to dissipated energy. It usually is derived for processes in which the dissipated energy is much smaller than the stored energy. Under these conditions, $Q \approx \omega_n/\delta$ (Feynman et al., 1963, p. 24-2), where ω_n is the fundamental frequency of the system, and δ is identical to the velocity-dependent damping parameter employed in mechanical engineering (Timoshenko et al., 1974). Consequently, because observations of active faults indicate that wide ranges in value of fault damping need to be considered, the measure of damping employed here is δ instead of Q*.

Active faults reveal that the styles of fault slip in the upper crust include typical earthquakes of seconds duration, slow earthquakes of minutes to tens of minutes duration (Sacks et al., 1981; Beroza and Jordan, 1990), accelerating–decelerating fault creep events, each of an hour or so duration, and fault creep events of individually longer duration (Wesson, 1988). Collectively, they indicate that velocity-dependent fault damping is a fundamental factor in several faulting processes because, in mechanical engineering studies, the larger the damping along a slip surface, the slower and smaller is the slip.

We therefore proceed as follows. First, several important features of thrust sheets of a foreland belt are discussed with respect to faulting, followed by a discussion of the equations for slip along an active segment of a decollement. Then the elastic stiffnesses of strata are derived for a thrust sheet, with constant and velocity-dependent damping parameters derived for the active segment of the basal decollement. To illustrate how these concepts can be incorporated in quantitative analyses of thrust belts, these concepts are applied to the Hogsback thrust sheet in the Kemmerer region of the Wyoming Salient.

* See Appendix for a Table of Symbols.

2. Some Deformational Features of a Foreland Belt

One of the important features of foreland belts is the flat–ramp–flat 'staircase' structure of the thrust faults, illustrated in Figure 1. The thrust sheet moves along the decollement and over the ramp to produce the ramp anticline shown. Subsequently, the ramp region undergoes complex deformation. An example is shown in Figure 2. During subsequent deformation, the basal decollement breaks through beneath the ramp region in some undetermined manner, extending the basal decollement towards the foreland, and then cuts obliquely upsection to form another ramp or related fault structure.

Thus, apart from complicating effects such as out-of-sequence thrusting (Boyer and Elliott, 1982), the basal decollement becomes younger in the foreland direction, with earlier-formed thrust sheets carried piggyback towards the foreland. Consequently, Figure 1 represents a stage that is replicated several times during the development of a foreland belt. The study of the external thrust sheets accordingly provides critical clues to the development of some of the more intensely deformed internal thrust sheets.

2.1. Comments on external low-angle thrusts

The preceding illustrates that faulting is a critical component in the development of foreland belts. Field data reported for several low-angle thrusts, however, are not consistent with some of the scaling relations reported for faults in general. For example, Elliott (1976) calculated that the width along-strike of external thrusts in the Southern Canadian Rockies foreland belt is roughly proportional to the maximum net slip, in contrast with the square-root scaling relation reported for faults in general: $W_T = aD^{1/2}$, where W_T = width, D = maximum displacement, and a is a constant (Walsh and Watterson, 1988; Scholz, 1989).

There is also no consistent increase of thickness of either the gouge zone or the fault zone with progressive fault slip, contrary to the scaling relation reported for faults in general, i.e., that $T_f = bD$, where T_f = thickness and b is a constant (Scholz, 1989). In

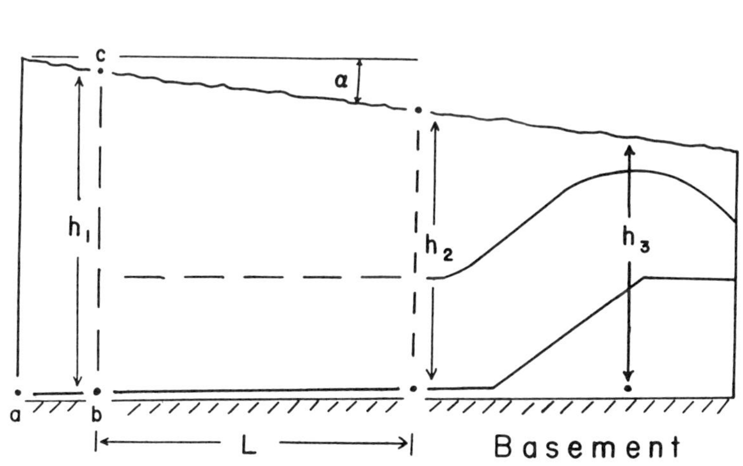

Figure 1. A developmental stage of a foreland belt. A temporally active thrust-belt segment is bounded by L along the basal decollement, vertical sections h_1 and h_2, the topographic slope, and width W perpendicular to the diagram.

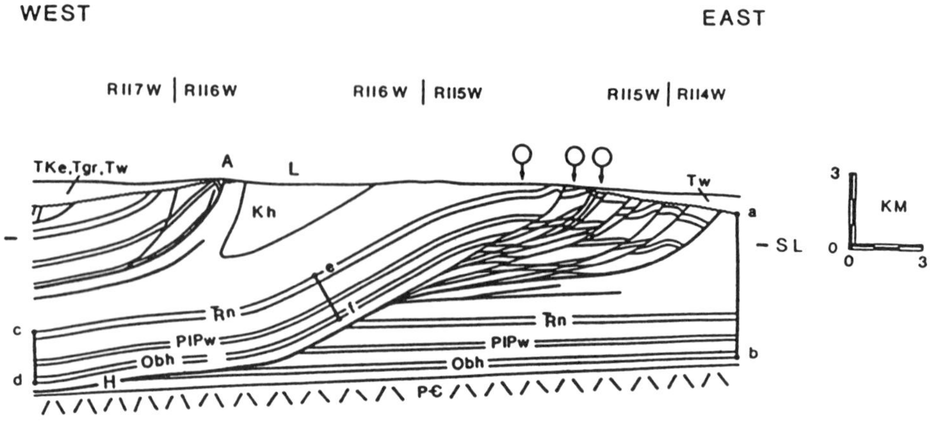

Figure 2. Balanced cross-section of the Absaroka (A) and Hogsback (H) thrust sheets 15 km north of Kemmerer in the Wyoming Salient (see figs. 1 and 2 of Delphia and Bombolakis, 1988, for location and detailed stratigraphy). The Lazeart syncline (L) is demarcated by a solid-line curve drawn along contact between the Frontier formation and the Cretaceous Hilliard clastic sequence (K_h). Open circles indicate locations of three deep wells close to section. Three key marker beds within the predominantly competent sequence are the Ordovician Bighorn Dolomite (O_{bh}), the Permo-Pennsylvanian Weber sandstone (PIP_W), and the Triassic Nugget sandstone ($\bar{R}_n$).

a number of cases, the low-angle thrusts possess (1) pervasive arrays of mesoscopic minor faults in wallrock adjacent to cataclasites, or (2) thin layers of pervasively cleaved rock along movement surfaces (Wojtal, in press). Features of several of these fault zones in the Southern Appalachians indicate cycles of work hardening and work softening (Wojtal and Mitra, 1986). Some of the fault zones thickened, whereas others thinned with progressive fault slip (Wojtal and Mitra, 1988). Recent studies indicate that episodes of wall rock shortening and elongation occur during increments of sheet motion via contractional and extensional mesoscopic faults in the wallrock (Platt and Leggett, 1986; Woodward et al., 1988; Wojtal, in press). These faults occur in the hanging walls of some thrusts, in the footwalls of some thrusts, and occasionally cut across the main faults. Several exposures in the Southern Appalachians show cataclasites without mesofaults, thereby indicating that cataclasite development does not depend solely on mesofaulting of the wallrock (Woodward et al., 1988). In a number of other examples, the fault contact is reported to be knife-sharp, independent of net slips ranging from meters to kilometers to a hundred kilometers (Price, 1988).

2.2. Inferred styles of fault slip along external low-angle thrusts

Different styles of fault slip have apparently occurred along external low-angle thrusts in foreland belts. One style is the rapid fault slip associated with typical earthquakes. For example, synthetic seismogram analyses indicate that several preferred low-angle nodal planes of earthquakes in the Himalayan foreland belt lie either on a basal decollement or on flats associated with the Main Boundary Thrust system (Baranowski et al., 1984). Similarly, south of Coalinga and east of the North Kettleman Hills Dome bordering the San Joaquin Valley, balanced cross-sections indicate that the M 5.5 Avenal earthquake is consistent with transitory low-angle thrusting averaging 0.5 m s^{-1} or so along a decollement (Namson and Davis, 1988).

The styles of fault slip also include episodic motion with velocities that may prove to be appreciably smaller than 0.5 m s^{-1}. Palynologic dating and structural analyses of syntectonic conglomerates and breccias in the Wyoming Salient clearly demonstrate that episodic motion of the Absaroka thrust sheet (A in Figure 2) had occurred (Lamerson, 1982). Episodic motion can theoretically result in various ways; e.g., from cyclic pore-pressure-induced rapid stick–slips, or as recurring slower slip due to work hardening and work softening processes within the fault zone, as well as from sequential imbricate faulting or buckle folding of strata just above the basal decollement. The Absaroka sheet is carried piggyback by the Hogsback sheet in Figure 2, but the normal stratigraphic sequence along c–d near the base of the Hogsback sheet exhibits no imbricate faulting or buckle folding within the resolution of high-quality seismic reflection profiles and other subsurface data of Lamerson (1982) and Dixon (1982) for a distance of almost 30 km west of c–d. Erosion kept pace with thrusting (Warner and Royse, 1987), and so fluctuations of boundary loads did not result from the fluctuations of temperature and confining pressure gradients usually hypothesized for rapid emplacement of thrust sheets. Consequently, the emplacement of the Hogsback sheet is discussed in a subsequent section in terms of variations of slip rate ranging from earthquakes to fault creep along the basal decollement.

There are several lines of evidence of fault creep in foreland belts. Apart from thrusting along evaporites, the evidence includes (1) pressure-solution-induced layer-parallel shortening of strata above a decollement (Engelder and Engelder, 1977; Geiser, 1988); (2) thrust-fault exposures of ancient pressure-solution fault slip (Elliott, 1976); and (3) near-surface fault creep measurements along the Buena Vista Hills splay thrust southeast of Coalinga (Wilt, 1958) and the Nahan splay thrust in the Himalayan foreland belt (Sinvhal et al., 1973).

Mineral-fibre slickensides and related features indicative of pressure-solution fault slip are reported to be widespread in portions of the Central Appalachians (Cloos, 1971), Southern Appalachians (Wojtal, in press), Western Alps (Durney and Ramsay, 1973), and the Canadian Rockies foreland belt (Elliott, 1976). In a theory developed by Elliott (1973, 1976), the slip velocity is linearly related to the surficial shear stress, and his equation indicates that the slip velocity may be relatively insensitive to the fractal geometry of the active fault segment.

In the case of the Buena Vista Hills oil field, the creeping splay thrust dips 20–25°, cutting Pleistocene Tulare gypsiferous sands and clays. It is not known to what extent oilfield operations have been a causative factor of the fault creep. The average slip rate, however, is very close to that reported for the Nahan thrust of shale over wet sandstone. In both cases, the average slip rates are in the range of 10^{-3}–10^{-4} cm s^{-1}, appreciably larger than the average geologic slip rates of thrust sheets. For example, estimated rates for the Wyoming Salient (Wiltschko and Dorr, 1983) are similar to the 10^{-7}–10^{-8} cm s^{-1} slip rates calculated by Elliott (1976) for external thrusts of the Canadian Rockies foreland belt. The average seismic slip rate of thrusting in the Himalayas is estimated to be 10^{-8}–10^{-9} cm s^{-1}, with a minimum uncertainty factor of 2 (Molnar and Deng, 1984). Therefore, all of the data discussed thus far indicate that we need to consider various styles of fault slip identified along active faults when analyzing faulting in the upper crust.

An important observation can be made that is consistent with very heavy fault damping. Hardly any cases of long-term fault creep have been documented in the upper continental crust since intensive monitoring of active faults was initiated in 1958, despite the fact that global positioning satellite data apparently have accuracies as small as or smaller than 1 cm. In fact, the most prominent example of 'long-term' fault creep, the creeping section of the central San Andreas, actually appears to be dominated by episodic fault creep events and small earthquakes (Wesson, 1988). Furthermore, the common assumption that afterslip is a form of fault creep that decreases slowly in a continuous manner is not strictly valid. The afterslip following the 1987 Superstition Hills earthquake sequence was found to consist of discrete fault creep events for which the recurrence interval became progressively longer, without any systematic change in the amplitude of the events (Bilham, 1989). And these events are similar in form to many episodic fault creep events along the central San Andreas.

Therefore, given the current difficulties we have in applying constitutive relations such as friction theories and flow laws to specific fault segments (see Scholz, 1989), these ongoing research attempts to characterize fault behavior need to be supplemented with semiempirical fault damping parameters analogous to those employed for slip surfaces in mechanical engineering. This approach is illustrated in preliminary terms for low-angle thrusts, as follows.

3. Fault-Slip Equations

A simpler version of the fault-slip equations derived by Bombolakis (1989a) is given by

$$X \approx \frac{(P - S)LW}{k_1 + k_2}[1 - \cos \omega_n t] \tag{1}$$

where X is horizontal displacement, L is the length of an intermittently active thrust-belt segment of width W in Figure 1, $(P - S)$ is the average stress drop along L, k_1 and k_2 are the composite elastic stiffnesses, respectively, of sedimentary packages along vertical sections h_1 and h_2, $\omega_n = \sqrt{(k_1 + k_2)/M}$ is the fundamental frequency of the thrust-belt segment of mass M, and t is time. The average net slip of a given event is achieved when $t = \pi/\omega_n$.

For this value of t, eq. (1) becomes identical in form to the net slip equations of traditional crack-growth models employed in earthquake seismology (Kanamori and Anderson, 1975):

$$U = \frac{\Delta\sigma\sqrt{A}}{C\mu} \tag{2}$$

where U is the average seismic net slip, $\Delta\sigma$ is the average stress drop, A is rupture area, μ is the rigidity modulus, and C is a dimensionless factor that depends on the shape of the spreading crack. $C \approx 1$ for several shapes. An analogous dimensionless factor in eq. (1) would be $C' = 1$ for the special case of $k_1 = k_2$ for $t = \pi/\omega_n$. L is proportional to $\sqrt{A}$ for unit width W, and S is the shear stress associated with constant damping along the fault. In this case, S is analogous to the residual strength in the slip-predictable model of earthquakes (Shimazaki and Nakata, 1980; Mogi, 1985). Therefore, the cross-correlation between eqs. (1) and (2) is excellent. But like the traditional crack-growth models of earthquake seismology, eq. (1) neglects velocity-dependent damping.

An important distinction between velocity-dependent damping and constant damping is that only the velocity-dependent damping effects the time duration of a discrete slip event (Den Hartog, 1956). Therefore, the predicted duration of seismic slip in eq. (1), $t = \pi/\omega_n$, should be appreciably smaller than the several seconds durations observed in typical earthquakes, assuming that velocity-dependent fault damping occurs during earthquake faulting. This hypothesis can be tested in a preliminary way with data obtained for the Hogsback thrust sheet. Palinspastic restoration of the structure in Figure 2 to the structural stage illustrated in Figure 1 indicates that h_2 was approximately 10 km during that stage of thrusting. The composite elastic stiffness of all the strata along vertical section h_2 was $k_2 \approx 3.7\ (W/l_0) \times 10^{14}\ \mathrm{N\,m^{-1}}$, where $l_0(\ll L)$ is a horizontal length perpendicular to width W (Bombolakis, 1989a). Stiffness k_1 was of the same order of magnitude as k_2. For an active thrust-belt segment in Figure 1, with rock density $\rho = 2500\ \mathrm{kg\,m^{-3}}$, $h_2 = 10$ km, $1\ \mathrm{km} \le L \le 30$ km, $10^{-3} \le l_0/L \le 10^{-2}$, and a topographic slope angle of a few degrees, the predicted time duration, $t = \pi/\omega_n$, ranges from only 1 s to a small fraction of a second. Consequently, velocity-dependent fault damping needs to be taken into account in eq. (1).

When velocity-dependent damping is incorporated in the derivation of eq. (1), we obtain (Bombolakis, 1989b)

$$X \approx \frac{(P-S)LW}{k_1 + k_2}\left[1 - e^{-(\delta/2)t}\left(\cos\omega_d t + \frac{\delta/2}{\omega_d}\sin\omega_d t\right)\right] \tag{3}$$

where δ is the damping parameter expressed in frequency units, $\omega_d = \sqrt{\omega_n^2 - \delta^2/4}$ is the damped fundamental frequency for $\delta/2 < \omega_n$, and $t = \pi/\omega_d$ is the duration of a net slip event. When δ is very small, eq. (3) reduces to eq. (1). Consequently, eq. (3) provides an approximate description of fast to slow rates of fault slip, depending on appropriate values of δ; i.e., for net slips associated with typical earthquakes and appropriate types of slow earthquakes and accelerating-decelerating fault creep events.

In contrast, when $\delta/2 > \omega_n$, we have the case of heavy damping (Timoshenko et al., 1974). The fault-slip equation for this situation (Bombolakis, 1990) is transformed to

$$X \approx \frac{(P-S)LW}{k_1 + k_2}\left[1 - e^{-(\delta/2)t}\left(\cosh\beta t + \frac{\delta/2}{\beta}\sinh\beta t\right)\right] \tag{4}$$

where $\beta = \sqrt{\delta^2/4 - \omega_n^2}$. Very heavy velocity-dependent damping occurs when $\delta \gg \omega_n$. For this case, $\delta/2 \approx \beta$, thereby causing the bracketed expression to approach zero because of a mathematical identity between hyperbolic and exponential functions. Thus, when $\delta \gg \omega_n$, $X \approx 0$ irrespective of time t. This means that no significant fault displacement can accumulate from recurring very slow fault creep of work-hardening/work-softening material. Instead, a potentially important effect of such events may be temporal thickening of the fault zone in accordance with conservation of energy principles. Heavy fault damping, accordingly, might be one of the basic reasons why so few examples of 'long-term' fault creep have been documented in the upper continental crust. It also is a physical condition that can preclude episodic fault-creep emplacement of thrust sheets at average geologic slip rates, as will be illustrated after the concepts of elastic stiffness and velocity-dependent damping are discussed.

4. Elastic stiffness

4.1. Individual stratigraphic unit

Carbonate twin lamellae measurements and paleostress analyses (Wiltschko et al., 1985, and included references; Craddock et al., 1988; Wojtal and Pershing, 1991) indicate that maximum principal axes had been subparallel to bedding in the direction of tectonic transport of a number of thrust sheets in the Wyoming Salient, the Southern Appalachians, and the Southern Canadian Rockies foreland belt. Here, we consider the case where the maximum principal stress has this orientation with respect to subhorizontal stratigraphic units of an external thrust sheet, assuming that isotropic elastic constants calculated from seismic reflection profiling data, and other field data, are applicable for estimates of elastic stiffness.

A representative portion of a sedimentary bed i is shown at depth H in Figure 3. Bed i has a stratigraphic thickness h_i, unit width W, and length l_0. For l_0 sufficiently small compared to L in Figure 1, the average state of stress is approximately homogeneous along l_0. The effective vertical stress $\bar{\sigma}_z$ under these conditions is

$$\bar{\sigma}_z \approx \rho g H(1 - \lambda) \tag{5}$$

where ρ is density, g is the gravitational acceleration, H is depth, and λ is the Hubbert–Rubey fluid pressure ratio (see Suppe, 1985). The effective principal stress in the direction of tectonic transport is designated by $\bar{\sigma}_x$, whereas $\bar{\sigma}_y$ is the effective principal stress in the strike direction. The sign convention followed here is that compressive stresses and compressive extensional strains are positive.

The elastic subhorizontal principal strains (Turcotte and Schubert, 1982) are

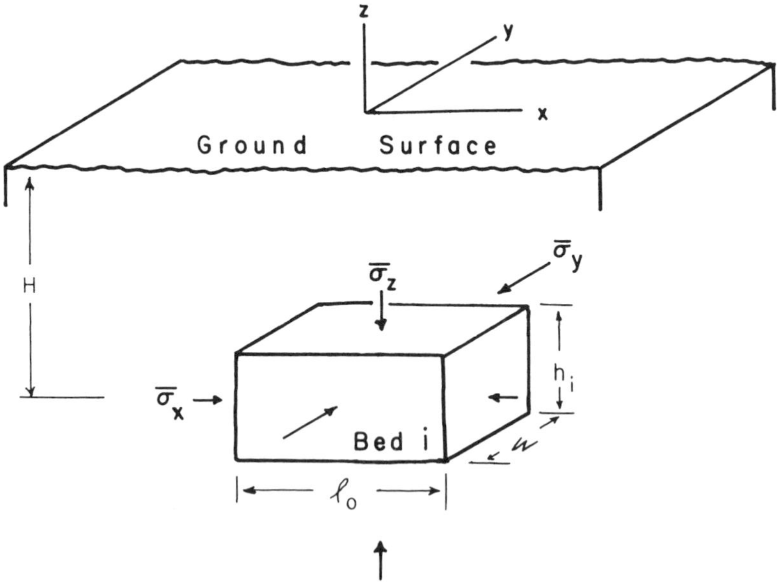

Figure 3. Subhorizontal 'bed i' at depth H in a thrust belt. Stratigraphic thickness is h_i, w is a unit width, and $l_0 \ll L$ of Figure 1. See text.

accordingly

$$\varepsilon_x \approx \frac{1}{E}[\bar{\sigma}_x - \nu(\bar{\sigma}_y + \bar{\sigma}_z)] - \alpha_T T \tag{6}$$

and

$$\varepsilon_y \approx \frac{1}{E}[\bar{\sigma}_y - \nu(\bar{\sigma}_x + \bar{\sigma}_z)] - \alpha_T T \tag{7}$$

where E is Young's modulus, ν is Poisson's ratio, α_T is the coefficient of thermal expansion, and T is temperature at depth H. The approximate signs are employed because E is usually greater parallel to bedding than perpendicular to bedding (Bombolakis, 1989a).

For the case of plane strain, $\varepsilon_y = 0$. The simultaneous solution of eqs. (5), (6), and (7) for this condition yields

$$\bar{\sigma}_x \approx \rho g H \frac{\nu(1-\lambda)}{1-\nu} + \frac{E\alpha_T}{1-\nu} T + \frac{E}{1-\nu^2}\varepsilon_x \tag{8}$$

If, for convenience, we regard one end of l_0 as being fixed, then the other end conceptually undergoes an elastic displacement x due to $\bar{\sigma}_x$. Consequently, $\varepsilon_x = x/l_0$. And if we multiply both sides of eq. (8) by the cross-sectional area $h_i W$, then eq. (8) is transformed to a spring force F. Since the spring constant $k = dF/dx$, the result of differentiation is

$$k_i \approx \frac{E_i}{1-\nu_i^2}\frac{h_i W}{l_0} \tag{9}$$

where k_i is the approximate spring constant of bed i in a stratigraphic sequence, and E_i and ν_i are the elastic constants of that particular bed.

4.2. Stratigraphic packages

From the standpoint of seismic stratigraphy, bed i is the interval velocity thickness over which E_i and ν_i are determined from seismic reflection profiling data (Thomsen, 1986). Therefore, for n beds or n interval velocity thicknesses, there are n equations of the form of eq. (9). These n equations enable a composite spring constant k_{ss} to be calculated for a sedimentary package, as follows.

In-situ stress measurements indicate that subhorizontal principal stresses in intact competent rock increase more or less linearly with depth (Haimson, 1977; McGarr and Gay, 1978). The fluctuations have been traced in some cases to differences in elastic constants of contrasting rock types (B.C. Haimson, personal communication). Consequently, the case considered here is that the elastic tectonic strain increases linearly with depth in a sedimentary package within a thrust sheet, as illustrated conceptually for the sedimentary package in Figure 4.

Each bed i in Figure 4 is subjected to a spring force $F_i = k_i x_i$ having its line of action passing near the midpoint thickness of bed i. F_i acts at a depth $z = h + d_i$, where d_i is the relative depth of bed i with respect to F_1 acting at $d_1 = 0$. Since the elastic displacement increases linearly with relative depth in Figure 4, $x_i = m'd_i + b$, where m' and b are constants. For $d_i = d_1 = 0$, we have $x_1 = b = l_0\varepsilon_0$, where ε_0 is the elastic strain in bed 1 over length l_0. Therefore, the resultant spring force F_R acting on the sedimentary

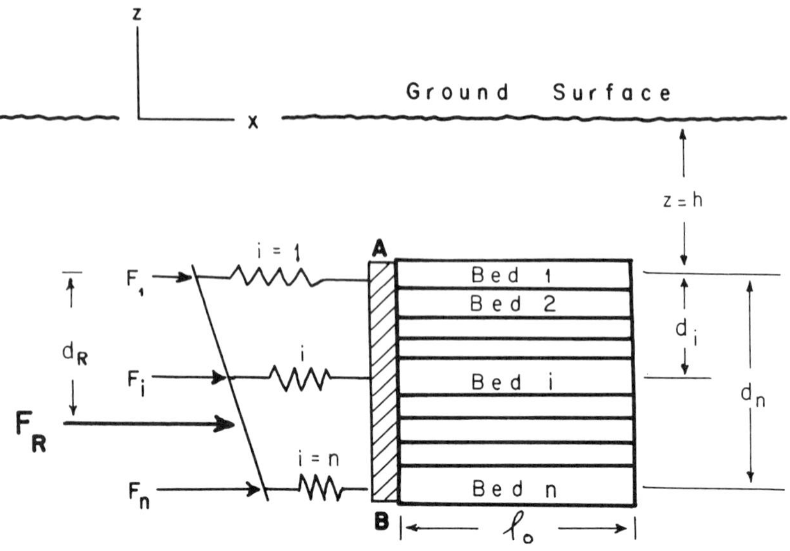

Figure 4. Subhorizontal sedimentary package at depth $z = h$, subjected to elastic tectonic strain under the condition of plane strain perpendicular to diagram. The elastic strain gradient is illustrated conceptually through progressive compression of each successive bed spring by bed force F_i. The relative depth of bed i is d_i, and the relative depth of the resultant tectonic force F_r acting on the sedimentary package is d_r. See text.

package is

$$F_R = \sum_{i=1}^{n} k_i(m' d_i + l_0 \varepsilon_0) = k_{ss} x_R \tag{10}$$

where k_{ss} is the composite spring constant of the sedimentary package, and x_R is the elastic displacement that would be produced by F_R acting on a rigid platen along $A - B$ of the sedimentary package.

Since F_R acts at relative depth d_R, we have

$$x_R = m' d_R + l_0 \varepsilon_0 \qquad \text{with} \qquad d_R = \sum_{i=1}^{n} M_i / F_R$$

in which the numerator is the sum of the moments of the F_i. Thus, choosing a moment center along the line of action of F_1,

$$d_R = \frac{\sum_{i=1}^{n} k_i x_i d_i}{\sum_{i=1}^{n} k_i x_i} \tag{11}$$

The composite spring constant per unit width W is accordingly

$$\frac{k_{ss}}{W} = \frac{1}{W} \frac{\sum_{i=1}^{n} k_i x_i}{m'\left[\sum_{i=1}^{n} k_i x_i d_i \Big/ \sum_{i=1}^{n} k_i x_i\right] + l_0 \varepsilon_0}$$

where $x_i = m'd_i + l_0\varepsilon_0$. If the numerator and denominator are divided by l_0, setting $m = m'/l_0$, then

$$\frac{k_{ss}}{W} = \frac{1}{W}\,\frac{\sum_{i=1}^{n} k_i(md_i + \varepsilon_0)}{m\left[\sum_{i=1}^{n} k_i(md_i + \varepsilon_0)d_i \Big/ \sum_{i=1}^{n} k_i(md_i + \varepsilon_0)\right] + \varepsilon_0} \qquad (12)$$

Equation (12) is identical to eq. (35) of Bombolakis (1989a), which was given without proof.

4.3. Application

The relation between the elastic stiffnesses of a loading system and laboratory specimens during rock deformation tests is one of the critical factors determining whether macroscopic failure occurs in an unstable or stable manner (Paterson, 1978; Cook, 1981). Brittle failure typically occurs unstably with sudden stress drops when the stiffness of the loading system is smaller than the stiffness of the test specimens, in contrast with the gradual failure induced with very stiff presses. Therefore, a comparison of the elastic stiffnesses of interactive stratigraphic packages in an upper crustal orogenic zone provides a means of evaluating whether fracturing and faulting should be seismic or aseismic. This comparison was made for the Hogsback thrust sheet in the Kemmerer region of the Wyoming Salient (Bombolakis, 1989a). It indicates that some of the imbrication of the Hogsback ramp anticline in Figure 2 was associated with seismic activity.

The Kemmerer region is one of the best-documented terrains of a foreland belt, largely because of the efforts of Lamerson (1982, 1985), supplemented by Delphia and Bombolakis (1988). The balanced cross-section in Figure 2 is based on extensive seismic profiling data and well data in addition to surface data. The three open circles represent three deep wells close to the cross-section, wells in which unusually extensive dip meter data were obtained. These data enabled the sequence of imbrication to be determined. Most of the imbricates had formed sequentially from east to west, thereby indicating that sequential locking of the imbricate faults had occurred during thrusting. The balanced cross-section accordingly was restored palinspastically to the stage of thrusting illustrated in Figure 1 (see fig. 8D of Delphia and Bombolakis, 1988) in order to calculate the elastic stiffnesses of the strata along vertical sections h_2 and h_3. These stiffnesses were calculated with eq. (9) from seismic reflection profiling data. The purpose of these stiffnesses is to estimate what the composite elastic stiffnesses were at h_2 and h_3 just prior to the imbrication of the ramp anticline. If the composite stiffness at h_3 is found to be greater than the composite stiffness at h_2, this relation indicates that some of the imbricates had developed seismically, a deduction consistent with sequential locking of the imbricate faults.

The composite stiffness is defined by eq. (12). The principal reason for the specific formulation of eq. (12) given here is that deformation profiles of thrust sheets (Sanderson, 1982) indicate that ε_x may increase nonlinearly with depth during several stages of thrust-sheet deformation. Equation (12) accordingly can be applied sequentially to a vertical sequence of sedimentary packages to evaluate the vertical variation of composite elastic stiffness with depth in a thrust sheet, in a manner analogous to that in which a mathematical curve is approximated by constructing tangents to the curve.

For preliminary estimates of the composite elastic spring constant, ε_0 can be interpreted as an 'average' value of the elastic tectonic strain, with $m = 0$. For these conditions, eq. (12) reduces to

$$\frac{k_{ss}}{W} = \frac{1}{W}\sum_{i=1}^{n} k_i \tag{13}$$

which is essentially identical to the composite elastic stiffness equation of Den Hartog (1956, p. 36). Therefore, using eqs. (9) and (13) for Hogsback strata along vertical sections h_2 and h_3 in Figure 1, the composite elastic stiffness of h_3 was found to be approximately 20% greater than that of h_2 (see table 2 of Bombolakis, 1989a). The strata along h_2 can be interpreted as the 'loading system' for the ramp region when recurring fault slip occurs along length L of the basal decollement. Consequently, the data indicate that seismic stress drops probably had occurred in the ramp region during imbrication of the Hogsback ramp anticline.

5. Estimate of Fault Damping along a Decollement

Equations (3) and (4) for fault slip illustrate that it would be especially helpful in trying to identify fault-slip styles of thrust-sheet emplacement if we could estimate the damping parameter δ from customary field data. This is the objective here.

The constant force component of damping in the fault-slip equations has a magnitude of SLW for the active thrust-belt segment in Figure 1. The velocity-dependent force component in mechanical engineering (Timoshenko et al., 1974) frequently is expressed as $C\dot{X}$, where C is the proportionality constant and $\dot{X}$ is the velocity of slip.

For these conditions, the average force associated with fault damping along L in Figure 1 is

$$C\dot{X}_{\mathrm{Avg}} + SLW = \tau_d LW \tag{14}$$

where τ_d is the average dynamic shear resistance associated with both components of fault damping, and $\dot{X}_{\mathrm{Avg}}$ is the average slip velocity during a fault-slip event. As shown by Timoshenko et al. (1974), the proportionality constant can be expressed as $C = \delta M$. From the geometry of the active thrust-belt segment in Figure 1, mass M is

$$M = \rho\left[\left(\frac{L}{2}\tan\alpha + h_2\right)LW\right] \tag{15}$$

where ρ is the average bulk density. Therefore,

$$\delta \approx \frac{(\tau_d - S)}{\rho\dot{X}_{\mathrm{Avg}}\left(\dfrac{L}{2}\tan\alpha + h_2\right)} \tag{16}$$

An important feature here is that δ is not sensitive to length L in Figure 1 because topographic slope $\tan\alpha$ is usually very small (e.g., see Suppe, 1985). Since ρ is a well-constrained parameter, the more important parameters required to estimate δ are the average slip velocity, the original depth h_2 of faulting at the leading edge of the thrust-belt segment, and $(\tau_d - S)$, which can be estimated by the fact that it has to be smaller than stress drop $(P - S)$ in eqs. (3) and (4). For example, the logarithmic mean of stress drops of typical earthquakes is close to 30 bar, whereas stress drops estimated by Wesson (1988) for many fault creep events are closer to 1 bar.

The implications of eqs. (3), (4), and (16) are illustrated by an analysis of the Hogsback thrust sheet in the Wyoming Salient (Bombolakis, 1992). Palinspastic analysis of Figure 2 indicates that h_2 in Figure 1 was close to 10 km during Hogsback thrusting. Section c–d in Figure 2 demarcates a predominantly competent sequence that includes the Ordovician Bighorn Dolomite, the Permo-Pennsylvanian Weber sandstone, and the Triassic Nugget sandstone. This sequence extends westward just above the basal decollement for 30 km with no evidence of significant deformation within the resolution of current subsurface data, until a major ramp of the Absaroka thrust is approached. Thus, there is a dramatic contrast in deformation east and west of c–d. The fundamental frequency for an active thrust-belt segment of the Hogsback sheet ranges from a few hertz for length L of 10 km to smaller values for greater lengths. The value of δ, however, is much larger than the fundamental frequency if we assume that the Hogsback sheet actually moved almost continuously (e.g., with very short recurrence intervals) at an average geologic slip rate of 10^{-7}–10^{-8} cm s^{-1}. A conservative estimate of $(\tau_d - S)$ is 0.1 bar, based on stress drops estimated by Wesson (1988) for fault-creep events along the central San Andreas. Equation (16) yields $\delta \approx 10^5$–10^6 Hz for these values. Therefore, the Hogsback sheet could not have been emplaced at an average geologic slip rate under these conditions. Instead, the Hogsback sheet was probably emplaced by intermittent modes of the more rapid forms of fault slip.

For example, suppose that fault damping can be approximated in some cases by Newtonian viscous damping. If the active fault segment in Figure 1 has a length L, width W, and an average 'cataclastic' thickness T_f, then

$$\delta \approx \frac{\eta}{\rho\left(\frac{L}{2}\tan\alpha + h_2\right)T_f} \tag{17}$$

where η is the coefficient of viscosity. The upper limit of δ is approximately 10^5–10^6 Hz for average geologic slip rates. A lower limit can be estimated with eq. (17) as follows. Suppose that fault creep along a–b of the basal decollement in Figure 1 induces a high pore pressure along $L = 10$ km, and that this pore pressure equals the lithostatic pressure corresponding to $h_2 = 10$ km when seismic slip is triggered along L. The lowest conceivable limit of δ would be determined by the viscosity of the pore water, but this limit would have to be appreciably smaller than the actual value of δ for a saturated cataclastic fault rock of $T_f = 1$ cm, even with internal high pore pressure. Standard Newtonian viscosities of water are in the centipoise range for laminar flow, whereas handbooks of physics and chemistry indicate that the eddy viscosity for turbulent flow can be greater by a factor of 10^5. Therefore, choosing a tentative value of $\eta = 10^3$ poise, the lower limit of δ is approximately 10^{-2} Hz if only damping due to viscosity of pore water is taken into account, and so the actual lower limit of δ would have to be appreciably greater than 10^{-2} Hz. A preliminary estimate can be made with eq. (3). The time duration of slip should be several seconds for a typical earthquake. The fundamental frequency of the thrust-belt segment specified here is only a few hertz. Therefore, the time duration equation, $t = \pi/\omega_d$, indicates that δ can be several hertz. Consequently, in this simplified example, eq. (17) indicates that the viscosity along the active segment of the basal decollement would be approximately 10^7 poise for this type of seismic slip. In contrast, the viscosity would be nearly 10^{12} poise for a fault-creep event occurring at an average geologic slip rate at 10 km depth.

6. Concluding Remarks

The difficulty we currently have in characterizing fault behavior of actual fault segments, using friction theories and flow laws, indicates that these attempts should be supplemented with elastic stiffnesses and semiempirical damping parameters that can be estimated from available field data. For example, mineral-fibre slickensides are usually considered indicative of slow fault creep. In contrast, mineral deposition from fluid-pressure fluctuations has been deduced for cyclic seismic slip events in the development of mineral deposits along several shear zones (Sibson et al., 1988). The elastic stiffness equations derived here provide a means of estimating whether deformation of external thrusts should be stable or unstable, and the velocity-dependent damping parameter derived here provides a means of estimating the rates of fault slip along the decollements. For a fast slip event, the velocity-dependent damping parameter can be almost equal to the fundamental frequency of the fast-moving portion of the fault block. But if this parameter is sufficiently larger than the fundamental frequency, then fault slip will proceed slowly at a rate determined by either the damped fundamental frequency of the system or by eq. (4).

The analysis of the Hogsback sheet illustrates potential modes of emplacement of external thrust sheets in a foreland belt. For each style of fault slip considered here, the increment of shear resistance associated with velocity-dependent damping is $(\tau_d - S)$, smaller than the stress drop $(P - S)$. For $(\tau_d - S)$ as small as 0.1 bar, data analysis for the Kemmerer region demonstrates that the 12 km or more of net slip could not have developed from episodes of fault creep occurring at rates within one or two orders of magnitude of average geologic slip rates, contrary to common assumptions for thrust sheets. Potential exceptions include thrusting along salt, and the LPS (layer-parallel shortening) thrusting described by Geiser (1988) for the New York plateau. Neither of these possibilities, however, applies to the Hogsback sheet in the Kemmerer region (Delphia and Bombolakis, 1988). Instead the Hogsback sheet was apparently emplaced by the more rapid forms of recurring fault slip (Bombolakis, 1992).

Fault-rock studies of several foreland thrusts in the Appalachians indicate that fault slip had involved cycles of work hardening and work softening (Wojtal and Mitra, 1986, 1988; Woodward et al., 1988). The analyses given here are consistent with this observation. Another important observation in recent years is that syntectonic extensional faulting is associated with contractional faulting in several thrust belts (e.g., Platt and Leggett, 1986; Wojtal and Mitra, 1986; Coward, 1988). Extensional and contractional faults in the wallrock bordering the decollement were associated with shortening and elongation episodes of several thrust sheets (Wojtal, in press). It is interesting, therefore, to note in Figure 1 that the sedimentary package along vertical section h_1 of the temporally active thrust-belt segment may undergo interludes of extension during dynamic slip along length L.

Acknowledgment

Helpful comments on a preliminary version of this paper were made by Brian Evans (M.I.T.), R.J. Martin III (New England Research, Inc.), Steve Wojtal (Oberlin), Nick Woodward (Baltimore, MD), and an anonymous reviewer. The review by Terry Engelder

(Pennsylvania State University) probed many points that seemed questionable or needed clarification. His efforts are much appreciated. Special thanks also are due to Jeffrey Townsend for the editing and arrangement of a technically difficult manuscript. Acknowledgment is made to the donors of the Petroleum Research Fund, administered by the ACS, for partial support of this research.

References

Baranowski, J., Armbruster, J., Seeber, L., and Molnar, P. (1984). Focal depths and fault plane solutions of earthquakes and active tectonics of the Himalayas. *J. Geophys. Res.* **89**, 6918–6928.

Beroza, G.C. and Jordan, T.H. (1990). Searching for slow and silent earthquakes using free oscillations. *J. Geophys. Res.* **95**, 2485–2510.

Bilham, R. (1989). Surface slip subsequent to the 24 November 1987 Superstition Hills, California, earthquake monitored by digital creepmeters. *Bull. Seismol. Soc. Am.* **79**, 424–450.

Bombolakis, E.G. (1989a). Thrust fault mechanics and dynamics during a developmental stage of a foreland belt. *J. Struct. Geol.* **11**, 439–455.

Bombolakis, E.G. (1989b). Fault slip and partitioning of elastic tectonic strains during dynamic low-angle thrusting. *Geol. Soc. Am. Abstr. with Programs* **20**, A134–A135.

Bombolakis, E.G. (1990). Emplacement of an external thrust sheet in a foreland belt. *Geol. Soc. Am. Abstr. with Programs* **22**, A224–A225.

Bombolakis, E.G. (1992). A developmental stage of a foreland belt. In *Thrust Tectonics* (ed. K.R. McClay). Chapman and Hall, London.

Boyer, S.E. and Elliott, D. (1982). Thrust systems. *Bull. Am. Assoc. Petrol. Geol.* **66**, 1196–1230.

Cloos, E. (1971). *Microtectonics Along the Western Edge of The Blue Ridge, Maryland.* The John Hopkins Press, Baltimore.

Cook, N.G.W. (1981). Stiff testing machines, stick slip sliding, and the stability of rock deformation. In *Mechanical Behavior of Crustal Rocks, The Handin Volume* (eds. N.L. Carter, M. Friedman, J.M. Logan and D.W. Stearns), pp. 193–202. American Geophysical Union Geophysical Monograph 24.

Coward, M. (1988). The Moine thrust and the Scottish Caledonides. In *Geometries and Mechanisms of Thrusting, with Special Reference to the Appalachians* (eds G. Mitra and S. Wojtal), pp. 1–16. Geological Society of America Special Paper 222.

Craddock, J.P., Kopania, A.A., and Wiltschko, D.V. (1988). Interaction between the northern Idaho-Wyoming thrust belt and bounding basement blocks, central western Wyoming. In *Interaction of the Rocky Mountain Foreland and the Cordilleran Thrust Belt* (eds. C.J. Schmidt and W.J. Perry, Jr.), pp. 333–351. Geological Society of America Memoir 171.

Delphia, J. and Bombolakis, E.G. (1988). Sequential development of a frontal ramp, imbricates, and a major fold in the Kemmerer region of the Wyoming thrust belt. In *Geometries and Mechanisms of Thrusting, with Special Reference to the Appalachians* (eds G. Mitra and S. Wojtal), pp. 207–222. Geological Society of America Special Paper 222.

Den Hartog, J.P. (1956). *Mechanical Vibrations.* McGraw-Hill, New York.

Dieterich, J.D. (1981). Constitutive properties of faults with simulated gouge. In *Mechanical Behavior of Crustal Rocks, The Handin Volume* (eds N.L. Carter, M. Friedman, J.M. Logan and D.W. Stearns), pp. 103–120. American Geophysical Union Geophysical Monograph 24.

Durney, D.W. and Ramsay, J.G. (1973). Incremental strains measured by syntectonic crystal growths. In *Gravity and Tectonics* (eds. A.D. Kees and R. Scholten). Wiley, New York.

Ekström, G. and England, P. (1989). Seismic strain rates in regions of distributed continental deformation. *J. Geophys. Res.* **94**, 10231–10257.

Elliott, D. (1973). Diffusion flow laws in metamorphic rocks. *Bull. Geol. Soc. Am.* **84**, 26455–26664.

Elliott, D. (1976). The energy balance and deformation mechanisms of thrust sheets. *Phil. Trans. R. Soc. Lond. A.* **283**, 289–312.

Engelder, T. and Engelder, R. (1977). Fossil distortion and decollement tectonics in the Appalachian Plateau. *Geology* **5**, 457–460.

Feynman, R.P., Leighton, R.B. and Sands, M. (1963). *The Feynman Lectures on Physics*, vol. 1. Addison-Wesley, Reading, Mass.

Geiser, P. (1988). Mechanisms of thrust propagation: some examples and implications for the analysis of overthrust terranes. *J. Struct. Geol.* **10**, 829–845.

Haimson, B.C. (1977). Crustal stress in the continental United States as derived from hydrofracturing tests. In *The Earth's Crust* (ed. J.C. Heacock), pp. 576–592. American Geophysical Union Geophysical Monograph 20.

Johnston, D.H. and Toksöz, M.N. (eds.) (1981). *Seismic Wave Attenuation*. Soc. Explor. Geophysicists, Geophys. Reprint Series No. 2.

Kanamori, H. and Anderson, D.L. (1975). Theoretical basis of some empirical relations in seismology. *Bull. Seismol. Soc. Am.* **65**, 1073–1095.

Lamerson, P.R. (1982). The Fossil Basin and its relationship to the Absaroka thrust system, Wyoming and Utah. In *Geologic Studies of the Cordilleran Thrust Belt* vol I (ed. R.B. Powers), pp. 279–340. Rocky Mountain Association of Geologists, Denver.

Lamerson, P.R. (1985). Fossil Basin and its relationship to Absaroka thrust system, Wyoming and Utah. *Bull. Am. Assoc. Petrol. Geol.* **69**, 853.

McGarr, A. and Gay, N.C. (1978). State of stress in the earth's crust. *Annu. Rev. Earth Planet. Sci.* **6**, 405–436.

Mogi, K. (1985). *Earthquake Prediction*. Academic Press, New York.

Molnar, P. and Deng, Q.-D. (1984). Faulting associated with large earthquakes and the average rate of deformation in central and eastern Asia. *J. Geophys. Res.* **89**, 6203–6228.

Namson, J. and Davis, T.L. (1988). Seismically active fold and thrust belf in the San Joaquin Valley, central California. *Bull. Geol. Soc. Am.* **100**, 257–273.

Paterson, M.S. (1978). *Experimental Rock Deformation – The Brittle Field*. Springer-Verlag, New York.

Platt, J.P. and Leggett, J.K. (1986). Stratal extension in thrust footwall, Makran accretionary prism: implications for thrust tectonics. *Bull. Am. Assoc. Petrol. Geol.* **70**, 191–203.

Price, R.A. (1988). The mechanical paradox of large overthrusts. *Bull. Geol. Soc. Am.* **100**, 1898–1908.

Ruina, A. (1983). Slip instability and state variable friction laws. *J. Geophys. Res.* **88**, 10359–10370.

Sacks, L.S., Linda, A.T., Snoke, J.A., and Suyehiro, S. (1981). A slow earthquake sequence following the Izu-Oshima earthquake of 1978. In *Earthquake Prediction, An International Review* (eds D.W. Simpson and P.G. Richards), pp. 617–628. American Geophysical Union, Maurice Ewing Series 4.

Sanderson, D.J. (1982). Models of strain variation in nappes and thrust sheets: a review. *Tectonophysics* **88**, 201–233.

Scholz, C.H. (1989). Mechanics of faulting. *Annu. Rev. Earth Planet. Sci.* **17**, 309–334.

Schimazaki, K. and Nakata, T. (1980). Time-predictable recurrence model for large earthquakes. *Geophys. Res. Lett.* **7**, 279–282.

Sibson, R.H., Robert, H., and Poulson, K.H. (1988). High-angle reverse faults, fluid-pressure cycling, and mesothermal gold–quartz deposits. *Geology* **16**, 551–555.

Suppe, J. (1985). *Principles of Structural Geology*. Prentice-Hall, Englewood Cliffs, N.J.

Sinvhal, H., Agrawal, P.N., King, G.C.P., and Guar, V.K. (1973). Interpretation of measured movement at a Himalayan (Nahan) thrust. *Geophys. J. R. Astron. Soc.* **34**, 203–210.

Thomsen, L. (1986). Weak elastic anisotropy. *Geophysics* **51**, 1954–1966.

Timoshenko, S., Young, D.H., and Weaver, W. (1974). *Vibration Problems in Engineering*. Wiley, New York.

Turcotte, D.L. and Schubert, G. (1982). *Geodynamics*. Wiley, New York.

Walsh, J.J. and Watterson, J. (1988). Analysis of the relationship between displacements and dimensions of faults. *J. Struct. Geol.* **10**, 238–247.

Warner, M.A. and Royse, F. (1987). Thrust faulting and hydrocarbon generation: discussion. *Bull. Am. Assoc. Petrol. Geol.* **71**, 882–886.

Wesson, R.L. (1988). Dynamics of fault creep. *J. Geophys. Res.* **93**, 8929–2951.

Wilt, J.W. (1958). Measured movement along the surface trace of an active thrust fault in the Buena Vista Hills, Kern County, California. *Bull. Seismol. Soc. Am.* **48**, 169–176.

Wiltschko, D.V. and Dorr, J.A. Jr. (1983). Timing of deformation in overthrust belt and foreland of Idaho, Wyoming and Utah. *Bull. Am. Assoc. Petrol. Geol.* **67**, 1304–1322.

Wiltschko, D.V., Medwedeff, D.A., and Millson, H.E. (1985). Distribution and mechanisms of strain within rocks on the northwest ramp of Pine Mountain block, southern Appalachian foreland: A field test of theory. *Bull. Geol. Soc. Am.* **96**, 426–435.

Wojtal, S. (in press). Shortening and elongation in foreland fold-thrust belt thrust zones. In *Structural Geology of Fold and Thrust Belts (Elliott Volume)* (eds. S. Mitra and G. Fisher). Johns Hopkins University Press, Baltimore.

Wojtal, S. and Mitra, G. (1986). Strain hardening and strain softening in fault zones from foreland thrusts. *Bull. Geol. Soc. Am.* **97**, 674–687.

Wojtal, S. and Mitra, G. (1988). Nature of deformation in some fault rocks from Appalachian thrusts. In *Geometries and Mechanisms of Thrusting, with Special Reference to the Appalachians* (eds. G. Mitra and S. Wojtal), pp. 17–33. *Geological Society of America Special Paper 222.*

Wojtal, S. and Pershing, J. (1991). Paleostress associated with faults of large offset. *J. Struct. Geol.* **13**, 49–62.

Woodward, N.B., Wojtal, S., Paul, J.B., and Zadins, Z.Z. (1988). Partitioning of deformation within several external thrust zones of the Appalachian orogen. *J. Geol.* **96**, 351–361.

Appendix

Table of Symbols

Symbol	*Definition*
α	Topographic slope angle in Figure 1
α_T	Coefficient of thermal expansion
β	Heavy-damping parameter of eq. (4)
$C\dot{X}$	Velocity-dependent damping force. ($C = \delta \cdot M$)
δ	Velocity-dependent damping parameter, in radians per second (1 Hertz $= 2\pi$/s)
D	Maximum net slip
d_i	Relative depth of bed i within a sedimentary package. See Figure 4
ε_0	$\varepsilon_x = \varepsilon_0$, the elastic strain in upper bed of sedimentary package
ε_x, ε_y	Subhorizontal elastic strains, with ε_x parallel to tectonic direction
η	Coefficient of viscosity, in poises
E, E_i	E = Young's modulus; E_i = Young's modulus of bed i in a sedimentary package
F, F_i, F_R	Spring forces, F_i = Spring force of bed i of a sedimentary package, F_R = resultant spring force acting on the sedimentary package
H	Depth to arbitrary stratum in Figure 3
h	Depth to top of a sedimentary package. See Figure 4
h_i	Thickness of bed i. See Figure 3
h_1, h_2, h_3	Total thicknesses of vertical sections indicated in Figure 1
k, k_i	Spring constants in Newtons per meter. k_i = spring constant of bed i of a sedimentary package
k_{ss}	Composite spring constant of a stratigraphic sequence
k_1, k_2, k_3	Composite spring constants of vertical sections in Figure 1
L	Length of intermittently active segment of basal decollement
l_0	$l_0 \ll L$, measured parallel to L at h_1, h_2, and h_3 in Figure 1
λ	Hubbert-Rubey fluid pressure ratio
M	Mass of active thrust block
V, V_i	Poisson's ratios. ν_i = Poisson's ratio of bed i of a sedimentary package
ω_n	Fundamental frequency in radians per second. (1 Hertz $= 2\pi$/s)
ω_d	Damped fundamental frequency in eq. (3)
$(P-S)$	Average stress drop along L in Figure 1

Symbol	*Definition*
P	Average peak shear strength
ρ	Density
S	Average residual shear strength
$\bar{\sigma}_x, \bar{\sigma}_y$	Subhorizontal effective compressive stresses, $\bar{\sigma}_x$ in tectonic direction
$\bar{\sigma}_z$	Effective overburden pressure
t	Time
T	Temperature
τ_d	Average basal shear stress associated with constant and velocity-dependent damping
$(\tau_d - S)$	Average basal shear stress associated with velocity-dependent fault damping

Symbol	*Definition*
T_f	Thickness of fault rock; e.g., gouge, cataclastic, etc.
W	Horizontal width of active segment of basal decollement, perpendicular to L. Also, a unit width if so specified
W_T	Total horizontal width of a fault, measured normal to maximum net slip
X	$X = (t)$, fault displacement during a fault-slip event
$\dot{X}$	$\dot{X} = \dot{X}(t)$, velocity of slip during a fault-slip event
$\dot{X}_{avg.}$	Average velocity of slip

Chapter 18

Brace–Goetze Strength Profiles, The Partitioning of Strike-slip and Thrust Faulting at Zones of Oblique Convergence, and the Stress–Heat Flow Paradox of the San Andreas Fault

Peter Molnar

Department of Earth, Atmospheric, and Planetary Sciences, Massachusetts Institute of Technology, Cambridge, MA 02139, U.S.A.

Abstract

The laboratory work of W.F. Brace, C.G. Goetze, and their colleagues provided much of the basis for the now common image of strength in the crust and upper mantle: a brittle upper crust obeying Byerlee's law, a weaker ductile lower crust, and a strong, but also ductile, uppermost mantle. Accordingly, for most plausible geotherms in continental crust and upper mantle, the strongest region may lie in the uppermost mantle. Although the lower crust is likely to be relatively weak, it couples the deformation of the upper mantle to that in the upper crust in most regions. Insofar as the strongest region lies in the uppermost mantle and deforms by ductile flow, the overall large-scale deformation of continential regions should be described best as that of a viscous continuum. Thus, the principal stresses and strain rates throughout the continental crust and upper mantle should be nearly perpendicular and parallel to the Earth's surface, and oblique-slip faulting should be a minor process. Insofar as blocks of upper crust rotate about vertical axes because they are coupled to the vorticity field of the underlying viscous substratum, sets of oblique-slip faults should not be stable under finite deformation. Consequently, with finite deformation, oblique convergence is likely to become partitioned into pure strike-slip parallel to the boundary between rigid plates or blocks, and pure convergence on thrust faults also with strikes parallel to the boundary. If, indeed, blocks of upper crust were effectively carried by the stronger, continuously deforming substratum, then orientations of stress in the upper crust would provide little insight into the dynamics of crustal deformation, much less insight than measurements of the full deformation gradient tensor or

FAULT MECHANICS AND TRANSPORT PROPERTIES OF ROCKS
ISBN 0-12-243780-2

velocity gradient tensor. Moreover, dissipative heating in the strong upper mantle could provide the source of excess heat flow in the Coast Ranges of California.

1. Introduction

In broad zones of oblique convergence between lithospheric plates or continental blocks, purely strike-slip displacements on vertical faults striking parallel to the zones commonly accommodate most, if not all, of the strike-slip component of relative plate or block motion. The convergent component results in thrust slip on inclined planes, but strike-slip components on these planes tend to be minor. Thus, oblique convergence is commonly partitioned into nearly pure strike-slip and nearly pure dip-slip faulting, and oblique-slip faulting is a relatively minor process (e.g., Fitch, 1972; Jarrard, 1986a; Mount and Suppe, 1987).

My main purpose here is to address the questions of why such partitioning occurs and conversely what it implies for the processes of deformation at depth in the earth. Much of this discussion is intended to be general, but I discuss the San Andreas fault as a specific example of such partitioning and as a feature about which much more is known than other such regions. Mount and Suppe (1987) and Zoback et al. (1987) inferred that the San Andreas fault zone is unusually weak and that its weakness accounts for the partitioning of slip. I do not contest that the San Andreas fault, like virtually all active faults, marks a zone of weakness between stronger, coherent blocks of crust, but I do dispute their inference that the weakness of the fault is necessary or sufficient for the partitioning of deformation.

An underlying theme of this chapter is that the continuous, ductile deformation of the upper mantle and lower crust is a vital ingredient not only in the partitioning of deformation at the surface (Richard and Cobbold, 1989), but also for understanding continental tectonics in general. Specifically, like England and Houseman (1986), Sonder and England (1986) and others, I think that, in general, resistance to deformation of continents is provided as much, if not more, by a strong uppermost mantle than by friction on faults in the upper crust. Moreover, I suspect that in most regions, the lower crust should be considered strong enough to couple the continuous deformation in the uppermost mantle to the brittle deformation and to the relative motion of blocks of upper crust, rather than being so weak that it should be thought of as decoupling them. As such, most faults in continental regions, including the San Andreas fault, might not penetrate the entire lithosphere as sharply defined discontinuities. Instead, they may diffuse into broad zones of shear in the lower crust and uppermost mantle over a width of a hundred kilometers or so. Accordingly, dissipative heating in the rapidly deforming but relatively cold uppermost mantle, at least beneath continents, could be large, and could account for the broad zone of high heat flow surrounding the San Andreas fault in California (Lachenbruch and Sass, 1973, 1980).

The crucial underpinning to the logic of this image derives from laboratory studies, a very large fraction of which were carried out on the 7th floor of the Green Building at M.I.T. in the laboratory of W.F. Brace. It seems appropriate to start with a summary of the implications of that work for strength in the crust and upper mantle, but some readers may choose to skip a review of what they already know. I then consider oblique slip faulting, diffuse deformation in continental regions, and some implications of continuous deformation of continental lithosphere for orientations and styles of deformation, culminating in a brief discussion of the laboratory experiments of Richard and Cobbold (1989). These discussions provide the basis for addressing the San Andreas fault and the stress–heat flow paradox and for a brief diatribe against undue emphasis on inferred orientations of stress in the upper crust of the earth.

2. Brace–Goetze Strength Profiles

Since Brace and Kohlstedt (1980) compared measurements of in-situ stress and geologic inferences of strength with those deduced from laboratory experiments, the conventional image of a profile of strength through continental crust and upper mantle has included (Figure 1): (1) a brittle upper crust in which strength increases with depth, (2) a ductile lower crust (in continents) in which strength decreases rapidly with depth, and (3) a ductile, but much stronger, upper mantle in which strength also decreases with depth. Here strength is taken to be the shear stress or stress difference needed to cause deformation at geologic strain rates. Such curves were first presented by the late C.G. Goetze in 1975, in classroom lectures at M.I.T., and they provided the basis for profiles of strength used by Bird (1978), Goetze and Evans (1979) (but only for old oceanic lithosphere), Molnar and Tapponnier (1981), Tapponnier and Francheteau (1978), and many subsequent studies. Although others, such as Sibson (1977) and Kirby (1980), also recognized this variation of strength with depth, I will refer to the qualitative form in Figure 1 as the 'Brace–Goetze strength profile.'

2.1. Brittle upper crust

The upper part of the strength profile, where brittle deformation occurs, is probably the best constrained. The strength there depends on 'Byerlee's law,' that the coefficient of friction, which determines the peak stress necessary to cause sliding on optimally oriented surfaces between parts of rock samples, is virtually identical and equal to 0.6 for most rocks (Byerlee, 1978). The rate at which strength increases with depth in the earth, therefore, should depend on that coefficient, on the pore fluid pressure, and on the orientations of the principal stresses. Because the vertical normal stress is a principal stress, frictional stresses on thrust faults, for which the vertical stress is the least compressive stress, should be larger at any depth than on normal faults, for which the vertical stress is the maximum compressive stress (e.g., McGarr, 1984).

The analyses of brittle deformation in Taiwan (Davis et al., 1983; Dahlen, 1984; Dahlen et al., 1984) and particularly the demonstration that the measured heat flow there matches that predicted by dissipation resulting from brittle deformation (Barr and Dahlen, 1988, 1989; Dahlen and Barr, 1989) constitute positive tests of the applicability of Byerlee's law to strength in the earth. The major questions about its applicability elsewhere concern the depth range over which it is appropriate and the level of pore fluid pressure in the earth.

Scholz (1990) emphasized that most renditions of Figure 1 have overlooked a fundamental aspect of the brittle part of the curve. Earthquake occurrence is commonly associated both with 'brittle' processes (e.g., Meissner and Strehlau, 1982; Sibson, 1982; Chen and Molnar, 1983) and with stick–slip deformation in the laboratory (Brace and Byerlee, 1966). Although both stick–slip and seismic slip seem to pass into stable sliding and aseismic slip at sufficiently high temperatures (Brace and Byerlee, 1970), the stresses

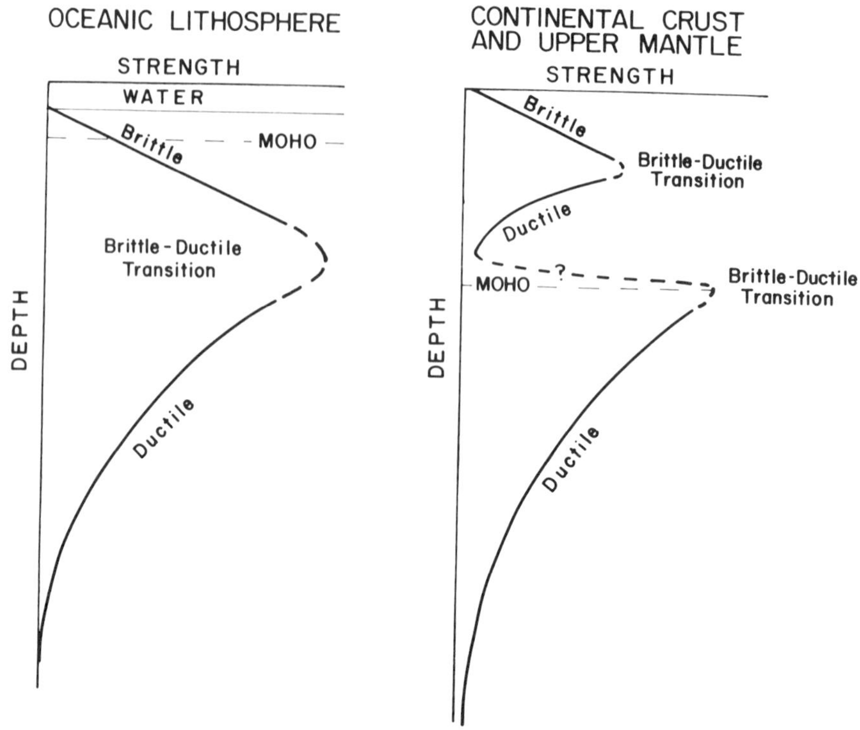

Figure 1. Brace–Goetze strength profiles for oceanic (left) and continental (right) lithosphere. In the upper part, strength, due to brittle friction, increases with depth. In lower continental crust, ductile processes govern the strength. In the upper mantle, plastic deformation occurs, and strength decreases with depth. In rare cases, earthquakes occur in the uppermost mantle, suggesting a second brittle ductile transition there (Chen and Molnar, 1983). No scales are shown to emphasize the qualitative nature of such profiles.

needed to cause high-temperature stable sliding continue to increase with increasing confining pressure (Stesky and Brace, 1973; Stesky et al., 1974). Hence, the 'brittle' part of Goetze's strength profile should continue well below the focal depth of the deepest crustal earthquake, a result clearly demonstrated also by Tse and Rice's (1986) numerical experiments on earthquakes as instabilities.

Scholz (1990) referred to this upper, 'brittle,' layer as the 'schizosphere', from the Greek 'schizo' meaning broken. Presumably, he also was punning on the split 'personality' of the upper crust, in which both seismic and aseismic slip occur on faults and at comparable levels of stress.

Probably the greatest difficulty in applying Byerlee's law to strength is ignorance of pore fluid pressures. For reference, note that in a strike-slip environment, if there were no fluid pressure, then the shear stress at a depth of 10 km would be about 160 MPa. For a pore fluid pressure 0.9 times the lithostatic pressure, it would be 16 MPa. Hence, Byerlee's law applied to dry friction provides an upper bound to measurements of stress in the uppermost crust (Brace and Kohlstedt, 1980; McGarr et al., 1982), but the vertical gradients in stress measured in situ are smaller than those predicted by dry friction (McGarr, 1980). In particular, McGarr's (1980) extrapolations indicate that maximum shear stresses are less than 100 MPa at a depth of 10 km, even for thrust faulting, and less than 100 MPa at the base of the seismogenic zone of the San Andreas fault at 14 km (McGarr et al., 1982). Estimates of the shear stress on the San Andreas fault based on heat flow measurements indicate shear stress of less than 20 MPa (Brune et al., 1969; Henyey and Wasserburg, 1971; Lachenbruch and Sass, 1973, 1980). As Berry (1973) reported pore fluid pressures reaching lithostatic pressures in the adjacent sedimentary rock, high fluid pressures appear to be a likely explanation of the shear stresses being lower than those predicted for dry environments by Byerlee's law.

Similarly, high fluid pressures seem to apply to regions of thrust faulting other than just the Coast Ranges of California. Davis et al. (1983) found that pore fluid pressures in accretionary prisms are commonly large fractions of the lithostatic pressure, and therefore that strengths are low. Barr and Dahlen's (1989) energy balance corroborates this for Taiwan. Moreover, low heat flow at subduction zones limits stresses on these major thrust faults to be ~ 100 MPa at depths of 30–60 km (Molnar and England, 1990). If pore fluid pressures were small, but Byerlee's law applied, such stresses, and hence strengths, should be of the order of 1 GPa.

In summary, Byerlee's law seems to apply to stresses in the upper crust, but pore fluid pressures seem to be sufficient to make shear stresses on faults lower than they would be if dry (see also Byerlee, 1990, and Rice, this volume, Chapter 20). Maximum shear stresses are likely to be about 100 MPa and almost surely less than 200 MPa in most regions.

2.2. Strength in the upper mantle

Let us consider next the mantle portion of the strength profile, before discussing the lower crust. Laboratory measurements of flow laws of dry olivine are commonly used for this, such as those of Evans and Goetze (1979), Goetze (1978), Kirby (1977, 1980, 1983) Kirby and Kronenberg (1987), Kohlstedt and Goetze (1974) and Weertman (1970). The high-temperature, low-stress part of the flow law is given by

$$\sigma_3 - \sigma_1 = \left(\frac{\dot{\varepsilon}}{A}\right)^{1/n} \exp\left(\frac{Q_a}{nR\theta}\right) \tag{1}$$

where $\sigma_3 - \sigma_1$ is differential stress and therefore twice the maximum shear stress, and $\dot{\varepsilon}$ is the resulting strain rate, A depends weakly on temperature but can be considered a constant here, Q_a is the activation enthalpy, $R = 8.3\ \mathrm{J\ K^{-1}\ mol^{-1}}$ is the gas constant, and θ is temperature in kelvin. Because of the weak dependence of $\sigma_3 - \sigma_1$ on $\dot{\varepsilon}$, and the limited range of geologically interesting strain rates, it is customary to treat $\sigma_3 - \sigma_1$ as a 'creep strength' (e.g., Weertman, 1970). Differences in such strengths for different plausible strain rates are small compared with the uncertainties in strength for any strain rate.

To illustrate this part of the strength profile, I use the values suggested by Goetze (1978; Goetze and Evans, 1979): $n \approx 3$, $A \approx 7 \times 10^{14}$ (for stresses in Pascals and strain rates in second^{-1}), and $Q_a = 520 \pm 60\ \mathrm{kJ\ mol^{-1}}$, and evaluate eq. (1) for $\dot{\varepsilon} = 3 \times 10^{-15}\ \mathrm{s^{-1}}$. First, the uncertainty in Q_a alone renders calculated strengths uncertain by 10 times at temperatures of about 700°C (Figure 2). Correspondingly, that uncertainty in Q_a is approximately equivalent to an uncertainty of about 100°C in temperature. Thus, a quantitatively useful application of (1) to specific levels in the mantle beneath continents is impossible, especially where a factor of 10 in the uncertainty in strength embraces a number of plausible, but mutually inconsistent, physical

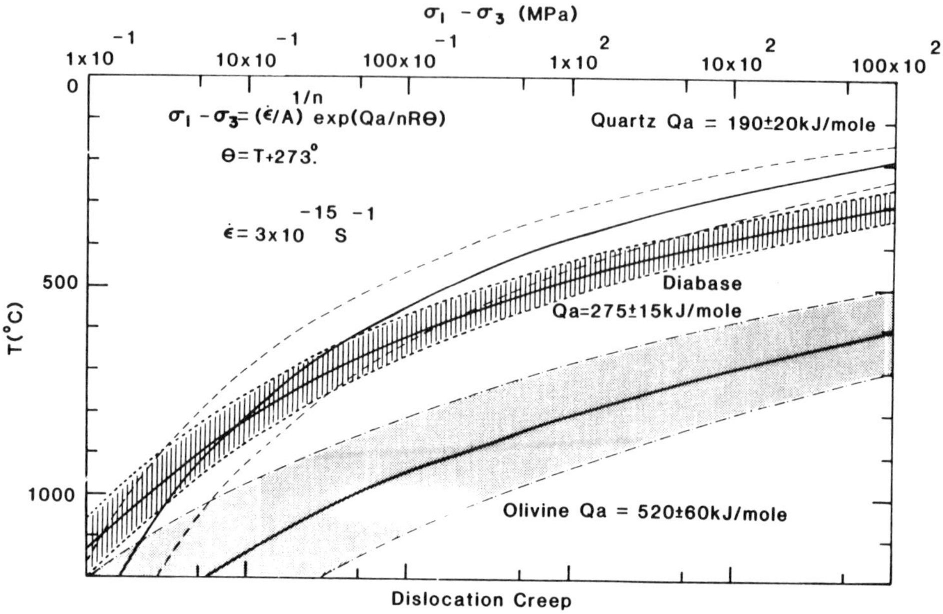

Figure 2. Flow laws of olivine (Goetze, 1978), diabase (Caristan, 1982), and quartz (Brace and Kohlstedt, 1980) appropriate for relatively low stress differences (eq. 1). Note the large uncertainties due only to measured uncertainties in activation enthalpies. (From Chen and Molnar, 1983.)

phenomena. If the quoted uncertainty in Q_a is not enough, subsequent studies of the effects of water and grain size on the flow of olivine make this uncertainty an underestimate (e.g., Chopra and Patterson, 1984; Mackwell et al., 1985; Karato et al., 1986; Goldsmith, 1987).

For temperatures below 600°C, eq. (1) yields creep strengths greater than 8.6 GPa, for a strain rate of $3 \times 10^{-15}\ s^{-1}$. Such a stress in the earth is absurd, and creep can occur by other processes at lower stresses than that described by (1). Evans and Goetze (1979) showed that experimental results obeyed a different theoretical relationship:

$$\sigma_3 - \sigma_1 = \sigma_p \left[1 - \sqrt{\frac{R\theta}{Q_a} \ln\left(\frac{\dot{\varepsilon}_0}{\dot{\varepsilon}}\right)} \right] \tag{2}$$

where $\sigma_p \sim 8.5$ GPa, $Q_a = 500 \pm 125$ kJ mol^{-1}, and $\dot{\varepsilon}_0 \approx 5.7 \times 10^{11}\ s^{-1}$ (Figure 3). Goetze (1978) suggested that for $\sigma_1 - \sigma_3 > 200$ MPa (maximum shear stress > 100 MPa), eq. (2) describes the strength in the mantle. Because the maximum strength of the brittle crust seems to be approximately 100 MPa, this transition is one that I consider further.

With a careful selection of activation enthalpies for (1) and (2), the two creep strengths can be matched at a common temperature that depends only weakly on strain rate (Goetze, 1978). For a strain rate of $7 \times 10^{-15}\ s^{-1}$ (corresponding, for instance, to 20 mm/year^{-1} of relative displacement distributed as simple shear over a zone 100 km wide), (1) and (2) are equal at about 750°C, but clearly, given the uncertainties in both estimates of Q_a, and in other properties of the mantle (e.g., water and grain size) that affect the strength, this number must be uncertain by at least 150°C. Thus, if the temperature at the Moho is less than approximately 750 ± 150°C, then, for a strain rate of

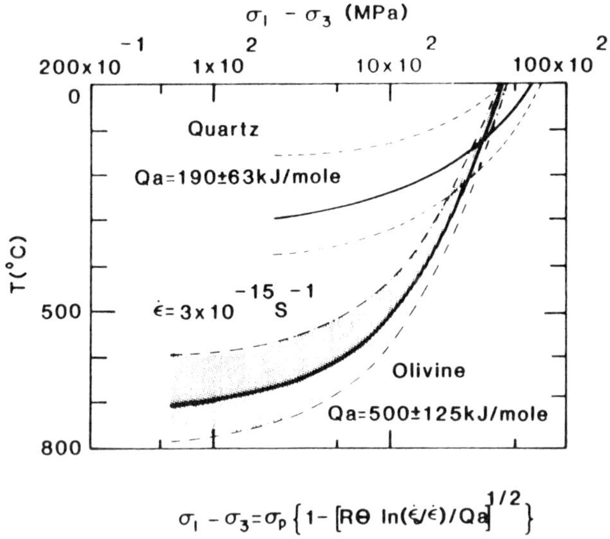

Figure 3. Flow laws of olivine and quartz (Evans and Goetze, 1979) for relatively large stress differences (eq. 2). (After Chen and Molnar, 1983.)

$7 \times 10^{-15}\,s^{-1}$, maximum shear stresses in the mantle are likely to reach 100 MPa. Correspondingly, temperatures must be greater than about $750 \pm 150°C$ for the mantle to be too weak to support stresses of about 100 MPa at geologic strain rates.

Given the enormous uncertainty in the temperature dependence of strength calculated from flow laws, Goetze and Evans (1979) inverted this dependence and used geologic constraints on strengths and strain rates to solve for temperatures in the Earth, Specifically, they compared flexural rigidities inferred from the flexures of lithospheric plates at deep-sea trenches with those calculated from Byerlee's law and from flow laws (eqs. 1 and 2). The free parameter in the latter estimate of the flexural rigidity is the temperature profile through the lithosphere. Goetze and Evans (1979) found that, in general, the temperature profiles required by the flexural rigidities were indistinguishable from those one would infer from the age of the oceanic lithosphere. Thus, we can say that the application of these flow laws to the mantle has been tested successfully, but we must not forget that the uncertainty in Q_a alone yields an uncertainty in creep strength of at least a factor of 10 at temperatures of 600–700°C (Figure 2). Moreover, unfortunately, Goetze and Evans's (1979) analysis of flexural rigidity does not provide a good test of the applicability of the low-temperature, high-strength flow law given by (2).

The uncertainties in the flow laws of olivine, in the range of their applicability, and in temperatures in the earth make it foolish to draw any firm conclusions about stresses or strengths in the mantle. Nevertheless, the olivine flow laws suggest one important property of the Brace–Goetze strength profile that is poorly appreciated; the strongest part of the crust and upper mantle commonly may lie in the uppermost mantle, and not in the upper crust where earthquakes occur on faults. Although this is recognized by scientists who study continental deformation as a whole (e.g., England and McKenzie, 1982; England and Houseman, 1986; Houseman and England, 1986; Sonder and England, 1986; Sonder et al., 1986; Vilotte et al., 1986), it has been largely ignored in considerations of deformation of the upper crust.

2.3. Lower crust

Aside from near the earth's surface and in the transition to the asthenosphere, the weakest part of the profile is probably in the lower crust. When temperatures are high enough that magmas are present, the lower crust might be orders of magnitude weaker than the upper crust and upper mantle (e.g., Gans, 1987).

Estimating the strength of this part of the Brace–Goetze strength profile in a quantitatively useful way, however, is fraught with several problems, and the suggestion of a minimum is strictly qualitative. Goetze himself, and many others after him, used laboratory measurements of the flow law of quartz to define the strength of the lower crust, but that flow law is very uncertain, especially given the possible role of water in the strength of quartz. Moreover, unlike the mantle, where olivine comprises 80% of the minerals, quartz is not especially common in the lower crust, where feldspars and pyroxenes prevail. The deformation of the lower crust might be governed by the flow of quartz, but only if quartz pervades the interstices between other minerals (e.g. Tullis and Yund, 1977). Because pyroxenes and feldspars seem to undergo substantial deformation only at temperatures higher than that for quartz, the quartz flow law is likely to underestimate the strength of the lower crust. The deformations of pyroxenes and feldspars, however, are not well enough understood to yield a reliable flow law. Figure 2 shows an extrapolation of Caristan's (1982) empirical flow law of Maryland diabase, an aggregate of feldspar and pyroxene, but this extrapolation is strictly qualitative, given that there is no reason for the deformation of an aggregate to be describable as a thermally activated process with a single activation enthalpy. Moreover, although it yields a strength intermediate between those of quartz and olivine (Figure 2), the uncertainties, due only to those of estimated activation enthalpies, overlap (e.g., Chen and Molnar, 1983). Finally, deformation of the crust is often associated with metamorphism and dehydration reactions. Thus, processes other than crystal plasticity probably contribute to (and may control) flow in the lower crust (Etheridge et al., 1980). Deformation might, for instance, be facilitated by high pore fluid pressures (Etheridge et al., 1980), by pressure solution (e.g., Rutter, 1983; Gratier, 1987), or by enhanced atomic diffusion at high pressure (Goldsmith, 1986, 1987, 1988). Thus, while it is easy to believe that the lower crust is relatively weak, quantitative evidence that demonstrates this remains elusive, if not illusive.

Perhaps the evidence most suggestive of a weak lower crust comes from geologic observations in regions of normal faulting (e.g., Jackson and McKenzie, 1983; Gans, 1987; Wernicke and Axen, 1988; Melosh and Williams, 1989; Block and Royden, 1990) and from analyses of the implications of the lack of an offset of the Moho in the Basin and Range Province (e.g., Froidevaux, 1986; Zuber et al., 1986). These studies show that a weak lower crust provides a mechanism for accommodating normal faulting in the upper crust, without the faults cutting through the lower crust and upper mantle (as envisioned, for instance, by Wernicke (1981, 1985)), and without requiring local isostatic compensation by variations in crustal thickness. Most of the studies cited above, however, refer to the Basin and Range Province, where horizontal extension and lithospheric thinning have created a relatively hot lower crust and upper mantle, and this region might not be typical.

A weak lower crust may constitute a zone of weakness along which crystalline nappes might detach from the underlying lower crust and upper mantle in convergent settings like the Himalaya (e.g., Chen and Molnar, 1983). The rapid underthrusting of cold material beneath such crystalline nappes, however, would transform an initially ductile shear zone into a fault zone, and the minimum in strength there would then persist for a reason quite different from that responsible for the initial localization of the fault.

Thus, a minimum in strength may not be a ubiquitous feature of the lower continental crust, and it might exist in only a few regions, such as regions of crustal extension where geotherms are especially steep. Even where such a minimum exists, its presence does not necessarily imply that the relative motion of blocks of upper crust occurs independently of deformation of the underlying upper mantle. Accordingly, it probably is a mistake to assume that, in general, the upper crust and upper mantle are 'detached' or 'decoupled' from one another at a weak layer in the crust. If the strongest region is the uppermost mantle, then viscous deformation in the lower crust is likely to couple the continuous deformation of the upper mantle to the relative motion of blocks of upper crust, instead of decoupling them.

3. The Relationship of Continuous Deformation at Depth to Faulting in the Upper Crust

3.1. Oblique convergence and oblique-slip faulting

At plate boundaries, the relative motion between plates is commonly orthogonal to the boundary, as at most spreading centers and subduction zones, or parallel to it, as along transform faults, but where relative plate motion is oblique to the boundary, slip on two fault systems commonly absorbs the obliquely oriented relative motion (e.g., Jarrard, 1986a, 1986b). Probably the best-known example where oblique convergence is partitioned into pure strike-slip faulting and largely thrust faulting on planes with parallel strikes is in Sumatra. Relative plate motion is oriented at about 45° to the strike of the subduction zone; subduction is oriented approximately 20° to the normal to the trench and volcanic arc, and rapid strike-slip faulting occurs parallel to it (Fitch, 1972; McCaffrey, 1991). Similarly, strike-slip faulting within the Aleutian Islands absorbs part of the strike-slip component of the Pacific–North American relative plate motion (Ekström and Engdahl, 1989; DeMets et al., 1990). The fault plane solution of one earthquake suggests that underthrusting occurs essentially perpendicular to the Hikurangi trough off the east coast of the North Island of New Zealand (e.g., Johnson and Molnar, 1972), and diffuse strike-slip shear within the island absorbs the component of Pacific–Australian motion parallel to the trench (Walcott, 1984).

The partitioning of oblique relative motion is also clear in intracontinental deformation. Faulting in central New Guinea seems to be largely either strike-slip or thrust on planes with similar strikes, but none of the slip vectors is parallel to the orientation of Pacific–Australian relative plate motions (Abers and McCaffrey, 1988; Ekström and England, 1989). Although less well constrained quantitatively, oblique convergence seems to be partitioned into pure strike-slip and thrust faulting in the Altai in western Mongolia (e.g., Tapponnier and Molnar, 1979), across the Venezuelan Andes (e.g., Kellogg and Bonini, 1982), and between the Kirthar and Sulaiman ranges, an active thin-skinned fold-and-thrust belt in Pakistan, and the Chaman strike-slip fault in Afghanistan (e.g., Molnar and Tapponnier, 1975).

Perhaps the best-studied example of such partitioning is along the San Andreas fault in Central California. The relative motion between the Pacific and North American plates includes a component of convergence of about 5 mm/year (normal to the San Andreas fault) (DeMets et al., 1990), and the active adjacent faulting and folding reflect a component of shortening perpendicular to the strike of that fault (e.g., Stein and King, 1984; Ekström and Dziewonski, 1985; Namson and Davis, 1988). Because deformation also occurs both west of the San Andreas fault and in the Basin and Range Province to the east of it, the rate of shortening perpendicular to the fault and absorbed by thrust faulting in Central California cannot yet be tightly constrained (Argus and Gordon, 1991). Nevertheless, the occurrence of orthogonal slip, consisting of components of strike-slip on the San Andreas strike-slip fault and of thrust-slip on faults with essentially the same strike, provides another example of the partitioning of oblique convergence into strike-slip and thrust slip (Mount and Suppe, 1987).

Finally, as discussed in more detail below, Richard and Cobbold (1989) observed such a partitioning of oblique convergence into pure strike-slip and thrust faulting in laboratory experiments using a sand layer overlying a viscous substratum, which in turn overlay mylar sheets that moved obliquely toward one another.

Although rare, oblique-slip faulting does occur in a few regions. In the Transverse Ranges of Southern California, large thrust and strike-slip components characterize the faulting associated with each of the 1952 Kern County (Gutenberg, 1955; Stein and Thatcher, 1981), the 1971 San Fernando (Whitcomb et al., 1973; Langston, 1978), the 1973 Point Mugu (Boore and Stierman, 1976), and the 1978 Santa Barbara (Corbett and Johnson, 1982; Ekström and Dziewonski, 1985) earthquakes. Oblique-slip faulting also occurs in northern Greece, where both normal and strike-slip components are present (Jackson and McKenzie, 1984). For both regions, paleomagnetic declinations measured in Late Cenozoic rocks (e.g., Luyendyk et al., 1985; Hornafius et al., 1986; Kissel and Laj, 1988) require large rotations about vertical axes, and other evidence suggests that such rotations continue (McKenzie and Jackson, 1983; Jackson and Molnar, 1990).

Rotations about vertical axes are likely where shear is distributed over a broad zone and where crustal blocks and their bounding faults are oriented obliquely to the orientation of regional simple shear. As the blocks rotate, so must the faults on their margins. As a result, after the blocks have rotated a sufficient amount, the orientations of the faults will become such that slip should cease. For continued deformation, new faults should form (e.g., Nur et al., 1986).

3.2. Stability of oblique-slip faults

McKenzie and Jackson (1983) showed that in zones of distributed deformation, where (1) the component of shortening is absorbed entirely by crustal thickening, and (2) the vertical vorticity is accommodated by rotations about vertical axes, the only stable orientations of faults are those on which oblique-slip does not occur. The first of these assumptions means that there is no net lateral translation (or extrusion) of material parallel to the belt, or that components of velocity parallel to the belt do not vary along-strike. The second assumption implies that vertical vorticity, resulting from distributed strike-slip shear, manifests itself as rotations of blocks about vertical axes. Thus, faults that bound crustal blocks are advected by the flow of material between bounding blocks that move with respect to one another.

In the case of a uniform deformation field between obliquely converging blocks (Figure 4), the strain rate and vorticity fields are uniform across the deforming zone.

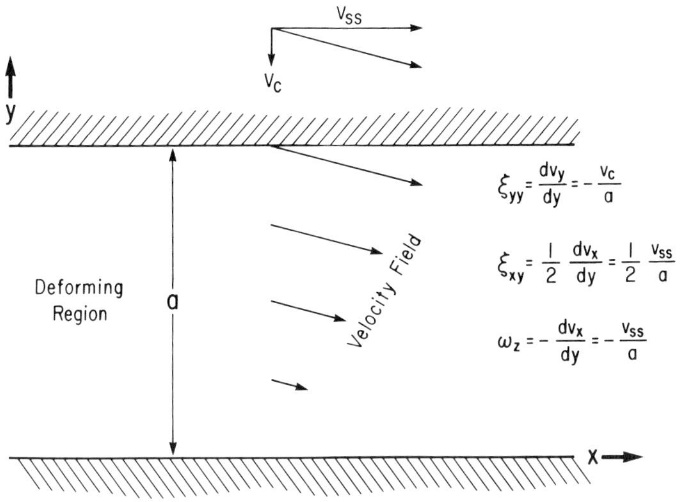

Figure 4. Map view of zone of simple oblique convergence, showing velocity field.

A rigid disk embedded in such a medium rotates only in response to vorticity, and at a rate given by half the vorticity (McKenzie and Jackson, 1983). Rigid objects with lower degrees of symmetry will respond in more complicated ways to the velocity field (Lamb, 1987), but if crustal blocks move in response to the vorticity field, then both the blocks and the faults bounding them must rotate.

By taking stability to mean that the orientations of and styles of faulting do not change with time, in the same sense that McKenzie and Morgan (1969) used for triple junctions in plate tectonics, McKenzie and Jackson (1983) showed that only faults striking parallel to the deforming zone are stable. Oblique-slip faulting could be stable if deformation were localized on one oblique-slip fault, and hence if deformation were *not* distributed. Thus, McKenzie and Jackson's (1983) analysis of stability does not categorically disallow major oblique-slip faults. Moreover, if deformation were distributed such that horizontal components of relative displacements of neighboring blocks were parallel to the orientation of relative motion between the bounding plates, such faults, including oblique-slip faults, could be stable. In such a case, there would be no rotation of blocks, and hence the blocks would not rotate in response to the vorticity field associated with an equivalent (or underlying) continuous deformation. Hence, McKenzie and Jackson's (1983) analysis indicates that (1) if deformation is distributed and (2) if the rotations of strong blocks give an average of the vorticity field, then the rotations of the faults bounding such blocks render such faults unstable. Oblique-slip faults in such a setting should be minor and transient.

The dependence on possibly restrictive assumptions makes it impossible to argue that oblique-slip faults are categorically unstable. Correspondingly, the value of McKenzie and Jackson's (1983) analysis is perhaps best appreciated if their argument is inverted. The rarity of major oblique-slip faulting except where rotations are evident suggests that the faults between blocks are advected with a flow that, in an average sense, reflects the vorticity field expected from the relative orientations of relative plate motion and the deforming zone between them. McKenzie (personal communication, 1990) insists that continuous deformation is not a prerequisite for rotations about vertical axes to reflect the average vorticity field, but clearly one way for them to do so is for them to be embedded in a continuum whose velocity gradient tensor is determined largely by kinematics: the velocity of relative plate motion and the orientations of the deforming zone.

3.3. Relative movements of crustal blocks or continuous deformation

The upper part of the Earth's crust is broken into blocks of various dimensions and separated by faults, which constitute zones of weakness. As discussed above, deformation is presumably continuous in the lower crust and upper mantle. Among the relevant questions of continental tectonics are: To what extent do these faults penetrate the lower crust and upper mantle? Is regional deformation best analyzed in terms of relative movements of blocks or as a continuous medium? To what extent is continental deformation driven and resisted by stress in the upper crust and friction on faults or by flow in the mantle and viscous deformation at depth?

Most attempts to examine the nature of faulting in a region in terms of the orientation of stress (e.g., Anderson, 1951) are, implicitly if not explicitly, based on the assumption that forces or stresses applied at lateral extremities of a region cause stresses within it. Most discussions of the orientation of stress in the crust based on geologic observations (e.g., Zoback and Zoback, 1980) make this assumption (e.g., Zoback et al., 1987). Similarly, most analyses of relative movements of crustal blocks in terms of stresses implicitly treat the substratum as stress-free (e.g., Nur et al., 1986). McGarr (1982), however, showed that the lateral variations in inferred orientations of stress (Zoback and Zoback, 1980), in general, require the existence of a basal shear stress in the intervening regions. If one views the crustal blocks as embedded in and carried by a viscous substratum, the significance of the orientations of stress in the upper crust becomes moot. McKenzie and Jackson's (1983) analysis of faulting ignores entirely the stress field and treats the orientations and senses of faulting as consequences of geometry, the presence of weakness (e.g., faults), and finite deformation in which vorticity plays a part. If the faults between the crustal blocks support lower shear stresses than those needed to deform the viscous substratum, a treatment of blocks carried by a viscous substratum is especially different from the one in which blocks are decoupled from the substratum.

In elastic or viscous homogeneous, isotropic media, the shear stresses are proportional to strains or strain rates. The components of shear stress and normal stress acting on the free surface must vanish and, therefore, shear strains and strain rates on horizontal planes must vanish at the surface too. Thus, the principal axes of strain rate should be vertical and horizontal, and oblique-slip faulting, for which the principal axes of stress and strain rate are not perpendicular to the earth's surface, should not occur (Anderson, 1951; Houseman and England, 1986; Richard and Cobbold, 1989).

For such orientations of principal axes, the strain and strain rate fields can be described as a superposition of elementary strain and strain rate fields associated with pure strike-slip, pure thrust, and pure normal faulting. Each of these three elementary strains can occur as a linear combination of two orthogonal elements. Thus, there are six independent elementary strain fields, and Houseman and England (1986, Appendix 1) demonstrated that the deformation of a continuous medium can be represented as combinations of pairs of these elementary fields. Ekström and England's (1989) analysis of faulting associated with earthquakes in several regions reveals such a partitioning of strain rate into pairs of elements similar to those predicted from the relative plate motion across the regions. Thus, the regional strain field associated with many earthquakes is consistent with the deformation predicted for a continuum, even if the individual earthquakes clearly rupture faults that separate crustal blocks.

Houseman and England's (1986) calculated orientations of strain rate fields are based on the idea that deformation can be described well as continuous deformation. Thus, one may view the observations of partitioning of faulting into pure strike-slip and dip-slip (summarized above) and of strain into corresponding elementary strain fields (Ekström and England, 1989) as tests of the idea that continuous deformation is an important element in the partitioning of oblique convergence. Richard and Cobbold's (1989) experiments provide an independent, additional case for the importance of continuous deformation at depth.

3.4. Richard and Cobbold's experiments

Richard and Cobbold (1989) carried out laboratory experiments using sand layers overlying mylar sheets and bounded laterally by flat pistons. The mylar sheets and the bounding pistons were displaced obliquely with respect to the linear boundary between the two mylar sheets. This boundary provided a discontinuity in displacement at the base of the layered medium. Richard and Cobbold (1989) performed two series of experiments with different ratios of strike-slip to convergent relative velocities of the mylar sheets (from 2:1 to 5:1). In one series, the sand layers overlay the mylar sheets directly, and oblique slip was common. In the other series, a layer of silicone, which deforms as a Newtonian viscous fluid, lay between the sand layers and the mylar sheets. In these latter experiments, oblique convergence was partitioned into pure strike-slip and pure dip-slip faulting on planes parallel to the discontinuity in the mylar sheet. Clearly, the viscous substratum seems to have played a vital role in the partitioning in these experiments.

Richard and Cobbold (1989) noted that the direct coupling of sand to the mylar sheets allows the movement of the sheets to impose a basal shear stress on the sand. As a result, at the base of the sand, the three principal stresses make oblique angles with respect to the vertical and horizontal, a condition necessary for oblique slip in a cohesionless material undergoing Coulomb friction. They inferred that the finite thickness of viscous silicone decoupled the sand layer from the mylar sheets, so that the principal stresses were nearly vertical and horizontal at the base of the sand. As noted above, this orientation should suppress oblique-slip faulting. I think that they are correct about the orientations of the principal stresses, but wrong in inferring that the viscous silicone decouples the sand from the mylar.

In Richard and Cobbold's (1989) experiments, the deformation was localized in a narrow zone parallel to and directly above the boundary between the mylar sheets. If the mylar sheets and overlying sand were decoupled from one another, deformation could have occurred anywhere in the sand. For this close juxtaposition of deformation in the sand over the discontinuity in the mylar, the relative displacements of the mylar sheets must have been coupled to the sand layer. In coupling them, the viscous silicone also smoothed the discontinuity in displacement between the mylar sheets into continuous deformation above it. The degree of smoothing presumably depends on both the viscosity of the silicone and the thickness of the layer of it.

Richard and Cobbold's (1989) results bear very directly on the origin of partitioning of oblique convergence along the San Andreas fault into pure strike-slip on that fault and into 'fault-normal compression' adjacent to it. Recall that Mount and Suppe (1987) and Zoback et al. (1987) inferred that such orthogonal compression (stress to them, but strain in the present context) results from a very low strength of the San Andreas fault. In Richard and Cobbold's (1989) experiments, however, the cohesionless sand contains no such localized weak zone; the strengths of both the strike-slip and the thrust faults should be comparable. Hence, although the attribution of such a partitioning to a weak fault might be correct for some faults, or partially so, it is not a unique explanation. (Note that I am not disputing here the possibiltiy of a weak San Andreas fault.)

3.5. Rate of energy release on oblique-slip faults

Michael (1990) showed that the rate of energy release on oblique-slip faults can be greater than that on pairs of strike-slip and dip-slip faults with parallel strikes, and he used this as an argument for why partitioning occurs. Initially, I began this work with a nearly identical analysis, but England and McKenzie (personal communications, 1989) persuaded me to abandon it. They argued that there is no known physical law that requires physical systems to choose a configuration in which energy dissipation is minimized. Michael (1990) justified his analysis by appealing to Hamilton's principle that systems choose configurations in which the time integral of the difference between kinetic energy and potential energy is minimized. This principle is known to yield Newton's laws of motion for conservative force systems. Michael (1990) assumed that the kinetic energy associated with faulting is converted again into potential energy, so as to conserve energy, and therefore, implicitly, to yield a conservative force system. Most of that kinetic energy, however, is dissipated as heat, either near the faults or as attenuated seismic waves. The earth is a notoriously inefficient thermal engine, and little of the earth's heat flow is converted into strain energy released by earthquakes. Thus, like England and McKenzie, I think Hamilton's principle does not apply, and Michael's (1990) argument is not, yet at least, based on sound physical principles.

4. The Stress–Heat Flow Paradox

Measurements of heat flow in Central California reveal two features that lie at the root of controversies over the state of stress in the earth. First, there is no measurable enhancement of conductive heat flow near the San Andreas fault, as there should be if significant shear heating were occurring along it (Brune et al., 1969; Henyey and Wasserburg, 1971; Lachenbruch and Sass, 1973, 1980). The uncertainties and the scatter in the measurements seem to permit average shear stresses of only 20 MPa on the fault. This is considered a paradox, because an application of Byerlee's law would yield shear stresses larger by an order of magnitude, unless pore fluid pressures attained a significant fraction of the lithostatic pressure. Byerlee (1990) and Rice (1992) have recently argued for low permeability in the fault zone and that high pore pressure is likely.

4.1. The broad zone of high heat flow

A more paradoxical, if less celebrated, aspect of the heat flow measurements is a broad zone of high heat flow in central California, some 200 km wide and with measurements averaging 30 mW m^{-2} higher than is typical for continental regions. Lachenbruch and Sass (1973) ascribed this high heat flow to viscous drag of the upper part of the lithosphere over its underlying substratum. Because such a substratum was presumed by many earth scientists to be weak, ascribing high heat flow to shear on it was not a widely embraced idea, especially because the shear stress must be large. Lachenbruch and Sass (1980), in fact, later suggested that this broad, high heat flow could alternatively be due to the passage of the Mendocino Triple Junction and the consequent emplacement of hot mantle material beneath the crust. I proceed here assuming that this alternative is not the explanation.

Recall that the rate of shear heating is given by the product of shear stress and rate of slip. Of the 48 mm/year of Pacific–North America motion (DeMets et al., 1990), suppose 40 mm/year are distributed beneath central California over a zone 200 km wide. Suppose that directly below the San Andreas fault, the upper crust on each side slides at about 20 mm/year with respect to the substratum, and that this rate diminishes to zero about 100 km from the fault. Then the average slip rate of the upper crust with respect to the substratum is about 10 mm/year, and heat generation at 30 mW m^{-2} would require an average shear stress of 100 MPa. Lachenbruch and Sass (1980) obtained a similar value with a more precise calculation.

This level of stress seemed prohibitively large to many investigators, but, if one accepts that the strongest part of continental lithosphere lies in the upper mantle, then shear stresses of 100 MPa are plausible. Moreover, if this possibility makes Lachenbruch and Sass's (1973) mechanism for generating high heat flow seem more plausible than before, it also permits another mechanism for generating heat by shear in the mantle: shear on vertical planes in the mantle.

The San Andreas fault is one element in a broad zone of right-lateral shear associated with the relative motion of the Pacific and North American plates. Right-lateral slip on the other faults and clockwise rotations of crustal blocks (e.g., Luyendyk et al., 1985; Hornafius et al., 1986) attest to such shear. If the upper crust undergoes right-lateral shear, so almost surely do the lower crust and uppermost mantle. Attempts using simple models of continuous deformation to account for the rates and magnitudes of rotation of parts of California (Sonder et al., 1986; Jackson and Molnar, 1990) are sufficiently successful to render very plausible the idea that the upper crust is coupled to a viscously deforming substratum. Thus, the uppermost mantle may deform as a continuum beneath the blocks of upper crust, sheared in a right-lateral sense.

If the upper mantle is indeed strong, as suggested above, then shear heating should occur within it. For 40 mm/year distributed over a zone 200 km wide below Central California, the average strain rate is $7 \times 10^{-15}\ s^{-1}$. If the average resisting stress over a zone 40 km thick were 100 MPa, the rate of shear heating per unit area parallel to the earth's surface would be roughly 30 mW m^{-2}, and, the entire heat flow anomaly could be explained by such shear heating.

Obviously, if the uppermost mantle is strong enough to support shear stresses of 100 MPa on horizontal planes, it must be able to do so on vertical planes as well. Conversely, if the mechanism imagined by Lachenbruch and Sass (1973, 1980) operates, there should be additional shear heating within the upper mantle due to shear on vertical planes. If such shear heating occurs, then it suggests that temperatures in the upper mantle are low enough to support stresses of 100 MPa at strain rates of about $7 \times 10^{-15}\ s^{-1}$.

Lachenbruch and Sass (1973) showed that temperatures at the Moho (at 30 km) beneath the Coast Ranges could be as low as about 550°C if the extra heat were produced in the upper crust and could be as high as about 850°C if it were produced in the mantle. These numbers are uncertain by at least 100°C, simply because of the uncertainty in the coefficient of thermal conductivity. Above, I argued that if the temperature in the mantle were lower than about 750°C, shear stresses of 100 MPa could be sustained at strain rates of $7 \times 10^{-15}\ s^{-1}$. Given the uncertainties assigned to both this temperature (at least 150°C) and that estimated by Lachenbruch and Sass (1973) for the mantle, one cannot reject the hypothesis that a shear stress of 100 MPa does characterize the upper mantle beneath the San Andreas fault. At the same time, the inference of such a level of stress in the uppermost mantle is clearly not firmly established.

4.2. Implications of gradients in stress

If the lower crust were weak enough to decouple the upper crust from the upper mantle, the level of shear stress on horizontal planes in the mantle would also be low and Lachenbruch and Sass's (1973, 1980) mechanism of heat generation would be ineffectual. Some strength in the lower crust is necessary for the deformation in the upper crust and upper mantle to be coupled, but if the lower crust were as strong as the upper mantle, a broad zone of deformation and heat generation in the upper mantle would be unlikely. We may use the equations of equilibrium and the breadth of the zone over which strike-slip and convergence is partitioned to examine the magnitudes of stress in the mantle. The existence of a shear stress on horizontal planes at depth, while such a component of stress must vanish at the earth's surface, requires a vertical gradient in that stress. From the equations of equilibrium, a vertical gradient in the shear stress on a horizontal plane requires a horizontal gradient in at least one component of stress on a vertical plane.

Consider the following right-handed coordinate system (Figure 5). Let the x-axis point northeast, perpendicular to the San Andreas fault, the y-axis southeast, parallel to it, and the z-axis vertically down. Let tension be positive, so that right-lateral shear τ_{xy} on the San Andreas is also positive. Movement of the earth's surface to the northwest

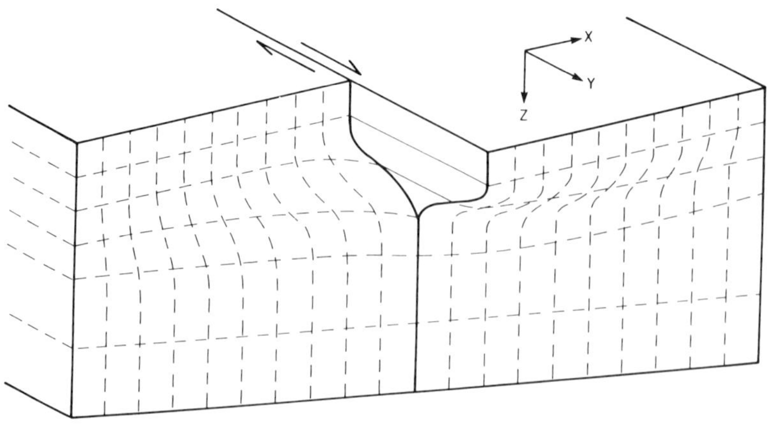

Figure 5. Simplified block diagram showing the displacement field at depth below a strike-slip fault. Strain at depth is relatively uniform and is localized at the fault at the surface.

or southwest with respect to the substratum corresponds to positive shear stresses τ_{yz} and τ_{xz}, respectively. From one of the three equations of equilibrium,

$$\frac{\partial \tau_{xy}}{\partial x} + \frac{\partial \tau_{yy}}{\partial y} + \frac{\partial \tau_{yz}}{\partial z} = 0 \tag{3}$$

Assuming a two dimensional configuration, we may ignore a y-dependence, corresponding to no important variation in stress along the San Andreas fault. Then $\partial \tau_{yy}/\partial y = 0$, and

$$\frac{\partial \tau_{xy}}{\partial x} + \frac{\partial \tau_{yz}}{\partial z} = 0 \tag{4}$$

Let us consider the effect of a weak lower crust for the simple case of pure strike-slip shear, with a velocity field $\mathbf{v} = (0, v_y, 0)$. For viscous deformation, $\tau_{xy} = 2\eta\, \partial v_y/\partial x$ and $\tau_{yz} = 2\eta\, \partial v_y/\partial z$, in which η is the viscosity. If η is constant, then (4) becomes

$$\frac{\partial^2 v_y}{\partial x^2} + \frac{\partial^2 v_y}{\partial z^2} = 0 \tag{5}$$

Hence, the gradient of v_y in the x-direction varies as rapidly as it does in the z-direction. For a case where v_y varied from 20 mm/year to nil over 10–20 km in the vertical, deformation might be localized near the San Andreas fault, not spread as much as 100 km from it. For power-law creep, with stress proportional to a power of strain rate, deformation can be yet more localized (e.g., Bell et al., 1977; England et al., 1985). If, however, η depends on depth, z, then (4) becomes

$$\eta(z)\frac{\partial^2 v_y}{\partial x^2} + \frac{\partial v_y}{\partial z}\frac{d\eta(z)}{dz} + \eta(z)\frac{\partial^2 v_y}{\partial z^2} = 0 \tag{6}$$

It is easy to see that for sufficiently large $d\eta/dz$, a steep gradient of v_y in the z-direction can be associated with a small gradient of v_y in the x-direction, and that shear on vertical planes can be spread over a wide area. Hence, a weak lower crust can couple a broad shear zone in the upper mantle to a narrow shear zone in the upper crust. This is apparent also in Li and Rice's (1987) more elegant calculations of *changes* in different components of stress during cycles of major earthquakes on the San Andreas fault. Recall that a broadening of shear zones with depth has been observed along deeply exhumed faults (e.g., Bak et al., 1975; Sørensen, 1983).

As the two blocks slip past one another, with shear stress on the San Andreas fault resisting that slip, shear between the upper mantle and the upper crust must develop; τ_{yz} has a finite, but not necessarily large, negative value, $-\tau_0$, at some depth z_0. Hence, $\partial \tau_{yz}/\partial z \approx -\tau_0/z_0$. From (4),

$$\frac{\partial \tau_{xy}}{\partial x} = \frac{\tau_0}{z_0} \tag{7}$$

Lachenbruch and Sass (1973) pointed out that such a configuration required an increase in the magnitude of the shear stress τ_{xy} from the San Andreas fault toward the northeast. The same applies toward the southwest, because the signs of both the vertical gradient in τ_{yz} and the horizontal gradient in τ_{xy} are opposite to those to the northeast.

Well aware of this argument, Zoback et al. (1987) used it in their discussion of a vanishingly small shear stress (τ_{xy}) on the San Andreas fault. For the shear stress on horizontal planes at the Moho ($z_0 = 30$ km) to reach $\tau_{yz} = -\tau_0 = -30$ MPa, the shear stress parallel to the San Andreas fault (τ_{xy}) should approach 100 MPa at a distance of 100 km from the fault. As Mount and Suppe (1987) and Zoback et al. (1987) pointed out, a zone of low strength, and hence low stress, along the San Andreas fault does not obviate relatively high shear stress (100 MPa) either on planes oriented obliquely to the plane of the San Andreas fault or far from the fault elsewhere within the lithosphere. Suppose that $\tau_0 \sim 100$ MPa at $z_0 = 30$ km, as Lachenbruch and Sass's (1973, 1980) explanation of heat flow requires. Then (7) implies that τ_{xy} increases by 300 MPa across the distance 100 km from the San Andreas fault. This requires what seems like a very large stress within the crust. Conversely, if shear stresses in the lithosphere did not exceed 100 MPa, and τ_{xy} attained this maximum value at 100 km from the fault, then (7) would set 30 MPa as the maximum value of τ_{yz} at a depth of 30 km. Of course, a maximum of 30 MPa for τ_{yz} at $z = 30$ km does not in any way forbid higher levels of τ_{xy}, shear stress on vertical planes parallel to the San Andreas fault, at that depth.

This argument applied to the anomalous heat flux of 30 mW m^{-2} (Lachenbruch and Sass, 1973, 1980) implies that only 10 mW m^{-2} could be produced by shear on horizontal planes. (As above, if we assume a mean rate of slip of the upper crust with respect to the lower crust of 10 mm/year, the product $\tau_{yz} \cdot v_y = 30$ MPa $\times$ 10 mm/year = 10 mW m^{-2}.) Thus the remaining 20 mW m^{-2} would seem to be produced by shear on vertical planes.

5. Some Comments on the Orientation of Stress

Knowledgeable readers cannot fail to notice that the discussion given above has paid little attention to inferences of orientations of stress in the earth. I do not doubt that in the regions undergoing folding and horizontal shortening adjacent to the San Andreas fault, the local orientation of the principal compressive stress is also orthogonal to the fault. What I doubt is that this information adds much to what is already known from the strain field or that it is relevant to the questions of partitioning of oblique convergence and the stress–heat flow paradox. In particular, I find the statement that 'the strength of the fault controls the orientation of the stress field and not the relative motion direction ($\sim$N35°W) of the Pacific and North American plates' (Zoback et al., 1987, p. 1108) to be misleading; at best it is only partially correct, and it is possibly totally false. It seems to me that, on the contrary, the inferred orientations of stress depend very much on the orientation of plate motion. Their statement is correct only in the sense that the fault is weak compared with the upper mantle, where continuous deformation results in a strain field that is subsequently used to infer orientations of stress.

First, if plate motion were obliquely divergent instead of convergent, thrust faulting and folding would be much less likely to occur perpendicular to the San Andreas fault than normal faulting, unless some other process pushed the Coast Ranges and the Sierra Nevada toward one another. Second, in Richard and Cobbold's (1989) experiments, the strike-slip fault is no weaker than the thrust faults adjacent to it. The same is likely for the San Andreas. Berry (1973) presented evidence that the folds adjacent to the San Andreas fault were associated with very high pore fluid pressures, in some cases in excess of the lithostatic pressure. Thus, I presume that neither the San Andreas fault nor the adjacent thrust faults support large stresses, where the word 'large' is referenced to the stresses that prevail in the underlying uppermost mantle.

When deformation on a large scale is considered, the entire lithosphere seems to behave as viscous continuum, with strain rate proportional to a power of stress difference (Sonder and England, 1986). With vertical and horizontal principal stresses, oblique slip is unlikely (Anderson, 1951; Houseman and England, 1986; Richard and Cobbold, 1989). In this sense, the stresses in blocks of upper crust merely take on the orientations necessary to accommodate the regional displacement field (McKenzie and Jackson, 1983), which is continuous at depth if not at the surface. The displacement field is governed largely by geometry – the relative orientations of the boundary and of relative motion between lithospheric plates (McKenzie and Jackson, 1983; England and Houseman, 1986; Houseman and England, 1986; Ekström and England, 1989) – and by the behavior of a continuous medium between essentially rigid blocks or plates. The stress field at the surface is complicated, in turn, by the local relative motions of blocks near their boundaries (e.g., Weldon and Springer, 1988).

6. Summary

I have tried to address two subjects, the partitioning of oblique convergence into pure dip-slip and pure strike-slip faulting and the stress–heat flow paradox. The fundamental assumptions include a basic image of strength and deformation in a profile of continental lithosphere. The upper crust consists of blocks bounded by relatively weak faults, and the strongest part of the crust and upper mantle seems to lie in uppermost mantle, where deformation is continuous and largely plastic. This continuous deformation is communicated and coupled to the blocks of the upper crust by continuous, ductile deformation in the lower crust. It should be clear that this image is drawn to a large extent from years of careful laboratory measurements, but geologic and geophysical observations of various kinds appear to be consistent with it.

With these assumptions, the partitioning of oblique convergence can be seen as the result of the principal stresses being essentially vertical or horizontal in the continental lithosphere (Houseman and England, 1986; Richard and Cobbold, 1989). The displacements at the earth's surface in continental regions result from the relative movements of crustal blocks that are carried by the deforming continuum beneath them. The rates, amounts, and orientations of deformation (the velocity gradient and deformation gradient tensors) in the underlying material depend on the orientation of relative displacements of neighboring, essentially rigid lithospheric plates, on the rate of that displacement, and on the orientation of the diffuse boundary between the plates. Accordingly, the particular style of deformation that one observes at the surface depends on the difference in the orientations of plate motion and of the diffuse boundary. The apparent, though presumably real, orientations of principal stresses in this upper crust are therefore *consequences* of this partitioning of strain and, as such, estimations of orientations of stress provide little additional information about the dynamic processes occurring in continental regions.

The strong uppermost mantle, which carries the crustal blocks above it, may provide the greatest resistance to deformation in continental regions. If so, shear heating within this layer is likely to be more important than in the upper crust above it. As noted above, I imagine but cannot demonstrate conclusively, that the higher heat flow in the Coast Ranges of California (Lachenbruch and Sass, 1973, 1980) results largely from such distributed shear in the mantle on vertical planes parallel to the San Andreas fault.

The least palatable aspect of the scenario presented here is probably the high strength of the upper mantle, which seems to be needed both for heat generation in California and to provide the ductile driving deformation at depth. At present, the evidence bearing on strength in the earth seems to be sufficiently uncertain to allow this assumption, but too uncertain to demonstrate wide applicability of it. Because upper mantle temperatures tend to be highest in regions of crustal extension, such as the Basin and Range Province or the Aegean region, the mantle there is less likely to be especially strong than in convergent settings. Thus, I have focused this analysis on zones of oblique convergence, where a strong upper mantle is most likely. Nevertheless, many of the arguments used here may apply also to areas of oblique extension. In any case, the image of crustal blocks separated by negligibly strong faults and overlying a homogeneous, isotropic continuum is an oversimplification, and the strengths of the faults surely are not negligible. If nothing else, then, perhaps this contribution will serve as an antidote to excessive attention focused on earthquakes and brittle faults in the upper crust (by both seismologists and funding agencies), at the expense of much needed understanding of deformation of the entire lithosphere.

Note Added in Proof

Since this paper was written, revised, and accepted, I discovered that Bird and Yuen (1979) argued much more elegantly and more convincingly than I that minimization of energy dissipation is not a reliable principle in tectonophysics. Also, since then, England and Molnar (1991) have discussed the heat flow in central California and made a more reasonable and more convincing case for shear heating at depth as the source of the extra heat.

Acknowledgments

I thank D. Davis, P. England, B. Evans, J. Jackson, A.H. Lachenbruch, R. McCaffrey, A. McGarr, D. McKenzie, R. Sibson, N. Sleep, and J. Stock for helpful comments on early drafts of the manuscript, and D. Frank for numerous retypings of it. In particular, England cleared up some misunderstandings about continuous deformation; he and McKenzie showed me that minimizing energy dissipation is a bankrupt approach. Jackson and McKenzie pointed out loopholes in my analysis of stability of faults; McCaffrey corrected some errors in my image of oblique-slip faulting; and Sleep showed me eq. (5) and its implications. This work was supported in part by NASA through grant NAG5-795.

References

Abers, G. and McCaffrey, R. (1988). Active deformation in the New Guinea fold-and-thrust belt: seismological evidence for strike-slip faulting and basement-involved thrusting. *J. Geophys. Res.* **93**, 13332–13534.

Anderson, E.M. (1951). *The Dynamics of Faulting and Dyke Formation with Applications to Britain.* Oliver and Boyd, Edinburgh.

Argus, D.F. and Gordon, R.G. (1991). Current Sierra Nevada–North America motion from very long baseline interferometry: Implications for the kinematics of the western United States. *Geology* **19**, 1085–1088.

Bak, J., Korstgaard, J., and Sørenson, K. (1975). A major shear zone within the Nassugtoqidian of west Greenland. *Tectonophysics* **27**, 191–209.

Barr, T.D. and Dahlen, F.A. (1988). Thermodynamic efficiency of brittle frictional mountain building. *Science* **242**, 749–752.

Barr, T.D. and Dahlen, F.A. (1989). Brittle frictional mountain building, 2. Thermal structure and heat budget. *J. Geophys. Res.* **94**, 3923–3947.

Barr, T.D. and Dahlen, F.A. (1990). Constraints on friction and stress in the Taiwan fold-and-thrust belt from heat flow and geochronology. *Geology* **18**, 111–115.

Bell, M.L., Crough, S.T., and Nur, A. (1977). Non-Newtonian antiplane flow and its application to plate deformation. *Tectonophysics* **39**, 501–514.

Berry, F.A.F. (1973). High fluid potentials in California Coast Ranges and their tectonic significance. *Am. Assoc. Petrol. Geol.* **57**, 1219–1249.

Bird, P. (1978). Initiation of intracontinental subduction in the Himalayas. *J. Geophys. Res.* **83**, 4975–4987.

Bird, P., and Yuen, D.A. (1979). The use of a minimum-energy dissipation principle in tectonophysics. *Earth Planet. Sci. Lett.* **45**, 214–217.

Block, L. and Royden, L.H. (1990). Core complex geometries and regional scale flow in the lower crust. *Tectonics* **9**, 557–567.

Boore, D.M. and Stierman, D.J. (1976). Source parameters of the Pt. Mugu, California, earthquake of February 21, 1973. *Bull. Seismol. Soc. Am.* **66**, 385–404.

Brace, W.F. and Byerlee, J.D. (1966). Stick-slip as a mechanism for earthquakes. *Science* **153**, 990–992.

Brace, W.F. and Byerlee, J.D. (1970). California earthquakes: why only shallow focus? *Science* **168**, 1573–1576.

Brace, W.F. and Kohlstedt, D.L. (1980). Limits on lithospheric stress imposed by laboratory experiments. *J. Geophys. Res.* **85**, 6248–6252.

Brune, J.N., Henyey, T.L., and Roy, R.F. (1969). Heat flow, stress, and rate of slip on the San Andreas fault, California. *J. Geophys. Res.* **74**, 3821–3827.

Byerlee, J.D. (1978). Friction in rocks. *Pure Appl. Geophysics* **116**, 615–626.

Byerlee, J. (1990). Friction, over pressure and fault normal compression. *Geophys. Res. Lett.* **17**, 2109–2112.

Byerlee, J.D. and Brace, W.F. (1970). Stick-slip, stable sliding, and earthquakes – effect of rock type, pressure, strain rate, and stiffness. *J. Geophys. Res.* **73**, 6031–6037.

Caristan, Y. (1982). The transition from high-temperature creep to fracture in Maryland diabase. *J. Geophys. Res.* **87**, 6781–6790.

Chen, W.-P. and Molnar, P. (1983). Focal depths of intracontinental and intraplate earthquakes and their implications for the thermal and mechanical properties of the lithosphere. *J. Geophys. Res.* **88**, 4183–4214.

Chopra, P.N. and Paterson, M.S. (1984). The role of water in the deformation of dunite, *J. Geophys. Res.* **89**, 7861–7876.

Corbett, E.J. and Johnson, C.E. (1982). The Santa Barbara, California, earthquake of 13 August 1978. *Bull. Seismol. Soc. Am.* **72**, 2201–2226.

Dahlen, F.A. (1988). Mechanical energy budget of a fold-and-thrust belt. *Nature* **331**, 335–337.

Dahlen, F.A. (1984). Non-cohesive Coulomb wedges: an exact solution. *J. Geophys. Res.* **89**, 10125–10133.

Dahlen, F.A. and Barr, T.D. (1989). Brittle frictional mountain building, 1. Deformation and mechanical energy budget. *J. Geophys. Res.* **94**, 3906–3922.

Dahlen, F.A., Suppe, J., and Davis, D. (1984). Mechanics of fold-and-thrust belts and accretionary wedges: cohesive Coulomb theory. *J. Geophys. Res.* **89**, 10087–10101.

Davis, D., Suppe, J., and Dahlen, A.F. (1983). Mechanics of fold and thrust belts and accretionary wedges. *J. Geophys. Res.* **88**, 1153–1172.

DeMets, C., Gordon, R.G., Argus, D.F., and Stein, S. (1990). Current plate motions. *Geophys. J. Int.* **101**, 425–478.

Ekström, G. and Dziewonski, A.M. (1985). Centroid-moment tensor solutions for 35 earthquakes in western North American (1977–1983). *Bull. Seismol. Soc. Am.* **75**, 23–39.

Ekström, G. and Engdahl, E.R. (1989). Earthquake source parameters and stress distribution in the Adak Island region of the central Aleutian Islands, Alaska. *J. Geophys. Res.* **94**, 15499–15519.

Ekström, G. and England, P.C. (1989). Seismic strain rates in regions of distributed continental deformation. *J. Geophys. Res.* **94**, 10231–10257.

England, P.C. and Houseman, G.A. (1986). Finite strain calculations of continental deformation 2. Comparison with the India–Asia collision. *J. Geophys. Res.* **91**, 3664–3676.

England, P.C. and McKenzie, D.P. (1982). A thin viscous sheet model for continental deformation. *Geophys. J. R. Astron. Soc.* **70**, 295–321.

England, P., and Molnar, P. (1991). Inferences of deviatoric stresses in actively deforming belts from simple physical models. *Phil. Trans. Roy. Soc. Lond.* A, **337**, 151–164.

England, P.C., Houseman, G.A., and Sonder, L. (1985). Length scales for continental deformation in convergent, divergent, and strike-slip environments: Analytical and approximate solutions for a thin viscous sheet model. *J. Geophys. Res.* **90**, 523–532.

Etheridge, M.A., Wall, V.J., Cox, S.F., and Vernon, R.H. (1980). High fluid pressures during regional metamorphism: implications for mass transfer and deformation mechanisms. *J. Geophys. Res.* **89**, 4344–4458.

Evans, B. and Goetze, C. (1979). The temperature variation of hardness of olivine and its implication for polycrystalline yield stress. *J. Geophys. Res.* **84**, 5505–5524.

Fitch, T.J. (1972). Plate convergence, transcurrent faults, and internal deformation adjacent to the southeast Asia and the western Pacific. *J. Geophys. Res.* **77**, 4432–4460.

Froidevaux, C. (1986). Basin and Range large-scale tectonics: constraints from gravity and reflection seismology. *J. Geophys. Res.* **91**, 3625–3632.

Gans, P.B. (1987). An open-system two layer crustal stretching model for the eastern Great Basin. *Tectonics* **6**, 1–12.

Goetze, C. (1978). The mechanisms of creep in olivine. *Phil. Trans. R. Soc. Lond. A*, **288**, 99–119.

Goetze, C. and Evans, B. (1979). Stress and temperature in the bending lithosphere as constrained by experimental rock mechanics. *Geophys. J. R. Astron. Soc.* **59**, 463–478.

Goldsmith, J.R. (1986). The role of hydrogen in promoting Al–Si interdiffusion in albite ($NaAlSi_3O_8$) at high pressures. *Earth Planet. Sci. Lett.* **80**, 135–138.

Goldsmith, J.R. (1987). Al/Si interdiffusion in albite: effect of pressure and the role of hydrogen. *Contrib. Mineral. Petrol.* **95**, 311–321.

Goldsmith, J.R. (1988). Enhanced Al/Si diffusion in $KAlSi_3O_8$ at high pressures: the effect of hydrogen. *J. Geol.* **96**, 109–124.

Gratier, J.-P. (1987). Pressure solution-deposition creep and associated tectonic differentiation in sedimentary rocks. In *Deformation of Sediments and Sedimentary Rocks* (ed. M.E. Jones and R.M.F. Preston), pp. 25–38. Geological Society of London Special Publication No. 29.

Gutenberg, B. (1955). The first motion on longitudinal and transverse waves of the main shock and the direction of slip. *Calif. Div. Mines. Geol. Bull.* **171**, 165–170.

Henyey, T.L. and Wasserburg, G.J. (1971). Heat flow near major strike-slip faults in California. *J. Geophys. Res.* **76**, 7924–7946.

Hornafius, J.S., Luyendyk, B.P., Terres, R.R., and Kamerling, M.J. (1986). Timing and extent of Neogene tectonic rotation in the western Transverse Ranges, California. *Geol. Soc. Am. Bull.* **97**, 1476–1487.

Houseman, G.A. and England, P.C. (1986). Finite strain calculations of continental deformation, 1. Methods and general results for convergent zones. *J. Geophys. Res.* **86**, 6115–6132.

Jackson, J. and McKenzie, D. (1983). The geometrical evolution of normal fault systems. *J. Struct. Geol.* **5**, 471–482.

Jackson, J. and McKenzie, D. (1984). Active tectonics of the Alpine-Himalayan belt between western Turkey and Pakistan. *Geophys. J. R. Astron. Soc.* **77**, 185–264.

Jackson, J. and McKenzie, D. (1988). The relationship between plate motions and seismic moment tensors, and the rates of active deformation in the Mediterranean and the Middle East. *Geophys. J.* **93**, 45–73.

Jackson, J., and Molnar, P. (1990). Active faulting and block rotations in the eastern Transverse Ranges, California. *J. Geophys. Res.* **95**, 22073–22087.

Jarrard, R.D. (1986a). Relations among subduction zone parameters. *Rev. Geophys.* **24**, 217–284.

Jarrard, R.D. (1986b). Terrane motion by strike-slip faulting of forearc slivers. *Geology* **14**, 780–783.

Johnson, T. and Molnar, P. (1972). Focal mechanisms and plate tectonics of the southwest Pacific. *J. Geophys. Res.* **77**, 5000–5032.

Karato, S.-I., Paterson, M.S., and FitzGerald, J.D. (1986). Rheology of synthetic olivine aggregates; influence of grain size and water. *J. Geophys. Res.* **91**, 8151–8176.

Kellogg, J.N. and Bonin, W.E. (1982). Subduction of the Caribbean plate and basement uplifts in the overriding South American plate. *Tectonics* **1**, 251–276.

Kirby, S.H. (1977). State of stress in the lithosphere: constraints provided by the experimental deformation of rocks. *Pure Appl. Geophys.* **115**, 261–274.

Kirby, S.H. (1980). Tectonic stress in the lithosphere: constraints provided by the experimental deformation of rocks. *J. Geophys. Res.* **85**, 6353–6363.

Kirby, S.H. (1983). Rheology of the lithosphere. *Rev. Geophys. Space Phys.* **21**, 1458–1487.

Kirby, S.H. and Kronenberg, A.K. (1987). Rheology of the lithosphere; selected topics. *Rev. Geophys.* **25**, 12198–12244.

Kissel, C. and Laj, C. (1988). The Tertiary geodynamical evolution of the Aegean arc: a paleomagnetic reconstruction. *Tectonophysics* **146**, 183–201.

Kohlstedt, D.L. and Goetze, C. (1974). Low-stress high-temperature creep in olivine single crystals. *J. Geophys. Res.* **79**, 2045–2051.

Lachenbruch, A.H. and Sass, J.H. (1973). Thermo-mechanical aspects of the San Andreas fault system. In *Proc. Conf. on Tectonic Problems of the San Andreas Fault System, Geological Sciences XIII*, pp. 192–205. School of Earth Sciences, Stanford University Publications.

Lachenbruch, A.H. and Sass, J.H. (1978). Models of an extending lithosphere and heat flow in the Basin and Range Province, in Cenozoic Tectonics and Regional Geophysics of the Western Cordillera. *Geol. Soc. Am. Mem.* **152**, 209–250.

Lachenbruch, A.H. and Sass, J.H. (1980). Heat flow and energetics of the San Andreas fault zone. *J. Geophys. Res.* **85**, 6185–6222.

Lamb, S.H. (1987). A model for tectonic rotations about a vertical axis. *Earth Planet. Sci. Lett.* **84**, 75–86.

Langston, C.A. (1978). The February 9, 1971 San Fernando earthquake: a study of source finiteness in teleseismic body waves. *Bull. Seismol. Soc. Am.* **68**, 1–29.

Li, V.C. and Rice, J.R. (1987). Crustal deformation in great California earthquake cycles. *J. Geophys. Res.* **92**, 11533–11551.

Luyendyk, B.P., Kamerling, M.J., Terres, R.R., and Hornafius, J.S. (1985). Simple shear of southern California during Neogene time suggested by paleomagnetic declinations. *J. Geophys. Res.* **90**, 12454–12466.

McCaffrey, R. (1991). Slip vectors and stretching of the Sumatran forearc. *Geology* **19**, 881–884.

Mackwell, S.J., Kohlstedt, D.L., and Paterson, M.S. (1985). The role of water in the deformation of olivine single crystals. *J. Geophys. Res.* **90**, 11319–11333.

McGarr, A. (1980). Some constraints on levels of shear stress in the crust from observations and theory. *J. Geophys. Res.* **85**, 6231–6238.

McGarr, A. (1982). Analysis of states of stress between provinces of constant stress. *J. Geophys. Res.* **87**, 9279–9288.

McGarr, A. (1984). Scaling of ground motion parameters, state of stress, and focal depth. *J. Geophys. Res.* **89**, 6969–6979.

McGarr, A., Zoback, M.D. and Hanks, T.C. (1982). Implications of an elastic analysis of in situ stress measurements near the San Andreas fault. *J. Geophys. Res.* **87**, 7797–7806.

McKenzie, D. and Jackson, J. (1983). The relationship between strain rates, crustal thickening, paleomagnetism, finite strain and fault movements within a deforming zone. *Earth Planet. Sci. Lett.* **65**, 182–202.

McKenzie, D.P. and Morgan, W.J. (1969). Evolution of triple junctions. *Nature* **224**, 125–133.

Meissner, R. and Strehlau, J. (1982). Limits of stresses in continental crusts and their relation to the depth-frequency distribution of shallow earthquakes. *Tectonics* **1**, 73–89.

Melosh, H.J. and Williams, C.A. (1989). Mechanics of graben formation in crustal rocks: a finite element analysis. *J. Geophys. Res.* **94**, 13961–13973.

Michael, A.J. (1990). Energy constraints on kinematic models of oblique faulting: Loma Prieta versus Parkfield-Coalinga. *Geophys. Res. Lett.* **17**, 1453–1456.

Molnar, P. and England, P. (1990). Temperatures, heat flux, and frictional stress near major thrust faults. *J. Geophys. Res.* **95**, 4833–4856.

Molnar, P. and Tapponnier, P. (1975). Cenozoic tectonics of Asia: Effects of a continental collision. *Science* **189**, 419–426.

Molnar, P. and Tapponnier, P. (1981). A possible dependence of tectonic strength on the age of the crust in Asia. *Earth Planet. Sci. Lett.* **52**, 107–114.

Mount, V.S. and Suppe, J. (1987). State of stress near the San Andreas fault: Implications for wrench tectonics. *Geology* **15**, 1143–1146.

Namson, J. and Davis, T. (1988). Seismically active fold and thrust belt in the San Joaquin Valley, central California. *Geol. Soc. Am. Bull.* **100**, 257–273.

Nur, A., Ron, H., and Scotti, O. (1986). Fault mechanics and the kinematics of block rotations. *Geology* **14**, 746–749.

Richard, P. and Cobbold, P. (1989). Structures en fleur positives et décrochements crustaux: modélisation analogique et interprétation mécanique. *C. R. Acad. Sci. Paris, Série II* **308**, 553–560.

Rutter, E.H. (1983). Pressure solution in nature, theory and experiment. *J. Geol. Soc. Lond.* **140**, 725–740.

Scholz, C.H. (1990). *The Mechanics of Earthquakes and Faulting*. Cambridge University Press.

Sibson, R.H. (1977). Fault rocks and fault mechanisms. *J. Geol. Soc. Lond.* **133**, 191–213.

Sibson, R.H. (1982). Fault zone models, heat flow, and the depth distribution of earthquakes in the continental crust of the United States. *Bull. Seismol. Soc. Am.* **72**, 151–163.

Sonder, L.J. and England, P. (1986). Vertical averages of rheology of the continental lithosphere: relation to thin sheet parameters. *Earth Planet. Sci. Lett.* **77**, 81–90.

Sonder, L.J., England, P.C., and Houseman, G.A. (1986). Continuum calculations in transcurrent environments. *J. Geophys. Res.* **91**, 4797–4810.

Sørenson, K. (1983). Growth and dynamics of the Nordre Strømfjord shear zone. *J. Geophys. Res.* **88**, 3419–3437.

Stein, R.S. and King, G.C.P. (1984). Seismic potential revealed by surface folding: 1983 Coalinga, California, Earthquake. *Science* **224**, 869–872.

Stein, R.S. and Thatcher, W. (1981). Seismic and aseismic deformation associated with the 1952 Kern County, California, earthquake and relationship to the Quaternary history of the White Wolf fault. *J. Geophys. Res.* **86**, 4913–4928.

Stesky, R.M. and Brace, W.F. (1973). Estimation of frictional stress on the San Andreas fault from laboratory measurements. *Proceedings of the Conference on the Tectonic Problems of the San Andreas Fault* (ed. R.L. Kovach and A. Nur), pp. 206–214. Stanford University Publications, Geol. Sci. Vol. XII.

Stesky, R.M., Brace, W.F., Riley, D., and Robin, P.-Y. (1974). Friction in faulted rock at high temperature and pressure. *Tectonophysics* **23**, 177–203.

Tapponnier, P. and Francheteau, J. (1978). Necking of the lithosphere and the mechanics of slowly accreting plate boundaries. *J. Geophys. Res.* **83**, 3955–3970.

Tapponnier, P. and Molnar, P. (1979). Active faulting and Cenozoic tectonics of the Tien Shan, Mongolia and Baykal Regions. *J. Geophys. Res.* **84**, 3425–3459.

Tse, S. and Rice, J.R. (1986). Crustal earthquake instability in relation to the depth variation of frictional slip processes. *J. Geophys. Res.* **91**, 9452–9472.

Tullis, J. and Yund, R.A. (1977). Experimental deformation of dry Westerly granite. *J. Geophys. Res.* **82**, 5705–5718.

Vilotte, J.P., Madariaga, R., Dagnières, M., and Zienkiewicz, O. (1986). Numerical study of continental collision: Influence of boundary forces and an initial stiff inclusion. *Geophys. J. R. Astron. Soc.* **84**, 279–310.

Walcott, R.I. (1984). The kinematics of the plate boundary zone through New Zealand: a comparison of short- and long-term deformations. *Geophys. J. R. Astron. Soc.* **79**, 613–633.

Weertman, J. (1970). The creep strength of the Earth's mantle. *Rev. Geophys. Space Phys.* **8**, 145–168.

Weldon II, R.J. and Springer, J.E. (1988). Active faulting near the Cajon Pass well, southern California: implications for the stress orientation near the San Andreas fault. *Geophys. Res. Lett.* **15**, 993–996.

Wernicke, B.P. (1981). Low-angle normal faults in the Basin and Range Province-nappe tectonics in an extending orogen. *Nature* **291**, 645–648.

Wernicke, B.P. (1985). Uniform-sense normal simple shear of the continental lithosphere. *Can. J. Earth Sci.* **22**, 108–125.

Wernicke, B.P. and Axen, G.J. (1988). On the role of isostasy in the evolution of normal fault systems. *Geology* **16**, 848–851.

Whitcomb, J., Allen, C., Garmany, J., and Hileman, J. (1973). San Fernando earthquake series 1971, focal mechanism and tectonics. *Rev. Geophys. Space Phys.* **11**, 693–730.

Zoback, M.L. and Zoback, M.D. (1980). State of stress in the conterminous United States. *J. Geophys. Res.* **85**, 6113–6156.

Zoback, M.D., Zoback, M.L., Mount, V.S., Suppe, J., Eaton, J.P., Healy, J.H., Oppenheimer, D., Reasenberg, P., Jones, L.M., Raleigh, C.B., Wong, I.G., Scotti, O., and Wentworth, C. (1987). New evidence on the state of stress of the San Andreas fault system. *Science* **238**, 1105–1111.

Zuber, M.T., Parmentier, E.M., and Fletcher, R.C. (1986). Extension of continental lithosphere: a model for two scales of Basin and Range deformation. *J. Geophys. Res.* **91**, 4826–4838.

Chapter 19
Hydraulic Pulses in the Earth's Crust

Amos Nur[1] *and Joseph Walder*[2]

[1]*Department of Geophysics, Stanford University, Stanford, CA 94305, U.S.A.*

[2]*U.S. Geological Survey, Cascades Volcano Observatory, 5400 MacArthur Blvd., Vancouver, WA 98661, U.S.A.*

Abstract

The analysis in this chapter attempts to quantify the coupling between pore space deformation and pore pressure in the earth's crust. Clearly, the state of free water deep in the crust is inseparable from the state of porosity, permeability, and pore pressure. If porosity and permeability are time-invariant, crustal pore pressure at depth must be close to hydrostatic. This is in conflict with evidence for ubiquitous high pore pressure at depth. Consequently crustal porosity, permeability, and water content must decrease with time, as pores close inelastically in the crust. When crustal permeability is low, and its rate of decrease with porosity is large, pore pressure P_p of trapped fluids must rise with time until spontaneous hydrofracturing, pulsed release of water, and the loss of some porosity occurs, followed by a drop in pore pressure and sealing of the system. After a P_p buildup period, another hydrofracturing episode occurs. Because inelastic pore deformation is enhanced by shear stress, this episodic process is most probable in tectonically active areas. Dimensional analysis shows that tens to hundreds of P_p episodes or 'burps,' each lasting 10^3–10^5 years, can occur before the crust dewaters in 10^6–10^7 years. Oceanic crust subduction may prolong this process to 10^8 years or more.

High pore pressure can explain deep crustal seismic reflectors and detachments, and large-scale horizontal fluid migration in the crust. Episodic pore pressure buildup and release in the crust can explain the formation of hydrothermal ore deposits and mineralized veins, and possibly some aspects of the mechanics of very large earthquakes.

1. Introduction

It was W.F. Brace's (1980) paper on crustal permeability that brought into focus one of the simplest yet most profound questions about the state of the earth's crust: what

FAULT MECHANICS AND TRANSPORT PROPERTIES OF ROCKS
ISBN 0-12-243780-2

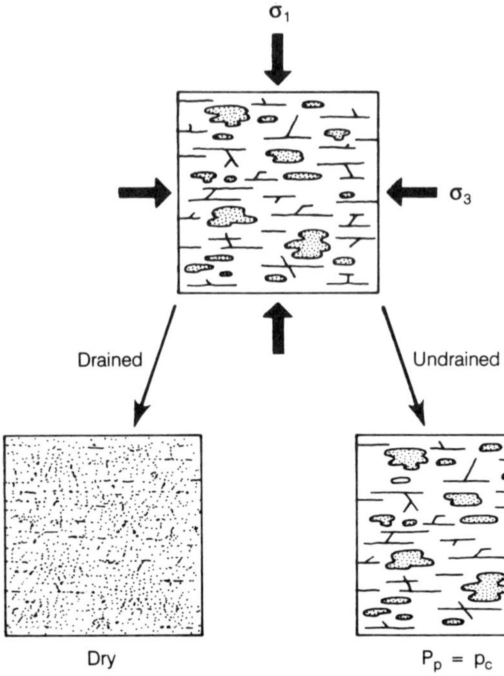

Figure 1. An initially stressed, water-saturated rock mass, with pore pressure P_p less than confining stress $P_c = \sigma_{ii}/3$, can either dry up and lose its porosity gradually if drainage is fast enough (lower left block), or, if drainage is too slow, pore pressure will with time reach the least confining stress.

is the depth of free water in the crust? This porosity is in turn inseparable from the water's ambient pore pressure P_p (Figure 1) which, as discussed below, is in turn inseparably linked to permeability.

In this chapter we attempt to answer quantitatively the three interlinked questions: (1) What is the depth and pressure of free water in the crust? (2) What are typical crustal porosity and permeability values? and (3) How do these parameters vary with time and over what timescale do they vary?

Because direct measurements of crustal hydrologic properties are confined to shallow depths, the answers to these questions can at present be based only on indirect evidence, some of which is listed in Table 1. For example, oxygen isotopic data (e.g., Taylor, 1977; Norton and Taylor, 1979) indicates water circulation down to 20 km. Deep zones of low electrical resistivity, inferred from crustal electromagnetic soundings, also suggest the presence of water (e.g., Nekut et al., 1977; Thompson et al., 1983). Seismic low-velocity zones (Feng and McEvilly, 1983) and low-seismic Q zones in the crust (Herrmann and Mitchell, 1975) are also consistent with pore pressure close to lithostatic (Nur and Simmons, 1969; Berry and Mair, 1977; Winkler and Nur, 1982). Reflections from deep crystal fault zones, especially subhorizontal detachments, may be due to zones of high P_p (Jones and Nur, 1984).

Fyfe et al. (1978) and Etheridge et al. (1984) reviewed geologic evidence for widespread free water with high P_p during low- to medium-grade regional metamorphism. The

Table 1. Possible indicators for the presence of free water in the earth's crust as suggested by the various referenced investigators. Also given are estimated depth values associated with these indicators

Indicator	Depth range (km)	Reference
Water table	0–2	Kozlovsky (1986)
Deep wells (i.e., the Kola Well)	to 12	Kozlovsky (1986)
Reservoir-induced seismicity	to 12	Bell and Nur (1978)
Crustal low-velocity zones	7–12	Berry and Mair (1977) Feng and McEvilly (1983) Jones and Nur (1984)
Crustal electrical conductivity zones	10–20	Nekut et al. (1977) Shankland and Ander (1983)
Oxygen isotopes	to 20	Taylor (1977)
Metamorphism	>20	Fyfe et al. (1978) Etheridge et al. (1984)
Crack healing and sealing	?	Richter and Simmons (1977) Sprunt and Nur (1976, 1977, 1979) Ramsay (1980) Smith and Evans (1984)
Formation of hydrothermal ore deposits	>5	Norton and Knight (1977) Cathles and Smith (1983)
Crustal seismic attenuation zones	7–15	Herrmann and Mitchell (1975) Winkler and Nur (1982)
Low stress on faults	0–10	Raleigh and Evernden (1982) Rice (this volume, Chapter 20)

ubiquity of mineralized fractures (Ramsay, 1980) also requires P_p to exceed the minimum principal stress during fracture formation and suggests widespread high P_p.

In order to maintain elevated P_p in the crust for geologically significant periods of time, the permeability of crustal rocks must be very low, down to 10^{-21} m^2 (1 ndarcy) or less (e.g., Bredehoeft and Hanshaw, 1968; Hanshaw and Bredehoeft, 1968). Such low permeability values are in remarkable disagreement with the conclusion of Brace (1980), who showed, on the basis of numerous geologic indicators, that fluid flow in the crust must be rapid and that crustal permeability must consequently be 10^{-5} m^2 (1 mdarcy) or higher to depths of 10 km or so. He concluded that P_p generally is not likely to exceed hydrostatic pressure at depth in the crust. This conclusion is in apparent conflict with the evidence for widespread, high crustal P_p.

In this chapter we model crustal hydraulics specifically to try to reconcile the two conflicting lines of evidence regarding crustal hydrology: ubiquitous water, high pore pressure, and the consequent low permeability on the one hand, and evidence for deep circulation and rapid flow on the other.

2. Time Dependence of Porosity and Permeability

The most common approach to the hydraulics of the crust is based on the (often unstated) assumption that porosity and permeability are time-invariant. Under this assumption, hydraulic permeability measured in situ is used to directly estimate typical ambient crustal permeability values. Brace's (1980; Brace and Walsh, 1984) extensive compilations thus lead to typical values of crustal permeabilities ranging from 100 darcy for shallow fractured rock masses, to 10^{-5} or 10^{-6} darcy at depth. With these permeabilities, crustal pore pressure can generally not be sustained above hydrostatic for geologically significant time.

The apparent contradiction between the indirect evidence for elevated crust P_p and the high permeability throughout the crust can therefore not be reconciled if porosity and permeability are assumed constant with time. However, if porosity, permeability, and consequently pore pressure, vary significantly with time, the two seemingly conflicting lines of evidence may be reconciled. There is in fact abundant, although indirect, evidence for episodic P_p pulses, associated with fluid flow during magmatic intrusions (Norton and Taylor, 1979), composite crack-filled veins, presumably episodically precipitated from solution (Ramsay, 1980), and isotopic variations across veins (Shemesh et al., 1992). For porosity, permeability, and pore pressure to vary with time, inelastic pore space variations are required. Tectonically or pore-pressure-induced fractures are the most likely mechanism for porosity to increase with time and pore pressure to decrease. The question that immediately arises is whether there is enough porosity deep in the crust, and whether it can decrease rapidly enough with time to produce large, lasting pore pressure increases. And how does permeability change through time to maintain P_p well in excess of hydrostatic?

Three obvious classes of mechanisms for porosity reduction, and consequently permeability reduction, are (1) plastic pore closure; (2) stress-induced dissolution and redeposition in the pores ('pressure solution') (e.g., Richter and Simmons, 1977; Sprunt and Nur, 1976, 1977, 1979; Padovani et al., 1982); and (3) crack healing (where mineral filling is the same as the host grain) and sealing (where crack filling is mineralogically different from the host grain).

Laboratory results strongly suggest that porosity reduction can be relatively fast (Smith and Evans, 1984) under mid-crustal conditions, perhaps especially so when rock is subject to deviatoric stress (Sprunt and Nur, 1976, 1977). However, two drastically different paths for the decrease of crustal porosity and permeability with time and the associated pore pressure changes can be envisioned, depending on the rate of porosity reduction and on the permeability (Figure 1).

Case 1. High crustal permeability. Owing to porosity reduction processes, the pore space at depth gradually decreases with time. If a permeable path exists to the earth's free surface, the fluid in the pore space will gradually be squeezed out so that P_p during this process will remain close to hydrostatic. Ultimately, this process leads to a dry, pore-free crust. Clearly, for this gradual dewatering at hydrostatic P_p to occur, the initial permeability of the rock system must be sufficiently high, and the rate of permeability decrease with decreasing porosity relatively small.

Case 2. Low permeability. In contrast, when porosity at depth decreases with time, but the hydraulic connectivity to the free surface is sufficiently low, the pore fluid cannot escape fast enough, and its pressure rises. However, this buildup of P_p is limited by the least normal stress (plus the tensile strength) of rock at depth. When P_p reaches this limiting stress level, natural hydrofracturing will occur which creates a short lived but significant hydraulic path. This in turn will cause a short episode of fluid expulsion, or

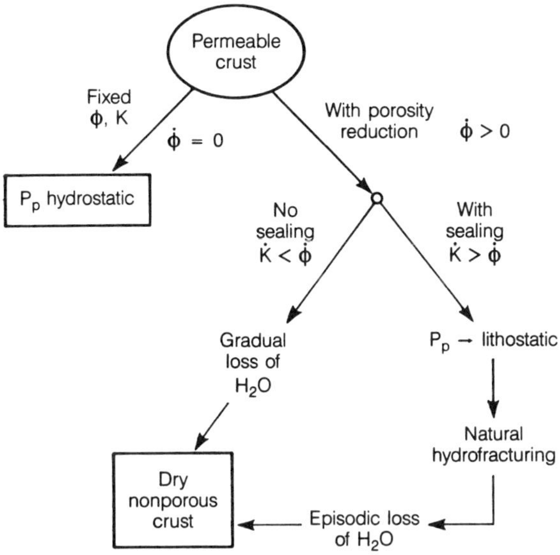

Figure 2. Different pore space/pore pressure paths for crustal rock mass depending on the rates of (1) porosity reduction $\dot{\phi}$ and (2) permeability reduction $\dot{K}$. When porosity ϕ and permeability K are fixed ($\dot{\phi} = 0$ and $\dot{K} = 0$), crustal pore pressure remains hydrostatic. It is much more likely, however, that porosity and hence permeability decrease with time. If permeability is low and its reduction rate is fast, episodic pore pressure buildup and release by repeated hydrofracturing is most likely, eventually leading to a dry, nonporous crust.

a pressure pulse, accompanied by a rapid partial pore pressure drop. Because of this drop, the hydrofracture closes, and consequently the system returns to its low prefac permeability state. Following that, as porosity reduction continues, P_p again builds up gradually to initiate another episode of hydrofracturing, fluid release, sealing, etc.

Thus porosity reduction in low-permeability systems involves episodic variation of pore pressure between lithostatic and less than lithostatic values in which each episode leads to a pulse of fluid release and rapid flow during a short period hydraulic fracturing and fluid release, followed by a long period of no or little flow, during which pore pressure recovers. The two different paths that pore pressure may follow in crustal rock is summarized in Figure 2.

3. Rates and Magnitudes of Pore Volume Strain and Pore Pressure Buildup

Next we try to determine quantitatively (1) whether actual crustal porosity reduction and its rate are hydrologically important, and (2) whether porosity reduction is typically gradual or episodic. Specifically, we need to somehow estimate four parameters: (1) the rate of porosity reduction ϕ required for P_p buildup; (2) the permeability values k and their rates of change with porosity, $dk/k\phi$, in situ; (3) the time required for pore pressure buildup and hence, the duration of the P_p cycles; and (4) the time required for the elimination of all connected porosity.

Walder and Nur (1984) have derived the following equation for pore pressure P_p:

$$\frac{1}{c}\frac{\partial P_p}{\partial t} = \Delta^2 P_p + \frac{\mu}{k}\dot{\phi} \tag{1}$$

where c is the hydraulic diffusivity $c = k/\mu\phi(\beta_f + \beta_\phi)$, β_f is pore fluid compressibility, β_ϕ is the *elastic* pore volume compressibility, μ is fluid viscosity, ϕ is porosity, $\dot{\phi} = \partial\phi/\partial t$ is the *inelastic* porosity reduction with time, $k = k(\phi)$ is hydraulic permeability, and t is time.

In the simple case of flow in the vertical direction, z, only, and using the scaling $P_p = P_0\bar{P}$, $t = t_0 \cdot \bar{t}$, and $z = h \cdot \bar{z}$, where P_0 is the maximum allowable P_p, $t_0 = h^2/c$, h is the depth to which porosity reduction occurs, and dimensionless variables are denoted by bars, eq. (1) is reduced to (Walder, 1984)

$$\frac{\partial \bar{p}}{\partial \bar{t}} = \frac{\partial^2 \bar{p}}{\partial \bar{z}^2} + \frac{\mu h}{\Delta\rho g}\left(\frac{\dot{\phi}}{k}\right)$$

It follows that the time for P_p buildup due to porosity reduction is controlled by the factor F,

$$F = \frac{\mu h}{\Delta\rho g}\frac{\dot{\phi}}{k} \tag{2}$$

where $\Delta\rho$ is the difference in density between rock and fluid. When $F < 1$, pore pressure remains close to hydrostatic during porosity reduction. In contrast, when $F \gg 1$, fluid pressure increases significantly above hydrostatic during porosity reduction. Furthermore, because F increases with depth h, induced high pore pressure is more likely at greater depth. This is further enhanced by the lower permeability k that is common at greater depth.

As an example, Walder and Nur (1984) considered a 10-km section of granitic crust undergoing porosity reduction, with the permeability-porosity relationship

$$k = k_0\left\{\frac{\phi^2 - \phi_c^2}{\phi_0^2 - \phi_c^2}\right\}$$

where k_0 and ϕ_0 are the initial permeability and porosity, respectively, and ϕ_c is the percolation threshold porosity at which flow can first occur. With $\phi_c = 2 \times 10^{-4} \cdot k_0 = 5 \times 10^{-20}\,\mathrm{m}^2$ (50 ndarcy), $\mu = 2 \times 10^{-4}$ Pa, $\Delta\rho = 1.7 \times 10^3\,\mathrm{kg\,m}^{-3}$, $F = 1$ when $\dot{\phi} = 4 \times 10^{-16}\,\mathrm{s}^{-1}$. In other words, this result suggests that were porosity reduction $\dot{\phi}$ to proceed throughout the 10-km-thick section at a rate lower than $4 \times 10^{-16}\,\mathrm{s}^{-1}$ or about 1% million years, pore pressure would remain hydrostatic. In contrast, when $\dot{\phi}$ is greater than $4 \times 10^{-16}\,\mathrm{s}^{-1}$, high pore pressure must develop, leading to pulsed, episodic releases.

Is the time required for pore pressure buildup to lithostatic fast enough to be geologically important? As mentioned above, the meager evidence which does exist indicates that porosity and permeability reduction can be geologically rapid if temperatures are sufficiently high. For example, Smith and Evans (1984) and Brantley et al. (1990) found that crack healing in quartz requires several hours at 600°C and several days at 400°C. Even at crustal depths of 5–12 km with temperatures between 100°C and 250°C, crack lifetimes increase to only hundreds of years. Furthermore, Sprunt and Nur (1977) found that porosity can be reduced by 30–40% in sandstones subject to moderate shear stress at 250°C in 2 weeks. The corresponding period of porosity reduction at 150°C, using (for lack of any more direct data) the Smith and Evans (1984) activation energy of 50–100 kJ mol^{-1}, would be 10^2–10^3 years.

The minimum time for pore pressure to reach lithostatic can be estimated for the extreme case of a hydraulically sealed rock system. Equation (1) for this case shows that the rate of pore pressure increase with time is simply proportional to the rate of relative porosity change with time:

$$\left.\frac{\Delta p}{\Delta t}\right|_{\text{no flow}} \cong \frac{\dot{\phi}}{\phi} \tag{3}$$

and the time τ_1 for pressure buildup to rupture, Δp, is

$$\tau_1 \cong \beta_f \phi \frac{\Delta P_p}{\dot{\phi}} \tag{4}$$

Taking $\Delta P_p = 200$ bar, $\beta_f = 5 \times 10^{-5}$ bar^{-1}, $\phi = 5 \times 10^{-3}$, and $\dot{\phi} = 3 \times 10^{-16}$ s^{-1}, we obtain $\Delta t \cong 10^{12}$ s, or 4×10^4 years. This buildup time can be much shorter, of course, if porosity reduction rate $\dot{\phi}$ is larger than 3×10^{-16} s^{-1}, when the system porosity ϕ is smaller than 5×10^{-3}, or when the pore pressure fluctuation is significantly less than 200 bar. For example for $\Delta p = 50$ bar, $\phi = 2 \times 10^{-3}$, and $\dot{\phi} = 10^{-15}$ s^{-1}, we find $\Delta t = 10^{10}$ s, or only 1500 years. we will assume therefore that typical value of time τ_1 required for P_p to increase say from hydrostatic to lithostatic pressure is on the order of $\tau_1 = 10^3$–10^5 years.

The porosity reduction during one cycle of pressure rise and decay for $\phi = 10^{-15}$ s^{-1} is around 3×10^{-4} (or 0.03%) for $\tau_1 = 10^4$ years, and 3×10^{-3} (0.3%) for $\tau_1 = 10^5$ years. This reduction during one episode of pore pressure buildup and hydraulic fracturing is therefore only a fraction of the total porosity. Consequently, the porosity reduction process in a low-permeability crust can involve a large number of episodes, as illustrated schematically in Figure 3.

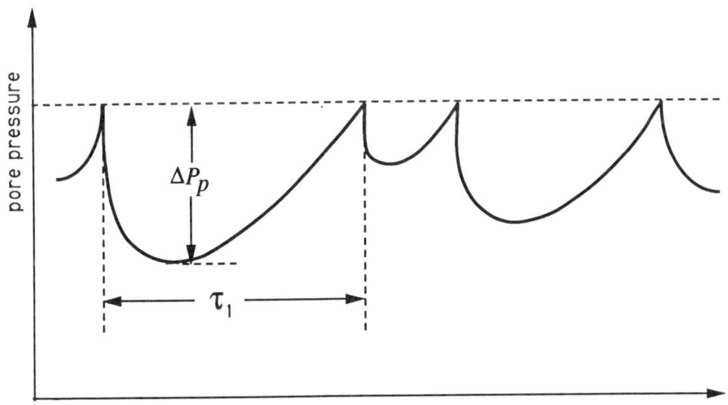

Figure 3. Idealized variations of pore pressure with time in a crust undergoing episodic pulses. Typical values for the duration of each episode are $\tau = 10^2$–10^4 years, and the pore pressure fluctuations $\Delta P_p = 100$–1000 atm. The time required to drain this kind of crust will be on the order of 10^5–10^7 years, involving tens to hundreds of episodes.

4. Geological Duration

The number of hydraulic pulses, involving P_p buildup, hydrofracturing, and release, can be estimated in two ways. Assuming that pulses will continue as long as porosity remains, we can define the duration $\tau_2 = \dot{\phi}^{-1}$. Thus for $\dot{\phi} = 10^{-16}\,s^{-1}$, $\tau_2 = 300 \times 10^6$ years; and for $\dot{\phi} = 10^{-15}\,s^{-1}$, $\tau_2 = 30 \times 10^6$ years. These estimates are probably unrealistically high, because porosity may lose its connectivity at some finite value, thus reducing the amount available for the reduction process considered here. But even if only 1/10 of the total porosity is involved, durations of $\tau_2 = 30 \times 10^6$ years (for $\dot{\phi} = 10^{-16}\,s^{-1}$) to $\tau_2 = 3 \times 10^6$ years (for $\dot{\phi} = 10^{-15}\,s^{-1}$) are expected.

Clearly, τ_2 is much longer than τ_1, the duration of a single pore pressure cycle. The number n of fluid release episodes is then simply

$$n = \frac{\tau_2}{\tau_1} \tag{5}$$

Taking $\tau_1 = 10^4–10^6$ years, and $\tau_2 = 10^7–10^8$ years, we find the range for $n = 10–10^4$ cycles.

The number of episodes can be estimated also from bounds on the magnitude of the pressure rise, ΔP_p during each cycle. Assuming that the pore fluid mass is conserved during the P_p buildup phase in each cycle (and is reduced only during the expulsion phase), we have

$$\frac{\Delta V_f}{V_f} = \beta_f \cdot \Delta p \tag{6}$$

where V_f is the pore fluid volume, ΔV_f is the change of V_f due to pore pressure Δp, β_f is the fluid's compressibility, and Δp is the induced P_p change during the buildup in one cycle. Neglecting the changes in fluid and grain density during the pore pressure cycle, we can write

$$\frac{\Delta \phi}{\phi} = \frac{\Delta V_f}{V_f} \tag{7}$$

where $\Delta \phi$ is the porosity change during a cycle. Combining eqs. (6) and (7) yields

$$\frac{\Delta \phi}{\phi} \approx \beta_f \cdot \Delta p \tag{8}$$

taking $\beta_f = 5 \times 10^{-5}\,\text{bar}^{-1}$ for water, eq. (8) indicates that pressure rises of $\Delta p = 100$, 300, and 1000 bar require 0.5%, 1.5% and 5% reduction of the initial porosity, respectively. Accordingly, if these ΔP_p values represent realistic P_p fluctuations, somewhere between 20 and 200 episodes are needed before porosity is eliminated and the episodes of fluid expulsion cease. This value is in reasonable agreement with the estimate n of eq. (5). It is also of the same order of magnitude as the number of crack seal layers observed in situ in some cases (e.g., Ramsay, 1980).

5. Water Replenished by Subduction

The duration τ_2 of crustal dewatering could be significantly prolonged if water were added to the crust, most likely from underlying subducting oceanic slabs (Figure 4) rich in water and/or hydrous minerals (i.e., McGeary et al., 1985).

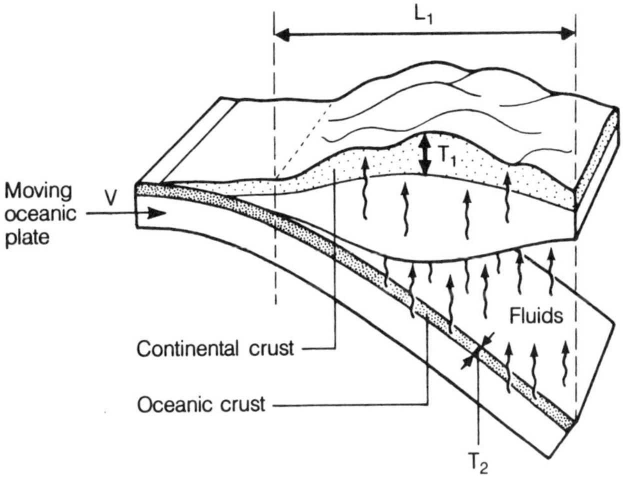

Figure 4. The addition of fluids from subducted lithosphere can prolong the lifespan of the restless, 'burping' overlying crust to be on the order of the duration of subduction – 10^8 years or more.

Table 2. The main time constants for the time-dependent hydraulic behavior of the crust

Duration of a P_p cycle	10^3–10^5 years
Time for crust to dry up	10^6–10^7 years
Duration of subduction	10^7–10^8 years

The amount of water added to the crust from a slab can be roughly estimated from the pore fluid fluxes involved. The volume V_c of fluid per unit width in the continental crust overlying subduction (Figure 4) is roughly $V_c = T_1 L_1 \phi_1$, where T_1 is the thickness of the fluid-bearing crust, L_1 is the length of the crust involved, and ϕ_1 is the average porosity over the thickness T_1. Taking $T_1 = 10$ km, porosity $\phi_1 = 1\%$, $L_1 = 300$ km, the volume of water per unit width in this crust is 30 km^2.

The volume V_0 of fluid subducted per unit width with the oceanic crust during the duration of subduction τ_3 is $V_0 = V\tau_3 T_2 \phi_2$, where V is subduction velocity, T_0 is the thickness of the porous oceanic crust, and ϕ_2 is its average porosity. Taking $T_2 = 5$ km, $T_2 = 1\%$, $\tau_3 = 10^8$ years, and $V = 5\,\text{cm}\,\text{y}^{-1}$, the volume of free water per unit width of subducted slab is on the order of 250 km^2, about 10 times V_c. This suggests that waters released from slabs at depth can add several times the initial pore fluid volume to the continental crust. Consequently, crustal 'burping' may last as long as typical subduction, so that $\tau_2 = \tau_3 = 10^8$ years, as summarized in Table 2.

6. Discussion and Conclusion

We have attempted to show that the question of the presence of free water deep in the crust is inseparable from the state of porosity, permeability and pore pressure in the crust, and especially their time dependence. If we *assume* that porosity and permeability

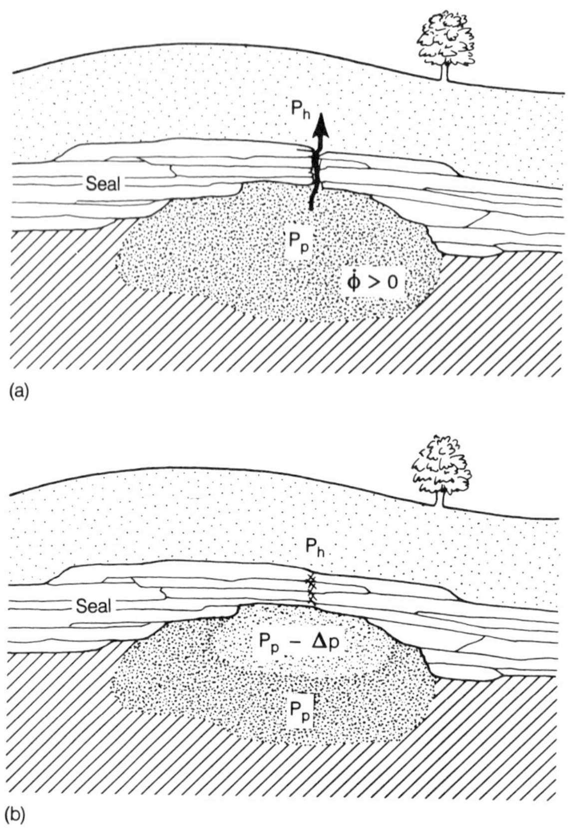

Figure 5. Schematic depiction of the process of natural hydrofracturing, local pore pressure reduction, hydrofracture sealing, and pore pressure buildup again.

are time-invariant, it follows that crustal pore pressure at depth must be generally hydrostatic. This conclusion is in conflict with geologic and laboratory evidence for rapid healing and sealing of flow paths, even at moderate temperatures.

It is therefore necessary to consider crustal hydrologic behavior which is time-varying (Figures 1 and 2). Our analysis suggests that crustal porosity, permeability, and water content in fact must decrease with time, as pores close inelastically at depth in the crust. When crustal permeabilities are high and the decrease of porosity with time is sufficiently small, the dewatering of the crust will be gradual, and pore pressure will remain near hydrostatic. However, as is more likely, if permeability is low and its rate of decrease with porosity or time is large, the pore pressure of the trapped fluid will rise rapidly. When the least principal stress is overcome, spontaneous hydrofracturing will occur. This will be accompanied by the pulsed release of water and the loss of some porosity, followed by a drop in pore pressure and sealing of the system (Figure 5). After a buildup period, another hydrofracturing episode will occur, and so on, as shown in Figure 3. Because inelastic pore deformation is enhanced or even driven by deviatoric stress, the episodic release process is most probable in tectonically active areas. Tens to hundreds of P_p buildup and release episodes, each with 10^3–10^5 years' duration, can occur before the crust dries up by dewatering in 10^6–10^7 years. Where oceanic lithosphere is being subducted, the process may last as long as subduction itself, on the order of 10^8 years.

Episodic pore-pressure buildup and release in the crust was suggested by Norton and Knight (1977) in conjunction with igneous instrusions, and by Cathles and Smith (1983) and Vrolijk (1987) for the formation of hydrothermal ore deposits and mineralized veins formed by periodic precipitation from brines (Ramsay, 1980). The analysis described in this chapter provides some of the quantitative aspects of the essential coupling between rock deformation and pore pressure involved in the formation of these geologic features.

Large earthquakes may also involve pore pressure fluctuations. It is often assumed that the repeat time of large earthquakes is controlled by the rate of tectonic strain accumulating on faults whose strength is constant. But it is now quite apparent that some sort of time-dependent fault strength is required, if only to reconcile laboratory rock failure results with in situ fault rupture (e.g., Dieterich, 1978). The increase of P_p towards lithostatic during a pore pressure cycle provides a simple mechanism for episodic fault weakening and failure. Unlike changes associated with elastic or dilatant deformation (e.g., Nur and Booker, 1971; Nur, 1972; Sibson et al., 1975, 1988; Rice, 1975, 1979; Rudnicki, 1977; McCaig, 1988; Byerlee, 1990), the pore pressure buildup due to the inelastic processes considered here is fairly insensitive to the details of the stress field accumulation around impending failure zones. Instead, we envision a robust process of sealing and trapping of the pore fluid in and around the fault zone, leading to the gradual buildup of P_p until the fault ruptures.

High pore pressure may be the cause of deep crustal seismic reflectors. Eaton (1980) and Jones and Nur (1984) on the basis of laboratory results (Nur and Simmons, 1969; Todd and Simmons, 1972) and field observation (Berry and Mair, 1977) suggested that pressurized crustal water trapped under permeability barriers could show up as seismic low-velocity zones. This possibility is especially intriguing in view of the growing evidence that these reflectors may be correlated with subhorizontal crustal detachment zones, which are mechanically very difficult to explain without high pore pressure being responsible for their low strength. Finally, Oliver (1986) has suggested that deep crustal fluids may migrate horizontally in the crust over very large distances away from subduction zones. The proposed process of high pore pressure development provides a mechanism which can drive fluids horizontally, sustain this drive, and account for the low-permeability barrier that is required to prevent the escape of pressurized brines upward.

Acknowledgments

The ideas and speculation presented in this chapter are the result of collaborations and discussions with many students and colleagues for over a decade or two. Many of these ideas here can be traced back in one form or another to work by Eve Sprunt, Terry Jones, Lee Bell, Ken Winkler, Sue McGeary, Mark Zoback, Bill Brace, Gene Simmons, Brian Evans, Jim Byerlee, Mike Etheridge, D. Norton, Rick Sibson, and others. We are grateful also for many important suggestions made by the two (anonymous) reviewers of this chapter. The funding for the research reported here was provided by the Division of Mathematical and Physical Sciences of the U.S. Department of Energy.

References

Bell, M.L. and Nur, A. (1978). Strength changes due to reservoir-induced pore pressure and stresses and application to Lake Oroville. *J. Geophys. Res.* **83**(B9), 4469–4483.

Berry, M.J. and Mair, J.A. (1977). The nature of the earth's crust in Canada. In *The Earth's Crust: Its Nature and Physical Properties* (ed. J.G. Heacock), pp. 319–348, Geophysical Monograph Series. American Geophysical Union, Washington, D.C.

Brantley, S.L., Evans, B., Hickman, S.H., and Crerar, D.A. (1990). Healing of microcracks in quartz: Implications for fluid flow. *Geology* **18**, 136–139.

Brace, W.F. (1980). Permeability of crystalline and argillaceous rocks. *Int. J. Rock Mech. Min. Sci. Geomech. Abstr.* **17**, 241–251.

Brace, W.F. and Walsh, J.B. (1984). The effect of pressure on porosity and the transport properties of rock. *J. Geophys. Res.* **89**(B11), 9425–9431.

Bredehoeft, J.D. and Hanshaw, B.B. (1968). On the maintenance of anomalous fluid pressures, I. Thick sedimentary sequences. *Geol. Soc. Am. Bull.* **79**, 1097–1106.

Bylerlee, J. (1990). Friction, overpressure and fault normal compression. *Geophys. Res. Lett.* **17**, 2109–2112.

Cathles, L.M. and Smith, A.T. (1983). Thermal constraints on the formation of Mississippi Valley-type lead-zinc deposits and their implications for episodic dewatering and deposit genesis. *Econ. Geol.* **78**, 983–1002.

Dieterich, J.H. (1978). Time-dependent friction and the mechanics of stick–slip. *Pure Appl. Geophys.* **116**, 790–806.

Eaton, G.P. (1980). Geophysical and geological characteristics of the crust of the Basin and range Province. In *Continental Tectonics* (eds. B.C. Burchfield, J.E. Oliver and L.T. Silver). National Academy of Science, Washington, D.C.

Etheridge, M.A., Cox, S.F., Wall, V.J., and Vernon, R.H. (1984). High fluid pressures during regional metamorphism and deformation: Implications for mass transport and deformation mechanisms. *J. Geophys. Res.* **89**, 4344–4358.

Feng, R. and McEvilly, T.V. (1983). Interpretation of seismic reflection profiling data for the structure of the San Andreas Fault zone. *Seismol. Soc. Am. Bull.* **73**(6), 1701–1720.

Fyfe, W.S., Price, N.J., and Thompson, A.B. (1978). *Fluids in the Earth's Crust.* Elsevier, Amsterdam.

Hanshaw, B.B. and Bredehoeft, J.D. (1968). On the Maintenance of anomalous fluid pressure, II. Source layer at depth. *Geol. Soc. Am. Bull.* **79**, 1107–1122.

Herrmann, R.B. and Mitchell, B.J. (1975). Statistical analysis and interpretation of surface-wave anelastic attenuation data for the stable interior of North America. *Seismol. Soc. Am. Bull.* **65**(5), 1115–1128.

Jones, T.D. and Nur, A. (1984). The nature of seismic reflections from deep crustal fault zones. *J. Geophys. Res.* **89**(B5), 3153–3171.

Kozlovsky, Ye. A. (ed.) (1986). *The Superdeep Well of the Kola Penisula*, Springer-Verlag, Berlin, 1–55.

McCaig, A.M. (1988). Deep fluid circulation in fault zones. *Geology* **16**, 867–870.

McGeary, S., Nur, A., and Ben-Avraham, Z. (1985). Spatial gaps in arc volcanism: the effect of collision or subduction of oceanic plateaus, *Tectonophysics* **119**, 195–221.

Nekut, A., Connerney, J.E.P., and Kuckes, A.F. (1977). Deep crustal electrical conductivity: Evidence for water in the lower crust. *Geophys. Res. Lett.* **4**, 239–242.

Norton, D. and Knight, J. (1977). Transport phenomena in hydrothermal systems: Cooling plutons. *Am. J. Sci.* **277**, 937–981.

Norton, D. and Taylor, H.P. Jr. (1979). Quantitative simulation of the hydrothermal systems of crystallizing magmas on the basis of transport theory and oxygen isotope data: An analysis of the Skaergaard intrusion. *J. Petrol.* **20**, 421–486.

Nur, A. (1972). Dilatancy, pore fluids, and premonitory variations of t_s/t_p travel times. *Seismol. Soc. Am. Bull.* **62**(5), 1217–1222.

Nur, A. and Booker, J.R. (1971). Aftershocks caused by pore fluid flow? *Science* **175**, 885–877.

Nur, A. and Simmons, G. (1969). The effect of saturation on velocity in low porosity rocks. *Earth Planet. Sci. Lett.* **7**, 183–193.

Oliver, J. (1986). Fluids expelled tectonically from orogenic belts: Their role in hydrocarbon migration and other geologic phenomena. *Geology* **14**, 99–102.

Padovani, E.R., Shirey, S.B., and Simmons, G. (1982). Characteristics of microcracks in amphibolite and granulite facies grade rocks from southeastern Pennsylvania. *J. Geophys. Res.* **87**, 8605–8630.

Raleigh, C.B. and Everndcn, J. (1982). Case for low deviatoric stress in the lithosphere. In *Mechanical Behavior of Crustal Rocks* (eds. N.L. Carter, M. Freidman, J.M. Logan, and D.W. Stearns), pp. 173–186, Geophysical Monograph Series. American Geophysical Union, Washington, D.C.

Ramsay, J.G. (1980). The crack-seal mechanism of rock deformation. *Nature* **284**, 135–139.

Rice, J.R. (1975). On the stability of dilatant hardening for saturated rock masses. *J. Geophys. Res.* **80**(11), 1531.

Rice, J.R. (1979). Earthquake precursory effects due to pore fluid stabilization of a weakening fault zone. *J. Geophys. Res.* **84**(B5), 2177.

Richter, D. and Simmons, G. (1977). Microcracks in crustal igneous rocks: Microscopy. In *The Earth's Crust: Its Nature and Physical Properties* (ed. J.G. Heacock), pp. 149–180, Geophysical Monograph Series. American Geophysical Union, Washington, D.C.

Rudnicki, J.W. (1977). The inception of faulting in a rock mass with a weakened zone. *J. Geophys. Res.* **82**(5), 844.

Shankland, T.J. and Ander, M.E. (1983). Electrical conductivity, temperature, and fluids in the lower crust. *J. Geophys. Res.* **88**, 9475–9484.

Shemesh, A., Ron, H., Erel, Y., Kolodny, Y. and Nur. A. (1992). Isotopic composition of vein calsite and its fluid inclusions: Implication to paleohydrological systems, tectonic events and vein formation processes. *Chem. Geol.* **94**(4).

Sibson, R.H., Moore, J.M., and Rankin, A.H. (1975). Seismic pumping – a hydrothermal fluid transport mechanism. *J. Geol. Soc. London* **131**, 653–659.

Sibson, R.H., Roberts, F., and Paulsen, K.H. (1988). High-angle reverse faults, fluid-pressure cycling, and mesothermal gold-quartz deposits. *Geology* **16**, 551–555.

Smith, D.L., and Evans, B. (1984). Diffusional crack healing in quartz. *J. Geophys. Res.* **89**, 4125–4135.

Sprunt, E.S. and Nur, A. (1976). Reduction of porosity by pressure solution: experimental verification. *Geology* **4**, 463–466.

Sprunt, E.S. and Nur, A. (1977). Destruction of porosity through pressure solution. *Geophysics* **42**, 726–741.

Sprunt, E.S. and Nur, A. (1979). Microcracking and healing in granites: New evidence from cathodoluminescence. *Science* **205**, 495–497.

Taylor, H.P. Jr. (1977). Water/rock interaction and the origin of H_2O in granitic batholiths. *J. Geol. Soc. London* **133**, 509–588.

Thompson, B.G., Nekut, A., and Kuckes, A.F. (1983). A deep crustal electromagnetic sounding in the Georgia Piedmont. *J. Geophys. Res.* **88**, 9461–9473.

Todd, T. and Simmons, G. (1972). Effect of pore pressure on the velocity of compressional waves in low-porosity rocks. *J. Geophys. Res.* **77**, 3731–3743.

Vrolijk, P. (1987). Tectonically driven fluid flow in the Kodiak accretionary complex, Alaska. *Geology* **15**, 466–469.

Walder, J.S. (1984). Coupling between fluid flow and deformation in porous crustal rocks. Ph.D. thesis, Stanford University.

Walder, J. and Nur, A. (1984). Porosity reduction and crustal pore pressure development. *J. Geophys. Res.* **89**(B13), 11539–11548.

Winkler, K.W. and Nur, A. (1982). Seismic attenuation: effects of pore fluids and frictional sliding. *Geophysics* **47**(1), 1–15.

Chapter 20

Fault Stress States, Pore Pressure Distributions, and the Weakness of the San Andreas Fault

James R. Rice

Division of Applied Sciences and Department of Earth and Planetary Sciences, Harvard University, Cambridge, MA 02138, U.S.A.

Abstract

The San Andreas Fault (SAF) is weak in an *absolute* sense, in that it moves under shear stresses far smaller than implied by the most obvious reading of laboratory friction results (Byerlee law with hydrostatic pore pressure and friction coefficient $f = 0.6$–0.9). It is also weak in a *relative* sense, in that the adjoining crust seems to be mechanically stronger; this is implied by the stress state there having a horizontal maximum principal direction that makes a steep angle to the trace of the SAF, much larger than the 25–30° angle (i.e., $45^\circ - 0.5 \arctan f$) expected from standard frictional failure considerations, and in the range of 60° to nearly 90°. It is shown that a maturely deformed fault zone which is weak relative to its surroundings, owing to inherent material strength and/or pore pressure differences, develops stresses within it which are distinct from those of its surroundings. Because of those stress differences, it is found that pore pressure distributions which are high, and near to the fault-normal compressive stress, within the fault zone, but which decrease with distance into the adjacent crust, are consistent with both the absolute and relative weakness of the SAF; the pore pressure in such distributions is less than the least principal stress at every point, so there is no hydraulic fracturing, even though the pressure in the fault zone may be greater than the least principal stress in the nearby crust. Such pore pressure distributions are shown to result from the following assumptions: (1) there is a supply of fluids near the ductile roots of crustal fault zones, where pore pressure must be nearly lithostatic; (2) active fault zones are far more permeable than the adjoining rock of the middle crust; and (3) fault permeability is a rapidly diminishing function of effective normal stress. Evidence in support of these assumptions is discussed. The resulting pore pressure distributions adjust significantly from hydrostatic, such that the effective normal stress, and hence also the brittle frictional strength, becomes approximately independent of depth along the fault zone. These assumptions also predict the possibility of diffusive surges of pore pressure that propagate upward along a fault in a slow wavelike manner.

FAULT MECHANICS AND TRANSPORT PROPERTIES OF ROCKS
ISBN 0-12-243780-2

1. Introduction

Evidence from the lack of a pronounced heat flow peak near the San Andreas Fault (SAF) show that it is weak in an *absolute* sense, in that it moves under shear stresses far smaller than the most obvious reading of laboratory friction results (Byerlee's law with hydrostatic pore pressure and friction coefficient $f = 0.6–0.9$) would seem to imply (Lachenbruch and Sass, 1980). That reading of laboratory results implies shear strength in the kilobar (10^8 Pa) range at 10 km depth, whereas the heat flow constraint suggests an average stress of order 100 bar or less over the seismogenic depth range.

More recently, results from various methods of inferring stress magnitudes and directions in the vicinity of the SAF (focal mechanism distributions, hydraulic fractures, wellbore breakouts) suggest that the fault is also weak in a *relative* sense, in that the adjoining crust seems to be mechanically stronger (Zoback et al., 1987). Relative weakness is implied since the stress state in the adjoining crust, driving the SAF, has a horizontal direction of maximum principal compressive stress σ_1 which makes a steep angle to the trace of the SAF, much larger than the 25–30° expected from the elementary frictional failure model (i.e., much larger than the angle $45° - \phi/2$, where $\phi = \arctan f$, with $f = 0.9–0.6$).

For example, Zoback et al. (1987) showed that several stress indicators suggest a maximum principal direction approaching 90° (fault-normal compression). They also note the compressional failures which take place in the nearby crust on faults striking subparallel to the SAF, and which are consistent with such a stress state; those failures accommodate the small amount of convergent plate motion which is present in addition to the primary strike-slip motion. Jones (1988) examined focal mechanisms of earthquakes on known faults within 10 km of the SAF and inferred that these, if regarded as being driven by a uniform stress field, required maximum stress directions in the range of 60–70° to the SAF, for the four of five regions studied in which the resulting σ_1 was horizontal. Similarly, along the strand of the Calaveras Fault active in the 1984 Morgan Hill earthquake, Oppenheimer et al. (1988) inferred a σ_1 direction in the adjoining crust of 60–80° at the time of the earthquake, for consistency of focal mechanisms of the aftershock sequence with the sum of that initial stress state plus the stress change associated with faulting. Such steep angles mean that the adjoining crust supports substantially more shear stress than does the SAF (or the Calaveras Fault).

It is likely that many major intraplate and interplate zones of active faulting are similarly weak. Fault weakness in continental thrust sheet emplacement is well known (Hubbert and Rubey, 1959), and recent works have emphasized the weakness, in both relative and absolute sense, of decollement faults beneath sediment-dominate accretionary prisms (Byrne and Fischer, 1990) and of transform faults along the ocean floor (Wilcock et al., 1990).

There seems to be no alternative to assuming that local stresses compatible with laboratory-measured friction properties are present in fault zones where seismic ruptures nucleate. Laboratory results suggest that classical friction concepts with f in the range 0.6–0.9 (Byerlee, 1978) describe the onset of sliding in most crustal rocks. Also, values of f in that range are known to be consistent with stress states on natural faults as inferred at approximately 2 km depth in the Rangely oilfield experiment (Raleigh et al., 1976), on faults intersected along the 3 km depth of the Cajon Pass borehole (Zoback and Healy, 1991), and generally on faults as studied at shallow depths by borehole stress measurement (e.g., Zoback et al., 1980; Hickman, 1991).

Explanations of fault weakness may thus be based on the assumption of presence at seismogenic depth of frictionally weak fault zone materials, like water-absorbing seismogenic depth of frictionally weak fault zone materials, like water-absorbing montmorillonite clays, although dewatering to illite and related frictional strengthening of such materials is likely to set in at temperatures above about 200°C (see, for example, Morrow et al., this volume, Chapter 3). They may also be based, more speculatively, on severe stress concentrations along the edges of isolated locked patches in a fault zone, owing to some mechanism of slow creep at low stress in neighboring fault zone material, provided that those locked patches slip-weaken very rapidly after instability begins (so that average stresses are small and stress drops, as conventionally inferred, appear to be far smaller than the actual kilobar range that would then be implied at the rupture-nucleating stress concentration site, if $f = 0.6$–0.9 and p is hydrostatic). Yet another possibility is that they may be based on elevated pore pressures in a fault zone having the usual frictional properties, $f = 0.6$–0.9. While it is possible that some combination of these explanations, and maybe others not yet elaborated, are sources of fault weakness, the aim here is to evaluate the pore pressure explanation.

It is shown that pore pressure distributions which are high and near to the normal stress within a major fault zone, but lower in the adjacent crust, are consistent with both the absolute and relative fault weakness discussed. Further, it is shown that such pore pressure distributions result from assumptions about crustal fluids and their motion which have extensive support from geochemical, petrologic and geophysical observations.

Since the work of Hubbert and Rubey (1959), high pore pressure has been appreciated widely as a plausible cause of what, here, has been called absolute fault weakness. Nevertheless, the more recent understanding of the *relative* weakness of major faults, signaled by the steep angle of σ_1, has led many to the conclusion that some explanation other than elevated pore pressure is required. This is because the pore pressure needed had seemed to be larger than the least principal stress. It is shown in the next section that such a conclusion is based on the tacit assumption that the stress tensor within the core of a mature fault zone like the SAF is fully identical to that in the nearby crust at the same depth. Instead, an analysis of the mechanics of a maturely deformed fault zone shows that these stress tensors differ in an essential way, with consequence that the pore pressure distributions required for relative fault weakness can meet, at every point, the condition that pore pressure be less than the least principal stress at that point, even though the pore pressure within the active fault core may be larger than the least principal stress prevailing a short distance away in the adjacent crust. (Independently, Byrne and Fisher (1990) have made similar mechanical arguments for a weak and overpressurized decollement beneath a sediment-dominated accretionary prism, and they also noted in their closing discussion the relevance of the concept of the SAF. Also, Byerlee (1990), in a study first presented at the same conference as was the present work, but published elsewhere and hence earlier, considers a clay fault gouge that is assumed to seal permanently a gradient of pore pressure perpendicular to itself. He comes to similar conclusions regarding stress states within and outside the fault core for the special case of fault-normal compression. In a recent work, Fournier (1990) addresses the fault weakness problem by assuming that fault zones, sealed by mineralization, can accommodate pore pressures significantly in excess of the least principal stress.)

2. Fault Stress States and Effects of Variations of Strength or Pore Pressure

It is shown here that the mechanics of a maturely deformed fault zone, which is weak relative to its surroundings owing to inherent material strength and/or pore pressure

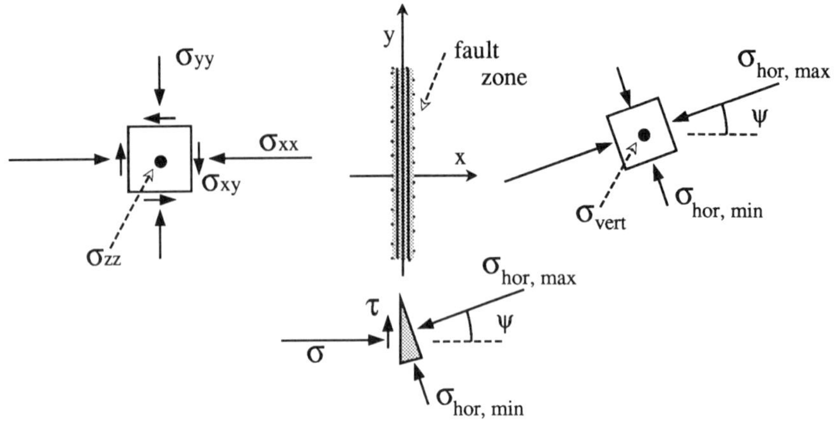

Figure 1. Stress state as seen in a horizontal plane through a vertical strike-slip fault zone. Axis *y* is horizontal and along the fault trace, *x* is perpendicular to the fault, and *z* is vertical.

differences, is such that a different stress state develops within the fault zone than in its surroundings. This result is applied to a fault zone such as the SAF and its adjacent crust, in the case for which both regions are assumed to have identical mechanical properties, including laboratory-like friction properties, but to sustain nonuniform pore pressure. It is found that pore pressure distributions which are high and near to the fault-normal compressive stress within the fault zone, but which decrease with distance into the adjacent crust, are consistent with both the absolute and relative weakness of the SAF. For example, stress distributions may exist for which a vertical fault such as the SAF is at a critical state for *strike-slip* failure, while the adjacent crust is at a critical state for *thrust* failure under a principal direction making an angle to the SAF that becomes *arbitrarily large* (but less than 90°) as pore pressure within the fault zone approaches the fault-normal compression there.

It is straightforward to see why high pore pressure has seemed incompatible with the steep σ_1 angle when such stress state differences are neglected. Figure 1 corresponds to a plane parallel to the earth's surface along a strike-slip zone like the SAF. The y-axis points along the fault trace in that plane and the x-axis is perpendicular to it; z is vertical. The direction of the maximum horizontal stress $\sigma_{\text{hor,max}}$ is assumed to make an angle ψ with the normal to the fault (this angle notation is used throughout, so that a steep $\sigma_{\text{hor,max}}$ direction corresponds to a small value of ψ, the $\sigma_{\text{hor,max}}$ direction being at angle $90^\circ - \psi$ with the fault trace). The minimum horizontal stress is denoted $\sigma_{\text{hor,min}}$. Cartesian stress components are also shown in Figure 1. If the stress state is to be understood as being critical for thrust failure in the nearby crust, as happens near the SAF, so that σ_1 is horizontal and σ_3 vertical, then $\sigma_{\text{hor,max}} = \sigma_1$ and $\sigma_{\text{hor,min}} = \sigma_2$.

The normal stress σ and shear stress τ acting along the fault (Figure 1) are

$$\sigma = \tfrac{1}{2}(\sigma_{\text{hor,max}} + \sigma_{\text{hor,min}}) + \tfrac{1}{2}(\sigma_{\text{hor,max}} - \sigma_{\text{hor,min}}) \cos 2\psi$$

$$\tau = \tfrac{1}{2}(\sigma_{\text{hor,max}} - \sigma_{\text{hor,min}}) \sin 2\psi$$

We assume that for slip along the fault, the friction condition

$$\tau = f(\sigma - p) = \tan\phi(\sigma - p)$$

must be met, where p is the local pore pressure and $f = 0.6$–0.9 (so that $\phi = 31$–$42°$). Solving for the value of p which just suffices to meet the friction condition then gives

$$p = \sigma_{\text{hor,min}} - (\sigma_{\text{hor,max}} - \sigma_{\text{hor,min}}) \cos\psi \sin(\psi - \phi)/\sin\phi$$

This equation shows that the required p will exceed $\sigma_{\text{hor,min}}$ whenever the maximum stress is at a sufficiently steep angle to the fault that the angle ψ of Figure 1 is less than ϕ (this observation has been made by Sibson, 1985). Further, if $\sigma_{\text{hor,min}}$ is σ_2, then p will exceed the vertical stress σ_3 under an even less stringent condition. This means that p will exceed $\sigma_{\text{hor,min}}$ whenever ψ is less than an angle of 30–40°, or whenever the angle $90° - \psi$, which $\sigma_{\text{hor,max}}$ makes with the fault direction, is greater than 50–60°. The steep angles mentioned above fall in the range of 60° to nearly 90°, and thus call for $p > \sigma_{\text{hor,min}}$. On this basis, it had been assumed that the steep angles could not be explained in terms of fault weakening by high pore pressure, since such pore pressures would lead to hydraulic fracture.

However, the argument just gone over is not strictly correct, at least if we regard the stress state with values $\sigma_{\text{hor,min}}$ and $\sigma_{\text{hor,max}}$, at angle ψ, as referring to the stress state in the nearby crust adjacent to a maturely deformed fault zone. This is because the stress state within a heavily deformed fault zone that is weakened either by having locally high pore pressure or by being made of inherently weak material will not be the same as the stress state in the nearby crust. Rather, there will develop larger principal stress magnitudes within the fault core than in the nearby crust, as shown below, when there is a small angle ψ in the nearby crust. Thus, even though one must expect the pore pressure to be higher within the fault core than in the nearby crust (as shown in the next two sections to follow from reasonable assumptions on the presence and motion of fluids in the crust), it develops that the pore pressure in the fault core is still less than the (locally elevated) least principal stress in the core, so that hydraulic fracture is not a problem. This is so even though the locally high pore pressure in the fault core is greater than the least principal stress prevailing at other locations, some distance away in the nearby crust.

If the stress state does vary near a fault zone, it is reasonable to think of the stresses at a given depth (in the vicinity of the fault zone and at locations not too near the earth's surface) as effectively varying only with the coordinate x, where x is measured perpendicular to the fault zone. For the strike-slip case in Figure 1, x is horizontal. Then, if we let the y-direction be along strike, and z vertical, it will be evident from conditions of mechanical equilibrium that the stress components σ_{xx} $(= \sigma)$, σ_{xy} $(= \tau)$, and σ_{xz} (assumed to vanish in the discussion above) must be essentially the same everywhere within the deforming fault zone as in the nearby crust outside it. There is no reason, based on equilibrium considerations, that σ_{yy}, σ_{zz}, and σ_{yz} (the latter also assumed above to vanish) should be essentially the same within the fault zone as outside it. Rather, the appropriate values of σ_{yy}, σ_{zz}, σ_{yz} are determined from mechanical constitutive relations, so as to assure that the material within the fault zone, deforming primarily if not exclusively by strike-slip shear, has essentially the same components of strain rate ε_{yy}, ε_{zz}, and ε_{yz} at points within the fault as outside. These are the components of strain rate governing distortions of planes lying parallel to the strike-slip fault plane. Just as the equilibrium conditions on the problem restrict σ_{xx}, σ_{xy}, and σ_{xz} to continuity, but leave σ_{yy}, σ_{zz}, and σ_{yz} unrestricted, so do the kinematical conditions on the problem restrict the strain rates ε_{yy}, ε_{zz}, and ε_{yz} to continuity but leave ε_{xx}, ε_{xy}, and ε_{xz} unrestricted;

here 'continuity' is to be interpreted as meaning having no strong variations as one passes along the x direction in the vicinity of the fault.

As is best illustrated by the simple example which follows, these considerations will require that if a maturely deformed fault zone is weak either because it is made of inherently weak material, or because of a pore pressure distribution which involves higher pressure within the fault zone than in the adjacent crust, then the equations of continuum mechanics generally preclude the possibility that all components of stress are the same within and outside the zone. In particular, the stress components σ_{yy} and σ_{zz}, for which variation with x is not constrained by equilibrium considerations, will be seen to vary significantly as we traverse the fault zone in the x-direction.

2.1. Simple example, soft ductile layer in hard ductile surroundings

A simple example is useful to clarify the significance of these remarks. It is done not for the case of primary interest here, involving a frictional material with sensitivity to pore pressure, but rather for the simpler case of classical rigid–plastic solids, for which the result will, hopefully, be found somewhat intuitive. Figure 2 shows a thin, soft ductile layer that is embedded within a harder ductile solid. Here, by 'ductile' is to be understood that the materials undergo volume-preserving plastic flow ($\varepsilon_{\gamma\gamma} = 0$) which is dependent, in the von Mises manner, on only the deviatoric stress components $s_{\alpha\beta}$ $(= \sigma_{\alpha\beta} - \frac{1}{3}\delta_{\alpha\beta}\sigma_{\gamma\gamma})$

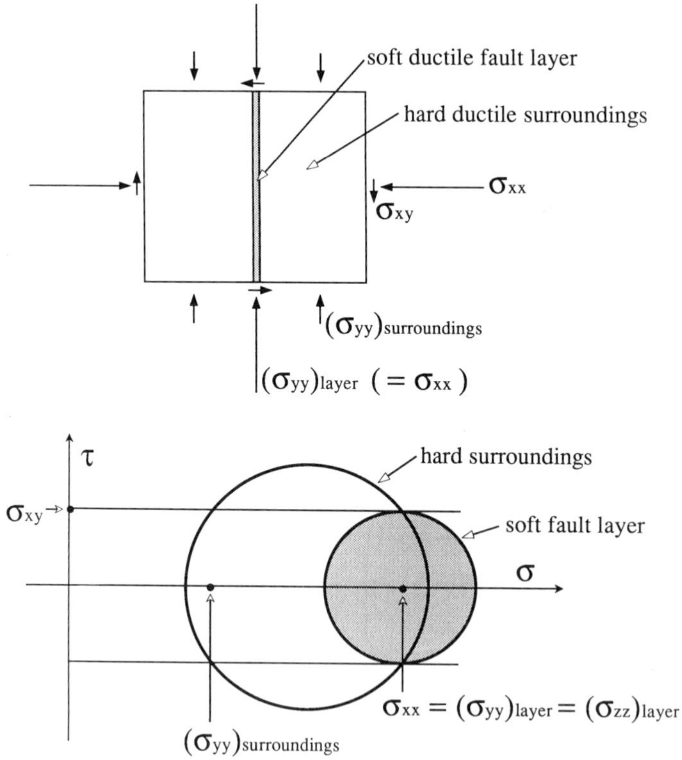

Figure 2. Stress states associated with the deformation of a soft ductile layer in harder ductile surroundings. The stress field induced in the layer is different from that in the surroundings.

and not on the mean normal stress ($\frac{1}{3}\sigma_{\gamma\gamma}$); here α, β, $\gamma = x, y, z$, with the usual summation convention on repeated indices. Further, the strain rates partition themselves among components so that the ratio of any $\varepsilon_{\alpha\beta}$ to $\varepsilon_{\gamma\delta}$ is the same as the ratio of $s_{\alpha\beta}$ to $s_{\gamma\delta}$.

The system is stressed with a large compression σ_{xx} acting perpendicular to the fault layer, but still of a magnitude well short of that to cause plastic flow in the harder solid, and with a shear stress σ_{xy} which suffices to cause the soft layer to flow plastically. Since the harder solid is nondeforming, all strain rates vanish within it and, since strain rate components ε_{yy}, ε_{zz}, and ε_{yz} must be the same within the fault layer as outside it, those same components vanish also within the layer. By incompressibility, $\varepsilon_{xx} = -(\varepsilon_{yy} + \varepsilon_{zz}) = 0$ additionally within the layer. Thus the soft fault layer can deform only in simple shear relative to its boundaries, and from the symmetry of the situation, ε_{xy} is the only nonzero strain rate within the layer. The constitutive relations thus require that all deviatoric stress components except s_{xy} ($= \sigma_{xy}$) vanish within the layer, and, in particular, this means that $\sigma_{yy} = \sigma_{zz} = \sigma_{xx}$ within the layer, where (from the required continuity of the set of stresses σ_{xx}, σ_{xy}, and σ_{xz}) the σ_{xx} here is the same within the layer as that applied to the harder solid outside it.

Thus the stress components σ_{yy} and σ_{zz} within the soft fault layer bear no relation whatever, in this example, to those stress components of the same subscripts which act on the harder solid at points remote from the fault zone. Mohr circles for the soft fault layer and for the adjoining hard solid are shown in Figure 2. Continuity of σ_{xx} ($= \sigma$) and σ_{xy} ($= \tau$) requires that both circles share this common (σ, τ) point of the Mohr plane. Although the soft layer sustains much less shear stress (smaller-diameter circle) than the harder surroundings, the principal stresses that develop within the soft layer, of which horizontal components are given by the σ-axis intercepts, are significantly larger than those for the stronger surroundings.

A few comments are in order before moving on from this simple example. First, reference was made in the introductory remarks to stress states being dissimilar between a 'maturely deformed' fault zone and its surroundings. A mature deformation state has been tacitly assumed in this example by adoption of a rigid-plastic model and solution for a more or less steady deformation rate field. If, instead, we imagined the system to be initially stress-free and then loaded it with σ_{xx} and σ_{xy} as discussed, the stresses σ_{yy} and σ_{zz} within the fault layer would not initially coincide with σ_{xx}, but rather would approach that value with ongoing plastic deformation; an elastic–plastic, rather than rigid-plastic, model would be needed to model such a transition.

Second, the case discussed thus far assumed that the harder solid did not deform. This is not such a good analog of the crust near the SAF, which accommodates some convergent deformation. We might therefore consider the case for which the stress state applied to the harder surroundings suffices to cause it to flow plastically, in a mode that must, in general, be expected to involve nonzero ε_{zz}, and perhaps ε_{yy}, as well as nonzero ε_{xx} (the convergent component) and ε_{xy} (nonzero since a finite σ_{xy} is sustained). The same strain rates ε_{zz} and ε_{yy} must be sustained within the soft layer, which, in addition, can be made (by application of suitable σ_{xy}) to take on any ε_{xy} value, not constrained kinematically by that outside. In the general case, all the $\varepsilon_{\alpha\beta}$ within the soft layer could be of comparable order, and there would be no need for, say, σ_{yy} within it to be close in value to σ_{xx}. However, to pursue our SAF analogy, the fault zone must be relatively thin compared to dimensions of the deforming hard block outside it, and the net convergent displacement accommodated is modest compared to the strike-slip motion. In such cases ε_{xy} within the soft fault layer must be much greater than any $\varepsilon_{\alpha\beta}$

in the harder solid outside, and in this case all $s_{\alpha\beta}$ except σ_{xy} within the soft layer will be small compared to σ_{xy}. Thus, for all practical purposes we recover the result $\sigma_{yy} = \sigma_{zz} = \sigma_{xx}$ within the layer, just as above for the case of a nondeforming outer region.

Third, there is no need to confine attention to a single soft layer. Instead one could consider an array of thin layers joined together, the array being still thin in aggregate, and embedded in a stronger outer solid. In the limit, this would represent a continuous variation of plastic strength, for example, weakest in the center of the fault zone and increasing as one moves away in the $+x$ or $-x$ directions. The principles of analysis are the same as above, with stress components σ_{xx}, σ_{xy}, σ_{xz}, and strain rate components ε_{yy}, ε_{zz}, ε_{yz}, being independent of x in the vicinity of the fault zone, and with the constitutive law obeyed everywhere. The details are somewhat complex, but it is easy to see that the nonconstrained stress and strain rate components will vary continuously with x, over the same portion of the x-axis as that over which the plastic strength varies. Also, the ratio of local strike-slip $(2\varepsilon_{xy})$ to convergent (ε_{xx}) deformation is position-dependent, and is largest at the weakest plane within the fault zone, where, if conditions as at the end of the previous paragraph apply, one will again have, approximately, $\sigma_{yy} = \sigma_{zz} = \sigma_{xx}$.

Finally, this analysis has been simplified by assuming a narrow fault region near which components of $\sigma_{\alpha\beta}$ and $\varepsilon_{\alpha\beta}$, if they vary at all, are assumed to vary only with distance x perpendicular to the fault. That is, $\partial/\partial x$ has been assumed to be of larger order than $\partial/\partial y$ or $\partial/\partial z$. Such will not generally be an acceptable approximation at shallow locations along the fault, at depths within a few fault zone widths of the free surface.

2.2. Brittle fault zone, variable pore pressure

Figure 3 shows stress states at a location within a brittle fault zone, presumed to represent the SAF, and in the adjacent, also brittle, crust. Pore pressure is imagined to be somewhat higher within the fault zone than outside, for reasons discussed in a subsequent section. The pore pressure varies continuously with x, from the maximum at, say, $x = 0$ in the fault core to a smaller value in the nearby crust at the same depth. In contrast to the simple example above, we assume here that the fault zone and the adjacent crust have identical strength properties, i.e., that both regions share the same friction coefficient f and obey the same law that 'failure' occurs when, and on the plane for which, the Amonton–Coulomb inequality $\tau < f(\sigma - p)$ is first violated. We assume that f is of the order of what is suggested by laboratory results, e.g., in the range from about 0.6 to 0.9. Hence, in this example, any difference in the stress and deformation response is rooted in the assumed variation in pore pressure. Figure 3 also emphasizes that not all stress components are required to be the same within the fault zone and in the adjacent crust.

Figure 4 has been prepared on the understanding, motivated by the previous subsection, that the local stress states within and near the brittle fault zone will adjust themselves to be compatible with the local deformation condition. The overall displacements accommodated are assumed to involve strike-slip along with some modest fault-normal convergence. This figure depicts coexisting stress states within the central core of a strike-slip fault at the most pressurized location, $x = 0$, and in the adjacent crust. The Mohr circles referring to horizontal stresses are shown as solid lines, and these intersect at the common point (σ, τ) of the Mohr plane, corresponding to the

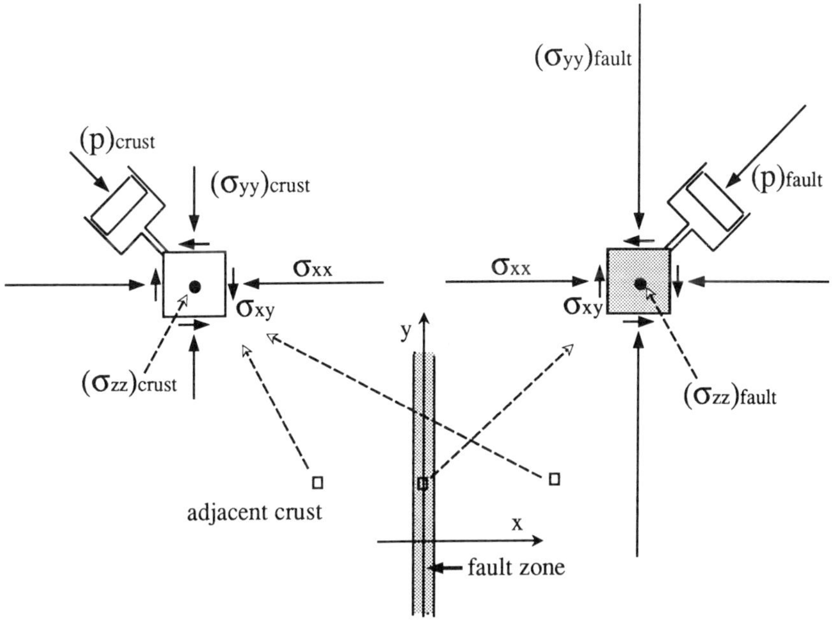

Figure 3. Stresses and pore pressure at locations within the core of the fault zone and in the adjacent crust. It is assumed that the pore pressure is high within the fault zone and diminishes somewhat with distance into the adjacent crust.

values of the fault-normal compressive stress σ_{xx} $(= \sigma)$ and fault-parallel shear stress σ_{xy} $(= \tau)$ depicted in Figure 3. Mohr circles involving vertical stress are shown as dashed lines in Figure 4. The largest Mohr circles in both regions meet the same Amonton–Coulomb failure condition, that $\tau = f(\sigma - p)$, where here τ and σ refer to the local stress values where the relevant Mohr circle has tangential contact with the limiting stress line, as illustrated. This happens on the plane $x = 0$ for the strike-slip failure within the fault core, and on a dipping thrust fault in the adjacent crust, trending subparallel to the SAF.

For the case illustrated, with friction coefficient $f = 0.7$ $(\phi = 35^\circ)$, the fault zone core is critically stressed for strike-slip failure, and the principal stress direction within it makes the (proper) shallow angle of 27.5° $(= 45^\circ - \phi/2)$ to the fault while, simultaneously, the adjacent crust is critically stressed for thrust failure under a steep principal direction at 72° $(= 90^\circ - \psi$, where $\psi = 18^\circ$ here) with the strike-slip fault. As will be evident from the construction, greater pore pressure within the fault, becoming closer to σ_{xx} (i.e., to the ordinate at which the two solid-line Mohr circles intersect), or closer to the local least principal stress in the fault core, would allow ψ to be as small as one would like.

Thus strike-slip failure, under conditions approaching fault-normal compression, is perfectly compatible with laboratory-range rock friction properties, so long as a mechanism exists for there being high pore pressures within the fault that grade to lower pressures in the immediately adjacent crust.

A range of values for the vertical stress within the fault core is possible in the construction of Figure 4, so long as that stress does not lie outside the range of the horizontal principal stresses there. If the fault core were to simultaneously meet conditions

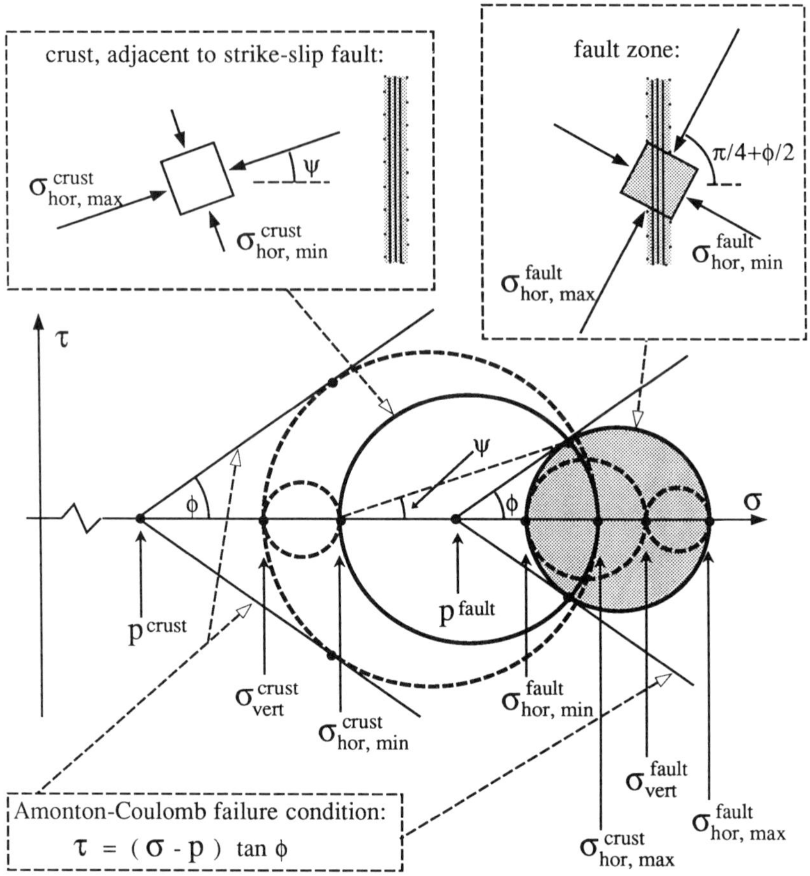

Figure 4. Example of failure under a stress state with maximum principal direction at a steep angle to a strike-slip fault. Construction shows stresses which are critical for strike-slip failure within the fault core and for thrust failure in the adjacent crust. Pore pressure and stresses are assumed to vary continuously, with distance x, between those of the two states shown. At each point the pore pressure is less than the least principal stress, although the pore pressure within the fault core is greater than the least principal stress in the adjacent crust. Drawn for $\phi = 35°$ ($\tan \phi = 0.7$) and $\psi = 18°$.

of strike-slip and thrust failure, as discussed in connection with the ductile layer example, then the vertical stress within the core would coincide with the least horizontal stress there. Unfortunately, the brittle case cannot be analyzed in the same precision as for the simple ductile layer, since failure involves stress drops rather than simple plastic flow at fixed stress, and, further, even in circumstances of stable frictional deformation, there is no generally agreed description of the combined stress state dependence of the flow process, analogous to the von Mises type formulation used for the ductile layer.

For such stress and pore pressure distributions as contemplated in connection with Figures 3 and 4, the pore pressure p is less than the least principal stress σ_3 at every point of the region, so there need be no hydraulic fracturing, even though p in the fault zone core may, in conditions like those illustrated, be greater than σ_3 at some distance away in the adjacent crust. Both p and σ_3 diminish with distance from the fault core, transitioning to values appropriate to the adjacent crust where σ_3 equals the overburden. At points near the fault core in that transition region, shear deformation of type γ_{xy}, less intense than in the core itself, is expected to induce some local elevation of σ_{yy} and σ_{zz} (and hence of σ_3), by the mechanism discussed above, so that $p < \sigma_3$ is met throughout the region of elevated p. These concepts seem to remove the difficulty some have had, based on the reluctance to accept fracture-inducing pressure levels, in reconciling the relative weakness of the SAF with elevated pore pressures. (Still, elevated pore pressures do sometimes lead to hydraulic cracking in seismogenic fault zones (Kerrich, 1986; Sibson, 1990), and the mechanism by which fluid transport is maintained in the crust, against depositional pore sealing processes, may involve pervasive hydraulic fracturing (Fyfe et al., 1978)).

We may conclude at this point that pore pressures which become close to the fault-normal compressive stress in the fault core, and which are somewhat lower in the adjacent crust, are consistent with both the absolute and relative weakness of major faults such as the SAF. Further, this is so even when the fault zone and nearby crust have identical strength properties, comparable to those found for rocks in the laboratory. The next two sections explain why it may be reasonable to expect such pore pressure distributions actually to be present in the crust.

3. Crustal Plumbing and Origin of Fault Zone Pore Pressure

Some assumptions about the supply of fluids in the crust and about the relative permeability of transport paths are stated below. All have substantial observational support. Fluid transport in the crust is complicated by a competition between depositional processes which seal-off permeability (e.g., Smith and Evans, 1984) and rupture processes which renew pore connectivity; the latter involve seismic and sometimes aseismic shear failures and, often, local tensile cracking driven by pressure build-up along clogged flow paths (Fyfe et al., 1978; Kerrich, 1986; Nur and Walder, 1990; this volume, Chapter 19; Sibson, 1990). It is thus a great simplification to treat the pore pressure distribution in the crust as resulting from solutions to the steady, time-independent equations for flow in porous media, as will be done here.

Nevertheless, within that simplification, relatively straightforward analysis (given in the next section) shows that if the assumptions to be stated hold good, then under a fairly wide variety of conditions there results a pore pressure distribution with the following characteristics: p is moderately close to the normal stress σ over nearly the entire fault depth, excepting a regime near the earth's surface where pressure variation with depth can be closer to hydrostatic, and p at sites only a few kilometers away from the fault zone, in the adjoining crust at seismogenic depths, is considerably lower. That is, of course, the type of pore pressure distribution which has been shown above to imply both absolute and relative fault weakness.

The assumptions are:

1. There is a source providing fluids deep in the crust near the ductile root of the fault zone, and pore pressures there approach lithostatic values (the second part of this assumption may be argued to follow from the first, since pore pressure within a source region in ductile rock must necessarily be near to lithostatic; otherwise, pore spaces would creep shut).

2. The core of a tectonically active fault zone, where pore connectivity is renewed by shearing, is much more permeable than the adjoining rock of the middle crust. (Another version of this assumption, sufficing either separately or in combination with that stated, is that the active core is much more permeable in directions along the plane of the core than perpendicular to it.)
3. Permeability along the fault zone is a strongly decreasing function of the effective normal stress, and hence is a strongly increasing function of pore pressure.

Because complete data are not always available from along the SAF, and because the phenomenon of weakness of major, crust-cutting faults is thought to be more general, observational data pertaining to these assumptions from many fault zones are reviewed here (in sequence 2, 1, 3).

3.1. Assumption 2

The assumption that fault zones are the preferred fluid permeation routes seems to be widely accepted on the basis of petrologic studies of exhumed fault zones. Such studies, involving compositional alterations along faults and fluid inclusions, are generally interpreted to show that large amounts of fluid (H_2O, CO_2) have moved through fault zones (Sibson et al., 1975; Beach, 1980; Newton et al., 1980; Etheridge et al., 1984; Kerrich, 1986; Parry and Bruhn, 1986; McCaig, 1988).

Geochemical tracers along *currently active* faults provide another basis for assumption 2. An elevated $^3He/^4He$ ratio for helium within crustal fluids is thought to be a definitive indicator of a mantle origin for that helium. In a general survey of helium in groundwater outflows in Western Europe, Oxburgh and O'Nions (1987) and O'Nions et al. (1989) show coincidence between areas where helium of mantle origin emerges and areas of currently active tectonics (mostly by extensional faulting in the area studied). This suggests that active fault zones are much better conduits for deep fluids than are stable areas of the crust and older, now inactive, fault zones, where groundwater emanations are not found to have elevated $^3He/^4He$ ratios. The helium is generally thought to occur in association with mantle-derived magmas emplaced into the middle or shallow crust, and this is thought to be most common in extensional tectonic regimes. Nagao et al. (1990) have recently given a preliminary report on the release of mantle-derived helium, and of CO_2 compatible with a mantle origin, in the strike-slip environment along the North Anatolian Fault in Turkey. Helium outflows of mantle origin, correlated with locations of active faulting, have been reported (Wakita et al., 1987) in a nonvolcanic forearc area of crustal seismicity in the Kinki District, Japan, and in association with the fault of the 1966 Matsushiro swarm (Wakita et al., 1978).

No general survey of helium emanations in groundwater outflows along and adjacent to the SAF has yet been reported. There are data on the Salton Trough area (Whelan et al., 1988; H. Craig, private communication, 1990) and the general occurrence of areas of elevated $^3He/^4He$ seems better explained in terms of shallow magmatic activity than fault location. There are, however, locations with elevated $^3He/^4He$ along the SAF, and such emanations as reported in the East Mesa area do not have an obvious shallow magma source. Irwin and Barnes (1980) point out that the areas of North America with greatest CO_2 discharge in spring outflow include the Franciscan rocks of the coastal ranges near the San Andreas fault. They reported that the $\delta^{13}C$ values along the SAF are compatible with a mantle origin, although carbon isotopes do not provide an unambiguous tag of origin. Thus, while the depth of origin of the CO_2 is not certain, its outflow in carbonic springs does support the notion of the fault as an effective transporter of fluids. While abundant, such springs certainly do not occur everywhere along the SAF. A speculation (R. Fournier, private communication, 1990) is that upwelling of CO_2 may be rather pervasive, but that clay gouge in the shallow reaches of the fault provides a seal so that massive leakages sufficient to feed carbonic springs are relatively rare events along-strike.

An additional insight on assumption 2 is provided by seismic inferences of fault zone properties. These have been carried out extensively for the New Madrid zone. Al-Shukri and Mitchell (1988) show that the *p*-wave velocity is reduced near the fault, and that inferred contours of lower velocity coincide well with the seismically most active part of the zone (and not, for example, with older, now inactive regions). This is shown in *p*-wave contours for both the 0–5 km and the 5–14 km depth ranges, spanning the depth of current seismicity there. Al-Shukri and Mitchell (1990) and Hamilton and Mooney (1990) show similar correlations with seismic attenuation. In studies of the response to blasts along refraction lines, the latter show that very weak arriving signals result for ray paths that traverse the currently active area, compared to the strength of signals for paths that avoid the fault or traverse now-inactive fault areas. While the interpretation of such results is not unambiguous, they strongly suggest that the currently active part of the fault is a region of increased porosity and crack content compared to the nearby crust, and hence would be fluid-infiltrated. The lack of such velocity and attenuation anomalies from now-inactive fault zones suggests that these have had their porosity filled in by depositional processes; this may be expected, as they now lack the essential effects of (shear) fracturing on renewal of pore connectivity and maintenance of permeability.

For the San Andreas system, Mooney and Ginzburg (1986) review evidence from seismic reflection studies, gravity (see also Wang et al., 1986), and other sources to conclude that the fault zone is a zone of reduced wave speeds and reduced density. They suggest that the zone of altered properties extends up to 3–4 km width along the creeping portion of the SAF to as little as 0.2 km or less along locked portions that fail in large earthquakes. Mooney and Ginzburg (1986) likewise suggest an interpretation of that zone of altered properties as one of increased porosity and fluid infiltration.

The interpretation to be made of assumption 2 is that the ratio $k_{\text{crust}}/k_{\text{fault}}$ is very much less than unity, where k_{crust} is the permeability of the middle crust outside the seismogenic fault zone, and k_{fault} the permeability within the fault zone. The smallness of the ratio is what controls the character of the predicted pore pressure distribution. Thus, for a ratio of 10^{-3}, it will not matter if $k_{\text{fault}} = 1\ \mu\text{d}$ (d = Darcy = $10^{-12}\ \text{m}^2$) and $k_{\text{crust}} = 1$ nd, or if $k_{\text{fault}} = 1$ md and $k_{\text{crust}} = 1\ \mu\text{d}$ (except that, judging from the results of Peacock (1990), fluid permeation would have negligible effect on the crustal temperature distribution in the former case but would perturb it somewhat, depending on the thickness of the permeable part of the fault zone, in the latter).

An alternate version was given for assumption 2 that has experimental support in that Arch and Maltman (1990) show, for sediments deformed in a laboratory shear apparatus, the development of distinctly anisotropic permeability that is several times larger in the direction perpendicular to the shear zone than along it. The amount of shear is small compared to that of a maturely deformed fault zone, in which the effect may be much larger. This alternate form of assumption 2 is also encouraged by the viewpoint that the core of a mature fault zone contains finely pulverised rock and sometimes appreciable clay components. These are not very permeable materials, and may form a very low-permeability lining to the walls of the highly deforming central portion of the fault zone. (In fact, Byerlee (1990) regards the entire width of the SAF fault zone as completely impermeable, in all directions, for pressure gradients below a critical threshold; this may be too extreme.) Thus there could be transport along the rupturing fault core, for reasons already discussed in connection with assumption 2 here, but little transport perpendicularly across the fault zone; this is equivalent at a more macroscopic scale to anisotropic permeability in the fault zone. It is again the smallness of a permeability ratio, now k_x/k_z, which matters in the assumption, rather than absolute values, where subscripts x and z denote permeability within the fault zone in directions respectively perpendicular and parallel to the plane of the fault core.

3.2. Assumption 1

The assumption on fluid presence near the ductile roots of fault zones has strongest evidence from magnetotelluric investigations. In these, the lower crust frequently shows a far smaller electrical resistivity than could be expected for dry rocks (Shankland and Ander, 1983; Haak and Hutton, 1986; Hyndman and Shearer, 1989; Madden et al., 1990). Further, areas that are tectonically active at present generally show the lowest values, typically less than 100 Ω m, assuming a 10 km conductive layer. Hyndman and Shearer (1989), following Shankland and Ander (1983), argue that the conductivity strongly suggests the presence of water in interconnected pore spaces in the lower crust. In view of the elevated temperatures there, the pore spaces are argued to have the form of quasiequilibrium tubular networks along grain triple junctions. Hyndman and Shearer (1989) infer that if the water is of salinity roughly like seawater, as fluid inclusion studies and the temperature dependence of resistivity (Shankland and Ander, 1983) suggest, that a porosity from about 0.5% to 3% is necessary in the lower crust for consistency with the low electrical resistivity.

Following earlier work correlating the inferred depth to the upper lid of the conductive layer with heat flow, Hyndman and Shearer (1989) note that the lid is in the temperature range of about 350–400°C, which reasonably coincides with temperatures at the lower limit of the seismogenic depth range as discussed in recent years by Brace and Kohlstedt (1980), Chen and Molnar, 1983), Sibson (1982) and Meissner and Strehlau (1982).

Thus the resistivity results imply that fluids are present at the ductile roots of major fault zones that traverse the crust, as in assumption 1. As mentioned there, the pore pressure in fluids long residing within ductile rocks must be lithostatic (or a modest amount above, to balance the forces stemming from surface energy that would otherwise cause sintering). Very interesting questions are raised by the presumed presence of fluids in the lower crust. Their exploration would take us too far afield here (see Gough, 1986; Hyndman and Shearer, 1989; Bailey, 1990), but they include the mechanisms of long-time sealing at the upper lid of the conductive layer, at least at positions away from its intersection by an active fault core, the reasons for the approximate coincidence found between the depth to the lid with that to the onset of lower crustal reflectors, which are usually also numerous in tectonically active areas, and the relation of both of those features to mid-crustal decollements.

Among possible sources for fluids near the ductile roots of fault zones is the release of volatiles, primordial or recycled, from the mantle or from underthrust crustal terrains. Newton (1989, 1990), O'Nions and Oxburgh (1988), and Sheppard (1989) provide recent reviews of the evidence on this. (Gold and Soter (1985) present a strong advocacy of the control of earthquake phenomena by deep fluid sources.) The use of $^3He/^4He$ ratios to tag the mantle origin of helium emerging in groundwaters, as discussed earlier in connection with assumption 2, may also hint at a rather broadscale process of volatile supply to the crust. The helium is thought to enter partial melts, including H_2O and CO_2, in the upper mantle, which are emplaced into or plated onto the lower crust. These hot fluids may induce further crustal melting, placing magma bodies into the middle to shallow crust. High $^3He/^4He$ ratios in groundwaters generally correlate with signs of such magmatism (e.g., Welhan et al., 1988), and perhaps because of that there is a certain tendency to assume that the underlying mantle supply of volatiles to the crust is limited to areas of active extensional tectonics. Instead, however, the process may be a more general one, with the extensional areas distinguished only by being regions with horizontal minimum principal stress direction, making it relatively easy for pressurized fluids to make their way upward in the crust. Crustal regions that are active in convergence may have rather similar mantle-fed fluid supply processes at their base, but (as Bailey, 1990, suggests) the fluids are then more readily trapped in horizontal reservoirs at the base of the brittle crust owing to the least principal stress being vertical, and do not lead to abundant surface geochemical emanations. The same argument may be made for a strike-slip regime with some convergence, like the SAF, where the last principal stress in the nearby crust is vertical.

A plausible driving process for the crustal faulting along the San Andreas system is that the crust is coupled, possibly by a weak horizon involving viscous flow in the hot lower crust or along a horizontal detachment, to a mechanically stronger upper mantle. That mantle region is undergoing a shear flow, essentially at steady rate on the interseismic time scale, which results from the Pacific/North America relative plate motion. Such a driving mechanism has been adopted in modeling stress accumulation in earthquake cycles by Li and Rice (1987) and has been discussed recently in terms of rheology and heat flow by Molnar (this volume, Chapter 18). The presumably broadscale shear flow in the upper mantle will result in a modest temperature elevation there, which Molnar (this volume, Chapter 18) suggests as the basis of the broad region of relatively high surface heat flows which encompasses the SAF. Partial melting process should be more active in this slightly warmer shear-flowing region than in adjacent upper mantle, and it should therefore be more effective in supplying volatiles to the base of the overlying crust. Thus there may be an inherent coupling between the flow processes in the upper mantle, which require faulting of the overlying cold crust, and the generation of fluid pressures, which allow that faulting to take place at low stresses.

Water-releasing metamorphic reactions initiated by shear heating of low-grade crustal rocks, as discussed by Scholz (1989, 1990), are also a possible source for highly pressurized fluids within a fault zone. This may be plausible along the SAF in terms of the mineralogy of the Franciscan accretionary complex forming the NE side of the SAF, NW of its intersection with the Garlock fault. Also, the development of deep fluids at high pore pressure along a fault by tectonic compression of adjacent, highly impermeable, fluid-bearing sedimentary rocks has been discussed by Berry (1973) and Irwin and Barnes (1975). Berry (1973) shows many examples of pore pressures well above hydrostatic, and sometimes approaching lithostatic, from borehole records in drilling for oil and gas in the Franciscan complex.

3.3. Assumption 3

The assumption that permeability k is a strongly decreasing function of effective normal stress, and hence a strongly increasing function of pore pressure, has support from laboratory studies as reported by Brace et al. (1968), Pratt et al. (1977), Brace (1978, 1980), and Huenges and Will (1989). Brace et al. (1968) find that k for intact, unfaulted Westerly granite is decreased from a few hundreds of nanodarcies to a few tens of nanodarcies as the effective confining pressure increases from 0 to 500 bar. Huenges and Will (1989) show that rocks of a given mineralogy with a low value of the bulk modulus at zero confining pressure show a much greater sensitivity of k to effective stress than do rocks of higher bulk modulus. Of course, low bulk modulus denotes a high density of open cracks. This suggests that to the extent that rocks of the heavily deformed fault core are extensively cracked, as suggested in the discussion of assumption 2, they will comply well with assumption 3 and show strong dependence of k on effective stress. Unfortunately, data defining the dependence of k on total stresses $\sigma_{\alpha\beta}$, p, rock type and fabric, and on permeation and temperature history are not as extensive as would be desirable.

We shall assume for the purpose of the simple analysis in the next section that k in the fault is a function only of the effective normal stress $\bar{\sigma} = \sigma - p$, where the total normal stress σ will be assumed to vary linearly with depth over the seismogenic range. Further, some specific results will be shown for the law $k = k_0 \exp(-\bar{\sigma}/\sigma^*)$, where k_0 is the permeability at zero effective stress and where σ^* is a constant. This form provides a tolerable, if very approximate, fit to the experimental results cited above for variations of $\bar{\sigma}$ between 0 and, say, 200–400 bar, when we take $\sigma^* = 50$–400 bar. The lower σ^* values (higher sensitivity to $\bar{\sigma}$) correlate with higher k_0, since they correspond with more highly cracked rocks as noted above.

Thus, for a given σ, k will be a strongly (approximately exponentially) increasing function of p. In the actual situation in a fault zone, where permeability is thought to result from a competition between depositional sealing and rupturing to renew pore connectivity, the strong increase of k with p that we assume here may be interpreted as an approximate proxy for saying that permeability will be negligible unless p is driven up toward σ (e.g., Fyfe et al., 1978). To accommodate such a phenomenon in a simple way, we may use the exponential relation for k just described, but with a yet smaller σ^* (now denoting the range of effective stress over which there is a transition from sealed, or rapidly resealed, to continuously permeable response).

4. Fault Zone and Crustal Pore Pressure Distributions

For simplicity, we model a vertical fault zone through the brittle seismogenic portion of the crust (of thickness $H \approx 15$ km) as a channel of uniform width w (Figure 5), small compared to H, and we start with an extreme form of assumption 2, to be relaxed later. That extreme form considers the adjacent crust to be so impermeable compared to the fault that the ratio $k_{\text{crust}}/k_{\text{fault}} \approx 0$, or, in the alternate/supplementary version of assumption 2, that $k_x/k_z \approx 0$.

For the extreme model described, we take the flow interval within the fault zone to extend vertically from $z = -H$ to the earth's surface at $z = 0$, ignoring the generally complex and permeable aquifer regime at shallow depths. Since there is negligible fluid loss from the fault zone to the crust over that flow interval, the mass rate of fluid upflow in the steady permeation state considered must be the same at every level z. Thus, ignoring variations in fluid mass density, the average (called q here) across the fault channel of the upward volumetric flow rate, q_z, must be independent of z. Letting k be

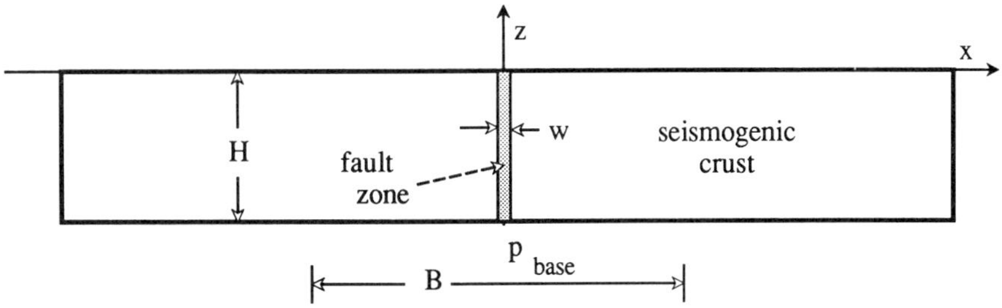

Figure 5. Cross-section of the seismogenic crust; B denotes the width over which elevated pressures are assumed to be present.

the average permeability at level z, letting γ_f be the weight density of the fluid, and noting that in the circumstances described p will be essentially independent of x within the channel, we must therefore have

$$q = -\frac{k}{\eta}\left(\frac{dp}{dz} + \gamma_f\right) = \text{constant (independent of } z)$$

(η is the viscosity of the permeating fluid, taken as constant here). This is to be solved subject to $p = 0$ at $z = 0$, and $p =$ a prescribed pressure, say, p_{base}, at $z = -H$. We will give k a chance to vary with z, as described shortly, but it is interesting to note now that if k is independent of z, then the solution is $p = -p_{\text{base}}z/H$. Thus, if normal stress σ on the fault varies linearly with depth, then the ratio p/σ is independent of depth. This shows that if p is a large fraction of σ at the base of the seismogenic zone, as in assumption 1, then according to this simple solution, p is the same large fraction of σ at every depth.

Now let k vary with z, but according to a law of the type $k = F(\bar{\sigma})$, where $F(\ )$ denotes a function which diminishes very rapidly with increasing values of its argument (shortly, we use the specific form $F(\bar{\sigma}) = k_0 \exp(-\bar{\sigma}/\sigma^*)$ discussed above). Let us write that $\sigma = -(\gamma + \gamma_f)z$ for the variation of σ with z, where γ is constant. If σ is approximately equal to the overburden, then $\gamma = \gamma_t - \gamma_f$ where γ_t is the total weight density of the fault zone material. For numerical illustrations we use $\gamma_f = 0.1$ bar m^{-1} ($= 10^4$ Pa m^{-1}) and $\gamma = 0.2$ bar m^{-1}, and also $H = 15$ km.

Thus, writing $p = \sigma - \bar{\sigma}$, we can rewrite the above equation for the flow rate as

$$\eta q = F(\bar{\sigma})\left(\frac{d\bar{\sigma}}{dz} + \gamma\right)$$

and this may be rearranged to

$$\gamma d(-z) = \frac{d\bar{\sigma}}{1 - \eta q/\gamma F(\bar{\sigma})}$$

Consider what happens as we integrate this equation from the surface, where $z = 0$ and $\bar{\sigma} = 0$, to deeper locations along the fault zone. In this integration, $d(-z) > 0$ and, for solutions of interest, one must conclude that the integration begins with $d\bar{\sigma} > 0$, which means that the (unknown) upflow $q < \gamma F(0)/\eta$ (the right side equals the flow rate due to a hydraulic gradient that is about twice in excess of hydrostatic in a solid with permeability equal to that of unstressed fault material). However, if $F(\bar{\sigma})$ diminishes to

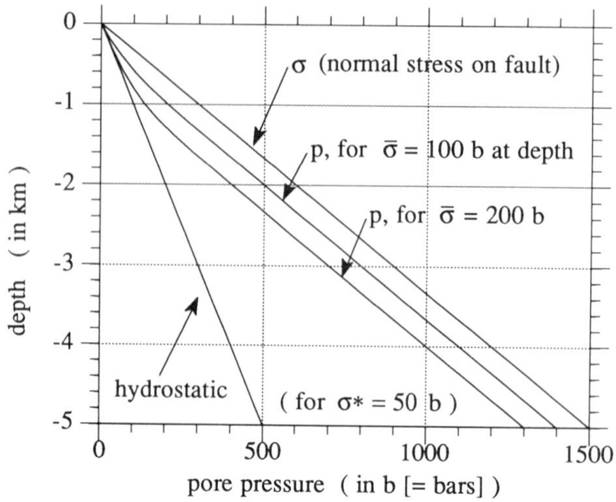

Figure 6. Pore pressure versus depth in fault zone for first 5 km of 15 km seismogenic thickness. The pore pressure distribution adjusts so that the effective normal stress, and thus also the permeability, becomes independent of depth. Based on permeability $k = k_0 \exp(-\bar{\sigma}/\sigma^*)$.

arbitrarily small values with increase of $\bar{\sigma}$, then the denominator in the integration will approach to a zero point as $\bar{\sigma}$ increases, and as that happens the ratio $d\bar{\sigma}/d(-z)$ must approach zero.

Thus the solution for $\bar{\sigma}$ approaches a constant value with increase of depth, that constant value being the solution of $\gamma F(\bar{\sigma}) = \eta q$; see Figure 6. Hence the permeability $k\ [= F(\bar{\sigma})]$ likewise approaches a constant (which legitimizes the assumption made in the simpler analysis above that k could be independent of depth). Thus, assuming that $\bar{\sigma}$ has effectively approached this constant asymptotic value (call that value $\bar{\sigma}_a$, which solves $\gamma F(\bar{\sigma}_a) = \eta q$) before the depth H is reached, the pressure toward the base of the brittle layer will be $p_{\text{base}} = (\gamma + \gamma_f)H - \bar{\sigma}_a$. That pressure is in excess of hydrostatic by the difference $\gamma H - \bar{\sigma}_a$, where γH is of order 3000 bar.

The upflow rate in the fault zone is given by

$$q = \gamma F[\bar{\sigma}_a]/\eta = \gamma F[(\gamma + \gamma_f)H - p_{\text{base}}]/\eta$$
$$\approx 6 \times 10^{14}/\text{year-m} \cdot F(4500\ \text{bar} - p_{\text{base}})$$

with $\eta = 10^{-3}$ Pa s. Thus, if p_{base} is hydrostatic, $p_{\text{base}} \approx 1500$ bar, the permeability function F will be reduced to a size that effectively prohibits flow. While the specifics of a supply mechanism for fluids at the base have not been described here, we might reasonably assume that if there is no venting flow q then p_{base} would not remain hydrostatic. Indeed, it is reasonable to assume that p_{base} increases until a steady state is reached, at which the venting up the fault zone at rate q just balances the supply rate. To have a nonnegligible flow rate, p_{base} will have to be driven upward in magnitude to within a small margin, perhaps within 50–300 bar, of the 4500 bar normal stress at the base.

For the simple exponential approximation $k = F(\bar{\sigma}) = k_0 \exp(-\bar{\sigma}/\sigma^*)$ described earlier, the integration discussed above may be carried out explicitly to give effective normal stress $\bar{\sigma}$ versus depth in the form

$$\bar{\sigma} = \sigma^* \ln\left(\frac{1}{\alpha + (1-\alpha)\exp(\gamma z/\sigma^*)}\right)$$

where $\alpha = q\eta/\gamma k_0$. This is seen to become asymptotic (as $\gamma z/\sigma^*$ becomes large negative) to the value $\bar{\sigma}_a = \sigma^* \ln(\gamma k_0/\eta q)$, and this value is effectively reached at a depth of about $2\sigma^*/\gamma$, which has the range 150 m to 1.5 km for $\sigma^* = 30$–300 bar. Further, the relation between base pressure and upflow rate is

$$q = \frac{\gamma k_0}{\eta}\exp\left(\frac{-\gamma H}{\sigma^*}\right)\left[\frac{\exp[(p_{\text{base}} - \gamma_f H)/\sigma^*] - 1}{1 - \exp(-\gamma H/\sigma^*)}\right]$$

and since $\gamma H/\sigma^* \approx 10$–100 for $\sigma^* = 30$–300 bar, this reduces, whenever p_{base} is at least a little in excess of hydrostatic, to

$$q = \frac{\gamma k_0}{\eta}\exp\left(-\frac{\sigma_{\text{base}} - p_{\text{base}}}{\sigma^*}\right)$$

Figure 6 shows plots of pore pressure p versus depth over the first 5 km for this solution, based on $\sigma^* = 50$ bar. Results are shown for two cases in which the base pressure is such as to make the deep, asymptotically approached, effective normal stress $\bar{\sigma}_a$ equal, respectively, to 100 bar and 200 bar. As seen, the pore pressure rapidly approaches a distribution in which it varies with depth just as does the normal stress σ, but with p reduced by a constant amount $(= \bar{\sigma}_a)$ from σ. Similar patterns have sometimes been seen in pressure logs from drilling for oil and gas in the Franciscan adjacent to the SAF, notably at Fort Bragg (Berry, 1973). In another context, Robin (1990) has also recently analyzed permeation with k dependent on effective stress.

Figure 7 shows the volumetric outflow rate per unit area within the fault channel, in millimeters of fluid discharged per year, in terms of the excess of p_{base} over hydrostatic at the bottom of the layer. Results are shown for values of σ^* ranging from 25 to 200 bar, and the curves have been drawn for $k_0 = 1$ md (and taking $\eta = 10^{-3}$ Pa s). In view of the earlier discussion, that relatively high k_0 value might be thought sensible only for the smaller σ^* values shown. It is easy to correct for other cases since the vertical axis in the plot is q times 1 md$/k_0$. Thus, if $k_0 = 1$ μd is taken as more suitable for the higher σ^* values shown, then the outflow rates plotted are reduced by 10^{-3}.

The response shown by this analysis, for the smaller σ^* values that are preferred for the reasons discussed earlier, reveals a substantial nonlinearity. The venting rate q is negligibly small unless p_{base} rises to become close in value to the lithostatic pressure at the base, and the venting increases very rapidly with p_{base} as that limit is approached, so that the system may be assumed to settle into an equilibrium at which venting balances deeper supply with p_{base} near to its limiting value.

This simple model suggests that the effective normal stress $\bar{\sigma}$ could be essentially independent of depth, within the seismogenic range of a fault zone above a highly pressurized root. This, in turn, implies that the limiting frictional strength, $\tau = f\bar{\sigma}$, may not have much variation with depth. Thus well-known plots of strength versus depth (Brace and Kohlstedt, 1980; Sibson, 1982; etc.), which are governed by brittle friction strength increasing linearly with depth at shallow depth and by ductile flow strength decreasing with depth as depth (and temperature) increase, may require some minor modifications in this case. A pore pressure distribution of the type shown here would deepen, compared to previous estimates, the depth at which frictional strength equals

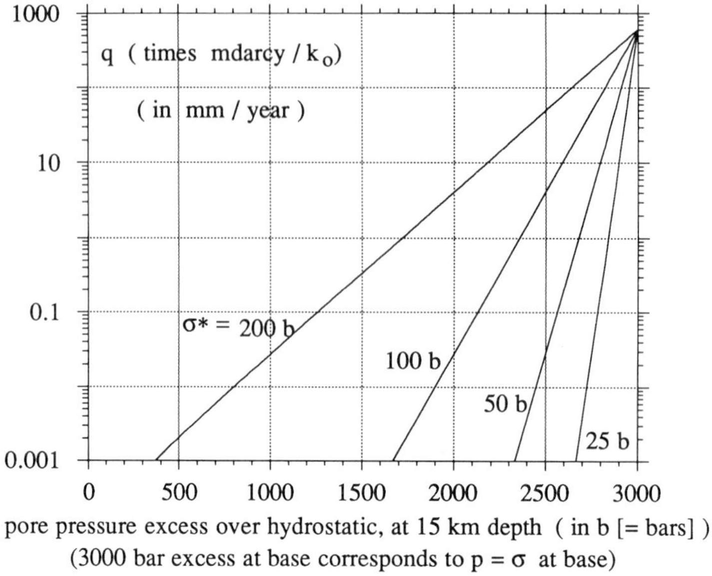

Figure 7. Relation between the fluid upflow rate (q) in the fault zone and the excess of actual pore pressure over hydrostatic at 15 km base of seismogenic zone. Curves drawn for 1 md zero-stress permeability, but can be adjusted to other values as indicated.

creep strength. (In fact, the actual depth of earthquake nucleation may more properly relate to the depth (and temperature) at which friction changes from the potentially unstable velocity-weakening type to the more inherently stable velocity-strengthening type. The relevance of that transition has been shown in earthquake cycle modeling based on rate- and state-dependent friction by Tse and Rice (1986), and foreshadowed by discussions of temperature effects on stick–slip by Brace and Byerlee (1970) and Brace (1972).)

These results show how the assumptions 1 to 3 (with an extreme version of 2) lead to a pore pressure distribution which is high, and near to the fault-normal compression, everywhere throughout the seismogenic zone. That suffices for absolute fault weakness. It remains to be seen that p will be somewhat reduced at points laterally away from the fault zone, in the adjacent crust, as required to explain relative weakness.

In fact, p can be defined in the adjacent crust, on the basis of the analyses just performed, once we specify a distribution of k there (which must be far smaller than k is for the fault, at least throughout the seismogenic layer, for validity of the analysis just described). The result depends on that k distribution and on boundary conditions along the base of layer considered in Figure 5. Rather than dealing with that limiting case directly, we shall consider two types of models where k_c ($= k_{\text{crust}}$) is far lower than k_f ($= k_{\text{fault}}$) but with k_c/k_f nonzero. In these cases the equation determining p is

$$\nabla \cdot [(k/\eta)\nabla(p + \gamma_f z)] = 0$$

where the excess (over hydrostatic) pressure, $p + \gamma_f z$, vanishes at the earth's surface, $z = 0$. At $z = -H$, we set $p + \gamma_f z = p_{\text{base}} - \gamma_f H$ over the interval $-B/2 < x < B/2$, so that B is the width of the pressurized zone (Figure 5). For $|x| > B/2$, we assume that

$p + \gamma_f z = 0$ (i.e., p is hydrostatic there, at $p = \gamma_f H$). Also, to simplify matters, we neglect any explicit dependence of k on p, adjusting very approximately for this by making k_f uniform in depth throughout the fault channel $-w/2 < x < w/2$, and also we continue to take η as constant.

In the first type of model considered, we take k_c to be uniform. This is probably very unrealistic, but will provide a contrast with the second type of model which, effectively, seals-off the lower crust, away from the fault zone, by having k_c decrease with depth to very small values compared to k_f.

For the first type of model (k_c = constant), the pressure distribution in the crust depends strongly on B. For $B \gg H$, the solution in the vicinity of the fault is simply $p = -p_{\text{base}} z/H$ and shows no attenuation of the elevated pressure with distance x from the fault. However, for B of the same order as H, the level contours of excess pore pressure depress significantly as one moves away from the fault zone, especially at typical earthquake nucleation depths (of order 10 km, or about $\frac{2}{3}H$) along the SAF, for cases considered here with $k_c \ll k_f$. Note that a difference of excess pressure between two points at the same depth is the same as the difference in total pore pressure between those points. Figure 8 shows such results (the view is of the half of the seismogenic layer, in the region $x > 0$ only) in the case for which $B = 1.52H$ and $k_c = 10^{-2} k_f$, with fault zone width $w = 0.05H$. The contour lines are equally spaced and the change in excess pore pressure from line to line corresponds to $1/10$ of the excess pore pressure at the base. These were calculated on the basis of the analogy between steady heat conduction and steady fluid permeation, using the finite-element program ABAQUS to solve for the temperature field in an analogously inhomogeneous medium. If the excess pressure is of order 3000 bar at the base, then the contour spacing is of order 300 bars excess pressure.

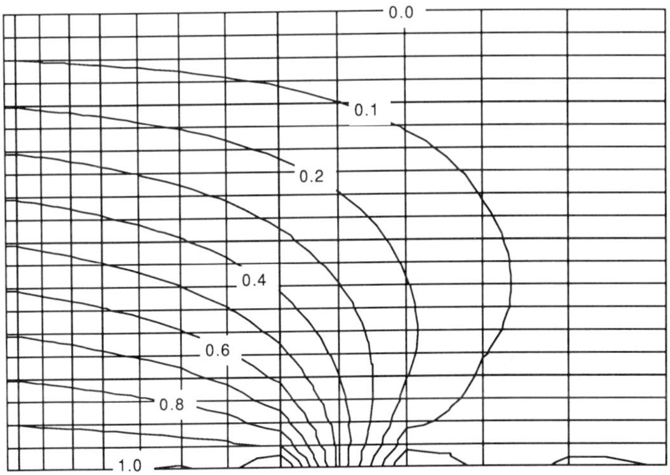

Figure 8. Contours of constant excess pore pressure, $p + \gamma_f z$, for the region of Figure 5 with $x > 0$, when the crustal permeability is uniform at $1/100$ of the fault zone permeability. Calculation done for fault width $w = 0.05H$ and pressurized base width $B = 1.52H$. Numbers on contours give local value of excess pore pressure as a fraction of $p_{\text{base}} - \gamma_f H$. Contour spacing of $1/10$ of the excess pore pressure at the base corresponds to a contour interval of about 300 bar when the base pressure is lithostatic.

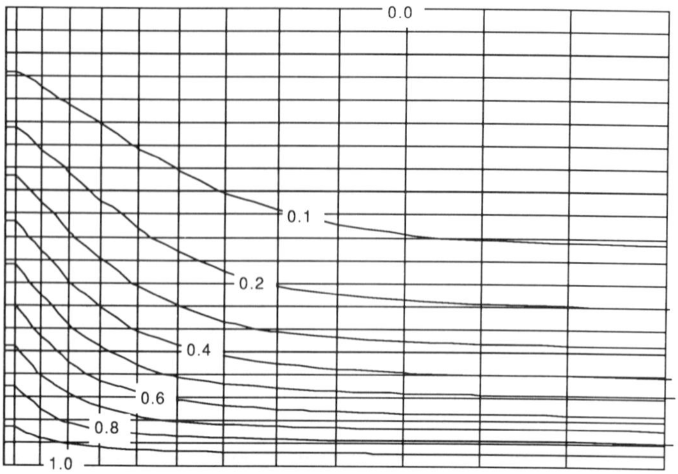

Figure 9. Contours of constant excess pore pressure, $p + \gamma_f z$, for the region of Figure 5 with $x > 0$, when the crustal permeability decreases exponentially with depth, from 1/10 of the fault zone permeability at the surface to 1/1000 of the fault zone permeability at the base. Calculation done for fault width $w = 0.05H$ and pressurized base width B extending over full width. Numbers on contours give local value of excess pore pressure as a fraction of $p_{\text{base}} - \gamma_f H$. Contour spacing of 1/10 of the excess pore pressure at the base corresponds to a contour interval of about 300 bar when the base pressure is lithostatic.

The second type of model examined significantly decreases the crustal permeability with depth. Now the results are insensitive to B, which can be taken as large as the overall width considered. Figure 9 shows results, again from ABAQUS via the steady heat conduction analogy, when the crustal permeability decreases with depth according to the relation $k_c = k_f 10^{-1+2z/H}$, so that $k_c = 10^{-1} k_f$ at the surface, $z = 0$, but k_c diminishes to $10^{-3} k_f$ at the base of the seismogenic layer, $z = -H$. The contours of excess pressure have the same spacing as above, and these are again seen to show higher p in the fault zone than at a small distance away, at the same depth, in the adjacent crust. Although the permeability of the middle depth range of the seismogenic crust is about the same here as in the previous example, the effect of distributing the low permeability toward the lower part of the seismogenic zone has a major effect on attenuating the excess pressure at points laterally away from the fault zone.

An alternative, or supplementary, version of assumption 2 was stated, for which horizontal permeability in the fault zone is far less than vertical. It is evident that such a property distribution will also attenuate the excess pore pressure at points laterally away from the fault zone.

At this point we may conclude that assumptions 1 to 3 lead to pore pressure distributions in the fault zone and in the nearby crust which are of the type noted earlier to be consistent with both the absolute and relative weakness of the SAF.

5. Vertically Propagating Pressure Pulses

For a simple analysis of time-dependent processes associated with the strongly nonlinear dependence of k on effective stress, we may return to the extreme form of assumption 2 adopted in the last section, in which mass outflow from the fault channel into the, presumably, much less permeable crust is neglected. Thus, if $m = m(z, t)$ is the average across the fault channel, at vertical position z, of the mass of fluid contained in unit volume of fault zone material, mass conservation requires that

$$\frac{\partial q}{\partial z} + \frac{\partial m}{\partial t} = 0$$

Although more accurate porous medium theories are possible, it is simplest to assume that m is a function only of the effective stress $\bar{\sigma} = \sigma - p$, that is, $m = G(\bar{\sigma})$. Thus with $k = F(\bar{\sigma})$ as before, we have

$$\frac{\partial}{\partial z}\left[-\frac{\rho_f}{\eta} F(\sigma - p)\left(\frac{\partial p}{\partial z} + \gamma_f\right)\right] + \frac{\partial}{\partial t}[G(\sigma - p)] = 0$$

for the equation controlling time dependence of pore pressure, where ρ_f is the mass density of fluid, regarded as constant here. Again, the normal stress σ is assumed in the form $\sigma = -(\gamma + \gamma_f)z$ as earlier, so that the previous equation can be rewritten as

$$\frac{\partial}{\partial z}\left[\frac{\rho_f}{\eta} F(\bar{\sigma})\left(\frac{\partial \bar{\sigma}}{\partial z} + \gamma\right)\right] + \frac{\partial}{\partial t}[G(\bar{\sigma})] = 0$$

If the fluids in the fault zone are mainly in cracklike pore spaces, as may be reasonable for a high pore pressure situation, it is reasonable to expect that k varies approximately with the cube of openings between the cracklike cavity walls (Poiseuille flow), but that m varies approximately with only the first power of those openings. In that case, and perhaps in others, it is reasonable to expect that $F(\bar{\sigma})$, which gives k, will be a much more rapidly varying function of $\bar{\sigma}$ than is $G(\bar{\sigma})$, which gives m. Formally, we may express this as saying that $F(\bar{\sigma})$ and $G(\bar{\sigma})$, both of which are assumed to decrease monotonically with increasing $\bar{\sigma}$, are also assumed to satisfy the inequality

$$\frac{F(\bar{\sigma}) - F(\bar{\sigma}_2)}{F(\bar{\sigma}_1) - F(\bar{\sigma}_2)} < \frac{G(\bar{\sigma}) - G(\bar{\sigma}_2)}{G(\bar{\sigma}_1) - G(\bar{\sigma}_2)} \qquad \text{for } \bar{\sigma}_1 < \bar{\sigma} < \bar{\sigma}_2$$

To make this inequality a little less abstract, suppose that F is proportional to h^3 and G to h, where h is some representative open width of cracklike, fluid-filled cavities in the fault zone, and h decreases with increasing $\bar{\sigma}$ so that $h_1 > h_2$ if $\bar{\sigma}_1 < \bar{\sigma}_2$. Then the above inequality is equivalent to the self-evident inequality

$$\frac{h^3 - h_2^3}{h_1^3 - h_2^3} \equiv \frac{(h - h_2)(h^2 + hh_2 + h_2^2)}{(h_1 - h_2)(h_1^2 + h_1h_2 + h_2^2)} < \frac{(h - h_2)}{(h_1 - h_2)} \qquad \text{for } h_1 > h > h_2$$

The above inequality in F and G is important, because it is the feature which suffices to give wavelike solutions for the upward propagation of a surge in pore pressure (decrease in $\bar{\sigma}$) due to an increase of pore pressure deep in the fault zone. Such deep increase might result from the shear rupture of a low-permeability barrier which has trapped fluid at more highly elevated pressures below.

Let us suppose then that the pore pressure distribution is initially in an asymptotic steady state with $\bar{\sigma}$ $(= \sigma - p)$ constant with respect to depth and $\partial p/\partial z = d\sigma/dz = -(\gamma + \gamma_f)$, as discussed earlier and shown in Figure 6. Let $\bar{\sigma}_0$ be that initial $\bar{\sigma}$ value. We seek to see when a solitary wave solution of form $\bar{\sigma}$ = function of $z - Vt$ could exist, which takes $\bar{\sigma}$ from its initial value $\bar{\sigma}_0$ to a final value $\bar{\sigma}_f$, as the wave passes by, and also to find the speed V of the wave. For such a wave field we can replace $\partial/\partial t$ by $-V\partial/\partial z$ in the above partial differential equation and then integrate it in z, from a position well ahead of the wavefront, where $\bar{\sigma} = \bar{\sigma}_0$, to one at a general state $\bar{\sigma}$ within the variable part of the wave, to one behind the wavefront where the new state $\bar{\sigma} = \bar{\sigma}_f$ has been established. Thus, fixing attention on any convenient moment of time and, to focus on the variation of $\bar{\sigma}$ with z, replacing $\partial\bar{\sigma}/\partial z$ with $d\bar{\sigma}/dz$, we have

$$\rho_f F(\bar{\sigma})\left(\frac{d\bar{\sigma}}{dz} + \gamma\right) - \eta V G(\bar{\sigma}) = \rho_f \gamma F(\bar{\sigma}_0) - \eta V G(\bar{\sigma}_0) = \rho_f \gamma F(\bar{\sigma}_f) - \eta V G(\bar{\sigma}_f)$$

Thus, if such a wave exists, the last equality of this expression shows that its velocity will be

$$V = \frac{\rho_f \gamma}{\eta} \frac{F(\bar{\sigma}_f) - F(\bar{\sigma}_0)}{G(\bar{\sigma}_f) - G(\bar{\sigma}_0)}$$

The monotonicity of F and G shows that this expression for V always yields a positive result. Hence, if such wavelike propagation of a pore pressure change can occur at all, it must involve an upwardly propagating pulse.

To see whether the wave can occur we substitute the expression for V into the expression before it, and then solve for $d\bar{\sigma}/dz$ to obtain

$$\frac{d\bar{\sigma}}{dz} = \frac{\gamma[F(\bar{\sigma}_f) - F(\bar{\sigma}_0)]}{F(\bar{\sigma})}\left[\frac{G(\bar{\sigma}) - G(\bar{\sigma}_0)}{G(\bar{\sigma}_f) - G(\bar{\sigma}_0)} - \frac{F(\bar{\sigma}) - F(\bar{\sigma}_0)}{F(\bar{\sigma}_f) - F(\bar{\sigma}_0)}\right]$$

By integrating in the positive direction in z, from the state of $\bar{\sigma} = \bar{\sigma}_f$ to that at $\bar{\sigma} = \bar{\sigma}_0$ (the corresponding z values in this type of wave analysis are at $-\infty$ and $+\infty$), this equation defines the variation in $\bar{\sigma}$ across the wavefront. A consistent result is obtained if z increases in the positive direction as $\bar{\sigma}$ is varied from $\bar{\sigma}_f$ to $\bar{\sigma}_0$, and such a result ($dz > 0$) is assured if the right-hand side of this equation is of the same sign as $\bar{\sigma}_0 - \bar{\sigma}_f$ when $\bar{\sigma}$ lies between $\bar{\sigma}_f$ and $\bar{\sigma}_0$. Because of the monotonic decrease of F with $\bar{\sigma}$, $[F(\bar{\sigma}_f) - F(\bar{\sigma}_0)]$ is always of that sign. The inequality involving F and G assumed earlier then assures us that the long bracketed set of terms on the right side of this equation also have the same sign as $\bar{\sigma}_0 - \bar{\sigma}_f$, at least when $\bar{\sigma}$ lies between $\bar{\sigma}_f$ and $\bar{\sigma}_0$. Thus the right-hand side is always positive when $\bar{\sigma}$ lies between $\bar{\sigma}_f$ and $\bar{\sigma}_0$, which therefore tells us that a wavelike solution can exist only when $\bar{\sigma}_0 - \bar{\sigma}_f > 0$.

Hence a wavelike solution exists only when it is a state with smaller $\bar{\sigma}$, or higher pore pressure, that is being propagated upward into the initial state. This will correspond to an upward surge of pore fluid.

The expression above for the velocity V of the surge reduces to a simple form when the decrease in $\bar{\sigma}$ is sufficiently strong that we may assume that $F(\bar{\sigma}_0)$ and $G(\bar{\sigma}_0)$ are, respectively, small compared to $F(\bar{\sigma}_f)$ and $G(\bar{\sigma}_f)$. In that limit the speed is

$$V = \frac{\rho_f \gamma k}{\eta m} = \frac{\gamma k}{\eta n}$$

where the k and m are those in the state behind the wavefront (that with $\bar{\sigma}$ denoted as $\bar{\sigma}_f$ above), and the latter form of the result arises when we write $m = \rho_f n$, where n is the fluid-filled porosity in the fault zone. Not unexpectedly, that latter term may be recognized as the average fluid particle velocity in a porous medium under an excess pressure gradient of γ.

To examine representative numerical values, if we assume $k = 1$ md, γ and η as before, and assume a porosity of $n = 0.01$, this gives $V = 60$ m/year. Values on the order of 1 km/year at not implausible for strong pressure surges which initiate additional microcracking, and result in a large increase in permeability to the 10–100 md range (they would, presumably, also involve a somewhat larger n). There may be a connection between the upwardly propagating surges, shown to result from the analysis here, and the unstably rising 'fluid domains' hypothesized by Gold and Soter (1985).

This analysis of pressure surges is an incomplete treatment of time effects, not only because it examines only a particular type of solution to the partial differential equation, but because it does not include explicit treatment of dissolution and deposition which, in aggregate, are thought to seal off permeability (Smith and Evans, 1984; Nur and Walder, 1990), and of shear rupturing processes which can renew pore connectivity and hence restore permeability. Time-dependent pore pressure has been emphasized by Gold and Soter (1985) and Sibson (1990) as a factor in addition to tectonic loading which could control the timing and aperiodicity of earthquakes.

6. Summary

It has been shown that a pore pressure distribution with p high and near to the fault-normal compression in a major fault zone such as the SAF, but somewhat lower in the adjacent crust, is consistent with the absolute and relative weakness of that major fault. This result remains valid for friction coefficients f in the typical range of 0.6–0.9 as for laboratory data on crustal rocks. Previously, such a pore pressure explanation has been regarded as invalid, since pore pressures within the fault zone would have to exceed the overburden there. However, it is shown here that a maturely deformed fault zone, weakened either by high pore pressure or by inherently weak material, will develop a stress state within its active core which is different from the state in the nearby crust. In particular, for the situation of a vertical strike-slip fault that is driven by stresses in the adjoining crust which have a maximum principal direction at a steep angle to the fault trace (and which are suitable for inducing thrust failures in the crust), all principal stress values in the active strike-slip fault core are elevated over those at the same depth in the adjoining crust. Because of that feature, high pore pressure distributions, as described, are possible with p less than the least principal stress at every point of the medium, even though p in the fault core may exceed the least principal stress σ_3 at some distance away in the crust (where, typically, adjacent to the SAF, σ_3 equals the overburden).

Three assumptions concerning the presence and motion of fluids in the crust were introduced. It was shown that these assumptions, when implemented as the basis for a steady-state fluid permeation analysis, lead to the type of pore pressure distributions which have been noted above to be consistent with both absolute and relative fault weakness. Briefly, the assumptions are that pore fluids are present near the ductile roots of fault zones, that active fault zones maintain a far higher permeability than do adjacent regions of the crust, and that permeability in the fault zone is a strongly decreasing function of the effective normal stress. This set of assumptions has support from a variety of sources, discussed in this chapter. These include laboratory rock permeability measurements, petrologic studies of exhumed fault zone rocks, geochemical characterization of groundwater outflows near faults, magnetotelluric inferences of electrical conductivity in the middle and lower crust, and seismic studies of wave speed and attenuation variations near active faults.

While the results of this chapter do not invalidate other possible explanations of weakness along major faults such as the SAF, they do show that elevated pore pressure is a plausible explanation. Deep drilling to the near vicinity of an active major fault zone, with measurement of stress state and pore pressure variations as the fault zone is approached, could provide a critical test of these ideas. If the pore pressure explanation of fault weakness is correct, then the traditional concept of frictional strength in the brittle crust varying in an approximately linear manner with depth must be replaced by one in which frictional strength is relatively uniform with respect to depth throughout the seismogenic zone.

Additional features of interest for earthquake phenomena follow from the assumptions made in the fluid permeation analysis. For example, it is shown that the time-dependent equations for the effective stress and pore pressure distributions without the fault zone have wavelike solutions corresponding to surges of pore pressure, and of pore fluid, which slowly make their way upward in a fault zone. Also, while the concept is not developed here, pore pressure distributions that result from a nonlinear, effective stress-dependent permeability provide new insights on the possible origins of seismic asperities. In addition to being due to variations in inherent frictional properties within the fault zone, such asperities may also result from variations in permeability properties (e.g., in properties analogous to k_0 and σ^* in the simple exponential model introduced here), or from fluctuations about the approximately linear variation of normal stress with depth.

It is a delight to dedicate this chapter to William F. Brace, whose work over many years has contributed richly to many of the areas discussed. The work was presented orally in June 1990 at a conference in honor of Brace.

Acknowledgment

Support for this study was provided by the USGS under grant 14-08-0001-G-1788 to Harvard. Also, the program ABAQUS was kindly made available under academic licence by Hibbitt, Karlsson and Sorensen, Inc. of Providence, RI. I am grateful to many individuals for discussing the concepts of this chapter. Robert Newton of Chicago posed some interesting questions to me in late 1988 which began the thinking leading to this work. Helpful discussions, verbally and/or by correspondence, were had also with Atilla Aydin, Fred Chester, Harmon Craig, Jack Hermance, Steve Hickman, William Irwin, Louise Kellogg, Geoff King, John Logan, Ted Madden, Brian Mitchell, Julie Morris, Kieth O'Nions, Alan Rubin, Selvyn Sacks, Chris Scholz, Tom Shankland, Brian Stewart, Terry Tullis, Hiroshi Wakita, Chi-Yuen Wang, Youxue Zhang, and Mark Zoback.

References

Al-Shukri, H.J. and Mitchell, B.J. (1988). Reduced seismic velocities in the source zone of New Madrid earthquakes. *Bull. Seismol. Soc. Am.* **78**, 1491–1509.

Al-Shukri, H.J. and Mitchell, B.J. (1990). Three-dimensional attenuation structure in and around the New Madrid seismic zone. *Bull. Seismol. Soc. Am.* **80**, 615–632.

Arch, J. and Maltman, A. (1990). Anisotropic permeability and tortuosity in deformed wet sediments. *J. Geophys. Res.* **95**, 9035–9045.

Bailey, R.C. (1990). Trapping of aqueous fluids in the deep crust. *Geophys. Res. Lett.* **17**, 1129–1132.

Beach, A. (1980). Retrogressive metamorphic processes in shear zones with special reference to the Lewisian complex. *J. Struct. Geol.* **2**(1/2), 257–263.

Berry, F.A.F. (1973). High fluid potentials in California Coast Ranges and their tectonic significance. *Bull. Assoc. Petrol. Geol.* **57**, 1219–1249.

Brace, W.F. (1978). A note on permeability changes in geologic material due to stress. *Pure Appl. Geophys.* **116**, 627–633.

Brace, W.F. (1972). Laboratory studies of stick-slip and their application to earthquakes. In *Forerunners of Strong Earthquakes* (eds. E.F. Savarensky and T. Rikitake) *Tectonophysics* **14**(3/4), 189–200.

Brace, W.F. (1980). Permeability of crystalline and argillaceous rocks. *Int. J. Rock Mech. Min. Sci.* **17**, 241–251.

Brace, W.F. and Byerlee, J.D. (1970). California earthquakes: why only shallow focus? *Science* **168**, 1573–1575.

Brace, W.F. and Kohlstedt, D.L. (1980). Limits on lithospheric stress imposed by laboratory experiments. *J. Geophys. Res.* **85**, 6248–6252.

Brace, W.F., Walsh, J.B., and Frangos, W.T. (1968). Permeability of granite under high pressure. *J. Geophys. Res.* **73**(6), 2225–2236.

Byerlee, J.D. (1978). Friction of rocks. *Pure Appl. Geophys.* **116**, 615–626.

Byerlee, J.D. (1990). Friction, overpressure and fault normal compression. *Geophys. Res. Lett.* **17**, 2109–2112.

Byrne, T. and Fisher, D. (1990). Evidence for a weak and overpressurized décollement beneath sediment-dominated accretionary prisms. *J. Geophys. Res.* **95**, 9081–9097.

Chen, W.-P. and Molnar, P. (1983). Focal depths of intracontinental and intraplate earthquakes and their implications for the thermal and mechanical properties of the lithosphere. *J. Geophys. Res.* **88**, 4183–4214.

Etheridge, M.A., Wall, V.J., Cox, S.F., and Vernon, R.H. (1984). High fluid pressures during regional metamorphism and deformation: implications for mass transport and deformation mechanisms. *J. Geophys. Res.* **89**, 4344–4358.

Fournier, R.L. (1990). A San Andreas Fault model in which maximum principal stress nearly normal to the fault is advantageous (abstract). *EOS, Trans. AGU* **71**, 1635.

Fyfe, W.S., Price, N.J., and Thompson, A.B. (1978). *Fluids in the Earth's Crust*, Developments in Geochemistry 1. Elsevier Scientific, New York.

Gold, T. and Soter, S. (1985). Fluid ascent through the solid lithosphere and its relation to earthquakes. *Pure Appl. Geophys.* **122**, 492–530.

Gough, D.I. (1986). Seismic reflectors, conductivity, water and stress in the continental crust. *Nature* **323**, 143–144.

Haak, V. and Hutton, R. (1986). Electrical resistivity in continental lower crust. In *The Nature of the Lower Continental Crust* (eds. J.B. Dawson et al.), pp. 35–49. Geological Society of London Special Publication No. 24.

Hamilton, R.M. and Mooney, W.D. (1990). Seismic-wave attenuation associated with crustal faults in the New Madrid seismic zone. *Science* **248**, 351–354.

Hickman, S. (1991). Stress in the lithosphere and the strength of active faults. *Rev. Geophys.*, Supplement, pp. 759–775, April.

Hubbert, M.K. and Rubey, W.W. (1959). Role of fluid pressure in the mechanics of overthrust faulting I: Mechanics of fluid-filled porous solids and its application to overthrust faulting. *Geol. Soc. Am. Bull.* **70**, 115–166.

Huenges, E. and Will, G. (1989). Permeability, bulk modulus and complex resistivity in crystalline rocks. In *Fluid Movements – Element Transport and the Composition of the Deep Crust* (ed. D. Bridgwater), pp. 361–375. Kluwer, Dordrecht.

Hyndman, R.D. and Shearer, P.M. (1989). Water in the lower continental crust: modelling magnetotelluric and seismic reflection results. *Geophys. J. Int.* **98**, 343–365.

Irwin, W.P. and Barnes, I. (1975). Effects of geologic structure and metamorphic fluids on seismic behavior of the San Andreas fault system in central and northern California. *Geology* **3**, 713–716.

Irwin, W.P. and Barnes, I. (1980). Tectonic relations of carbon dioxide discharges and earthquakes. *J. Geophys. Res.* **85**(B6), 3115–3121.

Jones, L.M. (1988). Focal mechanisms and the state of stress on the San Andreas fault in Southern California. *J. Geophys. Res.* **93**, 8869–8891.

Kerrich, R. (1986). Fluid infiltration into fault zones: chemical isotopic and mechanical effects. *Pure Appl. Geophys.* **124**, 225–268.

Lachenbruch, A.H. and Sass, J.H. (1980). Heat flow and energetics of the San Andreas fault zone. *J. Geophys. Res.* **85**, 6185–6222.

Li, V.C. and Rice, J.R. (1987). Crustal deformation in great California earthquake cycles. *J. Geophys. Res.* **92**, 11533–11551.

Madden, T., La Torraca, G., and Park, S. (1990). Resistivity variations around the Palmdale section of the San Andreas; unpublished manuscript.

McCaig, A.M. (1988). Deep fluid circulation in fault zones. *Geology* **16**, 867–870.

Meissner, R. and Strehlau, J. (1982). Limits of stress in continental crusts and their relation to the depth-frequency distribution of shallow earthquakes. *Tectonics* **1**, 73–89.

Mooney, W.D. and Ginzburg, A. (1986). Seismic measurements of the internal properties of fault zones. *Pure Appl. Geophys.* **124**, 141–157.

Nagao, K., Kita, I., Matsuda, J., and Ercan, T. (1990). Helium and carbon isotopic composition of gas and water samples from Turkey (abstract). *EOS, Trans. AGU* **71**(28), 850–851.

Newton, R.C. (1989). Metamorphic fluids in the deep crust. *Annu. Rev. Earth Planet. Sci.* **17**, 385–412.

Newton, R.C. (1990). Fluids and shear zones in the deep crust. *Tectonophysics* **182**, 21–37. [Issue on *The Nature of the Lower Continental Crust* (eds. D.M. Fountain and A. Boriani.)]

Newton, R.C., Smith, J.V., and Windley, B.F. (1980). Carbonic metamorphism, granulites and crustal growth. *Nature* **288**, 45–50.

Nur, A. and Walder, J. (1990). Time-dependent hydraulics of the earth's crust. In *The Role of Fluids in Crustal Processes*, pp. 113–127. National Academy Press, Washington, D.C.

O'Nions, R.K. and Oxburgh, E.R. (1988). Helium, volatile fluxes and the development of continental crust. *Earth Planet. Sci. Lett.* **90**, 331–347.

O'Nions, R.K., Griesshaber, E., and Oxburgh, E.R. (1989). Rocks that are too hot to handle. *Nature* **341**, 391.

Oppenheimer, D.H., Reasenberg, P.A., and Simpson, R. (1988). Fault plane solutions for the 1984 Morgan Hill earthquake sequence: evidence for the state of stress on the Calaveras fault. *J. Geophys. Res.* **93**, 9007–9026.

Oxburgh, E.R. and O'Nions, R.K. (1987). Helium loss, tectonics, and the terrestrial heat budget. *Science* **237**, 1583–1588.

Parry, W.T. and Bruhn, R.L. (1986). Pore fluid and seismogenic characteristics of fault rock at depth on the Wasatch fault, Utah. *J. Geophys. Res.* **91**, 730–734.

Peacock, S.M. (1990). Fluid processes in subduction zones. *Science* **248**, 329–337.

Pratt, H.R., Black, A.D., Brace, W.F., and Swolfs, H. (1977). Elastic and transport properties of an in-situ jointed granite. *Int. J. Rock Mech. Min. Sci.* **14**, 35–45.

Raleigh, C.B., Healy, J.H., and Bredehoeft, J.D. (1976). An experiment in earthquake control at Rangely, Colorado. *Science* **191**, 1230–1237.

Robin, P.-Y.F. (1990). Water retention within a brittle upper crust: a numerical simulation. Abstract of presentation at Brace Symposium: Fault Mechanics and Transport Properties of Rock, MIT, June 1990.

Scholz, C.H. (1989). Mechanics of faulting. *Annu. Rev. Earth Planet. Sci.* **17**, 309–334.

Scholz, C.H. (1990). *The Mechanics of Earthquakes and Faulting.* Cambridge University Press.

Shankland, T.J. and Ander, M.E. (1983). Electrical conductivity, temperatures, and fluids in the lower crust. *J. Geophys. Res.* **88**, 9475–9484.

Sheppard, S.M.R. (1989). The isotopic characterization of aqueous and leucogranitic crustal fluids. In *Fluid Movements – Element Transport and the Composition of the Deep Crust* (ed. D. Bridgwater), pp. 245–263. Kluwer, Dordrecht.

Sibson, R.H. (1982). Fault zone models, heat flow, and the depth distribution of earthquakes in the continental crust of the United States. *Bull. Seismol. Soc. Am.* **72**, 151–163.

Sibson, R.H. (1985). A note on fault reactivation. *J. Struct. Geol.* **7**, 751–754.

Sibson, R.H. (1990). Rupture nucleation on unfavorably oriented faults. *Bull. Seismol. Soc. Am.* **80**, 1580–1604.

Sibson, R.H., McMoore, J., and Rankine, A.H. (1975). Seismic pumping – A hydrothermal fluid transport mechanism. *J. Geol. Soc. London* **131**, 653–659.

Smith, D.L. and Evans, B. (1984). Diffusional crack healing in quartz. *J. Geophys. Res.* **89**, 4125–4135.

Tse, S.T. and Rice, J.R. (1986). Crustal earthquake instability in relation to the depth variation of frictional slip properties. *J. Geophys. Res.* **91**, 9452–9472.

Wakita, H., Fujii, N., Matsuo, S., Notsu, K., Nagao, K., Takaoka, N. (1978). 'Helium spots': caused by a diapiric magma from the upper mantle. *Science* **200**, 430–432.

Wakita, H., Sano, Y., and Mizoue, M. (1987). High ^{3}He emanation and seismic swarms observed in a nonvolcanic, forearc region. *J. Geophys. Res.* **92**, 12539–12546.

Wang, C.-Y., Rui, F., Zhengsheng, Y., and Xingjue, S. (1986). Gravity anomaly and density structure of the San Andreas fault zone. *Pure Appl. Geophys.* **124**, 127–140.

Welhan, J.A., Poreda, R.J., Rison, W., and Craig, H. (1988). Helium isotopes in geothermal and volcanic gases of the western United States, I. Regional variability and magmatic origin. *J. Volcanol. Geotherm. Res.* **34**, 185–199.

Wilcock, W.S.D., Purdy, G.M., and Solomon, S.C. (1990). Microearthquake evidence for extension across the Kane transform fault. *J. Geophys. Res.* **95**, 15439–15462.

Zoback, M.D. and Healy, J.H. (1991). In situ stress measurements to 3.5 km depth in the Cajon Pass scientific research borehole: implications for the mechanics of crustal faulting, *J. Geophys. Res.*, in press.

Zoback, M.D., Tsukahara, H., and Hickman, S. (1980). Stress measurements at depth in the vicinity of the San Andreas fault: implications for the magnitude of shear stress at depth. *J. Geophys. Res.* **85**, 6157–6173.

Zoback, M.D., Zoback, M.L., Mount, V.S., Suppe, J., Eaton, J.P., Healy, J.H., Oppenheimer, D., Reasenberg, P., Jones, L., Raleigh, C.B., Wong, I.G., Scotti, O., and Wentworth, C. (1987). New evidence on the state of stress on the San Andreas fault system. *Science* **238**, 1105–1111.

Chapter 21

Paradigms or Small Change in Earthquake Mechanics

C.H. Scholz

Lamont-Doherty Geological Observatory, and Department of Geological Sciences, Columbia University, Palisades, N.Y. 10964, U.S.A.

Abstract

The stick–slip theory of earthquakes is one of the greatest contributions of W.F. Brace, yet it has not achieved broad acceptance among seismologists. Instead there has been a long struggle between this 'new', though now 25-year-old, idea and the old seismological paradigm of the earthquake mechanism as a fracture phenomenon that produces a total stress drop. I review the history of this conflict and discuss various factors that prompted it.

1. Introduction

In the view of many, the crowning achievement in the career of W.F. Brace was the announcement, in Brace and Byerlee (1966), of the stick–slip theory of earthquakes. This constituted a new paradigm for a major earth process, with a potential influence that extended far beyond the confines of Brace's field of rock mechanics. Although this paradigm has been accepted into the canon of rock mechanics, it has not yet, 25 years later, been consensually accepted into the world view of seismologists, the group who would have been expected to be its principal users and beneficiaries. If the measure of completion of a scientific revolution is the near-universal acceptance of a new paradigm, then this one is certainly not over.

The purpose of this essay is to explore the nature of this scientific struggle for a new paradigm for the earthquake mechanism and some of the reasons for its lack of general acceptance among seismologists. I use as a basis the language and concepts of Kuhn's *The Structure of Scientific Revolutions* (1970), but I do not dogmatically adhere to Kuhn's model. Other historians and philosphers of science have taken somewhat different views: for a discussion see the concluding sections of LeGrand's (1988) treatment of the plate tectonics revolution. My being neither a profesional science historian nor a

FAULT MECHANICS AND TRANSPORT PROPERTIES OF ROCKS
ISBN 0-12-243780-2

nonparticipant weakens this account, since I am susceptible to charges of bias. On the other hand, the struggle for the stick–slip paradigm has been largely covert and waged through proxy issues to such an extent that the outsider is unlikely even to detect the presence of a struggle. This account may be of interest to the professional historian because the revolution in progress is qualitatively different from the cases usually discussed as examples. It was not created by a crisis, hence there is no great pressure for a quick resolution. It is still possible for two groups of scientists to continue to do normal science while developing in parallel two opposed paradigms. The struggle is largely between disciplines with different and incommensurate training and experience, and may be viewed as a struggle for dominance between disciplines on the one hand and as an educational disparity on the other.

The very nature of this struggle makes it extremely difficult to document in a way that satisfies either the scientist or the historian. One cannot find a case in which a seismologist has explicitly stated in print that he or she does not accept the stick–slip theory of earthquakes. Instead, what one finds is that, as a group, seismologists *act* as though that theory does not exist. The theory is shunned. The extent to which this shunning is by ignorance or by intent is difficult to ascertain. To the extent that ignorance or misunderstanding is a problem, this essay may serve a useful purpose in clarifying points of dispute.

2. The Old and New Paradigms

Both conceptualizations that are the topic of this essay lie within the overall rubric of the faulting theory of earthquakes, as developed by Gilbert (1884) and Koto (1893), and consolidated by Reid (1910). In this, the primary paradigm is that earthquakes result from a sudden slippage on a preexisting geologic fault. The cumulative offset or displacement of a fault is produced by a repetition of these sudden, earthquake-generating slips that occur at infrequent intervals on each active fault or fault segment. The central issue of the paradigms to be discussed here has to do with what causes the sudden slip that results in earthquakes, and secondary issues are the value of shear stress necessary to initiate the earthquake and the value of the shear stress that remains on the fault immediately following the earthquake. This question was addressed only by Gilbert, of these early workers, when he said that the earthquake occurs when the stress accumulating on the fault became large enough to overcome starting friction.

The old paradigm, which may be called the strength theory, dates from Tsuboi (1933). He was studying earth strain as determined by the differences between repeated geodetic measurements in Japan. He found that the largest strain measurement ever observed was obtained by comparing measurements just before and just after a large earthquake – that is, the change in strain brought about by the earthquake. He called this the 'ultimate strain,' and concluded that this is the maximum strain that the earth's crust could sustain without rupturing in an earthquake. He thus implicitly assumed that the earthquake dropped the stress to zero, apparently by an (unstated) analogy to tensile fracture. Although a justification for this assumption was never articulated by him, the concept of total stress drop in earthquakes achieved paradigm status and became a central part of the seismologic tradition. More recently, for example, Chinnery (1964) equated the strength of the Earth's crust with earthquake stress drops. Given the lack of any articulated model to justify this assumption, it is difficult even to speculate on why it was tacitly accepted. It was not even recognized as an assumption until the stick–slip theory was proposed as an alternative.

Brace and Byerlee's (1966) statement of the stick–slip theory of earthquakes marks the beginning of the application of rock mechanics to the earthquake mechanism problem. Stick–slip motion had been known as a common mode of frictional sliding since at least the time of Coulomb. In this mode of behavior, slip initiates with rapid unstable slip when static friction is overcome, and the frictional resistance drops to a lower dynamic value. Slip ceases when the stress drops to a value near the dynamic friction level. The cycle is repeated after a period in which the stress is recharged to the static friction value. When Brace and Byerlee noticed that this mode of frictional sliding was commonly observed in their laboratory rock friction experiments, it was not a great step for them to propose it as the general mechanism of earthquakes. In fact, a reexamination of the faulting theory of earthquakes, as stated by Gilbert and paraphrased above, shows it to be by definition a description of stick–slip frictional motion. That is, earthquakes are dynamic slip instabilities occurring on preexisting sliding surfaces that repeat following long periods in which no slip occurs. So this new paradigm was simply the application of a well-understood, even classical, phenomenon borrowed from one field and applied to a new manifestation in another field. This seemed hardly the stuff of controversy.

In the strength model, earthquakes occur because a strong part of the fault breaks: weak faults presumably have few or no earthquakes. In the stick–slip model, the seismicity of faults has not to do with strength but depends on the stability or instability of the frictional sliding properties of the fault. This stability depends on the rate at which friction fluctuates with slip or slip velocity as compared to the rate at which stress can be elastically unloaded by the material surrounding the fault.

The stick–slip hypothesis revealed the arbitrariness of the total stress drop assumption. By stating that the stress drop in an earthquake was a drop from static to dynamic friction and hence was some fraction of the total applied stress, it did not conflict with any known geophysical observation. Tsuboi's ultimate strain was simply a coseismic strain drop and nothing more. Since an earthquake stress drop can be measured in a variety of ways but total stress is not readily accessible to measurement, the one immediately pressing reason for accepting the new over the old paradigm was that the new was based on a physical mechanism with many well-known examples, whereas the old was based on an unsupported assumption. There was no crisis that prompted the introduction of this new paradigm; it was introduced from outside the field to which it was applied for the sole purpose of correcting what was viewed as an earlier misconception. It did not upset any significant body of seismological work; aside from Chinnery's paper no extensive literature had yet been devoted to inferring strength or stress from stress drop. It did not affect the way earthquakes are described, although, as will be discussed later, it did profoundly affect the way the cause of earthquakes would be viewed. The immediate damage to the field of seismology was slight and could be easily sustained. On the other hand, as will be brought out further in this narrative, the seismological study of earthquakes suffers cumulative damage from the retention of the strength paradigm. Because this paradigm is purely descriptive, it does not offer a framework upon which the mechanics of the process can be built and thereby does not allow the observations to be unified.

Why has the stick–slip hypothesis not entered the mainstream of seismology? One factor must be that the total stress drop concept was near the center of the earthquake paradigm for 35 years, so it was very deeply embedded in the consciousness of workers in that field. Neither did it help that the new concept was based on an understanding of friction that was a part of neither the training nor the experience of seismologists: in Kuhn's thesis, there is an incommensurability between the old and new concepts, and between the two fields that are promoting them. Finally, the practice of seismology was threatened in two ways: the primary seismological measurement of an earthquake, the stress drop, could no longer be seen as a measurement of a fundamental quantity and hence was demeaned in value; and secondly, the new paradigm, coming from the outside, challenged the status of the earthquake problem as the sole intellectual property of the seismologist.

3. Differences in the Training and Experience of the Disciplines

Seismology has always been regarded by the public according to its dictionary definition – the science of earthquakes – but the fact is that most seismologists do not study earthquakes. The field was born in the late nineteenth century for the purpose of studying earthquakes. It quickly became apparent that the answer to even the simplest questions about earthquakes, such as their location and depth, required an understanding of elastic wave propagation in the Earth. From this was born the great research program of seismology: the determination of the internal structure and constitution of the Earth from the propagation of seismic waves, in which earthquakes became merely the 'seismic source,' the clapper that rang the bell. The interest was in the bell, not the clapper. This earth-structure program was to become the most powerful and successful research effort in solid earth geophysics; it would place seismology at the top of the earth science hierarchy, and for most of the twentiety century it would attract the most able scientists in the earth sciences.

In the meantime, the study of earthquakes was relegated to the smaller and less prestigious subfield of earthquake seismology. By contrast to the study of earth structure, it was for a long time a mainly descriptive effort. The work consisted of the compilation of earthquake catalogs, the development of magnitude scales and improved location methods, and the description of individual prominent earthquakes. This situation persisted until the mid-1960s, when several remarkable developments occurred, initiated by Sykes' (1967) use of earthquake focal mechanisms to verify the transform fault hypothesis.

Quite aside from its impact on the seafloor spreading hypothesis, which is another story, Sykes' paper was notable in that it clinched the notion that earthquakes could be proved, from their radiation, to be the result of shear motion on a fault releasing a double couple, which had been a matter of controversy and confusion in seismology for 40 years. This development, which finally cemented the faulting hypothesis as a *general* mechanism for earthquakes, together with the appearance of Brace and Byerlee a year earlier, provided the ingredients for starting a new era. This was consolidated by the publication of Isacks, Oliver, and Sykes (1968), and a new research program for seismology was initiated; the elucidation of tectonics from the study of earthquake focal mechanisms. This new program revitalized earthquake seismology. Most of the central problems in earth structure had been solved by this time, and this new program offered opportunities attractive to a new generation of scientists, and so the field grew and became much more prominent within seismology as a whole.

Throughout these developments, the training of seismologists remained essentially unchanged. Because their prime tool is in the study of seismic waves, their training concentrates heavily on those branches of physics and mathematics relevant to the propagation of waves in elastic media. This is a broad and complex topic, the mastery and practice of which leaves little time for the exploration of distant topics, such as the mechanics of fracture and friction. Nor, in the way seismology viewed itself, was there any need to study these other topics. Even though the new program concentrated on the earthquake mechanism, the earthquake was still viewed as a seismic source, and hence was still entirely a problem of the inversion of seismic waves to discover the nature of this source. The modern theory of seismic sources was entirely couched in these terms (Aki and Richards, 1980). This background led to an entirely different, and incommensurate, conceptualization of the earthquake mechanism from that held by the rock mechanicists.

In contrast to seismology, rock mechanics is an obscure field. If known at all, it is most closely associated with mining engineering, a field in which its most familiar applications lay. The branch that concentrates on geological applications is tiny and has no professional organization or journal devoted to it. Of this branch, the modern subfield that specializes in brittle processes is recent in origin and is frequently associated with the name of W.F. Brace, who founded one of its major schools. Its intellectual roots are closest to materials science, in which the materials of study are rocks and the applications are to geologic processes. It views geologic processes, in particular deformation processes, at a general level in terms of the basic nature of the mechanisms by which rock can deform. Thus the formation of faults is viewed as a brittle fracture process, and continuing slip on faults, including that during earthquakes, is viewed as a frictional process. The aim is to understand why these natural phenomena exhibit the properties they do by experimenting with these same materials in the laboratory and isolating those aspects of the phenomena that are intrinsic properties of the materials. Any other aspects of the phenomena must be a consequence of the physical conditions, which can be evaluated by mathematical modeling. In the case of earthquakes, rock mechanicists approach the problem as an instability in rock friction and seek to understand why and under what conditions rock exhibits unstable (stick–slip) frictional behavior.

Seismology and rock mechanics differ, at the most basic level, in their conceptualization of even what is the nature of the earthquake mechanism problem. This may be illustrated by considering the ambiguous definition of an earthquake. Is it the shaking that causes damage and is felt by people and seismometers, or is it the rapid fault slip that causes the shaking? In terms of defining the earthquake mechanism problem, most seismologists accept the first and most rock mechanicists the second definition, and each recognizes the other's definition as a *secondary* problem. With such cross-purposes, it is not hard to see why these fields communicate largely by talking through each other.

The rock mechanicist sees the earthquake as a frictional slip instability that dynamically propagates over a fault surface. In the energy balance that governs this process there is one term that describes the portion of energy that is lost to elastic wave radiation. Because seismologists study elastic radiation, this is the only part of the process that is available to them to study. The size of this elastic radiation term is proportional to the stress drop, but the other terms depend on total stress. Because stress drop is found to be a small fraction of total stress in laboratory experiments, rock mechanicists usually neglect the radiation term in their analyses of the problem. On the other hand, if the total stress drop hypothesis is accepted, then the radiation term becomes dominant and all essential information regarding the earthquake can be determined from a study of its seismic waves. Thus the ratio of stress drop to stress, sometimes called the seismic efficiency, is at the center of this debate, because it determines the fraction of information that is accessible to the seismological method and, consequently, the importance of this field in the study of earthquakes.

I am struck by conversations with several older and very influential seismologists in which they expressed the belief that laboratory studies are not relevant to the earthquake problem. One stated the view that any experimental study can be duplicated with a computer model. It is not recognized that the frictional constitutive law that gives rise to stick–slip instabilities must first be *discovered* to be a property of the material before it can be written into a computer program to simulate the phenomenon. The first important question is: Why does fault slip produce earthquakes at all? The seismologist accepts the instability as the cause of the earthquake; the rock mechanicist investigates the cause of the instability. In the mind's eye of the rock mechanicist, the breaking of atomic bonds is at the root of the problem, in the seismologist's, it is seen as an appropriate Green's function.

An important misconception among seismologists is that laboratory studies are a form of scale modeling. This idea may stem from the popularity in the early 1960s of model seismology, in which various seismic wave propagation problems, then inaccessible to numerical modeling, were studied by means of small laboratory models. In contrast, scale modeling of strength has long been in disrepute, owing to the well-known scale dependence of this property (Hubbert, 1937). But rock mechanicists do not view experimental work as a scale model, but rather as an investigation of the fundamental properties of the material that define the geologic process. In this case, the property under study is the friction of rock. Because the subject of study is a material property, no 'model material' may be substituted. To most seismologists, the role of the material is not recognized, except insofar as it determines the elastic wave velocity. It is sufficient to treat earthquakes with a phenomenological description as dynamic slip between elastic half-spaces. To a seismologist, an earthquake is an event, not a component of a process. The process is less interesting because it is unobservable to the seismologist.

Hidden within the total stress drop concept, as will become clearer later, is a confusion between (fracture) strength and friction. If an earthquake is thought to be a strength problem, then there are severe scaling problems between the laboratory and geologic scales, which could be used to criticize the relevance of laboratory studies. These problems do not exist with friction, as stated in Amontons' first law: friction is independent of the size of the surfaces in contact. Understanding the reason for that law, together with Amontons' second law that friction is proportional to normal load and independent of the material, was a fundamental problem which eluded mechanicists for over two hundred years. Once grasped, however, it places considerable constraint on how one views this property. Because these topics are not part of the training and experience of the seismological tradition, it is inevitable that misunderstandings will persist between the fields.

4. The San Andreas Stress Issue as a Proxy in the Struggle

Because of the stick–slip mechanism is an old and well-established tenet of the mechanics of friction, it could not be challenged directly. What could be challenged was its applicability to the case of seismic faulting. Is the frictional strength of faults of the order of 100 MPa, as would be expected from laboratory measurements of coefficients of friction for rock, or of the order of 10 MPa, the value of earthquake stress drops? This question could not even be asked in the absence of the new hypothesis, so it is no coincidence that a heat flow project, which sought to answer it, was initiated almost immediately following the publication of Brace and Byerlee.

Because stress cannot readily be measured, this project sought to determine it indirectly by detecting the presence of a heat flow anomaly produced by frictional heating on the San Andreas fault in California. The result was totally negative. No heat flow anomaly of any size was detected (Brune, Henyey, and Roy, 1969). The errors in measurement were such that it could be stated that the mean shear stress on the fault had to be less than 20 MPa for a heat flow anomaly to go undetected. This result might have rung the death knell for the stick–slip hypothesis, but it did not, which underscores the difficulty that sometimes arises in making a negative argument convincing. The detection of a small heat flow anomaly that could be associated with a low level of stress would have been far more convincing. The lack of any anomaly suggested that something was amiss with the experiment. Surface heat flow measurements are fraught with difficulty, and the analysis required that all heat transmission was by conduction, so that any number of reasons could be suggested for a failure of the experiment. Much effort was expended on this project over the next decade: the number of heat flow measurements in California was increased tenfold and their precision increased, but still no anomaly could be detected (Lachenbruch and Sass, 1980). This increased the polarization between the 'low stress' people and the 'high stress' people, but did not force a consensus. To the 'high stressers,' the confined failure to find an anomaly had become even more suspicious. To them, the greatly increased quantity and quality of heat flow data was making it difficult to claim that the stress on the fault was even high enough to generate earthquake stress drops.

By this stage the issue had spread far beyond its initial jurisdiction. The value of the strength of the Earth's lithosphere was crucial to a far wider range of problems than the mechanism of earthquakes. During the 1970s, for example, a vigorous effort emerged to determine the driving forces of plate tectonics. The success of this effort depended on knowing the absolute value of at least some of the forces, one of the most important of which is the resistance to plate motion by friction on the plate boundaries. Workers in this area were not mere bystanders: some of the most actively promoted driving forces would be rendered insignificant if the resistive forces were as high as to be expected if conventional (laboratory) friction values applied to plate boundaries. A meeting was held in 1979 to debate, on a broader front, the issue of the strength of the Earth's lithosphere (Hanks and Raleigh, 1980). The polarization between low-stressers and high-stressers was now largely complete among those involved with the San Andreas controversy, but a new point of view began to emerge from those who were willing to concede that the San Andreas fault may be weak but that the lithosphere in general was not. That is, the San Andreas fault was beginning to be treated as a special, rather than a general, case.

Out of this was born what was to become one of most costly single experiments ever conducted in tectonics: the Cajon Pass deep drilling experiment. It was meant to settle this question once and for all by drilling to seismogenic depths next to the San Andreas fault in order to measure the stress and heat flow directly. Cajon Pass was a failure on almost all counts. Drilling only penetrated to 2.5 km, half the planned depth and too shallow to be considered in the seismogenic range. Heat flow at that depth was found to be very high, and was explained as a local anomaly (Lachenbruch and Sass, 1988). Stress measurements made in this borehole indicated left-lateral shear stresses, opposite to the direction of slip on the fault, and this result too will probably come to be viewed as a local anomaly.

On the theoretical front, there have been several belated attempts to justify the total stress drop hypothesis on the basis of several mechanisms, such as friction melting and vaporization of fault zone fluids by frictional heating, that could cause the residual friction to vanish after the initiation of dynamic slip (McKenzie and Brune, 1972; Sibson, 1973). Because these mechanisms offer no explanation for why the *initial* friction could be very low, they do not resolve the San Andreas heat flow paradox: if normal friction were present, these mechanisms would predict 100 MPa stress drops.

The many aspects of the San Andreas stress controversy have been reviewed by Scholz (1990) and Lachenbruch and McGarr (1990). Lachenbruch and McGarr (1990) sum up the current status with the words: 'efforts to determine the stresses that give rise to San Andreas fault slip, to date, have not led to anything resembling scientific consensus. The uncertainty results from widespread disagreement over the implications of different methods of assessing the stresses.' Direct measurements of in-situ stresses near the San Andreas fault sharply conflict with stresses inferred from the heat flow data: measurements of stresses in the vicinity of the fault show, in the top 1 km, a strong increase in shear stress with depth. This shear stress gradient is very similar to that observed elsewhere in both nontectonic (McGarr and Gay, 1978; Pine et al., 1983) and tectonic areas (Zoback and Healy, 1984). Both the orientation of these stresses and their extrapolation to mid-crustal depths would indicate a crustal averaged shear stress of about 50 MPa on the San Andreas fault, consistent with laboratory friction values.

Lachenbruch and McGarr discuss various loopholes that could resolve this dilemma. The most interesting of these loopholes for rock mechanicists to investigate are mechanisms that could explain how the fault could be very weak. Mechanisms of this type involve either very high pore pressure within the fault zone, or that the fault is lined with a material that has intrinsically low friction. For example, the first of these mechanisms is explored by J.R. Rice (this volume, Chapter 20) and the second by Byerlee (1990).

The issues of the strength of the San Andreas fault and whether the San Andreas is a general or special case of faults cannot, however, any longer be viewed as a test of the stick–slip hypothesis. Analysis of the stick–slip mechanism with the rate- and state-variable friction laws (Dieterich, 1979; Ruina, 1983) shows that the instability mechanism is not a function of the base friction, which determines the absolute stress level. The San Andreas stress controversy is a separate problem.

5. Development of the Two Paradigms along Separate Paths

The conceptualization of the earthquake mechanism has continued to develop over the last 20 years by two separate paths that follow the two paradigms. These two paths have been followed independently by the two disciplines with essentially no crosstalk between them. As both conceptualizations have become more developed it is possible to see more clearly the distinctions between them. So too, these developments have sought to expand the range of the conceptualizations as the competition between them changed to that of demonstrating which was more successful at solving the greatest number of puzzles.

During this period, the techniques of seismic waveform analysis were greatly improved, so that it became possible not only to measure the mean slip and stress drop in an earthquake but to estimate their distribution in both time and space over the rupture surface. These closely related parameters turn out to be highly variable within a single earthquake rupture. The resolution of this variability depends on the seismic band being observed. Long-wavelength radiation reveals the first-order variability, and shorter-wavelength radiation reveals higher orders of variability superimposed upon it.

This internal complexity of earthquakes has spawned a new generation of models. In these models, slip variations within earthquakes are described in terms of distributions of strong regions, called 'barriers' or 'asperities,' depending on the particular school one ascribes to. In these conceptualizations, the regions within the rupture with the highest concentration of moment release, and hence local stress drop, are thought to mark strong regions on the fault, the asperities. The asperities are thought to play a special role in the earthquake process, being strong points where earthquakes are initiated and terminated. The size of earthquakes that can occur on a fault is thought to reflect the size of asperities on it. In turn, the relative size of asperities on a fault is said to reflect the 'coupling' of the fault. This model has been particularly applied to subduction zones, and to the problem of 'seismic coupling' of subduction zone interfaces. Seismic coupling refers to the ratio of moment release produced by earthquakes to the total moment release, a ratio called the seismic coupling coefficient. This ratio is usually near unity for crustal faults but varies widely, from nearly unity to zero for subduction zones. In the asperity model, the subduction zones with the highest seismic coupling are those with the largest fraction of their interface comprised of asperities, and this because they have the highest 'coupling' (Ruff and Kanamori, 1980).

To what extent does this model reflect the old or the new paradigm? Although the term asperity is taken from the friction literature, this new usage seems to be a malappropriation. In the friction literature an asperity is a protrusion on a surface, and the real area of contact between surfaces consists of the sum of contacts at the tips of the asperities. The use of this term in this earthquake model is at most an allusion. The language of the asperity model shows that it is a strength model and not a friction model. In friction, strength is simply a function of normal stress, not an inherent property, and the seismic behavior depends on the stability of sliding, not the strength. The asperity model can be seen instead to be an outgrowth of the old paradigm. Although the total stress drop hypothesis is no longer overtly stated, and is certainly never mentioned with respect to the asperity model, it is clearly rooted in the idea that the earthquake stress drop reflects the strength of the fault. Thus in the asperity model the spatial distribution of slip and stress drop on the fault is interpreted as (one for one mapping of) the strength (asperity) distribution on the fault.

The stick–slip hypothesis has also developed. Experimentation showed that regular stick–slip in rock was due to a negative dependence of friction on slip velocity. This work led to the formulation of rate- and state-variable friction constitutive laws, compact rheological equations that are successful in predicting the wide variety of rock frictional behavior observed in the laboratory (Dieterich, 1979; Ruina, 1983). In their most compact form these friction laws contain only three variables: two velocity coefficients that are material properties, and a critical length which reflects the topography of the surface contact. A combination of these parameters together with a stiffness of the system determines the stability of slip and thus controls whether or not slip is stable or occurs by stick–slip oscillations. The rate parameters depend on the lithology and on temperature and hence predict that the seismic behavior of faults will be a function of these factors. The factors that determine the stability of slip do not affect the frictional strength of the fault, which is still a linear function of normal stress. The distinction between this conceptualization and that of strength models is thus clear: the seismic behavior of faults depends on the stability of frictional sliding and not on strength.

When explored with the use of numerical experiments, the rate- and state-variable friction laws predict a rich variety of behavior that could not have been expected from the laboratory studies from which they were derived. This predicted behavior is remarkably similar to a wide range of naturally occurring faulting phenomena that had previously been considered to be separate phenomena. As an example, in the model of Tse and Rice (1986), a fault is treated as an entirely frictional feature obeying a rate- and state-variable law in which the parameters were taken from laboratory measurements on granite. The model predicts a seismic cycle in which an upper, seismically slipping region overlies an aseismically slipping deeper layer. The seismic cycle contains immediate preseismic nucleation, coseismic slip, and a postseismic period of rapidly decaying enhanced deep slip succeeded by a long period of steady interseismic deformation: that is, all the elements of the cycle that are actually observed. Although other models, such as those containing a brittle–plastic transition and intracrustal asthenosphere, can also produce such features, none is as simple as this model, which contains only one parameter that was freely varied, the critical slip distance.

We may compare the usefulness of the problem-solving ability of the asperity model and the stick–slip model in the case of the seismic coupling of subduction zones, discussed earlier. Three types of behavior are observed on subduction zone interfaces: regions where the seismic coupling is total; those where slip is seismic only when triggered by seismic slip from an adjacent region and which otherwise slip aseismically; and regions where slip is always aseismic. In the asperity model the first type of region is called an asperity, the second a weak zone, and the third is not explicitly discussed (Lay, Kanamori, and Ruff, 1982). Analysis of the rate- and state-variable friction laws show that they predict three conditions of slip stability: unstable, in which slip is unstable under all conditions; conditionally stable, in which slip is stable under quasistatic loading and unstable under dynamic loading; and stable. Thus the three types of behavior described for subduction zones can, with the frictional model, be interpreted as marking regions in which these three stability conditions exist. The stability conditions depend, under similar geometric configurations, only on the slip velocity parameters in the friction law, which are material properties. Laboratory studies show that, for sediments, these parameters will lead to progressively more unstable behavior as the sediment becomes compacted and lithified. So one may interpret the variation of relative abundance of these stability states among subduction zones as due to differences of lithification of the interface sediments brought about by, among other things, differences in the normal stress supported by the interfaces (Scholz, 1990). This provides a definition of the meaning of 'coupling' which makes specific predictions in terms of variations in the local driving forces that give rise to these differences, predictions that can in principle be independently checked. Furthermore, because a specific model is defined, predictions can be made for the behavior of other subduction zones, which, for example, may have no historic record of seismicity. In contrast, the asperity model provides only names for these phenomena and so is descriptive but not predictive. From this standpoint it is not really a model, but mere wordplay.

6. Resolution of the Revolution

The struggle between these two paradigms has now entered a mature phase. The San Andreas stress issue, which once played a central, though proxy, role, has become beside the point and evolved a separate life of its own. The competition between the paradigms has become more centered on questions of aesthetics and puzzle-solving utility. With the development of the rate- and state-variable friction laws, the stick–slip hypothesis may now be stated in a compact system of equations which may be studied by rigorous means. These equations predict a rich variety of behavior which is observed in nature and provides a unifying explanation of them. When experimentally determined parameters are used in these laws, the ensuing models predict behavior which is in quantitative agreement with natural phenomena, providing further confidence in the applicability of the laboratory measurements. In contrast, the old strength paradigm has progressed no further than the 'asperity model,' which is entirely descriptive. If one paradigm has to 'win,' it is the one that is quantitative, rather than descriptive.

To be sure, among observational earthquake seismologists the asperity model remains the great fashion of the day. To them, the rate- and state-variable friction laws and the world view they embody is still esoteric. But there are no longer any challenges to these laws or the models derived from them. This body of work has now reached the point of becoming mainstream but still too advanced to have yet had much impact on routine practice. The asperity model's current popularity can be ascribed to its easy digestibility and the fact that it is vigorously promoted by some of the most influential earthquake seismologists.

During this struggle, seismology has maintained its scope as restricted to problems of elastic wave propagation, and in so doing has lost its hegemony over the study of earthquakes. Geology, rock mechanics, and tectonophysics in general, are now recognized as having a very significant role to play, and the study of earthquakes is now becoming known as earthquake science, rather than seismology. Although seismology's role in earthquake science is still great, if it does not embrace the new paradigm it is soon to be threatened with being consigned to the backseat role of purely descriptive functions. But since seismology will always be an important part of the study of earthquakes, this revolution will only be resolved when seismologists have embraced the new view, and this will only take place over a prolonged period as the education of a new generation of seismologists encompasses the discipline of rock mechanics.

In this essay I have tried to trace the history of the impact of Brace and Byerlee (1966) and to explain some of the reasons for the convolutions in this history. It is interesting to compare it with the paper of Brace and Kohlstedt (1980), which is probably the most cited of any of Brace's papers. In this paper, Brace and Kohlstedt recapitulated the earlier work of Goetze and Evans (1979) to formulate a strength profile for the continental lithosphere from the constraints of laboratory measurements. This strength model has become adopted into the canon of geomechanics, where it plays a central role in the modeling of many geologic phenomena. It is ironic that it is this paper, and not the Brace and Byerlee paper, that is most challenged by the San Andreas stress issue. It is also telling that this fact has never been mentioned until the publication of the paper by J.R. Rice in this volume (Chapter 20).

References

Aki, K. and Richards, P.G. (1980). *Quantitative Seismology, Theory and Methods*. W.H. Freeman, San Francisco.

Brace, W.F. and Byerlee, J.D. (1966). Stick-slip as a mechanism for earthquakes. *Science* **153**, 990–992.

Brace, W.F. and Kohlstedt, D. (1980). Limits on lithospheric stress imposed by laboratory experiments. *J. Geophys. Res.* **85**, 6248–6252.

Brune, J.N., Henyey, T., and Roy, R. (1969). Heat flow, stress, and rate of slip along the San Andreas fault, California. *J. Geophys. Res.* **74**, 3321–3327.

Byerlee, J. (1990). Friction, overpressure, and fault normal compression. *Geophys. Res. Lett.* **17**, 2109–2112.

Chinnery, M.A. (1964). The strength of the earth's crust under horizontal shear stress. *J. Geophys. Res.* **69**, 2085–2089.

Dieterich, J. (1979). Modelling of rock friction: 1. Experimental results and constitutive equations. *J. Geophys. Res.* **84**, 2161–2168.

Gilbert, G.K. (1884). A theory of earthquakes of the Great Basin, with a practical application. *Am. J. Sci.* **XXVII**, 49–54.

Goetze, C. and Evans, B. (1979). Stress and temperature in the bending lithosphere as constrained by experimental rock mechanics. *Geophys. J. R. Astron. Soc.* **59**, 463–478.

Hanks, T.C. and Raleigh, C.B. (eds) (1980). Stress in the lithosphere. *J. Geophys. Res.* **85**, 6083–6435.

Hubbert, M.K. (1937). Theory of scale models as applied to geology. *Bull. Geol. Soc. Am.* **48**, 1459–1520.

Isacks, B., Oliver, J., and Sykes, L. (1968). Seismology and the new global tectonics. *J. Geophys. Res.* **73**, 5855–5899.

Koto, B. (1893). On the cause of the great earthquake in central Japan, 1891. *J. Coll. Sci. Imp. Univ.* **5**(4), 294–353.

Kuhn, T.S. (1970). *The Structure of Scientific Revolutions* (rev. edn.). University of Chicago Press.

Lachenbruch, A. and McGarr, A. (1990). Stress and heat flow. In *The San Andreas Fault System, California* (ed. R.E. Wallace), pp. 261–278. USGS Professional Paper 1515.

Lachenbruch, A. and Sass, J. (1980). Heat flow and energetics of the San Andreas fault zone. *J. Geophys. Res.* **85**, 6185–6222.

Lachenbruch, A. and Sass, J. (1988). The stress heat-flow paradox and thermal results from Cajon Pass. *Geophys. Res. Lett.* **15**, 981–984.

Lay, T., Kanamori, H., and Ruff, L. (1982). The asperity model and the nature of large subduction zone earthquakes. *Earthquake Pred. Res.* **1**, 1–71.

LeGrand, H.E. (1988). *Drifting Continents and Shifting Theories*. Cambridge University Press.

McGarr, A. and Gay, N. (1978). State of stress in the earth's crust. *Annu. Rev. Earth Planet. Sci.* **6**, 405–436.

McKenzie, D. and Brune, J. (1972). Melting on fault planes during large earthquakes. *Geophys. J. R. Astron. Soc.* **29**, 65–78.

Pine, R., Ledingham, P., and Merrifield, C. (1983). In-situ stress measurements in the Carmenellis granite – II. Hydrofracture tests at Rosemanowis quarry to 2000 m. *Int. J. Rock. Mech. Min. Sci.* **20**, 63–72.

Reid, H.F. (1910). The mechanics of the earthquake. In *Rept. State Earthquake Inv. Comm., The California Earthquake of April 18, 1906*. Carnegie Institute, Washington D.C.

Ruina, A. (1983). Slip instability and state variable friction laws. *J. Geophys. Res.* **88**, 10359–10370.

Ruff, L. and Kanamori, H. (1980). Seismicity and the subduction process. *Phys. Earth. Planet. Int.* **23**, 240–252.

Scholz, C.H. (1990). *The Mechanics of Earthquakes and Faulting.* Cambridge University Press.

Sibson, R.H. (1973). Interactions between temperature and pore fluid pressure during an earthquake faulting and a mechanism for partial or total stress relief. *Nature* **243**, 66–68.

Sykes, L.R. (1967). Mechanism of earthquakes and nature of faulting at mid-oceanic ridges. *J. Geophys. Res.* **72**, 2131–2140.

Tse, S. and Rice, J.R. (1986). Crustal earthquake instability in relation to the depth variation of frictional properties. *J. Geophys. Res.* **91**, 9452–9472.

Tsuboi, C. (1933). Investigation of deformation of the crust by precise geodetic means. *Japan J. Astron. Geophys.* **10**, 93–248.

Zoback, M.D. and Healy, J. (1984). Friction, faulting, and in-situ stress. *Annales Geophysicae*, **2**, 689–698.

Index

International Geophysics Series

EDITED BY

Volume 1 BENO GUTENBERG. Physics of the Earth's Interior. 1959*

Volume 2 JOSEPH W. CHAMBERLAIN. Physics of the Aurora and Airglow. 1961*

Volume 3 S. K. RUNCORN (ed.). Continental Drift. 1962*

Volume 4 C. E. JUNGE. Air Chemistry and Radioactivity. 1963*

Volume 5 ROBERT G. FLEAGLE AND JOOST A. BUSINGER. An introduction to Atmospheric Physics. 1963*

Volume 6 L. DUFOUR AND R. DEFAY. Thermodynamics of Clouds. 1963*

Volume 7 H. U. ROLL. Physics of the Marine Atmosphere. 1965*

Volume 8 RICHARD A. CRAIG. The Upper Atmosphere: Meteorology and Physics. 1965*

Volume 9 WILLIS L. WEBB. Structure of the Stratosphere and Mesosphere. 1966*

Volume 10 MICHELE CAPUTO. The Gravity Field of the Earth from Classical and Modern Methods. 1967*

Volume 11 S. MATSUSHITA AND WALLACE H. CAMPBELL (eds.). Physics of Geomagnetic Phenomena. (In two volumes.) 1967*

Volume 12 K. YA. KONDRATYEV. Radiation in the Atmosphere. 1969

Volume 13 E. PALMÉN AND C. W. NEWTON. Atmospheric Circulation Systems: Their Structure and Physical Interpretation. 1969

Volume 14 HENRY RISHBETH AND OWEN K. GARRIOTT. Introduction to Ionospheric Physics. 1969*

Volume 15 C. S. RAMAGE. Monsoon Meteorology. 1971*

Volume 16 JAMES R. HOLTON. An Introduction to Dynamic Meteorology. 1972*

*Out of print.

Volume 17 K. C. Yeh and C. H. Liu. Theory of Ionospheric Waves. 1972

Volume 18 M. I. Budyko. Climate and Life. 1974

Volume 19 Melvin E. Stern. Ocean Circulation Physics. 1975

Volume 20 J. A. Jacobs. The Earth's Core. 1975*

Volume 21 David H. Miller. Water at the Surface of the Earth: An Introduction to Ecosystem Hydrodynamics. 1977

Volume 22 Joseph W. Chamberlain. Theory of Planetary Atmospheres: An Introduction to Their Physics and Chemistry. 1978*

Volume 23 James R. Holton. Introduction to Dynamic Meteorology, Second Edition. 1979

Volume 24 Arnett S. Dennis. Weather Modification by Cloud Seeding. 1980

Volume 25 Robert G. Fleagle and Joost A. Businger. An Introduction to Atmospheric Physics, Second Edition. 1980

Volume 26 Kuo-Nan Liou. An Introduction to Atmospheric Radiation. 1980

Volume 27 David H. Miller. Energy at the Surface of the Earth: An Introduction to the Energetics of Ecosystems. 1981

Volume 28 Helmut E. Landsberg. The Urban Climate. 1981

Volume 29 M. I. Budyko. The Earth's Climate: Past and Future. 1982

Volume 30 Adrian E. Gill. Atmosphere–Ocean Dynamics. 1982

Volume 31 Paolo Lanzano. Deformations of an Elastic Earth. 1982

Volume 32 Ronald T. Merrill and Michael W. McElhinny. The Earth's Magnetic Field: Its History, Origin and Planetary Perspective. 1983

Volume 33 John S. Lewis and Ronald G. Prinn. Planets and Their Atmospheres: Origin and Evolution. 1983

Volume 34 Rolf Meissner. The Continental Crust: A Geophysical Approach. 1986

Volume 35 M. U. Sagitov, B. Bodri, V. S. Nazarenko, and Kh. G. Tadzhidinov. Lunar Gravimetry. 1986

Volume 36 Joseph W. Chamberlain and Donald M. Hunten. Theory of Planetary Atmospheres: An Introduction to Their Physics and Chemistry, Second Edition. 1987

Volume 37 J. A. Jacobs. The Earth's Core, Second Edition. 1987

Volume 38 J. R. Apel. Principles of Ocean Physics. 1987

Volume 39 Martin A. Uman. The Lightning Discharge. 1987

Volume 40 David G. Andrews, James R. Holton, and Conway B. Leovy. Middle Atmosphere Dynamics. 1987

Volume 41 Peter Warneck. Chemistry of the Natural Atmosphere. 1988

Volume 42 S. Pal Arya. Introduction to Micrometeorology. 1988

Volume 43 Michael C. Kelley. The Earth's Ionosphere: Plasma Physics and Electrodynamics. 1989

Volume 44 William R. Cotton and Richard A. Anthes. Storm and Cloud Dynamics. 1989

Volume 45 William Menke. Geophysical Data Analysis: Discrete Inverse Theory. 1989

Volume 46 George Philander. El Niño, La Niña, and the Southern Oscillation. 1990

Volume 47 Robert A. Brown. Fluid Mechanics of the Atmosphere. 1991

Volume 48 J. Holton. An Introduction to Dynamic Meteorology, Third Edition. 1992

Volume 49 A. Kaufman. Geophysical Field Theory and Method, Vol. 1. 1992

Volume 50 S. Butcher, R. J. Charlson, G. H. Orians and G. V. Wolfe. Global Biogeochemical Cycles. 1992

Volume 51 B. Evans and T. Wong. Fault Mechanics and Transport Properties of Rocks. 1992